# PARTICLE PHYSICS

# PARTICLE PHYSICS

Duncan Carlsmith

PEARSON

Boston Columbus Indianapolis New York San Francisco Upper Saddle River
Amsterdam Cape Town Dubai London Madrid Milan Munich Paris Montreal Toronto
Delhi Mexico City São Paulo Sydney Hong Kong Seoul Singapore Taipei Tokyo

Executive Editor: Jim Smith
Assistant Editor: Steven Le
Development Manager: Laura Kenney
Managing Editor: Corinne Benson
Production Project Manager: Dorothy Cox
Production Management and Composition: Integra
Cover Designer: Mark Ong
Manufacturing Buyer: Dorothy Cox
Marketing Manager: Will Moore

Cover Image Credit: CERN

Credits and acknowledgments borrowed from other sources and reproduced, with permission, in this textbook appear on the appropriate page within the text.



**Library of Congress Cataloging-in-Publication Data**
Carlsmith, Duncan L.
Particle physics / Duncan Carlsmith.
p. cm.
Includes bibliographical references and index.
ISBN-13: 978-0-321-67689-4
ISBN-10: 0-321-67689-0
1. Particles (Nuclear physics)–Textbooks. I. Title.
QC777.C37 2013
539.7′2–dc23 2011050835

ISBN 10: 0-321-67689-0
ISBN 13: 978-0-321-67689-4

PEARSON

www.pearsonhighered.com 1 2 3 4 5 6 7 8 9 10—CRW—16 15 14 13 12

# Preface

This introduction to standard model elementary particle physics is accessible to students with a background in intermediate electromagnetic theory and non-relativistic quantum mechanics. It strives to balance experiment and theory and to present the subject in depth, but not exhaustively, starting from essential principles and following the dictum "as simple as possible but not simpler."

The text begins with the an overview of the standard model, a review of modern physics, and a smattering of history. The lengthy introduction is followed by special relativity and kinematics with applications of relativistic mechanics to accelerators. A chapter is devoted to the interactions of particles with matter and the principles of particle detection. Next, potential scattering calculations and the quantum theory of radiation are developed and used to jumpstart estimates of physical quantities in the relativistic regime. The successively more intricate gauge theories of electrodynamics, chromodynamics, and the electroweak model are then derived and applied. Along the way, Feynman rules which are the crux of the theories are developed in an intuitive way. The two component spinor description of fermions is emphasized and used in tree level calculations of seminal differential cross sections and transition rates including polarization.

This foundation is followed by a look at radiative corrections and renormalization. Aspects of the phenomenology of hadrons are then presented including discrete symmetries and conservation laws, the quark model, neutral meson mixing, and $CP$-violation. The final chapter introduces speculative notions of physics beyond the standard model, a status report on neutrino oscillations, and a look at particle physics in the context of cosmology.

One milestone for the student is to learn to calculate tree level processes without approximation. Such calculations are the *lingua franca* of the field. In reaching this goal, the student will hopefully appreciate the experimental basis for particle science, its connections with other fields of physics, and acquire a sense of future directions. Relatively simple exercises and some quite challenging problems are provided along with generous hints and references. A complete set of solutions is available to instructors for download at the Instructor Resource Center: www.pearsonhighered.com/irc.

## CHAPTER BY CHAPTER SUMMARY

Chapter 1 provides an orientation to particle physics and the standard model. It introduces quarks, leptons, and their gauge interactions through Feynman diagrams. Properties of leptons are illustrated with a brief history of discoveries. The quark model of hadrons is then presented. The $\hbar = c = 1$ convention is introduced and used throughout, with factors made explicit to ease calculations in some cases.

Chapter 2 provides some basic tools beginning with the special theory of relativity. Lorentz transformations, 4-vectors, and invariants are introduced followed by relativistic kinematics and mechanics. Relativity theory is followed by an introduction to accelerators, luminosity, and cross sections. Storage ring concepts are presented for students interested in contemporary colliding beam experiments. Some technical aspects of LHC operation are given for illustration.

Chapter 3 is an introduction to the interactions of particles with matter and to particle detection techniques. It pays homage to technologies and discoveries of historical interest and ends with an introduction to contemporary colliding beam detector complexes at the Large Hadron Collider.

Chapter 4 bridges the gap between non-relativistic quantum mechanics and quantum field theory, covering topics often glossed over in an introductory course in quantum mechanics. Fermi's golden rule is derived and applied to potential scattering to illustrate how matter waves probe subatomic length scales. Quasi-classical radiation theory is presented and then generalized to estimate the decay rates of the $W$ boson, the muon, and the neutron. Along the way, the relativistic propagator and phase space are encountered. This chapter ends with an orientation to relativistic quantum field theory and Feynman rules via a model of interacting scalar particles.

Chapter 5 introduces relativistic field theory through the electrodynamics of a charged scalar field. Relativistic field equations, Lagrangians, and $U(1)$ gauge symmetry are discussed. The Feynman rules for scalar quantum electrodynamics are found and tree level processes are calculated. The chapter concludes with the electrodynamics of charged vector bosons such as the $W^{\pm}$.

Chapter 6 concerns spinor electrodynamics. It presents the Dirac equation in four component and two component forms and makes contact with the non-relativistic limit and the leading corrections, connecting the student to a prior course in quantum mechanics. Feynman rules are derived and applied to 2-to-2 processes. Two component spinors are used in calculations including polarization. This chapter provides the core technology needed for tree level calculations in the standard model.

Chapter 7 begins with general non-Abelian gauge theory and then specializes to quantum chromodynamics. The Feynman rules for quantum chromodynamics are derived and applied. The color exchange interaction energy of a color composite is considered in detail and some subtleties related to charge conjugation and decay rate calculations are illustrated. Additionally, color confinement and de-confinement are introduced. Historically, QCD appeared after the theory of the electroweak interaction but, although it involves three rather than two dimensions, $SU(3)_C$ is simpler than broken $SU(2)_L \times U(1)_Y$ so treated first.

Chapter 8 presents the $SU(2)_L \times U(1)_Y$ model of electroweak interactions with the minimal Higgs model of spontaneous symmetry breaking. The theory is illustrated with calculations of muon $\beta$ decay, electroweak boson production and decay, top quark decay, and the properties of the standard model Higgs boson.

Chapter 9 concerns radiative corrections and renormalization. Bremsstrahlung and loop diagrams in QED are analyzed and the running coupling constant found. The Lamb shift and anomalous muon and electron magnetic moments are discussed. Color confinement and asymptotic freedom are then presented with a look towards grand unification. The chapter also provides an introduction to high order calculations of electroweak processes.

Chapter 10 discusses global symmetries ($P$, $C$, $T$, isospin), static properties of baryons and mesons in the quark model, bag models and the color hyperfine interaction, the parton model and parton distribution functions, and hadronic weak interaction phenomena including mixing and $CP$ violation in strange and bottom flavored mesons.

Chapter 11 introduces popular ideas related to physics beyond the standard model including grand unification and the minimal supersymmetric standard model. The possibility of two Higgs doublets motivated by left right symmetry restoration and by supersymmetry is considered. The Large Hadron Collider and the physics it may reveal are briefly described. Evidence for neutrino mass is reviewed along with the three flavor mixing model. The text concludes with an introduction to particle physics in cosmology, evidence for the $\Lambda$CDM model, and open questions.

## SUGGESTED USES

This text can be used in a one or two semester course. Chapters 1-4 and parts of Chapter 10 contain much of what one might find in a minimal introduction with simple mathematics for students of nuclear physics. The core field theory of the standard model is contained in Chapters 5-8. A short course with emphasis on theory might start with Chapter 1, omit Chapters 2 and 3 and much of Chapter 4, and proceed directly to the standard model Chapters 5-8. Chapter 9 is somewhat technical and can be left as extra reading. A two semester course might cover Chapters 1-6 in the first semester and Chapters 7-11 in the second semester. The following sections could be considered optional: Chapter 2.4 (beam stability), Chapter 2.6 (LHC accelerator design), Chapter 3.9 (LHC experiments), Chapter 4.12 (quantum field operators), Chapter 5.6-5.8 (vector electrodynamics), Chapter 6.14-16 (special QED topics), Chapter 7.6-7.9 (topics in QCD), Chapter 8.7-8.11 (topics in electroweak physics), Chapter 9 (renormalization). The topics in Chapters 10 and 11 can be selected as desired.

Each chapter offers some suggested reading and links to explore. Some of these suggestions could serve as assigned research topics. No attempt is made to provide a comprehensive list of references but many of the references provided have extensive bibliographies. Links to public electronic archives are given when available. To best take advantage of the references, the reader should learn how to access journals electronically through an institutional proxy server.

## ACKNOWLEDGEMENTS

I thank the University of Wisconsin-Madison and the Physics Department faculty for their support and encouragement and also CDF and CMS colleagues who have made working in this field a joy. I apologize to the many physicists whose beautiful experiments and theories could not be included.

## COVER ILLUSTRATION

The cover of this text represents a beam view of a $p\bar{p}$ collision at the Fermilab Tevatron operating at close to two TeV center of mass collision energy. The tracks are of particles produced via constituent quark antiquark annihilation as reconstructed from ionization trails in the Collider Detector Facility. The event is a candidate instance of the simultaneous creation of a $t$-quark and $\bar{t}$-quark.

## FEEDBACK

Please send comments and suggestions directly to the author at:
duncan@hep.wisc.edu

# Contents

# List of Figures

# List of Tables

# List of Problems

# CHAPTER 1

# Introduction

Contents



## 1.1 ■ MATTER IN A NUTSHELL

Particle physics is about the stuff called matter and energy. The building blocks are elementary spin 1/2 fermions of two kinds (see Table 1.1) : six flavors of quarks, each quark appearing in three so-called colors, and six matching flavors of colorless leptons. The quarks and leptons appear in three families of increasing mass. In addition, there are several fundamental spin 1 bosons associated with electromagnetic, weak, and color gauge fields, and a hypothetical spin 0 Higgs boson related to mass (Table 1.2). Their nature as described by the immensely successful standard model is the subject of this book. We begin with a preview of the strange concepts behind standard model quantum field theories, and some history of particle physics. In later chapters, we study in detail how particles are created and studied and we develop and learn to apply the standard model theory. We will also take a look at possible extensions demanded by present data.

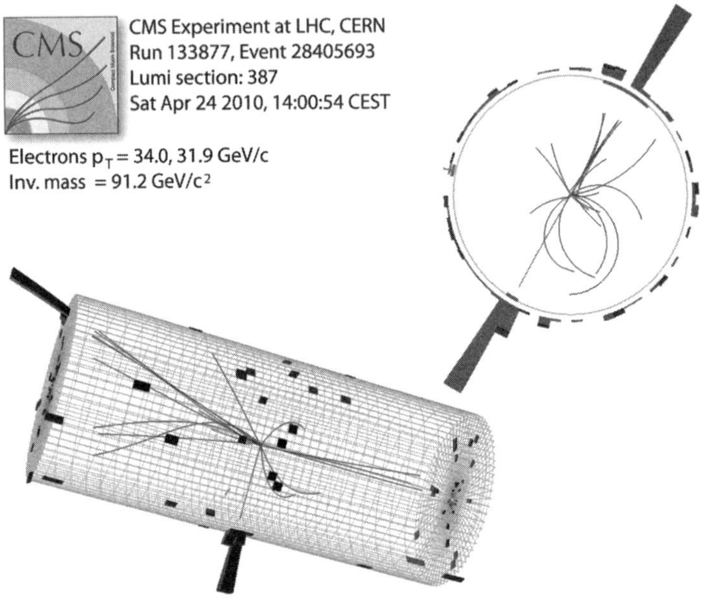


**FIGURE 1.1** A $Z$ boson in the CMS detector. Particle tracks and energy clusters are shown in an event display in a 7 TeV $pp$ collision at the center of the CMS detector at the LHC. A $Z$ weak boson decayed to an electron plus positron which emerge back to back transverse to the beams. Less energetic particles follow trajectories visibly bent by a solenoidal magnetic field.

| | Q | | | |
|---|---|---|---|---|
| Leptons | 0 | $\nu_e$ (<3 eV) | $\nu_\mu$ (<0.19 MeV) | $\nu_\tau$ (<18 MeV) |
| | $-1$ | $e^-$ (511 KeV) | $\mu^-$ (105 MeV) | $\tau^-$ (1.77 GeV) |
| Quarks | $+\frac{2}{3}$ | $u_a$ (1.5 − 3.3 MeV) | $c_a$ (1.27 ± 0.1 GeV) | $t_a$ (171 ± 2 GeV) |
| | $-\frac{1}{3}$ | $d_a$ (3.5 − 6 MeV) | $s_a$ (104 ± 30 MeV) | $b_a$ (4.2 ± 0.1 GeV) |

**TABLE 1.1** Elementary spin 1/2 fermions. Each quark appears in three colors ($a = r, g, b$). The electric charge $Q$ in units of $e$ is common to each row. Rest energy values in parentheses are from the Particle Data Group. Antiparticles such as $\bar{\nu}_e, e^+, \bar{u}, \bar{d}$ which have opposite charges and identical masses are not listed.

Normal atomic matter is composed of members of the lightest family: up and down quarks ($u$ and $d$) and electrons. The up and down quarks are confined in protons and neutrons (nucleons) with the composition

$$p_{uud} \quad \text{and} \quad n_{udd}. \tag{1.1}$$

| | | | | |
|---|---|---|---|---|
| Gauge bosons | $\gamma$ (0 eV) | $Z$ (91 GeV) | $W^{\pm}$ (80 GeV) | $g_{a\bar{b}}$ (0 eV) |
| Higgs boson | $H$ (> 115 GeV) | | | |

**TABLE 1.2** Elementary bosons. The photon $\gamma$ mediates the electromagnetic interaction. The massive neutral $Z$ boson and charged $W^{\pm}$ bosons mediate weak interactions. Eight gluons, often simply denoted by $g$, mediate color interactions. These gauge bosons have spin 1. The neutral spin 0 Higgs boson $H$ (sometimes called $h$) interacts with all massive particles.

Both have a radius of about 1 fm $= 10^{-15}$ m, a unit also called one fermi. Their masses are known to high precision:

$$\begin{aligned} m_p &= 938.272\,013(23)\ \mathrm{MeV\ c^{-2}} = 1.672\,621\,637(83) \times 10^{-27}\ \mathrm{kg} \\ m_n &= 939.565346(23)\ \mathrm{MeV\ c^{-2}} = 1.00\,137\,841\,887(58)\ \mathrm{m_p}. \end{aligned} \tag{1.2}$$

The nucleons are examples of bound states of three quarks known generically as baryons. The nucleus of an atom is roughly a spherical drop of nucleons with constant density and a radius of a few fm. In an atom, the nucleus is surrounded by a cloud of relatively light electrons with mass

$$m_e = 0.510\,998\,910(13)\ \mathrm{MeV\ c^{-2}} = 9.109\,382\,15(45) \times 10^{-31}\ \mathrm{kg}. \tag{1.3}$$

The electric force binds an electron to a proton and is characterized by the proton charge

$$e = 1.602\,176\,487(40) \times 10^{-19}\ \mathrm{C}. \tag{1.4}$$

In the neutral hydrogen atom, the electron moves within a radius about $10^{-10}$ m with a speed roughly 1% of light speed

$$c = 299\,792\,458\ \mathrm{m\ s^{-1}}. \tag{1.5}$$

Electromagnetic radiation is emitted and absorbed by accelerated charges such as the electrons in atoms. The quantum of free electromagnetic field energy is the photon, a massless ($\mathrm{m}_\gamma < 10^{-18}$ eV c$^{-2}$) particle that moves at light speed. The ordinary electrons, up-quarks, and down-quarks found in atoms today (along with a sea of photons, neutrinos, and dark particles) are relics of a high energy explosion at the beginning of time.

Particle physics (high energy physics) focuses on relativistic transformations of energy that reveal the extraordinary particles and fields. These include the muon $\mu^-$ and $\tau^-$ (essentially heavy electrons), the neutrinos $\nu_e$, $\nu_\mu$ and $\nu_\tau$ (neutral electron-like particles), the $c$ (charm), $s$ (strange), $t$ (top), and $b$ (bottom) quarks, the vector bosons $W^{\pm}$ and $Z$ (analogous to photons) which mediate weak interactions, and eight bi-colored gluons $g_{a\bar{b}}$ which are the quanta of the color force field. Every particle is found to be associated with an antiparticle of opposite charge but identical mass. The $e^+$, or positron, was the first antiparticle to be identified. The anti-electron-neutrino $\bar{\nu}_e$ is emitted in nuclear weak reactions. The antiproton is composed of antiquarks: $\bar{p} = \bar{u}\bar{u}\bar{d}$. Antimatter is best described as mass moving backwards in time. Six gluons carry net color and constitute three particle-antiparticle pairs (e.g. $\bar{g}_{r\bar{b}} = g_{b\bar{r}}$). The $Z$, $\gamma$, $H$ and the two color neutral gluons are identical with their antiparticles. In the standard model, matter and antimatter particles are generally created and destroyed together. As the early universe expanded and cooled, vast numbers of particle-antiparticle pairs annihilated to photons leaving the relatively small excess of matter (about one baryon per $10^9$ photons) seen today.

Quarks and gluons are found confined in color neutral composites called hadrons principally of two kinds: three quark ($qqq$) baryons, like the proton and neutron, and quark plus antiquark ($q\bar{q}$) mesons including the lightest hadronic subnuclear particles $\pi^-_{\bar{u}d}$ and $\pi^+_{u\bar{d}}$ called pions with mass

$$m_\pi = 139.57018 \pm 0.00035\text{MeV c}^{-2}. \tag{1.6}$$

Residual color interactions between hadrons in contact are often called strong interactions and are responsible for the binding of nucleons into nuclei and a plethora of other subnuclear phenomena.

Particle physics encompasses a wide range of energy and length scales. The smallest energies studied are represented by vacuum polarization corrections of order $10^{-9}$ eV in the electronic energy levels of hydrogen, and the $10^{-6}$ eV mass difference between the meson $K^0_{d\bar{s}}$ and its antiparticle the $\bar{K}^0_{s\bar{d}}$. Cosmic microwave background photons, which provide an image of the universe at fraction $3 \times 10^{-5}$ of its present age, have a spectrum of energies characterized by a temperature $T = 2.7$ K. Using the Boltzmann constant

$$k_B = 8.617\,343(15) \times 10^{-5}\text{ eV K}^{-1}, \tag{1.7}$$

we can estimate the typical relic photon energy $k_B T$ is about 1/4 meV. Neutrinos appear to have mass differences at this scale. Room temperature gaseous atoms have a thermal energy 100 times larger, about 1/38.68 eV. The energy scale of atomic and chemical physics, governed by the electron mass and charge, is much higher. To extract an electron from an atom requires an energy of order

$$1\text{ eV} \simeq 1.602 \times 10^{-19}\text{ J}.$$

Because this scale is much larger than the thermal scale under normal conditions, electric charges are generally bound in electrically neutral composites.

The residual color forces between nucleons result in a nuclear binding energy of about 1 MeV per nucleon, larger than the atomic scale by a factor of one million. This is the energy scale of nuclear physics. The raw color forces which bind quarks into color neutral nucleons are stronger still. To extract a quark from a nucleon requires an energy of several hundred MeV. We will denote this scale, governed by quantum chromodynamics, by

$$\Lambda_{QCD} = 0.2 \text{ GeV} \simeq m_\pi. \tag{1.8}$$

At this scale, the lightest strongly interacting quarks can materialize from energy in various forms and consequently hadronic physics is considerably more complex than atomic or nuclear physics. A further factor of one thousand brings us to the weak scale that we will denote by

$$v = 246 \text{ GeV}. \tag{1.9}$$

At the weak scale, the $W$ boson, $Z$ boson, and $H$ (if it exists) may be directly observed. Perhaps coincidentally, this weak scale is comparable to the rest energy $m_t c^2 = 171$ GeV of the heaviest known particle. It turns out that because the $W$ and $Z$ bosons are massive, the interactions in which they participate appear ephemeral for energies below the weak scale. Particle physics experiments today explore an energy scale of several TeV (1 TeV $= 10^{12}$ eV ) and may discover new particles and forces.

## 1.2 ■ STANDARD MODEL INTERACTIONS

The electromagnetic interaction is the prototype fundamental interaction in the standard model. In the quantum field theory we will study, the electromagnetic interactions of a charged particle derive from emission and absorption of photons and are represented by diagrams, invented by Richard Feynman, as shown in Figure 1.2. Feel free to combine such fundamental so-called first-order diagrams and to reverse the direction in time of a particle, interpreting it as an antiparticle. The diagrams shown in Figure 1.3 illustrate several important second-order quantum electrodynamic processes: (a) elastic scattering of two charged fermions by single photon exchange, (b) elastic scattering of a photon and an electron called Compton scattering, (c) particle-antiparticle fusion into a photon which decays to a particle-antiparticle pair of a different flavor, and (d) electron-positron annihilation into two photons.

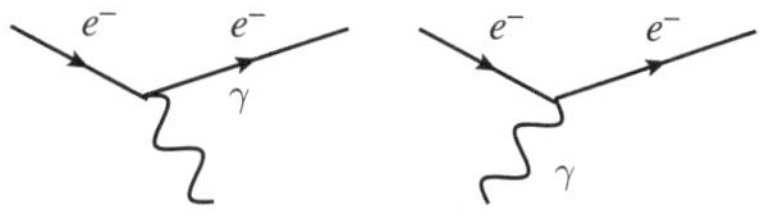


**FIGURE 1.2** Fundamental time-ordered diagrams for electrodynamics representing emission and absorption of a photon by an electron. Time increases to the right. Similar diagrams describe the electromagnetic interactions of other electrically charged particles.

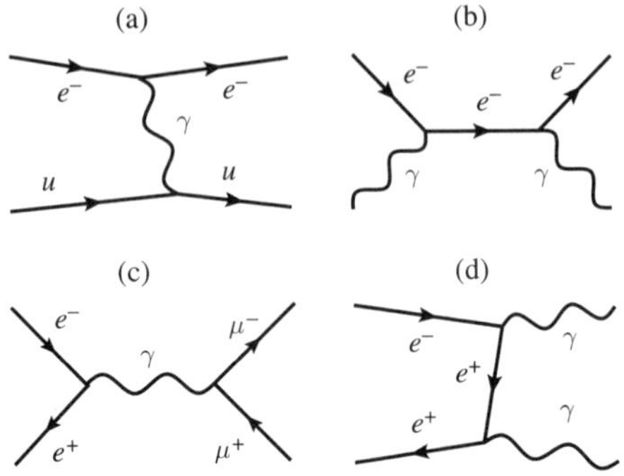


**FIGURE 1.3** Second order electromagnetic processes. Time increases to the right and arrows following charge backwards in time represent antiparticles. (a) Exchange of a single photon in $ue^- \to ue^-$ contributing to the binding and scattering of electrons and protons. (b) Compton scattering $\gamma e^- \to \gamma e^-$. (c) Annihilation of electron and positron to form a muon pair $e^+e^- \to \mu^+\mu^-$. (d) Annihilation of electron and positron resulting in two photons $e^+e^- \to \gamma\gamma$.

The complete set of first order standard model interactions appear in Figure 1.4 and some exemplary processes are listed in Table 1.3. Most of these processes entail a fermion radiating a gauge boson but others, such as $W^- \to W^-\gamma$ and $g \to gg$, are purely bosonic. Certain gauge bosons also couple to each other fundamentally four at time in second order but we will ignore these interactions for the moment.

A quick study of the fundamental processes reveals that only electrically charged particles interact with the photon. Neutral weak interactions mediated by the $Z$ boson affect both charged fermions and the electrically neutral neutrinos. Emission and absorption of a $W$ boson toggles the flavor of each lepton and quark flavor pair. While electrons participate in electromagnetic and weak interactions, neutrinos interact only weakly. Unlike the leptons, each quark flavor appears in three colors and undergoes color altering interactions mediated by gluons in addition to electromagnetic and weak interactions. Color interactions are much stronger than electromagnetic or weak interactions under normal conditions so dominate the physics of quarks.

The format of the table of quarks and leptons Table 1.1 reflects commonalities of electric and weak charges spanning the three families. Corresponding members of the three columns (also called generations) differ only in mass. Electric charge, weak charges, and color (quarks only) are common to each row. The pattern of fermions and the values of the masses are unexplained.

In particle physics, the notion of charge familiar from electromagnetism is generalized to include processes in which flavor or color are altered by boson emission. Only $W$ boson emission changes fermion flavor (called weak isospin) and it preserves lepton family ($e^- \to \nu_e W^-$) but links quark families. The weak interaction linking quarks of different family can be summarized succinctly by writing $d' \to uW^-$ where $d' = V_{ud}d + V_{us}d + V_{ub}b$ and $V$ is a unitary matrix

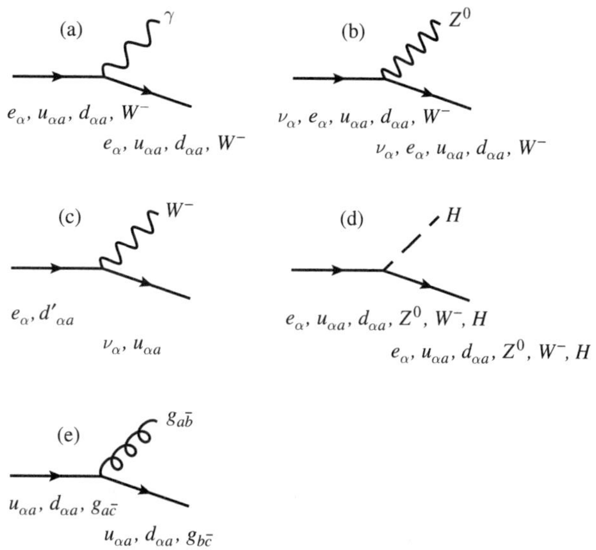


**FIGURE 1.4** Leading order standard model interactions. The index $\alpha = 1, 2, 3$ denotes the family, e.g. $\nu_2 = \nu_\mu$. The quark index $a = r, g, b$ denotes color. In each case, each particle in the initial state corresponds to a unique particle in the final state and the processes are sorted by the kind of boson emitted: (a) a photon, (b) a $Z$ boson, (c) a $W^-$, (d) a Higgs boson $H$, and (e) a gluon.

| | | | | |
|---|---|---|---|---|
| Electromagnetic | $e^- \to e^-\gamma$ | $u \to u\gamma$ | $d \to d\gamma$ | |
| Neutral weak | $\nu_e \to \nu_e Z$ | $e^- \to e^- Z$ | $u \to uZ$ | $d \to dZ$ |
| Charged weak | $e^- \to \nu_e W^-$ | $d' \to uW^-$ | | |
| Color | $u_r \to u_b g_{r\bar{b}}$ | $d_r \to d_b g_{r\bar{b}}$ | $g_{a\bar{c}} \to g_{a\bar{b}} g_{b\bar{c}}$ | |
| Higgs boson | $e \to eH$ | $u \to uH$ | $d \to dH$ | |
| Electroweak boson | $W^- \to W^-\gamma$ | $W^- \to W^- Z$ | $W^- \to W^- H$ | |

**TABLE 1.3** Standard model first order interactions. Exemplary electromagnetic, weak, and color interactions for first generation fermions. Some electroweak boson and Higgs boson interactions are also listed.

(determined empirically) called the Cabibbo-Kobayashi-Maskawa (CKM) matrix. The matrix element $V_{\alpha\beta}$ multiplies the weak charge in the transition $d_\beta \to u_\alpha W^-$. Otherwise, the charge governing $W^-$ emission is identical for all quarks and leptons. The magnitudes of the matrix elements are

$$|V_{\text{CKM}}| \equiv \begin{pmatrix} |V_{ud}| & |V_{us}| & |V_{ub}| \\ |V_{cd}| & |V_{cs}| & |V_{cb}| \\ |V_{td}| & |V_{ts}| & |V_{tb}| \end{pmatrix} = \begin{pmatrix} 0.97 & 0.22 & \sim 0.003 \\ 0.22 & 0.97 & \sim 0.04 \\ \sim 0.01 & 0.04 & 0.999 \end{pmatrix} \tag{1.10}$$

and the strength of the corresponding transitions are illustrated in Figure 1.5. In a similar vein, the collection of color changing gluon emission processes corresponds to a color charge common to all quarks and gluons multiplied by a matrix element $\lambda^c_{ab}$ for a process in which a quark of color $a$ emits a gluon of color $c$ and emerges with color $b$. Unlike the Cabibbo-Kobayashi-Maskawa, the $\lambda$ matrices are exactly predicted in standard model quantum chromodynamics. The form of these fundamental charge matrices will occupy us in Chapters 7 and 8. There are no weak interaction transitions like $\mu^- \rightarrow \nu_e W^-$ or $\mu^- \rightarrow e^- Z$ mixing lepton families so the total number of charged plus neutral leptons of each family is conserved but only the total number of quarks summed over families is conserved.

The still mysterious mass values of the fundamental fermions govern their stability. As we will see, a heavy particle will spontaneously decay to lighter particles through standard model interactions when such a decay is consistent with energy conservation. The decay rate is proportional to the square of the charges involved and to a factor called the phase space that we will study in chapter 4. The phase space factor is an increasing function of the energy released. Hence the $t$-quark, $b$-quark, $c$-quark, and $s$-quark all decay through $W$ emission to lighter quarks and ultimately to the $u$-quarks and $d$-quarks found in normal baryonic matter, while the $\mu^-$ decays to $\nu_\mu e^- \bar{\nu}_e$ and the $\tau^-$ decays to $\nu_\tau e^- \bar{\nu}_e$, $\nu_\tau \mu^- \bar{\nu}_\mu$, and $\nu_\tau d\bar{u}$ respectively. For this reason, matter in the universe today is composed of $u$-quarks, $d$-quarks, and electrons presumably immersed in a bath of weakly interacting invisible neutrinos.

In the standard model, the masses of the $W$ boson and $Z$ boson and also of the fermions derive from their interaction with a so-called vacuum condensate

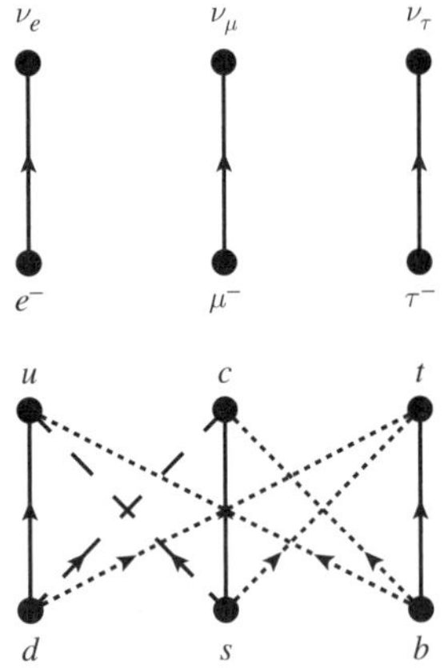


**FIGURE 1.5** Charged current transitions. Relative transition matrix elements corresponding to $W^-$ emission preserve lepton family and nearly preserve quark family. The amplitudes for quark transitions in decreasing order in magnitude are indicated by vertical solid arrows, diagonal arrows linking the first two families, and dotted arrows.

of the Higgs field. The mystery of the mass values is translated into a mystery concerning the origin of the strength of the interactions with the condensate. We haven't shown the static interaction with the condensate in our diagrams, only the coupling to the spin 0 Higgs field condensate wave excitation quantum $H$ called the Higgs boson. Interaction with the condensate is also responsible for intergeneration quark transitions $q_i \to q_j + W^-$ reflected in the Cabibbo-Kobayashi-Maskawa. Observation of the $H$ and elucidating the nature of the condensate are outstanding goals of present high energy experiments.

Some important second order weak processes are illustrated in Figure 1.6. The diagram in Figure 1.6 a) describes a neutrino scattering from a quark by $W$ boson exchange, a process which can convert the neutrino to a charged lepton. A $W$ boson exchange process like this is called a charged current process. Processes like $\nu_\mu e^- \to \nu_\mu e^-$ which take place through $Z$ boson exchange are called neutral current processes. The fusion of an electron and positron to form a $Z$ boson is shown in Figure 1.6 b). The Feynman diagram illustrates the decay $Z \to b\bar{b}$ but other decays such as $Z \to \mu^+\mu^-$ are possible. The diagram in Figure 1.6 c) illustrates the canonical weak interaction, neutron $\beta$-decay

$$n \to p + e^- + \bar{\nu}_e. \tag{1.11}$$

This decay derives from a quark weak interaction compounded with a leptonic weak interaction

$$d \to u + W^- \text{ plus } W^- \to e^- + \bar{\nu}_e \tag{1.12}$$

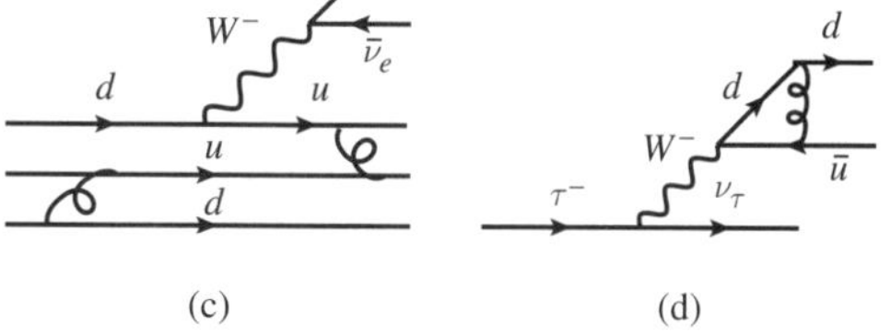

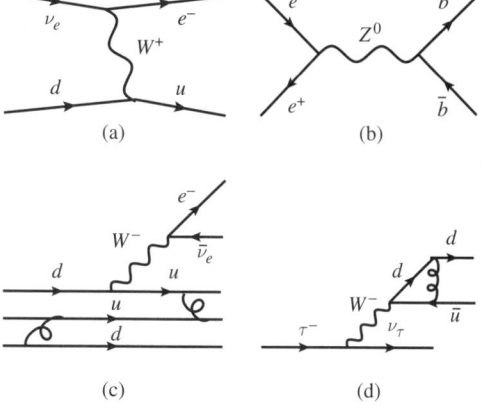

**FIGURE 1.6** Second order weak processes. (a) Neutrino quark scattering $\nu_e d \to e^- u$ through $W^+$ exchange in which a neutrino flavor converts to a charged lepton. (b) Electron and positron fusing to form a $Z$ boson which materializes as a $b\bar{b}$ pair. (c) Neutron $\beta$-decay $n \to pe^-\bar{\nu}_e$. (d) The decay $\tau^- \to \nu_\tau \pi^-_{d\bar{u}}$. In the last two diagrams, color interactions responsible for quark binding are indicated schematically.

with the $u$-quark remaining bound within the nucleon. The total energy released is about 1.3 MeV. This process and the related electron capture process

$$e^- + p \rightarrow n + \nu_e \tag{1.13}$$

are both critical in stabilizing atomic nuclei. A free neutron has a mean decay lifetime of about 15 minutes. The decay of a neutron in a nucleus is a source of so-called $\beta$ radioactivity leading to a more stable nucleus. Conservation of the energy of the entire system of interacting nucleons precludes neutron decay in a stable nucleus.

Neutron decay is but one example of charged current induced instability. In Figure 1.6 c), a diagram describing the weak decay of the $\tau^-$ to $\nu_\tau$ via $W^-$ emission is shown. The $W^-$ can decay to $e^-$ plus $\bar{\nu}_e$ as in neutron decay. The $\tau^-$ decay process shown indicates an alternate decay $W^- \rightarrow d\bar{u}$ and a schematic indication of the color interactions between the quarks which can lead to the formation of a $d\bar{u}$ bound state such as the $\pi^-$. The decay of the $\tau^-$ releases close to 1.6 GeV and its lifetime, about $0.3 \times 10^{-12}$ s, is much shorter than that for free neutron decay due to the larger energy release. Because $m_n < m_p + m_\pi$, a free neutron can not to a final state including a pion. Conservation of energy and of the number of quarks and leptons permits only $n \rightarrow pe^-\bar{\nu}_e$.

These examples may give you some sense of the power of the Feynman diagram language in describing fundamental processes. Given all fundamental diagrams and some considerations of energetics, you have the means to describe all possible particle processes in the standard model. Soon you will learn to calculate the rates for particle reactions precisely based on the standard model and relativistic kinematics. There is one more fact you should know. Weak interactions exhibit a polarization dependence not generally indicated by the diagrams. The $W^-$ couples only to a fermion with spin antiparallel to momentum and to an anti-fermion with spin parallel to momentum. (More precisely, the coupling constants are to left-handed fermions and right-handed anti-fermions as explained in more detail in Chapters 6 and 8.) The $Z$ boson couples unequally to the spin states of quarks and charged leptons and only to left-handed neutrinos. The two polarization states of a fermion have identical electric and color charges but different weak charges. It follows that right-handed neutrinos, if they exist, have no electroweak or color interactions. (They should still have gravitational interactions.) Mirror reflection symmetry and parity conservation (P) are violated in weak interactions. Symmetry under interchange of particles and antiparticles, called charge conjugation (C), and the associated conservation of charge conjugation parity are also violated. The origin of these fundamental asymmetries is mysterious.

## 1.3 ■ QUANTUM THEORY

Matter is governed by quantum mechanical principles welding the notions of wave and particle and the standard model emerged from much theoretical development culminating in quantum gauge field theory. In what follows, we review the origins of the quantum theory of matter and the theory of quantum electrodynamics. Along the way we will recall several concepts familiar from the study of nonrelativistic matter, introduce new notions important in relativistic quantum physics, attempt to de-mystify the language of Feynman diagrams, and provide a preview of standard model field equations and their quantum mechanical interpretation.

James Maxwell produced the first fundamental classical field theory and predicted electromagnetic waves moving at light speed. Maxwell's equations for the electric field **E** and magnetic field **B** in standard international units are

$$\begin{aligned}\nabla\cdot\mathbf{B} &= 0 \quad \nabla\times\mathbf{E}+\frac{\partial}{\partial t}\mathbf{B}=0\\ \nabla\cdot\mathbf{E} &= \epsilon_0^{-1}\rho \nabla\times\mathbf{B}-c^{-2}\frac{\partial}{\partial t}\mathbf{E}=\mu_0\mathbf{j}\end{aligned}\tag{1.14}$$

where $\rho$ and $\mathbf{j}$ are the electric charge and current densities. The two constants are

$$\mu_0 = 4\pi\times 10^{-7}\ \mathrm{N\,A^{-2}},\ \epsilon_0=\frac{1}{\mu_0 c^2}=8.854\ 187\ 817...\times 10^{-12}\ \mathrm{F\,m^{-1}}.\tag{1.15}$$

The homogeneous equations are identically satisfied by introducing scalar and vector potential fields $A^0$ and $\mathbf{A}$ defined by

$$\mathbf{E}=-\nabla A^0-\frac{\partial}{\partial t}\mathbf{A},\ \mathbf{B}=\nabla\times\mathbf{A}\tag{1.16}$$

up to a gauge transformation

$$A\equiv(A^0,\mathbf{A})\rightarrow\left(A^0+\frac{\partial}{\partial t}\chi,\mathbf{A}-\nabla\chi\right)\tag{1.17}$$

where $\chi(t,\mathbf{x})$ is an arbitrary scalar field. With the Lorenz gauge condition

$$\frac{\partial}{\partial t}A^0+\nabla\cdot\mathbf{A}=0,\tag{1.18}$$

the inhomogeneous Maxwell's equations reduce to the wave equations

$$\begin{aligned}\frac{1}{c^2}\frac{\partial^2}{\partial t^2}A^0-\nabla^2A^0&=\epsilon_0^{-1}\rho\\ \frac{1}{c^2}\frac{\partial^2}{\partial t^2}\mathbf{A}-\nabla^2\mathbf{A}&=\mu_0\mathbf{j}.\end{aligned}\tag{1.19}$$

Free electromagnetic plane waves have the form

$$\mathbf{A}\sim\boldsymbol{\epsilon}e^{-i(\omega t-\mathbf{k}\cdot\mathbf{x})}\tag{1.20}$$

where $\omega$ is the angular frequency, $\mathbf{k}$ is the wave vector, and $\boldsymbol{\epsilon}$ is a polarization vector. For wavelength $\lambda$, the magnitude of the wave vector is $|\mathbf{k}|=2\pi\lambda^{-1}$ and the angular frequency is given by $\omega=2\pi\nu=2\pi c\lambda^{-1}$. The invariance of light speed in Maxwell's equations under transformations between reference frames in relative motion, as verified by experiments, led Albert Einstein to the special theory of relativity and to relativistic mechanics according to which a free particle of mass $m$ and velocity $\mathbf{v}$ has energy and linear momentum given by

$$(E, \mathbf{p}) = (\gamma mc^2, \gamma m\mathbf{v}) \tag{1.21}$$

where $\gamma = (1 - \mathbf{v}^2/c^2)^{-1/2}$. These relationships are crucial in understanding elementary particle interactions.

Planck introduced the idea that electromagnetic radiation of frequency $\nu$ appears in quanta of energy

$$E_\gamma = h\nu = \hbar\omega \tag{1.22}$$

where h $= 6.626\,068\,96(33) \times 10^{-34}$ J s is Planck's constant and

$$\begin{aligned}\hbar = \frac{h}{2\pi} &= 1.054\,571\,628(53) \times 10^{-34}\ \text{J s}\\ &= 6.582\,118\,99(16) \times 10^{-22}\ \text{MeV s.}\end{aligned} \tag{1.23}$$

The kinematics of the photoelectric effect ($\gamma$ Atom $\rightarrow e^-$ Atom$^+$) and Compton scattering ($\gamma e^- \rightarrow \gamma e^-$) proved these quanta were massless particles (photons) of momentum

$$|\mathbf{p}|_\gamma = \hbar|\mathbf{k}| = \frac{h}{\lambda} = \frac{E_\gamma}{c}. \tag{1.24}$$

Subsequent experiments showed that the two circular polarization states of a classical electromagnetic wave correspond to photon spin angular momentum $\pm\hbar$ along the direction of motion.

De Broglie turned these notions around and suggested a massive particle of momentum $\mathbf{p}$ could be associated with a wave with wavelength related to momentum by

$$|\mathbf{p}| = \frac{h}{\lambda} \tag{1.25}$$

just as for photons. The de Broglie hypothesis was established by observations of diffraction of electrons with wavelength close to an atomic spacing ($\lambda \sim a_0$) from crystals. Subsequently Schrödinger developed a wave theory of nonrelativistic matter and a probability interpretation emerged. A free particle of mass $m$ is described by complex wave function $\psi$ satisfying the Schrödinger equation

$$i\hbar\frac{\partial}{\partial t}\psi = \frac{1}{2m}(-i\hbar\nabla)^2\psi \tag{1.26}$$

which is a transcription of $E = \mathbf{p}^2/2m$ with the operator equivalence

$$(E, \mathbf{p}) = \left(i\hbar\frac{\partial}{\partial t}, -i\hbar\nabla\right). \tag{1.27}$$

Plane wave solutions are

$$\psi = e^{-i(Et-\mathbf{p}\cdot\mathbf{x})/\hbar} \tag{1.28}$$

and the probability density and corresponding probability current density are

$$(\rho,\ \mathbf{j}) = \left(\psi^\dagger\psi,\ \psi^\dagger\left[\frac{i\hbar}{2m}\nabla\right]\psi + \left[\frac{i\hbar}{2m}\nabla\psi\right]^\dagger\psi\right). \tag{1.29}$$

The so-called minimal substitution

---

**Example 1.1. De Broglie wavelength of an electron.** Starting from rest, an electron accelerates in vacuum through an electrostatic potential difference $V$. For what value of $V$ will the electron wavelength be $\lambda = 10^{-10}$ m? Use $m_ec^2 = 0.511$ MeV and the expression $\hbar c = 197$ MeV fm.

Equating the kinetic energy gain to the potential energy difference times the charge and using the de Broglie hypothesis, we have

$$\begin{aligned}|q|V &= \frac{p^2}{2m_e} = \frac{1}{2m_ec^2}\left(\frac{2\pi\hbar c}{\lambda}\right)^2 \\ &= \frac{1}{2\times 0.511\text{ MeV}}\left(\frac{2\pi\times 197\text{ MeV fm}}{10^5\text{ fm}}\right)^2 = 150\text{ eV}\end{aligned} \tag{1.30}$$

or since $|q| = e$ the required voltage is 150 V.

---

$$(E,\mathbf{p}) \to \left(E - eA^0, \mathbf{p} - e\mathbf{A}\right) \tag{1.31}$$

in the free particle wave equation produces the Schrödinger equation (and current density) for a particle of charge $e$ interacting with an electromagnetic field

$$i\hbar\frac{\partial}{\partial t}\psi = \frac{1}{2m}\left(-i\hbar\nabla - e\mathbf{A}\right)^2\psi + eA^0\psi. \tag{1.32}$$

When the potential functions are subject to a gauge transformation, the wave function must be simultaneously transformed

$$\psi \to e^{-ie\chi}\psi \tag{1.33}$$

in order that the Schrödinger equation preserves its form. The electric charge density $e\rho$ and current density $e\mathbf{j}$ obtained through the minimal substitution are invariant under gauge transformation and serve in Maxwell's equations which also preserve their form. This covariance is called gauge symmetry.

Schrödinger's equation may be familiar in a special case. For an electron interacting with an electrostatic potential, the Schrödinger equation reduces to

$$i\hbar\frac{\partial}{\partial t}\psi = -\frac{\hbar^2}{2m_e}\nabla^2\psi + U\psi \equiv H\psi \tag{1.34}$$

where $U = eA^0$ is the potential energy and

$$H = \frac{\mathbf{p}^2}{2m} + U \tag{1.35}$$

is the Hamiltonian energy operator. The solutions describing bound and scattering electrons are the basis of nonrelativistic quantum mechanics in which the number of massive particles is conserved. Most notable are the bound state solutions for an electron in the Coulomb potential $A^0 = e(4\pi\epsilon_0 r)^{-1}$. The waves and energies can be expressed in spherical coordinates with integer indices $n > 0$, $0 \le l < n$, and $-l \le m \le +l$ as

$$\psi_{nlm}(t, \mathbf{x}) = R_{nl}(r)Y_{lm}(\theta, \phi)e^{-iE_nt/\hbar}, \; E_n = -\frac{E_0}{n^2}. \tag{1.36}$$

These solutions are eigenfunctions of the energy operator $H$, of the square of the orbital angular momentum operator

$$\mathbf{L} = \mathbf{x} \times \mathbf{p}, \tag{1.37}$$

and of its $z$-component $L_z$:

$$H\psi_{nlm} = E_n\psi_{nlm}, \; \mathbf{L}^2\psi_{nlm} = \hbar l(l+1)\psi_{nlm}, \; L_z\psi_{nlm} = \hbar m\psi_{nlm}. \tag{1.38}$$

Recall that the principal quantum number $n$ determines the number of nodes in the radial wave function $R_{nl}(r)$. The angular momentum quantum number $l$ is the number of nodes in $\theta$ in the spherical harmonic function $Y_{lm}$ and the orbital angular momentum has magnitude $L = \sqrt{l(l+1)}\hbar$. The magnetic quantum number $m$ counts the nodes in $\phi$ and is related to the $z$-component of orbital angular momentum by $L_z = m\hbar$. States with $l = 0$ are called $s$-wave and have non-vanishing wave function at the origin. States with $l = 1$ are called $p$-wave. The wave function of the lowest energy (ground) state in particular is

$$\psi_{100} = \frac{e^{-r/a_0}}{\sqrt{\pi a_0^3}} \tag{1.39}$$

where the Bohr radius, $a_0$, is

$$a_0 = \frac{4\pi\epsilon_0\hbar^2}{m_e e^2} = 0.529\,177\,208\,59(36) \times 10^{-10}\text{ m}. \tag{1.40}$$

The ground state energy is $-E_0$ where

$$E_0 = \frac{m_e e^4}{2(4\pi\epsilon_0)^2\hbar^2} = 13.605\,691\,93(34)\ \text{eV}. \tag{1.41}$$

For a charge $-e$ with mass $m_1$ bound to a charge $Ze$ with mass $m_2$, the electron mass $m_e$ in these expressions should be replaced by the reduced mass $\mu = m_1 m_2/(m_1 + m_2)$ and the quantity $e^2$ should be replaced by $Ze^2$.

The solutions describing the hydrogen atom exhibit very general features of quantum mechanics related to symmetry. A general multiparticle system is described by a wave function $\psi$ which is a solution to the abstract Schrödinger equation

$$i\hbar\frac{\partial}{\partial t}\psi = H\psi. \tag{1.42}$$

If $U$ is a time independent unitary transformation of states, then application of $U$ to both sides of the equation gives

$$i\hbar\frac{\partial}{\partial t}U\psi = UH\psi = UHU^{-1}U\psi. \tag{1.43}$$

If the Hamiltonian is unchanged by the transformation so that

$$UHU^{-1} = H \tag{1.44}$$

or $UH = HU$, then $U\psi$ is evidently also a solution. Put another way, $H$ evolves a transformed state into the transform of the evolved state. In particular, if $\psi$ is an eigenstate of $U$ so $U\psi = \lambda\psi$, we see that the eigenvalue $\lambda$ is unaltered by the time evolution and a conservation law is implied.

Consider now an energy eigenstate satisfying $H\psi = E\psi$. Then

$$HU\psi = EU\psi \tag{1.45}$$

so the states $U\psi$ obtained by transformation of any one energy eigenstate state $\psi$ have the same energy. The space of such states my be spanned by linearly independent states which form an energy degenerate "multiplet". If $U$ is a discrete symmetry like reflection in space satisfying $U^2 = 1$, any energy eigenstate and its transformed state form a pair of energy degenerate states

$$\psi_\pm = \frac{1}{\sqrt{2}}(\psi \pm U\psi) \tag{1.46}$$

with $U\psi_\pm = \pm\psi_\pm$. Suppose now $U$ is a member of a continuous group of transformations such as the rotation group, generated from the identity transformation by Hermitian generating transformations. If $H$ is unchanged by such transformations, the generating operators commute with $H$ so can be diagonalized along with $H$, the basis states being characterized by the eigenvalues of the generators as well as by energy. In the example of the hydrogen atom, symmetry under rotations generated by the angular momentum operators implies the energy eigenstates are multiplets of angular momentum.

The discrete transformation corresponding to space inversion $\mathbf{x} \to -\mathbf{x}$ (a mirror reflection followed by a rotation) is represented by the parity transformation $P$. Inversion corresponds to the change in spherical coordinates $\theta \to \pi - \theta$ and $\phi \to \phi + \pi$. The spherical harmonics satisfy

$$Y_{lm}(\pi - \theta, \phi + \pi) = (-1)^l Y_{lm}(\theta, \phi) \tag{1.47}$$

so the parity eigenvalue of $\psi_{nlm}$ is $(-1)^l$. The time reversal transformation, denoted by $T$, is represented by an anti-unitary transformation and $T$-symmetry relates the amplitudes for forward and reversed processes. In relativistic physics, a discrete internal symmetry called charge conjugation is important and it is denoted by $C$. Charge conjugation replaces particles by antiparticles and $C$-symmetry implies the energy eigenstates of a neutral multiparticle system have a conserved $C$-parity. In the standard model, we encounter other internal symmetries related to transformations of fermion flavor.

The wave mechanics of nonrelativistic Schrödinger theory blurs the classical notions of a particle properties. For example, a position coordinate $x$ of a particle can only be localized to a range $\Delta x$ if the wave function contains momentum components spanning a range $\Delta p_x$ limited by the uncertainty relation

$$\Delta p_x \Delta x \sim h. \tag{1.48}$$

In relativistic quantum mechanics, the notion of the coordinate is further blurred and no coordinate wave function may be consistently ascribed to a particle. The reason is ultimately that to localize a particle to within a range smaller than its Compton wavelength requires a probe particle with energy sufficient to create additional particles. The description of such an experiment demands a multiparticle approach from the start.

The creation and destruction of quantum particles in relativistic processes is described by quantum field theory. The quantum theory of a relativistic massive particle begins with the Klein-Gordon equation

$$\frac{1}{c^2}\frac{\partial^2}{\partial t^2}\phi - \nabla^2\phi + \left(\frac{m_\phi c}{\hbar}\right)^2 \phi = 0 \tag{1.49}$$

for a scalar field $\phi$. This equation follows from the relativistic relation between energy and momentum in the form

$$E^2 - \mathbf{p}^2 c^2 - m^2 c^4 \tag{1.50}$$

transcribed as an operator relation using Equation 1.27. Plane (de Broglie) wave solutions are

$$\phi \sim e^{\mp i(Et - \mathbf{p}\cdot\mathbf{x})/\hbar} \tag{1.51}$$

with $E = \sqrt{(\mathbf{p}c)^2 + (m_\phi c^2)^2}$ and it turns out these can be associated with emission and absorption of particles and antiparticles of mass $m_\phi$, relativistic energy $E$, and linear momentum $\mathbf{p}$.

An analogy is informative. The wave equation for the transverse displacement $y(t, z)$ of a string of mass density $\lambda$ and tension $T$ embedded in a restoring medium with spring constant per unit length $k$ is (see Figure 1.7)

$$\lambda \ddot{y} = T \frac{d^2 y}{dz^2} - ky. \tag{1.52}$$

If we write

$$y = \left(\frac{T}{\hbar c}\right)^{-1/2} \phi, \; c = \sqrt{T/\lambda}, \; k = T\left(\frac{m_\phi c}{\hbar}\right)^2, \tag{1.53}$$

we arrive at the Klein-Gordon equation in one space dimension. The energy quanta of wave excitations of such a mechanical system behave as bosons called phonons. The kinetic energy density $\lambda(\dot{y})^2/2$ and potential energy density $T(\partial_z y)^2/2 + ky^2/2$ may be generalized to three-dimensional space and the total energy density $\rho_E$ is then given by

$$(\hbar c)^{-1} \rho_E = \frac{1}{2}\left(\frac{1}{c}\frac{\partial \phi}{\partial t}\right)^2 + \frac{1}{2}(\nabla \phi)^2 + \frac{1}{2}\left(\frac{m_\phi c}{\hbar}\right)^2 \phi^2 \tag{1.54}$$

where $\phi$ has units of inverse length. That the effective mass for waves on a string is proportional to $\sqrt{k}$ and derives from an interaction with an external medium is a harbinger of the Higgs mechanism. In our mechanical model, the springs oppose displacement and try to reflect right-going waves as left-going waves and left-going waves as right-going waves resulting in wave dispersion. Similar effects appear in the propagation of light through a medium.

Extending the string analogy to displacements in two independent transverse dimensions, we can imagine two real relativistic scalar fields $\phi_1$ and $\phi_2$ with the

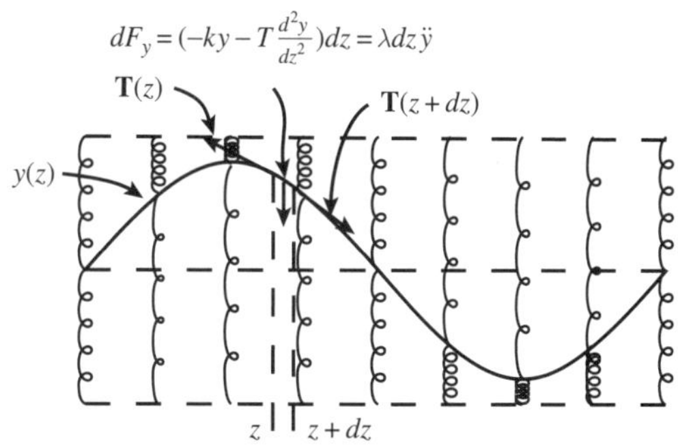


**FIGURE 1.7** Dynamics of a string. Waves on a string in a restoring medium satisfy the Klein-Gordon equation.

same mass corresponding to a homogeneous medium. These can be combined for convenience into a complex field

$$\psi = \frac{1}{\sqrt{2}}(\phi_1 + i\phi_2) \tag{1.55}$$

which represents a particle with a distinct antiparticle represented by the complex conjugate $\psi^\dagger$. In this case, the total energy density $\rho_E = \rho_1 + \rho_2$ is given by

$$(\hbar c)^{-1}\rho_E = \left|\frac{1}{c}\frac{\partial \psi}{\partial t}\right|^2 + |\nabla\psi|^2 + \left|\frac{mc}{\hbar}\right|^2 \psi^2. \tag{1.56}$$

With this definition, a normalized wave

$$\psi = \frac{(\hbar c)^{1/2}}{\sqrt{2E}} e^{-i(Et - \mathbf{p}\cdot\mathbf{x})/\hbar} \tag{1.57}$$

corresponds to one quantum of energy $E$ per unit volume.

A prototype model of interacting relativistic particles supposes a real field $\phi(t, \mathbf{x})$ and a complex field $\psi(t, \mathbf{x})$ and the coupled Klein-Gordon equations

$$\begin{aligned} \frac{1}{c^2}\frac{\partial^2}{\partial t^2}\phi - \nabla^2\phi &= -\tilde{m}_\phi^2\phi + g\psi^\dagger\psi = -\frac{\partial U}{\partial \phi} \\ \frac{1}{c^2}\frac{\partial^2}{\partial t^2}\psi - \nabla^2\psi &= -\tilde{m}_\psi^2\phi + g\phi\psi = -\frac{\partial U}{\partial \psi^\dagger}. \end{aligned} \tag{1.58}$$

Here $g$ is a coupling constant with dimensions of inverse length, and we write the Compton wavelengths as $\tilde{m}_\phi = m_\phi c/\hbar$ and $\tilde{m}_\psi = m_\psi c/\hbar$. As indicated, the terms on the right-hand-side postulated here may actually be derived from the potential energy density

$$U = \frac{1}{2}\tilde{m}_\phi^2\phi^2 + \tilde{m}_\psi^2\psi^\dagger\psi - g\psi^\dagger\psi\phi. \tag{1.59}$$

We will see in Chapter 6 that such wave equations are a generalization of Newton's second law

$$m\frac{d^2\mathbf{x}}{dt^2} = \mathbf{F} = -\frac{\partial U}{\partial \mathbf{x}} \tag{1.60}$$

to continuous variables such as $\phi(x)$ and $\psi(x)$.

The wave equation

$$\frac{1}{c^2}\frac{\partial^2}{\partial t^2}\psi - \nabla^2\psi + \tilde{m}_\psi^2\phi = g\phi\psi \tag{1.61}$$

is reminiscent of the Schrödinger equation for a charged particle with wave function $\psi$ and a potential energy function $g\phi$. The equation

$$\frac{1}{c^2}\frac{\partial^2}{\partial t^2}\phi - \nabla^2\phi + \tilde{m}_\phi^2\phi = g\psi^\dagger\psi \tag{1.62}$$

is reminiscent of the wave equation for the scalar potential in electrodynamics (Equation 1.19) with the addition of a mass term. The quantity $\rho = \psi^\dagger \psi$ may be interpreted as a number density and $g\rho$ as a charge density.

For a static point charge, the spherically symmetric solution to Equation 1.62 analogous to the Coulomb potential is the Yukawa potential

$$\phi = \frac{\mathrm{g}}{4\pi} \frac{\mathrm{e}^{-\tilde{\mathrm{m}}_\phi \mathrm{r}}}{\mathrm{r}}. \tag{1.63}$$

with an effective range equal to the Compton wavelength $\tilde{m}_\phi^{-1}$. A time dependent charge density may be shown to radiate $\phi$ field quanta if its frequency exceeds $m_\phi c^2 \hbar^{-1}$. Nuclear forces have an effective range $r \simeq 1.5$ fm and this motivated H. Yukawa to predict a quantum of a putative strong force field between nucleons with mass

$$m \simeq \frac{\hbar c}{r} = \frac{197 \text{ MeV fm}}{1.5 \text{ fm}} \simeq 130 \text{ MeV} \tag{1.64}$$

subsequently identified with the pion. The pion field indeed provides an approximate representation of the long range portion of nuclear interactions. Our model also captures an essential aspect of weak interactions which have a range $\hbar c/m_W \simeq \hbar c/m_Z$ much smaller than the radius of a nucleon.

---

**Example 1.2. Range of weak interactions.** Assuming a static weak $Z$ boson field is described by a Yukawa potential, estimate the range $\lambda = \hbar/m_Z c$ of the static weak interaction in fm. Use $m_Z$ mass in GeV and the expression $\hbar c = 0.197$ GeV-fm.

Assuming the weak analog of the electrostatic potential is suppressed by the exponential factor with characteristic length equal to the Compton wavelength of the $Z$ boson, we estimate a range $R$ as

$$R = \lambda_C = \frac{\hbar}{m_Z c} = \frac{\hbar c}{m_Z c^2} \simeq \frac{0.2 \text{ GeV fm}}{90 \text{ GeV}} = (1/450) \text{ fm}. \tag{1.65}$$

---

The quantum theory of fermions introduces several additional ideas. Foremost, any single electron appears in two states associated with intrinsic angular momentum (spin) of magnitude $\hbar/2$. A nonrelativistic spin 1/2 fermion of charge $e$ and mass $m$ is described by a 2-component complex spinor wave function

$$\psi = \begin{pmatrix} \psi_1 \\ \psi_2 \end{pmatrix} \tag{1.66}$$

governed by the Pauli equation

$$i\hbar \frac{\partial}{\partial t} \psi = \frac{[\boldsymbol{\sigma} \cdot (-i\hbar \nabla - e\mathbf{A})]^2}{2m} \psi + eA^0 \psi \tag{1.67}$$

where $\boldsymbol{\sigma}$ denotes a 3-vector of two-by-two Pauli matrices such as

$$\sigma_1 = \begin{pmatrix} 0 & 1 \\ 1 & 0 \end{pmatrix}, \; \sigma_2 = \begin{pmatrix} 0 & -i \\ i & 0 \end{pmatrix}, \; \sigma_3 = \begin{pmatrix} 1 & 0 \\ 0 & -1 \end{pmatrix} \tag{1.68}$$

that satisfy the algebraic conditions

$$\sigma_i \sigma_j = \delta_{ij} + i\epsilon_{ijk}\sigma_k \; ; \; \sigma_i^\dagger = \sigma_i \; ; \; \mathrm{tr}[\sigma_i] = 0. \tag{1.69}$$

Here, $i$ when not a subscript denotes $\sqrt{-1}$. Repeated indices are implicitly summed and Kroeneker's $\delta$ is defined as

$$\delta_{ij} = 1 \; (i = j), \; \delta_{ij} = 0 \; (i \neq j) \tag{1.70}$$

while the Levi-Civita symbol is defined as

$$\begin{aligned} \epsilon_{ijk} &= +1 \text{ if } (i, j, k = (1, 2, 3), (3, 1, 2), (2, 3, 1) \\ &= -1 \text{ if } (i, j, k = (1, 3, 2), (3, 2, 1), (2, 1, 3) \\ &= 0 \text{ otherwise.} \end{aligned} \tag{1.71}$$

Also, the symbol $\dagger$ denotes the Hermitian conjugate of a matrix meaning the transpose of the complex conjugate, and tr denotes the matrix trace. Introducing Dirac notation for the normalized eigenstates $\phi_\pm$ of $\sigma_z$, we write

$$|\uparrow> = \phi_+ = \begin{pmatrix} 1 \\ 0 \end{pmatrix}, \; |\downarrow> = \phi_- = \begin{pmatrix} 0 \\ 1 \end{pmatrix} \tag{1.72}$$

and a general spinor as

$$|\psi> = \psi_1 |\uparrow> + \psi_2 |\downarrow> . \tag{1.73}$$

The inner product of spinors is defined as a matrix multiplication

$$<\psi|\phi> \equiv \psi^\dagger \phi \tag{1.74}$$

and therefore we can write

$$<\psi|\psi> = (\,\psi_1^* \quad \psi_2^*\,) \begin{pmatrix} \psi_1 \\ \psi_2 \end{pmatrix} = |\psi_1|^2 + |\psi_2|^2. \tag{1.75}$$

The fields $\psi_1$ and $\psi_2$ are the probability amplitudes for the two spin states and, for a single conserved particle, the integral of the probability density $\rho = |\psi_1|^2 + |\psi_2|^2$ must be normalized to one.

Electron spin states are analogous to photon circular polarization states. However a photon carries spin angular momentum of magnitude $\hbar$, a fact represented mathematically by its association with a vector field, while an electron carries spin $\hbar/2$. Under a rotation of coordinates by an angle $\theta$ about a direction $\mathbf{n}$, the components of a spinor field are transformed according to the rule

$$\psi = \begin{pmatrix} \psi_1 \\ \psi_2 \end{pmatrix} \rightarrow \psi = e^{i\theta \mathbf{n} \cdot \mathbf{j}} \begin{pmatrix} \psi_1 \\ \psi_2 \end{pmatrix} \tag{1.76}$$

where the operator $\mathbf{j} = \mathbf{r} \times \mathbf{p} + \mathbf{s}$ is identified with total angular momentum and

$$\mathbf{s} = \frac{\hbar}{2}\boldsymbol{\sigma} \tag{1.77}$$

plays the role of an intrinsic angular momentum vector operator, the internal angular momentum direction represented by the orientation of the spinor in complex space. For an active rotation, the sign of $\theta$ is reversed. Consider for example a spinor that is initially a spin "up" eigenstate of $\sigma_z$. To actively rotate this spinor about the $y$-axis by a (polar) angle $\theta$, we use the identity

$$\begin{aligned} e^{ia\sigma_j} &\equiv 1 + (ia\sigma_j) + \frac{(ia\sigma_j)^2}{2!} + \ldots = \left(1 - \frac{a^2}{2} + \ldots\right) + i\sigma_j\left(a - \frac{a^3}{3!} + \ldots\right. \\ &= \cos a + i\sigma_j \sin a \end{aligned} \tag{1.78}$$

which follows from $\sigma_j^2 = 1$ and the series expansions for the exponential and trigonometric functions. Hence

$$\begin{aligned} e^{-i\frac{\theta}{2}\sigma_y}\begin{pmatrix}1\\0\end{pmatrix} &= \left(\cos\frac{\theta}{2} - i\sin\frac{\theta}{2}\sigma_y\right)\begin{pmatrix}1\\0\end{pmatrix} = \begin{pmatrix}\cos\frac{\theta}{2} & -\sin\frac{\theta}{2}\\ \sin\frac{\theta}{2} & \cos\frac{\theta}{2}\end{pmatrix}\begin{pmatrix}1\\0\end{pmatrix} \\ &= \begin{pmatrix}\cos\frac{\theta}{2}\\ \sin\frac{\theta}{2}\end{pmatrix}. \end{aligned} \tag{1.79}$$

An expanded form of Pauli's equation is

$$i\hbar\frac{\partial}{\partial t}\psi = \left[\frac{(-i\hbar\nabla - e\mathbf{A})^2}{2m} + eA^0\right]\psi - \boldsymbol{\mu}\cdot\mathbf{B}\psi \tag{1.80}$$

with $\mathbf{B} = \nabla \times \mathbf{A}$ and

$$\boldsymbol{\mu} = \frac{\mathrm{e}\hbar}{2\mathrm{m}}\boldsymbol{\sigma} = \frac{e}{m}\mathbf{s} \tag{1.81}$$

an intrinsic magnetic moment vector operator. The magnitude of the Pauli magnetic moment for an electron is called the Bohr magneton:

$$|\boldsymbol{\mu}_e| = \mu_B = \frac{e\hbar}{2m_e} = 5.788\,381\,7555(79) \times 10^{-11}\ \mathrm{MeV\ T^{-1}}. \tag{1.82}$$

The muon Pauli magnetic moment is a factor $m_e/m_\mu$ smaller.

In the absence of a magnetic field, the Pauli equation reduces to decoupled Schrödinger equations for $\psi_1$ and $\psi_2$ implying that the spin state is conserved. In light atoms, the relative velocities of the electrons and nuclei are small compared to $c$. Hence the magnetic fields due to motion $\mathbf{B} = q\mathbf{v} \times \mathbf{E}/c$ of the electrons and nuclei are small compared to their electric fields and the magnetic forces $\mathbf{F}_B = q\mathbf{v} \times \mathbf{B}$ are small compared to the electric forces. The spin and orbital angular momentum are, however, weakly correlated through the magnetic fields associated with orbital motion and magnetic moments and these correlations are approximately described by the Pauli equation.

By way of illustration of the spin degree of freedom, consider an electron or muon in uniform constant magnetic field. We can find solutions to Pauli's equation of the form

$$\psi(t, \mathbf{x}) = \psi_S(t, \mathbf{x})\psi_s(t) \tag{1.83}$$

where $\psi_S$ is a solution to the Schrödinger Equation (1.32) with $\int d\mathbf{x}\, |\psi_S|^2 = 1$ and

$$\psi_s(t) = \begin{pmatrix} a(t) \\ b(t) \end{pmatrix} = a(t)\begin{pmatrix} 1 \\ 0 \end{pmatrix} + b(t)\begin{pmatrix} 0 \\ 1 \end{pmatrix} \tag{1.84}$$

is a normalized spinor with $|a|^2 + |b|^2 = 1$. By substitution we find the equation of motion for the spinor

$$-i\frac{\partial}{\partial t}\psi_s = -\boldsymbol{\mu} \cdot \mathbf{B}\psi_s \tag{1.85}$$

and taking the $z$ axis along $\mathbf{B}$ we find the general solution

$$\psi_s = a_0 e^{-i\mu_B Bt}\begin{pmatrix} 1 \\ 0 \end{pmatrix} + b_0 e^{+i\mu_B Bt}\begin{pmatrix} 0 \\ 1 \end{pmatrix}. \tag{1.86}$$

The two spin states have energy $E = \pm\mu_B B$ and analysis reveals the average spin vector defined as

$$< \mathbf{s} > = \psi_s^\dagger \frac{\boldsymbol{\sigma}}{2}\psi_s \tag{1.87}$$

precesses around the direction of $\mathbf{B}$ as would the angular momentum of a classical system with magnetic moment vector along its internal angular momentum.

Spin precession is used to determine the magnetic moments of stable particles with high precision. In Chapter 6, we will derive relativistic corrections to the Pauli equation including the so-called spin-orbit interaction based on the Dirac theory and in Chapter 9 additional corrections due to multiparticle effects. These corrections explain the deviation of the effective magnetic moment of the electron from the Pauli value. The deviation is conventionally expressed by writing the magnetic moment of the electron as $\mu_e = (g_e/2)\mu_B$ where the value of the "g-factor" is given by $g_e/2 \simeq 1.00116$.

The next idea concerns the properties of identical particles. Two free particles are described in nonrelativistic quantum mechanics by a joint wave $\psi(t, \mathbf{x}_1, \mathbf{x}_2)$ that can be represented as a superposition of product states

$$\psi = \psi_a(1)\psi_b(2) \tag{1.88}$$

where $\psi_a(1) = \psi_a(\mathbf{x}_1)\eta_a(1)$ describes particle 1 in a state $a$ and $\psi_b(2) = \psi_b(\mathbf{x}_2)\eta_b(2)$ describes particle 2 in a state $b$. The wave functions of coexisting identical electrons are found to be always antisymmetric under exchange of particles. The electron is said to be a fermion. In contrast, the wave functions of coexisting photons are symmetric under interchange. The photon is said to be a boson. Generally, half integral spin objects are fermions and integral spin objects are bosons. For two identical particles corresponding to coordinates $\mathbf{x}_1$ and $\mathbf{x}_2$ and two states $\psi_a$ and $\psi_b$, the wave function must have the form

$$\psi(\mathbf{x}_1, \mathbf{x}_2) = \frac{1}{\sqrt{2}}\left[\psi_a(1)\psi_b(2) \mp \psi_b(1)\psi_a(2)\right] \tag{1.89}$$

where the negative sign applies to fermions and the positive sign to bosons. For fermions, the two states must be distinct or the wave function would vanish. This exclusion principle formulated by Pauli is essential to the electronic structure of atoms. Identical particle exchange symmetry must be imposed as an external constraint in nonrelativistic quantum mechanics. In quantum field theory, identical particles emerge naturally as quantum excitations of an underlying common quantum field and, as discussed in Chapter 4, the two kinds of exchange symmetry emerge from properties of the fundamental fields.

As an example of the Pauli exclusion principle, consider two electrons described by a joint wave function $\psi = \psi_S(t, \mathbf{x}_1, \mathbf{x}_2)\psi_s$ where the joint spin wave function $\psi_s$ transforms under rotations as

$$\psi_s \rightarrow e^{i\frac{\theta}{2}(\boldsymbol{\sigma}_1+\boldsymbol{\sigma}_2)\cdot\mathbf{n}}\psi_s. \tag{1.90}$$

In this expression, $\boldsymbol{\sigma}_1$ operates only on the wave function $\psi_1$ for the first particle and $\boldsymbol{\sigma}_2$ operates only on the wave function $\psi_2$ for the second particle in any product wave function $\psi_s$. The four outer products of the basis states for the two electrons form a basis for the spin states of the pair. More useful are groups of eigenstates of total spin $\mathbf{s} = \mathbf{s}_1 + \mathbf{s}_2$ that are transformed amongst themselves, in this case,

$$\begin{aligned}
|\mathrm{s} = 1, \mathrm{s_z} = +1 > &= |\uparrow\uparrow> \\
|\mathrm{s} = 1, \mathrm{s_z} = \ \ 0 > &= (|\uparrow\downarrow> + |\downarrow\uparrow>)/\sqrt{2} \\
|\mathrm{s} = 1, \mathrm{s_z} = -1 > &= |\downarrow\downarrow> \\
|\mathrm{s} = 0, \mathrm{s_z} = \ \ 0 > &= (|\uparrow\downarrow> - |\downarrow\uparrow>)/\sqrt{2}
\end{aligned} \tag{1.91}$$

The triplet of $s = 1$ states must have the same energy (be "degenerate") when $\mathbf{s}$ is conserved since then $\mathbf{s}^2$ commutes with the Hamiltonian. Being symmetric, a triplet spin wave function must accompany an antisymmetric space wave function. The singlet $s = 0$ state is antisymmetric and must accompany a symmetric wave function.

A relativistic electron is described by a relativistic spinor field obeying Dirac's generalization of the Pauli equation. Dirac's description of a free fermion includes a pair of 2-component spinors $\psi_L$ and $\psi_R$ called left-handed and right-handed coupled via the equations

$$i\hbar\frac{\partial}{\partial t}\psi_L - i\hbar c\boldsymbol{\sigma}\cdot\nabla\psi_L = mc^2\psi_R \tag{1.92}$$

$$i\hbar\frac{\partial}{\partial t}\psi_R + i\hbar c\boldsymbol{\sigma}\cdot\nabla\psi_R = mc^2\psi_L. \tag{1.93}$$

In case $m = 0$, these two equations decouple and, with $\mathbf{p}$ along the $z$-direction and $E = |\mathbf{p}|c$, the solutions are

$$\psi_L = u_L e^{-i(Et-\mathbf{p}\cdot\mathbf{x})/\hbar}\ ;\ u_L = \begin{pmatrix}0\\1\end{pmatrix} \tag{1.94}$$

$$\psi_R = u_R e^{-i(Et-\mathbf{p}\cdot\mathbf{x})/\hbar}\ ;\ u_R = \begin{pmatrix}1\\0\end{pmatrix}. \tag{1.95}$$

The solutions for $\mathbf{p}$ directed along an arbitrary direction may be found by rotation using Equation (1.76). These spinor amplitudes $u_L$ and $u_R$ represent respectively spin anti-parallel and parallel to the momentum. There are also solutions with $E = -|\mathbf{p}|c$ associated with antiparticles. When $m \neq 0$, it is customary to introduce a 4-component Dirac spinor

$$\psi = \begin{pmatrix}\phi_R\\\phi_L\end{pmatrix} \tag{1.96}$$

and write one Dirac equation in the Hamiltonian form

$$i\hbar\frac{\partial}{\partial t}\psi = \left(-i\hbar\nabla\cdot\boldsymbol{\alpha} + mc^2\alpha^0\right)\psi \equiv H_0\psi \tag{1.97}$$

where $(\alpha^0, \boldsymbol{\alpha})$ are four $4\times 4$ matrices. The relativistic charge and current densities turn out to be

$$j = e\left(\psi^\dagger\psi, \psi^\dagger\boldsymbol{\alpha}\psi\right) = e\left(\psi_R^\dagger\psi_R + \psi_L^\dagger\psi_L, \psi_R^\dagger\boldsymbol{\sigma}\psi_R - \psi_L^\dagger\boldsymbol{\sigma}\psi_L\right). \tag{1.98}$$

In Chapter 6, we will study the forms of the Dirac equation, its free particle solutions, and the interpretation of the current density in detail.

The Dirac equation for a fermion of charge $e$ interacting with an electromagnetic field is obtained by the gauge symmetry preserving operator substitutions $(E, \mathbf{p}) \rightarrow (E - eA^0, \mathbf{p} - e\mathbf{A})$. When the (gauge invariant) Dirac current density is used in Maxwell's equations, the fundamental equations of relativistic quantum electrodynamics maybe written as

$$\begin{aligned} i\hbar\frac{\partial}{\partial t}\psi - H_0\psi &= \left(eA^0 - e\boldsymbol{\alpha}\cdot\mathbf{A}\right)\psi \\ \frac{1}{c^2}\frac{\partial^2}{\partial t^2}A - \nabla^2 A &= e\left(\psi^\dagger\psi, \psi^\dagger\boldsymbol{\alpha}\psi\right) \end{aligned} \tag{1.99}$$

where $A = (A^0, \mathbf{A})$. An ultra-relativistic left-handed electron may be more simply described by a 2-component spinor field $e_L$ and the equations

$$\begin{aligned} i\hbar\frac{\partial}{\partial t}e_L - i\hbar c\boldsymbol{\sigma}\cdot\nabla e_L &= eA^0 e_L + ec\boldsymbol{\sigma}\cdot\mathbf{A}e_L \\ \frac{1}{c^2}\frac{\partial^2}{\partial t^2}A - \nabla^2 A &= e\left(e_L^\dagger e_L, -e_L^\dagger\boldsymbol{\sigma}e_L\right). \end{aligned} \tag{1.100}$$

The equations of quantum electrodynamics are the prototype for all standard model interactions of fermions with gauge bosons. The neutral current weak interaction is described by replacing the 4-vector field $A$ with a massive 4-vector field $Z$ and the electric charge $e$ with a weak charge $g_Z$ (different for $L$ and $R$ fermions). The charged current weak interaction is described by a complex 4-vector field $W$ with charge $g_W$ for $L$ fermions ($g_W = 0$ for $R$). For example, the weak coupling of the $e_L$ with a neutrino field $\nu_L$ in the limit $m_e = 0$ and $m_\nu = 0$ corresponds to the equations

$$\begin{aligned} i\hbar\frac{\partial}{\partial t}e_L - i\hbar c\boldsymbol{\sigma}\cdot\nabla e_L &= g_W W^0\nu_L + g_W c\boldsymbol{\sigma}\cdot\mathbf{W}\nu_L \\ \frac{1}{c^2}\frac{\partial^2}{\partial t^2}W - \nabla^2 W + \tilde{m}_W^2 W &= e\left(e_L^\dagger\nu_L\,,\, -e_L^\dagger\boldsymbol{\sigma}\nu_L\right). \end{aligned} \tag{1.101}$$

The field equations of chromodynamics are similar to those of the charged weak interaction with massless gluon fields $g_{a\bar{b}}$ coupling quark fields like $u_{L,a}$ and $u_{L,b}$ of the same flavor but different color. (Like electromagnetic interactions, color interactions have the same charge for $L$ and $R$).

While relativistic wave equations, especially the Dirac equation, appear to be simple generalizations of the Schrödinger equation, they can not generally be interpreted as describing a single conserved particle. Instead, they describe a field of identical particles. In quantum field theory, the field equations are interpreted as operator relations. We will use a simplified approach to quantum field theory, solving the field equations by successive approximation with a somewhat *ad hoc* quantum interpretation applicable to processes involving just a few particles. The idea can be illustrated with the equation for $A^0$ in quantum electrodynamics

$$\frac{1}{c^2}\frac{\partial^2}{\partial t^2}A^0 - \nabla^2 A^0 = e\psi^\dagger\psi. \tag{1.102}$$

Assume the product of a given normalized initial wave $\psi_1$ and a given normalized final wave $\psi_2$ may be used as an approximate (transition) charge density $e\psi_2^\dagger\psi_1$, fixed on the right-hand-side. The amplitude of the wave $A^0$ radiated by this source is proportional to the charge $e$ and may be found by standard means and the amplitude of a normalized photon wave component of this solution deduced. This number is interpreted as the quantum mechanical transition amplitude and its square governs the probability for radiation of one photon by the electron in making a transition from state $\psi_1$ to $\psi_2$.

The first order radiation process just described is captured by the diagram shown in Figure 1.8 (a). Alternatively, the product of an initial electron wave $\psi_1$ and a photon wave $A^0$ may be used to approximate the terms in Dirac's equation linear in the charge. These terms can be interpreted as a source for a scattered wave $\psi_2$ corresponding to a final electron state. The same diagram describing an initial and final fermion coupled to one photon applies with the photon absorbed rather than emitted as shown in Figure 1.8 (b). These two diagrams are just those in Figure 1.2.

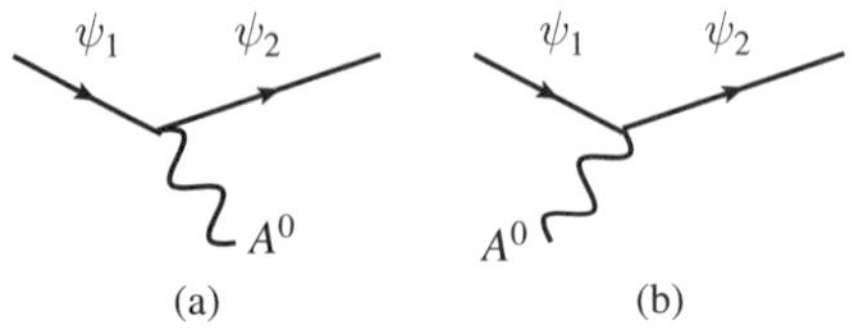


**FIGURE 1.8** Diagrams representing the solution to the equations of electrodynamics. (a) Emission of a radiation field time component $A^0$. (b) Absorption of a radiation field component $A^0$. In both cases, a charged particle makes a transition from state $\psi_1$ to state $\psi_2$.

The process of finding approximate solutions may be iterated. For example, given the approximate electromagnetic field associated with a transition of a muon from state $\psi_1$ to state $\psi_2$, we may use Dirac's equation to find an approximate solution describing the scattering of an electron in this field. The corresponding diagram appears in Figure 1.3 (a). In quantum language, one summarizes by saying the electron and muon have exchanged a quantum of electromagnetic energy, a photon, although, considered in detail, even this simple process entails an interaction of several multicomponent fields associated with electric and magnetic energy and intrinsic angular momentum. The iteration method leads to computations which can be simply codified in a set of rules associated with compound Feynman diagrams. The computations and rules must of course include all field components to represent the spin polarization states of the particles. The mathematical implementation of these ideas will be developed in detail later where we learn and apply gauge symmetry principles to specify the field equations of the standard model.

The amplitude for a particle to emit or absorb a photon, and in so doing make a transition from one state to another, is proportional to the charge of the particle. The transition probability is proportional to the square of the amplitude, so proportional to the electromagnetic fine structure constant

$$\alpha(0) = \frac{e^2}{4\pi\epsilon_0\hbar c} = 7.297\,352\,5376(50) \times 10^{-3} = 1/137.035\,999\,679(94). \tag{1.103}$$

The electromagnetic fine structure constant is a dimensionless number. Dimensionless coupling constants emerge naturally in gauge theories. We will see that standard model gauge theories predict relationships between electroweak coupling constants and between color coupling constants but do not specify the overall strength of such constants. It is an experimental fact that electromagnetic and weak coupling constants are small in magnitude at low energy. In contrast, color interactions are considerably stronger at achievable energies and are governed by the color interaction analog of the fine structure constant (pronounced "alpha ess")

$$\alpha_s(m_Z) = 0.1176(20). \tag{1.104}$$

Here the value for $\alpha_s$ is for center of mass energy near the weak scale.

From the preceding sketch of the theory of particle interactions, it should be clear that each diagrammatic vertex corresponding to an electromagnetic interaction contributes a factor $\alpha^{1/2}$ to the amplitude and increasingly complex diagrams connecting the same initial and final state represent a series expansion of a transition amplitude in powers of the coupling constants. Amplitudes corresponding to topologically distinct diagrams leading from the same initial state to the same final state must generally be summed to arrive at the complete amplitude.

For example, Figure 1.9 (a) shows the diagram for the leading order amplitude for $e^+e^- \to W^+W^-$ through an intermediate state $Z$ boson. Amplitudes described by diagrams with the $Z$ replaced by a photon or $H$ should be added to obtain the complete amplitude to second order in all couplings. Figure 1.9 (b) and (c) show diagrams corresponding to final state radiation amplitudes contributing to the distinct process $e^+e^- \to W^+W^-\gamma$. Initial state radiation amplitudes not shown will contribute to the total amplitude. The additional electromagnetic vertex indicated by a dot in these diagrams implies a factor $\alpha^{1/2}$. We can estimate that the rate for $e^+e^- \to W^+W^-\gamma$ is roughly a factor $\alpha$ lower than the rate for the distinct process $e^+e^- \to W^+W^-$, i.e. about one percent of $W^+W^-$ pairs will be accompanied by a photon. The diagram in Figure 1.9 (d) describes a leading order amplitude for $e^+e^- \to c\bar{c}$. The diagrams in Figure 1.9 (e) and (f) describe amplitudes including with additional internal bosons and fermions, amplitudes which must be added to that represented by (d) since the initial and final states are the same. These amplitudes have an additional factor of $\alpha$ from the two internal vertices.

The imaginative diagram in Figure 1.10 illustrates how complex a particle can appear when observed in detail. We concentrate on the simplest leading order descriptions of particle interactions. In Chapter 9 we consider some implications of higher order processes. But it must be pointed out up front that, when higher order processes are included, the effective coupling strength for any interaction is a function of process center of mass energy. In fact, at the weak energy scale $v \sim 250$ GeV, one finds the effective electromagnetic fine structure constant $\alpha(v) \sim 1/128$. For energies at the scale $\sim \Lambda_{QCD}$, the value is $\alpha_s(\Lambda_{QCD}) \simeq 1$. That $\alpha_s$ decreases

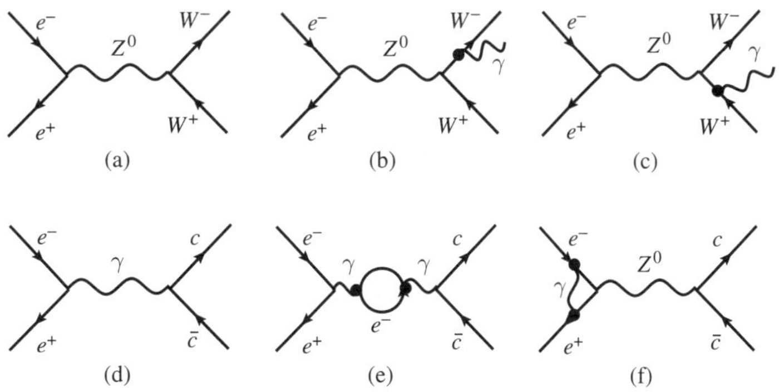


**FIGURE 1.9** Diagrams in electroweak physics.

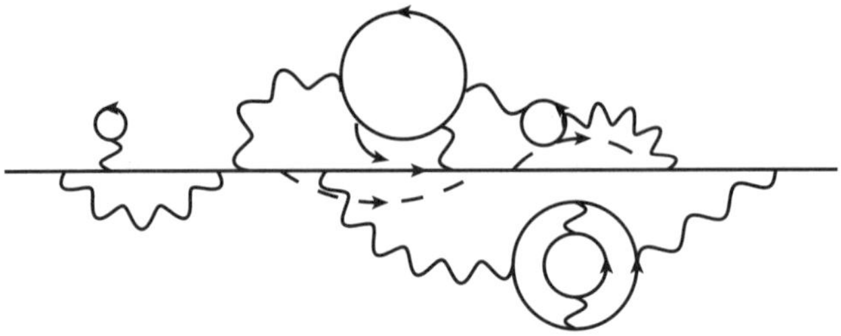

**FIGURE 1.10** Virtual particle cloud. High order diagrams may be said to represent a cloud of virtual particles potentially probed by an external particle. The energy dependence of such high order amplitudes renders the effective mass, charge, magnetic moment of a particle energy dependent.

when energy increases derives from color exchange by virtual gluons and has many ramifications. In particular, it implies that quarks and gluons behave increasingly like free particles in scattering experiments as the energy increases. Conversely, an expansion in couplings makes sense only if the couplings are small and, for strong interactions, will be reliable only at energies much higher than $\Lambda_{QCD}$. For lower energies, although color interaction diagrams provide a guide, the simplest diagrams can not serve as a foundation for calculations.

The complex array of particles, the unfamiliar concepts, and the novel mathematics introduced above are perhaps intimidating but with some experience will become second nature. As we familiarize ourselves with the theory and applications, it is important to bear in mind the scope of what is being described. The equations pertain over length scales from the astronomical to the smallest yet probed – about $10^{-19}$ m. A few essential fundamental constants like the mass and charge of the electron in the context of the standard model interactions are responsible for all of the structure observed within this range. In this limited sense, the standard model is a theory of everything. That the dynamics may be reduced to a few essential processes is simply astonishing!

That said, the universal gravitational interaction between particles is outside the scope of the standard model. Einstein's general theory of relativity, discussed in Chapter 11, provides a relativistic classical model of gravity linking energy to curvature of space and time. The linearized theory describes the gravitational field by a symmetric 4-tensor field corresponding to hypothetical massless spin two particles called gravitons. However, the quantum theory is fraught with difficulties undoubtably connected with an incomplete picture of the origin of rest energy. The gravitational interaction between elementary particles at achievable energies is so weak that quantum gravity is not accessible to experimental investigation at present. New phenomena including dark matter and neutrino oscillations are suggesting other ways that the standard model might be enlarged to encompass even a wider range of physics. While the standard model is perilously close to a theory of everything, it is known to be incomplete.

## 1.4 ■ REAL AND VIRTUAL PARTICLES

In our applications to scattering experiments, we will mostly be concerned with interactions connecting initial and final state particles (fields) which may be idealized as free of interaction and which correspond to the external lines in Feynman diagrams. Such so-called real particles correspond to free field wave functions which, when representing states of fixed momentum, have a relationship between wave vector and frequency fixed by free field wave equations. Such particles have a well defined mass. Virtual particles are represented by internal lines in diagrams and correspond to near fields and interaction (potential) energy. The use of the notion of a virtual particle requires care.

Four momentum conservation applies to all processes connecting free (real) particle states and can proscribe processes that appear to be represented by a valid Feynman diagram. The subject of relativistic kinematics will be discussed in Chapter 2. In Chapter 4, we will see how virtual particles in the states intermediate between initial and final states in a time ordered Feynman diagram correspond in perturbation theory to an apparent violation of energy conservation resulting from interaction energy. We will also see how for every diagram containing a virtual particle created at a time $t_1$ and destroyed at a time $t_2$, there is a diagram with a virtual antiparticle created at time $t_2$ and destroyed at time $t_1$. The sum of the corresponding amplitudes is Lorentz invariant and, for this sum, 4-momentum conservation applies. It is customary to draw just one time ordered Feynman diagram as a stand-in for all possible time ordered processes of the same topology. A virtual particle line then stands for all time orderings and, in calculations, energy and momentum are conserved at each vertex of a diagram while the virtual particle mass is permitted to take any value.

As an example of 4-momentum conservation in quantum processes, consider the case of an initial electron. A real electron may not convert to a real electron plus real photon nor to a real neutrino plus $W$ boson although the fundamental diagrams of electroweak theory suggest this is possible. In fact, it is easy to see in the rest frame of the initial electron that such processes would violate energy conservation. In the case of $e^- \to e^-\gamma$, the final recoiling electron would alone have more energy than the initial electron. In the case of $e^- \to W^-\nu_e$, the rest energy of the final $W$ boson alone exceeds the initial rest energy of the electron.

The only real processes represented by the fundamental diagrams are $W$ boson and $Z$ boson decay to fermion pairs excluding the $t$-quark. The allowed products of real $W$ boson decay are the pairs

$$W^- \to \begin{pmatrix} \bar{\nu}_e \\ e^- \end{pmatrix}, \begin{pmatrix} \bar{\nu}_\mu \\ \mu^- \end{pmatrix}, \begin{pmatrix} \bar{\nu}_\tau \\ \tau^- \end{pmatrix}, \begin{pmatrix} \bar{u} \\ d, s, b \end{pmatrix}, \begin{pmatrix} \bar{c} \\ d, s, b \end{pmatrix} \tag{1.105}$$

while a real $Z$ boson can decay to the lepton pairs

$$Z \to \begin{pmatrix} \bar{\nu}_e \\ \nu_e \end{pmatrix}, \begin{pmatrix} \bar{\nu}_\mu \\ \nu_\mu \end{pmatrix}, \begin{pmatrix} \bar{\nu}_\tau \\ \nu_\tau \end{pmatrix}, \begin{pmatrix} e^+ \\ e^- \end{pmatrix}, \begin{pmatrix} \mu^+ \\ \mu^- \end{pmatrix}, \begin{pmatrix} \tau^+ \\ \tau^- \end{pmatrix} \tag{1.106}$$

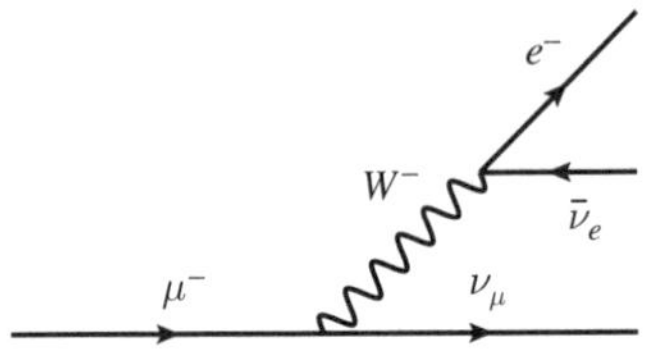


**FIGURE 1.11** The weak decay $\mu^- \to \nu_\mu e^- \bar{\nu}_e$. The virtual $W^-$ in the intermediate state should be regarded as having a mass equal to that of the $e^- \bar{\nu}_e$ system.

and to the quark pairs

$$Z \to \begin{pmatrix} \bar{u} \\ u \end{pmatrix}, \begin{pmatrix} \bar{d} \\ d \end{pmatrix}, \begin{pmatrix} \bar{c} \\ c \end{pmatrix}, \begin{pmatrix} \bar{s} \\ s \end{pmatrix}, \begin{pmatrix} \bar{b} \\ b \end{pmatrix} \tag{1.107}$$

A real $t$-quark can decay with real $W$ boson emission to $d$, $s$, or $b$:

$$t \to \begin{pmatrix} W^+ \\ d, s, b \end{pmatrix}. \tag{1.108}$$

If one of the three particles in a fundamental interaction is virtual, a process that would violate energy and momentum conservation for real particles may take place. Thus, in muon decay as described by the diagram in Figure 1.11, energy and momentum conservation prohibit the muon from radiating a real $W$ boson with invariant mass $m_W \simeq 80$ GeV c$^{-2}$ but a virtual $W$ boson with invariant mass of order the muon mass is permissible. On the other hand, weak decay to hadrons which is permitted for the $\tau^-$ is prohibited for $\mu^-$. In fact, $m_\mu = 105$ MeV c$^{-2}$ $<$ $m_\pi = 140$ MeV c$^{-2}$ and the pion is the lightest hadron composed of $\bar{u}d$ into which a $W^-$ can materialize. Hence the decay

$$\mu^- \to \nu_\mu + (W^- \to \bar{u}d \to \pi^-) \tag{1.109}$$

is disallowed. On the other hand $m_\tau > m_\pi$ and the analogous decay $\tau^- \to \nu_\tau + \pi^-$ is permitted. Both cases correspond to perfectly respectable diagrams in which a charged lepton emits a virtual $W^-$ which materializes as $d\bar{u}$ in a bound state. Kinematics dictates which process is viable.

## 1.5 ■ TRANSITION RATES IN PARTICLE PHYSICS

A general rule applies to the transformation rate of the state of any quantum system coupled to a large number of distinct quantum states. As described in Chapter 4, the transition rate denoted by $\Gamma$ and characteristic time $\tau$ for a process are given by Fermi's golden rule

$$\Gamma \equiv \frac{1}{\tau} \sim |M|^2 \rho. \tag{1.110}$$

Here, $M$ is a process dependent amplitude proportional to coupling constants and kinematic factors that may be deduced from the corresponding Feynman diagram. The density of states or phase space factor $\rho$ is the number of final quantum states over which the transition rate is integrated. In any fixed volume, the number of states is a rapidly increasing function of available energy. For this reason, the decay rate of an unstable particle is generally a rapidly increasing function of its rest energy.

The probability $P(t)$ of finding the system in its original state is consequently governed to high accuracy by a simple differential equation that implies an exponential distribution of decay times:

$$\frac{dP}{dt} = -\frac{1}{\tau}P \rightarrow P(t) = P_0 e^{-t/\tau}. \tag{1.111}$$

Here $\tau^{-1}$ is the characteristic transition rate and $\tau$ is the lifetime, a quantity generally subject to relativistic time dilation. When the initial system transforms to a collection of distinguishable final states each characterized by a partial transition rate $1/\tau_j$, we have

$$\frac{1}{\tau} = \sum_j \frac{1}{\tau_j}. \tag{1.112}$$

For example, the decay of the $W^-$ leads to $e^-\bar{\nu}_e$ and also to $\mu^-\bar{\nu}_\mu$ with a slightly different rate. Each process is called a decay mode and the fractional probability for a mode is called the branching fraction or branching ratio for that mode.

An unstable particle is by definition virtual as it inevitably suffers interactions. If its proper wave function decays exponentially in time,

$$\psi(t) = \psi_0 e^{(-iE_0+\frac{\Gamma}{2})t/\hbar}, \tag{1.113}$$

the Fourier transform gives the amplitude as a function of frequency

$$\psi(E) = \int_0^\infty dt\, \psi(t)e^{iEt/\hbar} = \frac{i\psi_0\hbar}{(E-E_0)+\frac{\Gamma}{2}} \tag{1.114}$$

which implies a (normalized) Breit-Wigner energy distribution

$$\frac{dP}{dE} = \frac{\Gamma/\pi}{(E-E_0)^2+\Gamma^2} \propto |\psi(E)|^2. \tag{1.115}$$

The mean value of the rest energy is $E_0$ and the quantity $\Gamma$ (also called the natural width) is the full width at half maximum of the distribution of the invariant mass of the particle systems which form the particle and into which it decays. The natural width is related to the proper lifetime $\tau$ by the uncertainty relation

$$\Gamma = \frac{\hbar}{\tau} \tag{1.116}$$

| | $n$ | $\mu^-$ | $\pi^-$ | $\pi^0$ | $J/\psi$ | $Z$ |
|---|---|---|---|---|---|---|
| $\tau$ (s) | 887 | $2 \times 10^{-6}$ | $2.6 \times 10^{-8}$ | $8.4 \times 10^{-17}$ | $0.7 \times 10^{-20}$ | $5 \times 10^{-25}$ |

**TABLE 1.4** Lifetimes of particles. The proper lifetimes of particles span a range of 28 orders of magnitude.

and the fractional uncertainty in rest energy can be defined as

$$\frac{\Delta E}{E} = \frac{\Gamma}{mc^2} = \frac{\hbar c}{(mc^2)(c\tau)}. \tag{1.117}$$

Since $\hbar c = 0.197$ GeV fm, a particle with a rest energy $mc^2$ of 1 GeV and a macroscopic proper decay length $c\tau$ of one meter will have a natural width of order $10^{-15}$ GeV, too small to measure directly. However, for relativistic energy $E = \gamma mc^2$, a particle will decay with mean travel distance $\gamma v\tau$ which may be macroscopic. The range of proper lifetimes encountered in particle physics is indicated by the examples in Table 1.4.

---

**Example 1.3. Charm quark decay.** What are the simplest modes of decay of the $c$-quark into fundamental fermions according to the standard model? If the lifetime of the $c$-quark is $\tau_c \simeq \tau_{D^+} \simeq 10^{-12}$ s, how far in mm does a $c$-quark typically travel at light speed (neglect time dilation), what is the uncertainty $\Delta E$ in its rest energy in eV, what is the ratio $\Delta E/k_B T$ at room temperature, and what is the fractional uncertainty $\Delta E/m_c c^2$ assuming $m_c c^2 = 1.25$ GeV?

The $c$-quark has electric charge $+(2/3)e$ and decays in the standard model by emission of a virtual $W^+$ to a lighter quark flavor ($s$ or $d$) with electric charge $-(1/3)e$. The virtual $W$ boson may decay to a lepton plus neutrino or quark plus antiquark of net charge $+e$ and mass smaller than $m_c - m_s$ or $m_c - m_d$ respectively. Inspection shows the pairs $e^+\nu_e$, $\mu^+\nu_\mu$, $u\bar{d}$ and $u\bar{s}$ are permitted. The distance a $c$-quark travels is $c\tau = (3 \times 10^8 \text{ ms}^{-1})(10^{-12} \text{ s}) = 0.3 \text{ mm} = 3 \times 10^{11}$ fm. We can estimate the uncertainty in its rest energy as

$$\Delta m_c c^2 = \frac{\hbar c}{c\tau} = \frac{0.2 \text{ GeV} - \text{fm}}{3 \times 10^{11} \text{ fm}} = 0.6 \times 10^{-3} \text{ eV} \tag{1.118}$$

which is 2.6% of room temperature thermal $k_B T = (1/39)$ eV. The fractional uncertainty is

$$\frac{\Delta m_c c^2}{m_c c^2} = 5 \times 10^{-13}. \tag{1.119}$$

To observe the natural width would require measurement of the energy and momentum of each decay product with fractional precision comparable to fractional uncertainty in the mass.

---

## 1.6 ■ LEPTONS

Leptons and quarks exhibit a tremendous variety of behavior. Each has a personality. To complete this introduction, we take a romp through the history of particle science in light of what we have learned of the standard model, beginning with the leptons.

The electron is said to have been discovered in 1897, when Thomson suggested a particle interpretation of cathode rays. Given the charge to mass ratio deduced from electromagnetic deflection, and the charge from Milikan's oil drop experiment, the mass of the electron was inferred and found to be much smaller than the mass of an atom. This was perhaps the first of many startling results in particle physics. Understanding the quantum nature of the electron, especially its spin and associated magnetism and later its relativistic behavior, coincided with the formulation of atomic theory, nonrelativistic quantum mechanics, and ultimately relativistic quantum field theory. The behavior of nonrelativistic electrons is familiar to the student of physics. Less familiar are the properties of the other fundamental particles in nature to which we now turn.

The muon was discovered in 1936 by Anderson and Neddermeyer. The muon is about two hundred times heavier than the electron and this makes a world of difference to the experimentalist. Unlike an electron, a muon plows through condensed matter as dense as uranium losing energy only gradually by collisions with electrons. The muon first appeared in cosmic ray showers as a mysterious penetrating charged particle leaving a long ionization trail. Cosmic ray induced showers and the process of ionization are considered in more detail in Chapter 3. The electron was amply sufficient to explain the properties of atoms and upon the discovery of the muon one physicist (Isidor Rabi) purportedly quipped: "Who ordered that?" First thought to be the quantum of the nuclear force field anticipated by Yukawa, the muon was found to be insensitive to strong nuclear forces and to behave like a heavy electron. The word lepton used today originally meant light weight particle and derived from the observation that neutrinos, the muon, and the electron are light compared to nucleons. While that meaning hardly applies to the $\tau^-$, the name stuck.

The cosmic ray induced flux of $\mu^+$ and $\mu^-$ at sea level is about 130 particles $s^{-1}\ m^{-2}$ with 1 GeV a typical energy. After stopping in matter, a $\mu^+$ is repelled by positively charged nuclei but attracts and can bind an electron to form a muonium atom $(\mu^+ e^-)_{\text{atom}}$. In contrast, a stopping $\mu^-$ may replace an electron in an atom forming a muonic atom or molecule. The properties of such exotic systems are those of normal matter scaled by the particle masses. In particular, the size of normal atoms is inversely proportional to the electron mass (see Equation 1.40) and a muonic hydrogen atom is about $m_\mu/m_e \simeq 205$ times smaller than an electronic hydrogen atom.

The muon decays to electron plus two neutrinos

$$\mu^- \to e^- + \bar{\nu}_e + \nu_\mu \tag{1.120}$$

with a lifetime $\tau_\mu = 2.3\ \mu\text{s}$, quite long compared to atomic time scales. The muon spin and its associated magnetic moment $e\hbar/(2m_\mu)$ can be determined from the radiation spectra of muonic atoms, and from the spin precession of free muons stopped or stored in a static magnetic field. As we will calculate in detail in Chapter 8, the direction of the electron in the characteristic muon decay is correlated with the muon spin direction. A precessing muon magnetic moment behaves like a precessing beacon of electrons and, from the precession rate, the magnetic moment has been measured with astounding precision. The muon lifetime augmented by time dilation permits a one GeV energy muon to travel over six kilometers at essentially light speed before decay. Particle physicists call stable any particle with a lifetime sufficient to be observed traveling over a macroscopic distance. At light speed, a relativistic particle such as the $\tau^-$ with lifetime of order one picosecond travels at least 0.3 mm and qualifies (barely) as stable.

In 1930, W. Pauli postulated the electron neutrino as a "desperate remedy" to account for missing energy and momentum in decays such as

$$
\begin{aligned}
{}^{137}_{55}\mathrm{Cs} &\rightarrow {}^{137}_{56}\mathrm{Ba} + e^- + \bar{\nu}_e \\
{}^{22}_{11}\mathrm{Na} &\rightarrow {}^{22}_{10}\mathrm{Ne} + e^+ + \nu_e
\end{aligned}
\tag{1.121}
$$

in which the atomic number $Z$ (the number of protons) of an element changes while the mass number $A$ (the number of nucleons) is unchanged. These processes could be ascribed to the weak interactions

$$
\begin{aligned}
n &\rightarrow p e^- \bar{\nu}_e \\
p &\rightarrow n e^+ \nu_e
\end{aligned}
\tag{1.122}
$$

after the discovery of the neutron in 1932 by James Chadwick and of the positron by Anderson in 1933. In the absence of the neutrino, conservation of energy and momentum in the rest frame of the decaying nucleus requires a unique value for the electron energy while a smooth spectrum of energies is observed. A 3-body decay with a neutral particle unobserved is in retrospect a natural explanation. The maximum electron energy permitted by energy and momentum conservation is sensitive to the electron neutrino mass $m_{\bar{\nu}_e}$ and the present upper bound $m_{\bar{\nu}_e} < 2.4$ eV is derived from processes like the tritium to helium decay,

$$
{}^3\mathrm{H} \rightarrow {}^3\mathrm{He} + e^- + \bar{\nu}_e. \tag{1.123}
$$

Note that since the neutron, proton, and electron are fermions, the neutrino must also be a fermion if angular momentum is conserved. Neutrinos carry no electric or color charge and interact only weakly with normal matter. Enrico Fermi used Pauli's hypothesis and gave the neutrino a name that stuck (neutrino means little neutral one) in developing a theory of the weak interaction before the neutrino was ever observed directly. The mean path length between interactions in normal matter of MeV energy electron neutrinos is of order $10^{15}$ m. Direct observation posed an incredible challenge. In 1956, in a telegram to Pauli, F. Reines and C. Cowen, Jr. announced the first observation of the inverse beta decay process

$$
\bar{\nu}_e + p \rightarrow n + e^+. \tag{1.124}
$$

The team's audacious experiment was located near the Savannah River nuclear reactor where uranium fission fragment decays produced an antineutrino flux of about $10^{13}$ $\text{cm}^{-2} \cdot \text{s}^{-1}$ with energy $\sim$1 MeV. A subsequent search by Davis and Harmer in 1959 for

$$\bar{\nu}_e + n \rightarrow p + e^- \tag{1.125}$$

in the form

$$\bar{\nu}_e + {}^{37}\text{Cl} \rightarrow {}^{37}\text{Ar} + e^- \tag{1.126}$$

gave a null result and indicated a distinction between neutrinos and antineutrinos.

Consider the result of the inverse $\beta$-decay in the Reines and Cowan experiment. An Mev energy positron loses energy through collisions in condensed matter and comes to rest, possibly forming a positronium $e^+e^-$ atom. The annihilation process

$$e^+ + e^- \rightarrow \gamma + \gamma \tag{1.127}$$

produces two 0.511 MeV gamma rays back to back in the rest frame of the atom. The gamma rays penetrate into matter but can transfer their energy to electrons by elastic Compton scattering. The high energy electrons thus produced, through subsequent collisions, leave a trail of excited and ionized atoms and the florescence of the excited atoms can result in observable flashes of light in sufficiently transparent and florescent material called scintillator. The experiment searched for coincident gamma rays producing simultaneous light flashes in tanks holding a total of 4200 liters of oil based scintillator and viewed by 330 light sensitive photomultiplier tubes. Cadmium chloride was added to primary target water tanks so that, after some rattling around, the neutron could be captured by a cadmium nucleus with additional gamma rays resulting from nuclear de-excitation giving confirming scintillation signals delayed by a few microseconds. Clever electronic fast timing and signal processing circuits were developed to record the electrical signals from the photomultipliers. The neutrino experiment illustrates some details of the interactions of particles with matter important in experimental particle physics. The interactions of particles with matter and particle detection techniques are the subject of Chapter 3.

As in nuclear $\beta$-decay, in muon decay $\mu^- \rightarrow e^-\bar{\nu}_e\nu_\mu$, the electron kinetic energy varies from zero up to a maximum value (nearly $m_\mu/2 = 52$ MeV) which suggests a 3-body decay with two missing neutral particles now established to be two distinct flavors of neutrino. Given the endpoint energy, an independent measurement of the muon mass, and a bound on the electron neutrino mass, an upper bound on the mass of the $\nu_\mu$ is inferred. The similarity of muon decay to neutron decay suggests a connection between these ephemeral phenomena - a universal weak interaction common to all quarks and leptons. Before the advent of high intensity muon beams made possible detailed studies of muon weak interactions with nuclei, the muon capture process

$$p + \mu^- \rightarrow n + \nu_\mu \tag{1.128}$$

was found to compete with muon decay in muonic atoms and provided early evidence of muon-quark weak interactions. The capture results from the exchange

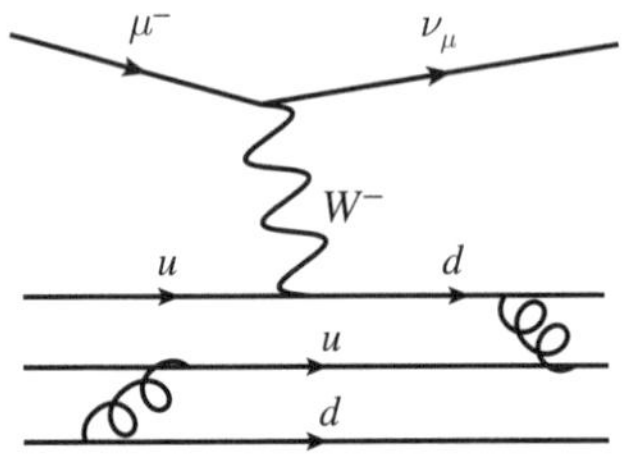


**FIGURE 1.12** Diagram describing muon capture in which the weak interaction $\mu^- u \to \nu_\mu d$ converts muonic hydrogen $(\mu^- p)_{\text{atom}}$ to the neutral particles $n$ and $\nu_\mu$.

of a $W$ boson converting the $\mu^-$ to $\nu_\mu$ and a $u$-quark to a $d$-quark as shown in Figure 1.12. Similar charged exchange processes like $\nu_\mu d \to \mu^- u$ are used to probe the structure of nucleons. By the way, electron capture occurs in unstable nuclei but electron capture in hydrogen is thankfully excluded by energy conservation and the accidental relationship $m_e < m_n - m_p$. Were the electron rest energy 1.3 MeV rather than 0.511 MeV, the normal electronic hydrogen atom would be unstable.

The discovery of the $\nu_\mu$ awaited a high intensity muon neutrino source. Muon neutrinos are not a by-product of nuclear decay in the standard model as they are produced in association with muons which have a rest mass much exceeding the nuclear energy scale. The most effective way to make muons and muon neutrinos is with a proton beam impacting a target producing charged pions and kaons which decay to muons. Beams of charged mesons are now routinely used to source neutrino beams and the appearance of muons in muon neutrino beam interactions was used to establish their existence.

At this point, so-called neutrino oscillations which lie outside the standard model must be mentioned. It is now established that a muon neutrino loses and regains its ability to produce a $\mu^-$ with a sinusoidal time dependence. There is evidence it oscillates to $\nu_\tau$ meaning it can produce $\tau^-$. Electron neutrinos also oscillate in flavor. Such oscillations indicate that the neutrinos of the standard model are quantum mechanical mixtures of mass eigenstates with mass squared differences $\Delta m^2 \simeq 8 \times 10^{-5}\ \text{eV}^2$ and $\Delta m^2 \simeq 3 \times 10^{-3}\ \text{eV}^2$. The mass values are not yet known but the sum of the neutrino masses is constrained by the standard cosmological model to be smaller than about 1 eV. The mass values are hence much smaller than bounds on neutrino mass from direct measurements. At an energy scale of one GeV, the typical threshold for experiments with hadrons, neutrino mass is irrelevant in kinematics and the oscillation lengths are large compared to the scale of all but dedicated experiments. Hence, neutrino mass effects can safely be neglected. In Chapter 11, we will return to this interesting topic.

The tau lepton was discovered by M. Perl and collaborators in 1975 in $e^+e^-$ collisions at center of mass energies above 3 GeV. A single virtual photon produced with such a mass can decay to the following pairs of charged particles:

$$e^+e^- \to \gamma \to e^+e^-,\ \mu^+\mu^-,\ \tau^+\tau^-,\ \text{or}\ q\bar{q} \tag{1.129}$$

where the quark flavors include $u$, $d$, $c$ and $s$, each in a choice of three colors. The threshold energy for $\tau^+\tau^-$ creation is close to that for $c\bar{c}$ production which was discovered essentially simultaneously causing great excitement.

The $\tau^-$ decays necessarily to a $\nu_\tau$ plus a virtual $W$ boson which may decay to a quark plus antiquark or to a lepton plus antilepton. Since the $\nu_\tau$ from the $\tau^-$ decay and $\bar{\nu}_\tau$ from the $\tau^+$ decay are essentially invisible, energy and momentum in the collision appear lost. If the $W^-$ decays to $d\bar{u}$, hadrons may be observed, e.g.

$$e^+e^- \to \tau^+\tau^- \to \bar{\nu}_\tau W^+ \nu_\tau W^- \to \bar{\nu}_\tau \pi^+_{u\bar{d}} \nu_\tau e^- \bar{\nu}_e. \tag{1.130}$$

The heavy $c$-quark decays necessarily to a virtual $W^+$ plus an $s$-quark or $d$-quark (principally the $s$-quark because $|V_{cd}^{CKM}| << |V_{cs}^{CKM}|$) and a leptonic decay such as $W^+ \to e^+\nu_e$ can also result in missing energy plus hadrons, e.g.

$$e^+e^- \to c\bar{c} \to sW^+\bar{s}W^- \to su\bar{d}\bar{s}e^-\bar{\nu}_e \to \bar{K}^0_{s\bar{d}} K^+_{u\bar{s}} e^- \nu_e. \tag{1.131}$$

The discovery of $\tau^+\tau^-$ production was based on events of the form

$$e^+e^- \to \tau^+\tau^- \to \bar{\nu}_\tau \nu_e e^+ \nu_\tau \bar{\nu}_\mu \mu^- \tag{1.132}$$

in which both the virtual $W^-$ and $W^+$ decay to lepton plus anti-lepton leading to charged leptons in association with missing energy and without additional hadrons. Figure 1.13 illustrates one of the first $\tau^+\tau^-$ heavy lepton pairs events identified. The $\tau^-$ is an integral charge spin 1/2 fermion without strong interactions which can

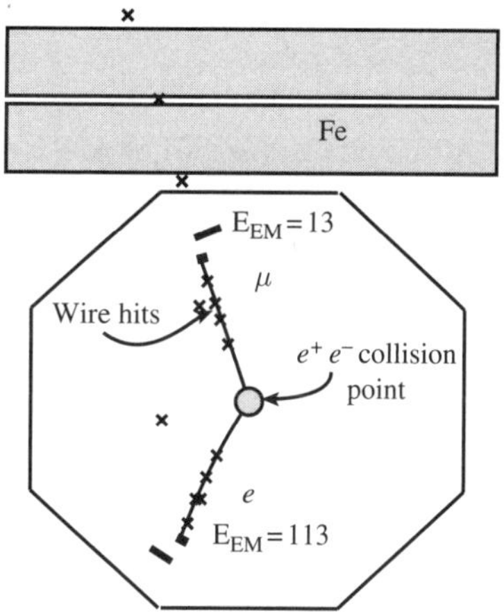


**FIGURE 1.13** Early $\tau^+\tau^-$ production event. A candidate for $e^+e^- \to \tau^+\tau^-$ is visualized. Decays such as $\tau^- \to e^-\nu_\tau\bar{\nu}_e$ lead to detectable $e^\pm\mu^\mp$ pairs plus invisible neutrinos. In this event, a muon moves upwards leaving ionization hits in a central cylindrical wire chamber. It deposits a small amount of energy in an external electromagnetic calorimeter and passes through steel plates leaving additional hits in interleaved wire chambers. An electron heads downwards and creates a shower of particles, its energy contained in the calorimeter. [after Martin L. Perl, SLAC-PUB-10150]

only be called a heavy lepton. Were the neutrino associated with $\tau^-$ decay either the $\nu_e$ or $\nu_\mu$, then one would expect $\tau^-$ production in the interactions of sufficiently high energy $\nu_e$ and $\nu_\mu$ with matter. Since (over short distances), $\nu_e$ and $\nu_\mu$ beams passing through matter do not produce $\tau^-$ leptons, the $\nu_\tau$ is a third neutrino flavor. The $\nu_\tau$ mass is constrained by the charged lepton energy spectrum in $\tau^- \to e^-\bar{\nu}_e\nu_\tau$ and $\tau^- \to \mu\bar{\nu}_\mu\nu_\tau$ to be $m_{\nu_\tau} < 24$ MeV. Direct observation of the $\nu_\tau$ requires a high intensity $\tau^-$ source. The observation of $\nu_\tau$ converting to $\tau^-$ in emulsion with $\nu_\tau$ sourced via a high intensity proton beam at Fermilab in the late 1990s was the last and perhaps too little recognized fundamental particle discovery. While the decay products in $Z \to \nu\bar{\nu}$ have never been observed directly, exactly three generations of light neutrinos are required to explain the natural width $\Gamma_Z$. Hence it is inferred that there are no additional fermion generations with light neutrinos and standard model interactions.

## 1.7 ■ QUARKS AND HADRONS

We now turn to the discovery of quarks. Recall that quarks are found bound by color forces in strongly interacting massive particles called hadrons. Such particles were classified as integral spin mesons and half integral spin baryons prior to the invention of the quark model. The quark model of hadrons was invented to explain the patterns of masses and of interactions of light hadrons composed of $u$-quarks, $d$-quarks and $s$-quarks and the model was readily extended to include heavier quark flavors.

In the naive nonrelativistic quark model, all hadrons have the base composition

$$\text{meson} = \mathrm{q\bar{q}} \text{ and baryon} = \mathrm{qqq} \text{ or } \mathrm{\bar{q}\bar{q}\bar{q}} . \tag{1.133}$$

All combinations of quark flavors in these two forms are observed - the color interaction does not distinguish flavor. The masses of hadrons composed entirely of the "light quarks" $u, d$ and $s$ are in the range 0.1–1.6 GeV. The masses of hadrons containing the "heavy quarks" $c$ and $b$ are approximately the sum of the masses of the constituent quarks. (The identification of hybrid combinations like $qq\bar{q}\bar{q}$ and $q\bar{q}g$ and pure gluonic states called glueballs is a topic of continued research.)

A meson is a $q\bar{q}$ quantum mechanical bound state analogous to the hydrogen atom. The two particle wave function for the ground state has zero orbital angular momentum about the center of mass ($s$-wave) and zero total spin angular momentum. Excited meson states of non-zero spin or orbital angular momentum can be identified as short-lived hadronic resonances in scattering experiments. The parity of the orbital wave function is $(-1)^l$ and a fermion and anti-fermion have opposite intrinsic parity (see Chapter 6) so the parity of a meson is $P = (-1)^{l+1}$. The ground state therefore has spin 0 and odd parity and such a particle is called a pseudoscalar. Parity is conserved in electromagnetic and color but not weak interactions so well respected in the strong interactions of hadrons. Establishing the parity and spin of hadrons was important in establishing the quark model and will be considered in Chapter 10.

The two particle wave function describing the pion, for example, is a spherically symmetric combination of product waves antisymmetric in spin of the form

$$\pi^+_{u\bar{d}} = \frac{1}{\sqrt{2}}\left(u^\uparrow \bar{d}^\downarrow - u^\downarrow \bar{d}^\uparrow\right) = \psi\left(|\mathbf{x}_u - \mathbf{x}_{\bar{d}}|\right)\frac{1}{\sqrt{2}}\left(\uparrow\downarrow - \downarrow\uparrow\right). \tag{1.134}$$

In case the $q$ and $\bar{q}$ have the same flavor, a meson may be an eigenstate of $C$. Since an exchange of the $q$ with the $\bar{q}$ implies a factor $(-1)^l$ from the orbital wave function and a symmetry factor $(-1)^s$ where $s$ is the total spin, the $C$ parity is $C = (-1)^{l+s}$. The three quarks in a ground state baryon have vanishing orbital angular momentum about their center of mass. The total spin angular momentum can be 3/2 or 1/2 and the ground state turns out to correspond to 1/2.

Color exchange underlies the stability of the peculiar $q\bar{q}$ and $qqq$ combinations. Baryon and meson wave functions are color neutral (technically color singlet) combinations of different quark color states such as

$$\begin{aligned}
\Omega^-_{sss} &= \frac{1}{\sqrt{6}}\left(s_r s_b s_g - s_r s_g s_b + s_g s_r s_b - s_g s_b s_r + s_b s_g s_r - s_b s_r s_g\right)\\
\pi^+_{u\bar{d}} &= \frac{1}{\sqrt{3}}\left(u_r \bar{d}_r + u_b \bar{d}_b + u_g \bar{d}_g\right).
\end{aligned} \tag{1.135}$$

The $\Omega^-_{sss}$ has spin 3/2 and its quark model wave function is symmetric in spin (because all three quark spins are aligned) and symmetric in space (because each of the three quarks is in an $s$-wave state about the center of mass) but antisymmetric in color and thus is consistent with the exclusion principle. A spin 1/2 color neutral triple strange ($sss$) ground state baryon is inconsistent with the exclusion principle and in fact not observed - there are eight spin-1/2 baryons and ten spin 3/2 baryons composed of $s$, $u$, and $d$. This peculiarity provided early evidence for the hidden color degree of freedom and motivated the development of quantum chromodynamics.

One significant fact that distinguishes the quark model of hadrons from the atomic model and from the standard model of nuclear structure is that quarks and gluons are not observed as free particles over macroscopic scales. Instead, they are masked through a process of energy materialization and color neutralization governed by quantum chromodynamics. The color electric fields joining color charges are compacted by nonlinear effects and form an effective connecting string of linear energy density $\sim$1 GeV/fm for charge separations exceeding about 1 fm. As illustrated in Figure 1.14, in high energy hadronic processes in which colored objects are produced, the color field spontaneously fragments into $q\bar{q}$ dipole pairs and recombination replaces primary quarks and gluons with color neutral hadrons. This fragmentation process indicates another significant difference between quark composites and atoms or nuclei. In the relativistic reactions of hadrons, the number and type of the constituents is not fixed. Color confinement was long an impediment to acceptance of the quark constituent model of hadrons. Quantum chromodynamics (QCD) provides a description of the spectrum of hadrons and an understanding of the origin of confinement.

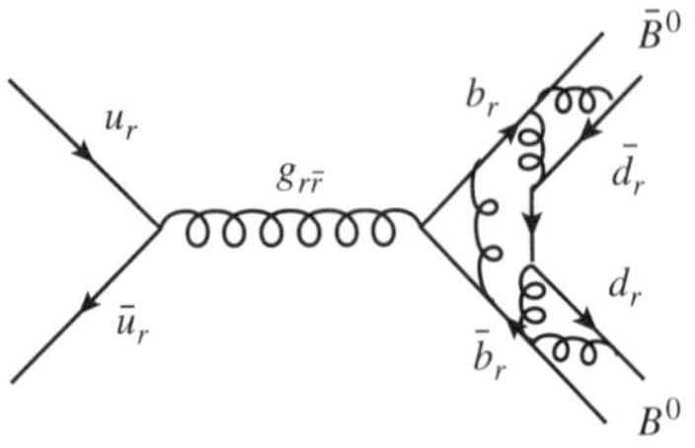


**FIGURE 1.14** Hadronization. An illustration of the hadronization of the quarks produced in $u\bar{u} \to g \to b\bar{b}$. Final state color interactions indicated schematically can result in a pair of long lived $B^0(d\bar{b})$ and $\bar{B}^0(\bar{d}b)$ mesons.

The original meson was the negatively charged pion $\pi^-_{d\bar{u}}$ and its antiparticle $\pi^+_{u\bar{d}}$. The name meson was invented to suggest an intermediate mass value, between $m_e$ and $m_p$. First seen during the 1940s in cosmic ray showers and distinguished from muons by its strong interactions with nuclei, a pion may appear when color field energy converts to light $q\bar{q}$ pairs. Suppose, for example, the color field of a proton spontaneously forms a $d\bar{d}$ pair. A little reorganization results in a process

$$p_{uud} \to ud(d\bar{d})u \to n_{udd}\pi^+_{\bar{d}u}. \tag{1.136}$$

It appears a pion has been radiated by a proton, converting it into a neutron. This kind of process explains how pions can appear as quanta of an effective nuclear strong force field.

Pions are the lightest hadrons and often the end product of energy transformations in hadronic interactions. Extraterrestrial (cosmic) protons striking nuclei in the upper atmosphere produce showers of pions by compounded strong interactions such as the one just described. The components of a cosmic shower are indicated in Figure 1.15. Charged pions decay through the process

$$\pi^-_{d\bar{u}} \to W^- \to \mu^-\bar{\nu}_\mu \tag{1.137}$$

with a mean proper lifetime $\tau_{\pi^-} = 2.6 \times 10^{-8}$ s and decay length $c\tau_{\pi^-} = 7.8$ m. Note that the pion lifetime is somewhat shorter than the lifetime $\tau_\mu = 2.3 \times 10^{-6}$ s for muon decay which also proceeds through a second order weak interaction: $\mu^- \to \nu_\mu W^- \to \nu_\mu e^- \bar{\nu}_e$. The pion decay rate is sensitive to the internal wave function of the pion as the $u$-quark and $d$-quark must fuse to form the virtual boson. Interestingly, the spin dependence of the weak interaction suppresses the alternate rare decay mode $\pi^- \to e^-\bar{\nu}_e$ which, due to the larger kinetic energy available to the decay products, would otherwise dominate. In the limit of vanishing electron mass, the electron and antineutrino spins coupling to the $W$ must combine to angular momentum of one while the $d$ and $\bar{u}$ spins in the pion sum to zero. Hence, angular momentum conservation suppresses the decay. This is just one example of the interplay of dynamical features and conservation laws at work in the behavior of hadrons that we will come to understand. The decay rate of the pion in particular is calculated in Chapter 10.

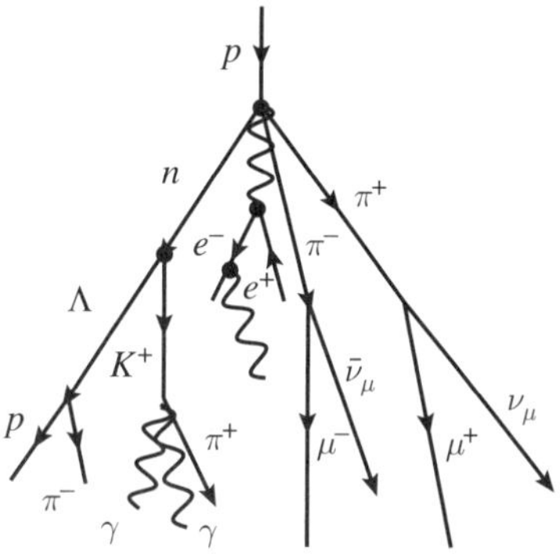


**FIGURE 1.15** Cosmic ray shower. A primary proton strikes a nucleus in the upper atmosphere creating pions and nuclear fragments which may subsequently interact strongly, multiplying the number of hadrons. Weak decays of shower particles result in penetrating muons and neutrinos. Photons from $\pi^0$ decay initiate additional electromagnetic showers.

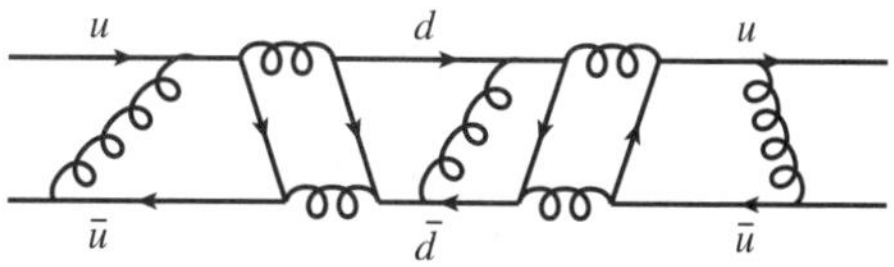


**FIGURE 1.16** Light quark meson mixing. The neutral pion contains both $u\bar{u}$ and $d\bar{d}$ components mixed through annihilation into gluons.

Pion decay is the principal source of long lived cosmic muons at sea level. Brought to rest by ionization energy loss in matter, a $\pi^+$ is repelled by nuclei. It may attract an electron forming an exotic atom $(\pi^+ e^-)_{\text{atom}}$ but ultimately decays through the weak interaction to $\mu^+ \nu_\mu$. In contrast, a $\pi^-$ brought to rest in matter is attracted by nuclei and may replace an electron forming a pionic atom such as pionic hydrogen $(\pi^- p)_{\text{atom}}$. Its proximity to the nucleus in the low principle quantum number atomic states permits the pion to be absorbed, the $\bar{u}$-quark in the pion combining with or annihilating with a $u$-quark in the nucleus, leading to processes such as

$$\pi^- p \to \pi^0 n, \gamma n. \tag{1.138}$$

The color interaction is flavor independent and hence $u\bar{u}$ and $d\bar{d}$ analogs of the charged pions are as readily produced in strong interactions as the $\pi^\pm$. However, these neutral states are flavor mixed through internal annihilation to gluons and behave quite differently from charged pions. The $u\bar{u}$ and $d\bar{d}$ mesons may be regarded as a two state quantum mechanical system and the eigenstates identified as the "pi-zero" meson with a rest energy of 135 MeV

$$\pi^0(135\ \text{MeV}) = \frac{1}{\sqrt{2}}\left(u\bar{u} - d\bar{d}\right) \tag{1.139}$$

and the much heavier "eta" meson

$$\eta^0(548\text{ MeV}) = \frac{1}{\sqrt{2}}\left(u\bar{u} + d\bar{d}\right). \tag{1.140}$$

Composed of particle and antiparticle of the same flavor like the ground state of positronium, the $\pi^0$ and $\eta^0$ decay rapidly by electromagnetic annihilation to two photons. The decay length $c\tau_{\pi^0} = 25$ nm is generally too short to observe directly except at extremely high energy. The $\eta$ decays to two photons but has sufficient rest energy to decay also to pions and has a decay length $c\tau_\eta = 1.6 \times 10^{-10}$ m.

The $\pi^0$ was discovered as the source of prompt gamma ray pairs emerging from hadronic reactions such as the charge exchange process

$$\pi^+ + n \to \pi^0 + p \to \gamma\gamma + p. \tag{1.141}$$

The decays of $\pi^0$ mesons add a gamma ray component to cosmic ray showers and the fusion of these real gamma rays with virtual gamma rays associated with the electric fields around nuclei convert the energy further to $e^+e^-$ pairs. This process contributes positrons and electrons to the mix of particles in cosmic ray showers.

Strong interactions between hadrons can produce an $s\bar{s}$ pair leading to associated production of pairs of strange hadrons a bit heavier than their $u$-quark and $d$-quark analogs. For example, the process $u\bar{u} \to g \to s\bar{s}$ underlies the hadronic reaction

$$\pi^+_{u\bar{d}} + n_{udd} \to K^+_{u\bar{s}} + \Lambda_{sud}. \tag{1.142}$$

Such processes in cosmic ray showers were the source of the first strange hadrons observed. Like all flavor numbers, strangeness (-1 for the $s$-quark and +1 for the $\bar{s}$-quark) is conserved in strong and electromagnetic interactions. The positively charged kaon $K^+(494)$ decays weakly ($\tau_{K^+} = 1.2 \times 10^{-8}$ s) through a strangeness changing fusion process analogous to $\pi^+$ decay resulting in the leptonic modes

$$K^+_{u\bar{s}} \to W^+ \to \mu^+\nu_\mu, e^+\nu_e \tag{1.143}$$

with 63% probability. Semi-leptonic ($K_{\mu 3}$ and $K_{e3}$) and all hadronic decay two and three particle modes include

$$K^+_{u\bar{s}} \to \pi^0_{u\bar{u}}W^+ \to \pi^0_{u\bar{u}}\left(\mu^+\nu_\mu, e^+\nu_e\right),\ \pi^0_{u\bar{u}}\,\pi^+_{u\bar{d}}. \tag{1.144}$$

The $\Lambda$(1115 MeV) can be thought of as a strange neutron, one of a group of strangeness bearing baryons. Generally, baryons heavier than the nucleons are called hyperons. The $\Lambda$ decays by the process $s \to W^-u$. The decay $W^- \to \bar{u}d$ or the absorption of the $W$ boson by a $d$-quark in association with spontaneous production of a light $q\bar{q}$ pair leads to

$$\Lambda_{sud} \to p_{uud}\,\pi^-_{\bar{u}d},\quad n_{udd}\,\pi^0_{\bar{u}u-\bar{d}d}. \tag{1.145}$$

with $\tau_\Lambda = 2.6 \times 10^{-10}$ s. A leptonic decay of the virtual $W$ boson resulting in $\Lambda \to pe^-\bar{\nu}_e$ analogous to $n \to pe^-\bar{\nu}_e$ is much less probable. It should be pointed out that the lifetime $\tau_\Lambda$ permits a $\Lambda$ to settle into a nucleus forming a so-called hypernucleus. As the $\Lambda$ is not excluded from the quantum states occupied by the nucleons, it is an interesting probe of nuclear forces.

The neutral kaons $K^0_{d\bar{s}}$ (497 MeV) and $\bar{K}^0_{\bar{d}s}$ exhibit a marvelous manifestation of weak interaction mixing. Like the analog $c$-quark and $b$-quark mesons, this pair of mesons mix through double $W$ boson exchange and a matter-antimatter oscillation phenomenon results. The energy eigenstates are (approximately) the symmetric and antisymmetric combinations "kay-short"

$$K_S \simeq \frac{1}{\sqrt{2}}\left(K^0 - \bar{K}^0\right), \tag{1.146}$$

which decays weakly to $\pi^+\pi^-$ or $\pi^0\pi^0$ with $c\tau_S = 2.6$ cm, and "kay-long"

$$K_L \simeq \frac{1}{\sqrt{2}}\left(K^0 + \bar{K}^0\right) \tag{1.147}$$

which decays weakly to three pions and to $\pi^\pm l^\mp \nu$ with $c\tau_L = 15$ m. These mesons are (nearly) eigenstates of $CP$ and the difference in the decay modes and lifetimes is a consequence of $CP$ conservation as described in Chapter 10. The weak energy difference between the $K_L$ and $K_S$ is only $m_L - m_S = 3.5 \times 10^{-6}$ eV and the corresponding oscillations between the two states can occur over macroscopic scales.

For example, a strongly produced $K^0$ is a quantum state that is initially part $K_L$ and part $K_S$ and the component amplitudes evolve independently. Since $K_L$ and $K_S$ are energy eigenstates, we can model the time evolution in the rest frame with the expression

$$K(t) = \frac{1}{\sqrt{2}}\left(K_S e^{-iE_S t/\hbar} + K_L e^{-iE_L t/\hbar}\right) \tag{1.148}$$

where $E_S = m_S c^2 - i\Gamma_S$ and $E_L = m_L c^2 - i\Gamma_L$ and express $K_S$ and $K_L$ in terms of $K_0$ and $\bar{K}_0$ to find the probability at time $t$ to observe a $K_0$ is

$$| < K^0|K(t) > |^2 = \left|\frac{1}{2}\left(e^{-im_S c^2 t/\hbar - t/2\tau_S} + e^{-im_L c^2 t/\hbar - t/2\tau_L}\right)\right|^2 \tag{1.149}$$

which has the form of a damped oscillation. Similarly, the probability for $\bar{K}^0$ oscillates. The oscillation can be observed via interactions with protons and neutrons which distinguish the $s\bar{d}$ and $d\bar{s}$ mesons, and by the lepton charge in semileptonic decays which tags whether an $s$-quark or $\bar{s}$-quark decayed. Recent additions to the observed meson data tables include the heavy quark combinations $B_c(\bar{b}c)$ and $B_s(\bar{b}s)$. Oscillations between neutral heavy meson states $B_d(\bar{b}d)$ and $\bar{B}_d$ and between $B_s$ and $\bar{B}_s$ entail off-diagonal elements of $V^{CKM}$ and are of great interest in understanding the Higgs mechanism in the standard model. These oscillations will be described in more detail in Chapter 10.

The first excited state of a $q\bar{q}$ system is an $s$-wave spin 1 vector meson. For example, the $\rho^+_{u\bar{d}}$ (770 MeV) corresponds to the three degenerate symmetric wave functions

$$\rho^+_{u\bar{d}} = u^\uparrow d^\uparrow, \; \frac{1}{\sqrt{2}}\left(u^\uparrow \bar{d}^\downarrow + u^\downarrow \bar{d}^\uparrow\right), \; u^\downarrow \bar{d}^\downarrow. \tag{1.150}$$

Such spin 1 states are generally heavier than their spin 0 counterparts and decay strongly because the energy level differences between spin 1 and spin 0 states exceeds the rest mass of the pion. For example, the $\rho^+$ decays strongly to two pions,

$$\rho^+_{u\bar{d}} \to \pi^+_{u\bar{d}}\pi^0_{u\bar{u}}, \tag{1.151}$$

through spontaneously formation of a $u\bar{u}$ (or $d\bar{d}$) pair from color field energy. Its proper lifetime is inferred from the width $\Gamma_{\rho^+} = 150$ MeV of the peak in two-pion invariant mass distributions in events in which the $\rho^+$ is produced and has the value

$$\tau_{\rho^+} = \frac{\hbar}{\Gamma_{\rho^+}} = \frac{6.6 \times 10^{-22} \text{ MeV s}}{150 \text{ MeV}} = 4 \times 10^{-24} \, s. \tag{1.152}$$

This value is comparable to a hadron diameter over light speed and typical of strong decay processes. The charged and neutral strange vector mesons also decay by pion emission, e.g. $K^{0*}(890) \to K^+\pi^-$ with $\Gamma = 50$ MeV. Notice that strangeness is conserved in this and all strong processes. Mass differences such as $m_\rho - m_\pi \sim$ 500 MeV reflect the spin-spin color magnetic interaction and are the hadronic analog of atomic hyperfine structure. As in atomic physics, such features of the spectroscopy of quark composites proved crucial in understanding their structure.

Heavy quark bound states are one proving ground for the chromodynamic theory of quark interactions. The heavy $c$-quark and $b$-quark first appeared as vector meson bound state resonances in $e^+e^-$ and hadronic interactions. Figure 1.17 shows the mass spectrum of $b\bar{b}$ mesons. It is similar in many ways to that of positronium reflecting the fact that the color Coulomb potential is similar to the electromagnetic Coulomb potential. The $n = 0$ ground state can be spin 0 or spin 1. The spin 1 state called the $\Upsilon(1S)$ is the spin triplet $l = 0$ state. The twelve 2P states with $n = 2$ are split into four multiplets through relativistic effects.

As we will see in Chapter 7, like its electromagnetic magnetic moment, the color magnetic moment of a quark is inversely proportional to its mass. The spin-spin interaction is consequently weaker for large quark mass. The spin 1 $\phi_{s\bar{s}}$ meson falls apart into two kaons bearing strangeness but the spin 1 $J/\psi_{c\bar{c}}$, the $\Upsilon_{b\bar{b}}$, and their $2s$ and $2p$ excitations $\phi'$, $\phi''$ and $\Upsilon'$, $\Upsilon''$ are tightly bound so unable to decay by strong interactions to heavy quark mesons such as the "open charm" $D^+_{c\bar{d}}$ plus $D^-$ and "open bottom" $B^-_{b\bar{u}}$ plus $B^+$. Hence, these states are unusually stable. Incidentally, in the 1970s, the $J/\psi$ was discovered decaying to lepton pairs in $e^+e^-$ collisions and independently in proton nucleus collisions so bears a double name. The existence of a $c$-quark had already been postulated to explain

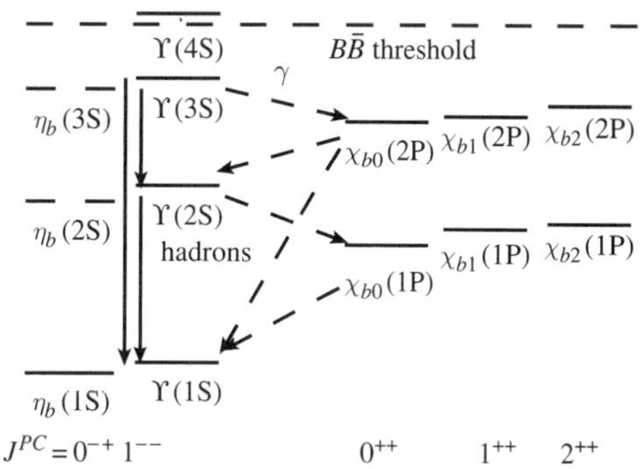


**FIGURE 1.17** Mass spectrum of $b\bar{b}$ mesons. The $J^{PC} = 1^{--}$ states called Υ have increasing principal quantum number, zero orbital angular momentum, and total spin 1. These are made in $e^+e^-$ collisions. Some electromagnetic and strong decay transitions are indicated by arrows.

the weak mixing of neutral kaons in which the $c$-quark participates as a virtual particle.

The discovery of the Υ and of open bottom quark flavored mesons in $e^+e^-$ and hadronic collision experiments shortly after the discovery of the $J/\psi$ and charmed particles, along with the discovery of the $\tau$ and $\nu_\tau$ leptons in the 1970s revealed a pattern of lepton and quark pairs and immediately suggested the existence of the top quark but it remained out of reach until 1995. As the standard model provides no basis for predicting the mass of a quark, early searches used ever more energetic $e^+e^-$ colliding beams to hunt for $t\bar{t}$ pair production. When $W$ bosons were produced in $p\bar{p}$ collisions, searches focused on decays of $W^+$ bosons to $t\bar{b}$, possible if $m_t < m_W - m_b$. Ultimately, the top quark pairs with $m_t = 175$ GeV $> m_W$ were found decaying to real $W$ bosons and $b$-quarks at the Fermilab Tevatron p$\bar{\text{p}}$ collider.

As illustrated in Figure 1.18, the $t\bar{t}$ pairs can be made through $u\bar{u}$ and $d\bar{d}$ annihilation and observed through cascade decays such as

$$t\bar{t} \to W^+bW^-\bar{b} \to \mu^+\nu_\mu be^-\bar{\nu}_e\bar{b} \tag{1.153}$$

with the $b$-quarks appearing in heavy long lived bottom flavored mesons and baryons. The $t$-quark, by the way, is too unstable to form a sensible hadron. Due to its large mass and consequent short lifetime for weak decay, its natural width $\Gamma_t$ is broader than the spectrum of bound states.

In the naive quark model, the lightest baryons $p_{uud}(938)$, $n_{udd}(939)$, and $\Lambda_{sud}(1115)$ are $s$-wave states with the three quark spins combined to form total spin 1/2. The magnetic moments of the constituent quarks add and result in nucleon magnetic moments $\mu_p = (g_p/2)\mu_N$ and $\mu_n = (g_n/2)\mu_N$ where the nuclear magneton is defined as

$$\mu_n \equiv \frac{e\hbar}{2m_p} = 3.152\,451\,2326(45) \times 10^{-14}\ \text{MeV T}^{-1} \tag{1.154}$$

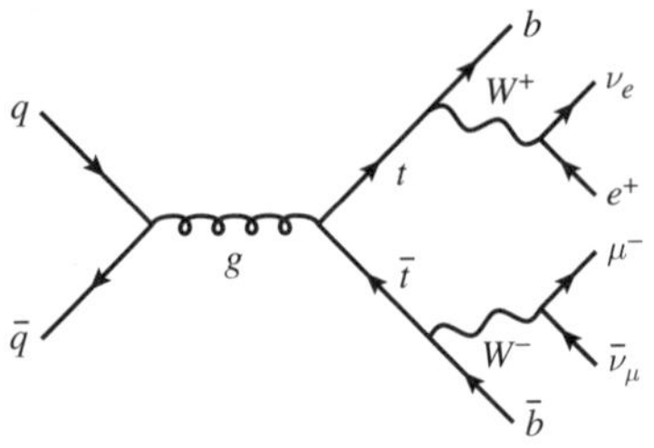


**FIGURE 1.18** Production of $t\bar{t}$ in a $p\bar{p}$ collider. The $t$-quark decays principally to $W^+b$. The $W^+$ and $W^-$ are most readily observed through their leptonic decays which lead to opposite sign charged lepton pairs ($e^+\mu^-$ are shown) and missing energy carried off by two neutrinos. The $b$ and $\bar{b}$ are observed as jets of particles each containing a $b$ or $\bar{b}$ flavored heavy hadron.

and $(g_p/2) = 2.792847356(23)$ while $(g_n/2) = -1.91304273(45)$. The origin of these "g-factors" in the quark model will be discussed in Chapter 10. Spin 3/2 baryons are roughly 300 MeV heavier - the color spin-spin interaction is again at work. Increased mass is also associated with increased strange quark content. Notably, the triple strange spin 3/2 $\Omega^-_{sss}(1672)$ was predicted by the quark model before it was observed. The wave function of the $s_z = +3/2$ spin state in the naive quark model is $\Omega^- = s^\uparrow s^\uparrow s^\uparrow$. This hadron was produced in 1964 (Figure 1.19) using a $K^-$ secondary beam and the strong associated production process

$$K^-_{s\bar{u}} + p_{uud} \to \Omega^-_{sss} + K^+_{u\bar{s}} + K^0_{d\bar{s}} \tag{1.155}$$

and observed to decay via a weak conversion of $s$-quark to $u$-quark $s \to u + W^- \to u\bar{u}d$ leading to

$$\Omega^-_{sss} \to \Xi^0_{sus}\pi^-_{d\bar{u}} \tag{1.156}$$

with the subsequent weak decays

$$\Xi^0_{sus} \to \Lambda_{sud}\pi^0_{d\bar{d}} \tag{1.157}$$

and

$$\Lambda_{sud} \to p_{uud}\pi^-_{d\bar{u}} \tag{1.158}$$

and the electromagnetic annihilation process $\pi^0 \to \gamma\gamma$. The discovery of the $\Omega^-$ at the expected mass vindicated the original three-flavor quark model of light hadrons. Charm and bottom flavored analogs of the light baryons that have been recently produced and identified in high energy colliders include the $\Lambda^0_b$(bud) and $\Sigma^-_b$ (bdd), the $\Xi^-_b$ (dsb), and the doubly strange $\Omega^-_b$(ssb). The $\Omega^-_b$ was discovered following the decay channel $\Omega^-_b \to J/\psi\Omega^-$ with $J/\psi \to \mu^+\mu^-$ and $\Omega^- \to \Lambda K^- \to (p\pi^-)K^-$. The $\Sigma^-_b$ was discovered in the decay sequence $\Sigma^-_b \to \Lambda^0_b\pi^-$ with $\Lambda^0_b \to \Lambda^+_c\pi^-$ and then $\Lambda^+_c \to pK^-\pi^+$.

The periodic table of fermions remains mysterious and motivates searches for substructure and for additional generations. The first generation of quarks and leptons might for example be three particle states of a charge -1/3 preon plus a

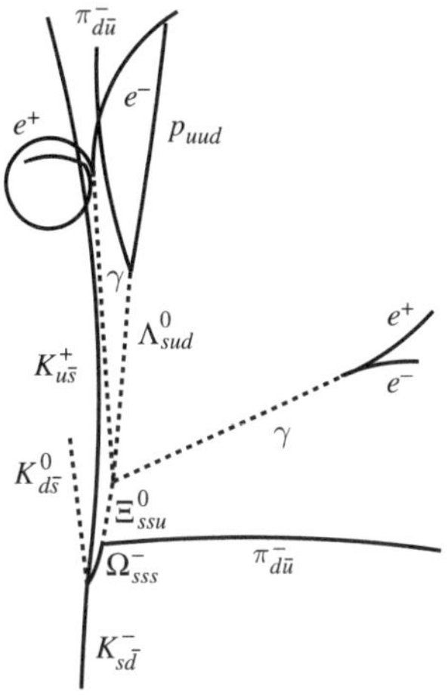


**FIGURE 1.19** Visualization of $\Omega^-(sss)$ decay in a bubble chamber. The $\Omega^-$ is produced in association with a $K^+$ and $K^0$ by a strong interaction of a $K^-$ with a proton in the liquid hydrogen. It decays by conversion of one $s$-quark to a $u$-quark producing a doubly strange baryon $\Xi^0$ plus $\pi^-$. A subsequent weak decay of the $\Xi^0$ to the singly strange baryon $\Lambda$ plus $\pi^0$ and the weak decay of the $\Lambda$ to proton are all evident. The ionization trails of charged particles are shown by solid lines. Neutral particles leave no ionization trail and their inferred paths are shown by dashed lines. The two photons from the electromagnetic decay $\pi^0 \to \gamma\gamma$ both converted to $e^+e^-$ pairs. [After V. E. Barnes *el al.*, Phys. Rev. Lett. **12**, 204 (1964)].

charge 0 preon bound by some new interaction. Other generations might then be excitations. Such ideas motivate searches for processes like $\mu^+ \to e^+\gamma, e^+\gamma\gamma$ and $\mu^+ \to e^+e^+e^-$ and for deviations from the point-like model for the scattering of quarks and leptons at high energy. The branching fractions for rare muon decay modes are presently known to be smaller than a few times $10^{-11}$ and no evidence has yet appeared for fermion compositeness from scattering experiments. Direct searches for fourth generation quarks and leptons have placed lower bounds for the masses of such states in the range of several hundred GeV while a fourth generation neutrino lighter than the $Z$ boson has also been ruled out. (See Chapter 8.) Our survey of leptons and quarks and their curious interactions ends here with just one of several fundamental open questions.

## 1.8 ■ BOSONS AND STANDARD MODEL INTERACTIONS

The photon was invented by Planck and Einstein to explain quantum aspects of electromagnetic radiation and it is surely familiar to the reader. Chapters 5 and 6 are devoted to the study of role of photons in quantum electrodynamics. The gluon was invented before it was discovered. The success of the quark model and other evidence for the color degree of freedom led to the theory of color dynamics which is described in Chapter 7. This theory called quantum chromodynamics

(QCD) predicted color gauge field radiation and was used to explain a variety of hadronic interaction phenomena including jets of particles associated with gluon hadronization.

The existence of the $W$ boson and $Z$ boson was predicted by the (now) standard model of the electroweak interaction. The $W$ boson was postulated as the boson mediating charged current weak interactions. The $Z$ boson was postulated on the basis of gauge symmetry described in Chapter 8. The first evidence for the $Z$ boson was the discovery of neutral current neutrino scattering in the large Gargemelle bubble chamber at CERN in 1973. Other hints emerged in parity violating asymmetries in $e^+e^-$ scattering. In 1983, the $W$ boson and $Z$ boson were directly produced via the constituent reactions $u\bar{d} \to W^+$ and $u\bar{u} \to Z$ in the European Organization for Nuclear Research (CERN) Super Proton Synchrotron, a $p\bar{p}$ collider operating at a center of mass energy of $\sqrt{s} = 630$ GeV. The $W$ boson and $Z$ boson have been extensively studied at the Fermi National Accelerator Laboratory (FNAL) Tevatron $p\bar{p}$ collider operating at a center of mass energy of close to 2 TeV, the CERN Large Electron Positron (LEP) $e^+e^-$ colliding beam operating just above the threshold for $W^+W^-$ production $\sqrt{s} = 160$ GeV, and the new CERN Large Hadron Collider (LHC) designed to produce $pp$ collisions at a center of mass energy of 14 TeV. The design of such machines is considered in Chapter 2.

The Higgs boson has not yet been observed. Its mass is expected to be close to the weak scale. Experiments at LEP set a lower limit $m_H > 114$ GeV on the mass of the Higgs boson. Experiments at the Fermilab Tevatron and CERN LHC are expected to discover or exclude the standard model Higgs boson. The physics of the minimal model Higgs boson is described in Chapter 8.

## 1.9 ■ UNITS

Particle physics is rife with humorous and peculiar conventions and nomenclature. As we will see, it is convenient in particle physics to set $\hbar$ and $c$ equal to one and to choose $\epsilon_0 = \mu_0 = 1$ so that Maxwell's equations have the Standard International form with fundamental constants $\epsilon_0$ and $\mu_0$ erased (Heaviside units). All quantities then may be expressed in units of a length such as 1 fm or an energy such as 1 eV. The charge of the proton in these natural units may be expressed as a dimensionless quantity: $e = \sqrt{4\pi\alpha} = 0.303$. This convention takes some getting used to and occasionally we will include factors such as $\hbar$ and c for clarity. In atomic physics it is handy to eliminate the factor of $4\pi$ from Coulomb's law. Accordingly, some authors use rationalized electromagnetic units in which charge is divided by a factor of $4\pi$. The factor $4\pi$ then appears in Maxwell's equations next to the charge and current densities and in electromagnetic energy but not in Coulomb's law and $\alpha = e^2/\hbar c$. Beware!

Factors of $\hbar$ and $c$ may be restored by dimensional analysis. It is useful to remember the approximate value

$$\hbar c = 197.326\,9631(49) \text{ MeV fm} \simeq 0.2 \text{ GeV fm} \tag{1.159}$$

and to use mass values in MeV or GeV units. For example, the quantity

$$a_0 = \frac{1}{m_e \alpha} \simeq \frac{137}{0.511\ \text{MeV}}(197\ \text{MeV fm}) = 0.529 \times 10^{-10}\ \text{m} \tag{1.160}$$

may be recognized as the Bohr radius of the hydrogen atom for infinite proton mass. The simple expression highlights the role of mass and charge. The cluttered full SI expression $a_0 = 4\pi\epsilon_0 m_e \hbar^2/m_e e^2$ does not contain the speed of light. The electron motion in hydrogen is nonrelativistic and the mean speed is $v = \alpha c$. The ground state kinetic energy is $K_0 = mv^2/2 = \alpha^2 m_e/2$ and the potential energy is twice this value and negative so

$$E_0 = -\frac{1}{2} m_e \alpha^2 = -13.6\ \text{eV}. \tag{1.161}$$

The ground state wave function by the way is $\psi = \psi(0)e^{-r/a_0}$ with density at the origin

$$|\psi(0)|^2 = \frac{1}{\pi a_0^3} = \frac{(\alpha m_e)^3}{\pi}. \tag{1.162}$$

For a $p\bar{p}$ atom, we would use the reduced mass $m_e \to \mu = m_p/2$.

Notice that were the proton and electron charges a bit larger, the mean speed of an electron in hydrogen would approach light speed, and the nonrelativistic Schrödinger theory and Bohr model would be inapplicable. Such is indeed the situation for color bound states, and for inner shell electrons in atoms with high atomic number Z for which $v \sim Z\alpha$. Notice, too, the role of the electron mass in setting the scale of atomic physics. The Bohr radius of muonic hydrogen $(\mu^- p)_{\text{atom}}$ is roughly a factor $m_\mu/m_e \sim 205$ smaller than that of normal hydrogen. If all electrons in normal matter were replaced with muons, the size of atoms and molecules would all be correspondingly reduced.

Gravity is characterized by Newton's constant

$$G_N = 6.674\,28(67) \times 10^{-11}\ \text{m}^3\ \text{kg}^{-1}\ \text{s}^{-2}. \tag{1.163}$$

By combining $G_N$ with other fundamental constants, we arrive at various intriguing Planck scales where quantum effects in gravitation are expected to be important : the Planck length

$$L_P = \sqrt{\frac{\hbar G_N}{c^3}} = 1.616\,252(81) \times 10^{-35}\ \text{m} \tag{1.164}$$

and the Planck mass

$$m_P = \sqrt{\frac{\hbar c}{G_N}} = 2.176\,44(11) \times 10^{-8}\ \text{kg}. \tag{1.165}$$

While the Planck mass is small on human scales, its rest energy equivalent is two billion joules. The Planck energy scale in engineering units is $E_P = m_P c^2 \simeq 1.2 \times 10^{19}$ GeV, far beyond the energy per particle in conceivable artificial particle accelerators.

## 1.10 FURTHER READING

Tables and reviews of particle properties and theories available at the Particle Data Group web site `http://pdg.lbl.gov/` are an essential supplement to this book. The fundamental constants found in this chapter come from the 2007 PDG compilation of NIST recommended values.

Particle physics is covered at an introductory level in Alessandro Bettini, *Introduction to Elementary Particle Physics*, Cambridge University Press (2008), M. G. Bowler, *Femtophysics*, Pergamon (1990), K. Gottfried and V. F. Weisskopf, *Concepts of Particle Physics*, Vol. I and II, Oxford (1986), D. Griffiths, *Introduction to Elementary Particles*, John Wiley & Sons, Inc. (1987), F. Halzen and A. D. Martin, *Quarks and Leptons*, John Wiley & Sons, Inc. (1984), Robert Mann, *Particle Physics and the Standard Model*, CRC Press (2010), D. H. Perkins, *Introduction to High Energy Physics*, Addison-Wesley (2000), Byron P. Roe, *Particle Physics at the New Millennium*, Springer (1996), A. Seiden, *Particle Physics: A comprehensive Introduction*, Pearson (2004).

Short texts emphasizing the standard model gauge theories include G. Kane, *Modern Elementary Particle Physics*, Addison-Wesley, Reading MA. (1993), W. N. Cottingham and D. A. Greenwood, *An Introduction to the Standard Model of Particle Physics*, Cambridge University Press (1998). Many text books are devoted to quantum field theory and specifically to the gauge theories. One especially readable example is Ian J.R. Aitchison and Anthony J.G. Hey, *Gauge Theories in Particle Physics*, Taylor & Francis (2002).

Observation of the electron neutrino is described in Cowan, C. L., Jr., F. Reines, F.B. Harrison, H. W. Kruse, and A. D. Mcquires, (1956). Detection of the Free neutrino: A confirmation., Science **124**, 103 (1956).

The distinction between electron neutrinos and antineutrinos was explored in R. Davis and D. S. Harmer, Bull. Am. Phys. Soc. **4**, 217 (1959) and the distinction between electron and muon neutrino reported in G. Danby et al., Phys. Rev. Lett. **9**, 36 (1962). The distinction between $\nu_\tau$ and $\nu_e$ and $\nu_\mu$ is described in N. Ushida et al., Phys. Rev. Lett. **57**, 2897 (1986).

The 1988 Nobel Prize in Physics was awarded jointly to Leon M. Lederman, Melvin Schwartz and Jack Steinberger "for the neutrino beam method and the demonstration of the doublet structure of the leptons through the discovery of the muon neutrino".

The 1969 Nobel Prize in Physics was awarded to Murray Gell-Mann for the development of the quark model of hadrons.

The 1976 Nobel Prize in Physics was awarded jointly to Burton Richter and Samuel Chao Chung Ting for the discovery of the charmed quark.

The discovery of the $\tau$ lepton is described in M. L. Perl *et al*, "Evidence for Anomalous Lepton Production in e+-e- Annihilation," Phys. Rev. Lett. **35**, 1489 (1975). Martin Perl was awarded the 1995 Nobel Prize in physics for this work along with Frederick Reines who was honored for his detection of the neutrino. The discovery of the $\tau$ neutrino is described in "Final tau-neutrino results from the DONuT experiment," DONuT Collaboration (K. Kodama et al.), Phys. Rev. **D78**, 052002 (2008), e-Print: arXiv:0711.0728

The observation of the doubly strange b-flavored baryon $\Omega_b^-$ is described in V. M. Abazov et al. (D0 Collaboration), Phys. Rev. Lett. **101**, 232002 (2008). See also T. Aaltonen, et al (CDF Collaboration), Phys. Rev. **D80**, 072003 (2009). The discovery of the $\Sigma_b$ (and $\Sigma_b^*$) is described in T. Aaltonen, et al (CDF Collaboration), Phys. Rev. Lett. **99**, 202001 (2007).

For a review of constituent models of quarks and leptons and gauge bosons, see Hidezumi Terazawa, "Subquark model of leptons and quarks," Phys. Rev. **D22**, 184 (1980).

The first observation of antihydrogen obtained by colliding a $\bar{p}$ beam with electron pairs in the Coulomb field of a nucleus is described in G. Baur *et al*, Phys. Lett. **B368**, 251 (1996). The first observation of antihydrogen by the ATHENA collaboration based on combining trapped positions and antiprotons is M. Amoretti *et al*, Nature **419**, 456 (2002).

## 1.11 ■ ONLINE RESOURCES

Some links to physics journals are

- Annual Reviews of Nuclear and Particle Science, `http://arjournals.annualreviews.org/loi/nucl`
- European Physics Journal C, `http://epjc.edpsciences.org/`
- Journal of High Energy Physics, `http://jhep.sissa.it/jhep/`
- Physical Review Letters, `http://prl.aps.org/`
- Physical Review D, `http://prd.aps.org/`
- Physics Letters B, `http://www.elsevier.com/wps/product/cws_home/505706/`
- Physics Reports, `http://www.elsevier.com/wps/product/cws_home/505703/`
- Reports on Progress in Physics, `http://www.iop.org/EJ/journal/RoPP/`
- Reviews of Modern Physics, `http://rmp.aps.org/`.

Electronic access to these journals may require an institutional or personal subscription. Most research papers and conference proceedings appear in the physics e-print archives `http://arxiv.org/archive/hep-xx/` with xx = ex for experiment, xx = ph for phenomenology, and xx = th for theory. The website provides a daily list of papers to download to computer or mobile device. Lectures by Nobel laureates provide perspectives on major discoveries with references. See `http://nobelprize.org/nobel_prizes/physics/laureates/xxxx/` where xxxx is the year.

## 1.12 ■ PROBLEMS

### Problem 1.1. Lepton numbers

Electron number is defined as the number of $e^-$ plus $\nu_e$ minus $e^+$ minus $\bar{\nu}_e$. Similar lepton numbers are defined for the $\mu^-$ and $\tau^-$ and all are conserved as illustrated by the absence of $\pi^0 \to \mu^\pm e^\mp$. [Ref: G. Danby *et al.*, Phys. Rev. Lett. **9**, 36 (1962).] Draw a leading order standard model electroweak Feynman diagram for each of the decays $\mu^- \to e^-\bar{\nu}_e\nu_\mu$ ($\Gamma_i/\Gamma \simeq 100\%$), $\mu^- \to e^-\bar{\nu}_e\nu_\mu\gamma$ ($\Gamma_i/\Gamma \simeq 1.4 \pm 0.04\%$), $\mu^- \to e^-\bar{\nu}_e\nu_\mu e^+e^-$ ($\Gamma_i/\Gamma \simeq (3.4 \pm 0.4) \times 10^{-5}$)) and show that these decays conserve electron and muon number. Show that the decays $\mu^- \to e^-\gamma$, $\mu^- \to e^-e^+e^-$, $\mu^- \to e^-\gamma\gamma$, $\tau^- \to e^-\mu^+\mu^-$, $\tau^- \to e^-\pi^+K^-$ and $\pi^0 \to \mu^\pm e^\mp$ violate lepton numbers. What are the experimental limits on the branching fractions for these examples of lepton number violating modes according to the Particle Data Group?

### Problem 1.2. The $\tau^-$ lepton

List the decay modes of the $\tau^-$ lepton which have a branching fraction larger than 1% according to the PDG and describe or draw a standard model interaction diagram explaining each mode containing three or fewer final state particles. What fraction of the $\tau^-$ decays correspond to one, three, and five final state charged particles?

### Problem 1.3. Quark and baryon number

By examining the fundamental diagrams, show that the standard model interactions conserve quark number defined as the number of quarks minus the number of antiquarks and then, assuming quark composites exist only as mesons and baryons or combinations of these particles (nuclei), show that the standard model interactions conserve baryon number, defined as the number of baryons ($qqq$) minus the number of antibaryons ($\bar{q}\bar{q}\bar{q}$). Show that in $\pi^- p$ collisions producing an $\bar{n}$, the minimal number of final state particles consistent with conservation of charge and baryon number is three and draw or describe a quark level standard model diagram for the process.

### Problem 1.4. Strangeness conservation

The number of $s$-quarks minus the number of s-quark is conserved in electromagnetic and strong interactions which produce or annihilate only a $q\bar{q}$ pair of the same flavor. Draw a Feynman diagram in terms of constituent quarks and fundamental interactions for the strong associated production process

$$\mathrm{p} + \mathrm{Nucleus} \to \Lambda + \mathrm{K}^+ + \mathrm{Nucleus}$$

which illustrates strangeness conservation in strong interactions. Draw quark level Feynman diagrams for the weak interaction decays $K^- \to \mu^- \bar{\nu}_\mu$ (63.5 %), $K^- \to \pi^- \pi^0$ (20.7 %) and $\Lambda \to p\pi^-$ (63.9 %) which violate strangeness conservation highlighting the role of the $W$ boson. (The branching ratio in parenthesis is the ratio of the partial decay width to the total decay width.)

### Problem 1.5. Photon mass

If the photon had mass $m_\gamma$, one might expect Maxwell's equations to be replaced by

$$\partial_t^2 A^0 - \nabla^2 A^0 + m_\gamma^2 A^0 = \rho \; ; \; \partial_t^2 \mathbf{A} - \nabla^2 \mathbf{A} + m_\gamma^2 \mathbf{A} = \mathbf{j}$$

with $\mathbf{B} = \nabla \times \mathbf{A}$ and $\mathbf{E} = -\nabla A^0 - \partial_t \mathbf{A}$. The Coulomb field $\mathbf{E}$ of a point charge and the dipole field $\mathbf{B}$ of a current loop would deviate from their $m_\gamma = 0$ form by a factor $e^{-m_\gamma r}$ as well as polynomial terms. If the Earth's dipole magnetic field is consistent with $m_\gamma = 0$ over a length scale $L$ of roughly 10,000 km to within $p = 5$ %, show that $\hbar/m_\gamma c \sim L/p$ implies the photon mass must be smaller than about $10^{-15}$ eV. Calculate the Compton wavelength $\hbar/m_\gamma c$ in astronomical units (1 a.u. $= 1.5 \times 10^{11}$ m) for a photon mass $m_\gamma = 10^{-17}$ eV. (Ref: A limit based on the Jovian field is found in L. Davis, A. S. Goldhaber, and M. M. Nieto, Phys. Rev. Lett. **35**, 1402 1405 (1975). A limit is of $7 \times 10^{-19}$ eV is given by Jun Luo, Liang-Cheng Tu, Zhong-Kun Hu, and En-Jie Luan, Phys. Rev. Lett. **90**, 081801 (2003). See also Alfred Scharff Goldhaber and Michael Martin Nieto, Rev. Mod. Phys. **82**, (2010), arXiv:0809.1003v5 [hep-ph].)

### Problem 1.6. Static weak interaction energy

Suppose a static weak potential between two fermions due to the $Z$ boson has the form

$$U_Z = -g^2 \frac{e^{-m_Z r}}{r}$$

with $g^2 \simeq \alpha$.
a) Suppose a nonrelativistic weak bound state is confined to a range $r = m_Z^{-1}$. Use the uncertainty principle to argue that such a state is possible only if $\mu g^2/m_Z > 1$ where $\mu$ is the reduced mass of the fermions. [Ref: F. Rogers, H. Graboske, Jr, D. Harwood, Phys. Rev **A1**,1577 (1970)]
b) Given $m_Z = 90$ GeV, what is the minimal $\mu$ in GeV? Given $\mu = m_e$, what maximal $m_Z$ could lead to a bound state?
c) Estimate the weak energy shift of the ground state of hydrogen by averaging the weak energy

$$\Delta E = \int d\mathbf{r}\; \psi^* U_Z \psi \simeq |\psi(0)|^2 \int d\mathbf{r}\; U_Z$$

using the unperturbed wave function $\psi = e^{-r/a_0}/\sqrt{\pi a_0^3}$ with $a_0 = 1/(\mu\alpha)$ and compare to the Rydberg $E_0 = \mu\alpha^2/2$. The tiny weak interaction is actually observable through its parity violation. See Marie-Anne Bouchiat and Claude Bouchiat, Rep. Prog. Phys. **60**, 1351 (1997).

---

**Problem 1.7. Light hadron decays**

---

Draw minimal compound fundamental diagrams showing the fundamental quarks, leptons and gauge bosons to describe the electromagnetic decay $\pi^0 \to \gamma\gamma$, the weak decay $\pi^- \to \mu^-\bar{\nu}_\mu$, and the strong decay $\rho^+(770) \to \pi^+\pi^0$. Find the lifetimes associated with these decay modes in the Particle Data Group tables of particle properties. Find the kinetic energy released by subtracting the rest masses of the final state particles from the rest mass of the initial state particle. Compare the lifetimes to those for $\eta \to \gamma\gamma$ (71%), $K^- \to \mu^-\bar{\nu}_\mu$ (63%), and $K^{*+}(890) \to K^+\pi^0$ $K^{\pi^0}$, $K^0\pi^+$ (100%) and show that in each case the lifetime decreases with the kinetic energy released. How does the CKM factor influence the $K^-$ decay rate?

---

**Problem 1.8. CKM matrix and heavy quark decay**

---

The amplitude for $W$ emission by a quark $q_i \to q_j + W^-$ is found to be proportional to a family independent weak charge times an element $V_{ij}$ of a nearly diagonal unitary matrix called the Cabibbo-Kobayashi-Maskawa (CKM) matrix. The inclusive decay rate $\Gamma(q_i \to q_j + X)$ where $X$ denotes other particles is proportional to the decay rate $q_i \to q_j$ is proportional to $|V_{ij}|^2$. From the empirical matrix elements and the quark mass values, what fraction of $t$-quark decays are to an $s$-quark? Argue that the dominant quark decay chain is $t \to b \to c \to s \to u$ where in each step a $W$ boson is emitted and every other step entailing an off-diagonal element of the CKM matrix is relatively slow. Were $V$ diagonal, which quarks would be stable?

---

**Problem 1.9. Rare decays of $t$ and $Z$**

---

Including the $WWZ$ interaction in addition to the gauge interactions of the fermions in the standard model, draw all leading order Feynman diagrams in terms of fundamental particles for the processes $t \to bW^+Z$ and $Z \to W^+\pi^-$. The total width of the $t$-quark is predicted to be about 1 GeV. If the initial and final state particles are real (not virtual), are these processes consistent with energy and momentum conservation?

---

**Problem 1.10. Fourth generation fermions**

---

Suppose a fourth fermion generation with a lepton pair $\nu_\omega(0)$ and $\omega^-(100)$ and a quark pair $x^{+2/3}(300)$ and $y^{-1/3}(50)$ where the masses in parenthesis are in GeV.

If the generation mixing through the weak interaction follows the observed pattern conserving lepton numbers and with a nearly diagonal quark mixing matrix, which fourth generation particle or particles would decay and which would be stable or relatively stable compared to the $t$-quark and why? Assume $m_t = 170$ GeV, $m_W = 80$ GeV, $m_Z = 90$ GeV, and $m_b = 5$ GeV and $m_c = 1.5$ GeV.

---

**Problem 1.11. Exotic atoms**

---

An antiproton ($\bar{p}$) brought to rest in liquid hydrogen by ionization energy loss may form a protonium atom $(p\bar{p})_{\text{atom}}$, a $p\bar{p}$ bound state analogous to the hydrogen atom. Other exotic atoms include positronium $(e^+e^-)_{\text{atom}}$, muonium $(\mu^+e^-)_{\text{atom}}$, pionium $(\pi^+e^-)_{\text{atom}}$, pionic hydrogen $(\pi^- p)_{\text{atom}}$, kaonic hydrogen $(K^- p)_{\text{atom}}$, and "pi-mu" atoms $(\pi^\pm\mu^\mp)_{\text{atom}}$.
a) Accounting for the reduced mass, give the Bohr model ground state radius $a_0$ in fm and energy $E_0$ in keV of each of these atoms.
b) In case of annihilation to two photons for positronium or protonium at rest, show that the ratio of photon energy to binding energy is $E_\gamma/|E_0| = 4/\alpha^2$ and that the ratio of the photon wavelength to the ground state Bohr radius is $\lambda_\gamma/a_0 = \alpha/2$.
c) Assume the capture of a $\pi^-$ in hydrogen to form $(\pi^- p)_{\text{atom}}$ starts at the hydrogen electronic Bohr radius. Show that the initial principle quantum number of the pion is $n \simeq 16$ and that in the cascade to the ground state, transitions from 4p, 3p, and 2p states to the 1s will source X-rays of energy approximately 3.04 keV, 2.88 keV and 2.43 keV.
d) The $(\pi^- p)_{\text{atom}}$ ground state is found to be shifted down from the Bohr model expectation by 7 eV and to have a full width at half maximum $\Gamma \simeq 1$ eV, both effects ascribed to strong interaction with the nucleus. What is the corresponding strong interaction lifetime and how does it compare to the free pion lifetime $\tau_{\pi^-} \sim 10^{-8}$ s?

---

**Problem 1.12. Muon catalyzed fusion**

---

Brought to rest in a liquid hydrogen target containing the isotopes deuterium D and tritium T, a $\mu^-$ can form diatomic exotic molecules such as $e^-\mu^-dd$ or $e^-\mu dt$ where $d$ and $t$ are the deuterium and tritium nuclei. Estimate the size of these molecules. [Hint: Consider first the $\mu^-dd$ ion.] Describe qualitatively the electronic excitation spectrum for such neutral molecules in gaseous form. The fusion reaction $t + d \to {}^4He + n + 17.6$ MeV can occur by quantum tunneling through the repulsive Coulomb barrier between $t$ and $d$. Under what conditions might muon fusion catalysis process provide a practical energy source? [Yu. V. Petriv, *Nature* **285**, 466 (1980); J. Rafelski and H. E. Rafelski, *Particle World*, Vol. 2, No. 1, 21 (1991).]

## Problem 1.13. Klein-Gordon equation for string

Consider displacements $y(z, t)$ from equilibrium of a string of linear mass density $\lambda$ under tension $T$ strung along the $z$-direction and embedded in a homogeneous elastic medium that gives a restoring force per unit length $f = -\kappa y$ with $\kappa$ a constant.
a) Apply Newton's second law to a small mass element to show that displacements of the string from equilibrium satisfy the one dimensional Klein-Gordon equation with $c^2 = T/\lambda$.
b) What is the general form of static solutions.
c) Show that the static shape coresponding to a knife-edge transverse force giving a fixed displacement at $z = 0$ in an infinitely long string is an exponential function. This is the one-dimensional analog of the Yukawa potential.
d) What is the minimal frequency of harmonic traveling waves?

## Problem 1.14. Yukawa potential

a) Show that the Yukawa potential $\phi = \frac{g}{4\pi}e^{-mr}/r$ is a solution to the Klein-Gordon equation

$$-\nabla^2\phi + m^2\phi = g\delta(\mathbf{x})$$

for a point source by expressing $\nabla^2\phi = (1/r)\partial_r^2(r\phi)$ in spherical coordinates and considering the point $\mathbf{x} = 0$ as in electrostatics.
b) Find the solution for a sphere of radius $R$ and constant particle density $\rho$ for $r < R$ and $r > R$. (Assume $\phi(r) = f(r)/r$, find an equation for f(r) and solve by imposing appropriate boundary conditions.)
c) Show that in the limit $m \to 0$, your solution for the sphere reduces to the usual electrostatic result derived from Gauss's Law while in the limit $mR \to \infty$ the solution approaches a spherical square well. This is expected since a test particle interacts only with matter within a range $\sim 1/m$ and, in the short range limit, the interior environment is uniform. As a model for the self-consistent field of a nucleus, this result suggests nucleons will behave as a confined free particle Fermi gas and be uniformly distributed. Limits on the contribution of a Yukawa-like component to gravity and non-gravitational interactions of macroscopic range and related particles including fat gravitons, chameleons, dilatons, ghost condensates and axions are described in E.G. Adelberger et al., Progress in Particle and Nuclear Physics **62**, 102 (2009).

## Problem 1.15. Pauli matrices

Show that the Pauli matrices satisfy the relations

$$\sigma_i\sigma_j = \delta_{ij} + i\epsilon_{ijk}\sigma_k \; ; \; \sigma_i^\dagger = \sigma_i \; ; \; \mathrm{Tr}[\sigma_i] = 0 \qquad (1.166)$$

and that if **a** and **b** are two vectors of operators which commute with the operators $\boldsymbol{\sigma}$ then

$$\mathbf{a}\cdot\boldsymbol{\sigma}\mathbf{b}\cdot\boldsymbol{\sigma} = \mathbf{a}\cdot\mathbf{b} + i\boldsymbol{\sigma}\cdot\mathbf{a}\times\mathbf{b}.$$

Use these results with $\mathbf{a} = a\mathbf{n}$ and $\mathbf{n}$ a unit vector to verify the equality of the Taylor series expansions for the relation

$$e^{i\mathbf{a}\cdot\boldsymbol{\sigma}} = \cos(a) + i\sin(a)\mathbf{n}\cdot\boldsymbol{\sigma}.$$

---

**Problem 1.16. Spin precession**

---

The equation of motion for the spin wave function of an electron in a uniform constant magnetic field is

$$-i\hbar\frac{d\phi}{dt} = -\boldsymbol{\mu}\cdot\mathbf{B}\phi = -\alpha\frac{e\hbar}{2m_e}\boldsymbol{\sigma}\cdot\mathbf{B}$$

where $\alpha = (g/2)$. The general solution is $\phi = a_0 e^{-i\omega t}\phi_+ + b_0 e^{+\omega t}\phi_-$ where $\omega = \alpha 2\mu B/\hbar = \alpha e B/m_e = \alpha\omega_c$ where $\omega_c$ is the cyclotron frequency and $\phi_\pm$ are solutions to the equation $\boldsymbol{\sigma}\cdot\mathbf{n}\phi_\pm = \pm\phi_\pm$. Define the average spin vector $<\mathbf{s}> = \phi^\dagger(\boldsymbol{\sigma}/2)\phi$ and show that

$$\frac{d}{dt} < \mathbf{s} > = \dot{\phi}^\dagger\mathbf{s}\phi + \phi^\dagger\mathbf{s}\dot{\phi} = \omega_c\mathbf{n}\times < \mathbf{s} > . \tag{1.167}$$

You may use the explicit solution choosing $\mathbf{n}$ along the $z$-axis, or substitute for $\dot{\phi}$ and $\dot{\phi}^\dagger$, express $\phi_\pm = \mp\boldsymbol{\sigma}\cdot\mathbf{n}$ and $\phi^\dagger_\pm = \mp\boldsymbol{\sigma}^\dagger\cdot\mathbf{n}$ and use the relations $\sigma_i\sigma_j = \delta_{ij} + i\epsilon_{ijk}\sigma_k$ and $\sigma_i^\dagger = \sigma$ to reduce the products of Pauli matrices. What is $\omega$ in $\mathrm{s}^{-1}$ for $|\mathbf{B}| = 1$ T?

---

**Problem 1.17. Planck scales**

---

Verify the numerical values of the following Planck numbers:

$$\begin{aligned}
L_P &= \sqrt{\frac{\hbar G_N}{c^3}} = 1.616\,252(81)\times 10^{-35}\ \mathrm{m}\\
m_P &= \sqrt{\frac{\hbar c}{G_N}} = 2.176\,44(11)\times 10^{-8}\ \mathrm{kg}\\
E_P = m_P c^2 &= \sqrt{\frac{\hbar c^5}{G_N}} = 1.9561\times 10^{9}\ \mathrm{J}\\
t_P = L_P c^{-1} &= \sqrt{\frac{\hbar G_N}{c^5}} = 5.391\,24(27)\times 10^{-44}\ \mathrm{s}\\
T_P = \frac{E_P}{k_B} &= \sqrt{\frac{\hbar c^5}{G_N k_B^2}} = 1.416\,785(71)\times 10^{32}\ \mathrm{K}
\end{aligned}$$

What is the ratio of water density to the Planck mass density $\rho_P = m_P/L_P^3$?

---

**Problem 1.18. Magnetars**

---

A magnetar is a neutron star with an abnormally large surface magnetic field $B \simeq 10^{10}$ T. Such strong fields lead to photon splitting, vacuum birefringence, and other strange effects such as squeezing a hydrogen atom into a needle shape.

a) Show the magnetic rest energy density $B^2/(2\mu^0 c^2)$ for a magnetar is over $10^4$ times the density of Pb.
b) Show that the energy difference $2\mu_B B$ between magnetic substates of an electron equals the electron rest mass at field strength

$$B_c = \frac{m_e^2 c^2}{\hbar e} = 4.4 \times 10^9 \text{ T}.$$

c) The classical radius of free gyration around the field lines is $r = p/(eB)$. Quantum mechanics requires in the ground state $p \sim \hbar/r$. Show that the implied radius is

$$r = \lambda_e \sqrt{B_c/B} = \alpha^{-1} r_e \sqrt{B_c/B}$$

where $\lambda_e = \hbar^2/(m_e c)$ and $r_e = \alpha/(m_e c^2)$. d) Show that this is smaller than the Bohr radius for magnetic fields exceeding $\alpha^2 B_c$. For such fields, atoms will be considerably distorted. [Ref: W. Rosner, G. Wunner, H. Herold, and H. Ruder, J. Phys. B: At. Mol. Phys. **17**, 29 (1984), Robert C. Duncan, "Physics in ultra-strong magnetic fields," Fifth Huntsville Gamma-Ray Burst Symposium,arXiv:astro-ph/0002442v1]

CHAPTER

# 2 Relativity, Accelerators, and Collisions

Relativity is essential in particle physics. In this chapter, we review the theory of special relativity and common notation for Lorentz covariant quantities. Some important aspects of relativistic kinematics are considered. We also take a look at the subject of accelerators governed by classical physics.

**Contents**



## 2.1 ■ SPECIAL RELATIVITY

Einstein's special theory of relativity is essential in understanding and describing particle physics. It derives from the observation that the laws of physics are the same in all frames of reference in uniform motion relative to the fixed stars (inertial reference frames) while the speed of light is independent of reference frame. Implicit is the assumption that space and time are homogeneous and isotropic. The theory teaches that time and length intervals are relative quantities - their values depend on frame of reference. The values of both the time coordinate $t$ and of the 3-space coordinates $\mathbf{x}$ of a localized event in one inertial frame of reference are related to their values $t'$ and $\mathbf{x}'$ in another inertial frame in relative motion by a Lorentz transformation. More generally, the Lorentz transformation relates the values of all physical quantities in one frame to those in another. In this chapter we study the implications of relativity for kinematics and classical mechanics.

The form of the Lorentz transformation may be found by considering the coordinates of a point on an expanding spherical pulse of light originating in

**FIGURE 2.1** Aerial view of the CERN Large Hadron Collider (LHC). The LHC is the world's largest and most powerful colliding beam accelerator. The LHC is located underground near Geneva, Switzerland. [CERN]

two frames of reference the origins of which coincide with the source of light. The coordinates of any point on the spherical light front satisfy

$$s^2 \equiv t^2 - \mathbf{x}^2 = t'^2 - \mathbf{x}'^2 = 0 \tag{2.1}$$

where we have set $c = 1$. This equivalence is consistent with an orthogonal rotation of the space coordinates such as

$$\begin{aligned} x' &= +x \cos\theta + y \sin\theta \\ y' &= -x' \sin\theta + y \cos\theta \end{aligned} \tag{2.2}$$

with $t' = t$. Such a transformation relates frames of reference with a common origin of space and time and at rest with respect to one another. Consider the case in which the two frames are similarly oriented and the second frame moves with velocity $v$ along the $x$-direction relative to the first frame as illustrated in Figure 2.2. The Lorentz transformation is

$$x' = \gamma(x - vt) = x \cosh\eta - t \sinh\eta \tag{2.3}$$

$$t' = \gamma(t - v)x) = t \cosh\eta - x \sinh\eta \tag{2.4}$$

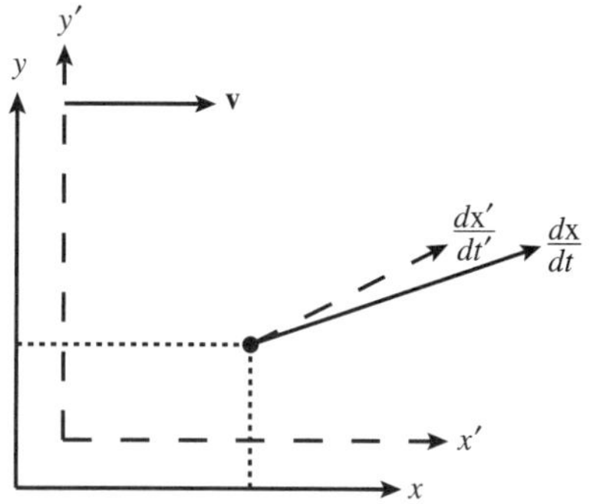


**FIGURE 2.2** Inertial frames of reference. The coordinates and velocity of a particle have different values relative to two inertial frames of reference with relative velocity **v**. The values are related by the Lorentz transformation.

with $y' = y$ and $z' = z$. Here

$$\gamma = \left(1 - v^2\right)^{-1/2} \tag{2.5}$$

and we introduce the quantity

$$\eta = \tanh^{-1}(v) \tag{2.6}$$

called the rapidity. This so-called boost transformation is mathematically equivalent to an orthogonal rotation in a plane with coordinates $x$ and $it$ by an imaginary angle $\theta = i\eta$. Correspondingly, a boost by rapidity $\eta_1$ followed by a boost by rapidity $\eta_2$ along the same spatial direction gives a boost with rapidity $\eta = \eta_1 + \eta_2$. The general Lorentz transformation of coordinates is a combination of a rotation and boost.

It is customary in particle physics to define contravariant (upper index) and covariant (lower index) 4-vector components as

$$x^\mu = (t, \mathbf{x}) \ , \ x_\mu = (t, -\mathbf{x}) \tag{2.7}$$

with the Lorentz index $\mu = 0,1,2,3$ so that if $\mathbf{x} = (x, y, z)$ then $x^0 = t$, $x^1 = x$, $x^2 = y$, and $x^3 = z$. The inner product of two 4-vectors is defined by implicit summation (contraction) over one covariant and one contravariant repeated index, e.g.

$$x^2 = \Sigma_{\mu=0}^{3} x_\mu x^\mu \equiv x_\mu x^\mu = t^2 - \mathbf{x}^2. \tag{2.8}$$

Lorentz indices may be raised or lowered with the diagonal numerical tensor

$$g_{\mu\nu} = g^{\mu\nu} = \begin{pmatrix} 1 & 0 & 0 & 0 \\ 0 & -1 & 0 & 0 \\ 0 & 0 & -1 & 0 \\ 0 & 0 & 0 & -1 \end{pmatrix} \tag{2.9}$$

Thus, with implicit summation, $x_\mu = g_{\mu\nu}x^\mu$ and $x^2 = g_{\mu\nu}x^\mu x^\nu$. The general Lorentz transformation has the form

$$x'^\mu = \Lambda^\mu_{\ \nu} x^\nu \tag{2.10}$$

where the matrix $\Lambda$ satisfies $\Lambda_{\mu\nu}\Lambda^\nu_{\ \sigma} = g_{\mu\sigma}$.

The Lorentz covariant gradient in this notation is

$$\partial_\mu = \frac{\partial}{\partial x^\mu} = (\partial_t, \nabla) \ ; \ \partial^\mu = \frac{\partial}{\partial x_\mu} = (\partial_t, -\nabla) \,. \tag{2.11}$$

Note very carefully the signs of the spatial components in these expressions. Covariant and contravariant 4-vector components transform in equivalent but mathematically distinct ways such that their inner product is invariant. The sign of the space components of the 4-gradient is opposite to the sign for the coordinate 4-vector ensuring, for example, that $\partial_\mu x^\mu$ is invariant.

The inner product of 4-vectors is invariant under Lorentz transformations. In particular, the interval $d\tau$ between two events with 4-vector difference $dx$, defined as

$$d\tau^2 = dx^2 = dt^2 - d\mathbf{x}^2, \tag{2.12}$$

is invariant. In a frame in which two events occur at the same point in space ($d\mathbf{x} = 0$), the interval $d\tau = dt$ may be identified as the proper time between the events and the relation between proper time and the time interval in a frame with relative velocity $v$ is

$$\frac{d\tau}{dt} = \sqrt{1 - \frac{d\mathbf{x}^2}{dt^2}} = \sqrt{1 - v^2} = \gamma^{-1}. \tag{2.13}$$

The trajectory of a particle is specified by four coordinate functions and the 4-velocity of a particle is defined as

$$v^\mu = \frac{dx^\mu}{d\tau} = \gamma(1, \mathbf{v}) \tag{2.14}$$

The 4-momentum $p$ of a particle of invariant mass $m$ and velocity $\mathbf{v} = |\mathbf{v}|\mathbf{n}$ with magnitude $|\mathbf{v}|$ along a unit vector $\mathbf{n}$ has components

$$p^\mu = (E, \mathbf{p}) = mv^\mu = (\gamma m, \gamma m\mathbf{v}) = m(\cosh\eta, \sinh\eta\ \mathbf{n}) \tag{2.15}$$

where $\eta = \tanh^{-1}(|\mathbf{v}|)$ is the rapidity associated with the boost to the rest frame of the particle. The relativistic total energy $E$ here can be regarded as a sum of rest energy and kinetic energy

$$E = \gamma m = m + K \tag{2.16}$$

with

$$K = E - m = (\gamma - 1)m. \tag{2.17}$$

When $v$ is small compared to one, we can make the approximation $\gamma \simeq 1 + v^2/2 + \ldots$ and find $K \simeq mv^2/2$. Similarly, the relativistic three momentum $\mathbf{p}$ reduces to the Newtonian expression $p = m\mathbf{v}$ when $v << 1$. The components of the 4-vectors $x$, $v$ and $p$ in different frames of reference are related by the same Lorentz transformation. By definition we have

$$p^2 = p_\mu p^\mu = E^2 - \mathbf{p}^2 = m^2 \tag{2.18}$$

which is invariant under Lorentz transformation.

## 2.2 ■ RELATIVISTIC KINEMATICS

Conservation of 4-momentum is a fundamental principle of physics. The total 4-momentum of a closed system is conserved even if the number or type of the constituents changes. In particular, in any transformation of a collection of initial free particles into a collection of final free particles, if $p_j$ is the 4-momentum of initial particle $j$ and $p'_k$ is the 4-momentum of final particle $k$, then:

$$p_{\text{initial}} = \Sigma_j p_j = \Sigma_k p'_k = p_{\text{final}} \tag{2.19}$$

where the equality holds for each 4-vector component. If the momenta of say two particles are $p_1$ and $p_2$ in some reference frame, the total initial momentum is $p_{\text{initial}} = p_1 + p_2$ and the value of each component of the 4-vector $p_{\text{initial}}$ is fixed. Since the components of $p_{initial}$ are each conserved, so is the square $p^2_{\text{initial}}$. As this is the square of a 4-vector, its value is independent of reference frame. It is called the invariant mass of the system.

If an interaction between the two particles results in, say, three free particles with total momentum $p_{\text{final}} = p_3 + p_4 + p_5$, then 4-momentum conservation requires all final components equal all initial components, $p_{\text{final}} = p_{\text{initial}}$, and consequently the final invariant mass is equal to the initial invariant mass

$$p^2_{\text{final}} = p^2_{\text{initial}}. \tag{2.20}$$

In other words, invariant (meaning the value is independent of frame) mass is conserved (meaning that value is not altered by any interaction).

In the fusion of two particles with momenta $p_1$ and $p_2$ to form a particle of momentum $p$ and mass $m$, the momentum vectors must satisfy

$$p_1 + p_2 = p \tag{2.21}$$

with $p_1^2 = m_1^2$, $p_2^2 = m_2^2$ and $p^2 = m^2$. Similarly in the decay of a particle of momentum $p$ into two particles with momenta $p_3$ and $p_4$, the 4-momentum vectors satisfy

$$p = p_3 + p_4 \tag{2.22}$$

with $p^2 = m^2$, $p_3^2 = m_3^2$ and $p_4^2 = m_4^2$. If two particles with momenta $p_1$ and $p_2$ interact leading to two particles of momenta $p_3$ and $p_4$, then

$$p_1 + p_2 = p_3 + p_4 \tag{2.23}$$

with $p_1^2 = m_1^2$, $p_2^2 = m_2^2$, $p_3^2 = m_3^2$, and $p_4^2 = m_4^2$. Four momentum conservation alone does not require the intermediate state to have a fixed invariant mass but if in fact the reaction proceeds through production and decay of a particle of invariant mass $m$, then we must also have

$$(p_1 + p_2)^2 = m^2 = (p_3 + p_4)^2 . \tag{2.24}$$

---

**Example 2.1. Limits to Dalitz decay of the $\pi^0$.** What is the approximate maximal number of electrons and positrons consistent with 4-momentum conservation that can be produced in the decay of a $\pi^0$ meson? Although consistent with energy and momentum conservation, why are such decays not observed?

Consider the decay in the rest frame of the $\pi^0$ where the total energy is $m_{\pi^0} =$ 135 MeV. Decays to $e^+e^-$ can result from high order virtual radiant photons each converting to a pair. If the final state particles are all nearly at rest, the available energy is optimally converted to rest mass. Since $m_e$=0.511 MeV, the maximum number of electrons plus positrons is about 260. In reality the $\pi^0$ decays most often to two photons. The decay to one real and one virtual photon that produces an $e^+e^-$ pair has a probability of roughly $\alpha = 1/137$. The decay to two $e^+e^-$ pairs has a probability of about $\alpha^2$. A decay to a cloud of stationary pairs has negligible probability.

---

The special frame of reference (defined up to a rotation) in which the total momentum of a particle system vanishes is called the center of mass or center of momentum frame. When a reaction proceeds through the production of an intermediate state particle of fixed mass, the center of mass frame is the rest frame of that particle but the center of mass frame is of more general utility. The momentum of a pair of particles for example can be expressed as

$$p = p_1 + p_2 = (E, \mathbf{p}) = \sqrt{s}\gamma(1, \mathbf{v}_{cm}) \tag{2.25}$$

where the invariant mass squared is

$$s \equiv p^2 = E^2 - \mathbf{p}^2. \tag{2.26}$$

Given the values of $p_1$ and $p_2$ in one frame, we may use the expression

$$v_{cm} = \frac{p}{\sqrt{s}} \tag{2.27}$$

and a corresponding Lorentz transformation to find the 4-momentum of each of the particles in the center of mass frame.

To express the magnitude of the energies and momenta of two particles in their center of mass in terms of invariants, we can take advantage of relativity and the following generally useful trick. If the momenta of two particles are $p_1$ and $p_2$ in some frame, the energy of 2 seen from the rest frame of 1 has a value all observers must agree upon and therefore it has a Lorentz invariant expression. The correct expression depends linearly on the momentum $p_2$ and on the velocity $v_1$ is readily guessed:

$$E_{21} = p_2 v_1 = \frac{p_2 p_1}{m_1} = \gamma_1 (E_2 - \mathbf{v}_1 \cdot \mathbf{p}_2) \tag{2.28}$$

which reduces to $E_2$ when $v_1 = (1, \mathbf{0})$. This expression is simply the Lorentz transformation of the time component of $p_2$. The magnitude of the momentum and of the relative velocity that follow are

$$\mathbf{p}_{21}^2 = E_{21}^2 - m_2^2 = \frac{(p_1 p_2)^2 - m_1^2 m_2^2}{m_1^2} \tag{2.29}$$

and

$$\mathbf{v}_{21}^2 = \frac{\mathbf{p}_{21}^2}{E_{21}^2} = \frac{(p_1 p_2)^2 - m_1^2 m_2^2}{(p_1 p_2)^2} = \mathbf{v}_{12}^2. \tag{2.30}$$

We can apply this idea to find the energy of say particle 2 in the center of mass of two particles with momenta $p_1$ and $p_2$ by replacing the 4-velocity of particle 1 with the 4-velocity of the center of mass:

$$E_{2,cm} = p_2 v_{cm} = p_2 \frac{p}{\sqrt{s}} = \frac{1}{2\sqrt{s}} \left(p^2 + p_2^2 - p_1^2\right) = \frac{1}{2\sqrt{s}} \left(s + m_2^2 - m_1^2\right) \tag{2.31}$$

where we used $2pp_2 = p^2 + p_2^2 - (p - p_2)^2$ and $p_1 = p - p_2$. Notice the result is manifestly invariant as it is expressed in terms of the invariant $s$ and the invariant masses $m_1$ and $m_2$. The magnitude of the momentum of either particle in the center of mass follows from $\mathbf{p}_{cm}^2 = E_{2,cm}^2 - m_2^2$ and is given by

$$|\mathbf{p}_{cm}| = \frac{1}{2\sqrt{s}} \left[\left(s - (m_1 + m_2)^2\right)\left(s - (m_1 - m_2)^2\right)\right]^{\frac{1}{2}}. \tag{2.32}$$

We will use these two kinematic formulae so often they are reproduced in Table 2.1 for convenience.

$$E_{1,cm} = \frac{1}{2\sqrt{s}}(s + m_1^2 - m_2^2)$$

$$E_{2,cm} = \frac{1}{2\sqrt{s}}(s + m_2^2 - m_1^2)$$

$$|\mathbf{p}_{cm}| = \frac{1}{2\sqrt{s}} \left[\left(s - (m_1 + m_2)^2\right)\left(s - (m_1 - m_2)^2\right)\right]^{\frac{1}{2}}$$

**TABLE 2.1** Energy and momentum of two particles in their center of mass frame.

**Example 2.2. Two-body decay of the top quark.** The decay of a top quark $t \to W^+ b$ produces a $W$ boson with a unique energy in the top quark rest frame. What is its approximate value? What is the corresponding speed of the $W$ boson in units of light speed? What is the speed of the $b$-quark in the $t$-quark rest frame?

If we take particle 1 to be the bottom quark and particle 2 to be the $W$ boson, we can apply the result just derived for $E_{2,cm}$ to find the $W$ boson energy in the center of mass of the pair which is the rest frame of the $t$-quark. Using $\sqrt{\mathrm{s}} = \mathrm{m_t}$ and round numbers $m_t$ =170 GeV, $\mathrm{m_W}$ = 80 GeV, and $m_b = 5$ GeV, we find

$$E_W = \frac{1}{2(170)}\left(170^2 + 80^2 - 5^2\right) \simeq 100 \text{ GeV} \tag{2.33}$$

implying $E_b = \mathrm{m_t} - E_W = 70$ GeV. The three momentum and speeds of the $W$ boson and $b$-quark are

$$|\mathbf{p}| = \left[100^2 - 80^2\right]^{1/2} = 70 \text{ GeV}\ ;\ \mathrm{v_W} = |\mathbf{p}|/E_W = 0.66\ \mathrm{c};\ \mathrm{v_b} = |\mathbf{p}|/E_b \simeq 1 \tag{2.34}$$

The $b$-quark momentum and energy are far in excess of its rest mass so the speed of the $b$-quark is close to $c$ with $\gamma \simeq p/m_b = 70/5 = 14$. In a laboratory frame moving with 4-velocity $v$ relative to the top quark, the energy of the $W$ boson could be found using the expression $E_{W,lab} = p_{W,cm} v$.

Now consider a three body decay of a mass $m_0$ to three masses $m_1$, $m_2$, and $m_3$. Four momentum conservation reads

$$p_0 = p_1 + p_2 + p_3 \tag{2.35}$$

and we can construct the invariants

$$\begin{aligned} s_{23} &= (p_0 - p_1)^2 = (p_2 + p_3)^2 \\ s_{13} &= (p_0 - p_2)^2 = (p_1 + p_3)^2 \\ s_{12} &= (p_0 - p_3)^2 = (p_3 + p_1)^2 \end{aligned} \tag{2.36}$$

which satisfy

$$s_{12} + s_{13} + s_{23} = m_0^2 + m_1^2 + m_2^2 + m_3^2. \tag{2.37}$$

The decay products must be coplanar so two angles or two invariants characterize the decay up to an overall rotation. Consider one of these in the rest frame of $m_0$:

$$s_{23} = (p_0 - p_1)^2 = m_0^2 - 2m_0 E_1 + m_1^2. \tag{2.38}$$

Since the minimum value of $E_1$ is $m_1$, the maximum mass of the others is

$$m_{23}^{\max} \equiv \sqrt{s_{23}^{\max}} = m_0 - m_1 \tag{2.39}$$

and this decay configuration leaves $m_1$ at rest while $m_2$ and $m_3$ recoil from each other in opposite directions. The minimum value of $s_{23}$ corresponds to a maximum value for $E_1$. The formula for the energy $E_1$ when $m_0$ decays to a mass $m_1$ and a mass $m_{23}$ also tells us that $E_1$ is maximal when $m_{23}$ is minimal. The minimal value is simply $m_{23}^{\min} = m_2 + m_3$. In this decay configuration, if $m_2 > 0$ and $m_3 > 0$, then $m_2$ and $m_3$ have vanishing relative velocity. If $m_2 > 0$ while $m_3 = 0$, then $E_3 = 0$. If both $m_2 = 0$ and $m_3 = 0$ then $m_{23} = 0$ implies $\mathbf{p}_2$ and $\mathbf{p}_3$ are parallel while these two particles may share the energy $m_0 - E_1$ in a variety of ways.

For a given value of $s_{23}$ between the limits just found, we can find the range of values of say $s_{12}$ if we think of the decay sequentially. Imagine $m_0$ decays to two masses $m_1$ and $m_{23}$ which recoil from each other with equal and opposite 3-momentum in the frame of $m_0$. The magnitude of the decay momentum in the frame of $m_0$ is fixed once $m_{23}$ is fixed along with $m_1$ and $m_0$. Then $m_{23}$ decays to $m_2$ and $m_3$ and in the frame of $m_{23}$ these two particles emerge back to back with momentum fixed by $m_{23}$, $m_2$, and $m_3$. If in the frame of $m_{23}$, the direction of $m_2$ is along the direction of $m_{23}$ so opposite to that of $m_1$, then $s_{12}$ will be maximal while if $m_2$ is directed opposite to $m_{23}$ then $s_{12}$ will be minimal.

We now calculate $s_{13}$ in the frame of $m_{23}$ where $\mathbf{p}_0 = \mathbf{p}_1$. (In this frame, a mass $m_0$ arrives, the mass $m_{23}$ appears at rest, and a mass $m_1$ heads off in the same direction as $m_0$ was headed.) The energy of $m_1$ viewed from the frame of $m_{23}$ is

$$E_{1,23} = p_1 v_{23} = p_1 \frac{p_{23}}{m_{23}} \tag{2.40}$$

where $p_{23} = p_2 + p_3$ with $p_{23}^2 = s_{23}$. Momentum conservation reads $p_0 = p_1 + p_{23}$ so

$$p_0^2 = p_1^2 + p_{23}^2 + 2p_1 p_{23} \tag{2.41}$$

and hence we have the invariant expressions for the energy and 3-momentum

$$\begin{aligned} E_{1,23} &= -\frac{s_{23} + m_1^2 - m_0^2}{2\sqrt{s_{23}}} \\ |\mathbf{p}_{1,23}| &= \sqrt{E_{1,23}^2 - m_1^2}. \end{aligned} \tag{2.42}$$

Notice that the energy is just the negative of the decay energy for the process $m_{23} \to m_0 + m_1$. The energy of $m_2$ in the $m_{23}$ frame can be found in terms of invariants using our two body decay formula again and its 3-momentum $p_{2,23}$ follows:

$$\begin{aligned} E_{2,23} &= \frac{1}{2\sqrt{s_{23}}}(s_{23} + m_2^2 - m_3^2) \\ |\mathbf{p}_{2,23}| &= \sqrt{E_{2,23}^2 - m_2^2}. \end{aligned} \tag{2.43}$$

Finally we construct the invariant mass in the $m_{23}$ frame as

$$s_{12} = (p_1 + p_2)^2 = m_1^2 + m_2^2 + 2(E_{1,23}E_{2,23} - |\mathbf{p}_{1,23}||\mathbf{p}_{2,23}|\cos\theta_{12,23}) \tag{2.44}$$

where $\theta_{12,23}$ is the angle between the momentum vectors of $\mathbf{p}_{1,23}$ and $\mathbf{p}_{2,23}$ in the frame of $m_{23}$. The values $\cos\theta_{12,23} = \pm 1$ determine the maximum and minimum values for $s_{12}$ for a given $s_{23}$. A plot of values of the pairs $s_{23}$ and $s_{12}$ is called a Dalitz plot and the boundaries of the region permitted by four momentum conservation are determined by Equation 2.44. Examples are shown in Figure 2.3.

We are often interested in an interaction of two initial particles with prescribed 4-momenta $p_1$ and $p_2$ leading to two final particles with momenta $p_3$ and $p_4$ satisfying

$$p_1 + p_2 = p_3 + p_4. \tag{2.45}$$

In the center of mass frame, the initial three momentum vectors are equal and opposite with some value $\mathbf{p}$ and three momentum conservation is satisfied if the final three momenta are also equal and opposite with some value $\mathbf{p}'$,

$$\mathbf{p}_1 = \mathbf{p} = -\mathbf{p}_2\ ;\ \mathbf{p}_3 = \mathbf{p}' = -\mathbf{p}_4 \tag{2.46}$$

The total energy $\sqrt{s}$ and the magnitude of $\mathbf{p}$ are determined by $\mathbf{p}_1$ and $\mathbf{p}_2$. The magnitude $|\mathbf{p}'|$ is determined by energy conservation in the center of mass

$$\sqrt{s} = E_3 + E_4 = \sqrt{m_3^2 + \mathbf{p}'^2} + \sqrt{m_4^2 + (-\mathbf{p}')^2}. \tag{2.47}$$

The only degree of freedom is the direction of $\mathbf{p}'$ relative to $\mathbf{p}$. Once this direction is specified, the 4-momentum vectors $p_3$ and $p_4$ in the laboratory frame can be found by Lorentz transformation from the center of mass frame to the laboratory frame.

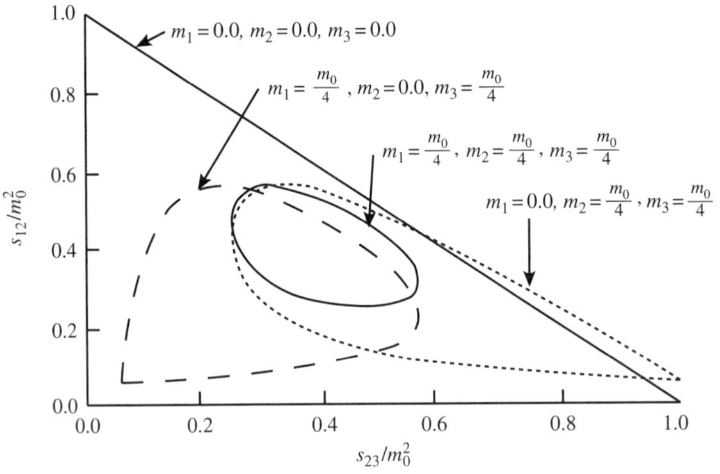


**FIGURE 2.3** Examples of Dalitz plot boundaries. The allowed regions are shown for $s_{12} = m_{12}^2$ and $s_{23} = m_{23}^2$ for a three body decay in which a mass $m_0$ produces masses $m_1$, $m_2$, and $m_3$.

In describing a two-to-two reaction, the values of quantities in the center of mass frame are Lorentz invariant - observers in different reference frame all agree on the values seen by an observer in the center of mass. It follows that these invariant aspects of the collision must have Lorentz invariant expressions in terms of inner products of the various 4-momentum vectors. It is convenient to use instead the Lorentz invariant Mandelstam variables (Figure 2.4) defined by

$$\begin{aligned} s &= (p_1 + p_2)^2 = m_1^2 + m_2^2 + 2p_1 p_2 = (p_3 + p_4)^2 = m_3^2 + m_4^2 + 2p_3 p_4 \\ t &= (p_3 - p_1)^2 = m_3^2 + m_1^2 - 2p_3 p_1 = (p_4 - p_2)^2 = m_4^2 + m_2^2 - 2p_4 p_2 \\ u &= (p_1 - p_4)^2 = m_1^2 + m_4^2 - 2p_1 p_4 = (p_3 - p_2)^2 = m_3^2 + m_2^2 - 2p_3 p_2. \end{aligned} \tag{2.48}$$

The quantity $s$ is the square of the energy in the center of mass. If the particles fuse into a single particle with momentum $p = p_1 + p_2$, then $s$ is the square of its mass. The quantity $t$ is called the square of the 4-momentum transfer. In the case particle 1 absorbs an exchanged particle and emerges as particle 3, the momentum of the exchanged particle is $p = p_3 - p_1$ and $t$ is the square of its mass. Inner products of the 4-momenta may all be expressed in terms of invariant masses and the Mandelstam variables using these expressions. Since

$$s + t + u = m_1^2 + m_2^2 + m_3^2 + m_4^2 \equiv h, \tag{2.49}$$

two Mandelstam variables suffice, say $s$ and $t$. These are equivalent to specifying to the total energy and the angle $\theta_{CM}$ between $\mathbf{p}_{CM}$ and $\mathbf{p}'_{CM}$ in the center of mass frame.

Suppose we prescribe the four momenta $p_1$ and $p_2$. Then we can compute $s = (p_1 + p_2)^2$ and then express the magnitudes of the energies and momenta of all four particles in the center of mass in terms of $s$ and the masses using the formula in Table 2.1. The four momentum transfer can be written in terms of the angle $\theta_{31,CM} \equiv \theta_{CM}$ between the momentum vectors $\mathbf{p}_{3,CM}$ and $\mathbf{p}_{1,CM}$ as

$$\begin{aligned} t &= (p_3 - p_1)^2 = (E_{3,CM} - E_{1,CM})^2 - (\mathbf{p}_{3,CM} - \mathbf{p}_{1,CM})^2 \\ &= (E_{3,CM} - E_{1,CM})^2 - |\mathbf{p}_{3,CM}|^2 - |\mathbf{p}_{1,CM}|^2 + |\mathbf{p}_{3,CM}||\mathbf{p}_{1,CM})| \cos\theta_{31,CM} \\ &= (E_{3,CM} - E_{1,CM})^2 - (|\mathbf{p}_{3,CM}| - |\mathbf{p}_{1,CM}|)^2 - 4|\mathbf{p}_{3,CM}||\mathbf{p}_{1,CM}| \sin^2 \frac{\theta_{CM}}{2} \end{aligned}$$

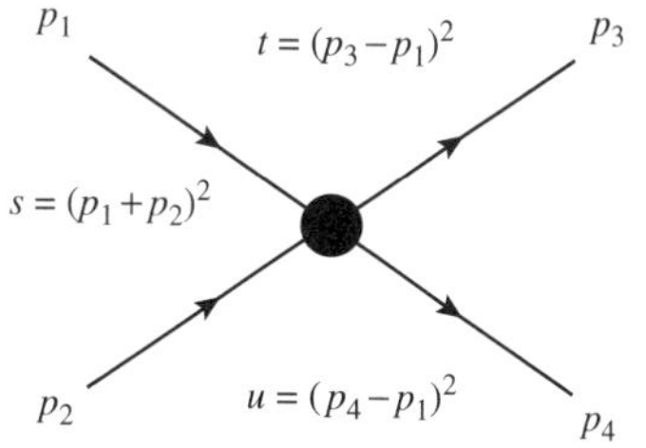


**FIGURE 2.4** Mandelstam variables. The squares of sum and differences of 4-momenta are used in the analysis of two-to-two reactions.

where we used $1 - \cos\theta = 2\sin^2(\theta/2)$. A similar expression applies for u. Since $\theta_{32,CM} \equiv \pi - \theta_{CM}$, we have

$$\begin{aligned} u &= (p_3 - p_2)^2 \\ &= (E_{3,CM} - E_{2,CM})^2 - (|\mathbf{p}_{3,CM}| - |\mathbf{p}_{2,CM}|)^2 - 4|\mathbf{p}_{3,CM}||\mathbf{p}_{2,CM}|\sin^2\frac{\theta_{CM}}{2}. \end{aligned}$$

In particular, for elastic scattering with $m_1 = m_3, m_2 = m_4$, and $|\mathbf{p}_{1,CM}| = |\mathbf{p}_{3,CM}|$, we have simply

$$t_{el} = -2\mathbf{p}_{CM}^2(1 - \cos\theta_{CM}) = -4\mathbf{p}_{CM}^2\sin^2\left(\frac{\theta_{CM}}{2}\right). \tag{2.50}$$

In the ultra-relativistic limit when all masses may be neglected, $|\mathbf{p}'| = |\mathbf{p}| = \sqrt{s}/2$ and

$$\begin{aligned} s_{rel} &= +2p_1p_2 = +2p_3p_4 = +4\mathbf{p}_{CM}^2 \\ t_{rel} &= -2p_1p_3 = -2p_2p_4 = -2\mathbf{p}_{CM}^2(1 - \cos\theta_{CM}) = -4\mathbf{p}_{CM}^2\sin^2\frac{\theta_{CM}}{2} \\ u_{rel} &= -2p_1p_4 = -2p_2p_3 = -2\mathbf{p}_{CM}^2(1 + \cos\theta_{CM}) = -4\mathbf{p}_{CM}^2\cos^2\frac{\theta_{CM}}{2} \end{aligned} \tag{2.51}$$

and we have $s + t + u = 0$.

In a two particle collision, the conserved and Lorentz invariant mass squared is, as we have seen,

$$s \equiv (p_1 + p_2)^2 = p_1^2 + p_2^2 + 2p_1p_2 = m_1^2 + m_2^2 + 2(E_1E_2 - \mathbf{p}_1 \cdot \mathbf{p}_2). \tag{2.52}$$

If particle 2 is at rest, then $\mathbf{p}_2 = 0$ and $E_2 = m_2$, and it follows that the beam energy $E_1$ to produce a mass $m = \sqrt{s}$ in striking a fixed target of mass $m_2$ is

$$E_1 = \frac{(s - m_1^2 - m_2^2)}{2m_2}. \tag{2.53}$$

For example, the minimal process to produce an antiproton in collisions of protons with hydrogen nuclei is

$$pp \to pp\bar{p}p \tag{2.54}$$

since quark and baryon number are conserved in the standard model. Since $m_{\bar{p}} = m_p$, this process requires a center of mass energy $\sqrt{s} > 4m_p$. At the threshold energy, the four nucleons are at rest in the center of mass. Additional energy is required if they are in motion in that frame. Hence the minimal beam energy is

$$E_1 = \frac{(4m_p)^2 - m_p^2 - m_p^2}{2m_p} = 7m_p = 6.57 \text{ GeV}. \tag{2.55}$$

Under threshold conditions, in the center of mass, each of the initial protons has energy $E_p = \sqrt{s}/2$ and a gamma factor $\gamma = E_p/m_p = \sqrt{s}/(2m_p) = 2$ so each initial proton has kinetic energy just equal to the rest energy of one of the produced particles. In the laboratory frame, the four final baryons all move with the velocity of the center of mass so each has total energy $E = \gamma m_p = 2m_p$.

Figure 2.5 illustrates the discovery at the Berkeley Bevatron cyclotron of the antiproton in collisions of 6.2 GeV protons with a copper target, taking advantage of nuclear Fermi motion to slightly decrease the threshold energy requirement. (Fermi motion refers to the kinetic energy of nucleons within the nucleus governed by the Fermi-Dirac distribution in a degenerate gas model of the nucleus.) The velocity of momentum selected antiprotons was measured by scintillator timing and Cherenkov techniques to establish the mass. The copper degraded the antiproton energy so that it came to rest in an emulsion film stack. An antiproton at rest is attracted to a nucleus composed of normal quarks. On contact, quarks and antiquarks react forming pions which carry away the energy equivalent of two nucleon masses.

In collisions with a fixed target, the beam energy $E$ to make a mass $m = \sqrt{s}$ much in excess of the initial particle masses is proportional to $m^2 = s$. In symmetric colliding beams, the beam energy to make the same mass need only be $\sqrt{s}/2$. For this reason, colliding beams are used to achieve the highest possible center of mass energies. The list of of stable particles which may be accelerated and collided includes electrons, protons, positrons, and antiprotons. Positrons and antiprotons are produced in collisions of high energy electrons and protons with stationary targets as described in Chapter 3. Briefly put, the energy of a relativistic electron is converted through collisions to high energy gamma rays which materialize as $e^+e^-$ pairs. Analogous color processes produce $q\bar{q}$ pairs which may combine to form antiprotons. Techniques have been developed to enable accumulation of order $10^{12}$ antimatter particles for injection into colliding beam machines.

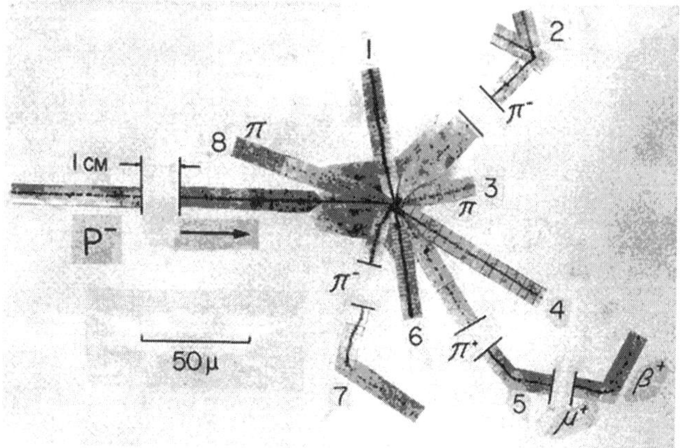


**FIGURE 2.5** Interaction of an antiproton with a nucleus in emulsion. Particles 1 and 4 are likely protons and particle 6 a $^3H$ nucleus. The visible energy release is 1300 MeV. [O. Chamberlain *et. al.*, Phys. Rev. **102**, 921 (1956)].

## 2.3 ■ RELATIVISTIC MECHANICS AND ACCELERATORS

We turn now to the subject of relativistic charged particle motion in classical mechanics. We will focus on motion subject to electric and magnetic fields which is important in understanding the classical aspects of the interaction of particles with matter and of artificial accelerators used to produce and study elementary particles.

The classical motion of a charged particle subject to an electromagnetic field is governed by the relativistic Lorentz force equation

$$\frac{d\mathbf{p}}{dt} = \mathbf{F} = e\mathbf{E} + e\mathbf{v} \times \mathbf{B} \tag{2.56}$$

with $\mathbf{p} = \gamma m\mathbf{v}$. Making $c$ explicit, $E = \sqrt{\mathbf{p}^2c^2 + m^2c^4}$ we have

$$dE = d\sqrt{\mathbf{p}^2c^2 + m^2c^4} = \frac{\mathbf{p} \cdot d\mathbf{p}c^2}{\sqrt{\mathbf{p}^2c^2 + m^2c^4}} = \mathbf{v} \cdot d\mathbf{p} = \mathbf{v} \cdot \mathbf{F}dt \tag{2.57}$$

from which follows the relationship between relativistic energy and the work $W$ done by the force:

$$dE = \mathbf{F} \cdot d\mathbf{x} \equiv dW. \tag{2.58}$$

It follows that the energy is not affected by a magnetic field and that the energy gain of a charge $q$ in dropping through an electrostatic potential difference $V$ is $\Delta E = qV$ as in nonrelativistic mechanics. Through the application of an external force, the energy and momentum of a particle may be increased without bound while the velocity nevertheless remains below light speed. An analytic expression for the motion in the case of acceleration of a charged particle by a uniform electric field is derived in a worked example.

The relativistic Lorentz force equation for motion of a charge $e$ in a magnetic field is

$$\frac{d\mathbf{p}}{dt} = e\mathbf{v} \times \mathbf{B}. \tag{2.59}$$

Since the magnetic force is orthogonal to the displacement, it does no work and therefore the particle energy is conserved. Since relativistic energy and momentum are related by $\mathbf{p} = E\mathbf{v}/c^2$ and since $E$ is constant, we can write

$$m_r\frac{d\mathbf{v}}{dt} = e\mathbf{v} \times \mathbf{B} \tag{2.60}$$

which has the form of the nonrelativistic expression $m\mathbf{a} = \mathbf{F}$ with the replacement $m \to m_r \equiv E/c^2$ in SI units. The general motion in a constant uniform magnetic field is a helix with constant velocity component along the field direction and uniform circular motion about an axis along the field direction. In case there is no velocity component along the field direction, uniform circular motion results and we may equate the centripetal components of mass times acceleration and magnetic force

$$m_r \frac{v^2}{r} = evB \tag{2.61}$$

and deduce the orbital radius

$$r = \frac{m_r v}{eB} = \frac{p}{eB}. \tag{2.62}$$

This expression is valid in SI units and has the engineering expression

$$r[\mathrm{m}] = \frac{p[\mathrm{GeV}]}{0.3B[\mathrm{T}]}. \tag{2.63}$$

The orbital period $T$ and cyclotron frequency $\omega$ are defined by

$$T = \frac{2\pi r}{v} = \frac{2\pi m_r}{eB}, \quad \omega = \frac{2\pi}{T} = \frac{eB}{m_r}. \tag{2.64}$$

While the cyclotron frequency of a nonrelativistic particle is independent of energy, in general it is inversely proportional to total energy.

---

**Example 2.3. Relativistic acceleration.** Starting from rest, an electron accelerates in a static electric field and passes through a potential difference of 50 kV. What is its energy in eV and what is its speed relative to light speed?

The initial energy is the rest energy $E_0 = m_e = 511$ keV. The energy gain is $\Delta E = \mathrm{qV} = \mathrm{e}(50\ \mathrm{kV}) = 50$ keV so the final energy is $E = E_0 + \Delta E = 561$ keV. The momentum is

$$\mathrm{p} = \sqrt{\mathrm{E}^2 - \mathrm{m}_\mathrm{e}^2} = \sqrt{(\mathrm{E} - \mathrm{m_e})(\mathrm{E} + \mathrm{m_e})} = \sqrt{(50\ \mathrm{keV})(1072\ \mathrm{keV})} = 231\ \mathrm{keV} \tag{2.65}$$

from which the speed is $v = p/E = 0.41$ in units of light speed.

---

**Example 2.4. Relativistic motion subject to a constant force.** Derive an expression for the relativistic speed of a charge $q$ as a function of time starting from rest subject to a constant force $F$.

The equation for the change in momentum may be integrated directly for a constant force

$$\frac{dp}{dt} = F \rightarrow p(t) = p(0) + \int_0^t dt\ F = p(0) + Ft = Ft \tag{2.66}$$

and this expression solved for the velocity as a function of time

$$p(t) = \frac{mv(t)}{\sqrt{1 - v(t)^2}} = Ft \rightarrow v(t) = \frac{Ft/m}{\sqrt{1 + (Ft/m)}} \qquad (2.67)$$

where factors of $c$ were suppressed. For $t << m/F$, the speed increases linearly with time while as $t \rightarrow \infty$ the speed approaches $c$.

---

In the Van de Graaf and Cockcroft-Walton electrostatic accelerators, mechanical and induction methods create a charge difference between two conductors. Charged particles injected into the electrostatic field between the conductors in such electrostatic accelerators can reach energies of order 1 MeV, sufficient to explore the properties of nuclei. Spontaneous spark discharges limit the electrostatic potential difference. To achieve higher energies, time dependent methods are required.

Linear accelerators make use of traveling or standing waves (radio frequency transverse magnetic waves in cavity strings) designed to keep bunches of particles in phase with an accelerating longitudinal electric field component. In high performance linear accelerators, frequencies range from 20 MHz to 10 GHz. The highest field strengths are obtained with superconducting Nb surfaces plated on copper, and power is supplied by vacuum tube triodes and tetrodes of up to 100 kW for frequencies less that 300 MHz and tuned microwave amplifiers called klystrons at higher frequencies. The achievable energy is a linear function of length and the largest accelerator of this kind was the Stanford Linear Accelerator with a length of 3.2 km and energy of 45 GeV.

The highest energy accelerators make repeated use of a linear accelerator. A synchrotron comprises a one dimensional lattice of dipole magnets arranged to circulate bunches of charged particles through a linear accelerator for repeated kicks of 1-10 MeV as illustrated in Figure 2.6. Throughout the lattice, the particles travel within a beam pipe evacuated to a pressure typically less the $10^{-12}$ atmosphere.

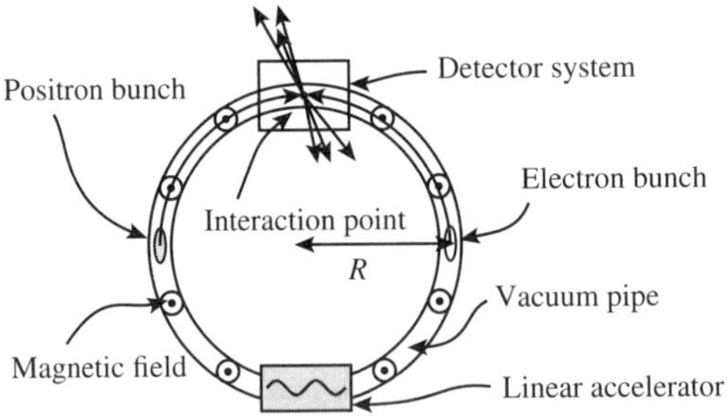


**FIGURE 2.6** Essentials of an $e^+e^-$ colliding beam synchrotron. Bunches of charged particles are guided around a ring of magnets and pass repeatedly through a linear accelerator. The magnetic field strength increases synchronously with increasing energy to maintain a fixed orbital radius $R$. Acceleration ceases when the maximum field strength is reached. In colliding beam mode, particle and antiparticle bunches circulate in an evacuated beam pipe in opposite directions and collide at the location of a detector system.

During acceleration, the magnetic field strength $B$ in the dipole magnets is synchronously ramped to fix the orbital radius. Additional quadrupole magnets focus and stabilize the beam while sextupole magnets compensate for the energy dependence of the focusing strength of the quadrupoles. The bunches are arranged so collisions occur at a fixed interaction point.

For a given maximum energy, the circumference and cost of a synchrotron is inversely proportional to the maximum magnetic field strength. In a room temperature electromagnet, a relatively small magnetic field produced by an electric current is used to induce a larger field in a ferromagnetic material. The maximum magnetic field strength of such an electromagnet is about 2 T, the saturation field of iron. The magnetic field that may be produced by a cooled current-bearing conductor is limited by Ohmic energy loss and it is difficult to produce a steady field of this magnitude without a ferromagnetic material. Superconducting high current density coils overcome the problem of heat generation permitting current generated fields in excess of 2 T but superconductivity may be maintained only below a critical field strength of about 10 T. For momentum p$\sim$ TeV and achievable field strength $B \simeq 10$ T, the radius of a synchrotron reaches kilometer dimensions.

The coils of a dipole and of a quadrupole magnet are illustrated in Figure 2.7. In high field strength superconducting magnets, the magnetic field is determined almost entirely by the distribution of current in the windings, surrounding steel providing the mechanical support. A uniform field within a cylinder requires a current density proportional to the cosine of the azimuthal angle and this is achieved approximately by a variable packing of a winding carrying constant current density. The 21 foot long coils of the Tevatron dipole magnet are made of superconducting niobium-titanium alloy wire embedded in a copper matrix that shunts current in the event of a local loss of superconductivity. There are 11 million wire-turns in a coil and 42,500 miles of wire carrying a total maximum current in excess of 4000 amperes. Liquid helium keeps the coils at 4.3 kelvin.

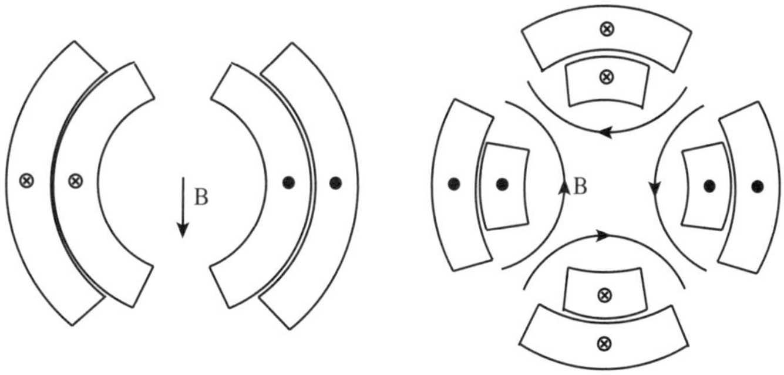


**FIGURE 2.7** Dipole and quadrupole magnet construction. In the dipole magnet, the current density of stacked coils approximates a $\cos\theta$ distribution which would produce an ideal uniform interior field.

**Example 2.5. Large electron positron collider.** An $e^+e^-$ synchrotron collider is designed to produce the $Z$ boson at rest. For field strength 0.1 tesla, what is the minimum orbital radius? If the dipole magnets occupy only 1/3 of the lattice, what is the minimum circumference?

The beam energy is half the rest mass of the $Z$ boson or about 45 GeV. The electron mass is negligible and the momentum in GeV/c has the same value. Hence we find

$$R[\mathrm{m}] = \frac{45\ \mathrm{GeV/c}}{0.3 \times 0.1\ \mathrm{T}} = 1500\ \mathrm{m} \tag{2.68}$$

This would be the radius of a ring composed entirely of dipole magnets. Allowing a factor three for other machine components, we arrive at a circumference $C = 2\pi R \times 3 = 28$ km.

---

Acceleration of particles to high energy typically occurs in stages. At Fermi National Accelerator Laboratory, the first stage of the Tevatron is a Cockroft-Walton electrostatic accelerator in which $H^-$ ions are accelerated to close to 1 MeV. The next stage is a 500 foot long linear accelerator that achieves 400 MeV. The ions are stripped of electrons in passing through a foil and the protons accelerated in the Booster synchrotron to an energy of 8 GeV and subsequently in the Main Injector to 150 GeV. The Main Injector also serves 120 GeV proton bunches to a target to manufacture antiprotons at the optimal energy with a yield of order $10^{-5}$ $\bar{p}$ per proton. The antiprotons are collected in the Debuncher ring that removes the bunch structure and are stored in the Accumulator. Antiprotons are reaccelerated in the Main Injector for insertion in the 1 km radius superconducting Tevatron which accelerates p and $\bar{p}$ simultaneously (in opposite directions) to close to 1 TeV.

The Lorentz force and the relativistic form of Newton's second law provide an incomplete description of particle motion as radiative reaction is neglected. An accelerated charge radiates electromagnetic energy and the associated energy loss in fact limits the reach of electron synchrotrons. For a particle of charge $e$, mass $m$, momentum $p$, in circular motion with radius $R$, the power loss to so-called synchrotron radiation is given by

$$\frac{dE}{dt} = -\frac{2\alpha}{3R^2}\left(\frac{p}{mc}\right)^4 (\hbar c^2) \tag{2.69}$$

and increases rapidly with energy. In a synchrotron of fixed radius, the maximum particle energy is reached when the energy increase in one pass through the accelerating section is radiated away in the next turn. This limitation motivates clashing linear accelerator designs to produce high center of mass energies. Compared with an electron of the same momentum, a proton radiates $(m_e/m_p)^4$ less energy and synchrotron radiation does not yet limit the reach of proton synchrotrons.

## 2.4 ■ BEAM STABILITY AND EMITTANCE

Successful operation of a cyclic accelerator or storage ring requires careful consideration of beam stability. The magnetic elements must guide a particle along an ideal equilibrium orbit and provide an effective restoring force to stabilize particles deviating from the ideal in position or momentum. The guidance system governs the transverse and longitudinal dimensions and energy spread of particle bunches and ultimately the rate of interactions in a colliding beam experiment. For this reason, and as an invitation to the field of accelerator physics, we look at some of the principles and parameters. It should be said that longitudinal and energy stability are also important but we will concentrate here on transverse motion.

While early synchrotrons used electric and magnetic fringe fields to provide restoring forces, modern synchrotrons use dedicated strong focusing elements - combinations of quadrupole magnets. The essential function of a quadrupole magnet can be described with reference to the schematic of the field lines shown in Figure 2.7. For particle velocity directed essentially down the axis of the magnet, the magnetic force $\mathbf{F} = q\mathbf{v} \times \mathbf{B}$ may be seen to restore particles towards the axis for small displacements along one transverse coordinate while amplifying displacements along the orthogonal coordinate. For one coordinate, the magnet acts as a focusing lens, while for the other coordinate, the magnet acts as a defocusing lens. In fact, for a quadrupole magnet with field gradient $B'$ and axis along the $z$-direction, the motion of a particle of charge $q$ and energy $E$ moving close to the axis is described by the equations (see problems)

$$\frac{d^2x}{dz^2} = -k^2 x; \ \frac{d^2y}{dz^2} = k^2 y \tag{2.70}$$

with $k^2 = qB'/Ev_z$ where $v_z$ is approximately constant. The solution for $k^2 > 0$ is

$$\begin{aligned} x(z) &= x_0(0)\cos(kz) + \frac{x'(0)}{k}\sin kz\,; \quad x' = \frac{dx}{dz} = \tan\theta_x \simeq \theta_x \\ y(z) &= y(0)\cosh(kz) + \frac{y'(0)}{k}\sinh kz\,; \ y' = \frac{dy}{dz} = \tan\theta_y \simeq \theta_y. \end{aligned} \tag{2.71}$$

A matrix expression for the effect of a focusing quadrupole of length $z$ is

$$\begin{pmatrix} x(z) \\ x'(z) \end{pmatrix} = \begin{pmatrix} \cos kz & \frac{1}{k}\sin kz \\ -k\sin kz & \cos kz \end{pmatrix} \begin{pmatrix} x(0) \\ x'(0) \end{pmatrix}. \tag{2.72}$$

Two quadrupole magnets rotated by 90° can serve to focus in both transverse coordinates. This is because a particle defocused in the first magnet arrives at the second with a larger displacement and encounters a stronger focusing force while a particle focused by the first magnet arrives at the second with a smaller displacement and encounters a smaller defocusing force. In the thin lens approximation, a

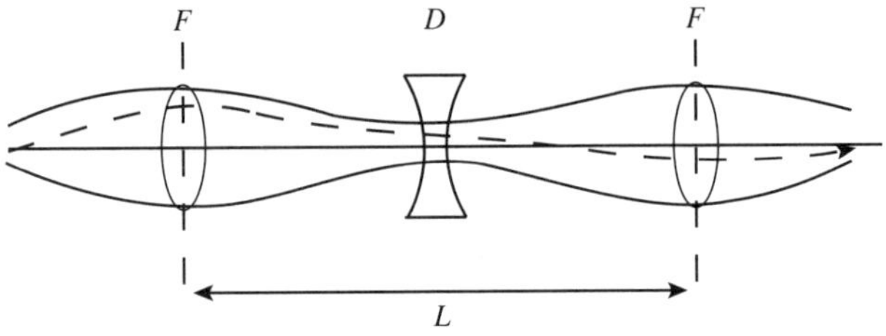


**FIGURE 2.8** FODO-cell. Schematic drawing of a symmetric FODO-cell in a lattice of alternate focusing and defocusing quadrupole magnets. The cell is defined as of a half of a focusing quadrupole magnet, a defocusing quadrupole magnet, and one half of a focusing quadrupole focusing magnet.

quadrupole doublet is focusing if separated by a gap of length $g$ such that each is within the focal length of the other and the focal length is found to be

$$f = -\frac{f_x f_y}{g}. \tag{2.73}$$

A synchrotron is typically constructed from a lattice of alternating focusing and defocusing quadrupole magnets with interleaving dipole bending magnets. The fundamental unit cell illustrated in Figure 2.8 containing one focusing plus one defocusing magnet is called a FODO-cell.

---

**Example 2.6. Magnetic focusing.** A relativistic particle of charge $e$ and momentum $p$ along the $z$-direction traverses a region of magnetic field $B_y$ over a short distance $L$. Find the angular deflection in the impulse approximation. If the field strength is proportional to transverse coordinate as in a quadrupole magnet $B_y = B'x$, show that the magnetic field acts as a lens and find the focal length. A quadrupole field has the form $B_y = B'x$ and $B_x = -B'y$ and focuses one transverse coordinate while defocusing the other. For $B' = 76\ \mathrm{T\ m^{-1}}$ over $L = 1.7$ m (Fermilab Tevatron quadrupole) and $p = 1000$ GeV/c, what is the focal length in m?

The Lorentz force equation gives in the impulse approximation

$$\frac{dp_x}{dt} = F_x = -ev_z B_y \rightarrow \Delta p_x = \int dt\, F = \int \frac{dx}{v} F = -eB_y L. \tag{2.74}$$

In the small angle approximation, the angular deflection is

$$\theta = \frac{\Delta p_x}{p} = -\frac{eB_y L}{p}. \tag{2.75}$$

If there is a field gradient, we can write this as $\theta = -x/f$ with inverse focal length $f^{-1} = eB'L/p$ which implies a parallel beam of particles is focused to a point a distance $f$ from the magnet. Notice that the focal length depends on energy so the beam suffers the equivalent of optical chromatic aberration. For the Tevatron quadrupole magnet,

$$f = \frac{p}{eB'L} = \frac{10^{12}\ \text{eV}\ 1.6 \times 10^{-19}\ \text{eV}^{-1}}{3 \times 10^{8}\ \text{m s}^{-1}\ 1.6 \times 10^{-19}\ \text{C}\ 76\ \text{T m}^{-1}\ 1.7\ \text{m}} = 25\ \text{m}. \quad (2.76)$$

---

We now consider a coasting particle in an ideal circular synchrotron. The deviation $x(s)$ from the ideal planar circular orbit in the vertical or radial coordinate is a function of path length $s$. To linear approximation, this coordinate function is governed by Hill's equation

$$\frac{d^2x}{ds^2} + k(s)x = 0 \quad (2.77)$$

where $k(s)$ is positive for a focusing region, 0 in drift regions, and negative for defocusing regions. Where $k$ is a constant, the general solution is

$$x(s) = x(0)\cos\left(\sqrt{k}s\right) + \left(x'(0)/\sqrt{k}\right)\sin\left(\sqrt{k}s\right) \quad (2.78)$$

If $k < 0$, the general solutions may be represented in terms of $\sinh(\sqrt{|k|}s)$ and $\cosh(\sqrt{|k|}s)$. The trajectory of a particle through a lattice of elements with constant $k$ may be constructed from piecewise solutions.

We now wish to know if a periodic system of magnets permits trajectories which remain close to the ideal orbit during repeated passes through the synchrotron. It is useful to write the solution for one cell in terms of a transfer matrix $\mathbf{M}$ operating on a two dimensional vector

$$\mathbf{x}(L) = \begin{pmatrix} x(L) \\ x'(L) \end{pmatrix} = \mathbf{M}\mathbf{x}(0) \quad (2.79)$$

where

$$x'(s) = dx/ds = \tan\theta_x \simeq \theta_x \quad (2.80)$$

is approximately the projected angle of the velocity vector. Now consider the eigenvectors $\mathbf{v}$ of $\mathbf{M}$ satisfying

$$\mathbf{M}\mathbf{v} = \lambda\mathbf{v}. \quad (2.81)$$

By combining the two equations, one finds solutions if $\lambda$ satisfies the characteristic equation

$$\lambda^2 - \lambda\,\text{tr}(\mathbf{M}) + |\mathbf{M}| = 0 \quad (2.82)$$

where $\text{tr}(\mathbf{M}) = M_{11} + M_{22}$ is the trace of $\mathbf{M}$ and $|\mathbf{M}| = M_{11}M_{22} - M_{12}M_{21}$ is the determinant. Since for our matrix $|\mathbf{M}|=1$, the solutions are

$$\lambda_{\pm} = \frac{\text{tr}(\mathbf{M})}{2} \pm \sqrt{\left(\frac{\text{tr}(\mathbf{M})}{2}\right)^2 - 1} \quad (2.83)$$

and therefore satisfy $\lambda_-\lambda_+ = 1$. We can express a general initial condition $\mathbf{x}_0$ in terms of the eigenvectors $\mathbf{v}_\pm$ as

$$\mathbf{x}_0 = a_+\mathbf{v}_+ + a_-\mathbf{v}_-. \tag{2.84}$$

The action of $N$ cells is then simply calculated:

$$\mathbf{x} = \mathbf{M}^N\mathbf{x}_0 = a_+\lambda_+^N\mathbf{v}_+ + a_-\lambda_-^N\mathbf{v}_-. \tag{2.85}$$

We can deduce that the orbit will be bounded only if the following stability condition is satisfied:

$$\left|\mathrm{tr}\frac{\mathbf{M}}{2}\right| \leq 1. \tag{2.86}$$

That this condition can be met in an idealized lattice is illustrated in a worked example.

---

**Example 2.7. Stability of FODO lattice.** Find the transfer matrix for a focusing magnet of length $L$, a defocusing magnet of length $L$, and the combination. What is the criterion for stability of a lattice of such magnet pairs?
For a focusing magnet of length $L$,

$$\mathbf{x}(L) = \begin{pmatrix} \cos(\sqrt{k}L) & \sin(\sqrt{k}L)/\sqrt{k} \\ -\sqrt{k}\sin\sqrt{k}L & \cos\sqrt{k}L \end{pmatrix} \begin{pmatrix} x(0) \\ x'(0) \end{pmatrix} = \mathbf{M}_1\mathbf{x}(0), \tag{2.87}$$

while, for a defocusing magnet of length L,

$$\mathbf{x}(L) = \begin{pmatrix} \cosh(\sqrt{k}L) & \sinh(\sqrt{k}L)/\sqrt{k} \\ \sqrt{k}\sinh\sqrt{k}L & \cosh\sqrt{k}L \end{pmatrix} \begin{pmatrix} x(0) \\ x'(0) \end{pmatrix} = \mathbf{M}_2\mathbf{x}(0). \tag{2.88}$$

A defocusing magnet followed by a focusing magnet both of length $L$ is described by the transfer matrix $\mathbf{M} = \mathbf{M_1M_2}$. The product matrix is

$$\mathbf{M} = \begin{pmatrix} \cos\alpha\cosh\alpha - \sin\alpha\sinh\alpha & (\cos\alpha\sinh\alpha + \sin\alpha\cosh\alpha)/\sqrt{|k|} \\ \sqrt{|k|}(\cos\alpha\sinh\alpha - \sin\alpha\cosh\alpha) & \cos\alpha\cosh\alpha - \sin\alpha\sinh\alpha \end{pmatrix} \tag{2.89}$$

where $\alpha = \sqrt{|k|}L$. The criterion for stability is

$$|\mathrm{tr}\mathbf{M}/2| = |\cosh\alpha\cos\alpha| < 1 \tag{2.90}$$

which translates to an operational condition for our two element cell

$$0 \leq \sqrt{k}L = \leq 1.86. \tag{2.91}$$

---

In a system designed for stable operation, trajectories of particles undergoing multiple passes are bounded and, as illustrated in Figure 2.9, in general exhibit an oscillation called a betatron oscillation. The number of oscillations per turn is

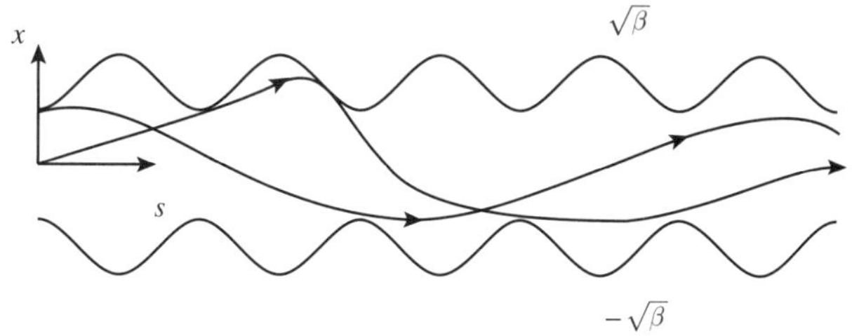


**FIGURE 2.9** Betatron oscillations. In a ring accelerator, each transverse coordinate $x$ of a particle oscillates as a function of path length $s$ within an envelope described by the so-called $\beta$-function. In the example illustrated, focusing quadrupoles are located at the peaks and defocusing magnets are located at the troughs of the $\beta$- function. A cosine-like trajectory starting at $s = 0$ and a sine-like trajectory starting at $s = 0$ are shown [After M Sands, "The Physics of Electron Storage Rings: An Introduction", SLAC-r-121, 35 (1970)]

known as the tune. The tune is generally chosen to be incommensurate with the orbital period so that imperfections in the accelerator, which appear as periodic forces in the equations of motion, do not generate a resonant response. It can be shown that the general bounded solution, if it exists, for Hill's equation with a periodic spring constant $k(s)$ has the form of

$$x(s) = A\sqrt{\beta}\cos(\Phi(s) + \delta) \tag{2.92}$$

where the amplitude $A$ and phase $\delta$ are determined by initial conditions while the positive definite and periodic $\beta$-function $\beta(s)$ derived from $k(s)$ characterizes the guide field. The $\beta$-function governs both the position dependent amplitude $A\sqrt{\beta}$ and the phase $\Phi(s) = \int_0^s ds\beta^{-1}(s)$. In case $k$ is constant, $\beta = 1/\sqrt{k}$. For a derivation of the general form, the reader should consult the suggestions for further reading at the end of this chapter. The $\beta$ function has dimensions of length and is the local wavelength of the oscillation. This guide function plays an important role in accelerator design. In particular, the physical size of the distribution of particle deviations from the ideal orbit increases and decreases as the $\beta$-function increases and decreases. Special insertions surrounding intersection points in colliding beam machines are designed to minimize the $\beta$-function at the collision point to increase the density of particles and the collision rate.

An important consideration in manipulations of beam bunch density is Liouville's theorem of Hamiltonian mechanics which requires conservation of the density of particle trajectories in phase space

$$\rho = \frac{dn}{d\mathbf{x}d\mathbf{p}}. \tag{2.93}$$

(Although usually Liouville's theorem is derived assuming no velocity dependent forces, the theorem is valid for motion in static electromagnetic fields.) Suppose we can neglect coupling between motions in the local coordinates orthogonal to the ideal orbit. Then the density of particles in the phase space variables $x$ and

$\theta_x = p_x/p = dx/ds = x'$ for fixed momentum magnitude is conserved. Consequently, if the density in coordinate $x$ is increased, the density in $x'$ must increase corresponding to larger angles. This, in turn, implies that if the transverse bunch dimensions are reduced at a collision point, they must suffer a relatively large excursion nearby and may exceed the aperture of the accelerator. Hence, there are limits to the bunch density that may be achieved and a high phase space density at injection to an accelerator is desirable.

The conserved phase space density of a bunch of particles is characterized by the product of the mean width of their distribution in $x$-coordinate times and the mean width of their $\theta_x$ distribution. If at some point a collection of particles is contained within an elliptical boundary of area $\pi\epsilon_x$ in $x - \theta_x$ space, then analysis shows the trajectories remain bounded by an ellipse as a function of position around the accelerator. Although the shape and orientation of the ellipse varies with $s$, its enclosed area is constant. The area of an ellipse enclosing 39% of the beam, $\epsilon_x$, is called the transverse emittance associated with $x$. For a round gaussian phase space probability distribution function

$$P(x, x') = \frac{e^{(x^2+x'^2)/(2\sigma^2)}}{2\pi\sigma^2}, \tag{2.94}$$

the probability of being within a radius $R$ is

$$P_R = \int_0^R r dr d\phi \, \frac{e^{-r^2/(2\sigma^2)}}{2\pi\sigma^2} = 1 - e^{-R^2/(2\sigma^2)} \tag{2.95}$$

and the ellipse is a circle of radius $\sigma$ and area $A = \pi\sigma^2$. In general

$$\epsilon_x = \pi\sigma_x\sigma_{x'}. \tag{2.96}$$

(Other definitions of the emittance of a beam are possible such as the area enclosing 95% of the beam corresponding to $R = \sqrt{(6)}\sigma$, with or without the factor of $\pi$.) During adiabatic acceleration, the density in position-momentum space is conserved. The normalized emittance, defined as the conserved position-momentum density per unit rest mass $\epsilon_n = \beta\gamma\epsilon_x$ remains constant, while the emittance decreases with energy.

By averaging over the phases $\delta$ of particles in a bunch assuming these are uncorrelated, the root mean square beam width $\sigma_x(s)$ can be expressed in terms of the emittance and $\beta(s)$ as

$$\sigma_x(s) = \sqrt{\beta(s)\epsilon_x} \tag{2.97}$$

and the root mean square angular divergence is

$$\sigma_{x'} = \sqrt{\frac{\epsilon_x(s)}{\beta(s)}}. \tag{2.98}$$

In a field free region where collisions occur, the $\beta$-function generally has a parabolic form centered on the collision point $s_0$

$$\beta(s) = \beta^* + \frac{(s - s_0)^2}{\beta^*} \tag{2.99}$$

where $\beta^*$ is the value at the collision point. The beam size increases linearly with distance for $s - s_0 >> \beta^*$ and must be accommodated by the aperture of the nearest quadrupole producing final focusing. For fixed emittance and a given aperture at a given distance, this aperture limit determines the minimum value $\beta^*$ and correspondingly the minimal beam size. For the LHC, the aperture is about 5 cm, the nominal emittance at 7 TeV is 0.503 nm, and $\beta^* = 55$ cm.

To achieve high density beams, accelerators are designed with low emittance sources and use a variety of stochastic cooling mechanisms to reduce the transverse and longitudinal emittance of captured particles. One method used in the Fermilab $p\bar{p}$ storage ring is to detect fluctuations in the bunch shape and apply damping electrical pulses. Another is to inject a "cold" electron beam co-moving with the protons, relying on inter-beam particle scattering to cool the stored beam. In the next section we will see how these accelerator parameters affect collision rates that are important to particle experiments.

## 2.5 ■ COLLISION RATES AND CROSS SECTIONS

Underlying the understanding of fundamental interactions are measurements of a myriad of reaction rates. A typical experiment entails a study of the processes resulting from the intersection of a beam of particles with another beam, or with fixed target of normal matter which, at energies much in excess of the atomic scale, may be regarded as a collection of stationary electrons and atomic nuclei. Provided the beam and target have sufficiently low density, the reactions may be understood in terms of two particle collision events characterized by an effective collision area called the cross section.

Figure 2.10 illustrates the idea of a cross section. Beam particles are shown encountering a collection of identical targets and are absorbed, transmitted, and scattered. We can see how absorption and scattering attenuate the beam and how, for a uniform flux of beam particles, the angular distribution of particles which scatter only once is characteristic of a single target. For a beam of hard spheres of radius $r_1$ encountering hard spheres of radius $r_2$, scattering occurs if the center of a beam sphere falls within a radius $r = r_1 + r_2$ of the center of a target sphere. The effective collision area per target particle $\sigma = \pi r^2$ governs the probability for scattering for a fixed target particle density and beam particle flux. Conversely, given knowledge of the target and beam overlap densities, from the reaction rate, one can calculate the cross section and determine the radius of the target sphere.

As illustrated in Figure 2.10, the interactions of beam particles with the targets can be of several sorts. All contribute to the effective total cross section. The effective cross section for any one process may be smaller or larger than what might be considered the geometrical extent of the target. In the case of light scattering from

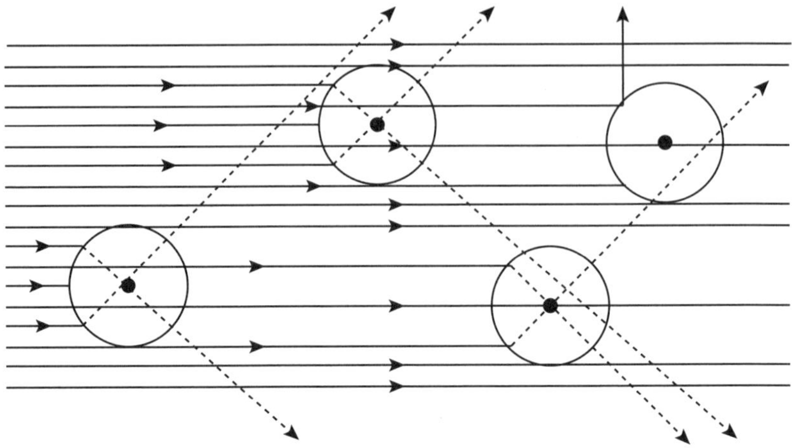

**FIGURE 2.10** Particle beam encountering a collection of identical targets. Scattering and absorption are illustrated. Each process is characterized by an effective cross sectional area per target.

glass spheres, for example, the particle representation fails when the wavelength of the light is comparable to the sphere radius and the scattering cross section is wavelength dependent. Even for wavelengths small compared to the sphere, since the refractive index depends on wavelength and reflection is polarization dependent, the distribution of scattered light depends on wavelength and polarization. In general, cross sections for individual particle scattering are energy and spin dependent.

The event rate in a collision experiment depends on both the experimental conditions and the process cross section $\sigma$ (m$^2$). Consider a beam of particle flux $j$ (particles m$^{-2}$ s$^{-1}$) and beam intensity $I$ (particles s$^{-1}$) impinging on a fixed target of number density $n$ (m$^{-3}$). In a distance $dx$ so thin that only two particle collisions are important, the beam intensity change or equivalently the reaction rate is

$$dI = -\int dA\ jn\sigma dx = -\frac{I}{\lambda}dx \tag{2.100}$$

with $\lambda = 1/n\sigma$ the mean free path. This expression implies the intensity falls exponentially with distance:

$$I(x) = I_0 e^{-x/\lambda} \tag{2.101}$$

with $I_0$ the initial intensity.

The unit of area called the barn is often used for cross sections and is defined as

$$1\text{ b} = 10^{-24}\text{ cm}^2 = 10^{-28}\text{ m}^2. \tag{2.102}$$

Hadrons react on contact and a typical geometric total interaction cross section is $\pi r^2 = 3 \times 10^{-26}$ cm$^2$ = 30 mb or $3 \times 10^{-26}$ cm$^2$. To get a feel for this number, consider that the mass density of liquid hydrogen is $\rho = 0.04$ gm cm$^{-3}$ and the mass of a proton $m_p = 1.67 \times 10^{-24}$ gm is essentially the mass of a hydrogen atom so the corresponding mean free path is

$$\lambda = \frac{m_p}{\rho\sigma} = \frac{1.67 \times 10^{-24}\text{gm}}{0.07 \text{ gm cm}^{-3}\ 3 \times 10^{-26}\text{cm}^2} = 8 \text{ m}. \tag{2.103}$$

We see that a typical hadronic interaction cross section corresponds to a macroscopic beam interaction and attenuation length.

The total cross section for various interactions of hadrons and photons is shown in Figure 2.11. The purely hadronic interaction cross sections vary considerably for center of mass energy below a few GeV (not shown), in the range of resonant production of hadrons. Above a few GeV, the cross sections increase only logarithmically with energy. As might be expected, the total cross section for photons interacting inelastically with protons is roughly a factor $\alpha$ smaller than the hadron proton cross sections, a reflection of the electromagnetic coupling of the photon to the quarks inside the proton in contrast to the strong coupling of incident quarks to the gluon fields of the target proton.

In a colliding beam machine, the event rate for a process of cross section $\sigma$ is given by

$$\dot{n} = L\sigma \tag{2.104}$$

where the quantity $L$ is called the instantaneous luminosity. The luminosity for a single collision of two bunches with particle densities $n_1$ and $n_2$ and bunch velocities $\mathbf{v}_1$ and $\mathbf{v}_2$ is in general

$$L = \int d\mathbf{x}\frac{dL}{d\mathbf{x}} = \int d\mathbf{x} \quad n_1(t,\mathbf{x})n_2(t,\mathbf{x})\sqrt{(\mathbf{v}_1 - \mathbf{v}_2)^2 - \frac{(\mathbf{v}_1 \times \mathbf{v}_2)^2}{c^2}}. \tag{2.105}$$

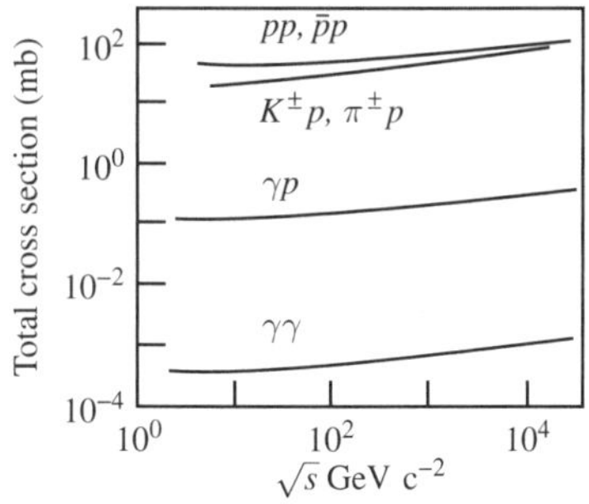


**FIGURE 2.11** Total cross sections for collisions of hadrons photons with protons as a function of center of mass energy. [C. Amsler *et al.* (Particle Data Group), Phys. Lett. **B667**,1 (2008)]

A subtlety in this expression in case the two beams are not collinear requires discussion. Consider the total number of events

$$n = \int dt\ L\sigma = \int dt d\mathbf{x} \quad n_1(t, \mathbf{x}) n_2(t, \mathbf{x}) \sqrt{(\mathbf{v}_1 - \mathbf{v}_2)^2 - \frac{(\mathbf{v}_1 \times \mathbf{v}_2)^2}{c^2}}\sigma. \tag{2.106}$$

This number is independent of reference frame as is the differential 4-volume $dx \equiv dt\ d\mathbf{x}$. One might define the cross section using a luminosity density

$$\frac{dL'}{d\mathbf{x}} = n_1 n_2 |\mathbf{v}_1 - \mathbf{v}_2| \tag{2.107}$$

which coincides with the definition above for collinear beams. This quantity and the cross section so defined would depend on frame of reference. That is not surprising since, if we think of cross section as a projected area, we expect the area to be subject to length contraction under Lorentz transformation in the plane of the projection. It is shown in an exercise that the velocity of beam 2 in the rest frame of beam 1 (a Lorentz invariant) is

$$v_{2,1} = \left[ \frac{(v_1 v_2)^2 - 1}{(v_1 v_2)^2} \right]^{\frac{1}{2}} = \frac{\left[ (\mathbf{v}_1 - \mathbf{v}_2)^2 - c^{-2} (\mathbf{v}_1 \times \mathbf{v}_2)^2 \right]^{\frac{1}{2}}}{1 - c^{-2} \mathbf{v}_1 \cdot \mathbf{v}_2} \tag{2.108}$$

where the second form follows by expansion of the 4-vector products. The denominator is $v_1 v_2 / (\gamma_1 \gamma_2)$ so the conventional luminosity density can be written as

$$\frac{dL}{d\mathbf{x}} = \frac{n_1}{\gamma_1} \frac{n_2}{\gamma_2} v_1 v_2 v_{2,1} = n_{1,1} n_{2,1} v_1 v_2 v_{2,1} \tag{2.109}$$

where $n_{1,1}$ and $n_{2,1}$ are the densities in the rest frame of beam 1. This expression is manifestly Lorentz invariant and it follows that, with this flux factor, the cross section is invariant under both longitudinal and transverse Lorentz transformations. It is already referred to a frame in which the two beams are collinear so has the contraction associated with a transverse boost already factored out.

For head on collisions of relativistic particles, the kinematic factor in the integral is $2c$. For $N_1$ particles per bunch in beam 1 and $N_2$ particles per bunch in beam 2, and $n_b$ bunches per beam with revolution frequency $f_{\text{rev}}$, the collision frequency at one location is $f = f_{\text{rev}} n_b$ and the luminosity is

$$L = f \frac{N_1 N_2}{A} \tag{2.110}$$

with $A$ the effective overlap area. To the experimentalist, aside from beam energy, the luminosity is the most important machine parameter. For particle bunches colliding head on with gaussian transverse profiles characterized by root mean square widths $\sigma_x$ and $\sigma_y$, the effective overlap area is

$$A = 4\pi \sigma_x \sigma_y. \tag{2.111}$$

As we have seen, the overlap area is determined by the effective focal length of the magnets surrounding the interaction point characterized by the value of $\beta(s)$ at the interaction point called $\beta^*$, and the transverse emittance values $\epsilon_x$ and $\epsilon_y$. To get a feel for the numbers, note that a representative collider luminosity of $10^{33}$ cm$^{-2}$ s$^{-1}$ implies a collision rate of $10^8$ $s^{-1}$ if the cross section is 0.1 b and for a process with a cross section of $0.3 \times 10^{-40}$ cm$^2$ one could expect one event per year (1 yr $\simeq \pi \times 10^7$ s). That one event in some $10^{15}$ interactions might just represent a fundamental discovery.

---

**Example 2.8. LHC event rate.** The Large Hadron Collider collides protons at a center of mass energy of 14 TeV. The design calls for 2808 bunches per beam of r.m.s. length 7.5 cm containing $1.15 \times 10^{11}$ protons with normalized transverse emittance $\epsilon_n = 3.75 \times 10^{-6}$ m at the injection energy of 0.45 TeV and $\epsilon_t = \epsilon_n/\gamma = 5.03 \times 10^{-10}$ m at full energy. The revolution frequency is 11 kHz and bunches collide every 25 ns with crossing angle 0.3 mrad at an interaction point where $\beta^* = 55$ cm. The luminosity is L $= 10^{34}$ cm$^{-2}$s$^{-1}$. If the total cross section for inelastic processes is $\sigma_{tot} = 80$ mb, how many inelastic collisions occur in one second of operation? If the cross section for a rare process producing an exotic particle X is calculated to be $\sigma_X = 1$ fb $= 10^{-15}$ b, how many X particles are produced in one year $= \pi \times 10^7$ s? What is the effective area of overlap of the beams in square microns at full energy?

Since 1 mb $= 10^{-27}$ cm$^2$, the event rate is

$$R = L\sigma_{tot} = 10^{34}\,\text{cm}^{-2}\text{s}^{-1}\; 80 \times 10^{-27}\,\text{cm}^2 = 8 \times 10^8\,\text{s}^{-1} \tag{2.112}$$

or 0.8 GHz. Since 1 fb $= 10^{-39}$ cm$^2$, the number of X particles per year is

$$N_X = L\sigma_X t \simeq 10^{34}\,\text{cm}^{-2}\text{s}^{-1}\; 10^{-39}\,\text{cm}^2\; 3 \times 10^7\,\text{s} = \;300. \tag{2.113}$$

The effective area is A $= 4\pi\sigma^2 = 3.2 \times 10^{-9}$ corresponding to an r.m.s. coordinate in each transverse direction of $\sigma = \sqrt{\beta^*\epsilon_t}$ =16 micron.

---

For bunches of root mean square length $\sigma_z$ crossing with small angle $\theta_y$, the effective area is reduced and it turns out for gaussian shaped bunches the effect can be represented by the replacement

$$\sigma_y \to \left[\sigma_y^2 + \sigma_z^2\theta_y^2\right]^{\frac{1}{2}}. \tag{2.114}$$

Such an expression applies to $pp$ colliders like the LHC. At the design energy of 7 TeV with $\beta^* = 0.55$ m, $\sigma_x = 16.6\ \mu$m, $\sigma_z = 7.55$ cm, and $\theta_y = 185$ mrad, the crossing angle correction factor is 0.84.

Detailed measurements of process cross sections underlie the theoretical edifice describing fundamental particles and their interactions. In a general two particle collision, many statistically independent processes may be identified. The distributions of directions and velocities and types of particles are measured and used to define differential cross sections as these bear important information about particle dynamics. An example differential cross section for the elastic collision of an electron with a proton is found with classical mechanics in a worked example. We will encounter many such differential cross sections.

---

**Example 2.9. Rutherford scattering.** Calculate the scattering angle $\theta$ as a function of impact parameter $b$ for Coulomb scattering of a particle of charge $e$, mass $m$, and velocity $v_0$ initially at large distance from a fixed charge $Ze$. Use this to find the effective differential area (differential cross section) for scattering into a range $\theta$ to $\theta + d\theta$.

Figure 2.12 shows the geometry of a collision and defines the impact parameter $b$. Since at large distances the potential energy vanishes and since energy and angular momentum are conserved, the impact parameters of the incident and exiting directions are equal. The direction of the momentum $\Delta p$ imparted to the charge $e$ is along the symmetry axis of the trajectory. The corresponding polar angle from that direction is called $\phi$ and the radial coordinate is $r$. Equating the momentum change to the time integral of the component of force along its direction gives the expression

$$\Delta p = \int dt\ F \rightarrow 2m\gamma v \sin\frac{\theta}{2} = \int dt\ k_e \frac{Ze^2}{r^2}\cos\phi = \frac{k_e Ze^2}{v_0 b}\int_{\phi_i}^{\phi_f} d\phi\ \cos\phi \tag{2.115}$$

where $k_e = (4\pi\epsilon_0)^{-1}$ and we used the angular momentum conservation relation $L = m\gamma r^2 d\phi/dt = m\gamma_0 v_0 b$ to make the replacement $r^{-2} \rightarrow (v_0 b)^{-1} d\phi/dt$ on the right-hand-side. The initial angle $\phi_i$ and final angle $\phi_f$ are related by $\phi_f = -\phi_i = (\pi - \theta)/2$ and the integral is $2\sin\phi_f = \cos\frac{\theta}{2}$ so solving for the impact parameter gives the scattering angle as a function of polar angle

$$b = a\cot\frac{\theta}{2} \tag{2.116}$$

where $a = k_e Ze^2/(m\gamma_0 v_0^2)$. Incident particles with impact parameters in a ring of area $d\sigma = 2\pi b db = 2\pi b(db/d\theta)d\theta$ are scattered into the angular range $d\theta$ about $\theta$. An element of solid angle is $d\Omega = \sin\theta d\theta d\phi = d\phi|d\cos\theta|$. Using cylindrical symmetry, the differential scattering cross section is

$$\frac{d\sigma}{d\Omega} = b\frac{db}{d\theta}\frac{d\theta}{d\cos\theta} = \frac{a^2}{2}\frac{\cos(\theta/2)}{\sin^3(\theta/2)}\frac{1}{\sin\theta}. \tag{2.117}$$

Using the identity $4\sin^2(\theta/2)\cos^2(\theta/2) = (1-\cos\theta)(1+\cos\theta) = \sin^2\theta$, and defining $E_k = m\gamma_0 v_0^2/2 = E(v/c)^2/2$, this may be written as

$$\frac{d\sigma}{d\Omega} = a^2\frac{1}{4\sin^4(\theta/2)} = \frac{\alpha^2 Z^2}{16E_k^2\sin^4\frac{\theta}{2}}. \tag{2.118}$$

We assumed the charge escapes the Coulomb field which is only valid in the relativistic Kepler problem if $L > \hbar(\alpha Z)^{-1}$.

---

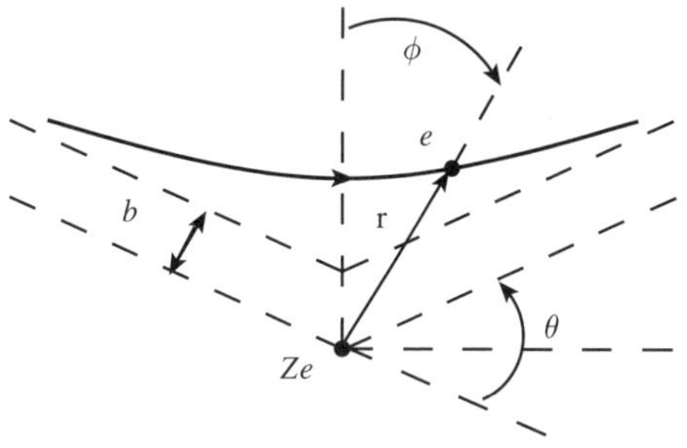


**FIGURE 2.12** Classical view of a collision of a charge $e$ with a fixed charge $Ze$. The impact parameter $b$ is the distance of closest approach for infinite incident momentum. The scattering angle relative to the incident direction is $\theta$. The angular coordinate relative to the scattering symmetry axis is $\phi$.

## 2.6 ■ THE LHC

The CERN Large Hadron Collider (LHC) is a proton-proton colliding beam machine with a design center of mass energy of 14 TeV. About three thousand bunches per ring, each bunch containing nominally $1.15 \times 10^{11}$ protons collide symmetrically at design interaction points every twenty-five ns to provide an average luminosity in excess of $10^{34}$ cm$^{-1}$ $s^{-1}$. Two general purpose detectors called CMS and ATLAS and two special purpose detectors ALICE and LHCb study the intense streams of matter produced in collisions. The CMS and ATLAS experiments use complementary designs to study the highest energy collisions and are described in detail in Chapter 3. The ALICE experiment is optimized to study lower energy collisions of nuclei and to examine signatures for the formation of a quark-gluon plasma state (see Chapter 9) while the LHCb experiment is designed to study the properties of bottom flavored matter and antimatter. In this section we take a look at the technology of the accelerator complex.

The LHC accelerator is assembled in a tunnel formerly home to the CERN Large Electron Positron storage ring (LEP) which studied electron positron collisions at center of mass energies up to twice the mass of the $W$ boson. The tunnel (Figure 2.13) is about 4 m in diameter and runs in a circle 27 kilometers in circumference at a depth of 50 to 175 meters, spanning the border of Switzerland and France near Geneva, Switzerland. The LHC storage ring is composed of eight arc sections each containing 23 cells and eight straight sections as shown in Figure 2.14. Two counter rotating beams collide at four intersection points. In the tunnel, there are 1232 dipole bending magnets 15 m in length and 392 focusing quadrupole magnets 5-7 m in length. The magnets are superconducting and use a twin aperture design to accommodate the two proton beams. Each dipole magnet (Figure 2.16) is 15 m in length and has an inductance of 0.11 H. At the nominal current of 12 kA, the field strength is 8.3 T and the stored energy is 7.1 MJ. The NiTn strand coils have an inner diameter of 56 mm and an

**FIGURE 2.13** Inside the Large Hadron Collider tunnel. A string of superconducting dipole magnets advances through the underground tunnel. [CERN]

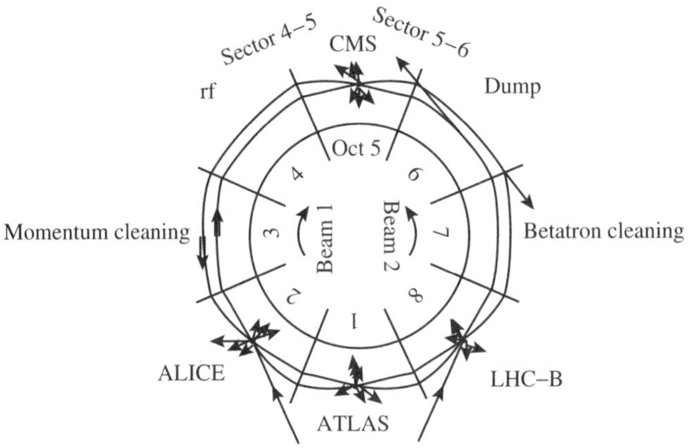


**FIGURE 2.14** Layout of the Large Hadron Collider. Shown are the locations of eight arc sections and eight straight sections, the injection and extraction points, collimation points called betatron cleaning stations, the radio frequency (RF) cavities for acceleration, and the four experimental halls.

outer diameter of 120 mm. The quadrupole magnet field gradient is 223 T m$^{-1}$. About 100 tons of liquid helium refrigerant maintains the magnet coil temperature at close to two degrees K. Each cell is cooled by approximately 1700 liters of superfluid helium. The magnets form a lattice of FODO cells (Figure 2.15) each containing focusing and defocusing quadrupole magnets which together provide a net focus in both transverse directions. At the collision points special focusing magnets reduce the beam separation from 194 mm over a distance of 200 m and guide the beam bunches through the crossing point. At crossing, the beams have transverse dimension of about 16 microns and crossing angle $3 \times 10^{-4}$ radians, sufficient to prevent simultaneous distributed collision of multiple bunches.

The proton injection complex begins with a duoplasmatron, a device delivering 300 mA of proton current at 92 keV. This beam is accelerated by a radio frequency quadrupole (RFQ) reaching 750 keV followed by a 50 MeV linear accelerator

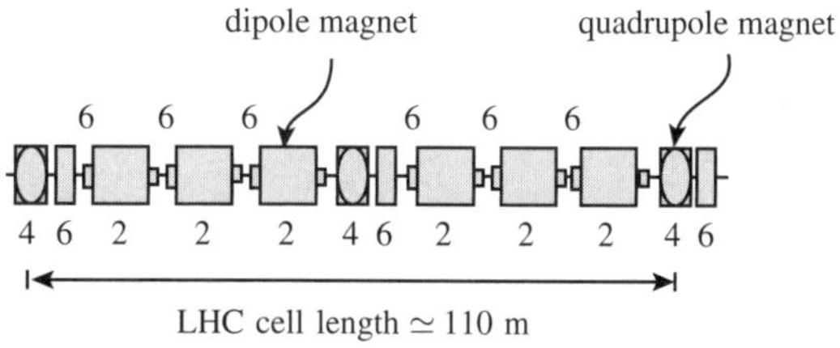


**FIGURE 2.15** Schematic of one cell of the LHC magnet lattice. Each cell contains six dipole bending magnets and two quadrupole focusing magnets. Additional corrector magnets are inserted between these essential elements.

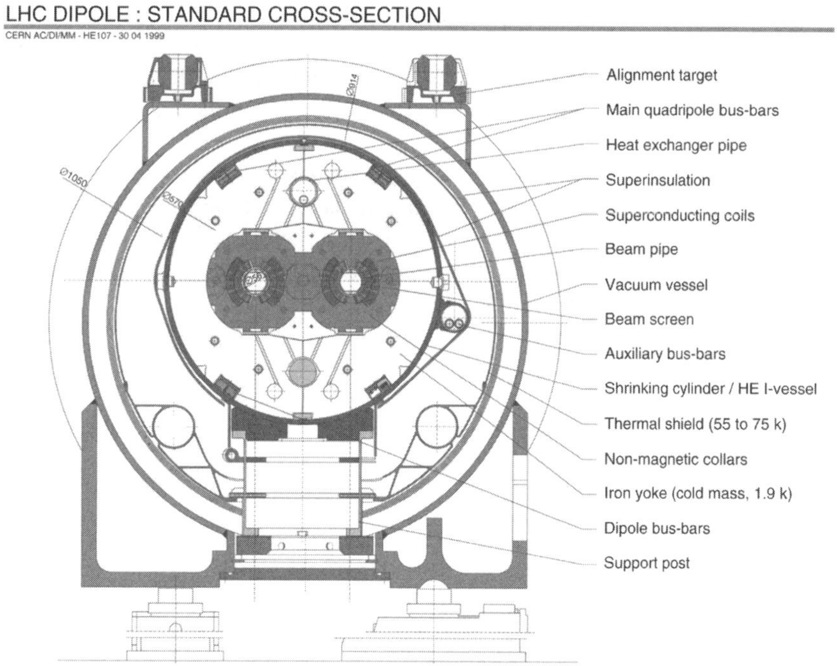


**FIGURE 2.16** Design of the dual bore superconducting LHC dipole magnet. [CERN Photolab, http://cdsweb.cern.ch/record/40524]

(LINAC 2) feeding the Proton Synchrotron Booster (PSB) which reaches 1.4 GeV. From the booster, the protons enter the Proton Synchrotron (PS) where their energy is brought to 26 GeV and subsequently enter the Super Proton Synchrotron (SPS) where their energy reaches 450 GeV. The final acceleration in the LHC to 7 TeV entails ramping the LHC magnetic field from about 0.5 to 8.3 T over a period of about 28 minutes. At full energy, the protons have a Lorentz factor $\gamma = E/m_p$ in excess of 7000 and circulate with a period of about 90 $\mu$s, colliding during a design storage time of 10-20 hours. The filling of the LHC entails moving batches of 72 full plus 12 empty bunches of 25 ns bunch spacing from the PS into the SPS. The SPS injection takes about four s per batch and four batches fill about 35% of the SPS. The acceleration to 450 GeV takes about eight s. Nine super-cycles of three batch SPS fills and three super-cycles of four batch fills produce 2808 proton bunches which are moved from the SPS into the LHC for each beam with a bunch structure

$$\begin{aligned} 3564 = &\{[(72+8e)\times 3+30e]\times 2+[(72+8e)\times 4+31e]\}\times 3 \\ &+\{[(72+8e)\times 3+30e]\times 3+81e\}. \end{aligned} \tag{2.119}$$

In this expression, "(72 + 8e)" denotes 72 bunches followed by eight empty bunches and so on. This bunch train structure is designed with four fold symmetry so at each experiment only full bunch pairs or empty bunch pairs collide. This filling process requires 4.3 minutes per beam.

The LHC is designed to also collide Pb ions with an energy of 2.8 TeV per nucleon and a luminosity of $10^{27}$ cm$^{-2}$ s$^{-1}$. The production of the ions starts with the electron cyclotron resonance (ECR) source. In such a source, a plasma of charged ions and electrons is confined in a magnetic trap. Electrons are heated to keV energies by a microwave field at the cyclotron frequency defined by the longitudinal magnetic field of the trap and collisions with the ions produces a distribution of ionization states. (The cross section for collisional ionization is approximately $\sigma \sim 10^{-16}\ \mathrm{cm}^2/Z^4$ for electron energy in excess of the ionization energy.) The resulting charge state distribution represents a balance between collisional ionization

$$e^- + \mathrm{Pb}^{N+} \rightarrow \mathrm{Pb}^{(N+1)+} + (2e^-) \tag{2.120}$$

and charge exchange

$$\mathrm{Pb}^{(N+1)+} + \mathrm{X}^0 \rightarrow \mathrm{Pb}^{N+} + \mathrm{X}^+. \tag{2.121}$$

The more highly charged ions have a longer confinement time as the trapping depends on charge as well as temperature. They are released as a pulse. The addition of lighter ions cools the heavier ions and lengthens their trapping time enhancing the production of higher charge states.

In the CERN injector complex, $\mathrm{Pb}^{27+}$ ions are accelerated by a radio frequency quadrupole (RFQ, see problem) and linear accelerator LINAC 3 to an energy of 4.2 MeV/nucleon. The ions are passed through a stripper to remove the remaining electrons. They are accumulated in the LEIR ring and subsequently injected into the PS with an energy of 73 MeV/nucleon. The design calls for the PS to accelerate four bunches of fully stripped $Pb^{82+}$ ions each containing $1.6 \times 10^8$ ions. These bunches are injected into the SPS at 5.11 GeV per nucleon and accelerated to 177 GeV/nucleon. Thirteen injections give a batch of 52 bunches separated by 125 ns. Three batches with a gap of seven bunches injected into the LHC form a bunch train. Four bunch trains are separated by eight empty bunches. The last batch has nine not 13 injections and the total is 608 bunches per LHC ring requiring 9.5 minutes per ring of filling time.

Many of the details just described are only illustrative and will evolve with experience but hopefully they indicate the complex orchestration of many technologies required in the successful operation of a high energy accelerator. At the design luminosity and energy, the LHC supplies about one billion *pp* collisions

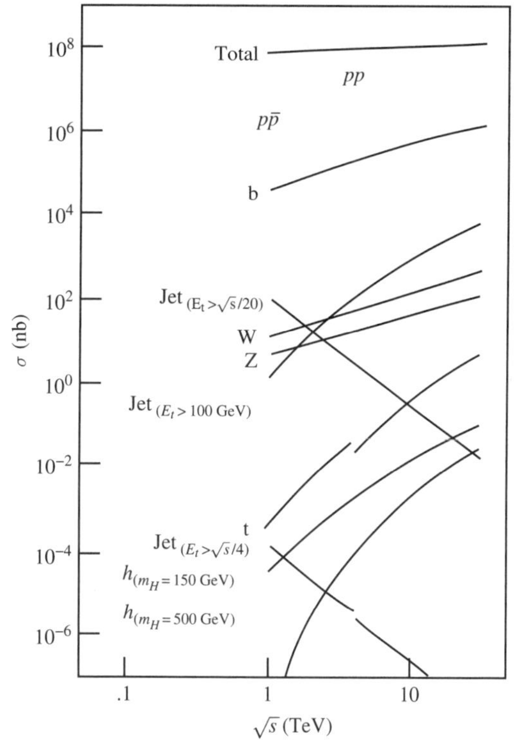


**FIGURE 2.17** Cross sections for particle production at high energy. Cross sections for production of heavy quarks, jets, electroweak gauge bosons, and the standard model Higgs boson in the center of mass energy range of the LHC (pp collisions) and for the Tevatron ($p\bar{p}$ collisions) are seen to increase with available energy. The fractional energy per constituent collision, reflected in the cross sections for jets with a fixed fraction of the total energy, decreases. [J. M. Campbell, J. W. Huston, W. J. Stirling, Rept. Prog. Phys. **70**, 89 (2007), arXiv:hep-ph/0611148v1]

per second. A typical data set of 10 $\text{fb}^{-1}$ obtained in one year of running at low luminosity corresponds to the production of order $10^{13}$ $b\bar{b}$ pairs, two trillion $W$ bosons, 500 million $Z$ bosons, 10 million $t\bar{t}$ pairs, and 20 thousand Higgs bosons (if $m_H = 150$ GeV.) The detectors systems used to study these collisions will be described in Chapter 3.

## 2.7 ■ FURTHER READING

The 1921 Nobel Prize in Physics was awarded to Albert Einstein "for his services to Theoretical Physics, and especially for his discovery of the law of the photoelectric effect."

The 1939 Nobel Prize in Physics was awarded to Ernest Lawrence "for the invention and development of the cyclotron and for results obtained with it, especially with regard to artificial radioactive elements."

The 1951 Nobel Prize in Physics was awarded jointly to Sir John Douglas Cockcroft and Ernest Thomas Sinton Walton "for their pioneer work on the transmutation of atomic nuclei by artificially accelerated atomic particles" using artificial accelerators.

The 1959 Nobel Prize in Physics was awarded jointly to Emilio Gino Segre and Owen Chamberlain "for their discovery of the antiproton."

Special relativity and relativistic mechanics are described in L. D. Landau and E. M. Lifshitz, *The Classical Theory of Fields*, Pergamon, Oxford (1975), Wolfgang Rindler *Essential Relativity*, Springer-Verlag (1977), and J. D. Jackson, *Classical Electrodynamics*, John Wiley and Sons, Inc., (1999). Antimatter in classical physics is discussed in John P. Costella, Bruce H. J. McKellar, Andrew A. Rawlinson, Am. J. Phys. **65**, 835 1(1997), arXiv:hep-ph/9704210v1.

Texts describing high energy particle accelerator design principles and applications include A. Chao and M. Tigner, eds, *Handbook of Accelerator Physics and Engineering*, World Scientific (1999), D. A. Edwards and M. J. Syphers, *An Introduction to the Physics of High Energy Accelerators*, John Wiley & Sons, Inc., (1993), and Helmut Wiedemann, *Particle accelerator physics*, Springer (2007). See also *CERN Accelerator School: 5th General Accelerator school*, CERN 94-01 (1994) at `http://preprints.cern.ch/cernrep/1994/94-01/94-01_v1.html` and `http://preprints.cern.ch/cernrep/1994/94-01/94-01_v2.html`, M. Sands, *The Physics of Electron Storage Rings: An Introduction*, SLAC-R-121, and E. D. Courant and H. S. Snyder, Annals of Physics **3**,1 (1958).

The problem of stability and storage of high energy beams are similar to those of storing and cooling atomic and molecular beams. For an introduction, see Wolfgang Paul, Electromagnetic Traps for Charged and Neutral Particles, Nobel Lecture, December 8, 1989.

The polarization of electron and proton beams and the generation of synchrotron radiation are two fascinating aspects of accelerator physics we have omitted to consider. An introduction to the classical mechanics of relativistic spinning particles is provided in J. D. Jackson's text. An extensive literature describes the kinematic effect called Thomas precession which appears in relativistic physics of spin 1/2 and spin 1 particles.. See for example John P. Costella, Bruce H. J. McKellar, and Andrew A. Rawlinson, Am. J. Phys. **69**, 837 (2001)

Just one scholarly article on beam polarization is S. R. Mane,"Electron-spin polarization in high-energy storage rings. I. Derivation of the equilibrium polarization," Phys. Rev. **A36**, 105119 (1987).

The design and construction of the LHC is described in Lyndon R. Evans (ed.), *The Large Hadron Collider: A Marvel of Technology*, CRC Press (2009) and in Lyndon Evans and Philip Bryant JINST **3**, S08001 (2008).

## 2.8 ■ PROBLEMS

### Problem 2.1. Velocity addition

The Lorentz transformation for aligned frames with relative velocity $v$ along the $x$-axis for two events is

$$dx' = \gamma(dx - vdt)\ ;\ dt' = \gamma(dt - vdx).$$

a) Divide these two to derive the formula

$$u' = \frac{u - v}{1 - uv}$$

for the velocity $u'$ of a particle moving along the $x$-direction with velocity $u$ as seen from a frame moving along the $x$-direction with velocity $v$.

b) Consider a collision of a proton of velocity $+u_p$ with a proton of velocity $-u_p$ in the laboratory frame. Use the velocity addition formula to show that the velocity of the second proton in the rest frame of the first is $u' = -2u_p/(1 + u_p^2)$ and its energy is $E' = (1 + u_p^2)/(1 - u_p^2)$. Show that the same expression for the energy $E'$ results from Lorentz transformation of the 4-momentum from the laboratory frame to the frame of the first proton.

c) Use the general transformation

$$\mathbf{x}' = \mathbf{x} - \gamma t\mathbf{v} + \frac{\gamma^2}{\gamma + 1}(\mathbf{v}\cdot\mathbf{x})\mathbf{v},\ t' = \gamma t - \gamma(\mathbf{v}\cdot\mathbf{x})$$

to derive the more general and intricate velocity addition rule

$$\mathbf{u}' = \frac{1}{\gamma(1 + \mathbf{u}\cdot\mathbf{v})}\left[\mathbf{u} - \gamma\mathbf{v} + \frac{\gamma^2}{\gamma + 1}(\mathbf{v}\cdot\mathbf{u})\mathbf{v}\right].$$

The rule is not symmetric in $\mathbf{u}$ and $\mathbf{v}$ reflecting the fact that boost transformations $L(\mathbf{u})$ and $L(\mathbf{v})$ along directions which are not co-linear do not commute. It can be shown that $L(\mathbf{v})L(\mathbf{u}) = L(\mathbf{u}')R(\mathbf{u}, \mathbf{v})$ where $R(\mathbf{u}, \mathbf{v})$ is a rotation which is related to Thomas precession in accelerated frames. See for example Sergei A. Klioner, arXiv:0803.1303v2 [astro-ph], and John P. Costella, Bruce H. J. McKellar, and

Andrew A. Rawlinson, Am. J. Phys. **69**, 837 (2001) and, for a different approach, John A. Rhodes and Mark D. Semon, Am. J. Phys. **72**, 943 (2004).

---

**Problem 2.2. Relative velocity**

---

Consider two particles with 4-velocity vectors $v_1$ and $v_2$. Show that in the rest frame of the first particle, the inner product of the 4-velocities reduces to

$$v_1 v_2 = \gamma(v)$$

where $v$ is the magnitude of the velocity $\mathbf{v}_{21}$ of the second particle relative to the first. Solve for $v$ in terms of the inner product and derive the expression

$$\mathbf{v}_{21}^2 = \frac{(p_1 p_2)^2 - m_1^2 m_2^2}{(p_1 p_2)^2} = \mathbf{v}_{12}^2.$$

---

**Problem 2.3. Decay momentum**

---

Starting with the expression $E_{2,cm} = \left(s + m_2^2 - m_1^2\right)/(2\sqrt{s})$, show that the momentum of either of two particles in their center of momentum frame is

$$p = \frac{1}{2\sqrt{s}}\left[\lambda\left(s, m_1^2, m_2^2\right)\right]^{1/2}$$

where $\lambda(a, b, c) \equiv a^2 + b^2 + c^2 - 2ab - 2ac - 2bc$. Find the value of the momentum for the decays $K^+ \to \mu^+ \nu_\mu$ and $K^+ \to \pi^+ \pi^0$ and compare to the values in the Particle Data Group tables of particle properties.

---

**Problem 2.4. Three-body decay kinematics**

---

Consider the decay $K_L \to 3\pi^0$ in the center of mass. Regard two of the pions as a system of mass $m_{23}$ and use the 2-body decay formula $E_1 = (m_K^2 - m_{23}^2 + m_1^2)/(2m_K)$ to show that the maximum energy of one $\pi^0$ in the kaon rest frame corresponds to the other two having equal velocity vectors and mass $m_{23} = 2m_\pi$. Next compute the two body decay momentum for this pair mass to show that the maximum pion momentum is

$$p = \frac{1}{2m_K}\left[\left(m_K^2 - (3m_\pi)^2\right)\left(m_K^2 - m_\pi^2\right)\right]^{1/2} = 139 \text{ MeV}.$$

---

**Problem 2.5. Dalitz plot boundaries**

---

Sketch the boundaries of the allowed region in the plane $s_{12}/m_0^2$ vs $s_{23}/m_0^2$ for the decay of a mass $m_0$ to masses $m_1$, $m_2$ and $m_3$ for the cases a) $m_1 = 0$, $m_2 = m_0/4$, $m_3 = 0$ and b) $m_1 = m_0/4$, $m_2 = 0$, $m_3 = 0$ in which two of the particles are massless.

## Problem 2.6. Neutron Decay kinematics

Consider the decay $n \Rightarrow pe^-\bar{\nu}$ of a neutron of relativistic kinetic energy $K = 500$ MeV in case, in the rest frame of the neutron, the decay leaves the proton at rest, the electron and antineutrino being emitted in opposite directions. The rest energies of these particles are $m_n = 939.565$ MeV, $m_p = 938.272$ MeV, $m_e = 0.511$ MeV, and $m_{\bar{\nu}} = 0.000$ MeV.

a) What is the speed $v$ in m/s of the neutron?

b) What is its maximum thickness along its direction of motion if at rest it is a sphere of radius 1 fm?

c) What is the total relativistic energy in MeV of the electron and antineutrino in the neutron rest frame? This energy is also the invariant mass of the pair since their total momentum vanishes in this special case.

d) By transforming the invariant mass of the electron plus antineutrino and the invariant mass of the proton, find the total relativistic energy in MeV of the electron plus antineutrino in the laboratory frame and the proton kinetic energy in the laboratory frame.

## Problem 2.7. Neutral pion mass

The mass of the $\pi^-$ is known from pionic atom emission spectra and $m_n$ can be deduced from the threshold photon energy for $\gamma d \to np$, given the deuteron and proton mass. Pion capture from rest in hydrogen results in (a) $\pi^- p \to n\pi^0$ and (b) $\pi^- p \to n\gamma$ with relative probability 1.56. Find $E_\pi^0$ and $E_\gamma$ for these reactions. The decay $\pi^0 \to \gamma\gamma$ is isotropic. Show that the lab energy of the $\gamma$ rays from the decay of the $\pi^0$ in reaction (a) are uniformly distributed between limits determined by the masses. Calculate the limits and sketch the total energy distribution over a range 0-150 MeV. From this distribution, the mass difference $m_{\pi^0} - m_{\pi^-} = 4.6$ MeV is deduced. [Ref: Crowe and Phillips, Phys. Rev. **96**, 470 (1954)].

## Problem 2.8. Energy transfer to an electron

Consider the elastic scattering of a mass $m$ of energy $E$ and momentum $p$ from a stationary electron. The velocity of the center of mass is $v = p/(E + m_e)$. Find the electron velocity in the center of mass, reverse it, and transform back to the laboratory and show that the maximum energy transfer to the electron is

$$\Delta E_{max} = 2m_e \frac{p^2}{s} = 2m_e \frac{p^2}{2Em_e + m^2 + m_e^2}.$$

Evaluate this for a proton with total energy of 10, 100, and 1000 GeV. What fraction of the proton energy is transferred if $E \to \infty$?

### Problem 2.9. Relativistic acceleration

Use the relativistic equation of motion $d\mathbf{p}/dt = \mathbf{F}$ to find the motion of a mass $m$ starting from rest subject to a constant force $F$. Show that the time $t$ to travel a distance $x$ is

$$t = \sqrt{2(x/a) + (x/c)^2}$$

where $a = F/m$. Explain the behavior of this expression for small and large $x$.

### Problem 2.10. Photon acceleration

An electron of energy $E_e = 50$ GeV collides head on with a photon of energy $E_\gamma = 1$ eV. Transform to the center of mass using $\gamma_{cm} = (E_e + E_\gamma)/\sqrt{s}$, reverse momenta, and transform back to show that the energy of photons scattered in the direction of the electron beam is approximately 22 GeV.

### Problem 2.11. Synchrotron energy loss

An electron of energy $E$ moves in a circular orbit in a uniform magnetic field $\mathbf{B}$. The electric field in the instantaneous rest frame is

$$\mathbf{E}' = \gamma(\mathbf{E} + \mathbf{v} \times \mathbf{B})$$

with $\gamma = E/m_e c^2$. Compute the radiated power $P$ in the instantaneous rest frame with the nonrelativistic Larmor formula $P = e^2 a^2/(6\pi\epsilon_0 c^3)$ and use the fact that $dE/dt$ is a Lorentz invariant to find the lab frame energy loss rate is

$$\frac{dE}{dt} = -\frac{2}{3}\frac{\alpha}{R^2}\left(\frac{p}{mc}\right)^4 .$$

### Problem 2.12. Relativistic Kepler motion

Relativistic motion in a Coulomb field is described by the equation

$$\frac{d\mathbf{p}}{dt} = \frac{d}{dt}\left(\frac{m\mathbf{v}}{\sqrt{1 - v^2}}\right) = -\alpha\frac{\hat{r}}{r^2}$$

where $\mathbf{p} = \gamma m \mathbf{v}$. a) Expand the time derivative and then take the dot product of both sides with $\mathbf{v}$ and integrate over time to derive conservation of the energy $E = \gamma m - \alpha/r$. b) Show that $\mathbf{L} = \mathbf{r} \times \mathbf{p}$ is conserved. c) In polar coordinates, $\mathbf{v} = \dot{r}\hat{r} + r\dot{\theta}\hat{\theta}$ and $L = \gamma m r^2 \dot{\theta}$. Put $s(\theta) = 1/r(\theta)$ and $\dot{r} = \dot{\theta} dr/d\theta$ and show that

$$(E + \alpha s)^2 = \mathbf{p}^2 + m^2 = L^2\left(\frac{ds}{d\theta}\right)^2 + L^2 s^2 + m^2$$

and then differentiate with respect to $\theta$ and divide by $ds/d\theta$ to find the orbit equation

$$\frac{d^2s}{d\theta^2} + \kappa^2 s = \frac{\alpha E}{L^2}$$

where $\kappa^2 = 1 - (\alpha/L)^2$. d) Verify the solutions of these two equations

$$s(\kappa^2 > 0) = \sqrt{\frac{E^2 - m^2\kappa^2}{L^2\kappa^4}} \cos\left[\sqrt{\kappa^2}(\theta - \theta_0)\right] + \frac{E\alpha}{L^2\kappa^2}$$
$$s(\kappa^2 = 0) = \frac{E}{2\alpha}(\theta - \theta_0)^2 + \frac{m^2 - E^2}{2E\alpha}$$
$$s(\kappa^2 < 0) = \sqrt{\frac{E^2 - m^2\kappa^2}{L^2\kappa^4}} \cosh\left[\sqrt{-\kappa^2}(\theta - \theta_0)\right] + \frac{E\alpha}{L^2\kappa^2}.$$

For $E > m$, unbound orbits with $\kappa^2 > 0$ may loop around the origin if $\kappa \sim 0$ (the number of loops may be found by setting $u = 0$ and solving for $\theta$). For $E > m$ and $\kappa < 0$, the orbits implant upon the origin. Thus positive energy solutions with $L < \alpha$ or impact parameter $b < \alpha/p$ do not escape. For $E < m$, the orbits with $\kappa^2 > 0$ are precessing ellipses with two turning points while for $\kappa < 0$ the orbits spiral out from and back to the origin. [Ref: T. H. Boyer, Am. J. Phys. **72**, 992-997 (2004), arXiv:physics/0405090v1]

---

### Problem 2.13. Electron cooling

A hot antiproton beam with mean energy $E_{\bar{p}} = 8.0$ GeV is cooled with a collinear cold beam of electrons. What mean electron energy is required for a speed equal to the mean antiproton speed?

---

### Problem 2.14. Van der Graaf accelerator

The Oak Ridge National Laboratory model 25 URC Van de Graaf accelerator comprises a high-voltage generator with a terminal at 25.5 MV inside a 33 m high cylindrical vessel containing $SF_6$ gas at 65-80 psi. Negative ions injected at the base and accelerated upwards pass through a stripper within the terminal producing ions with reversed charge. Their direction is reversed by a bending magnet, and they are again accelerated, downwards.

a) Assuming the terminal and vessel are spheres of radius 2.5 m and 5 m, estimate the electric field strength at the surface of the terminal.

b) The critical voltage $V_c$ for spark breakdown of a gas at absolute pressure $p$ over a distance $d$ is a function of the product $x = pd$. It has the value $V_c = V_{min} = 327$ V at $x_{min} = 0.567$ torr cm for air. For $x$ larger than this minimum value, $V_c = V_{min} x/x_{min}$. At smaller $x$, $V_c$ is large as the gas behaves increasingly like a vacuum. What is the critical electric field strength for air at 1 atm = 760 torr?

### Problem 2.15. Betatron

In the betatron, charged particles are accelerated between the pole tips of a dipole magnet while the magnet field strength is increased. The circulating electric field associated by Faraday's Law with the time dependent magnetic field increases the energy while the magnetic field confines. Show that the relativistic equations of motion in cylindrical coordinates for a charge $e$ with energy $E(t)$, radius $r(t)$ and azimuth $\phi(t)$ are

$$\begin{aligned}\dot{E}\dot{r} + E(\ddot{r} - r(\dot{\phi})^2) &= er\dot{\phi}B_z \\ \dot{E}r\dot{\phi} + E(r\ddot{\phi} + 2\dot{r}\dot{\phi}) &= eE_\phi + e(\dot{z}B_r - \dot{r}B_z) \\ \dot{E}\dot{z} + E\ddot{z} &= -er\dot{\phi}B_r\end{aligned}$$

where $E_\phi$ is the electric field strength and $B_r$ and $B_z$ are the components of the magnetic field strength. Show that a circular orbit of radius $R$ in the $x - y$ plane subject to a time varying magnetic field cylindrically symmetric about the $z$-axis requires the magnetic field strength must decrease with radius such that the enclosed magnetic flux satisfies $\int_A \mathbf{B}\cdot d\mathbf{a} = 2\pi R^2 B_z(R)$.

### Problem 2.16. Strong focusing

Suppose a particle of charge $q$ and energy $E$ passes through a quadrupole magnet with axis along the $z$-direction. Assume $v_z \sim$ constant and that the magnetic field has the form $\mathbf{B} = (B'y, B'x, 0)$. Show that

$$\frac{d\mathbf{p}}{dt} = Ev_z\frac{d\mathbf{v}}{dz} = q\mathbf{v}\times\mathbf{B} \rightarrow \frac{d^2x}{dz^2} = k_x^2 x\;;\; \frac{d^2y}{dz^2} = k_y^2 y$$

where $k_x^2 = qB'/(Ev_z) = -k_y^2$. Show that the transfer matrix for a field free region of length $z$ followed by a quadrupole of length $w$ is

$$M = \begin{pmatrix} \cos(k_x w) & \cos(k_x w)z + \frac{\sin(k_x w)}{k} \\ -k\sin(k_x w) & \cos(k_x w) - zk\sin(k_x w)\end{pmatrix}$$

and that the focal point lies a distance $f = k^{-1}\cot(k_x w)$ from the end of the magnet.

### Problem 2.17. Hill's equation

The deviation from the ideal reference orbit in an accelerator is described in linear approximation by Hill's equation

$$x'' + k(s)x = p$$

where $x(s)$ is a transverse coordinate expressed as a function of path length $s$ along the reference orbit, $x'' = d^2x/ds^2$, $k(s)$ is an effective force constant, and $p$ represents various forcing functions including nonlinearities. Suppose $C(s)$ and $S(s)$

are solutions to the homogeneous equation ($p = 0$) satisfying $C(0) = S'(0) = 1$ and $C'(0) = S(0) = 0$ where the prime denotes the derivative with respect to $s$. Show that the general solution is

$$x = Cx_0 + Sx_0' + S\int_0^s dt\ C(t)p(t) - C\int_0^s dt\ S(t)p(t)$$

by substitution to find $x'' + kx = (S'C - C'S)p$ and then show that $S'C - C'S$ is constant and has initial value unity. In the case of a storage ring or synchrotron that provides net stable beam guidance, such a pair of solutions are periodic and called cosine-like and sine-like solutions and the oscillations referred to as betatron oscillations.

---

**Problem 2.18. FODO lattice magnet tolerance**

---

Show that the transfer matrix for a focusing quadrupole followed by a defocusing quadrupole of unequal strength is

$$\mathbf{M}_{FD} = \begin{pmatrix} \cos\alpha_1\cosh\alpha_2 - \sqrt{k_2/k_1}\sin\alpha_1\sinh\alpha_2 & \cos\alpha_1\sinh\alpha_2/\sqrt{k_2} + \sin\alpha_1\cosh\alpha_2/\sqrt{k_1} \\ -\sqrt{k_1}\sin\alpha_1\cosh\alpha_2 + \sqrt{k_2}\cos\alpha_1\sinh\alpha_2 & -\sqrt{k_1/k_2}\sin\alpha_1\sinh\alpha_2 + \cos\alpha_1\cosh\alpha_2 \end{pmatrix}$$

and that the condition for beam stability is $-1 \le \mathrm{Tr}\mathbf{M}/2 \le +1$ where

$$\mathrm{Tr}\mathbf{M}/2 = \cos\alpha_1\cosh\alpha_2 + \sin\alpha_2\sinh\alpha_2\left[\sqrt{k_2/k_1} - \sqrt{k_1/k_2}\right]/2.$$

A lattice of FD pairs is also a lattice of DF pairs which implies a similar constraint with the labels 1 and 2 interchanged. Stability for both FD and DF ensures stability of both transverse coordinates. For the case of equal strength ($k_1 = k_2$) but different length, plot the curves $\cos\alpha_1\cosh\alpha_2 = \pm 1$ and $\cos\alpha_2\cosh\alpha_1 = \pm 1$ and show the enclosed stable region has a necktie shape starting from $\alpha_1 = \alpha_2 = 0$, permitting an increasing deviation from the line $\alpha_1 = \alpha_2$ until the maximal value $\alpha_1 = \alpha_2 = 1.86$ is reached. A intermediate nominal value permits the largest tolerance on magnet length and strength.

---

**Problem 2.19. Quadrupole combinations**

---

Two quadrupole magnets rotated by 90 degrees can focus both transverse coordinates of a beam but the focal points are generally at different locations. By a choice of parameters, the focal points may be made to coincide but the magnifications are generally different. Symmetric triplets or quartets (e.g. DF'D, DFFD or FDDF) can provide stigmatic focusing with the same magnification in both transverse coordinates. "Triplets" of 200 T m$^{-1}$ quadrupole magnets 23 m on either side of the intersection points are used to focus the colliding beams at the LHC.
a) For a drift space of length $d$ and a thin lens of focal length $f$, the transfer matrices are

$$\mathbf{d} = \begin{pmatrix} 1 & d \\ 0 & 1 \end{pmatrix} ; \ \mathbf{f} = \begin{pmatrix} 1 & 0 \\ -1/f & 1 \end{pmatrix}.$$

Show a thin lens between two drift spaces corresponds to a transfer matrix

$$\mathbf{M} = \mathbf{d}_1\mathbf{f}\mathbf{d}_2 = \begin{pmatrix} 1 - d_1/f & d_1 + d_2 - d_1 d_2/f \\ -1/f & 1 - d_2/f \end{pmatrix}.$$

An image is formed if $x_2 = M_{11}x_1 + M_{12}x_1'$ is independent of $x_1'$ so $M_{12} = 0$ which gives the thin lens equation $f^{-1} = d_1^{-1} + d_2^{-1}$ and in this case the magnification $\mu = x_2/x_1 = M_{11}$. In general, for optical element represented by a matrix $\mathbf{m}$ between two drift spaces with transfer matrix $\mathbf{M} = \mathbf{d}_1\mathbf{m}\mathbf{d}_2$, is

$$\mathbf{M} = \mathbf{d}_1\mathbf{m}\mathbf{d}_2 = \begin{pmatrix} m_{11} + d_1 m_{21} & (d_1 m_{21} + m_{11})d_2 + d_1 m_{22} + m_{12} \\ m_{21} & m_{22} + m_{21} d_2 \end{pmatrix}.$$

The condition for imaging is given by $M_{12} = m_{12} + m_{22}d_1 + m_{11}d_2 + m_{21}d_1d_2 = 0$. In this case, the magnification is $M_{11}$ and, since detM $= 1$, we have $M_{22} = 1/M_{11}$.

b) A quadrupole magnet of effective length $L$ may be represented by a thin lens at the center of a drift space of length $L$ with transfer matrix

$$\mathbf{F} = \begin{pmatrix} 1 - \frac{1}{2}kL & L - \frac{1}{4}kL^2 \\ -k & 1 - \frac{1}{2}kL \end{pmatrix}$$

with $k \equiv f^{-1}$. Show that the transfer matrix $\mathbf{M}_{FD}(k)$ for a defocusing quadrupole followed by focusing similar quadrupole in this approximation is

$$\mathbf{M}_{FD}(k) = \begin{pmatrix} 1 + kL - \frac{1}{2}k^2L^2 & 2L - \frac{1}{4}k^2L^3 \\ -k^2L & 1 - kL - \frac{1}{2}k^2L^2 \end{pmatrix}$$

and the matrix for a focusing quadrupole followed by a defocusing quadrupole is $\mathbf{M}_{DF}(k) = \mathbf{M}_{FD}(-k)$. Now construct $\mathbf{d}_1\mathbf{M}_{FD}\mathbf{d}_2$ and $\mathbf{d}_1\mathbf{M}_{DF}\mathbf{d}_2$ and show that if we require $d_1 = d_2$ for point to point imaging in both transverse coordinates then $\mu_{FD} \neq \mu_{DF}$.

---

### Problem 2.20. Linear accelerator

---

An Alvarez linear accelerator is constructed of copper tubes along a common axis with alternating voltage of frequency $f$ applied between neighbor tubes. Within a tube, the electric field vanishes. Between tube ends, an alternating electric field accelerates particles traveling along the axis. For a fixed voltage and frequency, the tube length must increase with distance to maintain the correct phase relationship between particle and field. Determine the required tube lengths for frequency $f$, accelerating voltage $V$, injection kinetic energy $E_0$, particle mass $m$, and charge $e$. Neglect the travel time across the gaps between tubes.

## Problem 2.21. Radio frequency quadrupole accelerator

A radio frequency quadrupole (RFQ) combines electrostatic focusing with acceleration. It is constructed with four parallel electrode pole tips along a z-axis driven by an alternating voltage of frequency $\omega$ creating an alternating quadrupole electric field in the $x - y$ plane. For electrode separation $d << c/\omega$, the electric field may be regarded as static and

$$E_x = E'_x x \sin(\omega t) \; ; \; E_y = -E'_y y \sin(\omega t)$$

where $E'_x$ and $E'_y$ are the field gradients along the two transverse directions. A particle moving along the axis sees alternately a focusing and defocusing force which can provide net focusing. The pole tips are machined so that the separation between electrode pairs is modulated resulting in a longitudinal field component and the modulation in $x$ and $y$ is 90 degrees out of phase. a) Show that if the transverse fields have the form $E'_x = E'_o(1 + \epsilon \sin(kz))$ and $E'_y = E'_o(1 - \epsilon \sin(kz))$ then Gauss's Law requires a longitudinal field

$$E_z = -2\epsilon \frac{E'_o}{k} \cos(kz) \sin(\omega t) = \epsilon \frac{E'_o}{k}(\sin(kz + \omega t) - \sin(kz - \omega t)).$$

b) If the backwards going wave represented by the first term is ignored, how must $k$ vary with $z$ to be consistent with constant axial acceleration?

## Problem 2.22. Synchrotron parameters

Consider building a 20 TeV on 20 TeV $pp$ collider.
a) For a maximum dipole magnet field strength of 6.6 Tesla, what is the minimum circumference? The actual circumference would have to accommodate focusing magnets, acceleration sections, injection and extraction mechanisms, detectors etc.
b) Assuming a peak luminosity of $10^{33}$ cm$^{-2}$s$^{-1}$ at an intersection region, a total $pp$ interaction cross section of 90 mb and 30 percent live time, how many collisions will be observed by a detector in one year?
c) For an energy gain of 5.26 MeV per pass through the radio frequency acceleration system, how many seconds are required to reach maximum beam energy?
d) For a fill of $10^{14}$ protons, find the beam current in ampere, total beam energy in MJ, and the synchrotron radiation power going into the liquid He refrigeration for the magnets.
e) What energy is required to achieve the same center of mass energy in the collision of a proton with a fixed hydrogen nucleus?

## Problem 2.23. Antiproton source

The Fermilab Tevatron collider produces about $4 \times 10^{13}$ antiprotons per week.
a) How many grams of antimatter are made per year and what is the total rest

energy of the antiprotons in joules. b) Supposing collisions of 1 TeV protons and 1 TeV antiprotons consume $4 \times 10^{13}$ antiprotons per week, what is the average power in watts corresponding to these collisions?

---

**Problem 2.24. Synchrotron radiation**

---

The power radiated by a relativistic electron in circular motion with radius $\rho = p/eB$ is

$$P = \frac{1}{6\pi\epsilon_0}\frac{e^4B^2\gamma^2}{m_e^2c}$$

and the spectrum of photons is characterized by the critical energy $u_c = \hbar\omega_c = \hbar(3/2)\gamma^3 c/\rho$. The number of photons emitted per second is $\dot{N} = (15\sqrt{3}/8)P/u_c$, the mean photon energy is $< u >= P/\dot{N} = 8u_c/(15\sqrt{3})$, the mean square energy is $< u^2 >= (11/27)u_c^2$, and the energy loss per turn is $U_s = e^2\gamma^4/(3\epsilon_0\rho)$. The energy damping time is $\tau_E \simeq TE/U_s$ where $T$ is the period of a revolution and the root mean square energy spread $\sigma_E$ resulting from fluctuations in the synchrotron radiation is approximately the root mean square photon energy times the square root of the number of photons emitted during a damping time: $\sigma_E \simeq \sqrt{\dot{N}\tau_E < u^2 >}$. For the Cornell Electron Storage Ring (CESR) $e^+e^-$ collider with particle energy $E$ =5.29 GeV and $\rho$ =98 m, show that $< u >\simeq 1$ keV, the energy loss per turn is $U_s$ =0.71 MeV, the number of photons radiated per turn is 2060, the number of turns for an electron to radiate all of its energy is $E/U_s \simeq 7300$, the damping time is $\tau_s$ =15 ms, and the energy spread is $\sigma_E \simeq$ MeV and compare the energy spread to the natural width of the $\Upsilon(1s)$.

---

**Problem 2.25. Fission cross section and critical mass**

---

The density of uranium is 19.6 g cm$^{-3}$. Enriched uranium contains U-235 which releases upon fission fast neutrons with mean energy 2 MeV. a) Estimate the mean free path in cm for such a fast neutron in pure U-235 to trigger a further fission if the fission cross section is the cross sectional area of the U-235 nucleus. b) Find the mass in kg of a sphere of U-235 of this radius. Criticality is reached when the probability a neutron induces a fission times the number of prompt neutrons released per fission (2.5 for U-235) equals one. For a bare sphere, the critical diameter is 17 cm and the critical mass is 52 kg.

---

**Problem 2.26. Hadron collision cross sections**

---

Total cross sections in mb ($10^{-27}$ cm$^2$) for 200 GeV ($\pi^+$, $\pi^-$, $K^-$, $K^+$, $p$, $\bar{p}$) interacting with protons/deuterons are (24, 24, 21, 20, 39, 42)/( 46, 47, 40, 39, 74, 78) and are dominated by inelastic short range strong processes. From these data, estimate the sizes of mesons and baryons, and assume a nuclear radius

$1.1\ A^{1/3}$ fm $\rightarrow \sigma_T \propto A^{2/3}$ to estimate their mean free paths in iron. Observe that mesons are a bit smaller than baryons and $q\bar{q}$ annihilation channels do not much affect the total cross sections.

---

**Problem 2.27. Neutron halos**

---

The cross section for interaction of neutron rich carbon isotopes at 40 MeV per nucleon with a stationary hydrogen target is measured to be about $754 \pm 22$ mb for mass number $A = 19$ and $1338 \pm 274$ mb for $A = 22$. Compare to the expectation for a solid sphere assuming root mean square matter radius $r = (1.2\,\text{fm})A^{1/3}$. The enhancement is ascribed to a halo of one or more loosely bound neutrons. See "Observation of a Large Reaction Cross Section in the Drip-Line Nucleus $^{22}$C", K. Tanaka et al., Phys. Rev. Lett. **104**, 062701 (2010).

---

**Problem 2.28. Rutherford scattering**

---

Using the Rutherford differential cross section for elastic scattering

$$\frac{d\sigma}{d\Omega} = a^2 \frac{1}{4\sin^4(\theta/2)}$$

with $a = k_e Ze^2/(mv^2)$, find the cross section in mb for a 1 MeV proton to be scattered into the backward hemisphere ($\cos\theta < 1$) by a carbon nucleus and find the back scattering probability for a 0.1 mm thickness graphite plate. Neglect the proton energy degradation due to collisions with electrons.

---

**Problem 2.29. Beam gas lifetime**

---

A stored proton beam is degraded by collisions with residual gas in the beam pipe. The beam gas loss rate is $dN/dt = -N/\tau$ with beam lifetime $\tau = 1/(nc\sigma)$ for $v = c$ and gas atomic number density $n$. Show that the molecular density for an ideal gas at pressure $p$ and temperature $T$ given by $n_{\text{mol}} = p/(k_B T)$ is numerically

$$n_{\text{mol}}[m^{-3}] = 9.66 \times 10^{24} \frac{p[\text{Torr}]}{T[\text{deg K}]}$$

and the lifetime is in engineering units

$$\tau[\text{hr}] = \frac{0.479\ T[\text{deg K}]}{p[\text{nTorr}]\sigma[\text{b}]}.$$

For the 8 GeV Fermilab antiproton accumulator with $N_2$ gas in the beam pipe at $T = 293$ deg K, the cross section for scattering leading to loss from the machine is $\sigma = \sigma_C + \sigma_N$ where $\sigma_C = 0.08$ b for Coulomb collisions and $\sigma_N = 0.4$ b for nuclear collisions. For $p = 1.7 \times 10^{-8}$ Torr, show that $\tau = 4$ hr.

### Problem 2.30. Fool's collider

A true muonium $(\mu^+\mu^-)_{\text{atom}}$ atomic state with principal quantum number $n$ could be produced in $e^+e^-$ collisions with $\sqrt{s} = 2m_\mu - \alpha^2 m_\mu/(4n^2)$.
a) Show that the required energy of a positron beam colliding with stationary electrons in a laboratory would be about 43 GeV.
b) By scaling the Bohr radius $R_\infty = (\alpha m_e)^{-1}$ of hydrogen, and accounting for length contraction, show the thickness along the direction of motion in the laboratory frame of a $(\mu^+\mu^-)_{\text{atom}}$ atom in its ground state so produced is $t_{lab} = 2R/\gamma = 4m_e^2/(m_\mu^3\alpha)$.
c) Show that in merging a positron beam with an electron beam both with energy 1 GeV to form true muonium, the small angle between the beams would need to be about 12 degrees. Other production methods are considered in Egil Holvik and Haakon A. Olsen, "Creation of relativistic fermionium in collisions of electrons with atoms," Phys. Rev. **D35** 2124 (1987).

### Problem 2.31. LHC event rates

At a luminosity of $L = 2 \times 10^{33}$ cm$^{-2}$s$^{-1}$, the LHC would be expected to produce the following event rates: weak bosons decaying to electrons $W^- \to e^-\bar{\nu}_e$ (40 Hz), $Z \to e^+e^-$ (4 Hz); heavy quarks $t\bar{t}$ (1.6 Hz), $b\bar{b}$ ($10^6$ Hz), Higgs boson for $m_H = 120$ GeV (0.08 Hz). a) What are the corresponding cross sections in fb? b) If the LHC is live one third of the time on average, how many events of each type occur during one calendar year? c) What would be the integral luminosity in fb$^{-1}$?

### Problem 2.32. LHC parameters

Assume an LHC dipole magnetic field is two continuous tubes of diameter $d = 200$ mm (a bit bigger than the outer coil diameter) and length 15 m, each tube containing a uniform magnetic field of strength of $B = 8$ T. The energy density in MKSC units is $\mathbf{B}^2/(2\mu_0)$.
a) What are the total stored magnetic linear energy density in MJ per meter of magnet, the energy stored in each 15 m length dipole, and the energy in GJ in a 27 km length ring?
b) If 1 GJ is the equivalent of 1/4 ton of TNT, what is the TNT equivalent of the LHC stored energy?
c) Assuming 3000 bunches per beam and $1 \times 10^{11}$ protons per bunch and using $m_p = 1.67 \times 10^{-27}$ kg, how many nanograms of hydrogen must be used to fill one beam and carry its 360 MJ?

### Problem 2.33. LHC multiple interactions

The estimated cross section $\sigma_X$ for pp collisions at 14 TeV to produce X for some exemplary process are $\sigma_{tot} = 10^8$ nb, $\sigma_b = 10^6$ nb, $\sigma_{jet}(E_t > 0.1\ TeV) = 10^3$ nb,

$\sigma_W = 200$ nb, $\sigma_Z = 50$ nb, $\sigma_{t\bar{t}} = 1$ nb, $\sigma_h(m_h = 150\ GeV) = 3 \times 10^{-2}$ nb, and $\sigma_H(m_h = 500\ GeV) = 4 \times 10^{-3}$ nb. Assume 2808 bunch collisions in each beam collide at the ATLAS detector, the bunch revolution frequency being 11 kHz providing an average luminosity of $L = 10^{34}$ cm$^{-2}$ s$^{-1}$.
a) What is the total number of interactions per second?
b) How many bunch crossings occur per second and what is the average number of interactions per bunch crossing?
c) What is the probability of a W being produced in any given bunch crossing?
d) If the mean number of events of a certain type in a bunch crossing is $\mu$, the probability of $N$ such events is given by the Poisson distribution $P(N, \mu) = \mu^N e^{-\mu}/N!$. What is the probability of two or more $W$ bosons being produced in distinct pp collisions in the same bunch crossing? How does this compare to the probability for producing a Higgs boson of mass 150 GeV that decays to two $W$ bosons in any crossing?

CHAPTER

# 3 Experiments

Particles are detected by their fundamental interactions - the physics and the detectors go hand in hand - but real detector operation also involves many fascinating details of physics on subatomic, atomic, molecular, and larger scales. Interesting experiments often require novel techniques and engineering acumen. We take a look at important principles behind particle experiments.

## Contents



## 3.1 ■ ENERGY DEPOSITION BY CHARGED PARTICLES

Experiments detect traces of energy left by long-lived particles in passing through normal matter, in particular the ionization trails created by charged particles. The ionization process can be largely understood in classical terms and we start our introduction to experiments with a simple classical model.

Suppose a charge $e$ with speed $v$ passes an electron with impact parameter $b$ and is only slightly deflected. Suppose also that the electron has insufficient time to move so may be regarded as stationary. The geometry is illustrated in Figure 3.2. The momentum transfer that follows from the relativistic force equation

$$\frac{d\mathbf{p}}{dt} = \mathbf{F} \rightarrow \Delta\mathbf{p} = \int dt\ \mathbf{F} \tag{3.1}$$

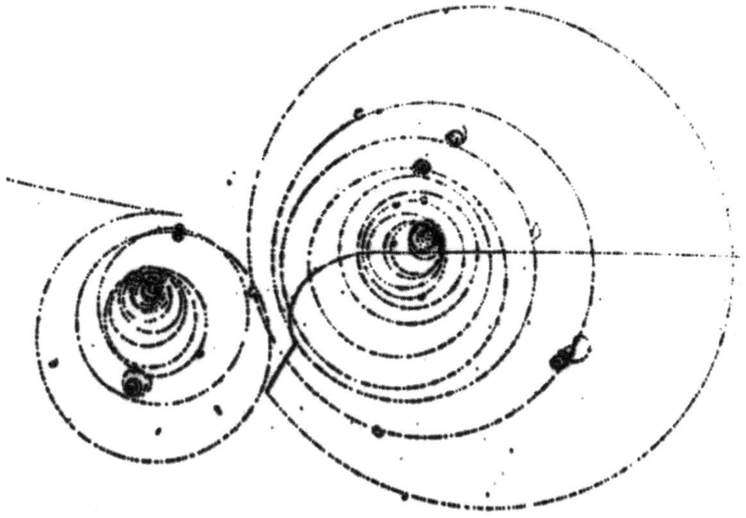

**FIGURE 3.1** Photograph of particle interactions in a bubble chamber. A positively charged kaon enters from the right and decays weakly in the center of the image to pions $K^+ \to \pi^+\pi^0$. The neutral pion decays promptly to two photons, one of which converts to $e^+e^-$ in the liquid. The positively charged pion decays to $\mu^+\nu_\mu$ and the $\mu^+$ decays to $e^+\bar{\nu}_e\bar{\nu}_\mu$. The positrons range out and annihilate to photons which are not observed. The invention of bubble chambers in 1952 revolutionized the field of particle physics. [CERN photo archive.]

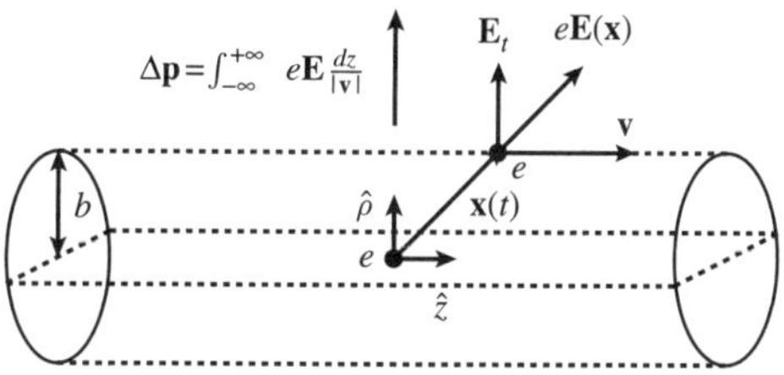


**FIGURE 3.2** Impulse in a Coulomb interaction. A charge $e$ with velocity $\mathbf{v}$ pass a stationary electron with impact parameter $b$. In the impulse approximation, time integral of the electric force $e\mathbf{E}$ determines the transverse momentum kick $\Delta\mathbf{p}$.

is determined by the transverse component $\mathbf{E}_t$ of the electric field of either charge and is readily found in this impulse approximation:

$$|\Delta p| = e\int_{-\infty}^{\infty} |\mathbf{E}_t|\frac{dz}{v} = \frac{e}{2\pi bv}\int_{cylinder} \mathbf{E}\cdot d\mathbf{A} = \frac{e^2}{2\pi bv} = \frac{2\alpha}{bv} \tag{3.2}$$

where we set $\hbar = c = \epsilon_0 = 1$, converted to a surface integral over a cylinder of radius $b$ and applied Gauss's Law. The energy transfer is

$$\Delta E = \frac{\Delta p^2}{2m_e} = \frac{2\,\alpha^2}{b^2v^2m_e}. \tag{3.3}$$

Notice that the slower the speed, the longer the interaction time and the larger the energy transfer. When $v$ approaches light speed, the energy transfer no longer depends on the energy of the incident particle. The impulse approximation used here provides only a guide. Close collisions corresponding to large energy transfer produce hard primary electrons called knock-ons or delta-rays. More distant collisions transfer just enough energy to excite atoms. At still larger impact parameter, the bulk response of the medium should be considered. The energy transfer to an atomic nucleus of atomic number $Z$ is smaller by a factor

$$\frac{Z^2 m_e}{m_N} \sim \frac{Z m_e}{m_p} \tag{3.4}$$

and may be neglected - a nucleus is much heavier than an electron and can absorb momentum but relatively little energy in the elastic collisions we will be considering.

---

**Example 3.1. Collisional ionization of hydrogen.** A relativistic muon passes an electron with an impact parameter equal to the Bohr radius $(\alpha m_e)^{-1}$. What is the energy transfer to the electron in units of the Rydberg energy $R = m_e\alpha^2/2$? What is the maximum impact parameter leading to ionization of hydrogen?

Using Equation 3.3 with $v = 1$, we have

$$\Delta E/R = \frac{2\alpha^2}{\left(\frac{1}{\alpha m_e}\right)^2 m_e} / \left(m_e\alpha^2/2\right) = 4\alpha^2. \tag{3.5}$$

Evidently the energy transfer is insufficient to ionize a hydrogen atom. The maximum impact parameter leading to ionization is $2\alpha \simeq 1/70$ of the Bohr radius in this classical model.

---

In amorphous matter of atomic number density $n$ and atomic number $Z$, the number of electrons with impact parameter in a range $db$ about $b$ in a distance $dx$ is $dN = Zndx2\pi bdb$. The average energy loss to electrons per unit length of track estimated by integrating over impact parameter is

$$\frac{dE}{dx} = \int \frac{dN}{dx} \Delta E(b) = \frac{4\pi\alpha^2 Zn}{m_e v^2} \ln \frac{b_{\max}}{b_{\min}}. \tag{3.6}$$

Approximate limits $b_{\max}$ and $b_{\min}$ on the range of impact parameters are arguably as follows. The maximum possible energy transfer permitted by 4-momentum conservation by a mass $M >> m_e$ to an electron at rest is

$$\Delta E_{\max} \simeq 2m_e\gamma^2 v^2 \tag{3.7}$$

where $\gamma = 1/\sqrt{1-v^2}$. (For nonrelativistic motion with $\gamma = 1$, the energy $2m_e v^2$ corresponds to reversing the electron speed in the center of mass so its speed is $2v$ in the laboratory frame.) This kinematic limit corresponds to an effective minimum impact parameter $b_{\min} = \alpha(\gamma m_e v^2)^{-1}$. The energy transfer may be expected to become inefficient for adiabatic collisions when the collision time $\Delta t \sim b_{\max}(\gamma v)^{-1}$ is comparable to an atomic orbital period $T = 2\pi/\omega$ where $\omega$ is some average atomic frequency. In a quantum mechanical description of target atoms, the ionization and excitation energies of the bound electrons will provide a lower bound to the energy transfer. The average energy loss is not very sensitive to the exact values of $b_{min}$ and $b_{max}$. A calculation including a quantum mechanical model of atomic excitation gives the ionization energy loss formula

$$\frac{dE}{dx} = \frac{4\pi\alpha^2 Zn}{m_e v^2} L \tag{3.8}$$

where

$$L = \ln\left[\frac{2m_e\gamma^2 v^2}{I}\right] - (v/c)^2 \tag{3.9}$$

is a slowly varying function of energy that depends on the average ionization energy $I$ ($I \simeq 10Z$ eV for a monatomic gas).

As illustrated in Figure 3.3, the energy loss rate drops with particle speed to a minimum value when $\gamma \simeq 1$, then slowly rises as $v$ approaches $c$. (The logarithmic rise is even less than predicted here due to long range dielectric screening.) A charged particle with an energy loss rate close to the minimum value is called a minimum ionization particle (MIP). For $\gamma \simeq 1$, and $Z$ =10 (Neon), $L \simeq \ln[1\text{ MeV}/0.1\text{ keV}] - 1 \simeq 8$. For hydrogen liquid, $I \simeq 22$ eV and $L \simeq 10$. In general, for mass density $\rho$ in g cm$^{-3}$, and atomic number density n, the electron number density is $Zn = \rho N_A Z/A$ where Avogadro's number $N_A = 6 \times 10^{23}\ g^{-1}$ is the number of nucleons per gm and $A$ the number of nucleons per atom essentially. Stable nuclei have $Z/A \sim 1/2$ so the energy loss per length measured in g cm$^{-2}$ for all relativistic particles of charge $\pm e$ is roughly material independent and the value is approximately

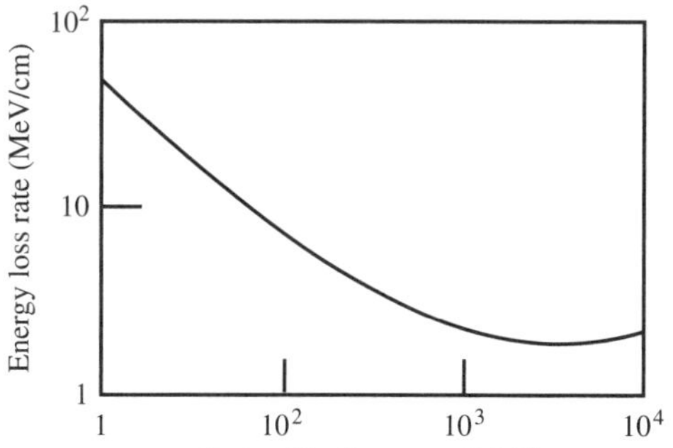


**FIGURE 3.3** Ionization energy loss rate of protons. The computed rate of energy deposition by protons in a plastic scintillator material of density $\rho \simeq 1$ gm cm$^{-3}$ is shown as a function of proton kinetic energy. The loss rate drops to a minimum when the proton speed approaches light speed.

$$\frac{1}{\rho}\frac{dE}{dx} = N_A \frac{Z}{A}\frac{4\pi\alpha^2}{m_e}(\hbar c)^2 L$$
$$= 6 \times 10^{23}\ \mathrm{g}^{-1}\frac{1}{2}\frac{12}{137^2}\left(197 \times 10^{-13}\ \mathrm{MeV\ cm}\right)^2 10$$
$$= 1.5\ \mathrm{MeV\ g^{-1}\ cm^2}. \tag{3.10}$$

Charged particles with velocities approaching light speed are therefore called "minimum ionizing.'

The average range of the particle of some initial kinetic energy $K$ may be found by integration

$$x = \int_0^K dK \left(\frac{dE}{dx}(K)\right)^{-1}. \tag{3.11}$$

For a relativistic particle, the average energy loss rate is approximately constant. Near the end of the range when the velocity becomes nonrelativistic, the energy loss per unit length and ionization density per unit length rise. A constant energy loss rate is often assumed for purposes of estimation. A 10 MeV kinetic energy $\alpha$ particle (charge $2e$) in air has a range of about 0.012 gm cm$^{-2}$ or about 10 cm and will be stopped by a thin plate of nearly any solid material. The ionization density is minimally of order 10-100 cm$^{-1}$ in gases and $10^3$ times as large in condensed matter and increases as a stopping particle approaches the end of its range. The velocity dependence of the ionization density in conjunction with an independent momentum measurement can be used to discriminate particles of similar charge. The energy loss and hence range of a charged particle is subject to fluctuations which limit the use of energy loss in particle velocity measurements. Descriptions of energy loss fluctuations called Landau fluctuations and other details may be found in the suggested reading.

---

**Example 3.2. Scintillation.** A $\mu^-$ of initial energy 1 GeV passes through 2 cm of scintillating plastic of density 1 gm cm$^{-3}$. If one percent of the deposited energy is converted to photons of energy 1 eV, about how many photons are produced?

The muon is relativistic so the ionization energy loss is approximately 1.5 MeV cm$^2$ g$^{-1}$ $\times$ 1 g cm$^{-3}$ $\times$ 2 cm $=$ 3 MeV, small compared to the energy itself. About $3 \times 10^6/100 = 30,000$ photons would be produced.

---

**Example 3.3. Muon penetration power.** Roughly how much steel can be penetrated by a muon of initial energy 13 GeV?

In solid iron the minimum ionization energy loss is

$$\frac{dE}{dx}|_{MI} = 1.5\ \mathrm{MeV\ cm^2\ g^{-1}} \times 7.8\ \mathrm{g\ cm^{-3}} = 13\ \mathrm{MeV\ cm^{-1}} \tag{3.12}$$

so a 13 GeV muon may penetrate about 10 m of steel.

---

## 3.2 ■ IONIZATION TRAIL DETECTORS

The trail of ionization left by the passage of a charged particle in matter is the basis for many detector technologies. The trail comprises primary positively charged ions and electrons, and additional ions and electrons produced as primary electrons come to rest. Such ion trails may be observed to nucleate visible tracks of grains in a chemically developed emulsion film, drops in a supersaturated vapor and bubbles in a superheated liquid. Emulsion grains have a diameter of $\sim 0.2\ \mu$m and emulsion based detection still provides the highest resolution images of particle tracks, but processing stacks of 600 $\mu$m thick emulsion films is extremely time consuming and such a detector, which integrates over time, is only suitable for special experiments. The positron was discovered in photographs of cosmic rays passing through a supersaturated vapor called a cloud chamber. Bubble chambers are ideally suited to observations of production and decay of weakly decaying light hadrons with a few GeV of energy. A liquid hydrogen bubble chamber (see Figure 3.4) operates at a temperature of 25-30 K with a pressure range of a few bar provided by a piston. With externally triggered depressurization, bubbles grow from ionization seeds to a diameter of $\sim 10^{-6}$ m in a few ms after which the liquid is illuminated and stereo photographs taken. The cycle time is at most tens of hertz.

Bubble chambers provide exquisite pictures of particle interactions. Figure 3.1 shows the decay

$$K^+ \to \pi^+\pi^0 \tag{3.13}$$

in a bubble chamber. The $K^+$ enters from the right and decays near the center of the image. The $\pi^0$ decays promptly ($c\tau_{\pi^0}$ = 25 nm) to $\gamma\gamma$ and one of the photons is observed converting to an $e^+e^-$ pair. The decay

$$\pi^+ \to \mu^+\nu_\mu \tag{3.14}$$

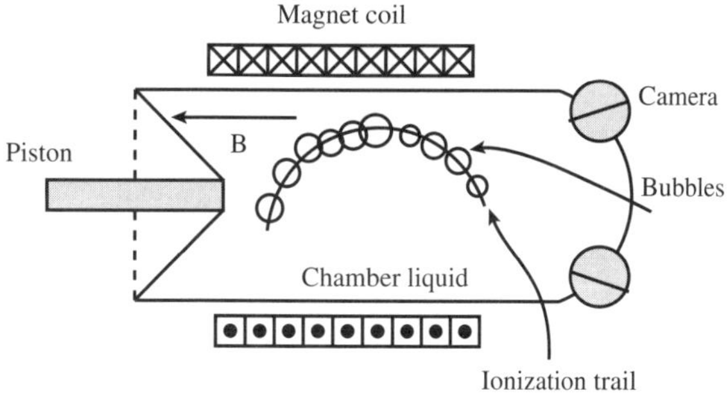


**FIGURE 3.4** Schematic drawing of a liquid hydrogen bubble chamber. The piston compresses the liquid within the chamber. When the pressure is released, ionization seeds bubble growth forming particle tracks.

followed by

$$\mu^+ \rightarrow e^+ \nu_e \bar{\nu}_\mu \tag{3.15}$$

produces additional tracks. The tightly curled tracks of several knock-on electrons are also evident in the image.

Emulsion, cloud, and bubble techniques have been superseded by faster technologies suitable for searches for rare process with high intensity beams and high interaction rates. High speed is obtained by collection of ionization in gases, liquids, and solid semiconductors, electronic amplification of the signal, and fast digital electronic information processing. However, low rate technology is still of use in specialized applications such as the search galactic dark matter particles producing recoiling nuclei in such experiments as COUPP, XENON, and PICASSO. (It has become a tradition to name particle physics experiments with an acronym and to supply a website. Information about these and other contemporary experiments is readily found online.)

Electrons and positive ions created in a gas or liquid by the passage of a charged particle, or through photo-absorption, may be accelerated in opposite directions by a macroscopic electric field. Such currents may be detected by external circuitry directly or by induction. In the Drude model of conduction, the average velocity $\bar{\mathbf{v}}$ of a charge $q$ moving in a medium subject to an electric field $\mathbf{E}$ is governed by

$$m\frac{d\bar{\mathbf{v}}}{dt} = q\mathbf{E} - \frac{1}{\tau}\bar{\mathbf{v}} \tag{3.16}$$

where $\tau$ in the mean time between collisions which randomize the direction of the velocity. This model implies a terminal velocity called the drift velocity

$$\bar{\mathbf{v}}_d = \frac{q}{m}\mathbf{E}\tau \tag{3.17}$$

which is inversely proportional to the particle mass $m$ if the mean time is governed by the thermal speed and target scattering center density. Thus, electrons drift considerably faster than ions. In practice, the Ohmic behavior predicted by the Drude model is respected by positive ions while the drift velocity of electrons in a gas is nonlinearly related to the electric field strength in part because the electron-atom scattering cross section is energy dependent.

The proportional wire chamber illustrated in Figure 3.5 is a device used to detect charged particles by sensing small ionization signals in a gas. A single detection cell of a proportional wire chamber contains a fine wire (0.05 mm diameter Au-plated W is typical) at positive potential. The electric field lines emanating from this wire trace out the collection region for the wire. Free electrons created by a passing charged particle drift along the electric field toward the wire at a speed determined by elastic collisions with gas molecules and the local electric field strength. Near the wire the electric field strength increases and in a mean free path an electron attains sufficient energy to ionize an atom by collision resulting in electron multiplication.

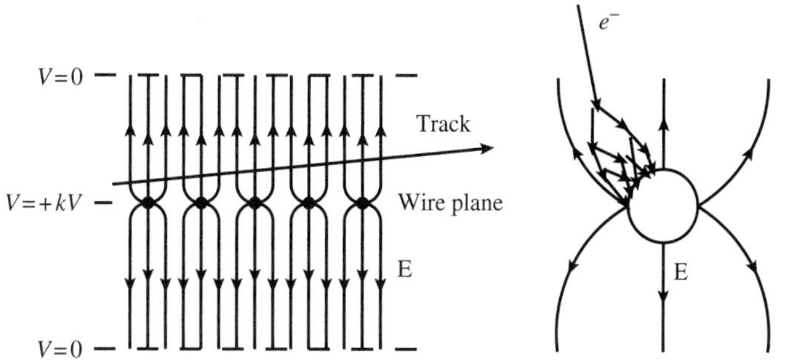


**FIGURE 3.5** Schematic of a multiwire proportional chamber. A plane of fine wires immersed in a gas collects electrons liberated by the passage of a charged particle. Near the wire surface, a drifting electron acquires sufficient energy between collisions to produce additional ionization.

Uncontrolled, cascade multiplication may lead to a plasma column, a streamer of ionization, and, if the streamer breaches between electrodes, a spark discharging the wire. A simple Geiger counter operates with a discharge signal. Photography and even acoustic detection of sparks and streamers generated by ionization was an early approach to the use of gas chambers in particle detection. In a streamer chamber, for example, gas is contained between two parallel plate electrodes several centimeters apart. A voltage of order 200 kV lasting a time of order 20 ns is generated by a Marx generator with a pulse shaping reflection line and applied to the gap to initiate nascent electrical breakdown.

In the proportional wire chamber, the electric field reaches the critical value for multiplication only within a few mean free paths of the wire surface where electron multiplication ends and the electrons enter and are trapped. No discharge of the wire takes place and the detector may run continuously. The total ionization produced is proportional to the number of electrons drifting in for multiplication factors less than roughly $10^6$. The wire behaves as a distributed extremely high gain high speed noiseless linear amplifier. With additional solid state electronic amplification, the signal due to a single drifting electron may be brought to a level sufficient for processing by room temperature solid state electronic circuitry.

The development of the signal is worth considering. When the drifting electrons arrive at the wire surface, the electrons and positively charged ions are still within a few microns of each other forming an electric dipole. Attraction to the positive ion charge pins the liberated electrons to the wire surface. But the positive ions are repelled by the large positive static charge on the high potential wire and drift (relatively slowly) away from the wire with a radial speed proportional to field strength so inversely proportional to radius. It follows that the ion current abruptly starts and decays following closely the form

$$I(t) = \frac{I_0}{\sqrt{1 + t/t_0}}. \tag{3.18}$$

To maintain electrostatic equilibrium within the cell, induced current proportional to the ion current flows in the wire and in surrounding electrodes. The induced current pulses may be detected by external electronics. The leading edge times of such pulses indicate the times at which each drifting electron arrives at the wire. The integral of the current is proportional to the number of ion pairs created in the gas.

The gas must not be electronegative or the drifting electrons will attach to gas molecules and no longer participate in multiplication. Stability often requires a gas component which absorbs ultraviolet photons emerging from the multiplication process which can otherwise lead to runaway discharges. A noble gas like argon mixed with a hydrocarbon like methane works well. In an electric field of $1\ \mathrm{kV\ cm^{-1}}$, a typical drift speed is

$$v_{\mathrm{drift}} \simeq 5 \times 10^6\ \mathrm{cm\ s^{-1}} \tag{3.19}$$

so an ionization trail may be collected over a distance as large as 1 m in 20 $\mu$s. A rate limitation appears only at charged particle fluxes of order $10^6\ \mathrm{cm^2\ s^{-1}}$ when the accumulation of drifting positive ions begins to influence the drift field and multiplication.

Detection of induced signals on segmented electrodes provides one method of localizing the original ionization. Timing can provide another. If the time at which a particle passed through the chamber is known from an external prompt detector such as a scintillator, and the difference in time between the prompt signal and the appearance of the amplified signal measured, with appropriate calibration of the drift velocity, the drift distance may be inferred from the drift time. The r.m.s. spatial resolution of such a drift chamber as small as

$$\sigma \simeq 0.15\ \mathrm{mm} \tag{3.20}$$

may be achieved.

Wire chambers may be constructed in various geometries. A plane of parallel wires separated by a few millimeters between planar electrodes forms a simple multiwire proportional chamber, each wire collecting ionization from its portion of the device. A long aluminum tube containing the gas, field shaping electrodes, and supports for the wire is a robust single cell drift detector module suitable for muon detection.

Figure 3.6 shows a reconstructed image of a high energy hadronic interaction observed with a collection of multiwire drift chambers. In the image, each point represents the reconstructed position of ionization collected by a single wire. This particular event, culled from millions by electronic data processing, represents one of the first observations of the weak production and decay of a $W$ boson. This kind of observation is only possible with fast electronic instrumentation.

High rate charged particle detection and fine resolution are obtained with silicon detectors. Lithographic techniques are used to create strip or pixel electrodes with

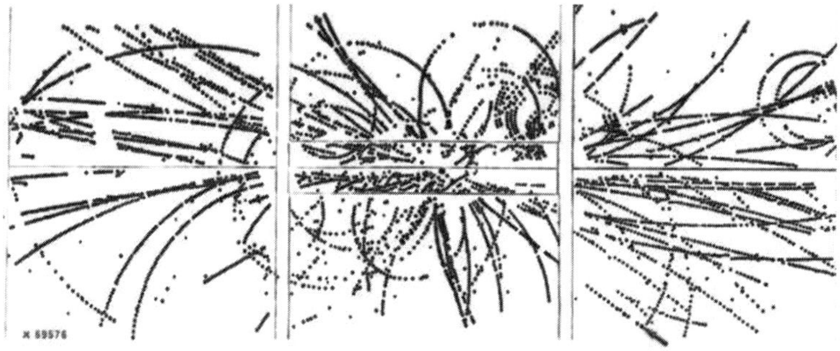


**FIGURE 3.6** Observation of $W$ boson production. One of the first observations of the production of a $W$ boson in the process $p\bar{p} \to W + X$ in the CERN UA1 detector central tracking system. Here X denotes additional particles such as pions produced in association with the fundamental fusion process $d\bar{u} \to W$. The decay $W \to e\nu$ produces an energetic electron corresponding to the straightest track. The neutrino is not observed directly. Its energy is inferred from 4-momentum conservation. [CERN Photo library. See G. Arnison *et al*, Physics Letters **122B** 103 (1983)]

feature size of about 50 $\mu$m on detector elements cut from doped crystal wafers a few inches in diameter. The passage of a charged particle through the wafer produces electron and hole pairs in a depletion region created by application of a bias voltage, and the subsequent current pulse is detected. High density integrated circuits are used for electronic signal amplification and processing.

The lifetimes of the $\tau$ and the $b$-quark and $c$-quark correspond to travel distances of a few millimeters. In colliding beam experiments, silicon detectors surrounding the beam pipe can pinpoint the origin of tracks seen in outer tracking chambers to tens of microns and readily distinguish prompt tracks from secondary tracks originating in heavy quark decay. In the very high luminosity environment of the LHC, gaseous wire chamber technology fails and large silicon only systems are used for inner tracking. The CMS silicon strip tracking detector comprises 10 million strips with a surface area of 200 m$^2$ requiring 80,000 readout chips. The detector elements are arranged in concentric barrels with forward and backward circular disks and surround the smaller pixel detector. A similar system at the ATLAS experiment comprises six million detector elements.

## 3.3 ■ LIGHT DETECTION

When a charged particle passes through a medium, energy is deposited in the form of atomic and molecular electronic excitation along the ionization trail. The de-excitation florescence that lies in the visible portion of the electromagnetic spectrum, called scintillation light, is detectable by photomultiplication. A useful scintillator must have a low attenuation length for scintillation light. Transparent plastic materials such as lucite doped with fluorescent organic molecules are available with response times of about 1 ns and are relatively inexpensive to fabricate in large volumes. One dopant is anthracene in which excitation of $\pi$-state electrons in its three benzene rings is a source of fast florescence and delayed so-called phosphorescence. Fast inorganic crystals such as NaI and BGO are more sensitive to energy deposition but more costly and delicate.

The photomultiplier tube (Figure 3.7) is a photosensitive detector capable of resolving single electromagnetic quanta. In its simplest form, a PMT comprises an evacuated glass cylinder containing typically a borosilicate glass window and a metallic structure serving as an electron multiplier. The interior of the entrance window is coated with a photosensitive low work function bi-alkali metal. An electron ejected by the photoelectric effect is accelerated to an energy of a few hundred eV and impacts a metal so-called dynode, electroplated to enhance secondary electron emission. An incident electron liberates several electrons. The secondary electrons are accelerated and impact subsequent dynodes further amplifying the signal. Ten such electron multiplication stages produce a measurable electronic pulse corresponding to a single photon after a delay of about 25 ns with a time jitter of a few hundred ps. The vacuum and the absence of magnetic fields are critical to the multiplication stages of a photomultiplier tube. A single stage photodiode is suitable to detect a large number of photons with unit gain in the presence of a magnetic field.

A photomultiplier followed by a solid state electronic amplifier is capable of producing a logic level pulse from a single photon. However typically only 0.1-0.2 of incident photons will produce the requisite initial photoelectron. The photomultiplier is suitable for detecting light from a single crystal or plastic scintillation channel corresponding to several photoelectrons. Special large and small area tubes and hybrid tubes with internally segmented multiplication channels and semiconductor secondary electron detection and amplification stages are commercially available, and the technology continues to be developed. One purely solid state photomultiplier with high quantum efficiency called a visible light photon counter (VLPC) is available but requires cryogenic operation temperature. With millimeter diameter single channel sensitive areas (pixels), these devices are well matched to

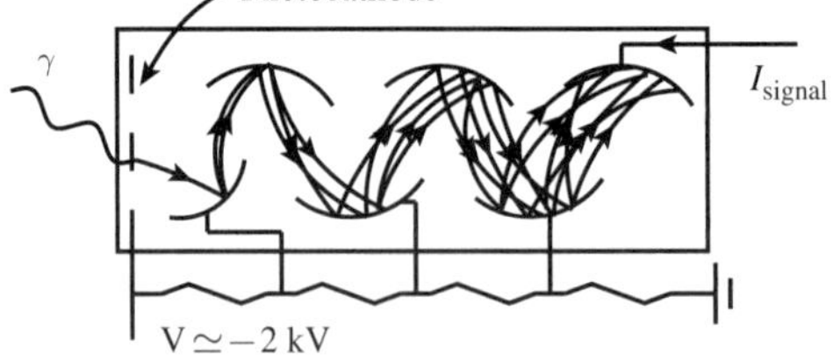


**FIGURE 3.7** Schematic of a photomultiplier tube. A photon enters an evacuated chamber and liberates an electron from a photocathode by the photoelectric effect. The electron is accelerated and impacts the first of several dynodes. Secondary electrons liberated by impact lead to multiplication.

tracking devices constructed of scintillating plastic fibers and have been deployed in the D0 experiment.

The wake field of a charge moving in a transparent medium appears as Cherenkov radiation. If the refractive index is $n$, and the particle speed exceeds the speed of light in the medium, $v > c/n$, Cherenkov radiation is produced along a cone with a velocity sensitive cone angle given by

$$\cos\theta_c = \frac{c}{nv}. \tag{3.21}$$

The refractive index of a gas is close to unity and somewhat adjustable. A volume of gas and suitable mirrors to direct Cherenkov radiation to a photomultiplier forms a fast counter for particles above a threshold speed. More elaborate Cherenkov light detector systems image the cone to measure the speed and direction. Cherenkov radiation has been used as the basis for particle detection in experiments to search for proton decay and to observe neutrino interactions in large volumes of water, to observe large cosmic ray showers in the atmosphere, and to study neutrino interactions in the ice at the South Pole. A somewhat related electromagnetic effect has also been put to use. If a charged particle passes through an interface between media, part of its associated electromagnetic field in a sense breaks off and appears as transition radiation.

## 3.4 ■ COULOMB SCATTERING

In the impulse approximation, the energy transfer in Coulomb scattering of a relativistic charged particle passing through matter as described by Equation 3.6 is inversely proportional to target particle mass. Scattering from nuclei contributes little to the energy loss in comparison to the energy loss to electrons on average. On the other hand, a close encounter with a heavy nucleus may result in substantial momentum transfer, significant deflection of the particle, and also substantial radiative energy loss called bremsstrahlung.

We consider first the deflection of a particle of momentum $p$ and charge $e$ by a nuclear charge $Ze$. The momentum impulse $\Delta p = 2\alpha Z/(bv)$ at impact parameter $b$ yields in the small angle approximation an angular deflection

$$\theta = \frac{\Delta p}{p} = \frac{2\alpha Z}{bvp} \tag{3.22}$$

where $\theta$ is the polar angle of the deflected particle relative to the incident particle direction. For small $\theta$, the projection of this deflection on a transverse direction $x$ is $\theta_x = \theta \cos\phi$ where $\phi$ is the azimuthal angle of the deflected particle (Figure 3.8). Coulomb interactions with different nuclei may be regarded as uncorrelated, except when a particle is directed exactly between planes of a single crystal. Uncorrelated multiple Coulomb scattering events result in an approximately Gaussian distribution of projected scattering angle. In a distance $dz$, the mean square deflection

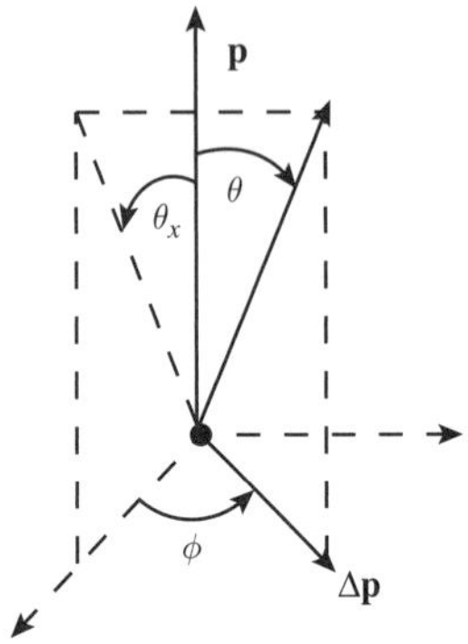


**FIGURE 3.8** Geometry of elastic Coulomb scattering. Multiple small angle collisions result in an approximately Gaussian distribution in the projected scattering angle $\theta_x$.

of a relativistic charged particle projected in a transverse direction averaged over azimuth and impact parameter for target number density $n$ is

$$\sigma_{\theta,x}^2 = \int bdb\, dz\, d\phi\, n \cos^2\phi \left(\frac{\Delta p}{p}\right)^2 \propto \left(\frac{\alpha Z}{p}\right)^2 dz. \tag{3.23}$$

The root mean square projected angle in passing through a length $z$ of material may be parameterized by the approximate formula

$$\sigma_{\theta,x} = \frac{p_0}{p}\sqrt{\frac{z}{L_r}} \tag{3.24}$$

and the corresponding displacement is $\sigma_x = z\sigma_\theta/\sqrt{3}$. Here

$$p_0 = \sqrt{\frac{3\pi}{2\alpha}} m_e[c] \simeq 14 \text{ MeV c}^{-1} \tag{3.25}$$

and we have introduced the electron radiation length defined by

$$L_r^{-1} \simeq \frac{4\alpha Z(Z+1) r_e^2 N_0}{A m_e^2} \ln \frac{183}{Z^{1/3}} \tag{3.26}$$

where $N_0 = 6.02 \times 10^{23}$ mole$^{-1}$, A is the mass number of the material traversed, and

$$r_e = \frac{\hbar c \alpha}{m_e c^2} = \frac{e^2}{4\pi m_e c^2} = 2.8 \text{ fm} \tag{3.27}$$

is the classical electron radius. The electron radiation length $L_r$ is usually expressed in gm cm$^{-2}$ and may be computed by expressing $A$ in gm mole$^{-1}$ and $r_e$ in cm. The radiation length in cm then follows by dividing by the mass density in gm cm$^{-3}$.

As described below, the radiation length is the mean length for a charged particle in passing through material to radiate $1/e$ of its energy by bremsstrahlung, a process which, like multiple Coulomb scattering, depends on the square of the atomic number of the material and also on the electron mass. Coulomb scattering of a heavy particle like a muon from nuclei has nothing to do with the electron mass but, with the definition of $p_0$, one material property (the *electron* radiation length), may be used to characterize both scattering and bremsstrahlung. The momentum dependence of the formula for $\sigma_{\theta,x}$ results from the fact that, at any impact parameter, the angular deflection is inversely proportional to momentum. The proportionality of the r.m.s. angle and position deflections to the square root of distance traversed in the material assumes many uncorrelated scattering events. Scattering disturbs the trajectory of a particle in an uncontrolled way, limiting measurements of initial position and momentum. For this reason, tracking detectors containing minimal material are desirable.

## 3.5 ■ BREMSSTRAHLUNG

The radiation length introduced above, $L_r$, characterizes the energy loss by electrons due to bremsstrahlung (braking radiation) associated with acceleration during Coulomb collisions with nuclei. Bremsstrahlung is illustrated by the diagram shown in Figure 3.9. Similar radiation processes affect all charged particles - any initial or final state charged particle may radiate a photon. Bremsstrahlung photons are emitted along the particle direction with mean angle $\theta \simeq m/E = 1/\gamma$ at high energy and with an energy spectrum inversely proportional to the photon energy. The process will be considered further in Chapter 9.

The average energy radiated by an electron in moving a distance $dx$ through material is

$$\frac{dE}{dx}_{rad} = -\frac{E}{L_r} \tag{3.28}$$

with $L_r$ given by Equation 3.26. The proportionality $1/L_r \propto \alpha^3 Z^2/m_e^2$ may be understood as follows. The acceleration in Coulomb scattering of an electron is

$$a \sim \frac{Ze^2}{m_e b^2} \tag{3.29}$$

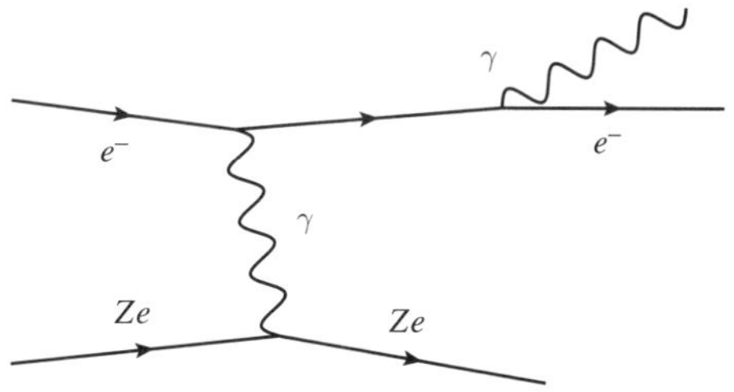


**FIGURE 3.9** Bremsstrahlung in Coulomb collisions. In collisions of an electron with a nucleus of charge $Ze$, radiation is generated.

for a time $t \sim b/v$ and, as long as the electron is nonrelativistic, the radiation power is given by the Larmor formula

$$P = \frac{2}{3}\alpha a^2. \tag{3.30}$$

The energy dependence of the mean energy loss rate derives from quantum mechanics. Classical theory predicts that in any glancing relativistic deflection, soft radiation is emitted with an energy spectrum inversely proportional to frequency $f$. The probability of a photon of energy $hf$ being radiated is constant but can not exceed the energy of the incident electron. Hence, the average radiated energy is proportional to the electron energy. The method of virtual photons arrives at the bremsstrahlung energy loss quasi-classically by associating a virtual photon density with the (screened) Coulomb field of the nucleus and averaging the Compton scattering from the electron for all impact parameters. The details of the interaction including screening of the nuclear charge by atomic electrons are embedded in the logarithmic dependence of the radiation length on atomic number.

With reference to Figure 3.9, in general, the amplitudes for radiation from the initial and final state of all charges must be summed. Only final state radiation from the electron is shown. The contribution from the much heavier nucleus is relatively small. This is justified by the classical Larmor formula since the accelerations of the incident electron and nucleus are in inverse proportion to their masses even though the nuclear charge can be much larger than the electron charge.

Bremsstrahlung is the dominant energy loss mechanism for relativistic electrons in matter. The ratio of radiative energy loss to ionization energy loss for electrons is

$$\frac{dE_{rad}}{dE_{ion}} = \frac{E}{E_c}. \tag{3.31}$$

The critical energy at which these two energy loss mechanisms are equal is $E_c =$ 800 MeV/Z. Bremsstrahlung from muons or pions is relatively rare ($\sigma \sim 1/m^2$) and the muon radiation length much larger than the electron radiation length in any material. Consequently a multiGeV muon can plough through hundreds of meters of material suffering little but ionization energy loss. This distinguishing behavior is the basis for muon identification. Only neutrinos have more penetrating power.

## 3.6 ■ PHOTON INTERACTIONS

Photons interact with charges in matter. Consider first the interaction with electrons. A free electron may not absorb a photon without violating energy and momentum conservation but elastic scattering is permitted. Classical nonrelativistic analysis gives the cross section for scattering of photons by free electrons. Suppose an electromagnetic wave of angular frequency $\omega$ and linear polarization with electric field along the $z$-direction. The equation of motion of an electron bound by a harmonic force is readily solved:

$$m_e\ddot{z} = -m_w\omega_0^2 + q_e E_0 \sin(\omega t) \rightarrow z(t) = \frac{q_e E_0}{m_e\omega^2\left(\frac{\omega_0^2}{\omega^2} - 1\right)} \sin(\omega t) \tag{3.32}$$

where we neglected the magnetic force. Larmor's radiation formula (Equation 3.30) from classical electromagnetic theory gives the power radiated. The time averaged power radiated by an electron that is accelerated by an electromagnetic wave with electric field $\mathbf{E}(t)$ in SI units is

$$P = \frac{e^2}{4\pi\epsilon_0}\frac{2}{3} < \mathbf{a}^2 >= \frac{1}{\left(\frac{\omega_0^2}{\omega^2} - 1\right)^2}\left(\hbar^2 c^6\right)\frac{8\pi}{3}\frac{\alpha^2}{(m_e c^2)^2} < \epsilon_0 \mathbf{E}^2 > . \tag{3.33}$$

The scattering cross section is obtained by dividing the radiated power by the incident energy flux (electric plus magnetic) $I =< \epsilon_0 \mathbf{E}^2 >$ and can be written as

$$\sigma = \frac{\omega^4}{\left(\omega^2 - \omega_0^2\right)^2}\sigma_T. \tag{3.34}$$

Here the Thomson scattering cross section is defined as

$$\sigma_T = \frac{8\pi}{3}\left(\frac{e^2}{4\pi\epsilon_0 m_e c^2}\right)^2 = \frac{8\pi}{3}r_e^2 = \frac{8\pi}{3}\left(\frac{\alpha\hbar}{m_e c}\right)^2 = \frac{8\pi}{3}(\alpha\lambda_C)$$
$$= 6.65 \times 10^{-25}\ \mathrm{cm}^2 \tag{3.35}$$

where $r_e = e^2/(4\pi\epsilon_0 m_e c^2)$ is the classical electron radius and $\lambda_C = \hbar/(m_e c)$ is the electron Compton wavelength. The Thomson cross section sets a scale for photon interactions with atomic electrons.

For photon frequencies much less than an atomic excitation frequency modeled by $\omega_0$, the elastic scattering is called Rayleigh scattering and the cross section is proportional to the fourth power of frequency. Photons with an energy equal to an atomic excitation frequency are strongly absorbed. As the photon energy increases, photoelectric effect processes become important. Just above threshold, the cross section for ejection of an innermost so-called K shell electron, for example, can be expressed in terms of the Thomson cross section

$$\sigma_{p.e.} = \sigma_T \alpha^4 Z^5 2\sqrt{2}\left(\frac{E}{m_e c^2}\right)^{-\frac{7}{2}} \tag{3.36}$$

where the K shell energy is approximately $E_{min} \simeq \frac{1}{2}m_e c^2\alpha^2 Z^2$. Only at energies large compared to atomic binding energies do photons interact with essentially free electrons by elastic Compton scattering. The cross section is $\sigma_{Compton} = \sigma_T$ for nonrelativistic scattering but decreases at high energy, approaching the form

$$\sigma_{Compton} \rightarrow \pi\frac{\alpha^2}{s}\left[\ln\frac{s}{m_e^2} + \frac{1}{2}\right] \tag{3.37}$$

where $s = m_e^2 + 2m_e E_\gamma$ is the square of the energy of the electron plus photon in the center of mass. The leading order Compton scattering diagram has two vertices corresponding photon absorption and emission - hence the factor $\alpha^2$. As described in Chapter 4, the factor $s^{-1}$ follows from dimensional analysis. The explanation of the origin of the logarithm is more involved. A practical consequence of the decrease in the cross section with energy is that the mean free path increases with energy.

Photons can scatter elastically from nuclei as well as from electrons. The amplitude is proportional to the atomic number $Z$ and the cross section to $Z^2$ but the nuclear mass, which replaces the electron mass in the Thomson formula, is proportional to $Zm_p$ and hence the ratio of the cross section for Compton scattering from the $Z$ electrons in an atom divided by the cross section for Compton scattering from the nucleus is roughly $m_p/m_e$ and, additionally, the energy loss to the nucleus is much smaller. Hence Compton scattering from nuclei is not important in determining the attenuation length of X-rays.

High energy photons are dominantly absorbed by production of $e^+e^-$ pairs. The attenuation length associated with pair production turns out to be about $(9/7)L_r$. As suggested by the diagram in Figure 3.10, pair production occurs through fusion with a virtual photon in the Coulomb field of a nucleus. The virtual photon density is proportional to $Z^2e^2$ so, like the cross section for bremsstrahlung, the cross section for pair production is proportional to the square of atomic number: $\sigma_{pair} \sim Z^2$. Above threshold, the cross section for the $\gamma\gamma \to e^+e^-$ process is $\alpha^2/s$ as for Compton scattering. In $\gamma Z \to e^+e^-Z$, a factor $Ze$ appears in the amplitude and there is an additional phase space factor as described in the Chapter 4 so the cross section rises to an energy independent limit

$$\sigma_{pair} = \frac{7}{6\pi} Z^2 \alpha \sigma_T \ln \frac{183}{Z^{\frac{1}{3}}} \sim \frac{Z^2\alpha^3}{m_e^2}. \tag{3.38}$$

The logarithm has to do with the screening of the electric field of the nucleus by atomic electrons. Both the electron and the positron tend to be produced at small angles ($\theta \sim m_e/E_e$) relative to the photon and the distribution of the fraction of the photon energy carried by either particle is fairly flat. To convert a photon to an

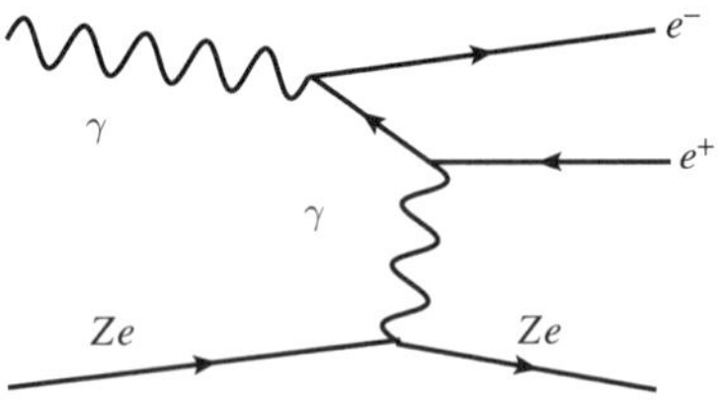


**FIGURE 3.10** Diagram describing pair production $\gamma(Ze) \to e^+e^-(Ze)$ in which an incident photon fuses with the electromagnetic field of a stationary nucleus of charge $Ze$. Another diagram has the electronic current reversed.

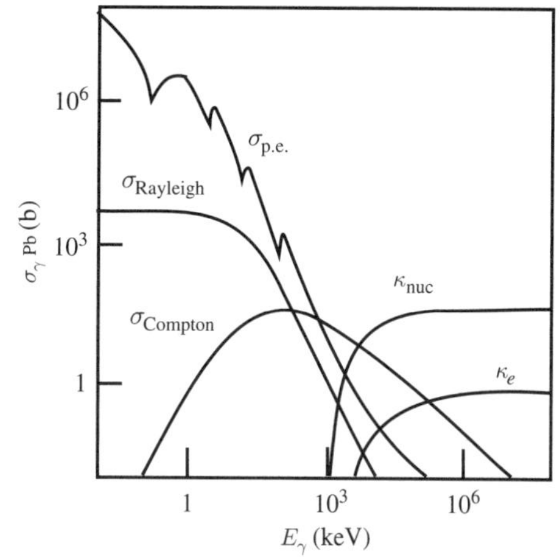


**FIGURE 3.11** Interactions of photons with matter. Cross sections as a function of photon energy $E_\gamma$ for photon interactions in Pb are shown for the photoelectric effect ($\sigma_{\text{p.e.}}$), coherent scattering from an atom ($\sigma_{\text{Rayleigh}}$), incoherent elastic scattering ($\sigma_{\text{Compton}}$), pair production due to nuclear field ($\kappa_{\text{nuc}}$), and pair production due to electrons ($\kappa_e$) [C. Amsler et al. (Particle Data Group), Phys. Lett. **B667**, 1 (2008) and 2009 partial update for the 2010 edition].

$e^+e^-$ pair, use lead, or uranium, if possible. The various contributions to the cross section for interactions of photons with matter are illustrated in Figure 3.11.

## 3.7 ■ ELECTROMAGNETIC SHOWERS

By compounded bremsstrahlung and pair production, a gamma ray or relativistic electron develops into a shower of soft electromagnetic particles as illustrated in Figure 3.12. The electrons and positrons lose energy by ionization while the positron rest energy is eventually converted to photons by annihilation with electrons in the material. Photons below 1 MeV lose energy by Compton scattering and eventually photoelectric processes of large cross section.

In a simple shower model, each electron or positron loses exactly half its energy to a photon in each radiation length, each photon produces an electron and a positron each carrying half the photon energy also in each radiation length, and all particles move in the incident direction. So the number of shower particles doubles every radiation length and the average particle energy is halved in each radiation length. The process ceases when the average energy reaches the critical energy which implies a depth

$$t \simeq L_r \ln E/E_c \tag{3.39}$$

which has a weak dependence on the initial energy E. In reality, the distance a particle moves before radiating or converting fluctuates as does the energy share given to each particle in each interaction and scattering causes the shower to

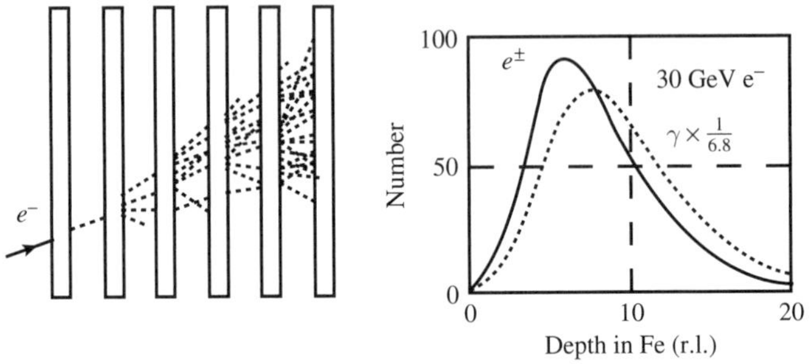


**FIGURE 3.12** Schematic picture of an electromagnetic shower. In the left illustration an electron enters a stack of metal plates. Bremsstrahlung and pair production result in multiplication of photons, electrons, and positrons which can leave ionization in detectors sandwiched between the plates. The mean number of electrons and photons as a function of depth in units of radiation length into a homogeneous material (Fe) is shown. [C. Amsler et al. (Particle Data Group), Phys. Lett. B667, 1 (2008) and 2009 partial update for the 2010 edition]

spread transversely. Figure 3.12 illustrates the average longitudinal development of particle density according to a detailed simulation.

The total ionization and visible radiation generated in an electromagnetic shower are proportional to the incident energy and permit "calorimetric" energy measurements. Calorimeters are often constructed as segmented sandwiches of high Z radiator/absorber (Pb, U) interleaved with scintillator or ionization detectors. Alternatives include high density transparent crystals and detectors of scintillation or Cherenkov light. Roughly twenty radiation lengths of material is required to contain most electromagnetic showers. Most of the energy is deposited in a thin core of order a radiation length in radius called the Molière radius. The position of the original particle may be determined from the energy distribution transverse to the particle direction with sufficiently fine transverse calorimeter segmentation. It may be more practical to insert a specialized fine grained detector for this function. For example, the ATLAS electromagnetic calorimeter at the LHC is constructed of lead sheets immersed in liquid argon. The 2 mm space between absorber sheets supports an electric field of magnitude 1000 V $mm^{-1}$ and is segmented transversely into ionization collection regions. In contrast, the CMS electromagnetic calorimeter is constructed of 76,000 scintillating lead tungstate crystals with photosensors - avalanche photodiodes in the barrel region and vacuum phototriodes in the forward region. The liquid argon option offers stability and unit gain, while the crystal option offers exceptional energy resolution.

## 3.8 ■ HADRON DETECTION

The interactions of both charged and neutral relativistic hadrons with matter are dominated by strong interaction processes. The collision of a relativistic hadron with a nucleus produces a profusion of hadrons including pions, protons, neutrons and nuclear fragments which, through subsequent collisions, multiply into a so-called hadronic shower. Hadronic showers are characterized by the hadronic absorption length of the material, the mean distance a hadron travels before colliding with a nucleus. The development of a hadronic shower is a complex process. Low energy hadrons interact strongly with nuclei via a variety of characteristic nuclear processes. Relativistic collisions can produce many flavors of stable and unstable mesons and baryons. Electromagnetic and weak decays of hadrons add leptons, neutrinos, and photons to the mix of shower particles. A schematic cosmic ray induced hadronic shower in the atmosphere is shown in Chapter 1, Figure 1.15.

In high energy experiments, it is important to measure the energy and momentum of photons and of neutral hadrons such as kaons and neutrons which leave no signal in tracking detectors, and this function is most simply performed by hadronic shower calorimeters. The hadronic absorption length is larger than the radiation length and hadronic showers are correspondingly more diffuse and penetrating than electromagnetic showers. A photon may be distinguished from a neutral hadron by as few as two longitudinal calorimeter segments - a front segment which senses mostly electromagnetic energy followed by a hadronic segment. The cross section for strong interaction with a nucleus is essentially geometrical and follows from the nuclear radius $R \simeq 1.1A^{1/3}$ fermi. In iron, for example, the radiation length is 1.76 cm, about one tenth of the hadronic interaction length. An electromagnetic segment of twenty radiation lengths and a total of 7-10 hadronic interaction lengths is a reasonable choice.

Fluctuations in shower development imply fluctuations in energy sampled by the calorimeter detection elements. The sampled shower fraction follows Poisson statistics, so the energy resolution improves with energy like $\sim 1/\sqrt{E}$ until limited by systematic effects. Hadronic showers include nuclear excitations, neutrons, muons, and neutrinos and exhibit larger fluctuations in detectable energy than electromagnetic showers. Hadronic showers also contain a fluctuating electromagnetic component. The charge exchange process $\pi^- p \rightarrow \pi^0 n$ for example will convert a pion into two photons which subsequently develop as electromagnetic showers. The proportionality constant between the observed signal and the initial particle energy is generally different for electromagnetic and hadronic showers degrading hadronic shower resolution although some tuning (compensation) can minimize the difference in response.

## 3.9 ■ LHC EXPERIMENTS

Two mammoth detectors, the Compact Muon Solenoid (CMS) and A Toroidal LHC ApparatuS (ATLAS) observe the *pp* collisions at the LHC. The design of CMS is shown in Figure 3.13 and the design of ATLAS is shown in Figure 3.15.

The principles of particle identification and momentum measurement used in these modern experiments are illustrated in Figure 3.14. The momentum and charge sign of each charged particle is determined by its deflection in a magnetic

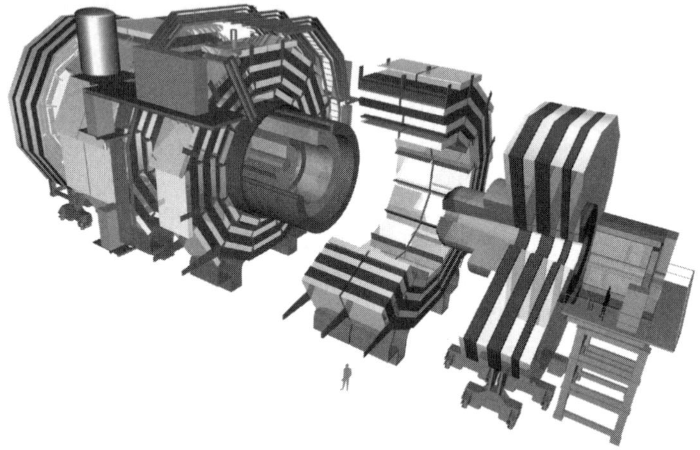

**FIGURE 3.13** The CMS detector. Designed to study high energy $pp$ collisions at the LHC, CMS features a large solenoidal central magnet surrounded by iron magnetic flux return instrumented for muon detection, and filled with calorimeters and tracking devices. [CMS collaboration]

field, the trajectory observed with silicon and wire chamber ionization tracking devices. A photon leaves no ionization trail and, like an electron, converts to an electromagnetic shower in a calorimeter. The shower energy and position determine the photon momentum. The absence of a track distinguishes the photon

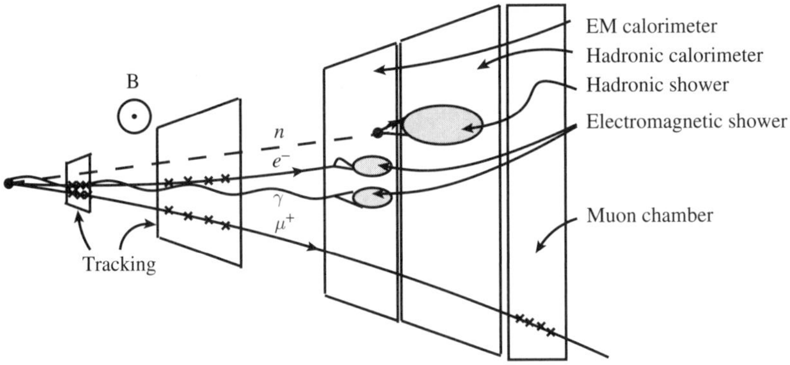


**FIGURE 3.14** Principles of particle detectors at colliders. Particle identification and momentum measurement in a general purpose detector system takes advantage of the different ways particles interact with matter. A photon or electron appears as an electromagnetic shower in an electromagnetic calorimeter. A hadron generates a hadronic shower in a backing calorimeter. A muon passes through both calorimeters leaving only a minimum ionizing trail of energy. Tracking devices detect the trajectories of charged particles. The momentum of a charged particle is determined from the curvature of its trajectory caused by a magnetic field.

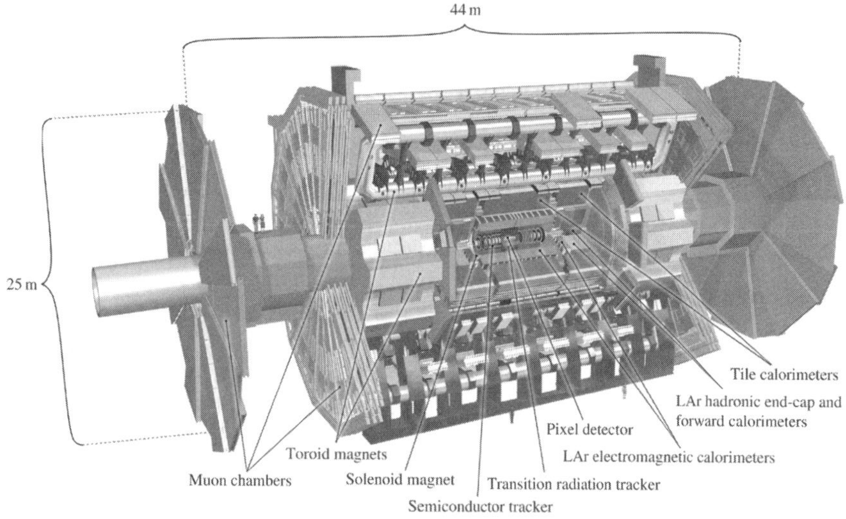


**FIGURE 3.15** Isometric view of the ATLAS detector at the LHC.

shower from an electron induced shower. A hadron such as a $\pi^+$ or $n$ has an interaction length longer than that of an electron or photon and creates a hadronic shower in a calorimeter placed behind the electromagnetic calorimeter. A muon leaves essentially only a trail of ionization in the calorimeters and is identified by dedicated muon tracking devices behind the hadron calorimeter.

The CMS detector comprises a large 3.8 T solenoid magnet with a superconducting coil of inductance of 14 H, nominal current 19 kA and stored energy 2.7 GJ. The coil encloses silicon tracking and an electromagnetic calorimeter. Outside the coil is a hadron calorimeter and an instrumented iron magnetic flux return used to identify and momentum analyze muons. The CMS interaction point is surrounded by a 13 layer barrel silicon tracking device. The innermost layers comprise 66 million pixels, each 100 $\mu$m $\times$ 150 $\mu$m in transverse dimension. The next layers are segmented into strips of width 180 $\mu$m, four layers of strips with length 10 cm inside six layers of strips of length 25 cm. The barrel device is complemented by disk shaped endcap silicon tracking. The CMS electromagnetic calorimeter (ECAL) surrounds the silicon tracking. It comprises dense lead tungstate ($PbWO_4$) crystals of radiation length just 9 mm and dimension 22 mm $\times$ 22 mm $\times$ 230 mm. There are 68.5 thousand crystals arrayed to point at the interaction region. Each crystal is backed by a silicon avalanche photodiodes capable of detecting light while immersed in a strong magnetic field. The hadron calorimeter (HCAL) is constructed of brass absorber plates and plastic scintillator tiles. The light in the tiles is collected by embedded fibers which absorb scintillation light and reradiate at longer wavelength. The fibers channel reradiated light by total internal reflection to hybrid photodiodes capable of operating in the magnetic field. Within the barrel magnetic flux return, proportional wire drift tubes (DT) track and identify muons escaping the hadronic calorimeter.

The endcap detectors are trapezoidal cathode strip chambers. These comprise 6 layer stacks of wire planes, each wire plane a multiwire proportional counter. The wires trace circles around the beam direction and electronic pulses on the wires measure the radial coordinate of muon tracks. Cathode strips of order 1 cm in width running in the radial direction between wire planes sense the orthogonal azimuthal coordinate. The centroid of the charge induced on several strips provides a measure of the azimuthal coordinate. Muons are deflected in the azimuthal direction by the CMS solenoidal magnetic field so this is the critical coordinate to measure in the endcaps to confirm the momentum of a corresponding interior track.

Both muon detection systems are supplemented with resistive plate chambers. A resistive plate chamber (RPC) comprises several gas filled gaps a few millimeters in thickness between bakelite panels. A potential difference across the gap is maintained that an ionization trail initiates a spark. The resistance of the panels lowers the electric field as current begins to flow and terminates the discharge. The location of the spark is determined digitally by the signals induced on conducting strips on the external surfaces of the resistive panels. A short gap is necessary for fast timing and multiple gaps are used to compensate for the small inefficiency of a single gap device. Resistive plate chambers are used extensively at the Large Hadron Collider experiments instead of scintillation counters because of their relatively low production cost, combined with a time resolution of about a nanosecond. The spatial resolution determined by the strip width is easily customized.

CMS deploys a complex electronic trigger system designed to select events from about 100 of the 40 million bunch crossings per second at design luminosity. The detector itself is continuously live and produces about 1 MB of data per crossing. All data is stored in a continuous pipeline for about one microsecond while the lowest level pattern recognition electronics recognizes signatures such as coincident signals in layered tracking devices indicating a track and clusters of energy in neighboring segments of calorimeter. Signals from parts of the detector separated by more than a bunch spacing in light travel time are assembled and jets, photons, muons and electrons crudely identified. About 50 kHz of data is passed to a collection of processors running event reconstruction programs which make the final event selection.

The ATLAS detector (Figure 3.15) is 44 m long and 25 m in diameter, the size dictated by the choice of air core toroidal external magnets used for muon momentum measurement. The ATLAS innermost detector contains 50 $\mu$m by 400 $\mu$m silicon pixels in three layers with 80 million readout channels. It is surrounded by a silicon strip tracker called the semi-conductor tracker (SCT) based on 80 $\mu$m by 12.6 cm strips in four layers constituting 6.2 million readout channels. Outside the SCT, a transition radiation tracker (TRT) is deployed. It is comprised of drift tubes (straws) 4 mm in diameter and up to 144 cm in length with roughly 200 $\mu$m spatial resolution. There are about 350 thousand straws containing a Xe rich gas mixture sensitive to X-rays through the photoelectric effect. The TRT is designed to detect, in addition to the primary ionization signals from penetrating particles, the X-ray radiation emitted when these particles make a transition between materials of different refractive index. The number of photons emitted by a charge $e$ in traversing an interface is proportional to the particle gamma factor times the fine structure constant ($N_\gamma \simeq \alpha\gamma$) and a count of these photons can be used to discriminate electrons from hadrons for momentum up to hundreds of GeV. The number of photons per interface is small. The TRT straws are embedded in plastic fibers 15 $\mu$m in diameter which provide many interfaces and the layers of straws provide on average 35 hits per track with a position resolution of 170 $\mu$m.

The ATLAS superconducting solenoid contains the TRT and SCT and has a field strength of 2 T. It is surrounded by an electromagnetic calorimeter and hadronic calorimeter. The electromagnetic calorimeter is made up of lead absorber plates sheathed with stainless steel, stacked in an accordion pattern, and immersed in liquid argon, all contained within a cryostat. Between the plates, copper electrodes mounted on insulating sheets create a static electric field in the gaps to drift and collect the ionization created in the argon by charged particles. Outside the electromagnetic calorimeter, the barrel hadronic calorimeter is constructed of steel absorber plates which serve to return the flux of the solenoid and scintillating plastic tiles. The ATLAS muon spectrometer begins at a radius of 4.25 m and extends to 11 m. A toroidal magnetic field is created in this region by air-core superconducting coils. About 370 thousand 30 mm diameter round cross section monitoring drift tubes (MDT) in three modules precisely measure the trajectories of muons in the central region. The open air geometry minimizes multiple Coulomb scattering and the momentum resolution at 1 TeV is better than 10%. In the forward and backward regions, muons are detected with cathode strip chambers. ATLAS deploys 355 thousand channels of RPCs to trigger on muons in the barrel region and thin gap multiwire chambers (which have a higher rate capability) in the endcaps.

This description of the detector technologies deployed by the CMS and ATLAS collaborations is incomplete. Numerous special purpose detectors are embedded within the major components and in the forward and backward directions. However it indicates the scope of the research and development required to construct a modern general purpose detector. Each component entails a custom design, extensive testing, and generally a custom fabrication facility. It is for this reason that a large international collaborative effort is behind each experiment.

## 3.10 ■ FURTHER READING

The 1958 Nobel Prize in Physics was awarded jointly to Pavel Alekseyevich Cherenkov, Ilja Mikhailovich Frank, and Igor Yevgenyevich Tamm "for the discovery and the interpretation of the Cherenkov effect."

The 1960 Nobel Prize in Physics was awarded to Donald A. Glaser for the invention of the bubble chamber.

The 1968 Nobel Prize in Physics was awarded to Luis Alvarez "for his decisive contributions to elementary particle physics, in particular the discovery of a large number of resonance states, made possible through his development of the technique of using the hydrogen bubble chamber and data analysis."

The 1992 Nobel Prize in Physics was awarded to George Charpak for the development of the multiwire proportional chamber.

A review of electronic detection techniques is T. J. Ferbel (ed.), *Experimental Techniques in High Energy Physics*, Addison-Wesley (1987). Elements of particle detection are also described in the texts A. Melissinos, *Experiments in Modern Physics*, Academic Press (1996), B. Rossi, *High Energy Particles*, Prentice-Hall (1952), J. D. Jackson, *Classical Electrodynamics*, John Wiley & Sons, Inc., (1999), and Richard Fernow, *Introduction to experimental particle physics*, Cambridge University Press (1986).

For theories and applications of statistics see A. G. Frodesen, O. Skjeggestad, and H. Tofte, *Probability and Statistics in Particle Physics*, Universitetsforlaget, Bergen (1979).

A brief history of experimental discoveries with reprints of original papers is R. H. Cahn and G. Goldhaber, *The experimental foundations of particle physics*, Cambridge University Press (1989).

For complete descriptions of the CMS, ATLAS, ALICE, and LHCb experiments, see the technical design reports (TDRs) at their websites. A book dedicated to technical aspects of the CMS and ATLAS experiments is Dan Green (ed.), *At the Leading Edge: The ATLAS and CMS LHC Experiments*, World Scientific (2010). Additional reviews are The CMS Collaboration *et al*, JINST **3** S08004 (2008), The ATLAS Collaboration *et al*, JINST **3** S08003 (2008), The Alice Collaboration *et al*, JINST **3** S08002 (2008), and The LHCb Collaboration *et al*, JINST **3** S08005 (2008).

## 3.11 ■ PROBLEMS

---

**Problem 3.1. Ionization range**

---

Starting with the expression for the range $R$ of a particle suffering ionization energy loss

$$R = -\int_E^0 \frac{dE}{dE/dx} = \int_E^0 \frac{dE}{f((v)}$$

and the expression $E = mv^2/2$, show that the range of a nonrelativistic particle of given initial speed is proportional to its mass $m$. Suppose a $K^+$ stops in liquid hydrogen and decays to $\pi^+\pi^0$ and that the range of the $\pi^+$ is $R$. At what distance from the stopping point would the ionization density on the track of the $K^+$ equal the initial ionization density of the $\pi^+$?

---

**Problem 3.2. Bubble chamber image analysis**

---

The schematic image in Figure 3.16 shows a $p\bar{p}$ annihilation producing $K^- K^0 \pi^+$ in a hydrogen bubble chamber immersed in a magnetic field. a) The $\bar{p}$ enters from

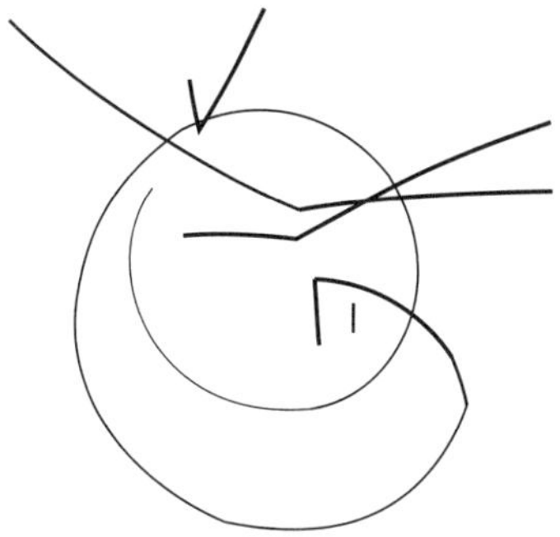

**FIGURE 3.16** Interaction of an antiproton in the 80 cm Saclay liquid hydrogen bubble chamber. Annihilation of the antiproton with a proton produces $K^- K^0 \pi^+$. [after CERN photo archive]

the right. Is the magnetic field directed into or out of the plane of the image? b) Provide a hypothetical reconstruction of all of the tracks and interactions in the image, describing the ultimate fate of each and every baryon, pion, kaon, muon, electron, and photon. What happened to the strange quark when the $K^-$ came to rest? What evidence is there that a $\Lambda$ was produced decaying to $p\pi^-$?

### Problem 3.3. Proton decay experiment

The proton is evidently long lived or we could not survive our own radioactivity. Proton decay experiments have largely been performed deep underground to reduce backgrounds from cosmic rays. Grand unified theories suggest the proton may in fact decay to, e.g. $e^+\pi^0$ with a lifetime estimated to exceed $10^{32}$ years. a) Calculate the mass in kiloton and the linear dimension in meters of a cube of water required to produce one decay in one year for such a lifetime. b) What would be the energy of the positron and what would be the opening angle of its Cherenkov light cone? c) About how far might the gamma rays from the decay $\pi^0 \to \gamma\gamma$ travel before converting to electron positron pairs?

### Problem 3.4. Pion stopping time

A $\pi^-$ of initial kinetic energy 100 MeV suffers ionization energy loss in liquid hydrogen. The density of the liquid is 70 kg m$^{-3}$. Estimate the time required for the pion to come to rest and show that it is short compared to the pion lifetime permitting pion capture.

### Problem 3.5. Ion drift velocity

Estimate the drift velocity $v_d = eE\tau/m$ of positive ions in a gaseous argon filled drift detector for electric field strength $E = 2000$ V cm$^{-1}$. Assume the mean time between collisions is $\tau = l/v$ with mean free path $l = 1/n\sigma$. Take the number density from the ideal gas law $n = P/k_BT$ with $T = 300$ K and $P = 10^5$ Pa and $k_B = 1.38 \times 10^{-23}$ J K$^{-1}$. Use the thermal speed $v = \sqrt{3k_BT/m}$ where $m = Am_p$ with mass number $A = 40$ and assume a hard sphere collision model with $\sigma = \pi d^2$ with argon diameter $d = 0.2$ nm. Compare $\mu \equiv v_d/E$ to the measured mobility (the drift speed per unit field strength) $\mu(\mathrm{Ar}^+) = 1.63\,\mathrm{cm}^2\,\mathrm{V}^{-1}\,\mathrm{s}^{-1}$. ["The Drift Velocities of Molecular and Atomic Ions in Helium, Neon, and Argon", John A. Hornbeck, Phys. Rev. **84**, 615(1951)]. What drift velocity would this model predict for electrons?

---

**Problem 3.6. Cosmic ray backgrounds**

---

A 1 km $\times$ 1 km $\times$ 10 m volume of purified water is instrumented with phototubes to search for $p \to e^+ + X$ where $X$ is not observable. A sea level experiment is subject to a cosmic ray muon flux of 100 Hz m$^{-2}$ at $E_\mu \simeq 2$ GeV. The entire surface of the volume is covered with scintillation counters each 1 m$^2$ in area and each costing 500 units of currency.
a) How efficient must each counter be at detecting crossing muons to veto essentially every cosmic muon entering from the top during one year of operation? b) How much did the veto system cost in currency units? c) If the experiment were underground, roughly how many meters of 3 g cm$^{-3}$ rock overburden would be required to range out 2 GeV muons by ionization energy loss before they reach the detector?

---

**Problem 3.7. Bremsstrahlung**

---

The differential cross section for bremsstrahlung by an electron of energy $E$ with emission of a photon of energy fraction $k = E_\gamma/E$ is

$$\frac{d\sigma}{dk} = \frac{A}{N_A X_0}\frac{1}{k}f(k)$$

where $f \simeq (4/3) - (4/3)k + k^2$, $X_0$ is the radiation length, $A$ is the mass number, and $N_A$ is Avogadro's number. a) Show that $\int_{k_0}^{1} dk\ f(k)/k = 5.3$ for $k_0 = 0.01$ and that for $N_2$ we have $A/N_A X_0 = 1.27$ b per molecule and the cross section for an energy loss exceeding 1% of the electron energy is about 6.5 b. b) In the LEP accelerator, a fractional energy loss in excess of about 1% would lead to loss of a beam particle. For beam pipe gas temperature and pressure $T = 300$ K and $P = 1$ nTorr$= 1.3 \times 10^{-7}$ Pa, show that the density of molecules is $\rho = 3.26 \times 10^{13}$ m$^{-3}$ and $\rho\sigma = 2.1 \times 10^{-14}$ m$^{-1}$. What is the corresponding mean lifetime in hr for a particle to stay in the machine?

---

**Problem 3.8. Compton scattering**

---

Consider electrons of energy $E$ =100 GeV in a room temperature beam pipe at LEP II. The number density of black body photons for temperature $T$ is

$\rho_\gamma = 8\pi\alpha(k_B T/(hc))^3$ with $\alpha = 2.404$ and the mean energy is $E_\gamma = 2.7k_B T$. a) Show that the a typical photon energy in the electron frame is $E^*_\gamma = E_\gamma E/m_e \simeq 6$ keV so the elastic scattering cross section in the electron frame is adequately described as Thomson scattering with cross section

$$\frac{d\sigma}{d\cos\theta} = \pi r_e^2 \left(1 + \cos^2\theta\right)$$

with $r_e = e^2/m_e = 2 \times 10^{-12}$ cm and total cross section $\sigma = \sigma_T = 0.6$ b. b) What is the mean lab energy of scattered photons in GeV? c) What is the corresponding mean scattering lifetime in hr? [Ref: Dehning *et al.*, Phys. Rev. Lett. **B249**,145 (1990)]

---

**Problem 3.9. Electron identification by synchrotron radiation**

---

A 100 GeV electron passes through a length $L = 1$ m of magnetic field $B = 1$ T perpendicular to the track. Calculate the deflection in space at the end of the magnet relative to infinite energy (no deflection). Synchrotron radiation is generated along the track and appears in a band of this width. Take the mean photon frequency as $\omega_s = 3\gamma^3/R$ where $R$ is the radius of curvature to estimate the total number of photons generated. How many would be generated by a muon of the same deflection? How is it that a classical continuous circular beam emits no synchrotron radiation?

---

**Problem 3.10. Thermal calorimeter**

---

Consider a 100 GeV electromagnetic shower contained in a room temperature Cu cylinder 20 radiation lengths in length by 2 radiation lengths in radius. (For Cu the electron radiation length is 1.43 cm.) Estimate the heat capacity $C_V$ using the atomic heat capacity $C_V^{\text{atom}} = 3k_B T$. Compare the deposited energy to the r.m.s. energy fluctuation in GeV of the detector

$$\sigma_E = \sqrt{k_B T^2 C_V}.$$

Calculate the energy fluctuation for an X-ray detector consisting of one cubic micron of silicon at temperature $T = 3$ K using the Debye model $C_V^{atom} = 234k_B(T/\Theta_D)^3$ with $\Theta_D = 645$ K.

---

**Problem 3.11. Millicharged particles**

---

Consider a particle with the mass of the electron but charge $10^{-3}e$. What would be its binding energy to a proton? The critical energy for an electron is the energy such that, per unit length, the energy loss by ionization is equal to that due to bremsstrahlung. It is about 800 MeV/Z for atomic number Z. What would be the critical energy of a millicharged electron? [Ref: A. A. Prinz *et al*, Phys. Rev. Lett. **81**. 1175 (1998), R. N. Mohapatra and I. Z. Rothstein, Phys. Lett. **B247**, 593 (1990)]

## Problem 3.12. Tracking resolution

The method of least squares fits a function $y(x, \mathbf{p})$ of a coordinate x and a vector of parameters $\mathbf{p}$ to a collection of independent measured values $y_i$ of similar variance at $n$ points $x_i$ by finding the parameter values which minimize the sum of squared deviations

$$\chi^2(\mathbf{p}) = \Sigma_i (y(x_i, \mathbf{p}) - y_i)^2/(n-1).$$

The optimal parameter values $\mathbf{p}_0$ are found by solving the equations $\partial s^2/\partial p_i = 0$. For a fit to a straight line of the form $y(x) = a + bx$, the method provides values for $a$ and $b$. Put $\bar{x} = \Sigma x_i/n$ and $\bar{y} = \Sigma y_i/n$ and define

$$\begin{aligned} s_{xx} &= \mathbf{x} \cdot \mathbf{x} - n\bar{x}^2 \equiv n\sigma_x^2 \\ s_{xy} &= \mathbf{x} \cdot \mathbf{y} - n\bar{x}\bar{y} \equiv n\sigma_{xy}^2 \\ s_{yy} &= \mathbf{y} \cdot \mathbf{y} - n\bar{y}^2 \equiv n\sigma_y^2. \end{aligned}$$

Then the estimates for $b$ and $a$ are $b = s_{xy}/s_{xx}$ and $a - \bar{y} - b\bar{x}$ and the variance of the measurements $y_i$ can be estimated from the squared deviations from the fit

$$s^2 = \chi^2(\mathbf{p}_0) = \frac{s_{yy}s_{xx} - s_{xy}^2}{s_{xx}(n-2)}.$$

The variances of the parameters are $\sigma_a^2 = s^2(n^{-1} + \bar{x}^2/s_{xx})$ and $\sigma_b^2 = s^2/s_{xx}$ and $a$ and $b$ are uncorrelated if $\bar{x} = 0$. The variance on an extrapolated point $y = y(x, \mathbf{p}_0)$ is

$$\sigma_y^2 = \frac{s^2}{n}\left[1 + \frac{(x-\bar{x})^2}{\sigma_x^2}\right].$$

Suppose a four layer silicon tracking device has layers separated by 4 cm, each providing a measurement of point on a straight track with error $\sigma = 20\ \mu$m. What is the error in millimeters on the track position extrapolated to a distance of one meter from the center of the device?

## Problem 3.13. Iron spectrometer

A relativistic $\mu^+$ passes through a 2 m thick slab of iron with magnetic field strength $B_y = 2$ T perpendicular to the muon momentum. The muon is deflected by the magnetic field through a small angle $\theta_x$ and by multiple Coulomb scattering. Suppose the entrance angle $\theta_x$(initial) is measured with two drift chambers separated by $\Delta z = 1$ m, each chamber having normally distributed position resolution with standard deviation $\sigma_x = 0.25$ mm. A similar system measures the exit angle $\theta_x$(final). If the momentum is estimated from the angular deflection, calculate the fractional momentum resolution $\sigma_p/p$ as a function of momentum. Neglect energy loss. At what momentum does the contribution from the chamber measurement error equal that from multiple Coulomb scattering?

## Problem 3.14. Proportional chamber

A grounded cylinder of length $L$ and inner radius $b = 1$ cm has an axial wire of radius $a = 20$ $\mu$m at voltage $V$ and is filled with a 1 atm mixture of argon and ethane. Electrons are multiplied near the wire where in a mean free path $l_{scat}$ the energy gain $eE(r)l_{scat}$ exceeds the ionization threshold $\epsilon^{\text{thres}}$. The multiplication has the form

$$\frac{\mathrm{dN}}{\mathrm{dr}} = \alpha(\mathrm{r})\mathrm{N} \to \mathrm{M} = \frac{\mathrm{N}}{\mathrm{N}_0} = \exp\left(\int_{\infty}^{\mathrm{a}} \mathrm{dr}\ \alpha(\mathrm{r})\right)$$

where, since there is multiplication only near the wire, the integration has been extended to infinity. Suppose the Townsend coefficient (ionization mean free path) has the form

$$\alpha = \frac{1}{\mathrm{l_{scat}}} \exp\left(\frac{-\epsilon^{\text{thres}}}{\mathrm{eE(r)l_{scat}}}\right).$$

Here $\mathrm{l_{scat}} = (\mathrm{n}\sigma)^{-1}$, where $n$ is the number density of molecules and $\sigma = 2.75 \times 10^{-16}$ cm$^2$ is an energy independent cross section for ionization by electron impact. The effective ionization threshold is $e\epsilon^{\text{thres}} = 16.7$ eV.

a) Determine the approximate voltage and wire charge per unit length $Q'$ required for a gain of $M = 1 \times 10^5$.

b) Multiplication leaves a cylindrical shell of ionic charge near the wire surface which drifts radially outwards along the electric field with mean speed $\mathrm{v} = \mu\mathrm{E(r)}$ with constant mobility $\mu$. Find the radius of the ionic charge as a function of time. In the absence of external connections, show the wire voltage change as the ions drift is approximately

$$v(t) = -\frac{q}{4\pi\epsilon_0 L} \ln\left[1 + \frac{\mu Q'}{\pi\epsilon_0 a^2} t\right].$$

c) If the wire is terminated in a high voltage blocking capacitance $C_b$ and a series resistance $R$ to the outer cylinder, the signal voltage across the resistor is $V_o = RC_t \frac{dv}{dt}$ with $C_t$ the tube capacitance if $C_b >> C_t$. Show that

$$\frac{V_o^{max}}{V} = R \frac{1}{[\ln(b/a)]^2} \frac{q\mu}{a^2}.$$

## Problem 3.15. Lead glass shower counter

A lead loaded transparent glass block absorbs high energy photons and electrons incident at one end. The other end is viewed by a photomultiplier tube which detects Cherenkov light, produced by the electrons and positrons in the electromagnetic shower, and transported by total internal reflection to the photocathode.

The glass has density $\rho = 4.5$ g cm$^{-3}$, refractive index $n = 1.7$ for visible light, and radiation length 2 cm.

a) If all of the incident energy is ultimately dissipated in ionization by minimum ionizing particles (pair produced $e^{\pm}$ and $e^-$ from Compton scattered gamma rays), what is the approximate sum of the lengths of all tracks in a 10 GeV shower?

b) The number of visible Cherenkov photons per unit track length and frequency interval is about 400 cm$^{-1}$. If the light transport efficiency is 10% and the photocathode efficiency is 10%, how many photoelectrons $N$ are produced in the photomultiplier on average per 10 GeV shower?

c) If nonlinearity and multiplication fluctuations are negligible, the charge produced by the photomultiplier tube is proportional to the number of photoelectrons which is proportional to the shower energy. Assuming the average number of photons $N(E)$ for a given incident energy is proportional to energy and follows Poisson statistics so (for large $N$) has variance $\sigma^2_{\mathrm{N(E)}} = N(E)$, calculate the relative energy resolution of the measurement in the form

$$\frac{\sigma_E}{E} = \frac{P}{\sqrt{E}}$$

where $P$ is in percent and $E$ in GeV. Notice that the resolution improves with increasing energy until limited by other systematic effects.

---

**Problem 3.16. MWPC efficiency**

---

Suppose the mean number $\bar{N}$ of ion clusters produced by a minimum ionizing particle in a gas is 30 cm$^{-1}$. The actual number fluctuates and the probability of $N$ clusters is

$$P(N) = \frac{\bar{N}^N e^{-\bar{N}}}{N!}.$$

What is the efficiency in percent for detecting tracks crossing a chamber with 3 mm thickness of gas, assuming the gain and electronic threshold are such that the cell is sensitive to a single electron produced in the gas?

---

**Problem 3.17. Berm design**

---

Design a rock pile to range out 500 GeV muons. Assume constant mean energy loss $dE/dx = 1.7$ MeV g$^{-1}$ cm$^{-2}$, a mass density $\rho = 3$ g cm$^{-3}$ and radiation length $L_{rad} = 20$ g cm$^{-2}$. Consider the requisite longitudinal and transverse dimensions of the berm. In determining the transverse size, derive an expression for the multiple scattering related displacement by integrating the variances in transverse position due to multiple scattering in small slabs along the beam direction, including the increase in r.m.s. multiple scattering angle as the particle energy decreases.

## Problem 3.18. Ultracold neutrons

An ultracold neutron (UCN) with temperature of a few mK may be coherently reflected by condensed matter which presents a potential barrier of a few hundred neV. Such a neutron may be piped around, behaving in some cases like optical radiation confined by total internal reflection, and in other cases like a classical particle. To produce ultracold neutrons, the low energy tail of the thermal energy distribution of neutrons at a nuclear reactor is selected, cooled by scattering from hydrogen atoms in a polymer, and further cooled through mechanical work on a fan turbine. Neutrons produced by proton beam collisions with a tungsten target have been cooled by injection into solid deuterium at a few degree K. With densities of several hundred per cubic centimeter, the $\beta$ decay angular distribution may be observed and the neutron lifetime ($\tau_n = 885 \pm 0.8$ s) measured.
a) Show that the critical speed to overcome the effective potential of $V = 210$ neV in iron is 6 m s$^{-1}$ and calculate the kinetic energy $K$ in neV of a neutron dropped from a height of 10 m with local gravitational acceleration g = 10 m s$^{-2}$. What is the equivalent temperature $T = E/k_B$? Compare the de Broglie wavelength of such a neutron to a typical interatomic spacing of $10^{-10}$ m.
b) Magnetic trapping and spin polarization filtering of such low energy neutrons is possible. Calculate the energies

$$V_{\text{tot}} = V \pm \mu_n B$$

in neV associated with the interaction of the neutron magnetic dipole moment with the magnetic field of strength $B = 2$ T inside a permanent magnet. The magnitude of neutron magnetic moment is

$$|\mu_\text{n}| = 1.91\ \text{e}/2\text{m}_\text{p} = 6.0 \times 10^{-14}\text{MeV T}^{-1}.$$

[Reference: V.K. Ignatovich, *The Physics of Ultracold Neutrons*, Clarendon Press (1990), Oxford, UK. See also R.W. Pattie, Jr. *et al* (UCNA Collaboration) "First Measurement of the Neutron $\beta$ Asymmetry with Ultracold Neutrons," PRL **102**, 012301 (2009).]

## Problem 3.19. Knock-on electrons

Show that the recoil kinetic energy $T$ of an electron knocked out of an atom by a fast primary particle through the Coulomb interaction is related to the center of mass and laboratory recoil angles $\theta_{cm}$ and $\theta$ by

$$T = \gamma^2\beta^2 [1 - \cos\theta_{cm}]\, m_e = \frac{2\beta^2\cos^2\theta}{1-\beta^2\cos^2\theta} m_e$$

where $\beta = v_{cm}/c$ and $\gamma = (1-\beta^2)^{1/2}$ with $v_{cm}$ the velocity of the center of mass of the collision. The electron is assumed free and at rest initially in the laboratory. From the lab angle and energy of the so-called knock-on secondary electron, the energy of the primary may be determined.

## Problem 3.20. Askaryan effect

The Askaryan effect is the coherent radio wavelength Cherenkov emission by the pancake of charged particles at the core of an electromagnetic shower. It can be used to detect EeV energy showers in the atmosphere or polar ice. For a track length $L$, an electron of speed $v$ radiates in a frequency range $\Delta\nu = \nu_2 - \nu_1$ an energy

$$E_A = \frac{\pi h}{c}\alpha L\left(1 - \frac{1}{n^2(v/c)^2}\right)\left[\nu_2^2 - \nu_1^2\right]$$

where $\alpha \simeq 1/137$ is the fine structure constant and $n$ is the refractive index of the shower medium.

Suppose the maximum number of charged particles in a shower of energy $E$ is $N \simeq E/(1\text{ GeV})$ with a $\sim$ 20% excess of electrons over positrons due to Compton scattering of photons within the shower so the net charge in the shower core is $Z = -0.2Ne$. If the transverse dimensions of the shower core are a few cm and are small compared to the wavelength of the emission, the emission may be regarded as coherent. Then the radiated energy is $(Z/e)^2 E_A$. Take $E = 10^{18}$ eV, $v \simeq c$, $L = 6$ m and $n = 1.8$ for ice, and $\nu_1 = 0.3$ GHz and $\nu_2 = 0.9$ GHz. a) Show that the radiated energy is about 6 nJ. b) The energy is emitted into a cone of polar angle $\theta_c = \cos^{-1}(c/(nv))$ with angular width $\Delta\theta_c = \frac{c}{\bar{\nu}L}\sin\theta_c$ with $\bar{\nu}$ the mean frequency and appears in a bandwidth limited detector during a time of order $\Delta t \simeq 1/\Delta\nu \simeq 1.6$ ns. Show that the energy flux is about $1.7 \times 10^{-8}$ J sr$^{-1}$ and, at an observation distance $d = 600$ km, the peak flux density is about 5 MJy where 1 Jy $= 10^{-26}$ W m$^{-2}$ Hz$^{-1}$. These parameters correspond to the observation range and threshold energy for the Antarctic Impulsive Transient Antenna (ANITA) with a detector noise level dominated by thermal emission from the ice of about 2 MJy.

## Problem 3.21. Transition radiation

When a particle of charge $e$ and energy $E = \gamma mc^2$ crosses a boundary between vacuum and a medium electron density $N_e$, transition radiation is emitted with mean energy $E_{tr} = \alpha\gamma\hbar\omega_p/3$ where $\alpha = e^2/(4\pi\epsilon_0\hbar c) \simeq 1/137$ and $\omega_p$ is the plasma frequency given by

$$\hbar\omega_p = \sqrt{4\pi N_e r_e^2 \frac{m_e c^2}{\alpha}}$$

with $r_e = e^2/(4\pi\epsilon_0 c^2)$ the classical electron radius. a) Show that this plasma energy can be expressed in terms of the Bohr radius $a_\infty = r_e/\alpha^2$ as

$$\hbar\omega_p = \sqrt{16\pi N_e a_\infty^3}\, E_0$$

where $E_0 = m_e c^2\alpha^2/2 = 13.6$ eV. Since the number of electrons within a volume of radius $a_\infty$ is of order the atomic number, we see that the plasma energy is 20-100 eV. b) If the typical transition radiation photon energy is $E_\gamma = \gamma\hbar\omega_p$, roughly how many interfaces are required to produce such an X-ray? If an interface is created every 15 $\mu$m in a dense collection of fibers of that diameter, what thickness of fiber material will be required per X-ray?

---

**Problem 3.22. Charmed meson mass measurement**

---

The lightest mesons containing a charm quark include the $D^+_{c\bar{d}}$, $D^0_{c\bar{u}}$, $\bar{D}^0_{\bar{c}u}$, and $D^-_{\bar{c}d}$. Each decays in a multitude of ways. For example, the $D^0$ decays to $K^-\pi^+$ with a probability (branching fraction) of 3.9%. The mean mass of the $D^0$ obtained from many measurements is $1864.84 \pm 0.17$ MeV. Suppose $D^0$ mesons are produced at rest, the $K^-$ and $\pi^+$ (identified by some means) each travel a distance $L = 1$ m in a uniform magnetic field of strength $B = 2$ T, their momenta are determined from the radii of curvature $R_\pi$ and $R_K$, and the mass $m_D$ is estimated from the sum $E_\pi + E_K$ of the measured energies. a) Use the formula $\sigma_f^2 = \Sigma_i \left(\frac{\partial f}{\partial x_i}\right)^2 \sigma_i^2$ for the variance of a function $f(\mathbf{x})$ of a vector of independent variables $x_i$ with individual variances $\sigma_i^2$ to express the error on the mass of the pair $\sigma_{m_D}$ in terms of the errors $\sigma_{R_\pi}$ and $\sigma_{R_K}$ on the track curvature radii. b) With what accuracy in percent must $R_\pi$ and $R_K$ be measured to achieve a mass error of 0.17 MeV? c) Suppose the standard error on the measured sagitta $s \simeq L^2/(8R)$ is 0.1 mm, and the error on the mean mass for $N$ events is $\sigma = \sigma_{m_D}/\sqrt{N}$. How many events are required to achieve an error of 0.17 MeV? d) The linear thermal expansion coefficient of aluminum is $\alpha = 23 \times 10^{-6}/°\mathrm{C}$ at room temperature. What temperature jump in K would increase the dimensions of a 1 m radius Al tracking detector support structure by 0.1 mm? e) The accuracy of an NMR probe measurement of a magnetic field strength is about $2 \times 10^{-5}$. The magnet current must be stabilized to this level or continuously monitored and variations in field strength along the track must be known to this level to render their contribution to the overall measurement error negligible. What is the systematic error in $m_D$ associated with a 20 ppm uncertainty in the field strength $B$?

---

**Problem 3.23. Charmed meson lifetime measurement**

---

The $D^+$ decays to $K_S\pi^+$ with a probability (branching fraction) of 1.45%. The mean mass of the $D^+$ obtained from many measurements is $1869.62 \pm 0.20$ MeV and the mean proper (rest frame) lifetime is $(1040 \pm 7) \times 10^{-15}$s. The typical time resolution of the fastest particle detector is of order 1 ns making a direct measurement of the decay time of $D^-$ impractical. Time dilation implies the decays of $D^-$ mesons with energy $E = \gamma m_D c^2$ have an exponential decay length distribution with mean decay length $L = vt = v\gamma\tau$ from which $\tau = Lm/p$ may be inferred. a) What is the mean decay length in millimeters of $D^+$ mesons of momentum 10 GeV? b) Neglecting the inaccuracy in decay length measurements, how many events must be measured to determine the proper lifetime with a relative precision of 7/1000? c) How many microns of decay length does this error correspond to? (For comparison, the point resolution for a silicon strip detector is about 20 $\mu$m.) d) How does the width $\Gamma = \hbar/\tau$ compare to the mass resolution $\sigma = 0.2$ MeV?

## Problem 3.24. Ionization cooling for muon collider

Ionization energy loss may be put to use to reduce the spread in transverse momentum of muons in a future $\mu^+\mu^-$ collider. The idea is to degrade the magnitude of the momentum while replenishing the longitudinal momentum with RF acceleration. Suppose an absorber of thickness $dz_{ab}$ reduces the mean energy of a muon of velocity $v$ by $dE = \epsilon'_{ab} dz_{ab}$.
a) Neglecting multiple Coulomb scattering and energy loss fluctuations (called straggling), show that the fractional change in mean transverse momentum for a relativistic muon is

$$\frac{dp_x}{p_x} = v^2 \frac{dE}{E}.$$

b) For acceleration energy gradient $\epsilon'_{acc}$, the restoration of the energy requires a distance $dz_{acc} = dz_{ab}\epsilon'_{ab}/\epsilon'_{acc}$. In total distance $dz = dz_{ab} + dz_{acc}$ we have

$$\frac{dp_x}{dz} = -p_x v^2 \frac{dE}{Edz} = -p_x/\lambda$$

where

$$\lambda = E\frac{dz_{ab}}{v^2 dE}\left(1 + \epsilon'_{ab}/\epsilon'_{ac}\right) = Ev^{-2}\left(\frac{1}{\epsilon'_{ab}} + \frac{1}{\epsilon'_{acc}}\right).$$

The ratio of the damping distance to mean decay distance for proper lifetime $\tau$ is

$$R = \frac{\lambda}{\gamma v \tau} = \frac{m_\mu v}{v^3 \tau}\left(\frac{1}{\epsilon'_{ab}} + \frac{1}{\epsilon'_{acc}}\right).$$

For $c\tau = 658$ m and $\epsilon'_{\rm acc} = 20$ MeV m$^{-1}$ and $\epsilon'_{\rm ab} = 30$ MeV m$^{-1}$ (liquid hydrogen) and $v \simeq c$, show that $R^{-1} \simeq 73$ so the muon lifetime does not limit use of this damping mechanism.
c) In the distance $dz$ the transverse momenta are rescaled by a common factor, $p_x \to (1-\alpha)p_x$ with $\alpha = dz/\lambda$. The variance of a distribution of transverse momentum changes by a cooling factor $(1-\alpha)^2 \simeq 1 - 2\alpha$ while multiple Coulomb scattering heats the transverse momentum distribution so

$$d < p_x^2 >= -2\frac{dz}{\lambda} < p_x^2 > + \left(\frac{E_s}{v}\right)^2 \frac{dz_{ab}}{X_0}$$

where $E_s \simeq 14$ MeV and $X_0$ is the radiation length. Show that the minimum value is

$$< p_x^2 >_{min} = E\frac{E_s^2}{2X_0\epsilon'_{ab}}.$$

CHAPTER

# 4 Quantum Physics

All elementary particles and interactions are governed by quantum mechanics. This chapter introduces perturbation theory in quantum mechanics with applications to scattering and multiparticle processes. It also provides a framework for understanding more sophisticated standard model interacting relativistic field theories.

## Contents



## 4.1 ■ SCHRÖDINGER'S EQUATION

The quantum theory of matter starts with Schrödinger's wave equation for a particle of mass $m$ subject to an external force described by a potential energy function $U(x)$:

$$i\frac{\partial}{\partial t}\psi(x) = \frac{(-i\nabla)^2}{2m}\psi(x) + U(x)\psi(x) \tag{4.1}$$

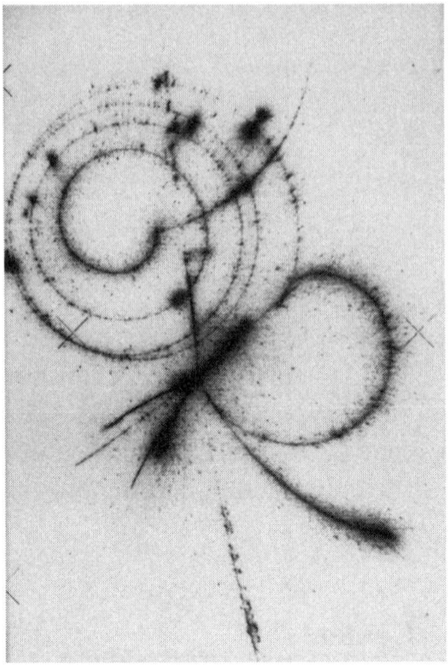

**FIGURE 4.1** Streamer chamber image of the decay chain $\pi^- \to \mu^- \bar{\nu}_\mu$ followed by $\mu^- \to e^- \nu_\mu \bar{\nu}_e$. The pion emerges from a hadronic collision, bends counterclockwise and decays. The muon spirals four times before decaying to the electron which exits the chamber. Understanding the nature of such small particles of matter led ultimately to the present standard model. [CERN Photo archive]

In this equation, $\psi(x)$ is a complex function of the space and time coordinates $x = (t, \mathbf{x})$ and $|\psi(x)|^2 d\mathbf{x}$ is the probability for finding the particle in a volume $d\mathbf{x}$ of space about coordinate vector $\mathbf{x}$ at time $t$. We have set $\hbar = 1$. When $U = 0$, solutions to Schrödinger's equation are superpositions of plane waves corresponding to momentum $\mathbf{p}$ and kinetic energy $E$

$$\psi(\mathbf{x}.t) = Ae^{-i(Et - \mathbf{p}\cdot\mathbf{x})} = Ae^{-ipx} \tag{4.2}$$

with $E = \mathbf{p}^2/(2m)$. A localized free particle is described by a superposition with flux density

$$\mathbf{j} = \frac{1}{2i}\left(\psi^\dagger \nabla \psi - \psi \nabla \psi^\dagger\right). \tag{4.3}$$

The Schrödinger equation for a charge $e$ interacting with an electromagnetic field is

$$i\frac{\partial}{\partial t}\psi = \left[\frac{(-i\nabla - e\mathbf{A})^2}{2m} + eA^0\right]\psi \tag{4.4}$$

where $A^0$ is the scalar potential and $\mathbf{A}$ is the vector potential.

The reader is assumed to be familiar with applications of the Schrödinger equation to describe nonrelativistic electrons and other stable particles in atoms, molecules, condensed matter, and perhaps atomic nuclei. The universal applicability of Schrödinger's equation to nonrelativistic matter suggests it is an appropriate starting point to develop a more general quantum framework.

## 4.2 ■ PERTURBATION THEORY

Analytic solutions to Schrödinger's equation exist for a few explicit forms of the potential energy function. Generally, for attractive potentials there exist a discrete spectrum of bound standing waves ($E < 0$) and a continuous spectrum of traveling scattering waves ($E > 0$). When an exact analytic solution may not be found, it may be possible to find a solution by successive approximation, treating a part of the problem as a perturbation to a solvable problem.

Let's write the Schrodinger equation as

$$i\frac{\partial}{\partial t}\psi = [H_0 + U]\,\psi \tag{4.5}$$

Here we have expressed the energy operator in terms of a part $H_0$ for which we know the solutions and a part $U$ the effect of which will be considered a small perturbation. In our application to single particle potential scattering, we will put $H_0 = -\nabla^2/(2m)$ for which plane waves constitute a complete set of solutions and let $U(\mathbf{x})$ be the entire potential energy function.

Suppose for simplicity that $U$ is constant in time. Expand a general solution $\psi$ in terms of energy eigenstates $\psi_m$ of $H_0$ as

$$\psi = \sum_m a_m \psi_m \tag{4.6}$$

with $\psi_m = \psi_m(\mathbf{x})e^{-iE_m t}$. Then multiplication by $\psi_n(\mathbf{x})e^{-iE_n t}$ and integration over $\mathbf{x}$ assuming the orthonormality condition

$$\int d\mathbf{x}\ \psi_n^\dagger(\mathbf{x})\psi_m(\mathbf{x}) = \delta_{nm} \tag{4.7}$$

leads to coupled equations for the amplitudes

$$\frac{da_n}{dt} = -i\sum_m a_m \int d\mathbf{x}\psi_n^\dagger U\psi_m = -i\sum a_m U_{nm} e^{iE_{nm}t}. \tag{4.8}$$

Here $E_{nm} = E_n - E_m$ and

$$U_{nm} = \int d\mathbf{x}\ \psi_n^\dagger(\mathbf{x})U\psi_m(\mathbf{x}) \tag{4.9}$$

with $\mathrm{d}\mathbf{x} = \mathrm{dx\,dy\,dz}$. What we have in mind here is an initial plane wave encountering a region of interaction (non-vanishing potential) resulting in a scattered wave which may be expressed as a superposition of plane waves.

Start with a pure state of index $i$ represented by initial amplitudes $a_m = \delta_{m,i}$. (We use the symbol $i$ to denote both $\sqrt{-1}$ and to denote the initial state index here.) With the amplitudes $a_m$ fixed on the right-hand-side as a first approximation, the amplitude $S_{fi}$ of a different state $f \neq i$ generated during a time interval $t$ follows by integration in first approximation and is

$$S_{fi} = -i \int dx\, \psi_f^{\dagger} U \psi_i = a_f = -i \int dt\, U_{fi} e^{iE_{fi}t} = -U_{fi} \frac{e^{iE_{fi}t} - 1}{E_{fi}} \tag{4.10}$$

where $dx = dt d\mathbf{x}$. If we write the differential equation for $a_f$ as

$$\frac{da_f}{dt} = -iU_{fi} e^{iE_{fi}t} a_i - i \sum_{n \neq i} U_{fn} e^{iE_{fn}t} a_n, \tag{4.11}$$

use the approximation $a_i = 1$ in the first term, and use the first approximation expression Equation 4.10 for the other amplitudes $a_n$, we find the equations for the amplitudes in the next approximation:

$$\frac{da_f}{dt} = -iU_{fi} e^{iE_{fi}t} - i \sum_{n \neq i} U_{fn} \frac{1}{E_{ni}} U_{ni} \left[ e^{iE_{fi}t} - e^{iE_{fn}t} \right]. \tag{4.12}$$

If the second term in brackets, which involves a jumble of frequencies, is dropped, integration over time yields

$$S_{fi} = a_f = +iM_{fi} F(E_{fi}) \tag{4.13}$$

with amplitude defined by

$$iM_{fi} = -i \left[ U_{fi} + \sum_{n \neq i} \frac{U_{fn} U_{ni}}{E_{in}} \right] \tag{4.14}$$

and an energy function

$$F(E) = \int dt\, e^{iEt} = \frac{e^{iEt} - 1}{iE}. \tag{4.15}$$

The amplitude $M_{fi}$ includes the two leading terms in an expansion where the first term represents the effect of the perturbing potential acting once, generating a transition from $i$ to $f$, and the second term is a sum over amplitudes for transitions from the initial state $i$ through intermediate states $n$ to the final state $f$, the perturbation acting twice. The series can be represented by the diagram shown in Figure 4.2.

The probability of a transition to a final state $f$ is

$$|S_{fi}|^2 = |M_{fi}|^2 |F(E_{fi})|^2 \tag{4.16}$$

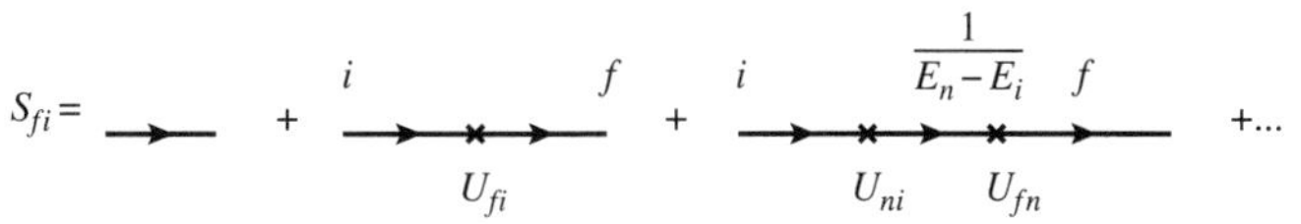


**FIGURE 4.2** General diagram describing perturbation theory. The amplitude $S_{fi}$ for a transition from an initial state $i$ to a final state $f$ generated by an interaction potential $U$ in perturbation theory is described by a series of terms. The first term from the left vanishes if $i$ and $f$ are orthogonal. The second term represents the effect of $U$ acting once. The second term represents transitions through intermediate states $n$ with $U$ acting twice.

where the square of the energy function is

$$|F(E)|^2 = t\frac{\sin^2(Et/2)}{t(E/2)^2} \simeq 2\pi\delta(E). \tag{4.17}$$

Here we observe that, for time intervals large compared to an energy level difference $E$, the factor $|F|^2$ is strongly peaked near $E = 0$ and approaches $2\pi t$ times a Dirac delta function enforcing energy conservation.

The Dirac $\delta$-function $\delta(x - a)$ is defined by the property

$$\int_{-\infty}^{+\infty} dx\ \delta(x - a) f(x) = f(a) \tag{4.18}$$

for integrable functions $f(x)$. It is sufficient that it has a peak about $x = a$, narrow compared to the scale of variation of the function $f(x)$, and itself have unit integral area. One representation of the Dirac $\delta$-function that we will use extensively is

$$\delta(x) = \frac{1}{2\pi}\int_{-\infty}^{+\infty} dk\ e^{ikx}. \tag{4.19}$$

Given the relations for a function $f(x)$ and its Fourier transform $f(k)$,

$$f(x) = \frac{1}{\sqrt{2\pi}}\int dk\ f(k)e^{-ikx}\ ;\ f(k) = \frac{1}{\sqrt{2\pi}}\int dx\ f(x)e^{+ikx}, \tag{4.20}$$

we can show that $\int dx\ \delta(x) f(x) = f(0)$. (The integrals implicitly cover the range $[-\infty, +\infty]$). In fact, if we insert the expression for $f(k)$ into the expression for $f(x)$ and interchange the order of integration, we find

$$f(x) = \frac{1}{\sqrt{2\pi}}\int dk\ f(k)e^{-ikx} = \frac{1}{2\pi}\int dkdx'\ e^{-ik(x-x')} f(x')$$

$$= \frac{1}{2\pi}\int dx'\ f(x')\int dk\ e^{-ik(x-x')} = \int dx'\ f(x')\delta(x - x') \tag{4.21}$$

where we identify the $\delta$-function representation above and observe it has the required property. The identification of $|F|^2$ with a delta function of energy times time is formally equivalent to writing

$$F = \int dt e^{iEt} = 2\pi\delta(E); \; |F|^2 = 2\pi\delta(E) \int dt e^{0t} = 2\pi\delta(E)t. \tag{4.22}$$

We now write the transition probability as

$$|S_{fi}|^2 = 2\pi |M_{fi}|^2 \delta(E_f - E_i)t. \tag{4.23}$$

The transition probability from $i$ to $f$ is proportional to time and the transition rate is defined as

$$\Gamma_{fi} = \frac{|S_{fi}|^2}{t} = 2\pi\delta(E_f - E_i)|M_{fi}|^2. \tag{4.24}$$

This formula is called Fermi's golden rule and applies to an initially pure state coupled to many final states. Feedback from the final states to the initial state was assumed incoherent and neglected. (When a finite number of nearly degenerate states are coupled by a perturbation, as is the case for meson and neutrino mixing, the feedback between these states must be treated explicitly.) Fermi's rule implies that the probability that a system is found in the initial state decays exponentially with time and that the total decay rate is the sum of decay rates to all final states. In our application to scattering with a continuous spectrum of plane waves, we will multiply the transition probability by a smooth density of final states and integrate over final energies absorbing the apparent singularity represented by the delta function. We will later construct a 4-dimensional version of this expression incorporating linear momentum conservation as well as energy conservation.

## 4.3 ■ POTENTIAL SCATTERING

We can use the preceding perturbation theory to calculate approximately the component amplitudes of a wave scattered by a region of fixed potential as illustrated in Figure 4.3. Suppose an initial plane wave

$$\psi_i = \frac{e^{-ip_i x}}{\sqrt{V}} \tag{4.25}$$

(representing one particle of momentum $p_i = (E_i, \mathbf{p}_i)$ in volume V) interacts with a static potential $U(\mathbf{x})$. A final state plane wave is

$$\psi_f = \frac{e^{-ip_f x}}{\sqrt{V}}. \tag{4.26}$$

Defining the 4-momentum transfer vector $q = p_f - p_i = (q^0, \mathbf{q})$, we can write

$$U_{fi} = \int d\mathbf{x}\, \psi_f^\dagger U(\mathbf{x})\psi_i = \frac{1}{V}\int d\mathbf{x}\, U(\mathbf{x})e^{-i\mathbf{q}\cdot\mathbf{x}} = \frac{1}{V}U(\mathbf{q}). \tag{4.27}$$

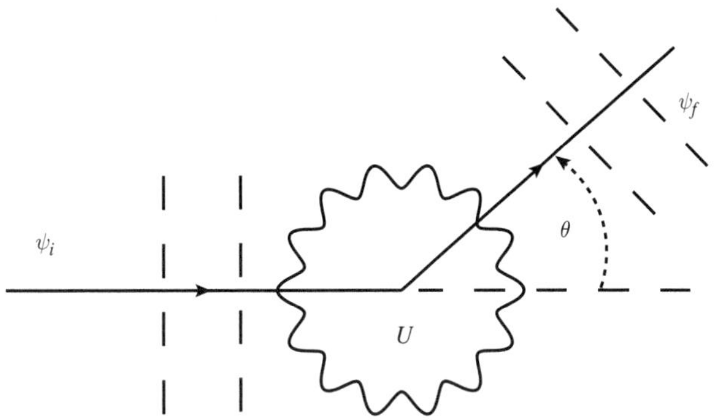


**FIGURE 4.3** Potential scattering in wave mechanics. A plane wave $\psi_i$ encounters a region of space where its potential energy function $U$ is nonvanishing. The interaction as described by wave mechanics results in a scattered wave with components $\psi_f$.

The quantity $U(\mathbf{q})$ here is the three dimensional Fourier transform of the potential function

$$U(\mathbf{q}) = \int d\mathbf{x}\, U(\mathbf{x}) e^{-i\mathbf{q}\cdot\mathbf{x}}. \tag{4.28}$$

We can regard $U(\mathbf{q})$ as the amount of plane wave of wave vector $\mathbf{q}$ in the potential function $U(\mathbf{x})$. If we were to write the potential function $U(\mathbf{x})$ as a sum of plane waves, and consider $U(\mathbf{x})\psi_i$ as a fixed source term in Schrödinger's equation, we would see that the appropriate $\mathbf{q}$ combines with $\mathbf{p}_i$ in the product $U\psi_i$ to source $\mathbf{p}_f = \mathbf{p}_i + \mathbf{q}$.

Using Fermi's golden rule, we next calculate the transition rate between plane wave states in the volume to be

$$\Gamma_{fi} = 2\pi \frac{|U(\mathbf{q})|^2}{V^2} \delta(q_0). \tag{4.29}$$

The waves in a large volume have a nearly continuous spectrum. The momenta of waves in a cube (for simplicity) of volume $V$ are of the form $\mathbf{p} = 2\pi V^{-1/3}\mathbf{n}$ with $\mathbf{n}$ a 3-vector of integers. So the number of momentum states $d\mathbf{n}_f$ with final momentum $\mathbf{p}_f$ in a range $d\mathbf{p}_f$ near $\mathbf{p}_f$ is

$$d\mathbf{n}_f = V \frac{d\mathbf{p}_f}{(2\pi)^3}. \tag{4.30}$$

Here $d\mathbf{n} = dn_x dn_y dn_z$. Multiplying the transition rate per state by $d\mathbf{n}_f$, we calculate the transition rate to states with $\mathbf{p}_f$ within the range $d\mathbf{p}_f$ to be

$$d\Gamma_{fi} = \Gamma_{fi} d\mathbf{n}_f = 2\pi \frac{|U(\mathbf{q})|^2}{V} \delta(q_0) \frac{d\mathbf{p}_f}{(2\pi)^3}. \tag{4.31}$$

Let's switch from Cartesian coordinates to spherical coordinates with the $z$-axis along the $\mathbf{p}_i$ direction and write the final momentum volume element as

$$d\mathbf{p}_f = \mathbf{p_f}^2 d|\mathbf{p_f}| d\Omega_f \tag{4.32}$$

where $d\Omega = \sin\theta\, d\theta\, d\phi = d\cos\theta\, d\phi$ is the differential solid angle subtended by $d\mathbf{p}_f$. (In writing for simplicity $\sin\theta d\theta = d\cos\theta$, we ignore a pesky minus sign.) Next, we switch integration variables from $|\mathbf{p}_f|$ to energy $E_f$ to integrate over the $\delta$-function and make the replacements

$$\delta(q_0) d\mathbf{p}_f = \delta(E_f - E_i)|\mathbf{p}_f|^2 \frac{d|\mathbf{p}_f|}{dE_f} dE_f d\Omega = |\mathbf{p}_f| E_f d\Omega = v_f E_f^2 d\Omega \tag{4.33}$$

where the relativistic form of the energy $E_f = \sqrt{\mathbf{p}_f^2 + m^2} = |\mathbf{p_f}|/v_f$ was used to write

$$\frac{d|\mathbf{p}_f|}{dE_f} = \frac{E_f}{|\mathbf{p}_f|} = \frac{1}{v_f}. \tag{4.34}$$

Energy conservation here implies $E_f = E_i \equiv E$, $|\mathbf{p}_f| = |\mathbf{p}_i| \equiv \mathbf{p}$, and $v_f = v_i \equiv v$. Using these expressions and dividing the transition rate density by the incident particle flux

$$j_i = \frac{v_i}{V} = \frac{|\mathbf{p}_i|}{E_i V}, \tag{4.35}$$

we define the differential cross section $d\sigma$ for scattering into angular range $d\Omega$ to be

$$d\sigma_{fi} = \frac{\Gamma_{fi} d\mathbf{n}_f}{j_i} = \frac{1}{(2\pi)^2} E^2 \frac{v_f}{v_i} |U(\mathbf{q})|^2 d\Omega = \frac{1}{2\pi} E^2 \,|U(\mathbf{q})|^2\, d\cos\theta \tag{4.36}$$

where the last equality applies if $U$ is independent of azimuthal angle $\phi$ so we may integrate over $d\phi$. The initial and final energy is fixed in this expression and the momentum transfer $\mathbf{q}$ is implicitly a function of scattering angle. Note that the dimensions of $U(\mathbf{q})$ are energy times volume and if we divide the right-hand-side by a factor of $(\hbar c)^4$, we obtain an area.

As an example, we find the differential cross section for elastic scattering for the Yukawa potential energy function,

$$U = \frac{g^2}{4\pi} \frac{e^{-mr}}{r}. \tag{4.37}$$

As we have seen in Chapter 1, such a potential energy function is associated with a field corresponding to a massive particle. With $g^2 = Ze^2$ and $m = 0$, the Yukawa potential energy represents the electrostatic energy of a charge $+e$ interacting with a nucleus of charge $Ze$. With $m \simeq Z^{\frac{1}{3}}/(1.4a_0)$, the potential energy models the electrostatic interaction of a charge $+e$ with the screened nuclear charge (assumed fixed) of a neutral atom. The Fourier transform of the Yukawa potential may be found using spherical coordinates and is

$$U(\mathbf{q}) = \int r^2 dr(2\pi) d\cos\theta \, \frac{g^2}{4\pi} \frac{e^{-mr}}{r} e^{-i|\mathbf{q}|r\cos\theta} = \frac{g^2}{\mathbf{q}^2 + m^2}. \tag{4.38}$$

Since $|\mathbf{p}_f| = |\mathbf{p}_i| \equiv |\mathbf{p}|$ and $\mathbf{q}^2 = (\mathbf{p}_f - \mathbf{p}_i)^2 = |\mathbf{p}|^2(1 - \cos\theta)$, we can write

$$d\mathbf{q}^2 = 2|\mathbf{p}|^2 d\cos\theta \tag{4.39}$$

and express the differential cross section as

$$\frac{d\sigma}{d\mathbf{q}^2} = \frac{1}{4\pi v^2} \frac{g^4}{(\mathbf{q}^2 + m^2)^2}. \tag{4.40}$$

For Coulomb scattering of an electron from a fixed point charge, we set $m = 0$ and $g^2 = Ze^2 = 4\pi\alpha Z$. We can also write the magnitude of the momentum transfer as

$$\mathbf{q}^2 = 2\mathbf{p}^2(1 - \cos\theta) = 4\mathbf{p}^2 \sin^2\frac{\theta}{2} \tag{4.41}$$

and use Equation 4.39 to express the cross section in terms of the scattering angle

$$\frac{d\sigma_R}{d\Omega} = \frac{Z^2\alpha^2}{4\mathbf{p}^2 v^2 \sin^4(\theta/2)}. \tag{4.42}$$

In the nonrelativistic limit, $|\mathbf{p}| = mv$ and this so-called Born approximation formula reproduces the Rutherford formula (see Example 2.9) deduced by analyzing all classical hyperbolic trajectories for all impact parameters. Rutherford used classical physics to understand the distribution of alpha particles scattered from nuclei thereby establishing the essential picture of an atom as a tiny massive nucleus surrounded by a cloud of electrons. In fact, he deduced an upper bound on the size of

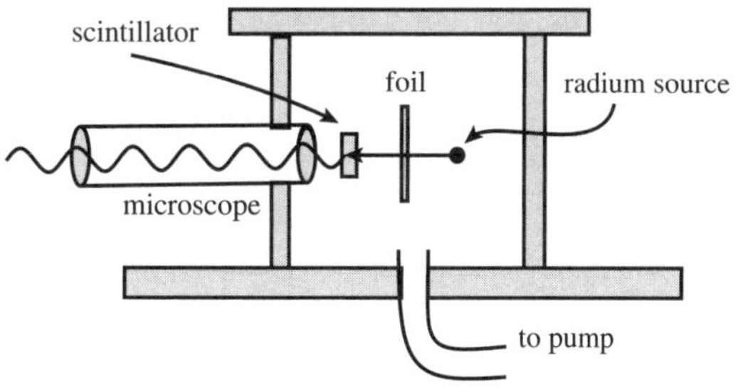


**FIGURE 4.4** Rutherford scattering apparatus. Alpha particles from a radium source in an evacuated container passed through a foil. A zinc-sulphide scintillator attached to a rotatable microscope served to detect deflected alpha particles. [after Dr. H. Geiger and E. Marsden, Philosophical Magazine Series 6, Volume 25, Number 148, April 1913.].

the nucleus from the highest momentum transfer collisions which, with sufficient energy, can reach the interior of a nucleus. H. Geiger and E. Marsden performed the measurements. A schematic of the apparatus is shown in Figure 4.4. It features all of the elements of a modern day particle physics experiment in miniature.

## 4.4 ■ THE STRUCTURE OF OBJECTS

Rutherford used the measured angular distribution of scattered particles to deduce the form of the potential within an atom. Another method would be the study of the energies of bound states which sample the potential in various ways and compare to the energy levels expected for various potentials, the quantum analog of how Newton used Kepler's Laws to deduce the form of the universal law of gravitation. While still cumbersome, you may agree that analysis of scattering is more direct and we next explore other applications of this method.

As we have seen, the cross section for elastic scattering from a single point source of potential energy reflects the strength and shape of the potential energy function $U$. Suppose the potential energy is a superposition of potential energy functions associated with some static spatial distribution of similar sources. In the first (linear) approximation in perturbation theory, the scattering amplitude is a coherent superposition of scattering amplitudes as illustrated in Figure 4.5. The total scattering amplitude from a spatial distribution of the sources follows from the replacements

$$U_{tot}(\mathbf{x}) = \sum_j g_j U(\mathbf{x} - \mathbf{x}_j) \rightarrow U_{tot}(\mathbf{q}) = U(\mathbf{q}) f(\mathbf{q}) \tag{4.43}$$

where $U(\mathbf{x})$ is the potential energy for a unit source charge and the structure factor

$$f(\mathbf{q}) = \sum_j g_j e^{-i\mathbf{q}\cdot\mathbf{x}_j} \tag{4.44}$$

characterizes the source distribution. The factors $g_i$ allow for different source charges. Correspondingly, the cross section $d\sigma$ for scattering from a distribution

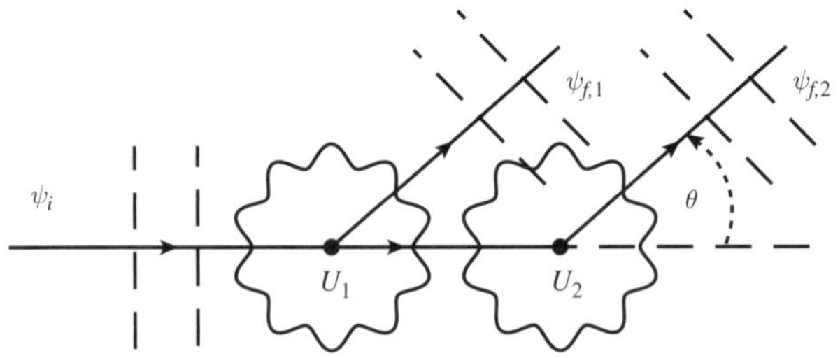


**FIGURE 4.5** Wave scattering as a probe of structure. Interference in the superposition of waves scattered by two sources of potential energy reflects the spatial distribution of the sources.

of sources is the cross section $d\sigma_0$ for a single source multiplied by the square of the structure factor

$$d\sigma = |f|^2 d\sigma_0. \tag{4.45}$$

A structure factor characterizes wave scattering in the linear approximation quite generally and may be familiar from the theory of x-ray scattering from crystals. When an incoming wave scatters from a distribution of scattering centers, the waves emerging from the scattering centers are coherent. The interference of the scattered waves at some point in space depends on the relative phases which in turn depend on the location of the scattering centers. For a continuous source density $\rho(\mathbf{x})$, the generalization of the structure factor is called a form factor and is proportional to the Fourier transform of the source density:

$$F(\mathbf{q}^2) = \int d\mathbf{x}\, \rho(\mathbf{x}) e^{-i\mathbf{q}\cdot\mathbf{x}}. \tag{4.46}$$

---

**Example 4.1. Form factor for a spherical charge distribution.** Expand the form factor of a general spherically symmetric charge distribution in powers of momentum transfer and interpret the first two nonvanishing terms.

For a spherically symmetric source, expansion in powers of $\mathbf{q} \cdot \mathbf{x}$ gives

$$F \simeq \int d\mathbf{x}\, \rho(\mathbf{x}) \left[ 1 - i\mathbf{q} \cdot \mathbf{x} - \frac{1}{2} (\mathbf{q} \cdot \mathbf{x})^2 + \cdots \right] \tag{4.47}$$

The first term is the total charge $Q = \int d\mathbf{x}\, \rho$. The second term is proportional to the first (dipole) moment and vanishes for a spherical charge distribution. The third term after integration over angles is $\mathbf{q}^2 < g\mathbf{x} >^2 /6$ where

$$< g\mathbf{x}^2 > = \int d\mathbf{x}\, \rho \mathbf{x}^2 \tag{4.48}$$

is the second moment of the source distribution. At low momentum transfer, the scattering is sensitive to the net charge because all the scattered waves interfere constructively as the phase differences are small. At high momentum transfer, the scattering from randomly placed constituents usually encountered in particle physics is incoherent so the form factor generally decreases with $|\mathbf{q}|$.

---

High energy particles have a short de Broglie wavelength and high momentum transfer collisions are used to resolve structure on small scales. To resolve a feature of size $L$ requires momentum transfer

$$|\mathbf{q}| \sim \frac{1}{L} = \frac{197 \text{ MeV/c}}{L/1 \text{ fm}}. \tag{4.49}$$

Elastic scattering from fixed nuclei of electrons with energies of order 1 GeV reveals that the electric charge of ground state nuclei is distributed with constant density over a sphere of radius

$$R \simeq (1.18 \text{ fm})\, A^{\frac{1}{3}} - 0.48 \text{ fm} \tag{4.50}$$

where $A$ is the mass number. More precisely, the radial dependence of the nuclear charge density can be described by a Fermi function

$$\rho(r) = \frac{\rho(0)}{1 + e^{(r-R)/a}}. \tag{4.51}$$

The density drops to half the central value $\rho(0)$ at $r = R$. The thickness of the surface is characterized by $a \simeq 0.54$ fm. Precise measurements reveal ripples in the density that can be understood as quantum mechanical bound state shell effects. Slight differences between the proton and neutron density distributions are also observed.

The Born approximation description of electromagnetic and weak elastic scattering of leptons from nuclei at moderate energies is successful when little energy is deposited in the nucleus. In general, probes which resolve the structure of a nucleus have energy sufficient to interact through resonant excitations and inelastic processes. The scattering of hadrons from nuclei is dominated by absorption through a wide range of channels. The effect of absorption on elastic scattering of strongly interacting particles may be approximately incorporated into potential scattering theory via the addition of an imaginary part to the potential energy function. In the simplest incarnation of such a model, the imaginary part dominates and scattering amplitudes calculated in such a model reduces to black sphere diffraction amplitudes governed by Huygens' principle perhaps familiar from classical optics.

According to Huygens' principle, the propagation of a wave can be represented by considering each small patch of area on a wave front as an isotropic wave source of strength proportional to the incident wave amplitude. The coherent superposition of such waves reproduces free wave propagation. (The method of Green's functions can be used to justify this principle.) If a wave $\psi = \psi_0 e^{-i(Et-pz)}$ encounters a circular aperture of radius $R$ in the plane $z = 0$, the wave beyond the aperture is given by an integral over the aperture area

$$\psi(t, \mathbf{x})|_{z>0} = \psi_0 \frac{e^{-i(Et-p|\mathbf{x}|)}}{|\mathbf{x}|} 2\pi R^2 \frac{J_1(pR\theta)}{pR\theta} \tag{4.52}$$

where $\theta = |\mathbf{x_t}|//|\mathbf{x}|$ is the scattering angle and $J_1(x)$ is a Bessel function. The quantity $J(x)/x$ equals 1/2 at $x = 0$ and has a first zero at $x = 3.83$.

When a plane wave encounters an absorbing disk of radius $R$, the wave that would have been generated by virtual sources within the aperture is deleted. Equivalently, the effect of the absorption is the addition to the incident wave of the negative of the wave generated by an open aperture. Hence the intensity of the scattered wave resulting from black sphere absorption is that of an aperture as given above. At a given beam energy, the angle determines the (transverse) momentum transfer $|\mathbf{q}|$. The scattering amplitude is proportional to the two dimensional Fourier transform of a constant density disk and the first minimum occurs when $|\mathbf{q}|R \simeq 3.83$.

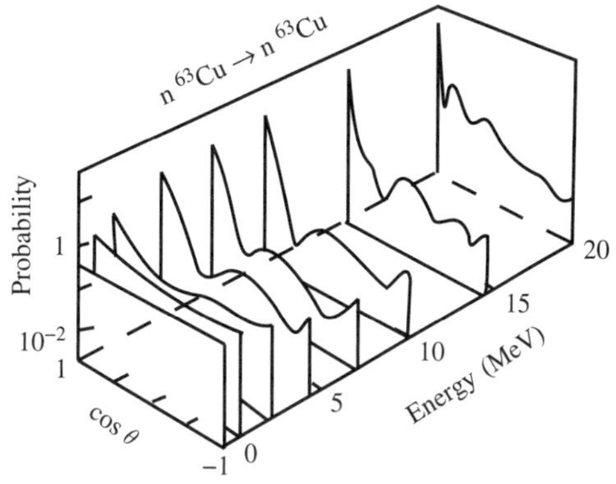


**FIGURE 4.6** Neutron scattering. The distribution of scattering angle $\theta$ of neutrons scattered from copper nuclei varies with energy. At low energy, the neutron wavelength is large compared to the nuclear radius and the scattered neutrons are isotropic. At energies of a few MeV, diffraction is evident. The neutron scattering data is from t2.lnal.gov.

The differential cross section for elastic scattering of neutrons from a medium sized nucleus as a function of angle and energy is shown in Figure 4.6 and illustrates nuclear diffraction. In *pp* elastic scattering, diffraction adds to the Coulomb scattering amplitude and leads to a dip in the differential cross section at momentum transfer $|\mathbf{q}| \sim 1/r_p$ where $r_p \sim 0.8$ fm. The diffraction pattern in *pp* and *pn* scattering reveals that nucleons have no sharp edge and detailed analysis is required to interpret the result.

The differential cross sections for electron, muon, and neutrino elastic scattering from hydrogen and deuterium nuclei are used to incisively study the charge distributions within individual nucleons. Relativistic energies are required and the cross section formulae are complicated by relativistic effects. The cross section for scattering of a relativistic spin 1/2 electron from a fixed spin 1/2 proton by single photon exchange is given by the Rosenbluth expression

$$\frac{d\sigma_{e^-p\to e^-p}}{d\Omega} = \frac{d\sigma_0}{d\Omega}\left[\left(G_E^2 + x\left(1+\kappa_p\right)^2 G_M^2\right)/(1+x) + 2x\left(1+\kappa_p\right)^2 G_M \tan^2(\theta/2)\right] \tag{4.53}$$

where $x = q^2/4m_p^2$ with $q^2 \equiv -Q^2$ the 4-momentum transfer, $\kappa_p$ is the anomalous magnetic moment of the proton, and

$$\frac{d\sigma_0}{d\Omega} = \frac{d\sigma_R}{d\Omega}\frac{\cos^2(\theta/2)}{\left[1+(2E/M)\sin^2(\theta/2)\right]}. \tag{4.54}$$

The differential cross section $d\sigma_0$ is the Rutherford differential cross section $d\sigma_R$ multiplied by an angular factor resulting from spin and divided by a factor associated with proton recoil. In Chapters 6 and 10 we will see how such formulas may be derived.

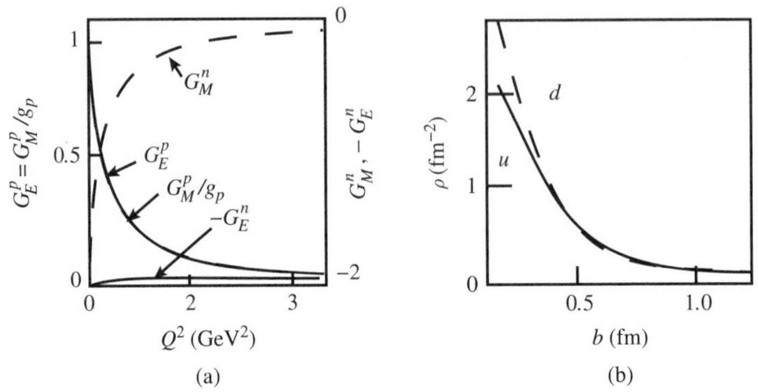


**FIGURE 4.7** Nucleon form factors and quark distributions. (a) The electric and magnetic form factors derived from elastic electron proton scattering as functions of 4-momentum transfer squared $Q^2$. As $Q^2$ approaches zero, $G_E$ approaches the nucleon charge and $G_M$ approaches the $g$-factor of the nucleon's magnetic moment. [Ref: B. Pasquini and S. Boffi, Phys. Rev. **D76**, 074011 (2007)] (b) The $u$-quark and $d$-quark transverse charge densities in the proton deduced from nucleon form factors. [Ref: Gerald A. Miller, Phys. Rev. Lett **99**, 112001 (2007)]

The two form factors $G_E$ and $G_M$ (often called $F_c$ and $F_m$) describe, loosely speaking, the charge and magnetic moment density distributions within the proton. The electric and magnetic form factors for both proton and neutron are shown in Figure 4.7 (a). By comparing the cross section for scattering from a proton to the cross section for scattering from a neutron, the form factors can be interpreted in term of $u$-quark and $d$-quark densities in the transverse direction shown in Figure 4.7 (b). It is found that, in contrast to heavy nuclei and as observed in diffraction scattering, the charge density within a nucleon is not uniform. The proton and neutron mean squared charge radii are given by

$$< q\mathbf{x} >_p^2 = (0.8768(69) \text{ fm})^2, \quad < q\mathbf{x} >_n^2 = -0.1161(22) \text{ fm}^2 \tag{4.55}$$

and in magnitude are comparable to the range of the strong interaction. As suggested by the charge radius squared for the neutron, it is also found that within a nucleon the $u$-quark and $d$-quark have slightly different distributions. The general shape of the distributions can be understood in the bag model of nucleons described in Chapters 9 and 10. Light quarks are relativistic and approximately confined to a sphere of radius 1 fm. The ground state wave functions peak near the origin.

Quarks and leptons are said to be structureless. The precise meaning of such a statement will become more clear when their interactions are described but, roughly, it means relativistic scattering models assuming canonical point-like interactions without intrinsic form factors suffice for their description at present. However in relativistic interactions, matter and energy are transformed, clouding the notion of a point particle. For example, as the center of mass energy is increased in photon electron collisions, the electron's electromagnetic field energy at short range is increasingly available in the form of virtual particles like $e^+e^-$ pairs. The electron appears as a multiparticle system described by calculable form factors. The description of interactions of point-like particles with hadrons demands additional form factors which characterize their process dependent structure.

## 4.5 ■ QUANTUM THEORY OF RADIATION

The perturbation theory we have applied to potential scattering is readily extended to time dependent perturbations. Let's start with the formula for the transition amplitude written in the form:

$$a_f = -i \int dx\, \psi_f^\dagger U(t, \mathbf{x}) \psi_i. \tag{4.56}$$

If the perturbation, which may be complex, has a harmonic component,

$$U(t, \mathbf{x}) = U(\mathbf{x}) e^{-i(\pm\omega)t}, \tag{4.57}$$

then the generalized Fermi golden rule for the transition rate is

$$\Gamma_{fi} = \frac{|S_{fi}|^2}{t} = 2\pi\delta(E_f - E_i \pm \omega)|M_{fi}|^2. \tag{4.58}$$

Positive and negative frequency components $\pm\omega$ correspond to absorption of energy and emission of energy respectively.

Now consider the description of an electron interacting with an electromagnetic field. A nonrelativistic electron is described by the Pauli equation

$$i\frac{\partial}{\partial t}\psi = \left[\frac{(-i\nabla - e\mathbf{A})^2}{2m} - \frac{e}{2m}\boldsymbol{\sigma} \cdot \nabla \times \mathbf{A} + eA^0\right]\psi \tag{4.59}$$

The electromagnetic field may be considered to have two parts: a fixed field $A_0^\mu$ due perhaps to a nucleus which includes a Coulomb potential and the magnetic field of the nucleus, and an external field $A^\mu$. Then we can write

$$i\frac{\partial}{\partial t}\psi = [H_0 + U]\psi \tag{4.60}$$

where $H_0$ is the Hamiltonian for an unperturbed atomic electron and the interaction with the external field is described by the potential energy operator

$$U = eA^0 - i\frac{e}{2m}\left[(\nabla \cdot \mathbf{A}) + 2\mathbf{A} \cdot \nabla - \frac{e}{2m}\boldsymbol{\sigma} \cdot \nabla \times \mathbf{A} + \frac{e^2}{2m}\mathbf{A}^2\right] \tag{4.61}$$

An example of a time dependent external field is that associated with a passing charge. By expressing the field in terms of its frequency components, we could use our extended golden rule to calculate the probability that an initial state electron

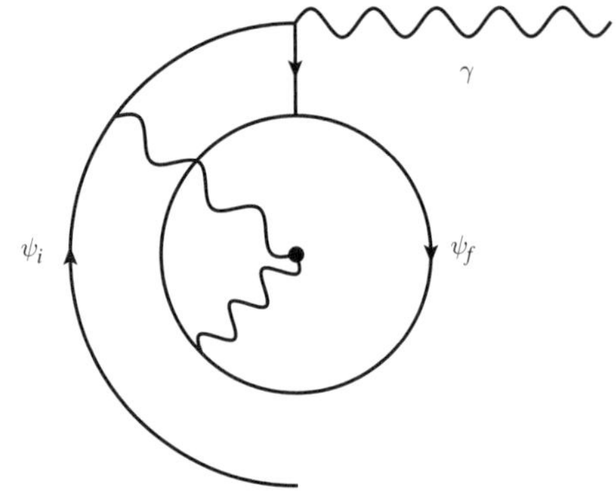


**FIGURE 4.8** Radiation from atoms. Diagram illustrating the emission of radiation when an atomic electron transitions between bound states $\psi_i$ and $\psi_f$. The internal photon lines represent the binding of the electron to the nucleus.

bound in an atom is induced to transition to different atomic states or into free particle states as a result of the passage of the external charge. By such means, detailed formulae for the energy loss of the passing particle may be developed. Here we focus on external radiation fields associated with free photons and with the emission of radiation illustrated in Figure 4.8.

A radiation field in a general gauge is described by a potential $A^\mu = (A^0, \mathbf{A})$ satisfying the free field Maxwell's equations. For a single photon of 4-momentum $p = (k, \mathbf{k})$, the field is proportional to

$$A^\mu = \epsilon^\mu e^{-i(k^0 t - \mathbf{k}\cdot\mathbf{x})} \tag{4.62}$$

where $\epsilon^\mu$ is a constant vector. When this $A^\mu$ is inserted into the perturbation operator $U$, the time dependence $e^{-ik^0 t}$ corresponds to emission of energy $k^0$ by the electron. The complex conjugate photon wave function with time dependence $e^{ik^0 t}$ corresponds to absorption of energy $k^0$. The time independent matrix element of the first term in the perturbation for emission is

$$M = -\int d\mathbf{x}\, \psi_f^\dagger e A^0 \psi_i = -e \int d\mathbf{x}\, \psi_f^\dagger \epsilon^0 e^{i\mathbf{k}\cdot\mathbf{x}} \psi_i = -e\epsilon^0 \rho_{fi}(\mathbf{k}) \tag{4.63}$$

with transition charge density

$$\rho_{fi}(\mathbf{k}) = \int d\mathbf{x}\, \psi_f^\dagger e^{i\mathbf{k}\cdot\mathbf{x}} \psi_i. \tag{4.64}$$

This matrix element is proportional to the coupling $e$ and to the components of the wave functions of the initial and final particles. We may interpret it as a part of the amplitude for emission of the photon while the electron makes a transition from the initial state to a final state. The matrix element with the complex conjugate photon wave function is an amplitude for absorption of the photon while the electron makes the inverse transition. Generally the terms linear in $A^\mu$ contribute to single photon emission and absorption in first order perturbation theory and to two photon processes in second order perturbation theory while the term quadratic in $\mathbf{A}$ is proportional to $e^{\pm 2ik^0 t}$ and contributes to two-photon processes.

In developing this idea, it is convenient to use the Coulomb gauge with $A^0 = 0$ and $\nabla \cdot \mathbf{A} = 0$ which permits us to express the radiation field in terms of $\mathbf{A}$ alone. A field corresponding to energy $\hbar\omega$ in volume V is

$$\mathbf{A} = \frac{1}{\sqrt{2\omega V}} \boldsymbol{\epsilon} e^{-i(\omega t - \mathbf{k}\cdot\mathbf{x})} = \mathbf{A}(\mathbf{x}) e^{-i\omega t} \tag{4.65}$$

where $\boldsymbol{\epsilon}$ is a polarization unit vector satisfying $\boldsymbol{\epsilon}^\dagger \cdot \boldsymbol{\epsilon} = 1$ and $\mathbf{k} \cdot \boldsymbol{\epsilon} = 0$, and the entire photon wave function has been normalized so that the time average electromagnetic energy in the volume V is

$$E = < \int d\mathbf{x}\, \frac{1}{2} \left(\mathbf{E}^2 + \mathbf{B}^2\right) >= \int d\mathbf{x}\; |\mathbf{E}|^2 = \int d\mathbf{x} \left| \frac{\partial \mathbf{A}}{\partial t} \right|^2 = \omega \tag{4.66}$$

consistent with Planck's relation $E = h\nu$ with $\hbar = 1$. We note that this normalization is just that for scalar boson de Broglie waves.

In Coulomb gauge, the matrix element describing a transition between nonrelativistic electron states $\psi_i \to \psi_f$ is

$$M = -U_{if} = -\int d\mathbf{x}\; \psi_f^\dagger(\mathbf{x}) \left[\mathbf{A} \cdot (e\mathbf{v}) - \frac{e}{2m} \boldsymbol{\sigma} \cdot \nabla \times \mathbf{A}\right] \psi_i(\mathbf{x}) \tag{4.67}$$

where we have set $A^0 = 0$, used the condition $\nabla \cdot \mathbf{A} = 0$, dropped the quadratic term, and written $\mathbf{v} = -i\nabla/m$. Note that the first term does not involve the electron spin while the second term does. For electron bound states in an atom of dimension $a_0$, we can make the estimate $|\nabla\psi| \sim |\psi/a_0|$ and, if $\lambda$ is the photon wavelength, assume $|\nabla \times \mathbf{A}| \sim |\mathbf{A}|/\lambda$. It follows that the spin changing term in the matrix element is roughly $a_0/\lambda$ smaller than the spin conserving term and, when $\lambda$ is large compared to the size of the atom, it may be neglected. In this approximation, spin is conserved in radiative transitions. The matrix element then reduces to a form which may be derived from the Schrödinger equation, namely

$$M_{fi} = -\int d\mathbf{x}\; \psi_f^\dagger(\mathbf{x})\, [\mathbf{A} \cdot (e\mathbf{v})]\, \psi_i = -\frac{1}{\sqrt{2\omega V}} \boldsymbol{\epsilon} \cdot \mathbf{j}_{fi}(\mathbf{k}) \tag{4.68}$$

with the dependence on the electronic state transition encapsulated in the quantity

$$\mathbf{j}_{fi}(\mathbf{k}) = \int d\mathbf{x}\; e^{i\mathbf{k}\cdot\mathbf{x}} \psi_f^\dagger(\mathbf{x})(e\mathbf{v})\psi_i(\mathbf{x}). \tag{4.69}$$

The quantity $\mathbf{j}_{fi}$ is seen to be the Fourier transform of a transition current, a sort of form factor vector. Given the initial and final wave functions, the transition current may be calculated and the matrix element for emission of a photon with a given photon polarization $\boldsymbol{\epsilon}$ and momentum $\mathbf{k}$ then found.

A photon corresponding to an optical frequency or lower has wavelength large compared to the scale of atomic electron wave functions. In atomic physics, it is therefore useful to expand the exponential $e^{i\mathbf{k}\cdot\mathbf{x}}$ since $\mathbf{k}\cdot\mathbf{x} << 1$ giving

$$\mathbf{j}_{fi}(\mathbf{k}) = \int d\mathbf{x}\ \left[1 + i\mathbf{k}\cdot\mathbf{x} - (1/2)\,(\mathbf{k}\cdot\mathbf{x})^2 + ...\right] \psi_f^{\dagger}(\mathbf{x})(e\mathbf{v})\psi_i(\mathbf{x}). \qquad (4.70)$$

The leading term produces a so-called electric dipole transition. If only this term is retained, the properties of the spherical part of the wave functions may be invoked to demonstrate that electric dipole transitions occur only between atomic states which differ in orbital angular momentum by one unit. When electric dipole transitions are forbidden, the next to leading term may still effect a so-called magnetic dipole transition but with an amplitude roughly a factor $kx \sim a_0/\lambda$ smaller and with a transition rate smaller by a factor $(a_0/\lambda)^2$. The spin dependent interaction term is of that order and should be considered in that case. The electric dipole approximation is of less utility in calculating radiation from hadronic bound states for which $\mathbf{k}\cdot\mathbf{x}$ is not small.

In calculating transition matrix elements in the electric dipole approximation, the interaction is conventionally simplified still further. The vector potential is set equal to zero at the location of the interaction and assumed to be constant in space over the radiating system. Instead of $e\mathbf{A}\cdot\mathbf{p}$, a scalar gauge invariant interaction

$$U = e\mathbf{E}\cdot\mathbf{r} \qquad (4.71)$$

is used with

$$\mathbf{E} = -\mathbf{i}\omega\boldsymbol{\epsilon}/\sqrt{\mathbf{2}\omega}. \qquad (4.72)$$

Since $(H_0\mathbf{r} - \mathbf{r}H_0)\psi = \mathbf{v}\psi$, the matrix element of the velocity operator is $\mathbf{v}_{fi} = (E_f - E_i)\mathbf{r}_{fi}$, and for first order transitions, $\omega = E_f - E_i$, so the two forms are equivalent. But for higher order processes, they are not.

In Fermi's golden rule above, we see the positive frequency corresponds to absorption of energy and the negative frequency to emission. The transition rate is proportional to a delta function representing energy conservation. In absorption processes, the transition rate may be integrated over the energy distribution of the initial photons with the finite width in energy of the initial and final electron states considered, and then divided by the photon flux to derive the absorption cross section. In the case of emission, the differential decay rate may be found by starting with the transition rate from an initial electron state $\psi_i$ to a final state $\psi_f$ plus a photon of fixed polarization $\boldsymbol{\epsilon}$ and momentum $\mathbf{k}$. The transition rate for final state photons with wave vectors in a range $d\mathbf{k}$ follows by multiplying by the corresponding number of photon states

$$d\mathbf{n}_\gamma = \frac{d\mathbf{k}}{(2\pi)^3 V}. \qquad (4.73)$$

The volume $V$ cancels out so it can be set equal to 1 and we have

$$\begin{aligned} d\Gamma_{i\to f+\gamma} &= 2\pi \left|M_{fi}\right|^2 \delta\left(E_f + k - E_i\right) d\mathbf{n}_\gamma \\ &= \frac{\left|M_{fi}\right|^2}{(2\pi)^2} \delta\left(E_f + k - E_i\right) k^2 dk\, d\Omega_{\mathbf{k}} \end{aligned} \tag{4.74}$$

where the photon momentum was expressed in spherical coordinates. We can then integrate over the $\delta$-function and set the photon energy to be $k = E_f - E_i$ and we find the differential decay rate for photon emission into a solid angle $d\Omega_{\mathbf{k}}$ is

$$\frac{d\Gamma_{i\to f+\gamma}}{d\Omega_{\mathbf{k}}} = (2\pi)^{-2} k^2 \left|M_{fi}\right|^2 = \frac{k}{8\pi^2} \left|\boldsymbol{\epsilon}\cdot\mathbf{j}_{fi}(\mathbf{k})\right|^2 . \tag{4.75}$$

The angular dependence of the decay rate is governed by the projection of the photon polarization vector onto the Fourier component of the transition current density.

The theory presented here is useful in describing radiative transitions in atomic, nuclear, and quark bound states and is illustrated in two worked examples. In what follows, we generalize the quantum theory of radiation to standard model processes and particles in a simplistic fashion and consider the implications of relativistic kinematics and antimatter.

---

**Example 4.2. Electric dipole radiation in hydrogen.** Calculate the lifetime for the radiative decay of the 2p electronic states of hydrogen with magnetic quantum number $\pm 1$ to the ground state

$$\psi_{2p,m=\pm1} = \frac{\mp 1}{8\left(\pi a_0^3\right)^{\frac{1}{2}}} \frac{[x \pm iy]}{a_0} e^{-r/(2a_0)} \to \psi_{1s} = \frac{1}{\left(\pi a_0^3\right)^{\frac{1}{2}}} e^{-r/a_0} \tag{4.76}$$

using the leading term in the expansion of the photon wave function.

We start by calculating the core transition current density for the upward transition as this is easier. Integration by parts shows that $\mathbf{j}_{fi} = \mathbf{j}_{if}^\dagger$. Inserting the wave functions, we have $\nabla\psi_{1s} = -a_0^{-1}(\nabla r)\psi_{1s} = -a_0^{-1}(\mathbf{r}/r)\psi_{1s}$ and

$$\begin{aligned} \mathbf{j}_{1s\to 2p}(\mathbf{k}) &= -i\frac{e}{m}\int d\mathbf{x}\, \psi_{2p}^\dagger(\mathbf{x})\nabla\psi_{1s}(\mathbf{x}) \\ &= -i\frac{e}{m}\frac{\pm 1}{8\pi a_0^5}\int d\mathbf{x}\, (x \mp iy)\left[\frac{x\hat{x} + y\hat{y} + z\hat{z}}{r}\right] e^{-3r/(2a_0)} \\ &= -i\frac{e}{m}\frac{\pm 1}{8\pi a_0^5}\int d\mathbf{x} \left[\frac{x^2\hat{x} \mp iy^2\hat{y}}{r}\right] e^{-3r/(2a_0)} \end{aligned} \tag{4.77}$$

where symmetry was used to drop terms with an integrand odd under reflection. We can next calculate

$$\int d\mathbf{x}\, \frac{x^2}{r} e^{-3r/(2a_0)} = \frac{1}{3}\int d\mathbf{x}\, \frac{x^2+y^2+z^2}{r} e^{-3r/(2a_0)}$$
$$= \frac{1}{3}\int r^3 dr d\Omega\, e^{-3r/(2a_0)} = 8\pi\left(\frac{2a_0}{3}\right)^4 \tag{4.78}$$

and arrive at the transition current density $\mathbf{j}_{2p\to 1s}(\mathbf{k}) = ie(m_e a_0)^{-1}(2/3)^4(\hat{x} \pm i\hat{y})$ and the differential decay rate

$$\frac{d\Gamma_{2p(\mathrm{H})\to 1s(\mathrm{H}),\gamma}}{d\Omega_{\mathbf{k}}} = \frac{k}{8\pi^2}\left|\boldsymbol{\epsilon}\cdot\mathbf{j}_{2p\to 1s}\right|^2 = \frac{\alpha k}{2\pi\,(m_e a_0)^2}\left(\frac{2}{3}\right)^8 \left|\boldsymbol{\epsilon}^\dagger\cdot(\hat{x}\pm i\hat{y})\right|^2. \tag{4.79}$$

Here $\boldsymbol{\epsilon}$ is a polarization vector perpendicular to $\mathbf{k}$. As we integrate over all directions of $\mathbf{k}$, $\boldsymbol{\epsilon}$ varies over all directions so, to find the decay rate integrated over all directions, we use spherical coordinates and the integral

$$\int d\Omega_{\mathbf{n}}\, \left|\mathbf{n}\cdot\hat{x} \pm i\mathbf{n}\cdot\hat{y}\right|^2 = \int d\cos\theta d\phi\, \left|\sin\theta\cos\phi \pm i\sin\theta\sin\phi\right|^2 = \frac{8\pi}{3}. \tag{4.80}$$

The decay rate is independent of spin state as it must be by rotational invariance. Using $a_0 = 1/m_e\alpha$, $k = E_{2p} - E_{1s} = (3/8)m_e\alpha^2$, and $\alpha = e^2/(4\pi)$, and multiplying by two to include the other polarization, we find

$$\Gamma^{tot}_{2p(\mathrm{H})\to 1s(\mathrm{H}),\gamma} = (2/3)^8\alpha^5 m_e = 0.627\times 10^9\ s^{-1} = 1/(1.6\ \mathrm{ns}). \tag{4.81}$$

---

**Example 4.3. Magnetic dipole radiation.** In an electron synchrotron, the electron magnetic moment interacts with the guiding magnetic field. Radiative decay from the higher energy spin state to the lower energy state results in transverse polarization of an initially unpolarized beam. Calculate the radiative lifetime in the electron rest frame and deduce the polarization time in the laboratory frame.

A relativistic electron in a uniform magnetic field $\mathbf{B}$ has orbital frequency $\omega = evB/p = \omega_c/\gamma$ where $\omega_c = eB/m$ is the cyclotron frequency. The magnetic field $\mathbf{B}'$ in the instantaneous rest frame of the electron is related to that in the laboratory frame by $\mathbf{B}' = \gamma\mathbf{B}$. The magnetic interaction energy operator is

$$U = -\boldsymbol{\mu}\cdot\mathbf{B} = \frac{e}{2m}\boldsymbol{\sigma}\cdot\nabla\times\mathbf{A}.$$

In the rest frame, the interaction with the static field produces two energy levels separated by $\omega_{12} = \frac{eB'}{m} = \gamma^2\omega$. The same magnetic interaction operator with $\mathbf{A} = (2k)^{-1/2}\boldsymbol{\epsilon} e^{i\mathbf{k}\cdot\mathbf{x}}$ accounts for radiation of a photon with polarization vector $\boldsymbol{\epsilon}$ and wave vector $\mathbf{k}$ per unit volume. The transition matrix element is

$$U_{fi} = \int d\mathbf{x}\, \psi^\dagger \frac{-e}{2m}\boldsymbol{\sigma}\cdot\left(-i\mathbf{k}\times\frac{\boldsymbol{\epsilon}^\dagger}{\sqrt{2k}}\right)e^{-i\mathbf{k}\cdot\mathbf{x}}\psi_i \simeq ik\frac{\sqrt{4\pi\alpha}}{\sqrt{2k}}\boldsymbol{\sigma}_{fi}\cdot\boldsymbol{\epsilon}' \tag{4.82}$$

where $\boldsymbol{\epsilon}' = \mathbf{k}\times\boldsymbol{\epsilon}/k$, $e^{i\mathbf{k}\cdot\mathbf{r}} \simeq 1$ and $\int d\mathbf{x}\, \psi^\dagger\psi = 1$. For magnetic field along a direction $\hat{\mathbf{z}}$, the transition matrix element is proportional to

$$\boldsymbol{\sigma}_{fi} = <\downarrow|\boldsymbol{\sigma}|\uparrow>= \hat{\mathbf{x}} - i\hat{\mathbf{y}} \tag{4.83}$$

and hence $U_{fi} \propto \sin\theta$ where $\theta$ is the polar angle of $\boldsymbol{\epsilon}'$. From Fermi's golden rule, the differential decay rate is

$$d\Gamma_{i\to f,\gamma} = |U_{fi}|^2 2\pi\delta(k-\omega_{12})\frac{k^2 dk}{(2\pi)^3}d\Omega. \tag{4.84}$$

After integration, including a factor of two for the two photon polarization states, the total transition rate in the rest frame $\Gamma^{rest}_{i\to f,\gamma}$ and the lifetime in the lab frame $\tau$ are found to be

$$\Gamma^{rest}_{i\to f,\gamma} = \frac{2}{3}\frac{\alpha}{m^2}\gamma^6\omega^3,\quad \tau^{-1} = \frac{2}{3}\frac{\alpha\gamma^5}{m^2\rho^3} \tag{4.85}$$

where the transition rate in the lab is reduced by one power of $\gamma$ by time dilation and the storage ring radius is $\rho = v/\omega$. Acceleration results in decoherence of the emission. Transitions to and from the many different orbital states near some central emitting state must be considered. A precise calculation using the Dirac equation yields $P(t) = P_0(1 - e^{-t/\tau})$ where

$$\tau^{-1} = \frac{5\sqrt{3}}{8}\frac{\alpha\gamma^5}{m^2\rho^3}$$

and $P_0 = 8/5\sqrt{3} = .9238$. [J.D. Jackson, Rev. Mod. Phys. **48**, 417 (1957)]. The spin precession rate $\Omega = [1 + \gamma(\frac{g}{2} - 1)]\omega$ is comparable to the orbital frequency.

## 4.6 ■ GENERALIZED RADIATION THEORY

The weak and color gauge interactions are similar to the electromagnetic interaction in the standard model. Let's look briefly ahead to the complete relativistic theory. A relativistic free fermion is described by the Dirac equation

$$i\frac{\partial}{\partial t}\psi_L - i\boldsymbol{\sigma}\cdot\nabla\psi_L = E\psi_L + \boldsymbol{\sigma}\cdot\mathbf{p}\psi_L = m\psi_R \tag{4.86}$$

$$i\frac{\partial}{\partial t}\psi_R + i\boldsymbol{\sigma}\cdot\nabla\psi_R = E\psi_R - \boldsymbol{\sigma}\cdot\mathbf{p}\psi_R = m\psi_L. \tag{4.87}$$

With the substitution $(E, \mathbf{p}) \to (E - eA^0, \mathbf{p} - e\mathbf{A})$, and writing $\psi_L$ as $e_L$ for an electron, we find the interaction of a left-handed fermion with an electromagnetic field is represented by

$$\begin{aligned} i\frac{\partial}{\partial t}e_L &= i\boldsymbol{\sigma}\cdot\nabla e_L + m\psi_R + e\left[A^0 + \boldsymbol{\sigma}\cdot\mathbf{A}\right]e_L \\ &= H_0 e_L + m e_R + U e_L \end{aligned} \tag{4.88}$$

where $H_0 = i\boldsymbol{\sigma}\cdot\nabla$ and the perturbation is identified as

$$U = e\phi_A \equiv e\left[A^0 + \boldsymbol{\sigma}\cdot\mathbf{A}\right]. \tag{4.89}$$

A similar equation applies to $\psi_R$. In the ultrarelativistic limit, the mass term may be neglected and the $L$ and $R$ fields behave as independent particles. The Hamiltonian structure of the equation is then similar to the Pauli equation but the spin independent and spin dependent interaction terms are of the same magnitude.

Denote the wave function for a left-handed electron by $e_L$ and a left-handed neutrino wave function by $\nu_L$. Introduce a real vector field $Z^\mu$ for the $Z$ boson, complex fields $W^\mu$ and $W^{\mu\dagger}$ for the $W^-$ and $W^+$, and color fields $g^\mu_{a\bar{b}}$ corresponding to gluons. Then an interaction of a left-handed electron with the neutral weak field is represented by

$$i\frac{\partial}{\partial t}e_L = i\boldsymbol{\sigma}\cdot\nabla e_L + m e_R + g_{Zee}[Z + \boldsymbol{\sigma}\cdot\mathbf{Z}]e_L \tag{4.90}$$

where $g_{Zee}$ is the weak charge. The charged current weak interaction is described by equations of a similar form:

$$i\frac{\partial}{\partial t}e_L = i\boldsymbol{\sigma}\cdot\nabla e_L + m e_R + g_{We\nu}\left[W^0 + \boldsymbol{\sigma}\cdot\mathbf{W}\right]\nu_L. \tag{4.91}$$

In the standard model, the charged current weak coupling applies to all three left-handed lepton pairs $e^-\nu_e, \mu^-\nu_\mu, \tau^-\nu_\tau$ and has the value

$$g_W \equiv \frac{g}{\sqrt{2}} = \frac{e}{\sqrt{2}\sin\theta_W} \simeq \sqrt{2}e$$

where $\theta_W$ is called the Weinberg angle and $\sin^2\theta_W \simeq 0.22$. For the $u$-quark and $d$-quark, the coupling is $g_{Wdu} = g_W V_{du}$ where $V_{du}$ is an element of the CKM matrix.

The quantum theory of radiation is readily generalized. The transition amplitude for absorption of a quantum of a field $V$ has the form

$$S_{iV\to f} = -i\int dx\ \psi_f^\dagger U \psi_i = -i\int dx\ \psi_f^\dagger g_V \phi_V \psi_i \tag{4.92}$$

where $g_V$ is the coupling constant and

$$\phi_V = V^0 \pm \boldsymbol{\sigma} \cdot \mathbf{V} \tag{4.93}$$

where the plus and minus sign apply to $L$ and $R$ fermions. The factor $\psi_i$ represents the initial state of the fermion and the factor $\psi_f^\dagger$ represents the final state of the fermion. The quantity $\phi_V$ contains the boson wave function $V^\mu$ sewn up in a two-by-two matrix. In emission, the boson wave function is replaced by $V^{\mu\dagger}$. Antiparticle wave functions have opposite phase and emerging antiparticles appear as entering particles. For example, the transition amplitude for a $\nu_e$ to absorb a $W^-$ and become an $e^-$ is

$$S_{\nu_e e^- \to W^-} = -i \int dx\; e_L^\dagger g_W \phi_W \nu_L \tag{4.94}$$

with $\phi_W = W^0 + \boldsymbol{\sigma} \cdot \mathbf{W}$. In pursuing this reasoning in this chapter, we will often make a simplification and consider only the time component of the vector fields and neglect the spins of the particles, in effect treating fermions as Lorentz scalars. This will allow us to focus on the generic aspects of relativistic transition rate calculations with minimal mathematical burden. In later chapters we will explore the ramifications of spin.

We now turn to the calculation of the decay rate for the process

$$W^- \to e^- \bar{\nu}_e \tag{4.95}$$

illustrated in Figure 4.9. We describe each particle by a plane wave in a 4-volume $VT$, writing

$$e_L = \phi_e \frac{e^{-ip_e x}}{\sqrt{V}}, \quad \bar{\nu}_L = \phi_\nu \frac{e^{ip_{\bar{\nu}} x}}{\sqrt{V}}, \quad W^\mu = \epsilon^\mu \frac{e^{-ip_W x}}{\sqrt{2E_W V}} \tag{4.96}$$

where $\phi_e$, $\phi_\nu$ and $\epsilon^\mu$ are constant spin wave functions. The fermion wave function normalization $V^{-1/2}$ corresponds to one particle per unit volume appropriate to the

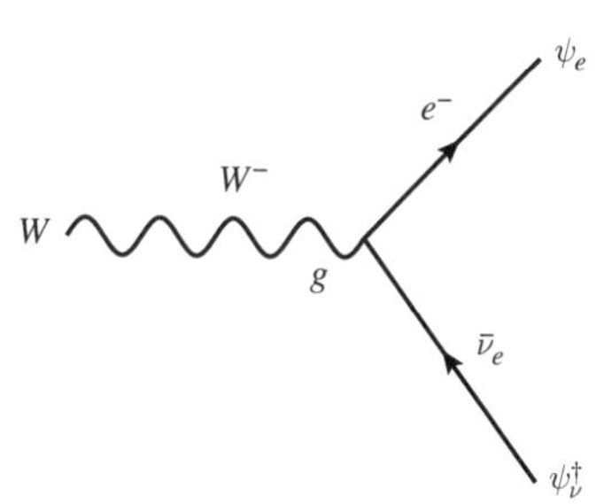


**FIGURE 4.9** Diagram for the decay $W^- \to e^- \bar{\nu}_e$. The diagram shows the factors which appear in the decay amplitude associated with the particles in a simple model which neglects spin polarization.

Schrödinger equation first order in time. (This normalization is also appropriate for the Dirac or Pauli equation.) The boson normalization $(2E_W V)^{-1/2}$ generalizes that used for photons and corresponds to one $W^-$ per unit volume. This different normalization emerges from the energy density associated with boson fields governed by wave equations which are second order in time. Notice that this normalization gives the boson wave function dimensions of inverse length like the Coulomb potential. We write the transition amplitude as

$$S_{W^-\to e^-\bar{\nu}_e} = -i \int dx\, e_L^\dagger g_W \phi_W \bar{\nu}_L = i M_{W^-\to e^-\bar{\nu}_e} I_p \tag{4.97}$$

where the essence of the interaction is in the matrix element

$$i M_{W^-\to e^-\bar{\nu}_e} = \frac{-i g_W}{\sqrt{2E_W}} \phi_e^\dagger \left[\epsilon^0 + \boldsymbol{\sigma}\cdot\boldsymbol{\epsilon}\right] \phi_{\bar{\nu}} \tag{4.98}$$

and

$$I_p \equiv \int dt d\mathbf{x} \left(\frac{e^{i p_e x}}{\sqrt{V}}\right)^\dagger \left(\frac{e^{i p_{\bar{\nu}} x}}{\sqrt{V}}\right)^\dagger \left(\frac{e^{i p_W x}}{\sqrt{V}}\right). \tag{4.99}$$

For large $T$ and $V$ we can convert the integrals of the exponential functions to a product of Dirac delta functions which enforce momentum and energy conservation and write

$$I_p = V^{-3/2}(2\pi)^4 \delta(p_e + p_{\bar{\nu}} - p_W) \tag{4.100}$$

where we define the 4-dimensional $\delta$-function as

$$\delta(p) = \delta(E)\delta(p_x)\delta(p_y)\delta(p_z). \tag{4.101}$$

The transition probability is $|S_{W^-\to e^-\bar{\nu}_e}|^2$. Let's generalize our result $\delta^2(E) = \delta(E)T$ in the derivation of Fermi's golden rule and interpret the square of the product of delta functions as

$$\delta^2(p) = (2\pi)^{-4} \int dt d\mathbf{x}\, e^{0x} \delta(p) = (2\pi)^{-4} T V. \tag{4.102}$$

Next, divide $|S_{W^-\to e^-\bar{\nu}_e}|^2$ by $T$ and multiply by a final state density factor

$$d\mathbf{n}_e = V \frac{d\mathbf{p}_e}{(2\pi)^3} \tag{4.103}$$

and by an additional factor

$$d\mathbf{n}_\nu = V \frac{d\mathbf{p}_{\bar{\nu}}}{(2\pi)^3}, \tag{4.104}$$

one for each final particle. We arrive at an expression for the decay rate to states with final particle momentum vectors $\mathbf{p}_e$ and $\mathbf{p}_{\bar{\nu}}$ in differential ranges in the form

$$d\Gamma_{W^-\to e^-\bar{\nu}_e} = \frac{\left|S_{W^-\to e^-\bar{\nu}_e}\right|^2}{T} V \frac{d\mathbf{p}_e}{(2\pi)^3} V \frac{d\mathbf{p}_{\bar{\nu}}}{(2\pi)^3} = \left|M_{W^-\to e^-\bar{\nu}_e}\right|^2 dPS. \quad (4.105)$$

Here the differential phase space is

$$dPS = (2\pi)^4 \delta(p_f - p_i) \frac{d\mathbf{p}_e}{(2\pi)^3} \frac{d\mathbf{p}_{\bar{\nu}}}{(2\pi)^3}. \quad (4.106)$$

Upon integration over $d\mathbf{p}_{\bar{\nu}}$, this reduces to

$$dPS = 2\pi \delta(E_W - E) \frac{d\mathbf{p}}{(2\pi)^3} \quad (4.107)$$

where, in the center of mass, $\mathbf{p} \equiv \mathbf{p}_e = -\mathbf{p}_{\bar{\nu}}$, $E = E_e + E_{\bar{\nu}}$, and the fermion energies are determined by the decay momentum: $E_e = \sqrt{\mathbf{p}^2 + m_e^2}$ and $E_{\bar{\nu}} = |\mathbf{p}|$.

We can simplify further by switching to spherical coordinates and writing $d\mathbf{p} = |\mathbf{p}|^2 d|\mathbf{p}| d\Omega$. Integration over $d\Omega$ produces

$$\begin{aligned} d\Gamma_{W^-\to e^-\bar{\nu}_e} &= \int d\Omega \left|M_{W^-\to e^-\bar{\nu}_e}\right|^2 \frac{|\mathbf{p}|^2 d|\mathbf{p}|}{(2\pi)^2} \\ &= 4\pi < \left|M_{W^-\to e^-\bar{\nu}_e}\right|^2 > \frac{|\mathbf{p}|^2\, d\,|\mathbf{p}|}{(2\pi)^2} \end{aligned} \quad (4.108)$$

where the angle average of $|M|^2$ is

$$< \left|M_{W^-\to e^-\bar{\nu}_e}\right|^2 > = \frac{1}{4\pi} \int d\Omega \left|M_{W^-\to e^-\bar{\nu}_e}\right|^2. \quad (4.109)$$

Next, the change of integration variables from $|\mathbf{p}|$ to $E$ using the expression $d|\mathbf{p}| = dE(d|\mathbf{p}|/dE)$ permits changing the integral over $d|\mathbf{p}|$ to an integral over $dE$ and elimination of the energy delta-function so we arrive at the form

$$\Gamma_{W^-\to e^-\bar{\nu}_e} = < |M_{W^-\to e^-\bar{\nu}_e}|^2 > \frac{1}{\pi} |\mathbf{p}|^2 \frac{d|\mathbf{p}|}{dE} = < |M_{W^-\to e^-\bar{\nu}_e}|^2 > PS. \quad (4.110)$$

Since $E = E_e + E_{\bar{\nu}}$ and $dE/d|\mathbf{p}| = v_e + v_{\bar{\nu}}$, the total 2-body phase space introduced here is

$$PS = \frac{1}{\pi} \frac{|\mathbf{p}|^2}{v_e + v_{\bar{\nu}}}. \quad (4.111)$$

If we choose to ignore the spin degrees of freedom with the simple approximation

$$M_{W^-\to e^-\bar{\nu}_e} = -\frac{g_W}{\sqrt{2E_W}}, \quad (4.112)$$

and neglect fermion masses, we have in the $W$ boson rest frame $E = 2|\mathbf{p}| = m_W$ and we estimate the proper decay rate to be

$$\Gamma_{W^- \to e^- \bar{\nu}_e} \simeq (2\pi) \frac{g_W^2}{2m_W} \frac{4\pi}{(2\pi)^3} \frac{m_W^2}{4} \frac{1}{2} = \frac{g_W^2}{16\pi} m_W . \qquad (4.113)$$

In hindsight, an expression of this form might have been guessed based on the physical idea of a heavy $W$ boson field. With $\hbar = c = 1$, a decay rate must have dimension of mass and, neglecting fermion masses, the $W$ boson mass is the only relevant parameter provided the weak charge is dimensionless, an assumption implicit in the preceding derivation. The further guess that the weak charge $g_W$ is comparable to the electromagnetic charge $e$ and universal would lead us to a more quantitative estimate of the $W$ boson lifetime including its decays to other lepton and quark pairs.

In the standard model, the $W$ boson wave function is proportional to a polarization 4-vector $\epsilon = (\epsilon_0, \epsilon)$ and the matrix element is

$$M = -\frac{g_W}{\sqrt{2E_W}} \left( \epsilon_0 \phi_e^\dagger \phi_{\bar{\nu}} + \boldsymbol{\epsilon} \cdot \phi_e^\dagger \boldsymbol{\sigma} \phi_{\bar{\nu}} \right) . \qquad (4.114)$$

The time component $\epsilon^0$ actually vanishes in the $W$ boson rest frame and the interaction energy density derives from the other terms. The leptonic current density $\mathbf{j} = \phi_e^\dagger \boldsymbol{\sigma} \phi_{\bar{\nu}}$ can be calculated to be a vector along the $\bar{\nu}$ direction. If the angle between the $W$ boson spin direction and $\mathbf{p}_{\bar{\nu}}$ is $\theta$, the projection of the $W$ boson spin on the lepton spinor current introduces a factor $\sin\theta$ into the matrix element and reduces the polar angle integral in the transition rate calculation by 2/3. The weak coupling in the standard model is $g_W = e/(\sqrt{2} \sin\theta_W) \simeq \sqrt{2}e$. The decay rate is

$$\begin{aligned} \Gamma_{W^- \to e^- \bar{\nu}} &= \frac{1}{\sin^2\theta_W} \frac{e^2}{4\pi} \frac{m_W}{12} \\ &= 4.5 \frac{1}{137} \frac{80 \text{ GeV}}{12} \frac{\left(c = 3 \times 10^{23} \text{ fm s}^{-1}\right)}{(\hbar c = .197 \text{ GeV fm})} \\ &= 3.4 \times 10^{23} \text{ s}^{-1} . \end{aligned} \qquad (4.115)$$

The weak charge $g_W$ is the same for all standard model fermions and the $W^-$ can decay to three lepton families and two quark families (in three colors) so the total decay rate $\Gamma_W$ is a factor of nine larger.

Real $W^-$ and $W^+$ bosons are produced pairs in $e^+e^-$ collisions as the decay products of virtual $Z$ boson and $\gamma$ bosons, in $p\bar{p}$ colliders in pairs as the decay products of virtual $Z$ boson and $\gamma$ bosons created by fusion of $u\bar{u}$ and $d\bar{d}$, singly by fusion of $d\bar{u}$ and $u\bar{d}$ quarks, and in the decay of top quarks. The $W$ bosons are most easily observed by leptonic decays such as $W^- \to e^- \bar{\nu}_e$ and $W^- \to \mu^- \bar{\nu}_\mu$. The decay length $c\tau_W = \hbar c / \Gamma_W$ is too short to directly observe. Its value is inferred from the width of the mass distribution of $W$ boson decay products and is consistent with a calculation which includes all possible $W$ boson decay modes.

**Example 4.4. Decay rate for $Z \to e^+e^-$.** Assume the $Z_0$ field is coupled to the electron with charge $g^L_{Zee} = -0.68e$ and $g^R_{Zee} = -0.53e$. Estimate the decay rate $\Gamma_{Z\to e^+e^-}$ in MeV and compare to the experimental value $\Gamma_{\mathrm{Z}\to\mathrm{e}^+\mathrm{e}^-} = \mathrm{BR}_{\mathrm{Z}\to\mathrm{e}^+\mathrm{e}^-}\Gamma^{\mathrm{tot}}_{\mathrm{Z}}$ $= (3.36\%)(2.49\ \mathrm{GeV}) = 83\ \mathrm{MeV}$.

The simple minded calculation for $W$ boson decay with $g_W = e$ gives

$$\Gamma_{Z\to e^+e^-} = \alpha m_Z/4 = (1/137)(91.2\ \mathrm{GeV})/4 = 166\ \mathrm{MeV}. \tag{4.116}$$

We should multiply by 2/3 for spin and include both L and R handed electrons by adding a factor $(g^L_{Zee}/e)^2 + (g^R_{Zee}/e)^2 \simeq 3/4$ which brings the naive estimate in line with experiment.

## 4.7 ■ INTERMEDIATE STATES AND PROPAGATOR

The process $\mu^- \to e^-\nu_\mu\bar{\nu}_e$ proceeds though intermediate states containing a virtual $W$ boson and can be used to illustrate the application of second order perturbation theory and, in even our simplified model, many features of relativistic processes. The symmetry between equations for matter and antimatter requires that we consider two coherent time ordered processes: a) $\mu^- \to \nu_\mu W^-$ followed by $W^- \to e^-\bar{\nu}_e$ and b) the spontaneous creation of $W^+e^-\bar{\nu}_e$ near a $\mu^-$ followed by $\mu^-W^+ \to \nu_\mu$. The amplitudes are represented by the two diagrams in Figure 4.10.

Take case a). The matrix element leading to an intermediate state containing a virtual $W$ boson of momentum $\mathbf{p}_W$ and energy $E_W = \sqrt{\mathbf{p}_W^2 + m_W^2}$ is

$$\phi_{ni} = \int d\mathbf{x}\, \psi_\nu^\dagger g_W \phi_W^\dagger \phi_\mu = g_W \frac{1}{\sqrt{2E_W}}(2\pi)^3\delta(\mathbf{p}_W + \mathbf{p}_\nu - \mathbf{p}_\mu) \tag{4.117}$$

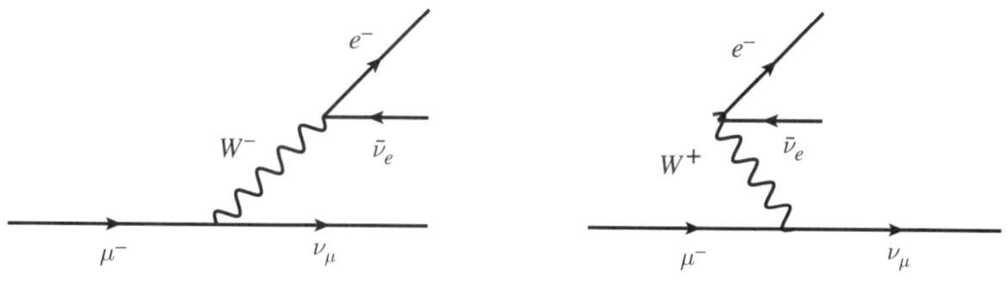


**FIGURE 4.10** Muon decay in second order Hamiltonian perturbation theory. The left diagram describes a transition path with an intermediate state containing $\nu_\mu W^-$. The right diagram describes a transition through an intermediate state containing $\mu^-W^+\bar{\nu}_e e^-$. Both amplitudes contribute to the transition $\mu^- \to \nu_\mu\bar{\nu}_e e^-$. The contribution of the $W$ boson to the sum is represented by the relativistic propagator.

where we have ignored spin and set the volume $V$ equal to one. (It cancels out in the end.) The matrix element from the intermediate state to the final state is

$$\phi_{fn} = \int d\mathbf{x}\, \psi_e^\dagger g_W \phi_W \psi_{\bar{\nu}}^\dagger = \frac{g_W}{\sqrt{2E_W}}(2\pi)^3 \delta(\mathbf{p}_W + \mathbf{p}_{\bar{\nu}} - \mathbf{p}_e). \tag{4.118}$$

The intermediate states are continuous and have density $d\mathbf{n}_W = d\mathbf{p}_W/(2\pi)^3$ so, with summation converted to integration, our second order perturbation theory formula Equation 4.14 takes the form

$$\begin{aligned} S_{\mu^- \to \nu_\mu \bar{\nu}_e e^-} &= -i \int d\mathbf{p}_W \frac{1}{(2\pi)^3} \phi_{fn} \frac{1}{E_\mu - E_\nu - E_W} \phi_{ni} 2\pi \delta(E_f - E_i) \\ &= -i \frac{g_W^2}{2E_W} \frac{1}{E_\mu - E_\nu - E_W} (2\pi)^4 \delta(p_f - p_i). \end{aligned} \tag{4.119}$$

Four momentum is ultimately conserved. Including case b) leads to a net amplitude

$$M_{\mu^- \to \nu_\mu \bar{\nu}_e e^-} = -\left( \frac{M_{W^- \to e^- \bar{\nu}} M_{\mu^- \to \nu \mu}}{E_\mu - E_a} + \frac{M_{\bar{\nu} \leftarrow \bar{W} \mu} M_{0 \to W^+ e^- \bar{\nu}}}{E_\mu - E_b} \right). \tag{4.120}$$

The energy of the $W$ boson in either case is given by $E_W^2 = m_W^2 + (\mathbf{p}_\mu - \mathbf{p}_\nu)^2$. The energies in the intermediate states are $E_a = E_W + E_\nu$ and $E_b = 2E_\mu + E_W - E_\nu$ where $E_\nu$ is the energy of the muon neutrino and we used energy conservation. When the matrix elements are all assumed to have the value $-g_W/\sqrt{2E_W}$, the energy differences in the denominators of the two amplitudes may be combined and expressed as

$$\frac{1}{E_\mu - E_a} + \frac{1}{E_\mu - E_b} = \frac{2E_W}{(E_\mu - E_\nu)^2 - E_W^2} = \frac{2E_W}{p_W^2 - m_W^2} \tag{4.121}$$

where $p_W^2 = (E_\mu - E_\nu)^2 - (\mathbf{p}_\mu - \mathbf{p}_\nu)^2$. The numerator $2E_W$ in this expression is cancelled by $W$ boson wave function normalization factors $(2E_W)^{-1/2}$ in the matrix elements $M$ and we arrive at the Lorentz invariant form

$$iM_{\mu^- \to \nu_\mu \bar{\nu}_e e^-} = (-ig_W)^2 \frac{i}{p_W^2 - m_W^2}. \tag{4.122}$$

Here $iM$ has been expressed in terms of a factor of $-ig_W$ associated with each vertex and a Lorentz invariant factor called a propagator

$$G(p_W) = \frac{i}{p_W^2 - m_W^2}. \tag{4.123}$$

The propagator arises from the sum over intermediate states with time ordering including both the emission and absorption of a virtual particle. As noted in

Chapter 1, a single Feynman diagram is conventionally used to denote the net amplitude summed over time ordering.

The propagator implies a resonant enhancement in the amplitude for $p_W^2 \sim m_W^2$ and appears singular. A higher order calculation in which the intermediate state $W$ boson is transformed to and from all of the particles to which it couples would eliminate the singularity. It turns out that the higher order effects can be approximately represented by the replacement

$$m^2 \to m^2 - im\Gamma \tag{4.124}$$

in the denominator of the lowest order propagator with $\Gamma$ the total decay rate or natural width leading to the expression

$$G(p_W) = \frac{i}{p_W^2 - m_W^2 - im\Gamma_W}. \tag{4.125}$$

The invariant mass $\sqrt{p_W^2} = \sqrt{(p_e + p_\nu)^2}$ of the virtual $W$ boson is determined by 4-momentum conservation and is generally not $m_W$. In the case of muon decay, its value is at most $m_\mu$ so $G(p_w)$ in this decay is effectively a constant. This reflects the fact that the interaction is short range.

## 4.8 ■ PHASE SPACE AND LIFETIMES

To calculate the decay rate of the muon, we consider next the density of final states. As in the previous calculation of 2-body decays of the $W$ boson, the number of wave states corresponding to a particle in the final state of momentum $\mathbf{p}$ in a range $\mathbf{dp}$ in a unit volume of space is $dn = d\mathbf{p}/(2\pi)^3$. The density of states or phase space for a general 3-body decay in the center of mass is

$$\begin{aligned} dPS &= (2\pi)^4 \delta\left(m - E_1(|\mathbf{p}_1|) - E_2(|\mathbf{p}_2|) - E_3(|\mathbf{p}_3)|\right) \\ &\quad \times \delta(\mathbf{p}_3 + \mathbf{p}_2 + \mathbf{p}_1) \frac{d\mathbf{p}_1}{(2\pi)^3} \frac{d\mathbf{p}_2}{(2\pi)^3} \frac{d\mathbf{p}_3}{(2\pi)^3} \\ &= 2\pi \delta\left(m - E_1\left(|\mathbf{p}_1|\right) - E_2\left(|\mathbf{p}_2|\right) - E_3\left(|\mathbf{p}_3)\,|\right)\right) \frac{d\mathbf{p}_1}{(2\pi)^3} \frac{d\mathbf{p}_2}{(2\pi)^3} \end{aligned} \tag{4.126}$$

where $m$ is the muon mass, $E_i = \sqrt{\mathbf{p}_i^2 + m_i^2}$, $\mathbf{p}_3 = -(\mathbf{p}_1 + \mathbf{p}_2)$, and we integrated over $d\mathbf{p}_3$. This expression may be further simplified. In spherical coordinates,

$$d\mathbf{p}_1 = \mathbf{p}_1^2 d|\mathbf{p}_1| d\Omega_1 = \mathbf{p}_1^2 d|\mathbf{p}_1| d\phi_1 d\cos\theta_1 \tag{4.127}$$

so integration over the angles $\theta_1$ and $\phi_1$ of particle 1 gives $d\Omega_1 \to 4\pi$. Now write $\mathbf{p}_2$ in spherical coordinates relative to the direction of particle 1 so

$$d\mathbf{p}_2 = \mathbf{p}_2^2 d|\mathbf{p}_2| d\Omega_{12} = \mathbf{p}_2^2 d|\mathbf{p}_2| d\phi_{12} d\cos\theta_{12} \tag{4.128}$$

and write $E_3$ as

$$E_3 = \sqrt{\mathbf{p}_3^2 + m_3^2} = \sqrt{\mathbf{p}_1^2 + \mathbf{p}_2^2 + 2|\mathbf{p}_1||\mathbf{p}_2| \cos\theta_{12} + m_3^2}. \quad (4.129)$$

Integration over angles $\phi_{12}$ and $\theta_{12}$ gives

$$dPS = \frac{1}{4\pi^3}\mathbf{p}_1^2 d\,|\mathbf{p}_1|\,\mathbf{p}_2^2 d\,|\mathbf{p}_2| \int_{-1}^{+1} d\cos\theta_{12}\,\delta\left(m - E_1 - E_2 - E_3\left(\cos\theta_{12}\right)\right). \quad (4.130)$$

The integral over the variable $x = \cos\theta_{12}$ can be converted to an integral over $E_3$

$$\begin{aligned} I &= \int_{-1}^{+1} dx\,\delta(m - E_1 - E_2 - E_3(x)) \\ &= \int_{E_3^-}^{E_3^+} dE_3 \frac{dx}{dE_3}\,\delta(m - E_1 - E_2 - E_3) \end{aligned} \quad (4.131)$$

where $E_3^{\pm} = [(|\mathbf{p}_1| \pm |\mathbf{p}_2|)^2 + m_3^2]^{\frac{1}{2}}$. Using $dE_3/dx = |\mathbf{p}_1||\mathbf{p}_2|/E_3$, one finds $I = E_3/|\mathbf{p}_1||\mathbf{p}_2|$ for values of $E_1$ and $E_2$ such that $E_3^- < m - E_1 - E_2 < E_3^+$ and otherwise I = 0. Then, since

$$\begin{aligned} d|\mathbf{p}_1| &= \frac{d|\mathbf{p}_1|}{dE_1}dE_1 = \frac{E_1}{|\mathbf{p}_1|}dE_1 \\ d|\mathbf{p}_2| &= \frac{d|\mathbf{p}_2|}{dE_2}dE_1 = \frac{E_2}{|\mathbf{p}_2|}dE_2, \end{aligned} \quad (4.132)$$

we arrive at the expression

$$dPS = 2(2\pi)^{-3}|\mathbf{p}_1||\mathbf{p}_2 E_3 d|\mathbf{p}_1|d|\mathbf{p}_2| = 2(2\pi)^{-3}E_1E_2E_3dE_1dE_2 \quad (4.133)$$

where $E_3 = m - E_1 - E_2$. The range of energies is restricted by

$$\sqrt{(|\mathbf{p}_1| - |\mathbf{p}_2|)^2 + m_3^2} < m - E_1 - E_2 < \sqrt{(|\mathbf{p}_1| + |\mathbf{p}_2|)^2 + m_3^2}. \quad (4.134)$$

The phase space distribution is a smooth function of the variables $E_1$ and $E_2$. The differential decay rate is the product of the phase space distribution and a function $|M|^2$ that represents the physics of the interaction. Correlations between particles appear in the differential decay distribution as a multiplicative factor. For example in the decay $t \to e_1^+ \nu_{e2} b_3$ through a virtual $W^+$, the propagator $G(p_W)$ associated with the $W$ boson can not be regarded as constant. In our model

$$d\Gamma_{t\to e^+\nu_e b} = |M|^2 dPS = \frac{g_W^4}{\left|p^2 - m_W^2 - im_W\Gamma_W\right|^2} 2(2\pi)^{-3}E_eE_\nu E_b dE_e dE_\nu. \quad (4.135)$$

where the invariant mass squared of the lepton pair $m_{e\nu}^2 = p^2$ is a function of the energies given by

$$p^2 = (p_e + p_\nu)^2 \simeq E_e E_\nu (1 - \cos\theta_{e\nu}) \qquad (4.136)$$

where the angle $\theta_{e\nu}(E_e, E_\nu)$ between the electron and neutrino must be expressed in terms of the energies $E_e$ and $E_\nu$, and the masses $m_b$, $m_e$, and $m_\nu$ have been neglected. Energy and momentum conservation in $t$-quark decay permit the virtual $W$ boson to have mass $m_W$ and qualify as real. As a consequence, the total decay rate is dominated by decays in which $m_{e\nu}$ is close to $m_W$ and the propagator $G(p_W)$ is large. Such correlations often appear in strong decays of hadrons which proceed dominantly through a superposition of intermediate state hadrons. Were $m_W$ slightly larger than $m_t - m_b$, the propagator would still vary significantly without exhibiting a peak. The correlation between the lepton energies in such a case might still be exploited to infer the presence of the high mass intermediate boson.

Returning to muon decay, when spin effects are neglected and $G(p_W)$ is approximated as a constant, the matrix element is independent of the momenta of the final particles. Hence, we can proceed to integrate the differential decay rate over all directions and energies of the final particles considering only the phase space. In general the bounding area for the integral is a complicated function of the masses. If we neglect all final state masses, we can find a relatively simple result. We have

$$|E_1 - E_2| < m - E_1 - E_2 < E_1 + E_2. \qquad (4.137)$$

The right inequality requires $E_2 > m/2 - E_1$ while the left requires $E_1 < m/2$ if $E_1 > E_2$ and $E_2 < m/2$ if $E_1 < E_2$ so as $E_1$ ranges from 0 to $m/2$, $E_2$ ranges from $m/2 - E_1$ to $m/2$. Hence

$$PS = \frac{2}{(2\pi)^3} \int_0^{m/2} dE_1 \int_{m/2-E_1}^{m/2} dE_2\, (m - E_1 - E_2) E_1 E_2 \propto m^5. \qquad (4.138)$$

With $g = e/(\sqrt{2}\sin\theta_W)$, including a sum over three intermediate $W$ boson polarization states with spin projection, the decay rate finally is given by the expression

$$\begin{aligned}
\Gamma_{\mu^- \to e^- \nu_\mu \bar{\nu}_e} &= \frac{1}{\tau_\mu} = |M|^2 PS = \frac{\alpha^2}{384\pi \sin^4\theta_W} \frac{m_\mu^5}{m_W^4} \\
&= \frac{137^{-2}}{384\pi 0.22^2} \frac{(0.105\ \text{GeV})^5}{(80\ \text{GeV})^4} \frac{c = 3 \times 10^{23}\ \text{fm s}^{-1}}{\hbar c = 0.197\ \text{GeV fm}} \\
&= (2.2\ \mu\text{s})^{-1} \qquad (4.139)
\end{aligned}$$

and this is the value that is observed. (In fact, it is from the muon lifetime that $g_W$ is usually determined.) The final formula for the decay rate contains a factor $g_W^2$ associated with each vertex in the diagram. The $W$ boson propagator contributes a factor $G^2(p_W) \simeq m_W^{-4}$. The factor $m_\mu^5$ from the phase space integration is therefore required by dimensional analysis. If the same weak interaction coupling constants govern

$$\tau \rightarrow \nu_\tau e \bar{\nu}_e$$

then the $\tau$ decay rate should be $(m_\tau/m_\mu)^5$ larger than the muon decay rate (and the lifetime correspondingly smaller) and it is.

The weak process that leads to neutron $\beta$-decay

$$n \rightarrow p e^- \bar{\nu}_e \tag{4.140}$$

is similar to muon decay but has a lifetime nine orders of magnitude larger. Let's calculate the decay rate supposing a similar interaction with the simplification $|M|^2 = (g_W/m_W)^4$. We will see that the lifetime difference is largely a consequence of phase space. Connecting these processes illustrates the universal nature of the weak interaction. It is too naive to replace the muon mass by the neutron mass in the muon decay rate formula which neglected the masses of the decay products. Therefore we revisit the 3-body phase space starting from the definition.

The phase space for neutron decay in the neutron rest frame is

$$dPS = \frac{2}{(2\pi)^3} E_\nu dE_\nu E_e dE_e (E_p = m_n - E_\nu - E_e) \tag{4.141}$$

with the restriction $E_p^- < E_p < E_p^+$ where $E_p^\pm = \sqrt{(E_\nu \pm p_e)^2 + m_p^2}$. For each inequality, squaring both sides and repeated application of $E^2 - p^2 = m^2$ allows one to solve for limits $E_\pm$ on the neutrino energy for given electron energy:

$$E_\pm = \frac{\left(m_n^2 - m_p^2 + m_e^2\right)/2 - m_n E_e}{m_n - E_e \mp p_e}. \tag{4.142}$$

With the definition

$$\epsilon = \frac{m_n^2 - m_p^2 + m_e^2}{2m_n} \simeq m_n - m_p, \tag{4.143}$$

we have

$$E_\pm = \frac{\epsilon - E_e}{1 - \left(\frac{E_e}{m_n} \pm \frac{p_e}{m_n}\right)} \simeq (\epsilon - E_e)\left(1 - \frac{E_e}{m_n} \mp \frac{p_e}{m_n}\right) \tag{4.144}$$

so $E_+ - E_- = (\epsilon - E_e)2p_e/m_n$ and $(E_+ + E_-)/2 = \epsilon - E_e$. Integration of the phase space gives

$$
\begin{aligned}
dPS &= \frac{2}{(2\pi)^3} E_e dE_e \int_{E_-}^{E_+} dE_\nu \; E_\nu (m_n - E_\nu - E_e) \\
&= \frac{2}{(2\pi)^3} E_e dE_e \left[ (m_n - E_e) \frac{\left(E_-^2 - E_+^2\right)}{2} - \frac{1}{3}\left(E_-^3 - E_+^3\right) \right] \\
&\simeq \frac{2}{(2\pi)^3} E_e dE_e m_n \frac{(E_- + E_+)(E_- - E_+)}{2} \\
&= \frac{1}{2\pi^3} E_e dE_e p_e (\epsilon - E_e)^2 
\end{aligned} \tag{4.145}
$$

where we retained only the leading order terms. We arrive at the decay rate as a function of electron energy

$$
\frac{d\Gamma_{n\to pe^-\bar{\nu}_e}}{dE_e} = \frac{g_W^4}{m_w^4} \frac{1}{2\pi^3} E_e p_e (\epsilon - E_e)^2 = \frac{g_W^4}{m_w^4} \frac{1}{2\pi^3} E_e p_e (\epsilon - E_e)^2. \tag{4.146}
$$

The electron energy spectrum rises from zero at vanishing electron energy and then falls as the electron energy approaches its maximum value $\epsilon$. This limit derives from the 2-body decay kinematics when the proton and neutrino have minimum invariant mass so the largest energy is released:

$$
E_e^{max} = \epsilon = \frac{m_n^2 - (m_p + m_\nu)^2 + m_e^2}{2m_n}. \tag{4.147}
$$

(The endpoint is decreased if $m_\nu \neq 0$ so provides a sensitive test of neutrino mass. Correction for the Coulomb attraction between the final state charged particles is especially important in establishing the endpoint.) By integrating over electron energy, we find the total decay rate is

$$
\Gamma_{n\to pe^-\bar{\nu}_e} = \frac{g_W^4}{m_w^4} \frac{1}{2\pi^3} \int_{m_e}^{\epsilon} dE_e \; E_e p_e (\epsilon - E_e)^2 = \frac{g_W^4}{m_w^4} \frac{1}{2\pi^3} \epsilon^5 I. \tag{4.148}
$$

With $y = m_e/\epsilon = 0.395$, the integral is

$$
I = \int_y^1 dx \; x(x^2 - y^2)^{1/2} (1 - x)^2 = 0.0157. \tag{4.149}
$$

Only the left-handed leptons participate in charged current weak interactions. If we assume only the left-handed component of the nucleons appear in the neutron decay matrix element, the spin averaged matrix element squared turns out to be

$$
\overline{|M_{n\to pe^-\bar{\nu}_e}|^2} = \frac{1}{2} \frac{g_W^4}{m_W^4} (1 - \mathbf{v}_p \cdot \mathbf{v}_e) \tag{4.150}
$$

and, since $|\mathbf{v}_p| << 1$, our result should be halved. Including the factor of 1/2, and writing $g^2 = 2\pi\alpha/sin^2\theta_w$, the decay rate is

$$
\Gamma_{n\to pe^-\bar{\nu}_e} = \frac{1}{\pi} \frac{\alpha^2}{\sin^4\theta_w} I \frac{\epsilon^5}{m_W^4} \simeq \frac{1}{1316} \text{ s}^{-1} \tag{4.151}
$$

while experiment gives $\tau_n = 887 \pm 2$ s.

How might we explain the discrepancy? The lifetime of a free neutron is actually difficult to measure. Neutrons decay within radioactive nuclei but are subject to a complicated environment. Ultracold (meV energy) neutrons can be confined in bottles long enough to observe their decay but nuclear absorption by the container walls leading to disappearance, also exponential in time, complicates the analysis. A spin polarized neutron can be confined in a magnetic trap and its decay studied. (The magnetic moment of the neutron is $\mu_{\rm n} = -60.3$ neV/T and, in a magnetic field gradient, the confinement force is $F = -\mu_n dB/dz$.) The experimental lifetime is inferred from all such techniques and is significantly different from our prediction.

One should check if the right-handed nucleons, unlike right-handed fundamental fermions, couple to the $W$ boson. This can be done by studying the spin and angular distributions of the electron in the decay of polarized neutrons and it turns out both $n_L$ and $n_R$ couple to the $W$ boson with comparable strength and this is a clue. The neutron $\beta$ decay derives from transitions of left-handed $d$-quarks to left-handed $u$-quarks but the neutron spin derives from the correlated spins of the three quarks $udd$. Careful studies indicate that the left handed $n \to p$ current that derives from the purely left-handed $d \to u$ currents is exactly like the $\nu_e \to e$ current except for a CKM matrix element $V_{ud} \sim 0.97$ while the right-handed $n \to p$ current is less accurately predicted. However, in any event, our naive decay rate must be multiplied by roughly two times $V_{ud}^2$. The inclusion of the right-handed component (and significant higher order radiative corrections) brings the spin averaged neutron decay rate calculation in line with experiment. [Ref: M. Faber et al., Phys. Rev. **C80**, 035503 (2009)]

## 4.9 ■ SCATTERING AND RESONANCES

In the picture we are developing of particle interactions, seemingly disparate processes like potential scattering and radiative decay are related. Neutrino electron scattering is a second order weak process that can be related to the weak muon decay. In a charged current ($W$ boson exchange) interaction, a muon neutrino may scatter from a stationary electron and emerge as a muon

$$\nu_\mu e^- \to \mu^- \nu_e. \tag{4.152}$$

The time ordered amplitudes for $\nu_\mu \to W^+\mu^-$ followed by $W^+e^- \to \nu_e$ and $e^- \to W^-\nu_e$ followed by $\nu_\mu W^- \to \mu^-$ are shown in Figure 4.11.

As in muon decay, the total amplitude $M$ for this $W$ boson exchange is a product of coupling constants times a propagator $G(p_W)$ corresponding to the virtual $W$ boson. The differential scattering cross section is the transition rate $|M|^2dPS$ for final states with density dPS divided by the relative flux of neutrinos impinging on electrons which, in the center of mass, is $j = v_{rel} = |v_\nu| + |v_e|$ so

$$d\sigma = \frac{d\Gamma}{j} = \frac{1}{v_{rel}}|M|^2 dPS. \tag{4.153}$$

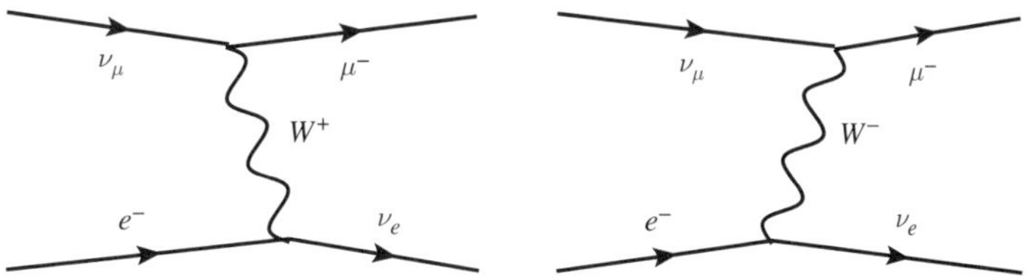


**FIGURE 4.11** Neutrino charged current interaction in second order Hamiltonian perturbation theory. The left diagram describes a transition path with an intermediate state containing $W^+$. The right diagram describes a transition through an intermediate state containing $W^-$. Both amplitudes contribute to the transition $\nu_\mu e^- \to \mu^- \nu_e$.

Here $v_{rel} = |\mathbf{v}_\nu - \mathbf{v}_e|$ is the relative speed as seen by an observer in the center of mass, not the speed of one of the particles as seen by an observer moving with the other particle. Neglecting spin, treating the phase space for the final two particles as in $W^-$ decay, and expressing it in terms of the polar angle $\theta$ of the final muon in the center of mass, we find

$$d\sigma = \frac{1}{v_{rel}} \left| \frac{g_W^2}{q^2 - m_W^2} \right|^2 \frac{1}{2\pi} |\mathbf{p}_\mu|^2 \frac{d|\mathbf{p}_\mu|}{dE} d\cos\theta \tag{4.154}$$

where $E = E_\nu + E_e = E'_\mu + E'_\nu$ is the total energy in the center of mass, $dE/dp = v'_\mu + v'_\nu$, and the square of the exchanged 4-momentum is $q$. The matrix element in this expression is $g_W^2 G(q)$. The 4-momentum can be expressed explicitly in terms of the angle $\theta$ as

$$q^2 = (p_\mu - p_{\nu_\mu})^2 = (p_{\nu_e} - p_e)^2 = m_\mu^2 - 2E_{\nu_\mu}(E_\mu - |\mathbf{p}_\mu| \cos\theta). \tag{4.155}$$

These results are easily generalized. The differential cross section for elastic scattering of a left-handed neutrino from a left-handed electron via $Z$ boson exchange shown in Figure 4.12 is simply estimated by the substitution $m_W \to m_Z$ and with coupling constants $g_W^2 \to g_{Z\nu\nu} g_{Zee}^L$ where $g_{Zee}^L$ is the neutral weak coupling of the left-handed electron, different from the right-handed coupling $g_{Zee}^R$. For scattering from unpolarized electrons, we should average over left-handed and right-handed electrons. In $Z$ boson exchange, the particle masses are not altered,

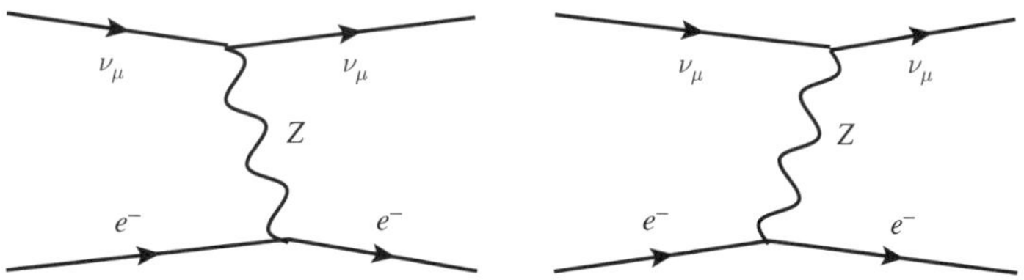


**FIGURE 4.12** Neutrino neutral current interaction $\nu_\mu e^- \to \nu_\mu e^-$ in second order Hamiltonian perturbation theory. The left diagram describes a transition amplitude with a $Z$ boson emitted by the $\nu_\mu$. The right diagram describes a transition amplitude with a $Z$ boson emitted by the $e^-$.

$q^2 = -\mathbf{q}^2$, and, if the target particle is extremely heavy, we can write $v_{rel} \to v$ and recover the Yukawa potential scattering result derived earlier in this chapter.

Returning to the charged current neutrino scattering process, if the neutrino energy is well in excess of the muon mass, we can neglect lepton masses and write $p = E/2$ and $v_{rel} = 2c$. In this limit, $dq^2 = 2E_\nu|\mathbf{p}_\mu|d\cos\theta$ and we find the Lorentz invariant expression for the differential cross section for ultra-relativistic weak scattering

$$\frac{d\sigma_{\nu_\mu e^- \to \mu^- \nu_e}}{dq^2} = \frac{g_W^4}{16\pi}\frac{1}{(q^2 - m_W^2)^2}. \tag{4.156}$$

The maximum momentum transfer squared is $q_m^2 = -(2p)^2 = -s$ where $s = (p_{\nu_\mu} + p_e)^2 \simeq 2m_e E^{lab}$ and integration over momentum transfer gives the total cross section

$$\sigma_{\nu_\mu e^- \to \mu^- \nu_e} = \int dq^2 \frac{d\sigma_{\nu_\mu e^- \to \mu^- \nu_e}}{dq^2} = \frac{g_W^4}{16\pi}\frac{1}{m_W^2}\frac{s}{s + m_W^2}. \tag{4.157}$$

We have $s = 2m_e E_{\text{lab}}$ with $E_{\text{lab}}$ the laboratory frame energy of the neutrino. Hence, for $s << m_W^2$ with spin effects included, we find the cross section increases linearly with the laboratory neutrino energy:

$$\begin{aligned}\sigma_{\nu_\mu e^- \to \mu^- \nu_e} &= \frac{\pi\alpha^2}{\sin^4\theta_W m_W^4} m_e E_{Lab} \\ &= \left(\frac{4.5}{137}\right)^2 \pi \frac{0.5 \times 10^{-3}\ \text{GeV}}{(80\ \text{GeV})^4}\frac{E_{lab}}{1\ \text{GeV}}\left(\hbar c = .197 \times 10^{-13}\ \text{GeV cm}\right)^2 \\ &= 1.6 \times 10^{-41}\ \text{cm}^2 \frac{E_{lab}}{1\ \text{GeV}}.\end{aligned} \tag{4.158}$$

The electron density in typical stuff is

$$\text{n}_\text{e} = \frac{5\ \text{gm cm}^{-3}}{1.7 \times 10^{-24}\ \text{gm/nucleon}}\frac{1\ \text{electron}}{2\ \text{nucleons}} = 1.5 \times 10^{24}\ \text{cm}^{-3} \tag{4.159}$$

so, even at $E_{lab} = 20$ GeV, the mean free path for a neutrino to interact by charge exchange with electrons in matter is of order $2 \times 10^{15}$ cm. This illustrates how feeble the weak interaction appears even at such an extremely relativistic energy. In contrast, the strong interaction between hadrons is roughly geometrical so the total cross section for hadronic interactions is approximately $\sigma_s \simeq (1\ \text{fm})^2 = 10^{-26}\ \text{cm}^2$ and hadronic mean free paths are about 20 cm.

Of course neutrinos interact with quarks as well as with electrons, and by neutral current $Z$ exchange in addition to charged current $W$ exchange. As discussed in Chapter 8, the neutral current cross section $\sigma_{\nu_\mu e^- \to \nu_\mu e^-}$ is smaller than the charged current cross section $\sigma_{\nu_\mu e^- \to \mu^- \nu_e}$ by a factor of ten by virtue of the couplings in the standard model. Quarks in a stationary nucleon have an energy $E_q \simeq 100$ MeV

(Fermi motion). For neutrino energy $E_{\text{lab}}$, the center of mass energy squared is of order $s = 2E_{\text{lab}}E_q$ and is about 200 times larger than in scattering from electrons. Also, a $W^-$ can be absorbed by two $u$-quarks in a proton and by the one $u$-quark in a neutron. The inclusive nucleon cross section $\sigma_{\nu_\mu N \to \mu^- X}$ for neutrino energies above 10 GeV is about 400 times $\sigma_{\nu_\mu e^- \to \mu^- \nu_e}$.

The amplitude for the process $e^+e^- \to \mu^+\mu^-$ through an intermediate $Z$ boson state is the sum of two amplitudes indicated by the diagrams in Figure 4.13. The matrix element neglecting spin has the form

$$-M_{e^+e^- \to \mu^+\mu^-} = \frac{g_{Zee}\, g_{Z\mu\mu}}{p^2 - m_Z^2 + i m_Z \Gamma_Z} \tag{4.160}$$

where the coupling constants $g_{Zee}$ and $g_{Z\mu\mu}$ depend on the fermion helicity even when spin is neglected. Generally, an intermediate state particle of 4-momentum square $p^2$ contributes to a transition amplitude a propagator $G(p)$ and, when $p^2 > 0$ (s-channel), as in this case, the amplitude is large at the value $p^2 = m^2$ (called a pole). Correspondingly, the cross section is said to exhibit a resonance.

Several different particles can appear in the intermediate state. For example, the entire amplitude for $e^+e^- \to \mu^+\mu^-$ in the standard model is a sum of amplitudes for formation of a virtual photon, $Z$ boson, and $H$ with the structure:

$$-M = \frac{e^2}{p^2 - m_\gamma^2} + \frac{g_{Zee}\, g_{Z\mu\mu}}{p^2 - m_Z{}^2 + i m_Z \Gamma_Z} + \frac{g_{eeH}\, g_{\mu\mu H}}{p^2 - m_H{}^2 + i m_H \Gamma_H} \tag{4.161}$$

The first amplitude has a pole at zero mass and the second at the mass of the $Z$ boson so the propagator for the photon dominates the amplitude at low mass and that for the $Z$ boson dominates near the $Z$ boson mass. It turns out that $g_{eeh} << g_{llZ} \sim e$ so the Higgs boson contribution is negligible. Consider the case of center of mass energy large compared to the muon mass but much smaller than the mass of the Z boson ($m_\mu^2 << s << m_Z^2$). Since $\sqrt{s}$ is large compared to $2m_\mu$, the phase space and flux factor are together proportional $s$ and the total cross section is $\sigma \sim \alpha^2 s^{-1}$. For $\sqrt{s} \simeq m_Z$, the cross section exhibits a resonant enhancement while for $\sqrt{s} >> m_Z$ the cross section is $\sigma \sim \alpha^2 s^{-1}(1 + (g_{Zee}/e)^2)^2$. Interference of the amplitudes for intermediate state photon and $Z$ boson is significant in the neighborhood of the $Z$ boson pole.

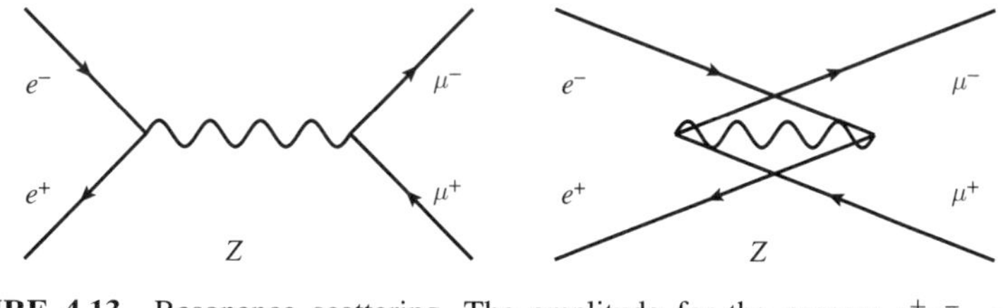


**FIGURE 4.13** Resonance scattering. The amplitude for the process $e^+e^- \to \mu^+\mu^-$ through a virtual $Z$ boson corresponds to two time ordered diagrams. The sum includes a propagator factor which is large if the mass of the $e^+e^-$ pair is equal to $m_Z$.

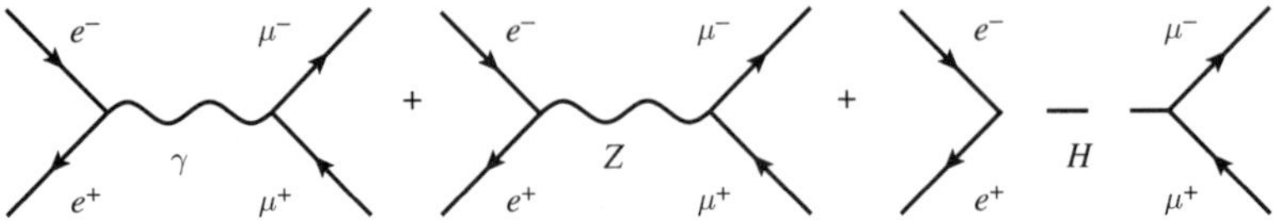


**FIGURE 4.14** Interference of resonances. Scattering through multiple resonances is illustrated by the process $e^+e^- \to \mu^+\mu^-$ through coherent virtual $Z$ boson, photon, and Higgs boson intermediate states. Near any resonance, one amplitude will dominate but in general the amplitudes may interfere.

It is informative to consider single resonance production generally. The amplitude for an intermediate state containing a particle $X$ of spin $s_X$ has the form

$$iM_{ab\to X\to cd} = \Sigma_\chi i M_{ab\to X_\chi} \frac{i}{p_X^2 - m_X^2 - im_X\Gamma_X} iM_{X_\chi\to cd} \tag{4.162}$$

where $p_X$ is the 4-momentum of $X$ and the summation is over the spin substates $\chi$ of particle $X$ with the spin states of particles $a$, $b$, $c$, and $d$ fixed. The matrix element $M_{X_s\to cd}$ contains a spin wave function $X_\chi(p_X)$ for particle $X$ while $M_{ab\to X_s}$ contains spin wave function $X_\chi^\dagger(p_X)$. The tensor product of the spin wave functions can be moved to the numerator of the propagator $G(p_X)$ for $X$ which becomes a sum over spin states $\Sigma_\chi X_\chi(p_X)X_\chi^\dagger(p_X)$:

$$G(p) = \frac{i\Sigma_\chi X_\chi(p_X)X_\chi^\dagger(p_X)}{p^2 - m_X^2 - im_X\Gamma_X}. \tag{4.163}$$

We will see instances of this later. Moreover, the transition rate is proportional to the square of the amplitude

$$|M^2| = \frac{\Sigma_{\chi,\xi} M_{ab\to X_\chi} M^\dagger_{ab\to X_\xi} M_{X_\xi\to cd} M^\dagger_{X_\chi\to cd}}{\left(s - m_X^2\right)^2 + m_X^2\Gamma_X^2}. \tag{4.164}$$

The spin states of initial and final particles are often not specified in experiments. When initial state spin indices $\alpha$ for $a$ and $\beta$ for $b$ and final state spin indices $\gamma$ for $c$ and $\delta$ for $d$ are summed, the squared matrix element can be written as

$$|M|^2 \propto \Sigma_{\alpha,\beta,\gamma,\delta,\chi,\xi} M_{a_\alpha b_\beta\to X_\chi} M^\dagger_{a_\alpha b_\beta\to X_\xi} M_{X_\xi\to c_\gamma d_\delta} M^\dagger_{X_\chi\to c_\gamma d_\delta}. \tag{4.165}$$

Now in the rest frame of $X$, the matrix element tensor $M_{X_\xi\to c_\gamma d_\delta}$ is a function of the angles of the decay momentum. Using rotational symmetry, it may be shown that after integration over the final decay angles and summation over final spins, the decay matrix element factors reduce to

$$\Sigma_{\gamma\delta} \int d\Omega_c M_{X_\xi\to c_\gamma d_\delta} M^\dagger_{X_\chi\to c_\gamma d_\delta} \propto \delta_{\chi,\xi} \Sigma_{\gamma\delta} |M_{X_\xi\to c_\gamma d_\delta}|^2 \propto \Gamma_{X\to cd} \tag{4.166}$$

where $\Gamma_{X\to cd}$ is the spin averaged decay rate to the final state particles. The summation over initial spins then permits us to write the production matrix element factors as

$$\Sigma_{\alpha\beta} M_{a_\alpha b_\beta \to X_\chi} M^\dagger_{a_\alpha b_\beta \to X_\chi} \propto \Gamma_{\bar{X}\to\bar{c}\bar{d}} = \Gamma_{X\to cd}. \tag{4.167}$$

The transition rate for $ab \to X \to cd$ then factorizes into a product of rates for production and decay of a spin $s_X$ object. If we assume time reversal invariance, valid for all but a few weak transitions between quarks, so that

$$\Sigma_{\alpha\beta} \left| M_{a_\alpha b_\beta \to X_\chi} \right|^2 = \Sigma_{\alpha\beta} \left| M_{X_\chi \to a_\alpha b_\beta} \right|^2, \tag{4.168}$$

then, upon multiplying $\overline{|M|^2}$ by the phase space and dividing by the flux, the cross section is found to be

$$\sigma_{ab\to cd} = \frac{4\pi(2s_X+1)}{(2s_a+1)(2s_b+1)p_{ab}^2} \frac{m_X \Gamma_{X\to ab} m_X \Gamma_{X\to cd}}{\left(s - m_X^2\right)^2 + m_X^2 \Gamma_X^2} \tag{4.169}$$

where $p_{ab}$ is the magnitude of the three momentum of $a$ or $b$ in the center of mass. This so-called relativistic Breit-Wigner form for resonance production can be distorted by interference between competing resonant amplitudes and by variation of the effective decay widths with mass $\sqrt{s}$ but is quite useful.

---

**Example 4.5. Event rate at a Z boson factory.** Use the general formula for resonance production to estimate the rate of $Z$ boson production and decay events at an $e^+e^-$ collider with a center of mass energy equal to the mass of the $Z$ boson and a luminosity of $\mathrm{L} = 10^{32}\ \mathrm{cm}^2\ \mathrm{s}^{-1}$. Assume for simplicity the $Z$ boson coupling to quark and lepton pairs in the standard model is independent of fermion flavor.

Suppose the interaction rate is dominated by virtual $Z$ production yielding events such as $e^+e^- \to Z \to \mu^+\mu^-$. Assume $\Gamma_{Z\to e^+e^-} = \Gamma_{Z\to\mu^+\mu^-}$ and $Z$ boson spin angular momentum $j = 1$. The cross section on resonance for $\mu^+\mu^-$ production is

$$\sigma_{e^+e^-\to\mu^+\mu^-} = \frac{3\pi}{(m_Z/2)^2} \left(\frac{\Gamma_{Z\to e^+e^-}}{\Gamma}\right)^2 = 12\pi \left(\frac{B_{ee}}{m_Z}\right)^2 \tag{4.170}$$

where $B_{ee} = \Gamma_{Z\to e^+e^-}/\Gamma$ is the fraction of $Z$ boson decays yielding electron-positron pairs. Kinematics permits the $Z$ boson to decay to $e^+e^-$, $\mu^+\mu^-$, $\tau^+\tau^-$, to $\nu_e\bar{\nu}_e$, $\nu_\mu\bar{\nu}_\mu$, $\nu_\tau\bar{\nu}_\tau$, and to $u\bar{u}$, $d\bar{d}$, $c\bar{c}$, $s\bar{s}$, and $b\bar{b}$ ($t\bar{t}$ is excluded by kinematics) in each of three colors. The masses of all of these fermions are negligible compared to $m_Z$ so if the coupling constants were identical, the branching fraction would be $B_{ee} = 1/21$ and the total cross section for $e^+e^-$ interactions would be $\sigma_{\mathrm{e^+e^-\to anything}} = \sigma_{\mathrm{e^+e^-}\to\mu^+\mu^-}/\mathrm{B_{ee}}$ or

$$\sigma_{\mathrm{e^+e^-\to anything}} = \frac{12\pi}{21} \frac{[\hbar c]}{\mathrm{m_Z^2}} = (1.8)\left(\frac{0.197\ \mathrm{GeV\ fm)}}{90\ \mathrm{GeV}}\right)^2 = 9\times 10^{-32}\ \mathrm{cm}^2 \tag{4.171}$$

implying an event rate of ten Hz. The coupling constants of the $Z$ boson to fermions are independent of generation but are actually different for $e^-$, $\nu_e$, u, and d, and $B_{ee} = 3.37\%$ so we should estimate $\sigma_{e^+e^- \to \mu^+\mu^-} = 2 \times 10^{-33}$ cm$^2$ and $\sigma_{\mathrm{e^+e^- \to anything}} = 60 \times 10^{-33}$ cm$^2$. (Of this 0.2 are to invisible neutrino pairs and 0.7 to hadrons.)

---

Discrete bound states appear as intermediate state resonances in relativistic physics. Consider for example $e^+e^- \to \mu^+\mu^-$ at $\sqrt{s} \sim 3$ GeV, a center of mass energy near threshold to make charm quark pairs. Below threshold, the $\mu^+\mu^-$ production process occurs via an intermediate state virtual photon. Near threshold, the virtual photon can decay to $c\bar{c}$ and such pairs, when produced essentially at rest, can bind into neutral charmonium mesons which subsequently annihilate to a virtual photon leading again to $\mu^+\mu^-$.

Among the higher order diagrams, the intermediate meson states correspond approximately to "ladder" diagrams in which, from the virtual photon, a $c\bar{c}$ pair of rails forms and binds via gluon exchange rungs, as illustrated in Figure 4.15. The sum of the corresponding amplitudes is anomalously large near the energy of a bound state. The evaluation of such a sum is a complex problem and can be cast in the form of an integral equation called the Bethe-Salpeter equation. The solution to the Bethe-Salpeter equation for a heavy fermion pair can be shown to be approximately equivalent to the solution of Dirac's generalization of Schrödinger's equation with relativistic corrections that may be understood quasi-classically. Exact solutions which include all multiparticle effects are not known.

Resonances produced via a virtual photon must have spin 1 (and odd parity) and $s$-wave states are dominant. Accessible states in $e^+e^-$ collisions are the $1s$ vector mesons composed of the quark plus antiquark pairs $u\bar{u}$, $d\bar{d}$, $s\bar{s}$, $c\bar{c}$, and $b\bar{b}$ called $\rho^0$, $\omega$, $\phi$, $J/\psi$ and $\Upsilon$, as well as $2s$ and $3s$ radial excitations. For the charmonium system, the $2s$ and $3s$ states are called the $\psi'$ and $\psi''$. For the $b\bar{b}$ system, these are called the $\Upsilon'$ and $\Upsilon''$. The appearance of bound state resonances is illustrated in Figure 4.16.

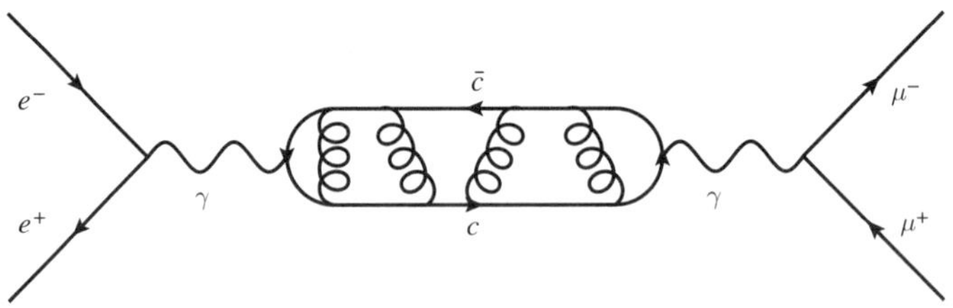


**FIGURE 4.15** Bound state resonances. Scattering through a bound state resonance is illustrated by the process $e^+e^- \to \mu^+\mu^-$ with a virtual $c\bar{c}$ meson in the intermediate state. The cross section for the process is enhanced when $m_{e^+e^-} \simeq m_R$ where $m_R$ is the resonance mass.

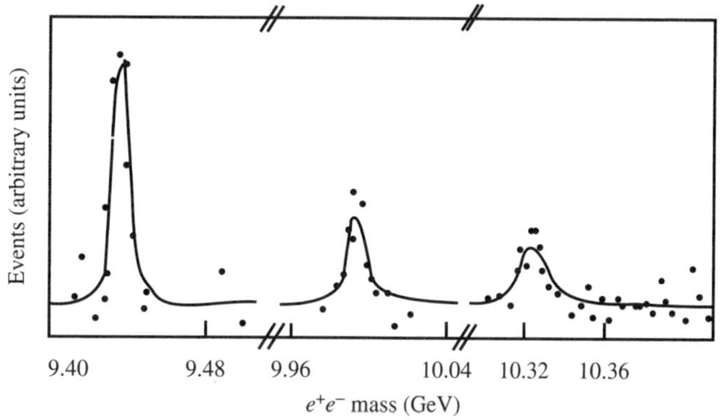


**FIGURE 4.16** Observation of the $b\bar{b}$ states $\Upsilon$, $\Upsilon'$, and $\Upsilon''$ at the Cornell Electron Storage Ring. The number of hadronic events normalized to the number of scattered $e^+e^-$ pairs is shown over several ranges of center of mass energy. [after T. Bohringer *et. al.*, Phys. Rev. Lett. **44**, 1111 (1980)].

We can get a feel for bound state resonance production as follows. Consider a $c\bar{c}$ bound state with nonrelativistic center of mass wave function $\psi(\mathbf{x})$ satisfying (for simplicity) Schrödinger's equation

$$\left[\frac{\mathbf{p}^2}{2m} + U\right]\psi(\mathbf{x}) = E\psi(\mathbf{x}) \tag{4.172}$$

where $\mathbf{x} = \mathbf{x}_c - \mathbf{x}_{\bar{c}}$, $m = m_c/2$ is the reduced mass, $U$ is the effective color averaged interaction potential, and $\chi$ is a spin wave function. For equal masses, the momentum operator is

$$\mathbf{p} = (m_{\bar{c}}\mathbf{p}_c - m_c\mathbf{p}_{\bar{c}})/(m_c + m_{\bar{c}}) = \frac{1}{2}(\mathbf{p}_c - \mathbf{p}_{\bar{c}}) \tag{4.173}$$

with $\mathbf{p}_c = -\mathbf{p}_{\bar{c}} = \mathbf{p}$. A particular space wave function is labeled by quantum numbers $\alpha = (n, l, m)$ specifying radial and angular factors. The $J/\psi$ corresponds to $(n, l, m_l) = (1, 0, 0)$ and a triplet of spin state combinations with $s = 1$ and $m = +1, 0, -1$. We can write the spin wave function as

$$\big|m> = \Sigma_{i\bar{i}} < i\bar{i}\big|m> \eta_i\eta_{\bar{i}}. \tag{4.174}$$

where $\eta_i$ is a spin wave function for the $c$ and $\eta_{\bar{i}}$ is a spin wave function for the $\bar{c}$. The amplitude for momentum $\mathbf{p}$ in the space wave function is

$$\psi(\mathbf{p}) = \int d\mathbf{x}\, e^{i\mathbf{p}\cdot\mathbf{x}}\psi(\mathbf{x}). \tag{4.175}$$

The rest frame states can be represented in terms of products of plane waves as

$$\psi_m(\mathbf{x}) = \Sigma_{i\bar{i}} < i\bar{i}|m> \int \frac{d\mathbf{p}}{(2\pi)^3}\,\psi(\mathbf{p})|\mathbf{p}, i> | -\mathbf{p}, \bar{i}> \tag{4.176}$$

where $|\mathbf{p}, i> = e^{i\mathbf{p}\cdot\mathbf{x}_c}\eta_i$ and $|-\mathbf{p}, \bar{i}> = e^{-i\mathbf{p}\cdot\mathbf{x}_{\bar{c}}}\eta_{\bar{i}}$ with spinors $\eta_i$ and $\eta_{\bar{i}}$. Now suppose $M_{ab\to|\mathbf{p},i>|-\mathbf{p},\bar{i}>}$ is the amplitude for $ab \to c\bar{c}$ in the center of mass with the $c$-quark having momentum $\mathbf{p}$ and spin state $i$ while the $\bar{c}$-quark has momentum $-\mathbf{p}$ and spin $\bar{i}$. The (not Lorentz invariant) amplitude for $ab \to J/\psi$ with $J/\psi$ spin state $m$ is

$$\begin{aligned} M_{ab\to\psi_m} &= \Sigma_{i\bar{i}} < m|i\bar{i}> \int \frac{d\mathbf{p}}{(2\pi)^3}\,\psi(\mathbf{p})M_{ab\to|\mathbf{p},i>|-\mathbf{p},\bar{i}>} \\ &= \psi(0)\Sigma_{i\bar{i}} < m|i\bar{i}> M_{ab\to|\mathbf{p},i>|-\mathbf{p},\bar{i}>}. \end{aligned} \tag{4.177}$$

In the second expression, the momentum dependence of $M_{ab\to|\mathbf{p},i>|-\mathbf{p},\bar{i}>}$ over the range included in the wave function was neglected and the integral reduced to give the value of the space wave function at the origin of coordinates. The production cross section is

$$\begin{aligned} \sigma_{ab\to\psi_m} &= \frac{|M_{ab\to\psi_m}|^2}{v_a+v_b}\delta(E_a+E_b-E_\psi)\delta(\mathbf{p}_a+\mathbf{p}_b-\mathbf{p}_\psi)\frac{d\mathbf{p}_\psi}{(2\pi)^3} \\ &= \frac{|M_{ab\to\psi_m}|^2}{v_a+v_b}\frac{\delta(\sqrt{s}-m_\psi)}{(2\pi)^3} \end{aligned} \tag{4.178}$$

where in the center of mass $E_\psi = m_\psi = m_c + m_{\bar{c}}$ and $\mathbf{p}_a = -\mathbf{p}_b$. The reverse process has differential decay rate

$$\frac{d\Gamma_{\psi_m\to ab}}{d\Omega_a} = \frac{|M_{ab\to\psi_m}|^2}{v_a+v_b}\frac{|\mathbf{p}_a|^2}{4\pi^2}. \tag{4.179}$$

Summed over spins of $a$ and $b$ and integrated over angles, the total decay rate is independent of $m$. Summation over all spin states in production and comparison to the decay rate gives the narrow resonance formula

$$\sigma = \frac{(2s_\psi+1)}{(2s_a+1)(2s_b+1)}\frac{\pi\Gamma}{|\mathbf{p}_a|^2}\delta\left(\sqrt{s}-M\right). \tag{4.180}$$

The instability of the intermediate state may be represented by replacing the delta function by a Breit-Wigner function and we return to the form given by Equation 4.169.

The Dalitz plot in Figure 4.17 illustrates the presence of resonances underlying the production of $\bar{K}^0\pi^-\pi^+$ in hadronic collisions. In the center of mass of a three particle system of fixed total mass, when the energies of two of the particles are specified, the energy of the third is determined. Equivalently, the invariant mass of two of the three pairs determines the third. For fixed total mass, the two energies or masses must lie within a closed region in the two dimensional energy or mass plane. The distribution of the two variables for three uncorrelated particles follows the phase space distribution $dPS \propto E_1E_2E_3dE_1dE_2$. When two of the particles result from decay of a resonance, their energies are correlated and their invariant mass distribution follows a Breit-Wigner form that corresponds to a vertical or horizontal band in the Dalitz plot. In the Dalitz plot shown in Figure 4.17, the presence of the vector mesons $K^*(890)$ and $\rho(770)$ appears.

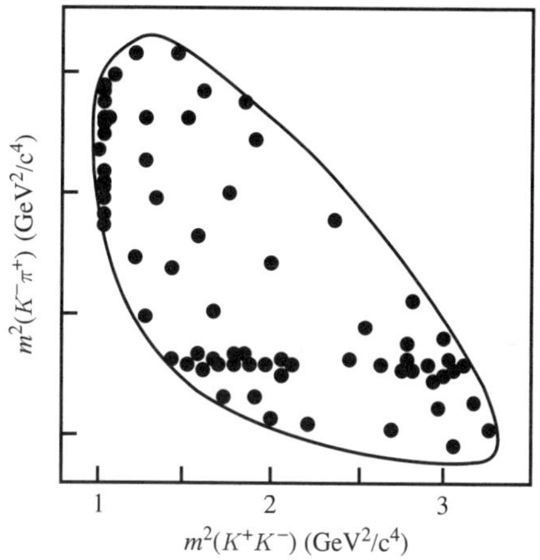


**FIGURE 4.17** Dalitz plot for decays $D_s \to K^+K^-\pi^+$. Each event is represented as a point in the plane of squared masses $m^2_{\bar{K}^+K^-}$ and $m^2_{K^-\pi^+}$. The boundary of the allowed region for the two two-particle mass values is shown. Enhancements corresponding to $\phi \to K^+K^-$ and $K^*(890) \to K^-\pi^+$ are observed. [after P. del Amo Sanchez *et al* (BaBar Collaboration), Phys. Rev. **D83**, 052001 (2011)]

## 4.10 ■ LORENTZ COVARIANT PERTURBATION THEORY

In this chapter, we have seen how Hamiltonian perturbation theory and a generalization of the theory of radiation leads to a description of relativistic processes. We now show how our results emerge from a manifestly covariant approach in the model of interacting scalar fields introduced in Chapter 1.

Our model of interacting relativistic fields supposes a complex Lorentz scalar charged field we will call $N$ with density $\rho = N^\dagger N$ coupled to a real scalar neutral field $\phi$. We have in mind the nucleon and pion fields of the original Yukawa model. The fields might instead represent a charged scalar electron $\tilde{e}$ or charged Higgs boson $H^+$ coupled to a neutral Higgs boson $H$ in present day supersymmetric extensions of the standard model. The Lorentz covariant field equations are (Equations 1.61 and 1.62)

$$\partial_\mu \partial^\mu \phi = -\frac{\partial U}{\partial \phi} \to \left(\partial_\mu \partial^\mu + m_\pi^2\right)\phi = N^\dagger g N \tag{4.181}$$

and

$$\partial_\mu \partial^\mu N = -\frac{\partial U}{\partial N^\dagger} \to \left(\partial_\mu \partial^\mu + m_N^2\right) N = \phi g N. \tag{4.182}$$

The first equation is the Klein-Gordon equation for the field $\phi$ with source density $N^\dagger N$ and implies nucleons source and sink the pion field. The second equation or its complex conjugate implies nucleons are scattered or bound by a pion field $\phi$. The energy density $L$ in Equation 1.59 has dimensions of energy per unit

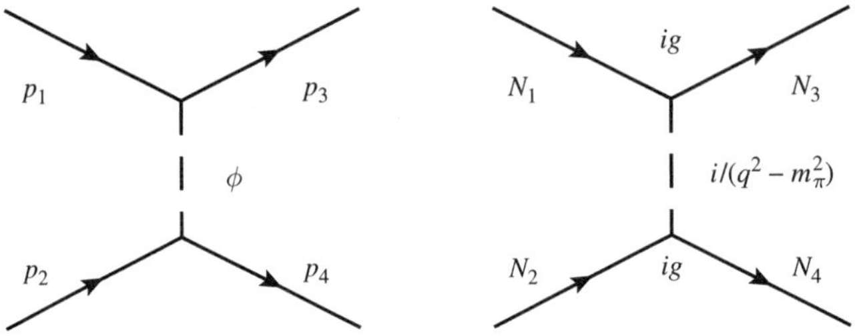


**FIGURE 4.18** Feynman diagram for scalar particle scattering. A single relativistic diagram describes scattering of scalar particles $N_1N_2 \to N_3N_4$ through an intermediate field $\phi$. The left diagram indicates the momenta of the particles. The right diagram shows the association of factors in the transition amplitude with parts of the diagram. Such associations are called Feynman rules.

volume or $m^4$ and, unlike the dimensionless electromagnetic charge, $g$ must have dimensions of mass like the fields here.

Consider the nucleon elastic scattering process described by the diagram in Figure 4.18

$$N_1N_2 \to N_3N_4 \tag{4.183}$$

for which the momenta of the particles are $p_1$, $p_2$, $p_3$, and $p_4$. Suppose de Broglie waves $N = N_1e^{-ip_1x}$ and $N^\dagger = N_3^\dagger e^{ip_3x}$ within some interaction 4-volume $VT$ correspond to scattering between nucleon states $N_1 \to N_3$ with momentum transfer $q = p_3 - p_1$. Using these waves in the interaction term in the first field equation above, an associated pion field may be found by inspection:

$$\left(\partial_\mu\partial^\mu + m_\pi^2\right)\phi = N_3^\dagger gN_1e^{iqx} \to \phi = \frac{i}{q^2 - m_\pi^2}N_3^\dagger(ig)N_1e^{iqx}. \tag{4.184}$$

Notice the appearance of the scalar field propagator

$$G(q) = \frac{i}{q^2 - m_\pi^2}. \tag{4.185}$$

Suppose next a wave $N = N_2e^{-ip_2x}$ interacts with this potential $\phi$ just found. Using the product of these two functions as an approximation to the interaction term on the right-hand-side of the second field equation, we have

$$\left(\partial_\mu\partial^\mu + m_N^2\right)N = gN_2\frac{i}{q^2 - m_\pi^2}N_3^\dagger(ig)N_1e^{i(p_3-p_2-p_1)x}. \tag{4.186}$$

For fixed space wave vector, this wave equation describes a simple driven harmonic oscillator. We are interested in radiated (escaping) waves and the crux of the solution is this: radiated waves are produced by resonant source components and the resonant response (like for a driven harmonic oscillators) is $\pi/2$ out of phase. The amount of $N_4 e^{-ip_4 x}$ sourced is consequently $\sqrt{-1}$ times the corresponding normalized Fourier component of the source. Equivalently, it is possible to construct the solution by multiplying by the propagator $G(p_3 - p_2 - p_1)$ and projecting out the amplitude for the positive energy wave for $t > 0$ (see for example Halzen and Martin). Hence we identify the transition amplitude as

$$S_{N_1 N_2 \to N_3 N_4} = i \int dx\ N_4^{\dagger} e^{ip_4 x} \left[ \frac{N_3^{\dagger} i g N_1}{q^2 - m_\pi^2} i g N_2 e^{i(p_3 - p_2 - p_1)x} \right]$$
$$= i M_{N_1 N_2 \to N_3 N_4} (2\pi)^4 \delta(p_4 + p_3 - p_1 - p_2) \tag{4.187}$$

where

$$i M_{N_1 N_2 \to N_3 N_4} = \left[ N_3^{\dagger}(ig) N_1 \right] \left[ \frac{i}{q^2 - m_\pi^2} \right] \left[ N_4^{\dagger}(ig) N_2 \right]. \tag{4.188}$$

The interaction volume was assumed large compared to relevant wavelength and period differences in writing the integral in terms of a 4-dimensional Dirac $\delta$-function. Since energy is conserved, for the case of elastic scattering in hand, $q^0 = 0$ and therefore $q^2 - m^2 = -\mathbf{q}^2 - m^2$ so $M_{N_1 N_2 \to N_3 N_4}$ is proportional to the spatial Fourier transform of a Yukawa potential expressed in Equation 4.38.

The amplitude $i M_{N_1 N_2 \to N_3 N_4}$ may be represented by a single momentum space diagram as shown in Figure 4.18 and constructed from factors associated with each element of the diagram. Coupling factors $N_3^{\dagger}(ig) N_1$ and $N_4^{\dagger}(ig) N_2$ are associated with each vertex and a propagator $G(q) = i/(q^2 - m^2)$ with the mediating field. The factors of $i$ keep track of phases. The single momentum space diagram represents both time orderings of the emission and absorption of the $\phi$ field. These rules are summarized in Table 4.1.

These (Feynman) rules permit calculation of the amplitudes of successive approximations by simple inspection of diagrams. For example, in the scattering of identical particles, the topologically distinct diagram with particle labels 3 and 4 interchanged corresponds to an amplitude with the momenta interchanged. This amplitude must be added (subtracted) in case the final bosons (fermions) are identical particles. Also we used $N(N^{\dagger})$ to destroy (create) nucleons. For antinucleon waves $Ne^{ipx}$, the roles are reversed. For example, a nucleon represented by $N \to N_1 e^{-ip_1 x}$ meeting an antinucleon represented by $N^{\dagger} \to [N_2 e^{ip_2 x}]^{\dagger}$

| Diagram element | Factor in amplitude |
|---|---|
| Ingoing or outgoing scalar | 1 |
| Ingoing or outgoing scalar antiparticle | 1 |
| Vertex for particles | $ig$ |
| Internal scalar line | $i/(p^2 - m^2)$ |

**TABLE 4.1** Feynman rules for a scalar field theory

combine into a source $gN_1N_2^\dagger e^{-ipx}$ for an intermediate pion field with 4-momentum $p = p_1 + p_2$. The amplitude corresponding to its formation and conversion back into a pair $N_3\bar{N}_4$ is

$$iM_{N_1\bar{N}_2\to N_2\bar{N}_4} = \left[N_2^\dagger(ig)N_1\right]\left[\frac{i}{p^2 - m_\pi^2 - im_\pi\Gamma_\pi}\right]\left[N_4^\dagger(ig)N_3\right]. \quad (4.189)$$

The addition $im_\pi\Gamma_\pi$ to the propagator denominator represents damping of the $\phi$ field by decay.

To relate to experiment a relativistic amplitude for a state transition $i \to f$ expressed in the form

$$S_{fi} = iM_{fi}(2\pi)^4\delta(p_f - p_i), \quad (4.190)$$

we proceed through now familiar steps. First, square $S_{fi}$ to get the transition probability and interpret the square of the $\delta$-function as

$$\delta(p)\delta(p) = \delta(p)\frac{1}{(2\pi)^4}\int d\mathbf{x}\, e^{i\mathbf{p}\cdot\mathbf{x}}|_{\mathbf{x}=0}\int dt\, e^{-iEt}|_{E=0} = \delta(p)\frac{1}{(2\pi)^4}VT. \quad (4.191)$$

The transition rate is therefore

$$\frac{|S_{fi}|^2}{T} = V|M_{fi}|^2(2\pi)^4\delta(p_f - p_i). \quad (4.192)$$

If each wave amplitude represents $2EV$ particles in the volume $V$, then conveniently $N^\dagger N = 1$ in our scalar wave model so the normalization factors may be set equal to unity in the matrix elements. Then the transition rate per particle follows dividing by $2E_jV$ for each particle $j$ in the initial or final state.

The joint transition rate for each final particle momentum $\mathbf{p}_k$ in a range $d\mathbf{p}_k$ about the value in the matrix element is obtained by multiplying the transition rate by the number

$$d\mathbf{n}_k = V\frac{d\mathbf{p}_k}{(2\pi)^3} \quad (4.193)$$

of distinguishable plane wave states in the volume $V$ corresponding to that momentum range. We arrive at a puffed up version of Fermi's golden rule:

$$d\Gamma = V\frac{|M_{fi}|^2}{\Pi_j(2E_jV)}\left[(2\pi)^4\delta(p_f - p_i)\Pi_k\frac{d\mathbf{p}_k}{2E_k(2\pi)^3}\right] \quad (4.194)$$

where the initial and final total 4-momentum vectors are $p_i = \Sigma_j p_j$ and $p_f = \Sigma_k p_k$. The factor in brackets defines the relativistic phase space density

$$dPS = (2\pi)^4\delta(p_f - p_i)\Pi_f\frac{d\mathbf{p}_f}{2E_f(2\pi)^3}. \quad (4.195)$$

The amplitude $M_{fi}$ defined with the relativistic wave function normalization and the relativistic phase space element are actually Lorentz invariant. It should be said that we have assumed that all possible final states are unoccupied. The exclusion principle may suppress transitions to occupied final states for fermions in a dense particle environment and correspondingly coherence may enhance transitions for bosons in a dense particle environment. We will not be considering such effects so do not attempt to quantify them here.

This expression applies to a transition for particles of fixed spin. When the spin states of the final particles are not measured, the observed rate is the sum over the final particles spin states which amounts to the replacement

$$|M_{fi}|^2 \rightarrow \Sigma_f |M_{fi}|^2 \tag{4.196}$$

where $\Sigma_f$ denotes summation over all spin states of all final particles. Such an expression would apply for example to a collision of a polarized electron and a polarized positron producing $\mu^+\mu^-$ the spins of which were not measured. When the initial state spins are randomly combined, the transition rate must be averaged over the distribution of initial spin states. If these spins are uncorrelated, the total transition rate to any one final state is given by the replacement

$$|M_{fi}|^2 \rightarrow < |M_{fi}|^2 >_i = \frac{1}{\Pi_i (2j_i + 1)} \Sigma_i |M_{fi}|^2 \tag{4.197}$$

where $2j_i + 1$ is the number of states for initial particle $i$, its spin being $j_i$ and $\Sigma_i$ denotes summation over all initial spin states. When neither initial nor final spins are specified and initial spins are uncorrelated, we require the replacement

$$|M_{fi}|^2 \rightarrow < \Sigma_f |M_{fi}|^2 >_i = \frac{1}{\Pi_i (2j_i + 1)} \Sigma_{f,i} |M_{fi}|^2 \equiv \overline{|M_{fi}|^2}. \tag{4.198}$$

This expression defines the spin averaged matrix element squared $\overline{|M_{fi}|^2}$. This doubly summed expression applies to collisions of unpolarized beams.

For one particle $a$ of mass $m_a$ in the initial state, factors of $V$ cancel and the differential decay rate is

$$d\Gamma_{a\rightarrow f} = \frac{|M_{a\rightarrow f}|^2}{2E} dPS = \frac{m_a}{E_a} \frac{|M^{\text{rest}}_{a\rightarrow f}|^2}{2m_a} dPS_{rest} = \gamma_a^{-1} d\Gamma^{\text{rest}}_{a\rightarrow f} \tag{4.199}$$

where $d\Gamma^{\text{rest}}_{a\rightarrow f}$ is the differential decay rate in the rest frame and $\gamma_a = E_a/m_a$ is the time dilation factor associated with changing to a frame in which the decaying particle has energy $E_a$. The total proper decay rate integrated over phase space and summed over all processes determines the proper lifetime

$$\tau = \frac{1}{\Gamma^{\text{rest}}}. \tag{4.200}$$

Since $M$, $m$, and $dPS$ are Lorentz invariant, the expression for $d\Gamma_{rest}$ is manifestly Lorentz invariant as it should be - the value of a proper quantity does not depend on frame of reference.

For two particles 1 and 2 in the initial state,

$$d\Gamma_{12\to f} = \frac{1}{2E_1 2E_2 V}|M_{12\to f}|^2 dPS. \tag{4.201}$$

The interaction cross section is defined as the interaction rate in the frame of particle 1 per unit volume density of particle 1 per unit flux of particle 2, obtained by evaluating $d\Gamma_{12\to f}$ in the frame of particle 1 and dividing by the flux $j = v_{21}/V$ where $v_{21}$ is the velocity of particle 2 relative to particle 1. An expression for $v_{21}$ follows from that for the energy of particle 2 seen from the frame of particle 1 (see Equation 2.28):

$$E_{21} = \frac{p_2 p_1}{m_1} = \frac{m_2}{\sqrt{1 - \mathbf{v}_{21}^2}} \to \mathbf{v}_{21}^2 = \frac{(p_1 p_2)^2 - m_1^2 m_2^2}{(p_1 p_2)^2}. \tag{4.202}$$

Since in the frame of particle 1, $p_1 = (E_1, 0)$, we have $p_1 p_2 = E_1 E_2$, and the differential cross section is

$$d\sigma_{12\to f} = \frac{|M_{12\to f}|^2}{4\left[(p_1 p_2)^2 - (m_1 m_2)^2\right]^{1/2}} dPS. \tag{4.203}$$

We can write this in a simpler but not manifestly covariant form using

$$\begin{aligned}\sqrt{(p_1 p_2)^2 - m_1^2 m_2^2} &= |\mathbf{p}_{cm}|\sqrt{s} = |\mathbf{p}_{cm}|\left(\frac{1}{E_{1,cm}} + \frac{1}{E_{2,cm}}\right) E_{1,cm} E_{2,cm} \\ &= (|\mathbf{v}_{1,cm}| + |\mathbf{v}_{2,cm}|) E_{1,cm} E_{2,cm}.\end{aligned} \tag{4.204}$$

If initial and final wave functions are normalized to one particle per unit volume, a convention we used in the development of the perturbation theory, a frame dependent amplitude

$$M'_{12\to f} = M_{12\to f} 4E_1 E_2 \Pi_f (2E_f) \tag{4.205}$$

is used together with the non Lorentz invariant phase space

$$dPS' = (2\pi)^4 \delta(p_f - p_i) \Pi_f \frac{d\mathbf{p}_f}{(2\pi)^3}. \tag{4.206}$$

We can express the differential cross section in the center of mass as

$$d\sigma_{12\to f} = \frac{|M'_{12\to f}|^2}{v_{1,cm} + v_{2,cm}} dPS'. \tag{4.207}$$

For two particles call them 3 and 4 in the final state, after integration over $\mathbf{p}_4$, the non Lorentz invariant phase space in the center of mass is

$$dPS' = (2\pi)^4 \delta(\sqrt{s} - E)\frac{d\mathbf{p}}{(2\pi)^6} = \frac{1}{(2\pi)^2} p^2 \frac{dp}{dE} d\Omega = \frac{d\Omega}{(2\pi)^2 v_f} p^2 \tag{4.208}$$

where $\mathbf{p} = \mathbf{p}_3 = -\mathbf{p}_4$, $E = E_3 + E_4$, and $E_{3,4} = \sqrt{m_{3,4}^2 + \mathbf{p}^2}$. We switched to spherical coordinates, $d\mathbf{p} = p^2 dp d\Omega$, and used $dE/dp = p/E_3 + p/E_4 = v_{3,cm} + v_{4,cm} \equiv v_f$. The Lorentz invariant 2-body phase space is

$$dPS = \frac{d\Omega}{16\pi^2} \frac{p^2}{p(E_3 + E_4)} = \frac{d\Omega}{16\pi^2} \frac{p}{\sqrt{s}}. \tag{4.209}$$

Let's assemble these results and calculate the decay rate for $\phi \to N\bar{N}$ in our scalar model. The Feynman diagram describes a $\phi$ field entering interaction and $N$ and $N^\dagger$ exiting. The invariant matrix element is $M_{\phi \to N\bar{N}} = g$, and the decay rate is

$$d\Gamma_{\phi \to N\bar{N}} = \frac{|M_{\phi \to N\bar{N}}|^2}{2m_\phi} \frac{d\Omega}{16\pi^2} \frac{p}{\sqrt{s}} \tag{4.210}$$

The center of mass energy is $\sqrt{s} = m_\phi$. The decay momentum is

$$|\mathbf{p}_{cm}| = \frac{1}{2\sqrt{s}} \left[\left(s - (m_N + m_{\bar{N}})^2\right)\left(s - (m_N - m_{\bar{N}})^2\right)\right]^{\frac{1}{2}} = \frac{1}{2}\left[m_\phi^2 - (2m_N)^2\right]^{\frac{1}{2}}. \tag{4.211}$$

Integration over the solid angle $d\Omega$ for the direction of the N gives a $4\pi$ and a total decay rate

$$\Gamma_{\phi \to N\bar{N}} = \frac{g^2}{2m_\phi} \frac{1}{4\pi} \frac{1}{2} \sqrt{1 - \left(\frac{2m_N}{m_\phi}\right)^2}. \tag{4.212}$$

This expression is valid only if $m_\phi > 2m_N$. Remember, $g$ in this model has dimensions of mass and so does $\Gamma_{\phi \to N\bar{N}}$. It is implicit that $m_\phi > 2m_N$.

## 4.11 ■ CROSS SECTION AND LIFETIME ESTIMATION

We will use Equation 4.194 many times, but for the moment, just notice the structure:

$$\text{Rate} = (\text{squared amplitude})\text{x}(\text{phase space}). \tag{4.213}$$

A single amplitude is proportional to a product of coupling constants and propagator factors. A propagator $G(p) \sim 1/(p^2 - m^2)$ implies a resonant enhancement if kinematics permits $p^2 \simeq m^2$ while otherwise the effects of a heavy virtual particle are suppressed. In low energy processes ($p^2 << m^2$), the propagator is effectively a constant. At high energies it decreases. On the other hand, the phase space rapidly increases with available energy. These facts and some dimensional analysis often suffice for rate estimation.

For example, the spin averaged cross section for $e^+e^- \to \mu^+\mu^-$ through a virtual photon in the limit $s >> m_\mu$ is

$$\sigma_{e^+e^-\to\mu^+\mu^-} = \frac{4\pi}{3}\frac{\alpha^2}{s}. \tag{4.214}$$

A factor $e$ corresponds to each vertex in the annihilation diagram and the center of mass energy squared is required to give the correct dimensions - it is the photon propagator ($G_\gamma \sim 1/s$) squared times a relativistic phase space factor ($\int dPS \sim$s). The numerical factor $4\pi/3$ is characteristic of the spin averaged equal left-handed and right-handed fermion couplings to the photon. The propagator replacement $s^{-2} \to (s - m_Z^2)^{-2}$ yields an estimate of the contribution of (neutral current) $Z$ boson exchange to elastic scattering and for $\sqrt{s} << m_Z$ the cross section is

$$\sigma_{e^-\mu^+\to e^-\mu^+} \sim \frac{4\pi}{3}\alpha^2\frac{s}{m_Z^4} \tag{4.215}$$

which is a factor $(\sqrt{s}/m_Z)^4$ smaller than the electromagnetic annihilation cross section. We have already estimated the cross section for the charge exchange interaction of neutrinos with matter. (See Equation 4.157.) Neutrinos only couple to stable matter via $Z$ boson or $W$ boson exchange. The minuscule cross section renders them almost invisible at low energy.

The proper top quark lifetime may be estimated with similar simple reasoning. The top quark decays to a $W$ boson plus (principally) a bottom quark. The coupling is of order $e$ and $|V_{tb}| \simeq 1$. If $m_W$ and $m_b$ are negligible, $m_t$ should set the scale. Dimensional analysis gives an expression similar to that we derived for the decay $W^- \to e^-\nu$, namely

$$\frac{1}{\tau_t} = \Gamma_t \simeq \alpha M_t = \frac{175\ \mathrm{GeV}(\mathrm{c} = 3\times10^{23}\ \mathrm{fm\ s^{-1}})}{137(\hbar\mathrm{c} = 0.2\ \mathrm{GeV\ fm})} \sim 2\times10^{24}\ \mathrm{s}^{-1}. \tag{4.216}$$

The $W$ boson mass is actually not negligible in the decay of the top quark so there is a phase space suppression and there are quite a few factors of $2\pi$ floating around, so we could be off by an order of magnitude.

The $b$-quark lifetime may similarly be estimated. The $b$-quark decays to a $c$-quark plus a virtual W$^-$ boson with $p_W^2 << m_W^2$. Kinematic constraints permit the virtual $W$ boson to decay to $e^-\bar{\nu}_e$, $\mu^-\bar{\nu}_\mu$, and $\tau^-\bar{\nu}_\tau$, or to one type of quark pair $d\bar{u}$ in three colors. Two vertices, a propagator $G_W \sim 1/m_W^2$, the CKM matrix element $|V_{bc}|$, and dimensional analysis imply

$$\frac{1}{\tau_b} \sim 6\alpha^2|V_{bc}|^2\frac{m_b^5}{m_W^4}. \tag{4.217}$$

The $b$-quark lifetime is actually $\tau_b \simeq 10^{-12}$ s. The lifetime for muon decay $\mu^- \to W^-\bar{\nu}_\mu \to (e^-\bar{\nu}_e)\nu_\mu$ should be longer by a factor

$$\frac{\tau_\tau}{\tau_b} = 6(m_b/m_\mu)^5|V_{bc}|^2 \simeq 10^6 \tag{4.218}$$

and in fact $\tau_\mu = 2 \times 10^{-6}$ s. A 50 GeV $b$-quark has a measurable mean decay length

$$l = v\gamma\tau_b = \frac{p}{m}(c\tau_b) \simeq 10(3 \times 10^{-2}\text{cm}) = 0.3 \text{ mm} \tag{4.219}$$

while at close to light speed a muon may travel hundreds of meters.

Returning to free neutron decay which is barely permitted by kinematics, we note that the mass difference $\epsilon = m_n - m_p = 1.3$ MeV governs the phase space. In fact, naively $\tau_n = (m_\mu/\epsilon)^5 \tau_\mu$ which is not that far from $\tau_n = 887$ s. (Thankfully $m_e + m_p < m_n$ so hydrogen is stable against inverse beta decay, one of several happy accidents permitting our familiar cold universe.) Our results for the phase suppression in neutron decay may be used together with CKM matrix elements to estimate the weak decay lifetimes of other hadrons. Standard model calculations which treat spin and phase space accurately provide a precise description of particle physics but the principle factors are on the table and naive analysis is often an extremely useful guide.

## 4.12 ■ FIELD OPERATORS AND PROPAGATORS

Let's look briefly at how our theory can be put on a more solid footing. Expand a complex scalar field $\phi = (\phi_1 + i\phi_2)/\sqrt{2}$ in volume $V$ in plane waves equivalent to standing wave modes

$$\phi = \int \frac{d\mathbf{p}}{(2\pi)^3} \left( a_\mathbf{p} \frac{e^{i\mathbf{p}\cdot\mathbf{x}}}{\sqrt{2E_\mathbf{p}V}} + b_\mathbf{p}^\dagger \frac{e^{-i\mathbf{p}\cdot\mathbf{x}}}{\sqrt{2E_\mathbf{p}V}} \right) \tag{4.220}$$

with amplitudes $a_\mathbf{p} \sim e^{-iE_\mathbf{p}t}$, $b_\mathbf{p} \sim e^{iE_\mathbf{p}t}$, and $E_\mathbf{p} = \sqrt{\mathbf{p}^2 + m^2}$. The energy is (no factor 1/2)

$$H = \int d\mathbf{x}\, |\partial_t\phi|^2 + |\nabla\phi|^2 + m^2|\phi^2| = \int \frac{d\mathbf{p}}{(2\pi)^3}\, E_\mathbf{p} \left( a_\mathbf{p}^\dagger a_\mathbf{p} + b_\mathbf{p} b_\mathbf{p}^\dagger \right). \tag{4.221}$$

Regard the field values at different $\mathbf{x}$, or equivalently the plane wave amplitudes, as independent quantum variables and the expansion of the field as an operator acting on a space of field states. If we treat each amplitude as a harmonic oscillator with natural frequency $E_\mathbf{p}$ and copy the operator formalism for the harmonic oscillator in nonrelativistic quantum mechanics, we can identify raising and lowering operators $a_\mathbf{p}^\dagger$ and $a_\mathbf{p}$ which increment and decrement the number of excitation quanta (create and destroy particles) and postulate the commutation relations

$$a_\mathbf{p} a_{\mathbf{p}'}^\dagger + a_{\mathbf{p}'}^\dagger a_\mathbf{p} = \delta_{\mathbf{p},\mathbf{p}'} \tag{4.222}$$

with all other operators commuting. If $|N_\mathbf{p}>$ denotes a field state containing $N_\mathbf{p}$ quanta of momentum $\mathbf{p}$, then

$$a_\mathbf{p}|N_\mathbf{p}> = \sqrt{N_\mathbf{p}}|N_\mathbf{p} - 1> = a_\mathbf{p}^\dagger|N_\mathbf{p} - 1>. \tag{4.223}$$

The operators $b_{\mathbf{p}}$ and $b_{\mathbf{p}}^{\dagger}$ describe antiparticle quanta, distinct if $\phi \neq \phi^{\dagger}$. The state empty of quanta $|0>$ is called the vacuum.

To describe the time evolution of a system including a quantum field, a Hamiltonian formalism may be used. The Hamiltonian for a free field is the energy operator

$$H = \int \frac{d\mathbf{p}}{(2\pi)^3} \, E_{\mathbf{p}} \left( a_{\mathbf{p}}^{\dagger} a_{\mathbf{p}} + b_{\mathbf{p}}^{\dagger} b_{\mathbf{p}} + 1 \right). \tag{4.224}$$

The vacuum energy is the sum of the ground state energies of all harmonic oscillators and infinite but may be ignored. The quantum mechanical Hamiltonian for a free electromagnetic radiation field in a volume $V$ may be similarly obtained by expanding the vector potential in modes and treating the amplitudes as boson operators. In Coulomb gauge, we may choose modes $\mathbf{A}_{\mathbf{p},\alpha} = \boldsymbol{\epsilon}_{\alpha} e^{-i(E_{\mathbf{p}}t - \mathbf{p}\cdot\mathbf{x})}$ with $E_{\mathbf{p}} = |\mathbf{p}|$ and $\alpha$ denoting the polarization. Computing $\mathbf{E} = \partial_t \mathbf{A}$ and $\mathbf{B} = \nabla \times \mathbf{A}$, one finds the energy

$$H_A = \frac{1}{2} \int d\mathbf{x} \, |\mathbf{E}|^2 + |\mathbf{B}|^2 = \Sigma_{\alpha} \int \frac{d\mathbf{p}}{(2\pi)^3} \, E_{\mathbf{p}} \left( a_{\mathbf{p}}^{\dagger} a_{\mathbf{p}} + \frac{1}{2} \right). \tag{4.225}$$

The total vacuum ground state energy is the sum of the infinite number of oscillator energies and is again embarrassingly but inconsequentially infinite.

A single nonrelativistic atomic electron interacting with a quantized electromagnetic radiation field is described by product states of the form $\psi = \psi_{\text{atom}} \psi_A$ where the photon wave function $\psi_A$ is a superposition of states with various excitation levels of each radiation field mode. The Hamiltonian for the system is

$$H = H_A + H_{\text{atom}} + H_{int} \tag{4.226}$$

where $H_{\text{atom}}$ is the Hamiltonian for the atom in the absence of the field and the interaction Hamiltonian is

$$H_{int} = -i\frac{e}{2m} \left[ (\nabla \cdot \mathbf{A}) + 2\mathbf{A} \cdot \nabla - \frac{e}{2m} \boldsymbol{\sigma} \cdot \nabla \times \mathbf{A} + \frac{e^2}{2m} \mathbf{A}^2 \right] \tag{4.227}$$

where $\mathbf{A}$ is a sum of raising and lowering operators. The matrix element of a lowering operator between a state with one photon of momentum $\mathbf{p}$ and a state with no photons is $< 0|a_{\mathbf{p}'\beta}|\mathbf{p}\alpha >= \delta(\mathbf{p} - \mathbf{p}')\delta_{\alpha\beta}$. The matrix element of the entire operator $\mathbf{A}$ projects out the wave function of the initial state photon.

Consider a transition between a $1s$ atomic state with one photon of momentum $\mathbf{p}$ and polarization $\alpha$ present to a $2s$ atomic state with no photon present. The transition matrix element for the system of the atom plus radiation field is found by taking the matrix element of $H_{int}$ between the product states of the atom and field. The field operator component proportional to $a_{\mathbf{p}\alpha}$ projects out the photon wave function (times a factor $\sqrt{N_{\mathbf{p}\alpha}}$ if there are $N_{\mathbf{p}\alpha}$ photons in the initial state) and we are left with

$$M_{fi} = \int d\mathbf{x} \, \psi_{\text{atom},f}^{\dagger} H_{int} \psi_{\text{atom},i} \tag{4.228}$$

where the operator **A** is replaced by the single photon wave function as we assumed for absorption. Similarly, the raising operators in **A** project into the conjugates of single photon wave functions in the matrix elements for transitions corresponding to photon emission, times a factor $\sqrt{N_{\mathbf{p}\alpha}+1}$ if there are $N_{\mathbf{p}\alpha}$ photons in the initial state. For transitions between states with at most one photon, this formalism exactly reproduces our naive application of time dependent perturbation theory to radiative transitions.

A further generalization treats both matter particles and radiation particles on an equal footing as quantum fields and assumes free quantum fields are coupled through an interaction energy density which is a product of the values of the fields and their derivatives. Consider an interaction energy between say three real fields of the form $V = g\int d\mathbf{x}\,\phi_3\phi_2\phi_1$. When expanded in terms of the mode operators, this implies matrix elements such as

$$\left(a^{\dagger}_{3,\mathbf{p}}a^{\dagger}_{2,\mathbf{q}}|0>\right)^{\dagger} V a^{\dagger}_{1,\mathbf{P}}|0> = g\delta(\mathbf{P}-\mathbf{p}-\mathbf{q}) \tag{4.229}$$

corresponding to the decay of a quantum of type 1 to a quantum of type 2 plus a quantum of type 3. If $\phi_2 = \phi_3$, the commutation relations guarantee the state with $\mathbf{p} \Leftrightarrow \mathbf{q}$ is produced with equal amplitude - states of two identical particles are symmetric with respect to interchange and scalar particles are bosons.

The canonical quantum theory of spin 1/2 fields is similar to the theory for boson fields - the field is expanded in harmonic plane waves the amplitudes of which are quantized. For example, the $R$ (+) and $L$ (-) 2-component spinor fields corresponding to a massless spin 1/2 particle have the expansion

$$\phi_{\pm} = \int \frac{d\mathbf{p}}{(2\pi)^3}\, a_{\mathbf{p},\pm}u_{\mathbf{p},\pm}\frac{e^{i\mathbf{p}\cdot\mathbf{x}}}{\sqrt{V}} + b^{\dagger}_{\mathbf{p},\pm}v_{\mathbf{p},\pm}\frac{e^{-i\mathbf{p}\cdot\mathbf{x}}}{\sqrt{V}}. \tag{4.230}$$

The free spinor fields satisfy $i\partial_t\phi_{\pm} = \pm i\nabla\cdot\sigma\phi_{\pm} = H_{\pm}\phi_{\pm}$ so $E_{\mathbf{p}}u_{\mathbf{p},\pm} = \pm\mathbf{p}\cdot\sigma u_{\mathbf{p},\pm}$ and $E_{\mathbf{p}}v_{\mathbf{p},\pm} = \pm\mathbf{p}\cdot\sigma v_{\mathbf{p},\pm}$ and for $\mathbf{p}$ along z, to within arbitrary phase factors, the polarization spinors are

$$u_+ = v_+ = \begin{pmatrix}1\\0\end{pmatrix}\ ;\ u_- = v_- = \begin{pmatrix}0\\1\end{pmatrix}. \tag{4.231}$$

Using $u^{\dagger}_{\pm}u_{\pm} = v^{\dagger}_{\pm}v_{\pm} = 1$ and $u^{\dagger}_{\mathbf{p},\pm}v_{-\mathbf{p},\pm} = 0$, the energy is found to be

$$H = \int d\mathbf{x}\,\phi^{\dagger}_{\pm}H_{\pm}\phi_{\pm} = \int d\mathbf{x}\,\phi^{\dagger}_{\pm}i\partial_t\phi_{\pm} = \int\frac{d\mathbf{p}}{(2\pi)^3}\,E_{\mathbf{p}}\left(a^{\dagger}_{\mathbf{p},\pm}a_{\mathbf{p},\pm} + b^{\dagger}_{\mathbf{p},\pm}b_{\mathbf{p},\pm} - 1\right). \tag{4.232}$$

As is readily checked, to obtain a sensible expression for the energy in terms of creation and destruction operators, we were forced to choose the so-called anti-commutation relations

$$a_{\mathbf{p},\alpha}a^{\dagger}_{\mathbf{p}',\beta} + a^{\dagger}_{\mathbf{p}',\beta}a_{\mathbf{p},\alpha} = b_{\mathbf{p},\alpha}b^{\dagger}_{\mathbf{p}',\beta} + b^{\dagger}_{\mathbf{p}',\beta}b_{\mathbf{p},\alpha} = \delta_{\mathbf{p}\mathbf{p}'}\delta_{\alpha\beta} \tag{4.233}$$

with $\alpha = \pm 1$ and $\beta = \pm 1$. (The vacuum energy of a fermion field is therefore negative and infinite and perhaps compensates the positive infinite vacuum energy of boson fields if boson and fermion degrees of freedom are equal in number as in supersymmetric theories.) It follows that $a_{\mathbf{p},\pm}a_{\mathbf{p},\pm} = a^{\dagger}_{\mathbf{p},\pm}a^{\dagger}_{\mathbf{p},\pm} = 0$ so it is not possible to create two quanta in the same state. Fermi statistics is thusly connected to positive energy being associated with both fermions and antifermions. This relationship between statistics and energy in relativistic quantum field theory is sometimes touted as an explanation of the Pauli exclusion principle.

The description of particle scattering using this quantized field theory and Hamiltonian perturbation theory, while not manifestly covariant under Lorentz transformations, may be shown to reproduce the rules we have deduced with an ad hoc derivation of Lorentz covariant transition matrix elements. In fact, our examples illustrate how Lorentz invariant amplitudes emerge from the inclusion of both particles and antiparticles in intermediate states. In particular, we have seen how the Lorentz covariant propagator emerges from combining "energy denominators" in second order "old fashioned perturbation theory." Related to this is a technical point concerning the propagator that should be mentioned.

The propagator is actually the Fourier transform a special kind of Green's function often used to describe the propagation of waves. Suppose $G(x, x')$ is a particular solution at field point $x$ to the Klein-Gordon equation with a four dimensional delta function source term at point $x'$:

$$(\partial_\mu \partial^\mu + m^2)G(x - x') = \delta(x - x'). \tag{4.234}$$

Since a general source term may be represented as

$$S(x) = \int dx' \delta(x - x')S(x'), \tag{4.235}$$

a solution to the inhomogeneous equation

$$\left(\partial_\mu \partial^\mu + m^2\right) \phi(x) = S(x) \tag{4.236}$$

may be constructed by superposition using the function $G(x, x')$:

$$\phi(x) = \phi_0(x) + \int dx' G(x - x')S(x'). \tag{4.237}$$

By substitution, it is easily seen that this is indeed a solution, provided $\phi_0(x)$ is a solution to the homogeneous equation with $S(x) = 0$. If the source term depends on a field, iterative replacement generates a series expression for the solution. For example, if $S(x) = gV(x)\phi$ with $g$ a constant, the first three terms are

$$\begin{aligned}\phi(x) &= \phi_0 + \int dx' G(x - x')gV(x')\{\phi_0(x') + \int dx'' G(x' - x'')gV(x'')\phi(x'')\\ &= \phi_0(x) + \int dx' G(x - x')gV(x')\phi_0(x')\\ &\quad + g^2 \int dx' dx'' G(x - x')V(x')G(x' - x'')\phi_0(x'').\end{aligned} \tag{4.238}$$

The first term represents no interaction. The term proportional to $g$ may be interpreted as the total amplitude for an initial particle to interact once at field points $x'$ and propagate to $x$. The term proportional to $g^2$ corresponds to the total amplitude for all double interactions in which the particle scatters at point $x''$, and propagates to and interacts at points $x'$ and propagates to $x$.

The propagator for the Klein-Gordon equation may be found by using the Fourier representations

$$G(x) = \frac{1}{(2\pi)^4}\int dp\, G(p)e^{-ipx}\ ;\ \delta(x) = \frac{1}{(2\pi)^4}\int dp\, e^{-ipx}. \tag{4.239}$$

Substitution into the defining equation gives

$$\left(\partial_\mu\partial^\mu + m^2\right)\left[\frac{1}{(2\pi)^4}\int dpG(p)e^{-ip(x-x')}\right] = \delta(x-x') = \frac{1}{(2\pi)^4}\int dpe^{-ip(x-x')} \tag{4.240}$$

from which we find $G(p)$ is just the propagator we identified in our ad hoc approach:

$$\left(-p^2+m^2\right)G(p) = 1 \rightarrow G(p) = -\frac{1}{p^2-m^2}. \tag{4.241}$$

Actually, the functions $G(x)$ and $G(p)$ are not completely defined by the previous expressions. Any solution to the homogeneous equation may be added to the propagator and it will still satisfy the defining inhomogeneous equation. Different propagators may be constructed representing different boundary conditions. Correspondingly, the Fourier transform $G(p)$ above is ill defined when $p^2 = m^2$. In quantum field theory of particles suffering localized interactions, we impose boundary conditions that in effect require matter waves propagate forwards in time while antimatter waves propagate backwards in time, and use the Feynman propagator defined by adding a small imaginary term to the invariant mass

$$m^2 \rightarrow (m - i\Gamma/2)^2 = m^2 - i\epsilon \tag{4.242}$$

with $\epsilon = m\Gamma$ and taking the limit of vanishingly small $\Gamma$. We can then write the denominator as

$$-p^2+m^2 = -p_0^2 + E^2 \rightarrow = -(p_0 - E_+)(p_0 - E_-) \tag{4.243}$$

where $E = \sqrt{\mathbf{p}^2+m^2}$ and $E_\pm = \pm E \mp i\epsilon$ and hence

$$G(x) = \frac{1}{(2\pi)^4}\int G(p)e^{-ipx} = \frac{1}{(2\pi)^4}\int dp^0 d\mathbf{p}\frac{-1}{(p_0-E_+)(p_0-E_-)}e^{-ip_0t}e^{+i\mathbf{p}\cdot\mathbf{x}}. \tag{4.244}$$

The integral over $p_0$ from $-\infty$ to $+\infty$ can be performed with the method of contour integration in complex variable analysis. The residue theorem states that counter clockwise line integral in the complex plane of a function with a simple pole of the form $g(z) = f(z)/(z - z_0)$ along a closed path enclosing the pole is

$$\int_C dz \frac{f(z)}{z - z_0} = 2\pi i f(z_0). \tag{4.245}$$

For $t > 0$, we extend the integral along the real axis to the plane of complex $p_0$ values, and choose a path along around a semicircle with negative imaginary part and infinite radius where the exponential factor $e^{itp_0}$ vanishes. The path enclosed the pole $E_+$ and (accounting for the path orientation) yields

$$\begin{aligned} G(x)(t > 0) &= \frac{1}{(2\pi)^4} \int d\mathbf{p} \left[ -2\pi i \frac{-e^{-ip_0 t}}{(p_0 - E_-)} \right] \Big|_{p_0 = E_+} e^{+i\mathbf{p}\cdot\mathbf{x}} \\ &= \frac{1}{(2\pi)^3} \int d\mathbf{p} \left[ i \frac{1}{(p_0 - E_-)} \right] \Big|_{p_0 = E_+} e^{+i\mathbf{p}\cdot\mathbf{x}} \\ &= i \frac{1}{(2\pi)^3} \int d\mathbf{p} e^{+iEt} \frac{e^{+i\mathbf{p}\cdot\mathbf{x}}}{2E}. \end{aligned} \tag{4.246}$$

For $t < 0$, we can close the path in the upper half plane and find

$$\begin{aligned} G(x)(t < 0) &= \frac{1}{(2\pi)^4} \int d\mathbf{p} \left[ +2\pi i \frac{-e^{-ip_0 t}}{(p_0 - E_+)} \right] \Big|_{p_0 = E_+} e^{+i\mathbf{p}\cdot\mathbf{x}} \\ &= \frac{1}{(2\pi)^3} \int d\mathbf{p} \left[ i \frac{-e^{-ip_0 t}}{(p_0 - E_+)} \right] \Big|_{p_0 = E_+} e^{+i\mathbf{p}\cdot\mathbf{x}} \\ &= -i \frac{1}{(2\pi)^3} \int d\mathbf{p} e^{-iEt} \frac{e^{+i\mathbf{p}\cdot\mathbf{x}}}{2E}. \end{aligned} \tag{4.247}$$

These expressions show that the Feynman propagator is a superposition of "positive energy" waves corresponding to particles propagating forwards in time and "negative energy" waves corresponding to antiparticles propagating backwards in time. It can therefore represent a particle propagating forwards in time to an interaction and onwards forwards in time or backwards in time as an antiparticle. This propagator and its interpretation coincide with our interpretation of such waves in the ad hoc approach.

## 4.13 ■ FURTHER READING

The Nobel Prize in Physics 1945 was awarded to Wolfgang Pauli "for the discovery of the Exclusion Principle, also called the Pauli Principle".

Quantum mechanics of nonrelativistic particles and systems, the theory of radiation and scattering, and an introduction to relativistic quantum mechanics are covered in many textbooks including A. S. Davydov, *Quantum Mechanics*, Pergamon (1964).

The theory of scattering of $\alpha$ and $\beta$ particles by matter and the structure of the atom is described in E. Rutherford, F.R.S., Philosophical Magazine, Series **6**, vol. 21, 669-688 (1911).

A review of nucleon electromagnetic form factors and comparison to unquenched lattice quantum chromodynamics calculations is C.F. Perdrisata, V. Punjabi, M. Vanderhaeghen, Progress in Particle and Nucl. Phys. **59**, 694 (2007).

A classic text on quantum electrodynamics covering purely electrodynamic phenomena using old fashioned time dependent perturbation theory as illustrated in this chapter is W. Heitler, *The Quantum Theory of Radiation, 3rd Ed.*, Oxford University Press (1954). The interaction $e\mathbf{E} \cdot \mathbf{r}$ instead of $e\mathbf{A} \cdot \mathbf{p}$ is discussed in Duckhwan Lee, Bull. Korean Chem. Soc. **6**, Vol. 20, 720 (1999) and in Willis Lamb, Jr., Phys. Rev. **85**, 259 (1959).

The equilibrium polarization of an electron beam in a storage ring is related to thermal radiation in an accelerated reference frame in W. G. Unruh, "Acceleration radiation for orbiting electrons," in *Quantum Aspects of Beam Physics*, Ed. Pisin Chen, World Scientific (1999), arXiv:hep-th/9804158.

An excellent text covering particle physics with the quantized field theory formalism is Otto Nachtmann, *Elementary Particle Physics: Concepts and Phenomena*, Springer-Verlag (1990).

Cross sections illustrating interferences between amplitudes for resonance production in $e^+e^- \rightarrow \pi^+\pi^-$ are given in L. M. Barkov et al., Nucl. Phys. **B256**, 365 (1985). The hadron production cross section in $e^+e^-$ near the $J\psi$ resonance is in A. M. Boyarski et al., Phys Rev. Lett. **34**, 1357 (1975).

A measurement of the neutron lifetime in a neutron storage ring making use of a sextupole magnetic field is given in W. Paul, F. Anton, L. Paul, S. Paul, and W. Mampe, Z. Phys. **C45**, 25 (1989).

Many texts describe the formalism of quantization of scalar, vector, and spinor field theories. The connection between spin and statistics that follows from requiring a positive energy for states in quantum field theory is described in W. Pauli, Phys. Rev. **58**, 716 (1940), in his Nobel Prize lecture, and in S. Weinberg, Phys. Rev. **133**, B1318 (1964). The requirement of invariance under the combination of symmetries $P$, $C$, and $T$ called the $PCT$ theorem also derived by Pauli appears in Niels Bohr and the Development of Physics: Essays Dedicated to Niels Bohr on the Occasion of His Seventieth Birthday. W. Pauli - editor, L. Rosenfeld - editor, V. Weisskopf - editor, McGraw-Hill, New York. (1955). Feynman provides another argument for the spin statistics connection using the space-time approach to quantum fields and properties of the propagators and closed loop diagrams. See R. P. Feynman, "Quantum Electrodynamics," Addison-Wesley (1961). A nonrelativistic spin statistics connection imperative is explored in a number of more recent papers.

## 4.14 ■ PROBLEMS

### Problem 4.1. Fermi's golden rule

The derivation of Fermi's golden rule starts with the expansion of the wave function $\psi = \sum_m a_m \psi_m$ with $\psi_m = \psi_m(\mathbf{x})e^{-iE_n t}$. Adopt the ansatz $a_i = e^{-i(\delta - i\Gamma/2)t}$ implying exponential decay. Here, $\delta$ is an energy shift and the total decay rate $\Gamma$ appears as an imaginary part of the energy. Show that, if $U_{ii} = 0$, the second order correction equation gives

$$\frac{i}{a_i}\frac{da_i}{dt} = \delta - i\Gamma/2 = -\sum_{n\neq i} \frac{|U_{ni}|^2}{E_i' - E_n}$$

with $E_i' = E_i + \delta - i\Gamma/2$ and then show that $\Gamma$ satisfies the implicit equation

$$\Gamma = \sum_{n\neq i} |U_{ni}|^2 \frac{\Gamma}{(E_i + \delta - E_n)^2 + \Gamma^2/4}.$$

For small $\delta$ and $\Gamma$, the normalized Cauchy-Lorentz distribution also known as the nonrelativistic Breit-Wigner function in this expression can be replaced by a Dirac $\delta$-function $f(E) = \pi^{-1}\Gamma/(E^2 + \Gamma^2) \to \delta(E)$ and we have reproduced Fermi's rule, summed over final states. For a discussion of the validity of the exponential decay law, see Mark Hillery, Phys. Rev. **A24**, 933 (1981).

### Problem 4.2. Proton form factor

Show that the form factor associated with an exponential charge distribution $\rho = \rho_0 e^{-r/R}$ is $F(q^2) = Q/(1 + \mathbf{q}^2 R^2)^2$ with $Q$ the total charge. Coulomb scattering of electrons from protons (corrected for recoil and magnetic effects) is consistent with this form with $R^{-2} = 0.71\ \text{GeV}^2$.

### Problem 4.3. Hydrogen form factor

Calculate the electric form factor and differential cross section for coherent potential scattering of a charge $e$ from the electron plus the proton in an entire hydrogen atom left in its ground state. What is the amplitude for scattering near the forward direction ($\theta = 0$)? Compare this amplitude to the case of an unscreened proton charge.

### Problem 4.4. Meson charge radii

The mean square charge radii of the stable mesons $\pi^+$, $K^+$, and $K^0$ have been measured by scattering from target electrons with the results $r_{\pi^+}^2 = 0.31 \pm 0.04$,

$r^2_{K^+} = 0.26 \pm 0.07$, $r^2_{K^0} = -(0.054 \pm 0.020)$ fm$^2$. In a nonrelativistic quark model, $\bar{r}^2_{\text{meson}} = < \Sigma_i e_i (\mathbf{r}_i - \mathbf{r}_{cm})^2 >$ where $e_i$ is the charge of the $i$-th quark, $\mathbf{r}_i$ the coordinate vector of the $i$-th quark, and the coordinate vector of the center of mass is

$$\mathbf{r}_{cm} = (m_q \mathbf{r}_q + m_{\bar{q}} \mathbf{r}_{\bar{q}})/(m_q + m_{\bar{q}}).$$

Express $r^2_{\text{meson}}$ in terms of the separation $\mathbf{r} = \mathbf{r}_q - \mathbf{r}_{\bar{q}}$ and assume effective masses $m_u = m_d = m_{\rho^0}/2$ and $m_s = m_\phi/2$ to compute $r^2_q >$ for these mesons in terms of quark antiquark separation $< \mathbf{r}^2 >$. One might expect a $u$-quark or $d$-quark to orbit a heavier $s$-quark and the charge radius of the $K^0$ to be negative. [E. Dally, Phys. Rev. Lett. **39**, 1176 (1977), A. Beretvas *et al.*, Tokyo Conference (1978) ; W. Molzon *et al.*, Phys. Rev. Lett. **41**, 1213 (1978)]

## Problem 4.5. Two-body phase space

Give an expression for the decay rate $\Gamma_{W^- \to e^- \bar{\nu}_e}$ using the simplified matrix element $M = g_W/\sqrt{2m_W}$ in the $W$ boson rest frame assuming $m_{\bar{\nu}} = 0$ but without neglecting the electron mass. Express your result in terms of $x = (m_e/m_W)^2$. For what value of electron mass would the decay rate be reduced by a factor of two?

## Problem 4.6. Muon capture

A $\mu^-$ stopped in liquid hydrogen rapidly cascades to the ground state of a muonic atom $(\mu^- p)_{\text{atom}}$. The atom is destabilized by muon decay and by the capture reaction $\mu^- p \to \nu_\mu n$. a) Manipulate $\sigma_{\nu_\mu e \to \mu \nu_e}$ to justify the estimate

$$\sigma_{\mu^- p \to \nu_\mu n} = \frac{1}{v^i_{rel}} \frac{g^4}{m_W^4} \frac{p^2}{\pi v^f_{rel}}$$

where $v^i_{rel}$ and $v^f_{rel}$ are the relative velocities of the particles in the initial and final states. b) The capture reaction rate is $\Gamma_{\mu^- p \to \nu_\mu n} = n v_{rel} \sigma_{\mu^- p \to \nu_\mu n}$ with $n = |\psi(0)|^2 \sim \alpha^2 m_\mu^3$. Show the capture rate is approximately

$$\Gamma_{\mu^- p \to \nu_\mu n} = \frac{1}{\pi^2} g^4 \alpha^3 \left(\frac{m_\mu}{m_W}\right)^4 m_\mu$$

or about $10^{-3}$ of the rate for decay $\mu^- \to \nu_\mu e^- \bar{\nu}_e$. The observed capture lifetime for the spin-singlet $1s$ state is about 1.6 ms. c) Muon decay leads to a delayed electron track while capture yields none. What is the neutron energy and range in liquid hydrogen? d) If the electron mass exceeded $m_n - m_p$, a neutral hydrogen would disappear and the neutron would be stable. Estimate the rate for muonium atom disappearance via

$$\mu^+ + e^- \to \nu_e + \bar{\nu}_\mu$$

and compare to the muon lifetime. Does the muonium atom generally disappear into invisible neutrinos or will a fast positron be observed? [Ref: G. Bardin *et al.*, Nucl. Phys. **A352**, 365 (1981)]

---

**Problem 4.7. Lifetime of the $\tau^-$**

---

Use the estimates of the $\mu^-$ and $b$-quark lifetimes to estimate the lifetime of the $\tau^-$ lepton in picoseconds summed over leptonic and hadronic (colored quark) decays. Compare your estimate to the experimental result in the particle data tables. Do the branching fractions of the leptonic and hadronic decays make sense?

---

**Problem 4.8. Kaon semi-leptonic decay**

---

One decay mode of the $K^-$ is the semi-leptonic process (called $Ke3$)

$$K^- \rightarrow \pi^0 e^- \bar{\nu}_e.$$

Estimate the lifetime for this weak decay mode by scaling from $\mu^- \rightarrow \nu_\mu e^- \bar{\nu}_e$ adjusting for the reduced phase space, the pion flavor wave function, and the CKM matrix element $V_{us}$ and compare to Particle Data Group lifetime $\tau_{K^-} = 1.24 \times 10^{-8}$s and branching fraction $\Gamma_{K\rightarrow\pi^0 e^- \bar{\nu}_e}/\Gamma_K = 5.1\%$.

---

**Problem 4.9. Simplified neutron decay phase space**

---

Assume in $n \rightarrow p\nu e$ that the proton is left stationary in the neutron rest frame - it recoils absorbing momentum with negligible kinetic energy. In this case, the differential phase space is that for a 2-body decay without a constraint on momentum, namely

$$dPS = 2\pi\delta(Q - E_\nu - E_e)\frac{d\mathbf{p}_e}{(2\pi)^3}\frac{d\mathbf{p}_\nu}{(2\pi)^3}$$

with $Q = m_n - m_p$ and $E_e = \sqrt{\mathbf{p}_e^2 + m_e^2}$. Integrate over angles and neutrino energy to derive the expression $dPS = dE_e E_e p_e (Q - E_e)^2 dE_e/(2\pi^3)$ and compare to the spectrum determined in the text from the 3-body phase space.

---

**Problem 4.10. Magnetic dipole radiation**

---

A muon with magnetic moment is at rest in a constant magnetic field of strength $B$. The matrix element for decay from the higher energy magnetic substate to the lower energy magnetic substate is

$$M_{fi} = \int d\mathbf{x}\psi^\dagger(\mathbf{x})\eta_f^\dagger\left[-\frac{e}{2m_\mu}\boldsymbol{\sigma}\cdot\nabla\times\mathbf{A}\right]\psi(\mathbf{x})\eta_i$$

where $\boldsymbol{\mu}_\mu = (e/2m_\mu)\boldsymbol{\sigma}$, $\psi$ is the muon space wave function, $\eta_i$ and $\eta_f$ denote initial and final muon spin wave functions, and $\mathbf{A} = (2E)^{-1/2}\epsilon_\alpha e^{i\mathbf{k}\cdot\mathbf{x}}$ is the wave function of the final photon. Assume the wavelength of the radiation is large compared to the space wave function of the muon and show that the decay rate is

$$\Gamma = \frac{2}{3\pi\epsilon_0}\frac{|\mu_\mu|^2\omega^3}{\hbar c^5} = \frac{16}{3\pi\epsilon_0}\frac{|\mu_\mu|^5 B^3}{\hbar^4 c^5}.$$

Find the lifetime for $B = 1$ T and compare it to the age of the universe. Scale your result to find the radiative lifetimes for a neutron and for an electron in a 1 T field.

---

**Problem 4.11. Vector meson radiative decay**

---

In the nonrelativistic quark model, the spin magnetic moment operator for a $q\bar{Q}$ system is

$$\boldsymbol{\mu} = \mu_q\boldsymbol{\sigma}_q + \mu_{\bar{Q}}\boldsymbol{\sigma}_{\bar{Q}}.$$

For an $L = 0$ meson, the diagonal matrix element describes the static magnetic moment (e.g. $|\boldsymbol{\mu}_{\rho^+}| = \mu_u + \mu_{\bar{d}}$) while the magnetic dipole radiative decay rate from the $s = 1$ vector state to the $s = 0$ (pseudo)scalar ground state is

$$\Gamma_{V\to P+\gamma} = \frac{2}{3\pi}k^3 \,|< V \mid \boldsymbol{\mu} \mid P >|^2$$

where $k$ is the photon energy. If the spatial wave functions are the same in the (pseudo)scalar and vector $q\bar{Q}$ mesons,

$$|< V \mid \boldsymbol{\mu} \mid P >| = \frac{1}{\sqrt{2}} \mid \mu_{\bar{Q}} - \mu_q \mid .$$

Assume magnetic moments $\mu_q = e_q/2m_q$ with effective masses $m_u = m_d = 363\ MeV$, $m_c = 1.5$ GeV to compute the rate for $\rho^- \to \pi^-\gamma$ and $J/\psi \to \eta_c\gamma$ and compare with the experimental results in the data tables. This calculation assumes a long wavelength approximation which is not really valid and neglects recoil effects.

---

**Problem 4.12. $Z'$ factory**

---

Suppose experiments discover a 10 TeV mass spin 1 neutral vector boson $Z'$ coupled exclusively to all standard model charged and neutral fermions as if they had identical charge. An $e^+e^-$ collider is built to produce it on resonance with a luminosity of $10^{35}$ cm$^{-1}$ s$^{-1}$. Assuming the collision energy spread is small compared to the natural width of the $Z'$, how many $e^+e^- \to Z' \to t\bar{t}$ events are produced in one year of continuous operation?

**Problem 4.13. Ice Cube neutrino detector**

The Ice Cube Neutrino Observatory is a cube of ice one km on a side instrumented with photodetectors at the South Pole. The top of the cube is at a depth of 1400 m. The inclusive cross section for muon neutrino charged current weak interactions on nucleons is

$$\sigma_{\nu_\mu N \to \mu^- X} = 0.67 \times 10^{-42}\ \mathrm{m}^2 \times \frac{E_\nu}{1\ \mathrm{GeV}}.$$

What is the probability a 10 GeV muon neutrino at the surface going straight down interacts within the detector?

**Problem 4.14. Natural width of the $W$ boson**

In the process $u\bar{d} \to W^+ \to e^+\nu_e$ at a $p\bar{p}$ collider, the longitudinal momentum of the $W^+$ is unknown and only the positron is observable. Suppose in the $W^+$ rest frame the angular distribution of the positron has the form

$$\frac{dN}{d\Omega} = f(\sin\theta)$$

where $\theta$ is the polar angle relative to the proton direction. Change variables to $p_t = (m_W/2)\sin\theta$ and show that

$$\frac{dN}{dp_t} = 2\pi f(2p_t/m_W)\frac{2p_t/m_W}{\sqrt{1-\left(\frac{2p_t}{m_W}\right)^2}}.$$

Since $p_t$ is invariant under longitudinal boost, this distribution applies the laboratory frame and exhibits a "Jacobian peak" at $p_t = m_W/2$. From the smearing of this peak as shown in Figure 4.19, the natural width $\Gamma_W$ can be measured.

**Problem 4.15. The decay $K_L \to \nu + (\pi^\mp\mu^\pm)_{\text{atom}}$**

The final states in a 3-body decay $a \to b + c + d$ include a 2-body subset in which $b$ and $c$ form an atom, $(bc)_{\text{atom}}$.

a) Show that the nonrelativistic wave function of such an atom with momentum $\mathbf{Q}$ has the form

$$\psi_{\text{atom}}(\mathbf{x}_\mathrm{b}, \mathbf{x}_\mathrm{c}) = \int \mathbf{dq}\ \psi_\mathrm{b}\left(\frac{\mathrm{m_b}}{\mathrm{m}}\mathbf{Q} + \mathbf{q}\right)\psi_\mathrm{c}\left(\frac{\mathrm{m_c}}{\mathrm{m}}\mathbf{Q} - \mathbf{q}\right)\psi(\mathbf{q})$$

where $m = m_b + m_c$ and $\psi(\mathbf{q}) = \int d\mathbf{x}\ e^{i\mathbf{q}\cdot\mathbf{x}}\psi(\mathbf{x})$ is the momentum space representation of the center of mass wave function $\psi(\mathbf{x})$ for the relative coordinate $\mathbf{x}_1 - \mathbf{x}_2$.

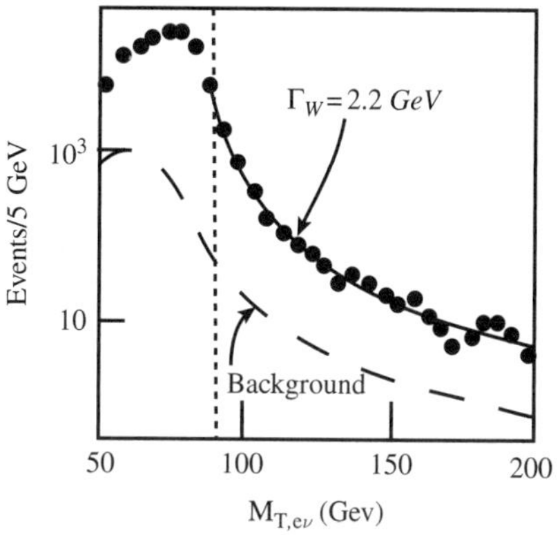


**FIGURE 4.19** Transverse mass distribution for $W \to e\nu$ events a $p\bar{p}$ collider. The estimated background is also shown. From the tail above the peak, the natural width of the W-boson is inferred. [CDF collaboration, Phys. Rev. Lett. **100**, 071801 (2008).].

b) For states normalized to one per unit volume, if $M'(\mathbf{p}_a, \mathbf{p}_b, \mathbf{p}_c, \mathbf{p}_d)$ is the decay amplitude for momenta $\mathbf{p}_a \to \mathbf{p}_b + \mathbf{p}_c + \mathbf{p}_d$, the amplitude to form the atom is

$$M'_{\text{atom}} = \int d\mathbf{q}\; M'(\mathbf{p}_a, \mathbf{q} + (m_b/m)\mathbf{Q}, -\mathbf{q} + (m_c/m)\mathbf{Q}, \mathbf{p}_d)\psi^*(\mathbf{q}).$$

Show that if the variation in $M'$ over the range of $\mathbf{q}$ is neglected, then

$$M'_{\text{atom}} = M(\mathbf{p}_a, (m_b/m)\mathbf{Q}, (m_c/m)\mathbf{Q}, \mathbf{p}_d)\psi^*(0)$$

where $\psi(0)$ is the atomic wave function at the origin $|\mathbf{x}_b - \mathbf{x}_c| = 0$. c) The Lorentz invariant amplitude is

$$M_{\text{atom}} = M'(\mathbf{p}_a, (m_b/m)\mathbf{Q}, (m_c/m)\mathbf{Q}, \mathbf{p}_d)\psi^*(0)\sqrt{2E_{\text{atom}}/(2E_b)(2E_c)}$$

where the wave function normalization factors evaluated in the atom frame amount to $(2\mu)^{-1/2}$ with $\mu = m_b m_c/(m_b + m_c)$ the reduced mass. The rate for the 3-body decay is $d\Gamma(a \to b, c, d) = |M|^2 dPS_3$ while the rate for the 2-body decay is $d\Gamma(a \to (bc), d) = |M_{\text{atom}}|^2 dPS_2 = |M|^2|\psi(0)|^2 dPS_2$. Supposing $|M|$ is constant, the branching ratio to atoms is

$$P_{\text{atom}} = |\psi(0)|^2 PS_2/PS_3.$$

The branching fraction $\Gamma(K_L \to \pi^{\pi}\mu^{\mp}\nu) = (27.04 \pm 0.07\ \%$. Use these factors to estimate the branching fraction for $K_L \to (\pi^{\mp}\mu^{\pm})_{\text{atom}}\nu$ summed over all atomic states and compare to the measured value $(1.05 \pm 0.11) \times 10^{-7}$. [ S. H. Aronson *et al.*, "Measurement of the rate of formation of pi-mu atoms in $K_L^0$ decay," Phys. Rev. **D33**, 3180 (1986).]

## Problem 4.16. Neutron decay to hydrogen

The neutron decay $n \to (pe^-)_{\text{atom}} \bar{\nu}_e$ produces a hydrogen atom of kinetic energy 326.5 eV plus an antineutrino of energy 783 keV in the neutron rest frame. a) Verify these values. b) Estimate the branching ratio using the formula

$$\frac{\Gamma_{n \to H + \bar{\nu}_e}}{\Gamma_{n \to p + e^- \bar{\nu}_e}} = |\psi(0)|^2 PS_2 / PS_3$$

where $\psi(0) = (n^3 \pi a_0^3)^{-1/2}$ is the wave function at the origin squared,

$$PS_2 = p^2 (dp/dE^{tot})/\pi = \frac{p^2}{\pi(v_H + v_\nu)}$$

is the phase space for the 2-body decay for which the momentum is $p$, and $PS_3 = \epsilon^5 I/(2\pi)^3$ is the 3-body phase space with $I = 0.0157$ and $\epsilon = m_n - m_p = 1.29$ MeV. A professional calculation gives $4 \times 10^{-6}$. [W. Schott *et al.*, "An experiment for the measurement of the bound-beta decay of the free neutron," Eur. Phys. J. **A30**, 603 (2006) and M. Faber et al.,"On Continuum-State and Bound-State $\beta^-$ Decay Rates of the Neutron," Phys. Rev. **C80**, 035503 (2009).]

## Problem 4.17. Phase space in $K_{\mu 3}$ and $K_{e3}$ decays

The decays $K_L \to \pi^- \mu^+ \nu_\mu$ (27.0%) and $K_L \to \pi^- e^+ \nu_e$ (40.5 %) both proceed via a $s \to uW^-$ transition yielding $K_L \to \pi^+ W^-$. Phase space gives the lepton energy spectrum

$$dPS = \frac{1}{4\pi^3} E_l dE_l \left[ (m_K - E_l) \frac{E_-^2 - E_+^2}{2} + \frac{1}{3} \left( E_-^3 - E_+^3 \right) \right]$$

with $E_\pm = (m_K^2 - m_\pi^2 + m_l^2)/2 - m_K E_l)/(m_K - E_l \mp p_l)$. a) Integrate the spectrum numerically for the two decays and compare the phase space ratio to the ratio of branching ratios. b) Estimate the phase space ratio assuming a two body phase space proportional to $p^2$ with $p$ the maximum neutrino energy. c) Estimate the phase space ratio assume $PS \sim \epsilon^5$ using $\epsilon = m_K - m_\pi - m_l$ and also $\epsilon = E_e^{\max}$.

## Problem 4.18. Scattering of atoms

Suppose a bound state of two particles with total mass $m = m_1 + m_2$ is described by a two particle wave function $\psi(\mathbf{x}_1, \mathbf{x}_2) = e^{i\mathbf{p}\cdot\mathbf{r}} \phi(\mathbf{x})$ with $\mathbf{x}_{cm} = (m_1 \mathbf{x}_1 + m_2 \mathbf{x}_2)/m$, $\mathbf{r} = \mathbf{x}_2 - \mathbf{x}_1$. The plane wave represents the motion of the center of mass with atom momentum $\mathbf{p}$. The internal wave function $\phi$ depends on the relative coordinates $\mathbf{r}$. If the particles carry equal and opposite charge $e$, the matrix element for scattering in an external potential $U$with an internal transition $i \to f$ is

$$S_{fi} = -ie2\pi\delta(E_f - E_i) \int d\mathbf{x}_1 d\mathbf{x}_2 \, \psi_f^\dagger [U(\mathbf{x}_1) - U(\mathbf{x}_2)] \, \psi_i.$$

a) Show that this may be written as $S_{fi} = -2\pi i \delta(E_f - E_i) M_{fi}$ with

$$M_{fi} = eU(\mathbf{q})[F_{fi}(-m_2\mathbf{q}/m) - F_{fi}(m_1\mathbf{q}/m)]$$

with the Fourier transform of the potential and transition density

$$U(\mathbf{q}) = \int d\mathbf{x}\, e^{i\mathbf{q}\cdot\mathbf{x}} U(\mathbf{x}) \; , \; F_{fi}(\mathbf{q}) = \int d\mathbf{x}\, e^{i\mathbf{q}\cdot\mathbf{x}} \phi_f^\dagger(\mathbf{x})\phi_i(\mathbf{x}).$$

b) Show that $M_{fi} = 0$ for $\mathbf{q} = 0$ and that, for $m_1 = m_2$, the amplitude for elastic scattering vanishes. c) The total cross sections summed over elastic, excitation, and ionization final states for scattering of tiny relativistic exotic atoms in their ground state from a Cu atom target, are calculated to be

$$\begin{aligned}
\sigma^{tot}(e^+e^-)_{\text{atom}} &= 1.0 \times 10^{-18}\ \text{cm}^2 \\
\sigma^{tot}(e^+\mu^-)_{\text{atom}} &= 8.2 \times 10^{-19}\ \text{cm}^2 \\
\sigma^{tot}(\mu^+\mu^-)_{\text{atom}} &= 12.5 \times 10^{-21}\ \text{cm}^2 \\
\sigma^{tot}(\pi^+K^-)_{\text{atom}} &= 6.5 \times 10^{-22}\ \text{cm}^2.
\end{aligned}$$

The density of Cu is $\rho = 8.96$ gm cm$^{-3}$ and the mass number is $A = 63.5$. Find and compare the mean free paths of a positronium atom and of a $(\pi^+K^-)_{\text{atom}}$ in Cu. [Ref: Stanislaw Mrowczynski, "Interactions of elementary atoms with matter," Phys. Rev.**A33**, 1549 (1986).]

---

### Problem 4.19. Low energy neutrino scattering

---

Use the spin independent amplitude $M = g_{eeZ}g_{\nu\nu Z}/m_Z^2$ to calculate the cross section for a massless muon neutrino with energy $E_\nu$ to scatter elastically from a stationary electron and show that $\sigma_{\nu_\mu e^- \to \nu_\mu e^-} \propto E_\nu^2$ for $E_\nu << m_e$ while $\sigma_{\nu_\mu e^- \to \nu_\mu e^-} \propto E_\nu m_e$ for $m_e << E_\nu << m_Z^2/(2m_e)$.

---

### Problem 4.20. Casimir-Lifshitz force

---

A long range attractive Van der Waals force appears between two metal plates with small separation and has an exploding literature connected with nanotechnology. The force was first derived by Casimir [H. B. G. Casimir, Proc. K. Ned. Akad. Wet. **51**, 793 (1948)] from the dependence on plate separation of the zero temperature vacuum energy for infinite conductivity, suggesting (erroneously) that the vacuum alone holds energy. A much more detailed description was formulated by Lifshitz [E. M. Lifshitz, Zh. Eksp. Teor. Fiz. **29**, 94 (1956)] who recognized the implicit role of charges in the walls in Casimir's calculation and showed the force could be attractive or repulsive depending on the material properties. For a review and connections to Hawking radiation, see S.K. Lamoreaux, Phys. Rev. Lett.**79**, 5(1997) and for massive vector fields see Lee Peng Teo, Phys. Rev. **D82**, 105002 (2010).

Using the ground state energy $\hbar\omega/2$ of each mode, show that the quantum ground state energy $U(b)$ per unit area between plates of separation $b$ and area $L^2$

with $L >> b$ for a massless scalar field subject to Dirichlet boundary conditions $\phi = 0$ appropriate for perfect conductors has the form

$$U(b) - U_0 = \frac{\hbar c}{16\pi^2} \int dk_x dk_y \left( \Sigma_{n=-\infty}^{+\infty} \sqrt{k_x^2 + k_y^2 + \left(\frac{\pi n}{b}\right)^2} \right.$$

$$\left. - \int_{-\infty}^{+\infty} dn \sqrt{k_x^2 + k_y^2 + \left(\frac{\pi n}{b}\right)^2} \right)$$

$$= \frac{\hbar c}{8\pi} \int_0^\infty dx \left( \left( \int_0^\infty dn \sqrt{x + \left(\frac{n\pi}{b}\right)^2} - \Sigma_{n=0}^{\infty} \sqrt{x + \left(\frac{n\pi}{b}\right)^2} + \frac{1}{2}\sqrt{x} \right) \right.$$

where the transverse degrees of freedom are treated as continuous and the energy for infinite $b$ is subtracted. Write this as

$$\Delta U(b) = -\frac{\hbar c}{8\pi} [ \int_0^\infty dn\ f(n) - \Sigma_{n=0}^{\infty} f(n) + \frac{1}{2} f(0)$$

with

$$f(n) \equiv \int_0^{x_m} dx \sqrt{x + \left(\frac{n\pi}{b}\right)^2}$$

where $\sqrt{x_m}c$ is some cut-off frequency such as the plasma frequency of the metal. Use the Euler-Maclauring formula

$$\int_0^\infty dn\ f(n) - \Sigma_{n=0}^{\infty} f(n) = -\frac{1}{2} f(0) + \frac{1}{6 \times 2!} f^{(1)}(0) + \frac{1}{30 \times 4!} f(3)(0) + \cdots$$

to find for large $x_m b^2$ that only the term proportional to the third derivative $f^{(3)}(0)$ contributes so the energy and force are

$$U_{\text{EM}} = 2\Delta U(b) = \frac{\pi^2 \hbar c}{720 b^3};\ F = \frac{dU_{\text{EM}}}{db} = -\frac{\pi^2 \hbar c}{240 b^4}$$

where we multiplied by two to (crudely) account for polarization of the electromagnetic field. This derivation is from F. J. Yndurain, “Relativistic Quantum Mechanics and Introduction to Field Theory,” Springer-Verlag (1996).

CHAPTER

# 5 Electrodynamics of Bosons

The theory of quantum electrodynamics for relativistic charged bosons introduces a number of key concepts. The scalar case is generalized to describe the electrodynamics of charged vector bosons such as the $W^{\pm}$. We derive the field equations and calculate illustrative fundamental processes.

## Contents



## 5.1 ■ FIELD EQUATIONS OF SCALAR ELECTRODYNAMICS

In this chapter we develop the quantum field theory of spin 0 charged particles interacting with an electromagnetic field and the generalization to the electrodynamics of spin 1 particles. Along the way, we will encounter the Lagrangian formulation of quantum field theories, the simplest gauge theory, covariant perturbation theory, and formulas for the cross sections for a variety of relativistic processes. There are no known elementary charged scalar particles so the theory is something of a laboratory. The theory of scalar electrodynamics does apply to charged mesons such the $\pi^{\pm}$ and $K^{\pm}$ for momentum transfers which do not resolve their structure. It also applies to proposed supersymmetric scalar electrons and muons, the selectron $\tilde{e}^{-}$ and smuon $\tilde{\mu}^{-}$, and to charged Higgs particles which appear in theories extending the standard model.

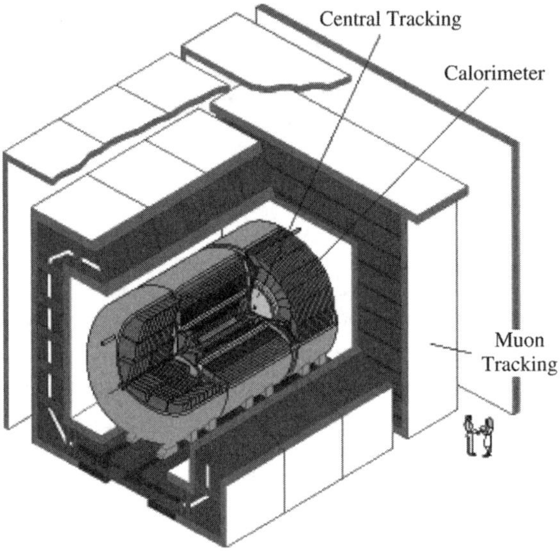


**FIGURE 5.1** Schematic of the D0 experiment at the Fermilab Tevatron. Weighing 5000 tons, the detector features a scintillating fiber tracking system, a stable liquid argon based calorimeter and extensive external muon detectors. The quantum electrodynamics of $W^{\pm}$ weak bosons has been studied at the Tevatron by the D0 and CDF collaborations. [D0 public picture gallery]

A pedestrian derivation of the fundamental relativistic field equations requires little more than de Broglie's idea of relativistic matter waves expressed by the Klein-Gordon equation, a generalization of the electromagnetic interaction in Schrödinger theory, and an appropriate incorporation of Maxwell's equations. Start with the Klein-Gordon wave equation 1.49 for a complex Lorentz scalar field $\phi(x) = (\phi_1 + i\phi_2)/\sqrt{2}$. In terms of the 4-momentum operator

$$p_\mu = i\partial_\mu = (i\partial_t, i\nabla)\,, \tag{5.1}$$

the free particle wave equation is

$$\left(-p_\mu p^\mu + m^2\right)\phi = \left(\partial_\mu\partial^\mu + m^2\right)\phi = \left(\frac{\partial^2}{\partial t^2} - \nabla^2 + m^2\right)\phi = 0. \tag{5.2}$$

Plane wave solutions are

$$\phi = \frac{1}{\sqrt{2E}}e^{\mp ipx} = \frac{1}{\sqrt{2E}}e^{\mp i(Et - \mathbf{p}\cdot\mathbf{x})} \tag{5.3}$$

with $E = \sqrt{\mathbf{p}^2 + m^2}$ and can be associated with a charged scalar particle and its antiparticle. As described in Chapter 1, the free particle Schrödinger equation is converted into an equation describing a particle of charge $e$ ($e = -1.602 \times 10^{-19}$ C for an electron) subject to an electromagnetic potential by the substitution

$$p^\mu = (E, \mathbf{p}) \to p^\mu - eA^\mu = \left(E - eA^0, \mathbf{p} - e\mathbf{A}\right) \tag{5.4}$$

where the potential functions are written as a 4-vector

$$A^\mu = \left(A^0, \mathbf{A}\right). \tag{5.5}$$

The substitution rule is equivalent to the replacement of the ordinary 4-derivative $\partial^\mu$ with the so-called gauge covariant derivative operator $D^\mu$ defined by

$$\partial^\mu \to D^\mu \equiv \partial^\mu + ieA^\mu. \tag{5.6}$$

The result of this substitution in the Klein-Gordon equation is simply

$$\left[D_\mu D^\mu + m^2\right]\phi = 0. \tag{5.7}$$

When terms describing interactions are moved to the right-hand side of the equation, the equation takes the explicit form

$$\partial_\mu \partial^\mu \phi + m^2 \phi = -ie\left[\partial_\mu \left(A^\mu \phi\right) + A^\mu \partial_\mu \phi\right] + e^2 A_\mu A^\mu \phi. \tag{5.8}$$

Exact solutions for $\phi$ may be found for a Coulomb potential. We will not derive them but simply observe that, for fixed point nucleus with atomic number $Z$, the energy spectrum of bound states can be written as

$$E_{n,l} = -\frac{m}{2}\left(\frac{Z\alpha}{n}\right)^2 \left[1 + \left(\frac{Z\alpha}{n}\right)^2 \left(\frac{n}{l + \frac{1}{2}} - \frac{3}{4} - \cdots\right]\right. \tag{5.9}$$

where terms of order $(Z\alpha)^6$ and higher have been neglected. Such a spectrum applies to pionic and kaonic atoms for which the meson is well within the radius of the innermost electron. Notice that states with different values of orbital quantum number $l$ for the same principal quantum number $n$ are not degenerate.

Gauge invariance lies behind the substitution rule. As described below, the electromagnetic potential $A^\mu$ may be subject to a gauge transformation

$$A^\mu \to A^\mu + \partial^\mu \chi \tag{5.10}$$

where $\chi$ is an arbitrary scalar function. Combined with a position dependent phase transformation

$$\phi(x) \to U(x)\phi(x) = e^{-ie\chi(x)}\phi(x), \tag{5.11}$$

the transformation given by Equation 5.10 leaves Equation 5.8 unchanged except for an inessential phase factor. In fact, when both $A^\mu$ and $\phi$ are transformed, we find the covariant derivative transforms in a simple homogeneous way:

$$D_\mu \phi \to \left(\partial_\mu - ieA_\mu - ie\partial_\mu \chi\right) e^{ie\theta}\phi = e^{ie\chi} D_\mu \phi \equiv U D_\mu \phi. \tag{5.12}$$

Equivalently, we can write the simultaneous transformation as

$$\begin{aligned}\phi &\to U\phi \\ D_\mu &\to U D_\mu U^{-1}\end{aligned} \tag{5.13}$$

and so we find that $D_\mu \phi$ transforms like $\phi$:

$$D_\mu \phi \to \left(U D_\mu U^{-1}\right)(U\phi) = U D_\mu \phi. \tag{5.14}$$

It follows that the entire Klein-Gordon equation is transformed as follows:

$$\left[D_\mu D^\mu + m^2\right]\phi = 0 \to U\left[D_\mu D^\mu + m^2\right]\phi = 0. \tag{5.15}$$

Since the factor $U$ may be cancelled, the transformed equation is equivalent to the original. The equation is said to be gauge covariant.

To arrive at an interacting theory, we need to specify how currents associated with the charged particles generate potential functions $A^\mu$ so turn to Maxwell's equations. Recall how the homogeneous Maxwell equations in classical electrodynamics

$$\nabla \cdot \mathbf{B} = 0\,, \quad \nabla \times \mathbf{E} - \partial_t \mathbf{B} = 0 \tag{5.16}$$

permit the the electric and magnetic fields to be expressed in terms of potentials. Since the divergence of a curl vanishes,

$$\nabla \cdot \mathbf{B} = 0 \to \mathbf{B} = \nabla \times \mathbf{A} \tag{5.17}$$

and, since the curl of a gradient vanishes,

$$0 = \nabla \times \mathbf{E} + \partial_t \mathbf{B} = \nabla \times (\mathbf{E} + \partial_t \mathbf{A}) \to \mathbf{E} = -\nabla A^0 - \partial_t \mathbf{A}. \tag{5.18}$$

The electric and magnetic fields are actually components of a Lorentz covariant field strength tensor $F^{\mu\nu}$ expressible in terms of the 4-vector of potential functions

$$F^{\mu\nu} = \partial^\mu A^\nu - \partial^\nu A^\mu = \begin{pmatrix} 0 & -E_x & -E_y & -E_z \\ E_x & 0 & -B_z & B_y \\ E_y & B_z & 0 & -B_x \\ E_z & -B_y & B_x & 0 \end{pmatrix}. \tag{5.19}$$

By definition, the components of $A^\mu$ in different frames of reference are related like the components of $x^\mu$ and the components of a 2-index tensor like $F^{\mu\nu}$ transform like the products of 4-vector components. These expressions define the Lorentz transformations of $\mathbf{E}$ and $\mathbf{B}$. The charge density $\rho$ and 3-space current density $\mathbf{j}$ which appear in the inhomogeneous Maxwell equations are components of the 4-current density

$$j^\mu = (\rho, \mathbf{j}) \tag{5.20}$$

which transforms as a 4-vector. The inhomogeneous equations in Lorentz covariant form are (with $\epsilon_0$, $\mu_0$, and $c$ equal to 1)

$$\partial_\nu F^{\nu\mu} = \partial_\nu \partial^\nu A^\mu - \partial^\mu(\partial_\nu A^\nu) = j^\mu \sim \begin{pmatrix} \nabla \cdot \mathbf{E} = \rho \\ \nabla \times \mathbf{B} - \partial_t \mathbf{E} = \mathbf{j} \end{pmatrix}. \tag{5.21}$$

For example, the time component is Gauss's Law

$$\partial_0 F^{00} + \partial_1 F^{10} + \partial_2 F^{20} + \partial_3 F^{30} = j^0 \equiv \nabla \cdot \mathbf{E} = \rho. \tag{5.22}$$

Current conservation is built in. A 4-divergence of 5.21 gives

$$\partial_\nu \partial_\mu F^{\mu\nu} = \partial_\nu j^\nu. \tag{5.23}$$

The left side vanishes identically since $\partial_\mu \partial_\nu = \partial_\nu \partial_\mu$ while $F^{\mu\nu} = -F^{\nu\mu}$ so we must have

$$\partial_\nu j^\nu = 0. \tag{5.24}$$

This equation is called the equation of continuity for charge and is usually written as

$$\partial_t \rho = -\nabla \cdot \mathbf{j}. \tag{5.25}$$

Integrating this equation over a volume and converting the right-hand side to a surface integral gives

$$\frac{dQ}{dt} \equiv \frac{d}{dt} \int d\mathbf{x}\, \rho = -\int d\mathbf{x}\, \nabla \cdot \mathbf{j} = -\int d\mathbf{a} \cdot \mathbf{j} \tag{5.26}$$

which implies the rate of change of charge $Q$ in the volume is equal to the net flux into the volume. Maxwell's equations do not specify $j^\mu$. In classical physics, current is associated with conserved particles in motion subject to the Lorentz force and charge conservation is assumed. There would be inconsistency if the current density was not conserved. When particles bearing charge are created and destroyed, charge conservation is a more delicate and interesting matter. The Klein-Gordon equation plays the role of the Lorentz force equation for quantum scalar particles.

The field strength tensor $F^{\mu\nu}$ is invariant under the gauge transformation

$$A^\mu \to A^\mu + \partial^\mu \chi \tag{5.27}$$

for any scalar field $\chi$. In fact, the identity $\partial^\mu \partial^\nu \chi = \partial^\nu \partial^\mu \chi$ implies

$$F^{\mu\nu} \to \partial^\mu (A^\nu + \partial^\nu \chi) - \partial^\nu (A^\mu + \partial^\mu \chi) = F^{\mu\nu}. \tag{5.28}$$

By virtue of gauge invariance, the Lorentz condition $\partial_\mu A^\mu = 0$ may be imposed by choosing a gauge transformation with $\chi$ a solution to the Poisson equation

$$\partial_\mu \partial^\mu \chi = -\partial_\mu A^\mu. \tag{5.29}$$

Given the Lorentz condition, the inhomogeneous Maxwell equations (see Equation 5.21) reduce to four independent massless wave equations

$$\frac{\partial^2}{\partial t^2}A^\mu - \nabla^2 A^\mu = j^\mu. \tag{5.30}$$

The photon is massless but we should point out here that a massive vector field satisfies the Proca equation

$$\partial_\nu F^{\nu\mu} + m_A^2 A^\mu = j^\mu \tag{5.31}$$

and the 4-divergence of this equation gives

$$m_A^2 \partial_\mu A^\mu = \partial_\mu j^\mu. \tag{5.32}$$

This equality implies that, if the current is conserved, the Lorentz condition is required as an external condition leaving three degrees of freedom. The mass term violates gauge invariance and is excluded from electrodynamics. We also note that the classical electromagnetic field is real corresponding to the fact that the photon is electrically neutral. A complex vector field $A^\mu$ corresponding to a charged vector particle will be considered below.

In Maxwell's equations, the current density is gauge invariant. To complete the gauge covariant theory of a scalar field $\phi$ interacting with the vector field $A^\mu$, we must find a conserved gauge invariant current density associated with $\phi$ to serve as a source and sink for the field $A^\mu$. For a free field $\phi$, the current density

$$j^\mu = ie\left(\phi^\dagger \partial_\mu \phi - \phi\left(\partial_\mu \phi\right)^\dagger\right) \tag{5.33}$$

is the natural generalization of the nonrelativistic probability current density. The divergence of this current is

$$\partial_\mu j^\mu = ie\left(\partial_\mu \phi^\dagger \partial^\mu \phi - \partial_\mu \phi \partial^\mu \phi^\dagger\right) + ie\left(\phi^\dagger \partial_\mu \partial^\mu \phi - \phi \partial_\mu \partial^\mu \phi^\dagger\right). \tag{5.34}$$

The first two terms cancel and the free field equation and its conjugate imply the next two also cancel so $\partial_\mu j^\mu = 0$. However, as may be easily verified, this current is not conserved when $\phi$ interacts with $A^\mu$ and obeys Equation 5.8.

A reasonable guess for a conserved current in the interacting case follows from the substitution $\partial_\mu \to D_\mu$ in the free field current density with care taken to obtain a real result:

$$j^\mu = ie\left[\phi^\dagger D^\mu \phi - \phi\left(D^\mu \phi\right)^\dagger\right] = ie\left[\phi^\dagger \partial^\mu \phi - \phi \partial^\mu \phi^\dagger\right] - 2e^2 \phi^\dagger \phi A^\mu. \tag{5.35}$$

It is left as an exercise to show that Equation 5.7 and its complex conjugate imply $\partial_\mu j^\mu = 0$ identically. The curious term proportional to $e^2$ in the current density is essential to charge conservation. Collecting our results and moving interaction terms to the right-hand side of the equations, we arrive at the coupled equations

$$\partial_\mu \partial^\mu \phi + m^2 \phi = -ie\left[\partial_\mu \left(A^\mu \phi\right) + A^\mu \partial_\mu \phi\right] + e^2 A_\mu A^\mu \phi \tag{5.36}$$

$$\partial_\nu \partial^\nu A^\mu - \partial^\mu (\partial_\nu A^\nu) = ie\left[\phi^\dagger \partial^\mu \phi - \phi \partial^\mu \phi^\dagger\right] - 2e^2 \phi^\dagger \phi A^\mu \tag{5.37}$$

which form the basis for scalar electrodynamics.

The Lorentz covariance of these equations turns out to imply conservation of energy and of linear and angular momentum. The gauge covariance is connected to charge conservation. The equations are invariant under the parity inversion $\mathbf{x} \to -\mathbf{x}$ combined with

$$\mathbf{A}(\mathbf{x}) \to -\mathbf{A}(-\mathbf{x}) \text{ and } \phi(\mathbf{x}) \to \pm\phi(-\mathbf{x}). \tag{5.38}$$

Inversion symmetry permits the intrinsic parity of the $\phi$ to be positive or negative. The equations are also invariant under the charge conjugation ($C$) transformation

$$\phi \to \phi^\dagger \text{ and } A^\mu \to -A^\mu \tag{5.39}$$

which replaces charged particles with antiparticles reversing electromagnetic potentials. In this sense, a photon is an odd eigenstate of $C$ while a charged particle is not a $C$ eigenstate.

## 5.2 ■ LAGRANGIAN FIELD THEORY

Wave equations for interacting fields are most easily derived from a scalar function $L$ (Lagrangian) of the field components and their (first) derivatives using, for each field component $\phi$, the Euler-Lagrange equation

$$\partial_\mu \frac{\partial L}{\partial\left[\partial_\mu \phi(x)\right]} = \frac{\partial L}{\partial \phi(x)}. \tag{5.40}$$

The Euler-Lagrange equation is $dp/dt = F$ for fields. It follows from a principle of stationary action as does a similar equation in the Lagrangian mechanics of a classical system of a few degrees of freedom. The action $S$ is defined as the integral over space-time of the Lagrangian. For independent small variations $\delta\phi$ and $\delta\partial_\mu\phi$ of the values of the field and of its derivatives at each point in space-time, the variation of the action is

$$\delta S \equiv \delta \int dx\; L = \int dx\; \frac{\partial L}{\partial \phi}\delta\phi + \frac{\partial L}{\partial\left(\partial_\mu \phi\right)}\delta\left(\partial_\mu \phi\right) \tag{5.41}$$

$$= \int dx \left[\frac{\partial L}{\partial \phi} - \partial_\mu \frac{\partial L}{\partial\left(\partial_\mu \phi\right)}\right]\delta\phi + \int dx\; \partial_\mu \left[\frac{\partial L}{\partial\left(\partial_\mu \phi\right)}\delta\phi\right]. \tag{5.42}$$

The gradient resulting from the 4-dimensional integration by parts may be dropped assuming the fields vanish at $\infty$. Stationary action $\delta S = 0$ for arbitrary functions $\delta\phi(x)$ requires Euler's equation.

An example or two may make sense of the Euler-Lagrange equations. For a free real scalar field, the Lagrangian is

$$L_\phi = \frac{1}{2}\partial_\mu \phi \partial^\mu \phi - \frac{1}{2} m^2 \phi^2. \tag{5.43}$$

From this, one finds the derivatives

$$\frac{\partial L}{\partial\left(\partial_\mu \phi\right)}=\partial^\mu \phi\ ;\ \frac{\partial L}{\partial \phi}=-m^2 \phi \tag{5.44}$$

and the Euler-Lagrange equation is the Klein-Gordon equation

$$\partial_\mu \partial^\mu \phi+m^2 \phi=0. \tag{5.45}$$

Notice in this example that because the Lagrangian is a Lorentz scalar, its derivative with respect to a 4-vector like $\partial_\mu \phi$ is a 4-vector with index in the opposite position upstairs or downstairs while its derivative with respect to a scalar is a scalar. The Euler-Lagrange equation is therefore Lorentz covariant. The Lagrangian for a pair of equal mass scalar fields $\phi_1$ and $\phi_2$ can be expressed in terms of a complex scalar field $\phi=(\phi_1+i\phi_2)/\sqrt{2}$ and has the form

$$\begin{aligned} L=L_{\phi_1}+L_{\phi_2} &= \frac{1}{2}\partial_\mu \phi_1 \partial^\mu \phi_1-\frac{1}{2} m^2 \phi_1^2+\frac{1}{2} \partial_\mu \phi_2 \partial^\mu \phi_2-\frac{1}{2} m^2 \phi_2^2 \\ &= \partial_\mu \phi^{\dagger} \partial^\mu \phi-m^2 \phi^{\dagger} \phi. \end{aligned} \tag{5.46}$$

Independent variations of $\phi_1$ and $\phi_2$ or of $\phi$ and $\phi^\dagger$ treated as independent variables yield a wave equation for each. A term in the Lagrangian which is a product of different field components leads to coupled field equations, e.g.

$$L_{12}=\frac{1}{2} \partial_\mu \phi_1 \partial^\mu \phi_1+\frac{1}{2} \partial_\mu \phi_2 \partial^\mu \phi_2-V\left(\phi_1, \phi_2\right) \rightarrow\left(\begin{array}{l}\partial_\mu \partial^\mu \phi_1=-\frac{\partial V}{\partial \phi_1} \\ \partial_\mu \partial^\mu \phi_2=-\frac{\partial V}{\partial \phi_2}\end{array}\right). \tag{5.47}$$

The wave equation for a free real vector field

$$\partial_\mu \partial^\mu A^\nu-\partial_\mu \partial^\nu A^\mu+m^2 A^\nu=0 \tag{5.48}$$

results from the Lagrangian

$$L_A=-\frac{1}{4} F_{\mu \nu} F^{\mu \nu}+\frac{1}{2} m^2 A_\mu A^\mu \tag{5.49}$$

where $F^{\mu\nu}=\partial^\mu A^\nu-\partial^\nu A^\mu$ and the space components of the mass term $m\mathbf{A}^2/2$ appear in $L_A$ with sign as for the scalar field and that dictates the minus sign in front of the first term in order to arrive at sensible wave equations. To derive the wave equations, write

$$\frac{\partial F_{\mu \nu}}{\partial\left(\partial_i A_j\right)}=\delta_{\mu i} \delta_{\nu j}-\delta_{\nu i} \delta_{\mu j} \tag{5.50}$$

and then

$$\begin{aligned} -\frac{\partial L}{\partial\left(\partial_\rho A_\sigma\right)} &= \frac{1}{4} \frac{\partial F_{\mu \nu} F^{\mu \nu}}{\partial\left(\partial_\rho A_\sigma\right)}=\frac{1}{2} F^{\mu \nu} \frac{\partial F_{\mu \nu}}{\partial\left(\partial_\rho A_\sigma\right)} \\ &= \frac{1}{2} F^{\mu \nu}\left(\delta_{\mu \rho} \delta_{\nu \sigma}-\delta_{\nu \rho} \delta_{\mu \sigma}\right)=F^{\rho \sigma}. \end{aligned} \tag{5.51}$$

The Euler-Lagrange equation for the vector field is therefore

$$-\partial_\rho \frac{\partial L}{\partial\left(\partial_\rho A_\sigma\right)}+\frac{\partial L}{\partial A_\sigma}=\partial_\rho F^{\rho\sigma}+m^2 A^\sigma=0 \tag{5.52}$$

and, for $m=0$, has the form of Maxwell's equations for a free electromagnetic field.

If the Lagrangian is invariant under a transformation of the fields, the corresponding equations are covariant and a general connection called Noether's theorem between symmetries and conservation laws can be established. For each continuous symmetry generated from small changes, there corresponds a locally conserved charge. In particular, for a transformation of the form

$$\phi_i \rightarrow \phi_i+\delta\phi_i=\left[1+\theta T_{ij}\right]\phi_j, \tag{5.53}$$

where $\theta$ is some parameter and $T$ is some matrix, the change in $L$ is

$$\begin{aligned} \delta L &= \frac{\partial L}{\partial\phi_i}\delta\phi_i+\frac{\partial L}{\partial\left(\partial_\mu\phi_i\right)}\partial_\mu\left(\delta\phi_i\right) \\ &= \left[\frac{\partial L}{\partial\phi_i}-\partial_\mu\frac{\partial L}{\partial\left(\partial_\mu\phi_i\right)}\right]\delta\phi_i+\partial_\mu\left(\frac{\partial L}{\partial\left(\partial_\mu\phi_i\right)}\delta\phi_i\right) \end{aligned} \tag{5.54}$$

Invariance of $L$ implies $\delta L=0$ and, since the term in square brackets vanishes for fields satisfying the Euler-Lagrange equation, we deduce a local conservation law in the form of a current the 4-divergence of which vanishes:

$$\partial_\mu\left[\frac{\partial L}{\partial\left(\partial_\mu\phi_i\right)}\delta\phi_i\right]=\partial_\mu j^\mu=0. \tag{5.55}$$

To see the connection between the condition

$$\partial_\mu j^\mu=\partial_t j^0+\nabla\cdot\mathbf{j}=0 \tag{5.56}$$

and a conserved quantity, integrate over a closed volume in space and apply the divergence theorem to convert the space part to an outward pointing surface integral:

$$\int d\mathbf{x}\,\partial_\mu j^\mu=0\rightarrow\partial_t\int d\mathbf{x}\,j^0=-\int d\mathbf{a}\cdot\mathbf{j}. \tag{5.57}$$

We can identify $j^0$ as a density of some sort of charge and $\mathbf{j}$ as an associated three current density and the charge in the arbitrary volume changes only by virtue of flowing from place to place.

When $\theta$ is independent of position, the current may be taken to be

$$j^\mu=\frac{\partial L}{\partial\left(\partial_\mu\phi_i\right)}T_{ij}\phi_j \tag{5.58}$$

For example, if $V = V(\phi_1^2 + \phi_2^2)$ in Equation 5.47, then $L_{12}$ is symmetric under a kind of flavor rotation

$$\begin{pmatrix} \phi_1 \\ \phi_2 \end{pmatrix}' = \begin{pmatrix} \cos\theta & \sin\theta \\ -\sin\theta & \cos\theta \end{pmatrix} \begin{pmatrix} \phi_1 \\ \phi_2 \end{pmatrix}. \tag{5.59}$$

Expanding for small theta, we have

$$T = \begin{pmatrix} 0 & 1 \\ -1 & 0 \end{pmatrix} \tag{5.60}$$

and find a conserved current density

$$j^\mu = (\partial^\mu \phi_1)\,\phi_2 - (\partial^\mu \phi_2)\,\phi_1 = i\left[\phi^\dagger \partial^\mu \phi - \phi \partial^\mu \phi^\dagger\right] \tag{5.61}$$

where $\phi = [\phi_1 + i\phi_2]/\sqrt{2}$. For the Lagrangian $L_{12} = |\partial_\mu \phi|^2 - V(|\phi|^2)$, the symmetry is a phase symmetry of the complex field. The derivation of energy-momentum conservation and angular momentum conservation laws from translational and rotational symmetry of the action may be found in most field theory text books. Turning this result around, if we start with a Lorentz scalar Lagrangian invariant under translation and rotation, the resulting dynamical equations are guaranteed to conserve energy and momentum and angular momentum.

The power of the Lagrangian formulation may be illustrated with a derivation of the coupled equations of scalar electrodynamics. Start with the free field Lagrangian for the free electromagnetic field and a complex scalar field:

$$L_0 = L_A + L_\phi = -\frac{1}{4} F_{\mu\nu} F^{\mu\nu} + \left(\partial_\mu \phi\right)^\dagger (\partial^\mu \phi) - m^2 \phi^\dagger \phi. \tag{5.62}$$

The first term $L_A$ is invariant under gauge transformations $A^\mu \to A^\mu - \partial^\mu \chi$. The remainder is invariant under a position independent (global) phase transformation. The substitution $\partial_\mu \to D_\mu = \partial_\mu + ieA_\mu$ leads to a Lagrangian coupling the fields $A^\mu$ and $\phi$

$$L = -\frac{1}{4} F_{\mu\nu} F^{\mu\nu} + \left(D_\mu \phi\right)^\dagger (D^\mu \phi) - m^2 \phi^\dagger \phi = L_0 + L_{int} \tag{5.63}$$

where the interaction corresponds to the following terms

$$L_{int} = -e\left(\phi^\dagger i \partial_\mu \phi - \phi i \partial_\mu \phi^\dagger\right) A^\mu + e^2 A^2 \phi^\dagger \phi. \tag{5.64}$$

From this Lagrangian, the equations for both the scalar field and the electromagnetic field of scalar electrodynamics follow. Notice we were careful to write $\partial_\mu \phi^\dagger \partial^\mu \phi$ as $(\partial_\mu \phi)^\dagger \partial^\mu \phi$ in order to obtain a real Lagrangian after the conversion of partial derivatives to gauge covariant derivatives.

The Lagrangian is invariant under the simultaneous transformations

$$\begin{aligned} \phi(x) &\to e^{ie\theta(x)} \phi(x) \equiv U\phi \\ A^\mu &\to A^\mu - \partial^\mu \theta. \end{aligned} \tag{5.65}$$

For such a transformation, $F^{\mu\nu}$ and $\phi^\dagger\phi$ are invariant while

$$D_\mu\phi \to \left(D_\mu - ie\partial_\mu\theta\right)e^{ie\theta}\phi = e^{ie\theta}D_\mu\phi = UD_\mu\phi \tag{5.66}$$

implies the kinetic plus interaction term is invariant:

$$\left(D_\mu\phi\right)^\dagger D_\mu\phi \to (U\phi)^\dagger UD^\mu\phi = \left(D_\mu\phi\right)^\dagger U^\dagger UD^\mu\phi = \left(D_\mu\phi\right)^\dagger D_\mu\phi. \tag{5.67}$$

We have unearthed here a technique for constructing a set of field equations consistent not just with conservation of energy momentum and angular momentum, but also with a dynamical symmetry, by simply writing down a Lagrangian invariant under that symmetry. Next we have deduced that the presence of the symmetry is linked to a conservation law. The electric current density associated with the gauge symmetry in the Lagrangian for scalar electrodynamics is readily found to be the conserved electric current density guessed above. Conservation of electric charge is connected with symmetry under local U(1) phase transformations. In Chapter 7, we will study the general theory of gauge symmetry. Incidentally, one can show that $\partial \to D_\mu = \partial_\mu + ieA_\mu$ in the free field Lagrangian leads to same equations as obtained substituting $\partial_\mu \to D_\mu$ in the free field equations. (Consider field variations with $A$ fixed so $\delta\partial_\mu\phi = \delta D_\mu\phi - ieA_\mu\delta\phi$.)

## 5.3 ■ VECTOR FIELD PLANE WAVES

The preceding sections introduced 4-vector fields and it is important to detail the properties of free field solutions. A vector field plane wave corresponding to a free particle of momentum $p$ has the form

$$A^\mu = \epsilon^\mu e^{\pm ipx} \tag{5.68}$$

where the polarization 4-vector $\epsilon$ plays the role of a spin wave function. For a wave representing $2EV$ particles in volume V, $\epsilon$ may be chosen to satisfy the normalization condition

$$\epsilon^\dagger \cdot \epsilon = -1. \tag{5.69}$$

A wave corresponding to one particle per unit volume is then

$$A^\mu = \frac{1}{\sqrt{2E}}\epsilon^\mu e^{\pm ipx}. \tag{5.70}$$

The normalization $\epsilon^2 = -1$ is a matter of convenience. The magnitude of the normalization appears consistent with that for scalar particles given both satisfy the Klein-Gordon equation.

The Proca equation for a massive vector field

$$\partial_\mu\left(\partial^\mu A^\nu - \partial^\nu A^\mu\right) + m^2 A^\nu = 0 \tag{5.71}$$

is modeled after Maxwell's equations. The mass term actually implies a constraint that reduces the number of degrees of freedom from four to the three that correspond to a spin 1 particle (see Equation 5.32). In fact, if we take the 4-divergence of this equation, then we find

$$\partial_\nu \partial_\mu \left(\partial^\mu A^\nu - \partial^\nu A^\mu\right) + m^2 \partial_\nu A^\nu = 0 \tag{5.72}$$

and the first term vanishes identically so the Lorenz condition must be satisfied. The Lorentz condition $\partial_\mu A^\mu = 0$ requires for a plane wave that $p \cdot \epsilon = 0$ which implies that in the rest frame of a massive particle, $\epsilon^0 = 0$. So the polarization vector in the rest frame has the form $\epsilon = (0, \mathbf{e})$. The three vector $\mathbf{e}$ can be expressed in terms of a basis of three orthonormal polarization vectors $\mathbf{e}_\alpha$ with $\alpha = 1, 2, 3$ satisfying the orthonormality condition

$$\mathbf{e}_\alpha^\dagger \cdot \mathbf{e}_\beta = \delta_{\alpha\beta}. \tag{5.73}$$

A Cartesian basis such as

$$\mathbf{e}_1 = \hat{x}, \ \mathbf{e}_2 = \hat{y}, \ \text{and } \mathbf{e_3} = \hat{\mathbf{y}} \tag{5.74}$$

corresponds to linear polarization states. Alternative helicity basis states illustrated in Figure 5.2 are defined as eigenstates of the spin operator $\mathbf{s}$ along the direction $\mathbf{n}$ of momentum $\mathbf{p}$ and are solutions to the equations

$$\mathbf{s} \cdot \mathbf{n}\mathbf{e}_\alpha = \alpha \mathbf{e}_\alpha \tag{5.75}$$

Here the three $3 \times 3$ spin matrices $\mathbf{s}^\alpha$ (i = 1,2,3) for spin 1 (corresponding to the three $2 \times 2$ Pauli matrices $\sigma^\alpha$ for spin 1/2 ) obey the algebraic conditions

$$\mathbf{s}^\alpha \mathbf{s}^\beta - \mathbf{s}^\beta \mathbf{s}^\alpha = i\epsilon^{\alpha\beta\gamma} \mathbf{s}^\gamma \tag{5.76}$$

and may be represented by

$$[\mathbf{s}^\alpha]^{\rho\tau} = -i\epsilon_{\alpha\rho\tau} \tag{5.77}$$

with $\epsilon_{\alpha\beta\gamma}$ the completely antisymmetric 3-tensor. For momentum along the $z$-direction, explicit solutions to

$$\mathbf{s} \cdot \hat{\mathbf{z}}\mathbf{e}_\alpha = \alpha \mathbf{e}_\alpha \tag{5.78}$$

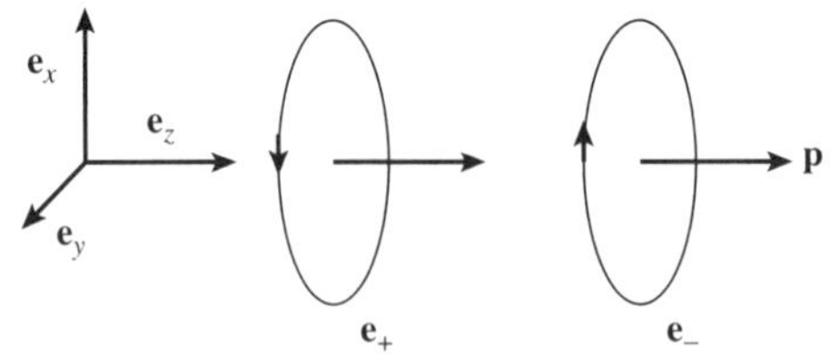


**FIGURE 5.2** Helicity states of a photon moving along a direction $\mathbf{e}_z$. The polarization vectors $\mathbf{e}_\pm = (\mathbf{e}_x \pm i\mathbf{e}_y)/\sqrt{2}$ represent spin angular momentum along the direction of momentum $\mathbf{s} \cdot \mathbf{p}/|\mathbf{p}| = \pm\hbar$.

are two transverse states with $\alpha = \pm 1$

$$\mathbf{e}_{\pm} = [1, \pm i, 0]/\sqrt{2} \tag{5.79}$$

and a longitudinal state

$$\mathbf{e}_0 = [0, 0, 1] \tag{5.80}$$

with $\alpha = 0$. Solutions for spin $+1$, 0 and $-1$ along an arbitrary rest frame direction $\mathbf{n}$ may be found by rotation of these three states. The solutions associated with a wave of momentum $p = \gamma m(1, \mathbf{v})$ may be found by Lorentz transformation from the rest frame. For example, for a wave in the $z$-direction, the boosted form of the polarization 4-vectors are

$$\epsilon_{+1} = (0, 1, i, 0) \ ; \ \epsilon_0 = [\gamma v, 0, 0, \gamma] \ ; \ \epsilon_{-1} = [0, 1, -i, 0]\,. \tag{5.81}$$

For a massless vector field such as the electromagnetic field, the Lorenz condition $\partial_\mu A^\mu = 0$ is optional and the particle and polarization interpretation is subtle. The transverse helicity states $(0, \mathbf{e}_\pm)$ correspond to classical circularly polarized radiation waves and to photons with spin $\pm 1$ along the momentum, while the longitudinal polarization $(0, \mathbf{e}_0)$ and scalar $(1, 0)$ in a general gauge conspire to represent a Coulomb interaction. In fact, if instead of the covariant Lorenz gauge condition $\partial_\mu A^\mu = 0$ that is required of a massive vector field, we use the frame dependent Coulomb gauge condition

$$\nabla \cdot \mathbf{A} = 0, \tag{5.82}$$

then $\mathbf{p} \cdot \mathbf{e} = 0$ for plane waves and the inhomogeneous Maxwell equations reduce to a wave equation (Ampere's Law)

$$\partial_t^2 \mathbf{A} - \nabla^2 \mathbf{A} = \mathbf{j} + \nabla \partial_t A^0 \tag{5.83}$$

plus an equation (Gauss's Law) through which $A^0$ is determined instantaneously by the charge density:

$$-\nabla^2 A^0 = \rho. \tag{5.84}$$

In this gauge, $A^0$ represents an instantaneous Coulomb interaction which should first be found and used in the wave equation and that equation then turns out to involve only transverse polarization states. In Lorenz gauge, as we will see, we must allow four (two transverse plus longitudinal and scalar) polarization states with negative normalization of the scalar state for intermediate state photons. But free photons in initial and final states are of the transverse variety only.

Anticipating the Feynman rules derived below, another difference between massless and massive vector particles should be pointed out. The Lorentz invariant amplitude $M$ for a process involving an external massive vector particle with polarization vector $\epsilon$ will have the form $M = \epsilon^\dagger J$ or $\epsilon J^\dagger$ where $J$ is some 4-vector. The transition probability summed over polarization states is therefore of the form

$$\sum_i |M|^2 = \sum_i \epsilon_i^\mu \epsilon_i^{\nu\dagger} J_\mu^\dagger J_\nu = \rho^{\mu\nu} J_\mu^\dagger J_\nu \tag{5.85}$$

where we introduce the polarization tensor

$$\rho^{\mu\nu} \equiv \sum_{i=1}^{3} \epsilon_i^\mu \epsilon_i^{\nu\dagger}. \tag{5.86}$$

The sum over polarization in intermediate states implies this polarization tensor also appears in the numerator of the propagator for the vector particle. It is a Lorentz 4-tensor and arguably must have the tensor structure

$$\rho^{\mu\nu} = a g^{\mu\nu} + b\frac{p^\mu p^\nu}{p^2} \tag{5.87}$$

where $a$ and $b$ are constants since the sum treats all orientations of the polarization vector equally. The inner product with $p$ and the Lorentz condition $\epsilon \cdot p = 0$ imply $a = -b$. Then the trace is $\rho_\mu^\mu = -3$ so we find that

$$\rho^{\mu\nu} = -g^{\mu\nu} + \frac{p^\mu p^\nu}{p^2}. \tag{5.88}$$

We define the propagator for a vector boson as

$$G_V^{\mu\nu}(p) = -i\frac{g_{\mu\nu} - \frac{p^\mu p^\nu}{p^2}}{p^2 - m_V^2}. \tag{5.89}$$

The second term appears problematic in the limit of zero mass but for photons does not contribute as photons interact with conserved currents - the photon is always radiated or absorbed by a current $J \sim e^{ipx}$ and $\partial J = 0 \rightarrow pJ = 0$, a fact connected with gauge invariance. Hence the second term in $\rho$ may be disregarded for photons and the factor $-g^{\mu\nu}$ thus effectively represents the sum over photon polarizations in the numerator of the photon propagator

$$G_\gamma^{\mu\nu}(p) = \frac{-ig^{\mu\nu}}{p^2}. \tag{5.90}$$

If, for example, $\mathbf{p}$ is along the 3-direction, current conservation implies

$$pJ = 0 \rightarrow p^0 J^0 - |\mathbf{p}| J^3 = 0 \rightarrow J^0 = J^3 \tag{5.91}$$

so using two transverse polarizations along the 1-direction and 2-direction, we have

$$\begin{aligned}\sum_{i=1,2} |M|^2 &= \sum_i \epsilon_i^\mu \epsilon_i^{\nu\dagger} J_\mu^\dagger J_\nu = |J^1|^2 + |J^2|^2 \\ &= |J^1|^2 + |J^2|^2 + |J^3|^2 - |J^0|^2 = -g_{\mu\nu} J^\mu J^\nu.\end{aligned} \tag{5.92}$$

Summed over the two transverse real photon polarizations, a squared matrix element is $-J^\dagger J$ as if the sum included longitudinal and scalar external photons as well. This trick applies to an entire gauge invariant amplitude. Current conservation implies that, if we replace $\epsilon$ by the photon momentum in a matrix element, it must vanish.

In summary, a massive vector field has plane wave solutions with three linearly independent polarization states corresponding to spin 1. A vector field mass term is inconsistent with gauge invariance and excluded from electrodynamics. Free photons are massless and have only two transverse polarization states. In intermediate states, longitudinal and scalar photons must be included to provide a complete representation of the electrodynamic interaction in Lorenz gauge.

## 5.4 ■ FEYNMAN RULES FOR SCALAR ELECTRODYNAMICS

Covariant Feynman rules emerge from a perturbative evaluation of the field equations

$$\partial_\mu \partial^\mu \phi + m^2 \phi = -ie\left[\partial_\mu \left(A^\mu \phi\right) + A^\mu \partial_\mu \phi\right] + e^2 A_\mu A^\mu \phi \tag{5.93}$$

$$\partial_\nu \partial^\nu A^\mu - \partial^\mu \left(\partial_\nu A^\nu\right) = ie\left[\phi^\dagger \partial^\mu \phi - \phi \partial^\mu \phi^\dagger\right] - 2e^2 \phi^\dagger \phi A^\mu \tag{5.94}$$

together with the conjugates. We follow the model of purely scalar field interactions in Chapter 4.

Suppose a photon described by $A^\mu = \epsilon^\mu e^{-ikx}$ encounters a charged scalar particle described by $\phi_1 = e^{-ip_1 x}$. To order $e$, these waves in the first term on the right-hand side of Equation 5.93, constitute a source for $\phi$ so we write the approximate equation

$$\partial_\mu \partial^\mu \phi + m^2 \phi = -ie\left[\partial_\mu \left(A^\mu \phi\right) + A^\mu \partial_\mu \phi\right] = -e\left[\left(p_1 + k\right) + p_1\right] \epsilon e^{-i(k+p_1)x}. \tag{5.95}$$

The amount of $\phi_2 = e^{-ip_2 x}$ produced is the projection

$$S_{\phi_1 A \to \phi_2} = i \int dx\, \phi_2^\dagger \left[-e\left[\left(p_1 + k\right) + p_1\right] e^{-i(k+p_1)x} = iM \left(2\pi\right)^4 \delta\left(p_1 + k - p_2\right) \tag{5.96}$$

where the amplitude for absorption is

$$iM_{\phi_1 A \to \phi_2} = \left(-ie\right)\left(p_2 + p_1\right)\epsilon. \tag{5.97}$$

Given an incoming wave $\phi = \phi_1$ and an exiting wave $\phi^\dagger = \phi_2^\dagger$ in the first term of Equation 5.93, the amplitude for a photon $A = \epsilon e^{-ikx}$ to be radiated is the projection (with a minus sign since $\epsilon^\dagger \epsilon = -1$)

$$S_{\phi_1 \to \phi_2 A} = -i \int dx\, A^\dagger \left[ie\left[\phi^\dagger \partial^\mu \phi - \phi \partial^\mu \phi^\dagger\right] = iM \left(2\pi\right)^4 \delta\left(p_1 - k - p_2\right) \tag{5.98}$$

with emission amplitude

$$iM_{\phi_1 \to \phi_2 A} = (-ie)\epsilon^\dagger (p_2 + p_1) . \tag{5.99}$$

The transition current density $j = e(p_1 + p_2)$ appears as in nonrelativistic quantum mechanics.

Second order processes are represented by elastic scattering of a scalar electron $\tilde{e}^-$ of initial momentum $p_1$ and final momentum $p_3$ and scalar muon $\tilde{\mu}^-$ of initial momentum $p_2$ and final momentum $p_4$

$$\tilde{e}_1 \tilde{\mu}_2 \to \tilde{e}_3 \tilde{\mu}_4 \tag{5.100}$$

by single photon exchange as illustrated in Figure 5.3. To describe this process, suppose two fields $\phi_e$ and $\phi_\mu$ and both associated currents appearing in Maxwell's equations. The current density for charge $e$ is

$$j_\nu = ie\left(\phi^\dagger \partial_\nu \phi - \phi \partial_\nu \phi^\dagger\right) - 2e^2 \phi^\dagger \phi A_\nu . \tag{5.101}$$

Put $\phi \to \phi_1 = e^{-ip_1 x}$ and $\phi^\dagger \to \phi_3^\dagger = e^{ip_3 x}$ and neglect the term proportional to $e^2$. The selectron transition current is

$$j_{13} = ie\left(\phi_3^\dagger \partial \phi_1 - \phi_1 \partial \phi_3^\dagger\right) \tag{5.102}$$

The associated electromagnetic field may be found using Equation 5.94 and the Lorentz gauge condition $\partial_\mu A^\mu$:

$$\partial_\nu \partial^\nu A^\mu = j_{13} = e\left(p_1^\mu + p_3^\mu\right) e^{iqx} \to A^\mu = \frac{-1}{q^2} e\left(p_1^\mu + p_3^\mu\right) e^{iqx} \tag{5.103}$$

with $q = p_3 - p_1$ the momentum transfer. To order $e$, Equation 5.93 for the smuon scalar field is

$$\left(\partial_\mu \partial^\mu + m^2\right) \phi = -ie\left(\partial_\mu A^\mu + A^\mu \partial_\mu\right) \phi \tag{5.104}$$

where the derivatives act on everything to their right. Use the approximate electromagnetic field and the muon field $\phi_2 = e^{-ip_2 x}$ on the right-hand side and treat

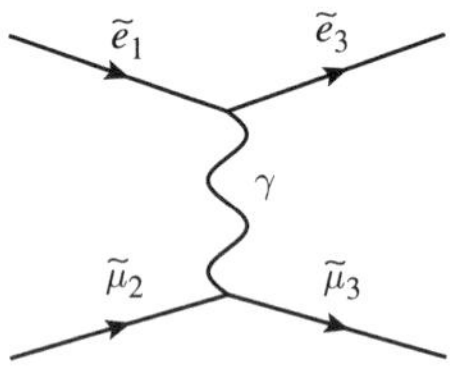


**FIGURE 5.3** Feynman diagram for smuon-selectron elastic scattering. The process $\tilde{e}^- \tilde{\mu}^- \to \tilde{e}^- \tilde{\mu}^-$ by single photon exchange is described by a single relativistic diagram. The relativistic diagram includes both time orderings of emission and absorption of the intermediate state photon.

it as a fixed source. The amount of $\phi_4 = e^{-ip_4 x}$ generated is the corresponding projection of the source

$$S = i \int dx\, \phi_4^\dagger (-ie) \left(\partial_\mu A^\mu + A^\mu \partial_\mu\right) \phi_2. \tag{5.105}$$

The first term is $\partial_\mu (\phi_4^\dagger A^\mu \phi_2) - (\partial_\mu \phi_4^\dagger) A^\mu \phi_2$. Integration by parts assuming fields vanish at infinity (our plane waves are properly speaking a stand-in for wave packets) gives the symmetrical form

$$S = i \int dx\, (-ie) \left(\phi_4^\dagger \partial_\mu \phi_2 - \phi_2 \partial_\mu \phi_4^\dagger\right) A^\mu = -i \int dx\, j_{24} A \tag{5.106}$$

which evaluates to

$$S_{\tilde{e}\tilde{\mu}\to\tilde{e}\tilde{\mu}} = i M_{\tilde{e}\tilde{\mu}\to\tilde{e}\tilde{\mu}} (2\pi)^4 \delta\left(p_4 + p_3 - p_2 - p_1\right) \tag{5.107}$$

where

$$i M_{\tilde{e}\tilde{\mu}\to\tilde{e}\tilde{\mu}} = [(-ie)(p_4 + p_2)^\mu] \left[\frac{-ig_{\mu\nu}}{q^2}\right] [(-ie)(p_1 + p_3)^\nu]. \tag{5.108}$$

By comparing the preceding expressions with the diagram in Figure 5.3, we can see that each vertex is associated with a factor $-ie$ and a momentum 4-vector representing the transition current. We call this combination a vertex factor

$$-ie\,(p_1 + p_2). \tag{5.109}$$

An intermediate state photon is associated with the photon propagator

$$G_\gamma^{\mu\nu}(q) = \frac{-ig^{\mu\nu}}{q^2} \tag{5.110}$$

The photon propagator is similar to the propagator or a massive scalar particle

$$G(q) = \frac{i}{q^2 - m^2}. \tag{5.111}$$

found in the previous chapter. As discussed in the preceding section (see Equation 5.90), in the photon propagator, the numerator tensor $-g_{\mu\nu}$ represents the sum over exchanged photon polarization states

$$\Sigma_i \epsilon_i^\mu \epsilon_i^\nu = -g^{\mu\nu} \tag{5.112}$$

reflecting the sum over intermediate states in Hamiltonian perturbation theory. In general, we must superpose all amplitudes which lead from one initial state to one final state. In particular, for scattering of identical particles such as $\tilde{e}^-\tilde{e}^- \to \tilde{e}^-\tilde{e}^-$, we should add another similar amplitude with final state particles interchanged to obtain the total scattering amplitude:

$$i M_{\tilde{e}\tilde{e}\to\tilde{e}\tilde{e}} = \frac{e^2}{q^2} \left[(p_1 + p_3)(p_2 + p_4) + (p_1 + p_4)(p_2 + p_3)\right]. \tag{5.113}$$

Let's consider the nonrelativistic limit of our result for $\tilde{e}\tilde{\mu}$ scattering. Neglecting the three momenta relative to the masses, we have

$$(p_4 + p_2)(p_3 + p_1) = 4m_{\tilde{e}} m_{\tilde{\mu}} \tag{5.114}$$

and the $\tilde{e}\tilde{\mu}$ scattering amplitude is

$$i M^{\text{nonrelativistic}}_{\tilde{e}\tilde{\mu}\to\tilde{e}\tilde{\mu}} = -i4m_{\tilde{e}} m_{\tilde{\mu}} \frac{e^2}{\mathbf{q}^2}. \tag{5.115}$$

The quantity $e^2/\mathbf{q}^2$ is the Fourier transform of the Coulomb potential $e^2/r$. Compare to the Born approximation for scattering by a potential in nonrelativistic quantum mechanics,

$$S^{\text{nonrelativistic}}_{fi} = -i \int dx\, \psi^\dagger_f V(x) \psi_i. \tag{5.116}$$

The factor $2m_{\tilde{e}} 2m_{\tilde{\mu}}$ results from our normalization. The free particle electric current density for our plane waves $e^{\mp ipx}$ is

$$j_\mu = e\left[\phi^\dagger i\partial_\mu \phi - i\phi\partial_\mu \phi^\dagger\right] = \pm e2p_\mu \tag{5.117}$$

so corresponds to number density $\rho = 2E$ or $\rho = 2m$ in the nonrelativistic limit. Complex conjugate waves correspond to antiparticles. So the amplitude for $\tilde{e}^-_1 \tilde{\mu}^+_2 \to \tilde{e}^-_3 \tilde{\mu}^+_4$ is obtained by the reversal of the sign of $p_2$ and of $p_4$ (or the exchange $p_2 \Leftrightarrow -p_4$) which amounts to a sign change of the overall amplitude representing the attractive rather than repulsive Coulomb interaction.

There are more interactions buried in these equations. Consider the process of selectron pair annihilation to form a smuon pair

$$\tilde{e}^-_1 \tilde{e}^+_2 \to \tilde{\mu}^-_4 \tilde{\mu}^+_3 \tag{5.118}$$

illustrated by the diagram in Figure 5.4. Substitute $\phi_1 = e^{-ip_1x}$ and $\phi_2 = e^{ip_2x}$ to write the selectron transition current $j^\nu_{12}$. The corresponding electromagnetic field found from Maxwell's equations (Equation 5.14) is

$$A^\nu = -\frac{j^\nu_{12}}{(p_1 + p_2)^2} \tag{5.119}$$

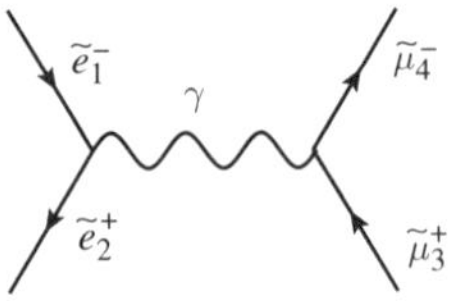


**FIGURE 5.4** Smuon pair production. A single diagram describes scalar electron and positron annihilating to form scalar muon and anti-muon $\tilde{e}^-\tilde{e}^+ \to \tilde{\mu}^-\tilde{\mu}^+$ through an intermediate photon state.

and couples to the smuon transition current $j_{34}$ constructed from $\phi_3 = e^{ip_3x}$ and $\phi_4 = e^{-ip_4x}$. The amplitude may be derived from the $\tilde{e}^-\tilde{\mu}^-$ scattering amplitude by the exchange $p_3 \Leftrightarrow -p_2$:

$$iM_{\tilde{e}^-\tilde{e}^+\to\tilde{\mu}^-\tilde{\mu}^+} = [(-ie)(-p_3+p_4)^\mu]\left[\frac{-ig_{\mu\nu}}{(p_1-p_2)^2}\right][(-ie)(p_1-p_2)^\nu]. \tag{5.120}$$

The factors corresponding to the currents and the photon propagator are already familiar. The diagram is that describing the scattering channel except external legs corresponding to antiparticles are associated with negative momenta. The legs have been "crossed" from outgoing to ingoing and the momentum dependence of the amplitudes for crossed processes are related. Note that in the scattering of identical particles there are two distinct diagrams corresponding to $\tilde{e}_1^-\tilde{e}_2^+ \to \tilde{e}_3^-\tilde{e}_4^+$ and the total amplitude is

$$M_{\tilde{e}^-\tilde{e}^+\to\tilde{e}^-\tilde{e}^+} = e^2\left[\frac{(p_1+p_3)(-p_4-p_2)}{(p_3-p_1)^2} + \frac{(p_1-p_2)(p_3-p_4)}{(p_1+p_2)^2}\right] \tag{5.121}$$

where the first term represents photon exchange and the second an annihilation amplitude that vanishes in the nonrelativistic limit.

We can understand the signs of the momenta in these expressions in classical terms. The essential photon and emission amplitudes given by Equations 5.97 and 5.99 have the form $M = A_\mu j_{12}^\mu$ where the transition current for a selectron is $j_{12} = e(p_1+p_2)$. In the center of mass of the pair, we have $p_1 = (E, \mathbf{p})$ and $p_2 = (E, -\mathbf{p})$ and

$$j_{12}^{cm} = e\,(p_1+p_2)^{cm} = (\rho, \mathbf{j}) = e\,[2E, \mathbf{0}]. \tag{5.122}$$

In this frame, we have one selectron moving along the direction $\mathbf{p}$ superposed on another selectron with opposite velocity. The charge density is the sum and the current densities cancel. The 4-current in a general frame is related by Lorentz transformation and has non-vanishing three current density along the average of the directions of the individual current densities. In the case of pair production, the transition current is $j_{12} = e(p_1 - p_2)$. In the center of mass of the pair, the selectron 4-momentum is $p_1 = (E, \mathbf{p})$ and the spositron momentum is $p_2 = (E, -\mathbf{p})$. The transition current is

$$j_{12}^{cm} = e\,(p_1+p_2)^{cm} = (\rho, \mathbf{j}) = e\,[0, 2\mathbf{p}]. \tag{5.123}$$

In this frame, we have one selectron moving along the direction $\mathbf{p}$ superposed on a spositron with opposite velocity. The charge densities cancel and the space current densities add like in a wire bearing electrical current.

New twists appear in scalar Compton scattering

$$\tilde{e}_1^-\gamma_2 \to \tilde{e}_3^-\gamma_4 \tag{5.124}$$

illustrated in Figure 5.5. With reference to Figure 5.5 (a), start with a selectron wave function $\phi_1 = e^{-ip_1x}$ interacting with a photon wave function $A_2 = \epsilon_2 e^{-ip_2x}$

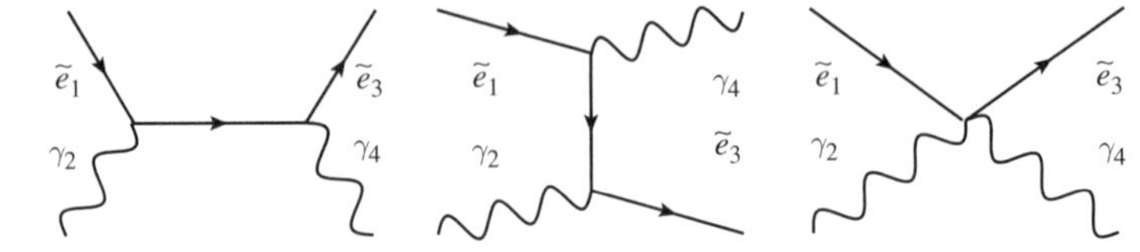


**FIGURE 5.5** Diagrams describing scalar electron Compton scattering $\tilde{e}^-\gamma \to \tilde{e}^-\gamma$. The total amplitude is the sum of amplitudes corresponding to three second order diagrams.

where the polarization vector satisfies $\epsilon_2 p_2 = 0$ (Lorenz gauge). An intermediate state selectron field satisfies

$$\left[\partial_\mu \partial^\mu + m^2\right]\phi = -2ieA_2\partial\phi_1 \to \phi = \frac{-1}{p^2 - m^2}(-ie)2A_2\partial\phi_1 \tag{5.125}$$

where $p = p_1 + p_2$. A selectron transition current

$$j^\mu = ie\left[\phi_3^\dagger \partial^\mu \phi - \phi^\dagger \partial^\mu \phi_3\right] \tag{5.126}$$

corresponding to the transition to $\phi_3 = e^{-ip_3x}$ is a source of electromagnetic field in Maxwell's equations. The projection onto $A_4 = \epsilon_4 e^{-ip_4x}$ of this current gives amplitude

$$S^{(a)}_{\tilde{e}^-\gamma\to\tilde{e}^-\gamma} = -i\int dx\; A_4^\dagger j = iM_{\tilde{e}^-\gamma\to\tilde{e}^-\gamma}(2\pi)^4\delta\left(p_4 + p_3 - p_1 - p_2\right) \tag{5.127}$$

where, again, the negative phase in the projection corresponds to the photon normalization $\epsilon^\dagger\epsilon = -1$, and the matrix element is

$$\begin{aligned} iM^{(a)}_{\tilde{e}^-\gamma\to\tilde{e}^-\gamma} &= \left[\epsilon_4^\dagger\,(-ie)\,(p + p_3)\right]\left[\frac{i}{p^2 - m^2}\right]\left[(-ie)\,(p_1 + p)\,\epsilon_2\right] \\ &= -i4e^2\frac{(\epsilon_4^\dagger p_3)\,(\epsilon_2 p_1)}{p^2 - m^2}. \end{aligned} \tag{5.128}$$

The conditions $\epsilon_4 p_4 = \epsilon_2 p_2 = 0$ and $p_1 + p_2 = p_3 + p_4$ were used to simplify the standard expression in the last step. The current densities have been written in a form consistent with previous results. The propagator for the intermediate state selectron appears and the external photons are represented by their polarization vectors. The shuffling of momenta makes the vertex factors consistent with those in the photon exchange process.

This is not quite the end of the story. We should also include the exchange contribution of Figure 5.5 (b) with matrix element

$$\begin{aligned} iM^{(b)}_{\tilde{e}^-\gamma\to\tilde{e}^-\gamma} &= \left[\epsilon_4^\dagger\,(-ie)\,(p_1 + q)\right]\left[\frac{i}{q^2 - m^2}\right]\left[(-ie)\,(q + p_3)\,\epsilon_2\right] \\ &= -i4e^2\frac{(\epsilon_4^\dagger p_1)\,(\epsilon_2 p_3)}{q^2 - m^2} \end{aligned} \tag{5.129}$$

with $q = p_4 - p_1 = p_2 - p_3$, and a contribution from the neglected effective potential $e^2A^2\phi\phi^\dagger$

$$iM^{(c)}_{\tilde{e}^-\gamma\to\tilde{e}^-\gamma} = 2ie^2\epsilon_4^\dagger\epsilon_2 \tag{5.130}$$

represented by Figure 5.5 (c). A factor $-i$ in this last expression comes from projecting the final photon when using Equation 5.94. Equivalently, using Equation 5.93, a factor $+i$ comes from projecting the final $e^-$ and a factor of two from treating the factors $A$ as independent. Crossed versions of these diagrams describe $\tilde{e}^+\tilde{e}^- \Leftrightarrow \gamma\gamma$ and complete the set of fundamental processes of order $e^2$. Beware: the vertex with two photons and two charged particles is essential in scalar electrodynamics and in vector boson electrodynamics but disappears in the quantum electrodynamics of spin 1/2 particles as we will see. Hence no such vertex appears in the fundamental processes involving quarks and leptons interacting with photons. The fundamental diagrams of scalar electrodynamics appear in Figure 5.6 and the essential Feynman rules are given in Table 5.1.

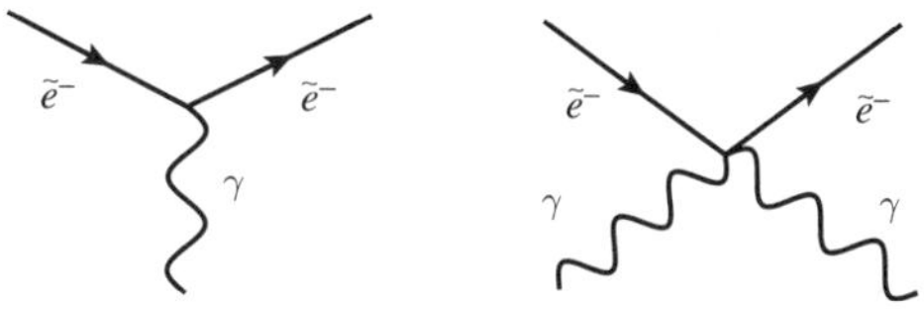


**FIGURE 5.6** Two and three particle fundamental diagrams in scalar electrodynamics. The diagram to the right is second order in the $\tilde{e}^-$ electric charge.

| Diagram element | Factor in amplitude |
|---|---|
| Ingoing or outgoing scalar | 1 |
| Ingoing or outgoing scalar antiparticle | 1 |
| Ingoing photon line | $\epsilon$ |
| Outgoing photon line | $\epsilon^\dagger$ |
| Vertex for particles | $-ie(p_1 + p_2)$ |
| Internal photon line | $-ig_{\mu\nu}/q^2$ |
| Internal scalar line | $i/(p^2 - m^2)$ |

**TABLE 5.1** Feynman rules for scalar electrodynamics.

## 5.5 ■ FEYNMAN RULES FROM LAGRANGIAN

While the perturbative analysis of the field equations provides an intuitive derivation of Feynman rules, there is a simpler and more general way to discover the vertex factors. The couplings in the field equations of scalar electrodynamics derive from the interaction Lagrangian

$$L_{int} = -e\left[\phi^\dagger i\partial_\mu\phi - \phi i\partial_\mu\phi^\dagger\right]A^\mu + e^2A^2\phi^\dagger\phi. \tag{5.131}$$

Each term in $L_{int}$ corresponds to a fundamental diagram and to an interaction term in the field equations. If we identify an interaction term in the Lagrangian with the negative of a potential energy, we can write the generalized transition matrix in Hamiltonian perturbation theory as

$$\begin{aligned} S_{fi} &= \delta_{fi} - i\int dx U_{fi} + \cdots = \delta_{fi} + i\int dx L_{int,fi} + \cdots \\ &= \delta_{fi} + iM_{fi}(2\pi)^4\delta\left(p_f - p_i\right). \end{aligned} \tag{5.132}$$

The matrix element $iM$ of the fundamental diagram may be found by inserting plane waves into $iL_{int}$, converting derivatives to momenta, and then erasing the exponential factors which integrate to the factor $(2\pi)^4\delta(p_f - p_i)$.

For example, inserting $\phi_1 = e^{-ip_1x}$, $\phi^\dagger = e^{+p_2x}$ and $A^\mu = \epsilon^\mu e^{-ip_3x}$ into the first term in the interaction Lagrangian for scalar electrodynamics gives

$$iM = -ie(p_1 + p_2)_\mu\epsilon^\mu \tag{5.133}$$

which is the product of the vertex and the photon polarization vector factors found from the field equations. The vertex factors derived in this way must be supplemented by the rule that a virtual particle is associated with a propagator factor $G(p) = i/[p^2 - m^2]$ times a sum over spin states of the virtual particle.

## 5.6 ■ SCATTERING CROSS SECTIONS

To relate amplitudes to experiment, we start with the cross section formula (see Equation 4.203)

$$d\sigma = \frac{|M_{fi}|^2}{4\sqrt{(p_1p_2)^2 - m_1^2m_2^2}}\left[(2\pi)^4\,\delta\left(p_f - p_i\right)\Pi_f\frac{d\mathbf{p}_f}{2E_f(2\pi)^3}\right] \tag{5.134}$$

and retrace the simplification steps of the preceding chapter using Mandelstam variables. For two final particles, call them 3 and 4, we can simplify the phase space factor in the usual way. In the center of mass, defining $p' = |\mathbf{p}_3| = |\mathbf{p}_4|$, $E = E_3 + E_4$ and $d\Omega \equiv d\Omega_3$, we have

$$dPS = \frac{(2\pi)^4}{2E_32E_4}\frac{1}{(2\pi)^6}\delta(E - \sqrt{s})p'^2\frac{dp'}{dE}d\Omega dE = \frac{1}{16\pi^2}\frac{p'}{\sqrt{s}}d\Omega \tag{5.135}$$

where we used

$$dE/dp' = dE_3/dp' + dE_4/dp' = p'\left(\frac{1}{E_3} + \frac{1}{E_4}\right) = p'\frac{\sqrt{s}}{E_3E_4}. \tag{5.136}$$

In the center of mass, $\sqrt{(p_1p_2)^2 - m_1^2m_2^2} = p\sqrt{s}$ with $p = |\mathbf{p}_1| = |\mathbf{p}_2|$ so

$$d\sigma = \frac{1}{64\pi^2}|M_{fi}|^2\frac{p'}{ps}d\Omega. \tag{5.137}$$

The 4-momentum transfer squared is

$$t = (p_3 - p_1)^2 = m_3^2 + m_1^2 - 2E_3E_1 + 2|\mathbf{p}_3||\mathbf{p}_1|\cos\theta. \tag{5.138}$$

In the center of mass, energies are fixed so $dt = 2|\mathbf{p}_3||\mathbf{p}_1|d\cos\theta$ and hence

$$d\Omega = -d\cos\theta d\phi = d(-t)d\phi/2pp' \tag{5.139}$$

so we can write down an expression for the differential cross section in terms of $t$ and the azimuth of the outgoing particles:

$$d\sigma = \frac{1}{64\pi^2}\frac{p'}{ps}|M_{fi}|^2 d\Omega = \frac{|M_{fi}|^2}{128\pi^2 p^2 s}dtd\phi = \frac{|M_{fi}|^2}{64\pi p^2 s}dt \tag{5.140}$$

where the direction of integration over $dt$ is usually obvious so the minus sign dropped, and if the squared amplitude is independent of azimuth, integration over azimuth gives the second expression.

Consider first $\tilde{e}^-\tilde{\mu}^-$ elastic scattering. The matrix element is

$$M_{\tilde{e}^-\tilde{\mu}^-\to\tilde{e}^-\tilde{\mu}^-} = \frac{e^2}{t}(p_1 + p_3)(p_2 + p_4) = e^2\frac{(s-u)}{t} \simeq e^2\frac{3+\cos\theta}{\cos\theta - 1} \tag{5.141}$$

where the last expression applies in the ultra-relativistic limit. The differential cross section is the same for the four combinations $\tilde{e}^\pm\tilde{\mu}^\mp$ and, expressed in terms of four momentum transfer, is

$$\frac{d\sigma_{\tilde{e}\tilde{\mu}\to\tilde{e}\tilde{\mu}}}{dt} = \frac{|M|^2}{64\pi p^2 s} = \frac{\pi\alpha^2}{4p^2 s}\frac{(s-u)^2}{t^2} \tag{5.142}$$

where $\alpha = e^2/4\pi$. We may also write the cross section in terms of scattering angle using $s + u + t = h = 2m_{\tilde{e}} + 2m_{\tilde{\mu}}$ and $t = 4\mathbf{p}^2\sin^2\theta/2$ as

$$\begin{aligned}\frac{d\sigma_{\tilde{e}\tilde{\mu}\to\tilde{e}\tilde{\mu}}}{d\Omega} &= \frac{1}{64\pi^2}\frac{p'}{ps}|M|^2 \\ &= \frac{\alpha^2}{16}\frac{s}{\mathbf{p}^4}\frac{1}{\sin^4(\frac{\theta}{2})}\left(1 - \frac{h-t}{2s}\right) \\ &= \simeq \frac{1}{4s}\alpha^2\left[\frac{3+\cos\theta}{1-\cos\theta}\right]^2\end{aligned} \tag{5.143}$$

where the last expression applies in the ultrarelativistic limit. The differential cross section becomes infinite as $\theta \to 0$ corresponding to small momentum transfer (large impact parameter) and the total cross section is infinite. In reality, one can not measure momentum transfers smaller than some experimental resolution and charges are ultimately screened.

Consider next the scalar pair production process $\tilde{e}_1^-\tilde{e}_2^+ \to \tilde{\mu}_3^-\tilde{\mu}_4^+$. The amplitude is without relativistic approximation

$$M_{\tilde{e}^-\tilde{e}^+\to\tilde{\mu}^-\tilde{\mu}^+} = e^2\frac{(p_4 - p_3)(p_1 - p_2)}{(p_1 + p_2)^2} = e^2\frac{t-u}{s} = \frac{e^2}{s}4\mathbf{p}_3\cdot\mathbf{p}_1. \tag{5.144}$$

The differential cross section in the center of mass is

$$d\sigma_{\tilde{e}^-\tilde{e}^+\to\tilde{\mu}^-\tilde{\mu}^+} = \frac{|M|^2}{64\pi^2 s}\frac{|\mathbf{p}'|}{|\mathbf{p}|}d\Omega = 4\left(\frac{e^2}{4\pi}\right)^2\frac{p'^3 p}{s^3}\cos^2\theta d\Omega$$
$$= \frac{1}{4}\frac{\alpha^2}{s}vv'^3\cos^2\theta d\Omega \tag{5.145}$$

and the total cross section is

$$\sigma_{\tilde{e}^-\tilde{e}^+\to\tilde{\mu}^-\tilde{\mu}^+} = \int\frac{d\sigma_{\tilde{e}^-\tilde{e}^+\to\tilde{\mu}^-\tilde{\mu}^+}}{d\Omega}d\phi d\cos\theta = \frac{\pi}{3}\frac{\alpha^2}{s}vv'^3 \tag{5.146}$$

where $v = p/E = 2p/\sqrt{s}$ and $v = p'/E' = 2p'/\sqrt{s}$ are the velocities of the electron and muon. The factor $\alpha^2$ comes from squaring the couplings, the factor $vv'^3$ describes the kinematic threshold and well above threshold the only scale to give the proper dimensions is $s$. The angular dependence can be understood as follows. Take the $z$-axis along $\mathbf{p}$ in the rest frame of the virtual photon. The $z$-component of the angular momentum of the initial particles and hence of the photon vanishes so the decay amplitude is proportional to $Y_{10} \sim \cos\theta$.

We now turn to scalar Compton scattering. The matrix element is

$$M_{\gamma\tilde{e}\to\gamma\tilde{e}} = M^{(a)}_{\gamma\tilde{e}\to\gamma\tilde{e}} + M^{(b)}_{\gamma\tilde{e}\to\gamma\tilde{e}} + M^{(c)}_{\gamma\tilde{e}\to\gamma\tilde{e}}$$
$$-e^2\frac{\epsilon_4^\dagger(2p_3+p_4)\epsilon_2(2p_1+p_2)}{p^2-m^2} - e^2\frac{\epsilon_4^\dagger(2p_1-p_4)\epsilon_2(2p_3-p_1)}{q^2-m^2}$$
$$+2e^2\epsilon_4^\dagger\epsilon_2$$
$$= 4e^2\frac{\epsilon_4^\dagger p_3\epsilon_2 p_1}{p^2-m^2} - 4e^2\frac{\epsilon_4^\dagger p_1\epsilon_2 p_3}{q^2-m^2} + 2e^2\epsilon_4^\dagger\epsilon_2 \tag{5.147}$$

where, in arriving at the last expression, the Lorentz conditions $\epsilon_4 p_4 = 0$ and $\epsilon_3 p_3 = 0$ were used. It is simplest to evaluate this in the rest frame of the initial selectron where $p_2$ has only a time component. We choose transverse photon polarization vectors so $\epsilon_4$ and $\epsilon_3$ only have space components. Then the first two terms vanish. Suppose the initial photon momentum is along the 3 axis and scatters through angle $\theta$ in the 1-3 plane. Pick initial linear polarization vectors $\epsilon_2 = (0, \mathbf{e}_1)$ in the scattering plane or $\epsilon_2 = (0, \mathbf{e}_2)$ perpendicular to the scattering plane, and final linear polarization vectors $\epsilon_4^\dagger = (0, \cos\theta, 0, -\sin\theta)$ or $\epsilon_4^\dagger = (0, \mathbf{e}_2)$. The amplitude $M^{(c)} = 2e^2\epsilon_4^\dagger\epsilon_2$ is maximal for initial and final polarization perpendicular to the scattering plane.

Averaged over initial polarizations and summed over final polarizations, the squared amplitude is readily found to be

$$\overline{|M_{\gamma\tilde{e}\to\gamma\tilde{e}}|^2} \equiv \frac{1}{2}\Sigma_f|M_{\gamma\tilde{e}\to\gamma\tilde{e}}|^2 = 2e^4(1+\cos^2\theta) \tag{5.148}$$

and, since $p^2 s = (s-m^2)2/4$, the differential cross section is

$$\frac{d\sigma_{\gamma\tilde{e}\to\gamma\tilde{e}}}{dt} = \frac{\overline{|M_{\gamma\tilde{e}\to\gamma\tilde{e}}|^2}}{64\pi p^2 s} = \frac{e^4}{8\pi}\frac{1+\cos^2\theta}{(s-m^2)^2}. \tag{5.149}$$

The center of mass energy squared evaluated in frame of the initial selectron is $s = m^2 - 2E_2 m$ and the momentum transfer is

$$t = (p_4 - p_2)^2 = 2E_4 E_2 (\cos\theta - 1) . \tag{5.150}$$

The energy $E_4$ of the scattered photon depends on the scattering angle. The relationship between the two, used by Compton to establish that photons behave as massless particles and carry momentum $p = E/c$, follows from 4-momentum conservation. If the final state selectron 4-momentum is (E,**p**), and **p** has angle $\phi$ relative to the initial photon direction, energy conservation requires

$$m + E_2 = E_4 + E, \tag{5.151}$$

and 3-momentum conservation requires

$$E_2 = E_4 \cos\theta + p\cos\phi,\ E_4 \sin\theta = p\sin\phi \tag{5.152}$$

and hence

$$\cos\phi^2 = 1 - \left(\frac{E_4}{p}\right)^2 \sin^2\theta. \tag{5.153}$$

So we find that

$$(E_2 - E_4\cos\theta)^2 = p^2\cos^2\phi = p^2 - E_4^2\sin^2\theta. \tag{5.154}$$

Substituting $p^2 = E^2 - m^2 = (m + E_2 - E_4)^2 - m^2$ in this expression and simplifying yields

$$E_4 = \frac{E_2}{1 + (E_2/m)(1 - \cos\theta)}. \tag{5.155}$$

By combining Equation 5.150 and Equation 5.155, one finds

$$t = \frac{2E_2^2(\cos\theta - 1)}{1 + \frac{E_2}{m}(1 - \cos\theta)} \tag{5.156}$$

from which we can derive

$$\frac{dt}{d\cos\theta} = \frac{2E_2^2}{\left[1 + \frac{E_2}{m}(1 - \cos\theta)\right]^2} \tag{5.157}$$

and the differential cross section in the rest frame of the initial selectron (the laboratory frame in Compton's experiments) is

$$\frac{d\sigma_{\gamma\tilde{e}\to\gamma\tilde{e}}}{d\Omega_{Lab}} = \frac{1}{2\pi}\frac{d\sigma_{\gamma\tilde{e}\to\gamma\tilde{e}}}{dt}\frac{dt}{d\cos\theta} = \frac{\alpha^2}{2m^2}\frac{1 + \cos^2\theta}{[1 + (E_2/m)(1 - \cos\theta)]^2}. \tag{5.158}$$

For $E_2 << m$, the cross section reduces to the classical Thomson result

$$\frac{d\sigma_{\gamma\tilde{e}\to\gamma\tilde{e}}}{d\Omega_{Lab}} = \frac{\alpha^2}{2m^2}(1 + \cos^2\theta) \tag{5.159}$$

which is maximal for forward and backward scattering but varies by only a factor of two.

For $E_2 >> m$, the shape of the distribution of scattering angles differs significantly from the nonrelativistic limit. The differential cross section is strongly peaked in the forward direction. This may be seen by considering the cross section ratios

$$\frac{d\sigma\,(\theta = \pi)}{d\sigma\,(\theta = 0)} = \frac{1}{\left(1 + \frac{2E_2}{m}\right)^2};\ \frac{d\sigma\,\left(\theta = \frac{\pi}{2}\right)}{d\sigma\,(\theta = 0)} = \frac{1}{2\left(1 + \frac{E_2}{m}\right)^2} \tag{5.160}$$

and is illustrated in Figure 5.7. At high energy, photons are scattered principally with angles $\theta < \sqrt{2m/E_2}$. It follows that the total cross section decreases with energy and the typical energy loss per collision is small. The total cross section can be found by integration and is

$$\sigma_{\gamma\tilde{e}\to\gamma\tilde{e}} = \int d\Omega_{Lab} \frac{d\sigma_{\gamma\tilde{e}\to\gamma\tilde{e}}}{d\Omega_{Lab}} = \frac{8\pi}{3}\frac{\alpha^2}{s} f(s/m^2) \tag{5.161}$$

where $f(s/m^2)$ is a slowly varying function of energy

$$f(x) = (3/2)(x + 1)(x^2 - 2x\ln(x) - 1)/(x - 1)^3 \tag{5.162}$$

and $x = s/m^2 = 1 + 2E_2/m$ in the rest frame of the initial selectron.

The choice of the lab frame simplifies the evaluation of the Compton scattering cross section above. In a general frame, there are three contributions to the amplitude each of which depends on the initial and final photon polarizations. The expression

$$\sum_{i=1,2} |M|^2 = \sum_i \epsilon_i^{\mu} \epsilon_i^{\nu\dagger} J_{\mu}^{\dagger} J_{\nu} = -g_{\mu\nu} J^{\mu} J^{\nu} \tag{5.163}$$

can be used to evaluate a cross section for a process with an external photon summed over photon polarization. In Compton scattering, there are two external

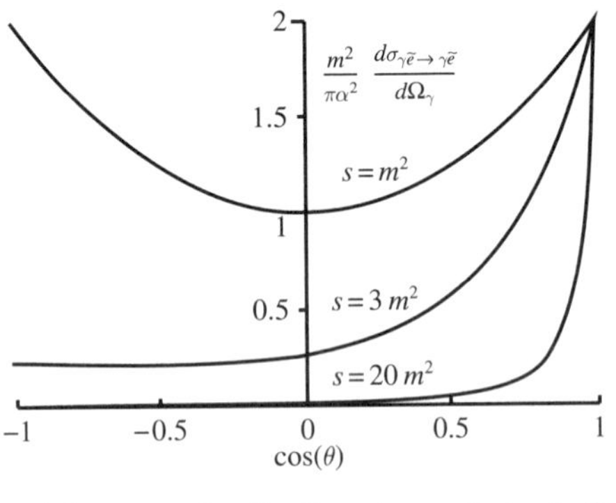


**FIGURE 5.7** Differential cross section for Compton scattering in the rest frame of a scalar target particle of mass $m$ for various values of center of mass energy $s$.

photons and we have to apply the trick twice to find the cross section summed over final photon polarization and averaged over initial photon polarization. We have $M = \epsilon_4^{\mu\dagger} \epsilon_3^{\nu} T_{\mu\nu}$ with

$$T_{\mu\nu} = -e^2 \frac{(2p_3 + p_4)_\mu (2p_1 + p_2)_\nu}{p^2 - m^2} - e^2 \frac{(2p_1 - p_4)_\mu (2p_3 - p_1)_\nu}{q^2 - m^2} + 2e^2 g_{\mu\nu} \tag{5.164}$$

and we must calculate $\Sigma|M|^2 = T^{\mu\nu} T_{\mu\nu}$. The result can be cast in the form

$$\Sigma|M|^2 = 4e^4 \left( m^4 \left[ \frac{1}{p_1 p_4} - \frac{1}{p_1 p_2} \right]^2 - 2m^2 \left[ \frac{1}{p_1 p_4} - \frac{1}{p_1 p_2} \right] + 2 \right). \tag{5.165}$$

There is a subtle point here. We must not impose $p_4 \epsilon_4 = 0$ and $p_2 \epsilon_2 = 0$ to simplify the matrix element prior to applying the trick, in this case of its double application. We must have $M = 0$ with $\epsilon_4 \to p_4$ and $\epsilon_2 \to p_2$ independently or an incorrect answer results. Nor can we apply the trick to the matrix element simplified in the lab frame to $M = 2e^2 \epsilon_4^\dagger \epsilon_2^\dagger$ as this part of the amplitude is not gauge invariant by itself and the conditions for the trick are not satisfied. To use this simplified matrix element, we must perform the sum over the two transverse real photon states explicitly.

To evaluate Equation 5.165 in the laboratory frame, we write

$$\left[ \frac{1}{p_1 p_4} - \frac{1}{p_1 p_2} \right] = m^{-1} \left[ \frac{1}{E_4} - \frac{1}{E_2} \right] = m^{-2} (1 - \cos\theta) \tag{5.166}$$

where we used Equation 5.155 in the form $E_4{}^{-1} - E_2{}^{-1} = m^{-1}(1 - \cos\theta)$. From this, we find $\Sigma|M|^2 = 1 + \cos^2\theta$ as before. The expressions

$$\begin{aligned} s &= (p_1 + p_2)^2 = m^2 + 2p_1 p_2 = (p_3 + p_4)^2 = m^2 + 2p_3 p_4 \\ t &= (p_4 - p_2)^2 = -2p_4 p_2 = -2p_4(p_4 + p_3 - p_1) = 2(p_1 p_4 - p_3 p_4) \end{aligned} \tag{5.167}$$

may be used to write

$$p_1 p_2 = \frac{s - m^2}{2}; \quad p_1 p_4 = \frac{t + s - m^2}{2} \tag{5.168}$$

and to express $\Sigma|M|^2$ in terms of invariants. The invariants may be expressed in terms of the center of mass scattering angle to find the scattering cross section in the center of mass.

## 5.7 ■ VECTOR ELECTRODYNAMICS

Two degenerate real vector fields $W_1^\mu$ and $W_2^\mu$ or a complex vector field are required to describe a charged spin 1 particle like the $W$ boson. The free particle Lagrangian is

$$L_0 = -\frac{1}{4}F_1^{\mu\nu}F_{1\mu\nu} - \frac{1}{4}F_2{}^{\mu\nu}F_{2\mu\nu} + \frac{1}{2}m_W^2 W_1{}^{\mu}W_{1\mu} + \frac{1}{2}m_W W_2{}^{\mu}W_{2\mu} \quad (5.169)$$

where the field strength tensor for $W_1$ is

$$F_1{}^{\mu\nu} = \partial^\mu W_1{}^\nu - \partial^\nu W_1{}^\mu \quad (5.170)$$

and the field strength tensor for $W_2$ is

$$F_2^{\mu\nu} = \partial^\mu W_2{}^\nu - \partial^\nu W_2{}^\mu. \quad (5.171)$$

Define the complex vector field

$$W^\mu = \frac{1}{\sqrt{2}}(W_1^\mu + iW_2^\mu) \quad (5.172)$$

which will be associated with a vector particle of charge $e$. (For the $W^-$, the charge is $e = -1|e|$.) The corresponding field strength tensor is

$$F_W{}^{\mu\nu} = \frac{1}{\sqrt{2}}\left(F_1^{\mu\nu} + iF_2^{\mu\nu}\right). \quad (5.173)$$

Then the free field Lagrangian can be written in the more compact form

$$L_0 = -\frac{1}{2}F_{W\,\mu\nu}^{\dagger}F_W{}^{\mu\nu} + m_W^2 W_\mu{}^\dagger W^\mu = -\frac{1}{2}F_W{}^\dagger F_W + m_W^2 W^\dagger W. \quad (5.174)$$

Now if the field strength tensor associated with the electromagnetic potential $A^\mu$ is

$$F_A{}^{\mu\nu} = \partial^\mu A^\nu - \partial^\nu A^\mu, \quad (5.175)$$

then the Lagrangian describing free $W$ bosons and photons is

$$L_0 = -\frac{1}{2}F_W{}^{\dagger}_{\mu\nu}F_W^{\mu\nu} + m_W W_\mu^\dagger W^\mu - \frac{1}{4}F_{A\mu\nu}F_A^{\mu\nu}. \quad (5.176)$$

Conversion of normal to gauge covariant derivatives, $\partial^\mu \to D^\mu = \partial^\mu + ieA^\mu$, leads to a gauge invariant Lagrangian describing the quantum electrodynamics of a vector boson. We modify the field strength for the $W$ boson to make it gauge covariant

$$F_W^{\mu\nu} \to F_W^A \equiv (\partial^\mu + ieA^\mu)W^\nu - (\partial^\nu + ieA^\nu)W^\mu = F_W^{\mu\nu} + ie(A^\mu W^\nu - A^\nu W^\mu) \quad (5.177)$$

and correspondingly

$$\begin{aligned} F_W{}^{\dagger}_{\mu\nu} \to F_W^{A\,\dagger} &\equiv (\partial_\mu - ieA_\mu)W_\nu^\dagger - (\partial_\nu - ieA_\nu)W_\mu^\dagger \\ &= F_W{}^{\dagger}_{\mu\nu} - ie\left(A_\mu W_\nu^\dagger - A_\nu W_\mu^\dagger\right). \end{aligned} \quad (5.178)$$

Making these substitutions in $L_0$ and collecting terms, we find

$$L = L_0 + L_{AWW} + L_{AAWW} \tag{5.179}$$

where the three groups of terms are defined as

$$\begin{aligned} L_0 &= -\frac{1}{2}F^{\dagger}_{W\,\mu\nu}F_W{}^{\mu\nu} + m_W^2 W^{\dagger}{}_{\mu}W^{\mu} - \frac{1}{4}F_{A\mu\nu}F_A{}^{\mu\nu} \\ L_{AWW} &= -ieA^{\mu}\left[W^{\dagger}{}_{\nu}F_W{}^{\mu\nu} - F^{\dagger}_{W\,\mu\nu}W_{\nu}\right] \\ L_{AAWW} &= -e^2\left(A_{\mu}A^{\mu}W^{\dagger\nu}W_{\nu} - A_{\mu}W^{\dagger\mu}A_{\nu}W^{\nu}\right). \end{aligned} \tag{5.180}$$

The first group is just the Lagrangian for free fields. The interaction terms linear and quadratic in $A_\mu$ and hence linear and quadratic in charge have been denoted by $L_{AWW}$ and $L_{AAWW}$. The linear terms describe interactions between one photon and two $W$ bosons, interactions such as radiation of a photon by a $W$ boson and the decay of a virtual photon into $W^+W^-$. The quadratic terms describe interactions between two photons and two $W$ bosons and contribute to second order processes including Compton scattering of a photon by a $W$ boson. The terms correspond to the three-particle and four-particle vertex diagrams shown in Figure 5.8. These are familiar from our study of scalar electrodynamics.

In the standard model of the electroweak interaction, an additional gauge invariant term

$$L_a = -ie\kappa W^{\dagger}_{\mu}W_{\nu}F_A^{\mu\nu} \tag{5.181}$$

appears with $\kappa = 1$. Such a term implies the magnetic moment of the $W$ boson is

$$\boldsymbol{\mu}_W = g_W \frac{e}{m_W}\mathbf{s}_W \tag{5.182}$$

with $g_W = 1 + \kappa = 2$ and $\mathbf{s}_W$ the spin operator given by Equation 5.77. The term proportional to $\kappa$ appears anomalous in the context of vector electrodynamics generated by the minimal coupling substitution of derivatives to gauge covariant derivatives and we point out its origin here. The discussion may make more sense later after our study of the electroweak model. Also, as will be described in Chapter 8, only in the context of the electroweak gauge theory with spontaneous symmetry breaking is vector electrodynamics with a massive vector boson renormalizable.

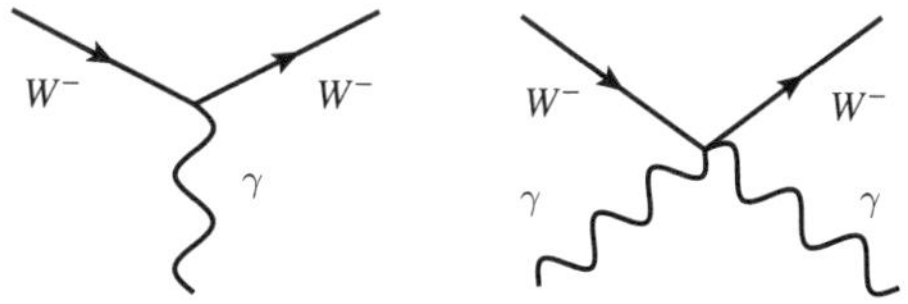


**FIGURE 5.8** Two and three particle fundamental diagrams in vector electrodynamics. The diagrams are similar to those for scalar electrodynamics. The two photon term is second order in the $W^-$ electric charge.

That said, the interactions described by bare vector field electrodynamics augmented with additional effective interactions and Lorentz invariant form factors provides a model for the interactions of massive charged spin 1 bosons such as the $\rho^-$.

The electroweak standard model contains a 3-vector of equivalent vector bosons $\mathbf{W} = (W_1, W_2, W_3)$. The component fields interact with each other in a manner prescribed by gauge symmetry. The weak interaction $W^-$ boson is identified as $W = (W_1 - iW_2)/\sqrt{2}$ with $W^\dagger = (W_1 + iW_2)/\sqrt{2}$ representing the $W^+$. The third field $W_3$ mixes with a minimally coupled fourth vector boson $B$. The massless photon field $A$ and the massive $Z$ boson are identified as orthogonal linear combinations of $W_3$ and $B$. The rotational symmetry relates the weak couplings of the $W_1$, $W_2$, and $W_3$. The mixing relates the weak and electromagnetic couplings. A term of the form $L_a$ for the $W_3$ is required in the pure $W_1, W_2, W_3$ gauge theory and, since the photon is part $W_3$ and part $B$, the anomalous term is induced in the electrodynamics of the $W$ boson. In sum, while the term proportional to $\kappa$ appears anomalous from the viewpoint of electrodynamics alone, in the context of the standard model, the value $\kappa = 1$ is predicted and deviations of the form $\kappa = 1 + \Delta\kappa$ are deemed anomalous.

Given the Lagrangian, we can derive the field equations for the electrodynamics of vector bosons and use these to calculate quantum interactions, generalizing the example of scalar electrodynamics. We start with the Euler-Lagrange equation for the electromagnetic field

$$\partial_\nu \frac{\partial L}{\partial(\partial_\nu A^\mu)} = \frac{\partial L}{\partial A^\mu}. \tag{5.183}$$

The derivatives $\partial_\nu A^\mu$ appear only in $L_0$ and $L_a$ and we can write

$$\partial_\nu \partial^\nu A^\mu - \partial^\mu \partial_\nu A^\nu = j_W^\mu \tag{5.184}$$

where the current density associated with the W can be identified as

$$\begin{aligned} j_W^\mu &= -\frac{\partial L_{AAW}}{\partial A_\mu} - \frac{\partial L_{AAWW}}{\partial A_\mu} + \partial_\nu \frac{\partial L_a}{\partial_\nu A_\mu} \\ &= \frac{1}{2}\left[\frac{\partial F_W^{A\dagger}}{\partial A^\mu} F_W^A + F_W^{A\dagger} \frac{\partial F_W^A}{\partial A^\mu}\right] + ie\kappa\partial_\nu\left[W^{\dagger\mu} W^\nu - W^{\dagger\nu} W^\mu\right] \\ &\quad -ie\left[W_\nu^\dagger F_W^{A\,\mu\nu} + F_{W\,\mu\nu}^{A\dagger} W_\nu\right] + ie\kappa\left[\left(\partial_\nu W^\dagger \mu\right) W^\nu - W^{\dagger\nu}\partial_\nu W^\mu\right] \end{aligned} \tag{5.185}$$

or finally

$$\begin{aligned} j_W^\mu &= ie\left[W_\nu^\dagger F_W{}^{\mu\nu} - F_{W\,\mu\nu}^\dagger W_\nu\right] + ie\kappa\left[\left(\partial_\nu W^{\dagger\mu}\right) W^\nu - W^{\dagger\nu}\partial_\nu W^\mu\right] \\ &\quad + e^2\left(2A^\mu W^{\dagger\nu} W_\nu - W^{\dagger\mu} A_\nu W^\nu - A_\nu W^{\dagger\nu} W^\mu\right). \end{aligned} \tag{5.186}$$

The constraint $\partial_\mu W^\mu$ was imposed in simplifying the anomalous term. The current density $j_W$ depends on $A$ as in scalar electrodynamics. Under an electrodynamic gauge transformation in which $A \to A + \partial\chi$ and $W \to UW$, we have $DW \to UDW$ and $F_W^A \to UF_W^A$ and the Lagrangian is invariant. The expression for $j_W$ in terms of the gauge covariant field strength $F_W^A$ shows that this current density is gauge invariant. The Euler-Lagrange equation for the $W$ boson field has the gauge covariant form

$$D_\mu \left(D^\mu W^\nu - D^\nu W^\mu\right) + m_W^2 W^\nu = -\frac{\partial L_a}{\partial W_\mu^\dagger} = ie\kappa W_\nu F_A^{\mu\nu} \tag{5.187}$$

which may be compared to the corresponding equation for a charged scalar field

$$D_\mu D^\mu \phi + m_\phi^2 \phi = 0. \tag{5.188}$$

In expanded form, we have for a $W$ the equation

$$\begin{aligned}\partial_\mu \partial^\mu W^\nu + m_W^2 W^\nu = &-ie\left[\partial_\mu \left(A^\nu W^\mu\right) - \partial_\mu \left(A^\mu W^\nu\right)\right] - ieA_\mu \left[\partial^\mu W^\nu - \partial^\nu W^\mu\right] \\ &+ie\kappa W_\nu \left(\partial^\mu A^\nu - \partial^\nu A^\mu\right) \\ &+e^2 A_\mu \left[A^\mu W^\nu - A^\nu W^\mu\right]\end{aligned} \tag{5.189}$$

while for a scalar particle of charge $e$ the expanded form is

$$\partial_\mu \partial^\mu \phi + m^2 \phi = -ie\left[\partial_\mu \left(A^\mu \phi\right) + A^\mu \partial_\mu \phi\right] + e^2 A_\mu A^\mu \phi. \tag{5.190}$$

Aside from the anomalous magnetic moment term added by hand, the vector field exhibits a more complex current as could be anticipated given its magnetic moment.

## 5.8 ■ FEYNMAN RULES FOR VECTOR QED

Following the example of scalar electrodynamics, the field equations of vector electrodynamics may be analyzed and scattering amplitudes and Feynman rules derived. However, the Feynman rules for vertices in vector field electrodynamics may be found most simply by inserting plane waves corresponding to incoming and outgoing particles into the Lagrangian interaction terms, taking the derivatives, then erasing the exponential plane wave factors and adding a factor $i = \sqrt{-1}$. The terms linear in $e$ together give vertex factor proportional to $e$ for a $WWA$ three particle vertex. The terms quadratic in $e$ together give the vertex factor proportional to $e^2$ for the $WWAA$ four particle vertex and, in the latter case, the two associations of two participating photons with each factor $A$ and the two associations of participating $W$ bosons with each factor $W$ must be counted.

For example, an interaction Lagrangian term responsible for photon emission is

$$L_{AWW} = -ieA_\mu \left[W_\nu^\dagger F_W^{\mu\nu} - F_W^{\dagger\ \mu\nu}\right] \tag{5.191}$$

Consider a photon emission process $W \to W + A$. Insert into this Lagrangian the waves

$$W \to \epsilon_1 e^{-ip_1 x}\ ,\ W^\dagger \to \left(\epsilon_2 e^{ip_2 x}\right)^\dagger\ ,\ A \to \epsilon_A e^{i(p_2 - p_1)x}. \tag{5.192}$$

After erasing the exponentials, we find

$$F_W^{\mu\nu} = i\left(p_1^\mu \epsilon_1^\nu - p_1^\nu \epsilon_1^\mu\right) \; ; \; F_W^{\dagger\,\mu\nu} = -i\left(p_2^\mu \epsilon_2^{\nu\dagger} - p_2^\nu \epsilon_2^{\dagger\mu}\right) \tag{5.193}$$

and, adding a factor $i$, we arrive at the amplitude

$$S_{WWA} = (i)(-ie)\epsilon_{A\,\mu}^{\dagger}\left[\epsilon_2^\dagger i\left(p_1^\mu \epsilon_1^\nu - p_1^\nu \epsilon_1^\mu\right) - (-i)\left(p_2^\mu \epsilon_2^{\nu\dagger} - p_2^\nu \epsilon_2^{\dagger\mu}\right)\epsilon_{1\nu}\right] \tag{5.194}$$

or more compactly

$$S_{WWA} = iM = \epsilon_{A\,\mu}^{\dagger}\left\{(-ie)\left[(p_1+p_2)^\mu\left(\epsilon_2^\dagger\epsilon_1\right) - \epsilon_1^\mu\left(\epsilon_2^\dagger p_1\right) - \epsilon_2^{\dagger\mu}\left(p_2\epsilon_1\right)\right]\right\}. \tag{5.195}$$

The interaction term $L_a = -ie\kappa W_\mu^\dagger W_\nu F_A^{\mu\nu}$ adds an additional amplitude and, given that $p_2\epsilon_2^\dagger = 0$ and $p_1\epsilon_1 = 0$, it may be written as

$$\begin{aligned} S_\kappa &= iM = i\,(-ie)\,\kappa\epsilon_{2\,\mu}^{\dagger}\epsilon_{1\nu}\left[i\,(p_2-p_1)^\mu\,\epsilon_A^\nu - i\,(p_2-p_1)^\nu\,\epsilon_A^\mu\right] \\ &= ie\epsilon_{2\,\mu}^{\dagger}\epsilon_{1\nu}\left[(-\kappa p_1)^\mu\,\epsilon_A^\nu - (\kappa p_2)^\nu\,\epsilon_A^\mu\right] \\ &= ie\epsilon_A^\mu\epsilon_{2\,\mu}^{\dagger}\epsilon_{1\nu}\left[-(\kappa p_2)^\nu\,(-\kappa p_1)^\mu - (\kappa p_2)^\mu\right]. \end{aligned} \tag{5.196}$$

The total amplitude $S = S_{WWA} + S_\kappa$ can be written as

$$S = iM = A_\mu j_W^\mu \tag{5.197}$$

where the current associated with the $W$ boson is

$$j_W^\mu = \left\{(-ie)\left[(p_1+p_2)^\mu\left(\epsilon_2^\dagger\epsilon_1\right) - \epsilon_1^\mu\left(\epsilon_2^\dagger(1+\kappa)\,p_1\right) - \epsilon_2^{\dagger\mu}\left((1+\kappa)\,p_2\epsilon_1\right)\right]\right\}. \tag{5.198}$$

The factor $\epsilon_\mu^A$ stands generally for the electromagnetic field and, for a virtual photon, would be replaced by a propagator $A^\mu \to G_\gamma^{\mu\nu} = -g^{\mu\nu}/q^2$. The vector $j^\mu$ is the electromagnetic transition current and its first term is

$$j_1^\mu = (-ie)\left[(p_1+p_2)^\mu\left(\epsilon_2^\dagger\epsilon_1\right)\right]. \tag{5.199}$$

This term may be recognized as the transition current for a scalar charged particle, multiplied by the projection of the spin state of the exiting $W$ boson onto the spin state of the incoming $W$ boson. The second and third terms are relativistic (magnetic) in origin. In fact, recall the condition $p_\mu\epsilon^\mu = 0$. In the nonrelativistic limit, the polarization vectors have vanishing time component while the 4-momenta have vanishing space components. Hence the second and third terms in the current vanish while the first term gives

$$M = eA^0(2m_W)\,(\boldsymbol{\epsilon}_2\cdot\boldsymbol{\epsilon}_1). \tag{5.200}$$

Here the factor $2m_W$ arises from the normalization of the waves. If $A^0$ were a component of an electrostatic field, the scattering amplitude represented by $M$

would be exactly that for a scalar particle with the polarization vector projection representing spin conservation.

The $W$ boson field propagator may be unearthed as follows. Generally a $W$ boson field is described by the Proca equation

$$\partial_\mu\left(\partial^\mu W^\nu - \partial^\nu W^\mu\right) + m_W^2 W^\nu = j^\nu \tag{5.201}$$

Taking the 4-divergence of both sides gives

$$\partial_\nu W^\nu = \partial_\nu j^\nu / m_W^2 \tag{5.202}$$

so we can write the Proca equation in the form of the Klein-Gordon equation

$$\partial_\nu \partial^\nu W^\mu + m^2 W^\mu = \left(g^{\mu\nu} + \frac{\partial^\mu \partial^\nu}{m_W^2}\right) j_\nu \tag{5.203}$$

and, from the examples of the scalar and photon propagator, the propagator for a massive vector field may be identified as

$$G_W^{\mu\nu} = -i\frac{g^{\mu\nu} - p^\mu p^\nu / m_W^2}{p^2 - m_W^2}. \tag{5.204}$$

This is the result stated in Equation 5.89 and the numerator in this expression is the tensor

$$\rho_W^{\mu\nu} = \Sigma_{i=0}^{3} \epsilon_i^\mu \epsilon_i^{\nu\dagger} \tag{5.205}$$

which represents the sum over free particle polarization states. In low energy limit when the components of $p^\mu$ are small compared to $m_W$, we have $p^2 << m_W^2$ and $p^\mu p^\nu / m_W^2 << 1$ and the $W$ boson propagator reduces to $i g^{\mu\nu}/m_W^2$. The Feynman rules for vector electrodynamics are summarized in Table 5.2.

As an example of vector electrodynamics, consider the process of annihilation of two charged scalar particles into a virtual photon which materializes as a $W^+W^-$ pair as illustrated in Figure 5.9. The virtual photon is realistically made through

| Diagram element | Factor in amplitude |
|---|---|
| Ingoing vector | $\epsilon$ |
| Outgoing vector | $\epsilon^\dagger$ |
| Ingoing vector antiparticle | $\epsilon^\dagger$ |
| Outgoing vector antiparticle | $\epsilon$ |
| Vertex for particles | various |
| Internal vector line | $-i(g^{\mu\nu} - p^\mu p^\nu / m^2)/(p^2 - m^2)$ |

**TABLE 5.2** Feynman rules for vector electrodynamics. The vertex factors for the various interaction terms are not listed.

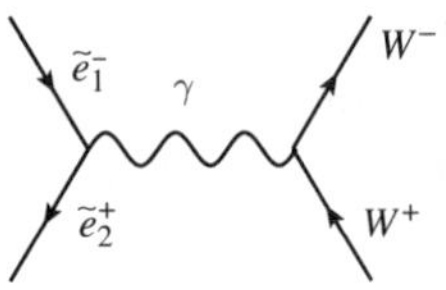


**FIGURE 5.9** Feynman diagram for $\tilde{e}^+\tilde{e}^- \to W^+W^-$.

fusion of $e^+e^-$, $u\bar{u}$ or $d\bar{d}$, processes considered in the next chapter. Here we use the machinery of scalar electrodynamics and vector electrodynamics to describe a hypothetical process. Combining the matrix element for the scalar $\tilde{e}^+\tilde{e}^-$ annihilation, the photon propagator, and the matrix element for $W^+W^-$ pair creation, we can immediately write down the matrix element in the form

$$iM_{\tilde{e}_1^-\tilde{e}_2^+\to W_3^-W_4^+} = j_e^\mu \left[\frac{-ig_{\mu\nu}}{(p_1+p_2)^2}\right] j_W^\nu . \tag{5.206}$$

The scalar current producing the photon is

$$j_e = (-ie)\,(p_1 - p_2)\,. \tag{5.207}$$

The current associated with $W$ pair production is that for photon emission by a $W^-$ (see Equation 5.195) with the relabeling of the exiting $W^-$ via $p_2 \to p_3$ and $\epsilon_2^\dagger \to \epsilon_3^\dagger$ while the incoming $W^-$ is replaced by an outgoing $W^+$ implying $p_1 \to -p_4$ and $\epsilon_1 \to \epsilon_4$, namely

$$j_W^\mu = ie\left[(p_3 - p_4)\left(\epsilon_4\epsilon_3^\dagger\right) + \left(\epsilon_3^\dagger p_4\right)\epsilon_4^\dagger - \epsilon_3^\dagger\,(p_3\epsilon_4)\right]. \tag{5.208}$$

There are three possible polarization states for the $W^-$ and three possible polarization states for the $W^+$ so nine combinations of amplitudes that may be computed. In electron positron annihilation to form $W^+W^-$, there are four possible initial polarization states so in total there are 36 transition rates predicted. Obviously, the analysis is somewhat complex and we will not pursue it further.

## 5.9 ■ VECTOR BOSON QED AT COLLIDERS

The intermediate vector bosons $W^\pm$ of the weak interaction have been produced in $p\bar{p}$ and $e^+e^-$ colliders in sufficient numbers to study the associated electromagnetic radiation. At a $p\bar{p}$ collider, a single $W^+$ is produced principally via the weak fusion of $u$-quark and $\bar{d}$-quark. Photons may be radiated by the initial state quarks, the intermediate state $W$ boson, and the final state charged lepton. A gluon from an initial state quark which happens to produce a narrow jet (a single $\pi^0$ for example) may fake a single photon signature. Production of a $Z$ boson which decays to $e^+e^-$ or $\mu^+\mu^-$ with one of the leptons fortuitously lost can fake the lepton plus missing energy signature used to identify the $W$ boson. The $Z$ boson is neutral and does not radiate in the standard model but a $Z$ boson which fakes a $W$ boson may nevertheless be accompanied by an additional photon from initial quarks and final leptons. The sum of the simulated contributions from each of these processes compares well to the data. Similar processes are under study at the LHC.

Production of $W^+W^-$ pairs at the LEP $e^+e^-$ collider provides another glimpse of vector electrodynamics. There are three principal leading order contributions to the amplitude corresponding to fusion into a virtual photon or $Z$ boson (fusion into a Higgs boson is negligible) and neutrino exchange:

$$e^+e^- \to \gamma \to W^+W^- \tag{5.209}$$

$$e^+e^- \to Z \to W^+W^- \tag{5.210}$$

and

$$e^+e^- \to W^+\bar{\nu}_e e^- \to W^+W^-. \tag{5.211}$$

These processes will be described in the context of electroweak unification.

If the $W$ boson is composite like the proton and neutron, then it should have a magnetic moment along its rest frame spin direction that differs from the value implicit in the structureless field theory. (The origin of the nucleon magnetic moments in their quark constituents is discussed in Chapter 10.) More generally, the electromagnetic interaction of the $W$ boson would be characterized by electric and magnetic structure factors. The $W$ boson does not live long enough to permit observation of a static magnetic moment as is possible for nucleons through spin precession in a magnetic field. Nor is it possible to scatter photons from a target of $W$ bosons. However, the radiation of a photon is the inverse of absorption, and the internal electromagnetic radiation from the $W$ boson observed at colliders is sensitive to $W$ boson structure especially for high transverse energy. Collider experiments place ever increasing limits on anomalous standard model electromagnetic couplings and on the existence of alternate effective Lagrangian interaction terms with tensor structure different from the ones we have derived in gauge theory. This sort of structure search extends to the $Z$ boson and to include $WWZ$ interactions described later.

## 5.10 ■ FURTHER READING

The relativistic scalar wave equation was introduced by W. Gordon, Z. Phys. **40**, 117 (1926) and O. Klein, Z. Phys. **41**, 407 (1927).

The use of a scalar field to describe nuclear forces is described in H. Yukawa, Proc. Phys. Math. Soc. Jpn. **17**, 48 (1935).

The description of massive charged spin 1 particles was introduced in A. Proca, J. Phys. Radium **7**, 347(1936), C. R. Acad. Sci. Paris **202**, 1366 (1936), C. R. Acad. Sci. Paris **202**, 1490 (1936), and J. Phys. Radium **9**, 61 (1938). The use of the modified Proca equation including the anomalous magnetic moment so that

$g = 2$ for a $W^-$ was studied by H. C. Corben and J. Schwinger [Phys.Rev. **58**, 953 (1940)] who found exact solutions for the Coulomb problem and identified states in which a vector particle will collapse to the origin. For a more recent analysis of this problem and its relationship to renormalizability, see Victor Flambaum and Michael Kuchiev, Mod. Phys. Lett. **A22**,2971 (2007), arXiv:hep-ph/0609194v1, and Phys. Rev. Lett. **98**, 181805 (2007).

Relativistic corrections to the nonrelativistic Schrödinger equation for spin 0, spin 1/2, and spin 1 particles are discussed in L. L. Foldy and S. A. Wouthuysen, Rev. Mod. Phys. **21**, 400 (1949) and K. M. Case, Phys. Rev. **95**, 1323 (1954).

## 5.11 ■ PROBLEMS

### Problem 5.1. Klein paradox

A charged scalar field subject to a constant potential $V = qA^0$ is described by the equation

$$\left(i\frac{\partial}{\partial t} - V\right)^2 \phi + \nabla^2\phi - m^2\phi = 0$$

and the solutions are combinations of free particle plane waves $\phi = e^{-ipx}$ with $p^2 = (E-V)^2 - m^2$. Suppose a potential barrier described by $V = 0$ for $z < 0$ and $V = V_0$ (a constant) for $z > 0$ and look for solutions

$$\begin{aligned}\phi(z<0) &= e^{-iEt+pz} + re^{-iEt-pz}\\ \phi(z>0) &= te^{-iEt+p'z}\end{aligned}$$

with $p' = \alpha\sqrt{(E-V_0)^2 - m^2}$ with $\alpha = \pm 1$. For a weak potential ($V_0 < E - m$), choose $\alpha = +1$. For an intermediate potential ($E - m < V_0 < E + m$), choose $p'$ to be negative and imaginary. For a strong potential ($V_0 > E + m$), compute the group velocity and argue that one must choose $\alpha = -1$ assuming there is no source of particle at $z = +\infty$. By requiring continuity of $\phi$ and $\partial_z\phi$ at $z = 0$, show that $t = 2p/(p + p')$ and $r = (p - p')/(p + p')$, compute the transmission coefficient $T$ and reflection coefficient $R$ and show that for a strong potential $R > 1$ and $T < 0$ and interpret in terms of particle and antiparticle pairs released from the barrier. [N. Dombey and A. Calogeracos, Phys. Rep. **315**, 41 (1999)]

### Problem 5.2. Classical electrodynamics

The electromagnetic field strength tensor and its dual are defined by

$$F^{\mu\nu} = \partial^\mu A^\nu - \partial^\nu A^\mu\ ;\ F_d^{\mu\nu} = \frac{1}{2}\epsilon^{\mu\nu\rho\sigma}F_{\rho\sigma}$$

where $\epsilon^{\mu\nu\rho\sigma}$ is the 4-dimensional completely antisymmetric numerical tensor with $\epsilon^{0123} = 1$ and $\epsilon^{\alpha\beta\delta\gamma} = +1(-1)$ for even (odd) permutations of the indices and $\epsilon^{\alpha\beta\delta\gamma} = 0$ otherwise. Note that $\epsilon^{\mu\nu\rho\sigma} = -\epsilon_{\mu\nu\rho\sigma}$ and that some authors define $\epsilon_{0123} = +1$. The components of $F_d$ are obtained from $F$ by the interchange $\mathbf{E} \to \mathbf{B}$ and $\mathbf{B} \to -\mathbf{E}$

$$F^{\mu\nu} = \begin{pmatrix} 0 & -E_1 & -E_2 & -E_3 \\ E_1 & 0 & -B_3 & B_2 \\ E_2 & B_3 & 0 & -B_1 \\ E_3 & -B_2 & B_1 & 0 \end{pmatrix} \; ; \; F_d^{\mu\nu} = \begin{pmatrix} 0 & -B_1 & -B_2 & -B_3 \\ B_1 & 0 & E_3 & -E_2 \\ B_2 & -E_3 & 0 & E_1 \\ B_3 & E_2 & -E_1 & 0 \end{pmatrix} .$$

a) Consider the Lagrangian

$$L = -\frac{1}{4} F^{\mu\nu} F_{\mu\nu} - j_\mu A^\mu$$

with current density $j$ independent of $A$ and of the derivatives $\partial^\mu A^\nu$. Show that the Euler equations for $A^\nu$ are $\partial_\mu F^{\mu\nu} = j^\nu$.
b) Next write the component equations in terms of $\mathbf{E}$ and $\mathbf{B}$ and verify they are the inhomogeneous Maxwell equations (Gauss's Law and Ampere's Law).
c) Next show that $\partial_\mu F_d^{\mu\nu} = 0$ is satisfied by virtue of $\partial_\mu \partial_\nu = \partial_\nu \partial_\mu$ and represents the homogeneous equations $\nabla \cdot \mathbf{B} = 0$ and Faraday's Law.
d) Finally show the Lorentz force equation has the covariant form

$$\frac{dp^\mu}{d\tau} = eF^{\mu\nu} v_\nu$$

where $d\tau = \sqrt{dt^2 - d\mathbf{x}^2} = dt/\gamma$ is the proper time along the particle path and $v^\mu = dx^\mu/d\tau$ is the four velocity of the particle parametrized in terms of proper time.

---

**Problem 5.3. Relativistic atom in a magnetic field**

---

As an application of the Lorentz transformation of the electromagnetic field, consider the DIRAC experiment which expects to see study pionium ($\pi^+\pi^-$) atoms of energy 5-6 GeV. Ground state atoms may be excited to atomic $p$ states with lifetimes of 11 ps by a Stark electric field effect in their rest frame. Consider first a neutral anti-hydrogen atom ($e^+\bar{p}$) with speed $v$ moving perpendicular magnetic field of strength $|\mathbf{B}| = 4$ T. a) Show that the Lorentz transformations imply that in the atom's rest frame, the electric field has strength $|\mathbf{E}| = (v/c)\gamma c|\mathbf{B}|$. b) The ground state will be ionized if the external force on the positron at about a Bohr radius $a_0$ balances the Coulomb force of the $\bar{p}$. Show that this condition can be written as $|\mathbf{E}| = 2V_0/a_0$ with $V_0 = 13.6$ Volts and find the energy of the atom in GeV for which the condition is satisfied. c) Now consider an exotic atom of reduced mass $m = m_\pi/2$. We characterize an achievable field strength by the magnetization density of a saturated ferromagnet $M = \mu_{Bohr}/V_a$ with $\mu_{Bohr} = e/2m_e = \sqrt{\pi\alpha}/m_e$ and atomic volume $V_a = (3a_0)^3 = 27/(m_e\alpha)^3$. Show that the ionization threshold velocity in such a field strength is given by $v\gamma = [27/(4\pi\alpha)](m/m_e)^2$.

## Problem 5.4. Equations of scalar electrodynamics

The Lagrangian for a charged scalar field interacting with an electromagnetic field is

$$L = (D_\mu\phi)^\dagger D^\mu\phi - m^2\phi^\dagger\phi - \frac{1}{4}F^{\mu\nu}F_{\mu\nu}$$

where $D^\mu = \partial^\mu + ieA^\mu$ and $F_{\mu\nu} = \partial_\mu A_\nu - \partial_\nu A_\mu$. Show that a) the Euler-Lagrange equation for $\phi$ is the Klein-Gordon equation with $\partial^\mu \to D^\mu$, and b) the Euler-Lagrange equations for $A^\nu$ are Maxwell's equations $\partial_\mu F^{\mu\nu} = j^\nu$ where $j^\nu = ie[\phi^\dagger D^\nu\phi - (D^\nu\phi)^\dagger\phi]$. Maxwell's equations require the 4-divergence of the current density vanish ($\partial_\mu j^\mu = 0$). Show that the divergence can be written in terms of covariant derivatives as

$$\begin{aligned}\partial_\mu j^\mu &= ie\partial_\mu\left[\phi^\dagger D^\nu\phi - (D^\nu\phi)^\dagger\phi\right]\\ &= \left(\left(D_\mu - ieA_\mu\right)\phi\right)^\dagger D^\nu\phi + \phi^\dagger\left(D_\mu - ieA_\mu\right)(D^\mu\phi)\\ &\quad - \left(\left(D_\mu - ieA_\mu\right)(D^\mu\phi)\right)^\dagger\phi - (D^\mu\phi)^\dagger\left(D_\mu - ieA_\mu\right)\phi\end{aligned}$$

by taking the gradient inside the conjugation and only then expressing it in terms of the gauge covariant derivative. Expand this form and cancel terms to show that the divergence indeed vanishes identically provided $\phi$ is a solution to the gauge covariant equation $D_\mu D^\mu\phi + m^2\phi = 0$.

## Problem 5.5. Axions and Higgslets

Relic neutral scalar particles called axions could be a component of cold dark matter. The axion field $\phi$ motivated by quantum chromodynamics is expected to be coupled through a virtual quark loop to the electromagnetic field with Lagrangian

$$L = -\frac{1}{4}F_{\mu\nu}F^{\mu\nu} + \frac{1}{2}\partial_\mu\phi\partial^\mu\phi - \frac{1}{2}m^2\phi^2 + \frac{g}{4}\phi F_d^{\mu\nu}F_{\mu\nu}$$

where $F_d^{\mu\nu} = \epsilon^{\mu\nu\rho\sigma}F_{\rho\sigma}/2$ and $F_d^{\mu\nu}F_{\mu\nu} = -4\mathbf{E}\cdot\mathbf{B}$. (Caution: Our convention is $\epsilon^{0123} = +1$ not $\epsilon_{0123} = +1$ and the sign of $F_d$ and $g$ depend on this convention.) a) Show that the Euler field equation for $\phi$ is

$$\partial_\mu\partial^\mu\phi + m^2\phi = -g\mathbf{E}\cdot\mathbf{B}$$

so superposed **E** and **B** fields form a source density for $\phi$. b) Noting that $\partial_\mu F_d^{\mu\nu} = 0$, show that Maxwell's equations become

$$\partial_\mu F^{\mu\nu} = g\partial_\mu(\phi F_d^{\mu\nu}) = g(\partial_\mu\phi)F_d^{\mu\nu}$$

or in component form

$$\begin{aligned}
\nabla \times \mathbf{B} - \partial_t \mathbf{E} &= \mathbf{j} - g\partial_t \phi \mathbf{B} - g\nabla\phi \times \mathbf{E} \\
\nabla \cdot \mathbf{E} &= \rho + g\mathbf{B} \cdot \nabla\phi \\
\nabla \times \mathbf{E} + \partial_t \mathbf{B} &= 0 \\
\nabla \cdot \mathbf{B} &= 0
\end{aligned}$$

where $j^\mu = (\rho, \mathbf{j})$ are any other electromagnetic current densities. Evidently, like the $\pi^0$ and Higgs boson, the $\phi$ couples to two photons and the overlap of a gradient of $\phi$ with an electromagnetic field is a source and sink of electromagnetic field. Therefore we can anticipate two photon decay $\phi \to \gamma\gamma$ and related processes such as $\gamma\phi \to \gamma$ and $\gamma\gamma \to \phi$ and mixing of the classical field $\mathbf{E}$, $\mathbf{B}$ and $\phi$. c) Deduce the Feynman rules and the covariant form of the vertex factor by inserting photon plane waves into the interaction term in the Lagrangian and erasing the exponential wave factors and show that the two photon decay rate is

$$\Gamma_{A\to\gamma\gamma} = \frac{g^2}{64\pi} m_\phi^3.$$

There are a number of limits on light mass ($m_\phi \approx$ eV) nearly stable particles of this type. Recent experiments include ADMX, GammeV, CAST, and the Tokyo axion helioscope. [P. Sikivie, Phys. Rev. Lett. **51**, 1415 (1983), J. Moody and F. Wilczek, Phys. Rev. **D30**, 130 (1984); L Maiani et al, Phys. Lett. **B175**, 359 (1986), K. Zioutas et al. (CAST Collaboration) Phys. Rev. Lett. **94**, 121301 (2005). See also "Identification of dark matter 2008," Proceedings of Science, `http://pos.sissa.it`. A compilation of limits is published by the Particle Data Group and a general review is Georg. G. Raffelt, Phys. Rep. **198**, 1 (1990).]

## Problem 5.6. Relativistic selectron potential scattering

Consider the cross section

$$\frac{d\sigma_{\tilde{e}_1\tilde{\mu}_2\to\tilde{e}_3\tilde{\mu}_4}}{d\Omega} = \frac{e^4}{64\pi^2}\frac{p'}{ps}\left|\frac{s-u}{t}\right|^2$$

in the center of mass frame in the limit $m_{\tilde{\mu}} \to \infty$. Show that $t = -4\mathbf{p}^2 \sin^2\frac{\theta}{2}$ and $s - u \to 4E_{\tilde{e}}m_{\tilde{\mu}}$ and hence

$$\frac{d\sigma_{\tilde{e}_1\tilde{\mu}_2\to\tilde{e}_3\tilde{\mu}_4}}{d\Omega} = \frac{\alpha^2}{4m_{\tilde{e}}^2\gamma^2 v^4}\frac{1}{\sin^4(\frac{\theta}{2})}$$

which is the relativistic classical cross section.

**Problem 5.7. Scalar electron muon scattering**

The cross section for $\tilde{e}_1\tilde{\mu}_2 \to \tilde{e}_3\tilde{\mu}_4$ scattering by photon exchange is

$$\frac{d\sigma}{dt} = \frac{\pi\alpha^2}{4p^2 s}\frac{(s-u)^2}{t^2}.$$

We wish to write the cross section in the frame in which $\tilde{\mu}_2$ is at rest. For simplicity, assume $m_{\tilde{e}}$ is negligible. Then

$$t = (p_3 - p_1)^2 = -2E_1E_3x = (p_4 - p_3)^2 = 2m_{\tilde{\mu}}\left(m_{\tilde{\mu}} - E_4\right)$$

where $x = 1 - \cos\theta = 2\sin^2(\theta/2)$ where $\theta$ is the deflection angle of the selectron. a) Express $s, u$, and $p^2 s$ in the rest frame of the smuon and use $m_{\tilde{\mu}} - E_4 = E_3 - E_1$ and solve for the energy of the scattered selectron as a function of angle to find $E_3 = E_1/(1 + (E_1/m_{\tilde{\mu}})x)$ so we can express $t$ in terms of angle

$$t = -2E_1E_3x = -2E_1^2x/\left[1 + \left(E_1/m_{\tilde{\mu}}\right)x\right].$$

b) Next show that $u = (p_3 - p_2)^2 = m_{\tilde{\mu}}^2 - 2E_3m_{\tilde{\mu}}$ and $s = (p_1 + p_2)^2 = m_{\tilde{\mu}}^2 + 2E_1m_{\tilde{\mu}}$ so

$$\frac{s-u}{t} = -\frac{m_{\tilde{\mu}}}{x}\left(\frac{E_1 + E_3}{E_1E_3}\right) = -\frac{2m_{\tilde{\mu}}}{E_1x}\left(1 + \frac{E_1}{2m_{\tilde{\mu}}}x\right).$$

c) Finally assemble these results to show that

$$\frac{d\sigma}{d\Omega} = \frac{1}{2\pi}\frac{dt}{d\cos\theta}\frac{d\sigma}{dt} = \frac{\alpha^2}{4E_1^2\sin^4\left(\frac{\theta}{2}\right)}\left(\frac{1 + \frac{E_1}{m_{\tilde{\mu}}}\sin^2\left(\frac{\theta}{2}\right)}{1 + \frac{2E_1}{m_{\tilde{\mu}}}\sin^2\left(\frac{\theta}{2}\right)}\right)^2$$

which has the relativistic Rutherford form (with $v \simeq 1$ since we neglected $m_{\tilde{e}}$) times a "recoil" factor in braces which in this case approaches unity when $m_{\tilde{\mu}} >> E_1$. For a relativistic spin 1/2 particle scattering from a Coulomb field (Mott scattering), this factor is $\gamma^{-2} + v^2\cos^2(\theta/2)$ and the Rutherford form is valid only in the nonrelativistic limit.

**Problem 5.8. Scalar Bhabba scattering**

a) Show that the amplitude for scalar charged particle scattering $\tilde{e}^+\tilde{e}^- \to \tilde{e}^+\tilde{e}^-$ is

$$M_{\tilde{e}^+\tilde{e}^-\to\tilde{e}^+\tilde{e}^-} = e^2\left(\frac{u-s}{t} + \frac{u-t}{s}\right).$$

b) For elastic scattering of identical mass particles, show that $t = -(s - 4m^2)S$ and $u = -(s - 4m^2)(1 - S)$ where $S = \sin^2(\theta^{CM}/2)$. c) Use the relations of part b) to show that the differential cross section in the center of mass is

$$\frac{d\sigma_{\tilde{e}^+\tilde{e}^-\to\tilde{e}^+\tilde{e}^-}}{d\Omega} = \frac{\alpha^2}{4s^3}\frac{1}{S^2}\left[\frac{s^2}{s-4m^2} + s\left(1-2S+2S^2\right) + 4m^2 S\,(1-2S)\right]^2.$$

---

**Problem 5.9. Scalar Compton scattering**

---

Verify that the squared amplitude for Compton scattering of a photon from a charged scalar particle such as $\gamma_2\pi^+ \to \gamma_4\pi^+$ in the rest frame of the initial charged particle, averaged over initial polarizations and summed over final polarizations, is as described in the text

$$\overline{|M|^2} \equiv \frac{1}{2}\sum |M|^2 = 2e^4\left(1+\cos^2\theta\right)$$

where $\theta$ is the scattering angle of the photon. Choose linear polarization vectors in and out of the scattering plane. For $\theta = \pi/2$, with initial polarization vector in the scattering plane, show that the amplitude vanishes for any final polarization.

---

**Problem 5.10. Compton scattering kinematics**

---

For scattering of a photon of 4-momentum $k$ and an electron of momentum $p$, show that squaring the 4-momentum conservation relation $p + k - k' = p'$ requires $pk - pk' - kk' = 0$. Evaluate in the rest frame of $p$ to derive the Compton relation between the final photon energy and angle.

---

**Problem 5.11. Scalar pair production**

---

Consider the process $\gamma\gamma \to \pi^+\pi^-$. a) Show that the invariant amplitude is

$$M = e^2\left[-\frac{\epsilon_1 p_+ \epsilon_2 p_-}{p_+ p_1} - \frac{\epsilon_2 p_+ \epsilon_1 p_-}{p_+ p_2} + \epsilon_1\epsilon_2\right]$$

where $p_1$ and $p_2$ and $\epsilon_1$ and $\epsilon_2$ are the photon 4-momentum and polarization vectors and $p_+$ and $p_-$ are the pion 4-momentum vectors. b) Express the 4-momentum vectors in the center of mass as $p_1 = k(1, \hat{\mathbf{z}})$, $p_2 = k(1, -\hat{\mathbf{z}})$, $p_+ = E(1, \beta\hat{\mathbf{n}})$, $p_- = E(1, -\beta\hat{\mathbf{n}})$, $\epsilon_1 = (0, \boldsymbol{\epsilon}_1)$, and $\epsilon_2 = (0, \boldsymbol{\epsilon}_2)$ and show that

$$M/e^2 = 2\beta^2\frac{\boldsymbol{\epsilon}_1\cdot\hat{\mathbf{n}}\boldsymbol{\epsilon}_2\cdot\hat{\mathbf{n}}}{1-(\beta\hat{\mathbf{n}}\cdot\hat{\mathbf{z}})^2} - \boldsymbol{\epsilon}_1\cdot\boldsymbol{\epsilon}_2.$$

c) Evaluate $M$ for the four combinations of polarization vectors $\boldsymbol{\epsilon}_1 = \hat{\mathbf{x}}$, $\boldsymbol{\epsilon}_1 = \hat{\mathbf{y}}$ and $\boldsymbol{\epsilon}_2 = \hat{\mathbf{x}}$, $\boldsymbol{\epsilon}_2 = \hat{\mathbf{y}}$ and show that the average is $\Sigma|M|^2/4 = (e^4/2)f(\theta)$ where

$$f(\theta) = 1 - \frac{2\beta^2\sin^2\theta}{1-\beta^2\cos^2\beta} + \frac{2\beta^4\sin^4\theta}{(1-\beta^2\cos^2\theta)^2}$$

where $\hat{\mathbf{n}} \cdot \hat{\mathbf{z}} = \cos\theta$. d) Plot the right-hand side as a function of $\cos\theta$ for the values $\beta^2 = 0.1, 0.5, 0.9$, and $0.99$ to see how the angular distribution evolves with energy. The cross section $\sigma = \int d\Omega f(\theta)\alpha^2/(2s)$ is compared to data in J. F. Donoghue and B. R. Holstein, Phys Rev. **D48**, 137 (1992).

---

## Problem 5.12. Pionium annihilation

---

The invariant matrix element for $\tilde{e}^+\tilde{e}^- \leftrightarrow \gamma\gamma$ summed over photon polarizations is

$$\Sigma|M|^2 = 2e^4\left[1 - \frac{2\beta^2\sin^2\theta}{1-\beta^2\cos^2\beta} + \frac{2\beta^4\sin^4\theta}{\left(1-\beta^2\cos^2\theta\right)^2}\right].$$

where $\theta$ is the angle between one photon and one of the pions in the center of mass and $\beta$ is the speed of either pion in the center of mass. The cross section is

$$\begin{aligned}\sigma_{\tilde{e}^+\tilde{e}^-\leftrightarrow\gamma\gamma} &= \int d\Omega\,\frac{1}{64\pi^2}\frac{k}{ps}\Sigma|M|^2\\ &= \frac{\pi\alpha^2}{\beta^2 s}\left[4\beta - 2\beta^3 + \left(1-\beta^4\right)\ln\frac{1-\beta}{1+\beta}\right].\end{aligned}$$

Use this expression and $\tau^{-1} = |\psi(0)|^2 v_{rel}\sigma$ to compute the lifetime $\tau$ for the decay $(\pi^+\pi^-)_{\text{atom}} \to \gamma\gamma$ for the atomic ground state of pionium and show that it is long compared to the lifetime of about $2.82 \pm 0.3$ fs for the charge exchange process $\pi^+\pi^- \to \pi^0\pi^0$ leading to four photons, a process similar to the decay $(\pi^-K^+)_{\text{atom}} \to \pi^0 K^0$. [ Nemenov, Leonid, "Lifetime measurement of $\pi^+\pi^-$ and $\pi K$ atoms to test low-energy QCD," CERN-SPSC-2009-013 ; SPSC-SR-044.]

---

## Problem 5.13. Coulomb production

---

The lifetime of the $\pi^0$ can be inferred from the cross section $\sigma_{\gamma Z\to\pi^0 Z}$ for fusion of a photon with a virtual photon in the Coulomb field of a nucleus Z, the inverse of the decay $\pi^0 \to \gamma\gamma$. The interaction Lagrangian is assumed to have the form $L = -j_N^\mu A_\mu + g_{\pi^0\gamma\gamma}\phi F^{\mu\nu}\tilde{F}_{\mu\nu}$ where $j_N^\mu$ is the electromagnetic current of the nucleus, $A^\mu$ is the electromagnetic potential, $\phi$ is a pseudoscalar pion field, $F^{\mu\nu}$ is the field strength tensor and $\tilde{F}^{\mu\nu}$ its dual, and $g_{\pi^0\gamma\gamma}$ is a form factor. From the second term (appropriate to a pseudoscalar meson decay), $\Gamma_{\pi^0\to\gamma\gamma}$ may be calculated. By combining both terms, $\sigma_{\gamma\text{N}\to\pi^0\text{N}}$ may be computed. Both are proportional to $g_{\pi^0\gamma\gamma}$ which may be regarded as constant. Assuming a fixed point nucleus and therefore $E_\pi = E_\gamma \equiv k >> m_\pi$, one finds

$$\frac{d\sigma_{\gamma Z\to\pi^0 Z}}{dt} = \frac{\pi}{k^2}\frac{d\sigma_{\gamma Z\to\pi^0 Z}}{d\Omega_\pi} = 8\pi\alpha Z^2\frac{\Gamma_{\pi^0\to\gamma\gamma}}{m_{\pi^0}^3}\frac{t-t_0}{t^2}.$$

Here $Z$ is the atomic number and $\alpha \simeq 1/137$. The four momentum transfer is

$$t = (p_\pi - p_\gamma)^2 \simeq -\mathbf{q}^2 + t_0$$

with transverse momentum transfer $\mathbf{q}^2 = 4kp_\pi \sin^2(\theta/2) \simeq (k\theta)^2$. The minimum value is $t_0 \simeq -(m_\pi^2/(2k))^2 \equiv -q_L^2$ with $q_L$ the longitudinal momentum transfer to the nucleus. Consider $E_\gamma = 5.5$ GeV and $Z = 6$. a) Use $E_{\pi^0} = E_\gamma$ to derive the expression for $t_0$. b) Compute $q_L$ and the recoil energy $E_L = q_L^2/(2Am_p)$. c) Compute the angle $\theta_{max}$ in radians at which the cross section is a maximum and, at $\theta_{max}$, the transverse momentum $|\mathbf{q}|$, the energy transfer $\mathbf{q}^2/(2Am_p)$, and $d\sigma/d\Omega$ in mb str$^{-1}$. d) Compute the angle $\theta \simeq 1/(E_\gamma R_N)$ for radius $R_N = 1.1A^{1/3}$ fm at which the nuclear form factor cuts off production. Integrate the cross section up to this value and find the total cross section in mb. [ A. Halprin, C. M. Andersen, and H. Primakoff, Phys. Rev. **152**, 1295 (1956), A. Bernstein, Proceedings of the 6th International Workshop on Chiral Dynamics, CD09 July 6-10, 2009]

# CHAPTER 6

# Electrodynamics of Fermions

Free fermions are described by spinor field equations. Gauge symmetry determines the form of electrodynamics. The Feynman rules for spinor QED are derived and applied.

## Contents



## 6.1 ■ DIRAC EQUATION FOR A FREE PARTICLE

Paul Dirac came up with a first-order matrix wave equation which describes a relativistic spin-1/2 particle. Dirac was motivated to generalize quantum mechanics to incorporate the special theory of relativity. His equation explained the doubling of stationary states of atomic electrons, already tentatively ascribed to spin, and predicted the existence of the anti-electron. Ultimately, the Dirac theory of electronic

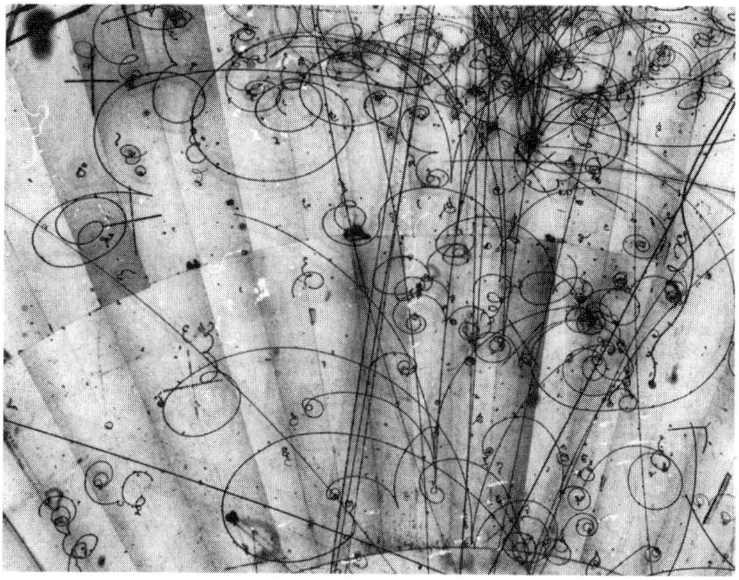

**FIGURE 6.1** Image from Fermilab 15-ft bubble chamber experiment E632 showing several instances of $e^+e^-$ pair production. A magnetic field causes electrons to curl clockwise and positrons anti-clockwise. [G.T. Jones, Birmingham University/ Fermi National Accelerator Laboratory, POW 980318]

matter waves interacting with an electromagnetic field led to the development of the first full fledged quantum field theory known as quantum electrodynamics.

The Dirac equation can be written in terms of the momentum operator $p_\mu = i\partial_\mu$ and a 4-vector of matrices $\gamma^\mu = (\gamma^0, \boldsymbol{\gamma}) = g^{\mu\nu}\gamma_\nu$ as

$$\gamma^\mu p_\mu \psi = m\psi \tag{6.1}$$

with $m$ a constant that will be identified with the particle mass. If we apply $\gamma_\nu p^\nu$ to both sides, we find

$$\gamma_\mu \gamma_\nu p^\mu p^\nu \psi = m\gamma_\mu p^\mu \psi = m^2\psi. \tag{6.2}$$

Since

$$\gamma_\mu \gamma_\nu = \frac{1}{2}(\gamma_\mu\gamma_\nu + \gamma_\nu\gamma_\mu) + \frac{1}{2}(\gamma_\mu\gamma_\nu - \gamma_\nu\gamma_\mu) \tag{6.3}$$

while $(\gamma_\mu\gamma_\nu - \gamma_\nu\gamma_\mu)p^\mu p^\nu$ vanishes identically, we have

$$\frac{1}{2}(\gamma_\mu\gamma_\nu + \gamma_\nu\gamma_\mu)p^\mu p^\nu \psi = m^2\psi. \tag{6.4}$$

In order that $\psi$ satisfy the Klein-Gordon wave equation

$$(p^2 - m^2)\psi = 0, \tag{6.5}$$

we require the $\gamma$ matrices satisfy

$$\gamma_\mu \gamma_\nu + \gamma_\nu \gamma_\mu = 2g_{\mu\nu} \tag{6.6}$$

implying for $\mu \neq \nu$ that $\gamma_\mu \gamma_\nu = -\gamma_\nu \gamma_\mu$ and

$$1 = \gamma_0^2 = -\gamma_1^2 = -\gamma_2^2 = -\gamma_3^2. \tag{6.7}$$

It now follows that there are de Broglie wave solutions of the form

$$\psi = ue^{\mp ipx} \tag{6.8}$$

where $u$ is a constant spinor amplitude and the 4-momentum $p^\mu = (E, \mathbf{p})$ satisfies $E = \sqrt{\mathbf{p}^2 + m^2}$. Solutions of the form $ue^{-ipx}$ will be associated with matter and solutions of the form $ue^{+ipx}$ with anti-matter.

If we write Dirac's equation as

$$i\gamma_0 \frac{\partial}{\partial t}\psi + i\boldsymbol{\gamma} \cdot \nabla \psi = m\psi \tag{6.9}$$

and multiply by $\gamma_0$ using $\gamma_0^2 = 1$, we find the Hamiltonian form

$$i\partial_t \psi = H\psi = \gamma_0 \boldsymbol{\gamma} \cdot \mathbf{p} + m\gamma_0 = \boldsymbol{\alpha} \cdot \mathbf{p} + m\alpha^0. \tag{6.10}$$

Here we define

$$\alpha^0 \equiv \gamma^0, \ \boldsymbol{\alpha} \equiv \gamma^0 \boldsymbol{\gamma}. \tag{6.11}$$

These four matrices satisfy

$$\alpha_j^2 = 1, \ \alpha_i \alpha_j = -\alpha_j \alpha_i (\text{for } i \neq j. \tag{6.12}$$

If $H$ is required to be Hermitian ($H^\dagger = H$) to ensure conservation of the free particle current as in Schrodinger theory, then $\gamma_0 = \gamma_0^\dagger$ and

$$\boldsymbol{\alpha} = \gamma_0 \gamma = \gamma^\dagger \gamma_0 \tag{6.13}$$

or, equivalently, we have an additional constraint on the $\gamma$ matrices that we can write as

$$\gamma_\mu^\dagger = \gamma_0 \gamma_\mu \gamma_0. \tag{6.14}$$

Dirac found that a minimal solution to the constraints on the $\gamma$ matrices requires $4 \times 4$ matrices acting on a 4-component wave function called a spinor. The standard representation of the $\gamma$ matrices is

$$\gamma^0 = \begin{pmatrix} 1 & 0 \\ 0 & -1 \end{pmatrix}, \ \boldsymbol{\gamma} = \begin{pmatrix} 0 & \boldsymbol{\sigma} \\ -\boldsymbol{\sigma} & 0 \end{pmatrix}, \ \boldsymbol{\alpha} = \begin{pmatrix} 0 & \boldsymbol{\sigma} \\ \boldsymbol{\sigma} & 0 \end{pmatrix} \tag{6.15}$$

where $\boldsymbol{\sigma}$ denotes the Pauli matrices (Equation 1.68). It will be useful in what follows to define the quantity

$$\gamma^5 = -i\gamma^0\gamma^1\gamma^2\gamma^3 = -\frac{i}{4!}\epsilon^{\mu\nu\rho\sigma}\gamma_\mu\gamma_\nu\gamma_\rho\gamma_\sigma = \begin{pmatrix} 0 & -1 \\ -1 & 0 \end{pmatrix} \tag{6.16}$$

where $\epsilon^{\mu\nu\rho\sigma} = +1/-1$ if the indices are an even/odd permutation of 0123 and $\epsilon^{\mu\nu\rho\sigma} = 0$ otherwise. (The reader is cautioned that other authors define $\gamma^5$ with a different phase.) We also define

$$\sigma_{\mu\nu} = \frac{1}{2}(\gamma_\mu\gamma_\nu - \gamma_\nu\gamma_\mu). \tag{6.17}$$

With reference to Equation 5.19, just as $\mathbf{E}$ and $\mathbf{B}$ can be identified as the components of an antisymmetric 4-tensor $F^{\mu\nu} \sim (-\mathbf{E}, \mathbf{B})$, the antisymmetric tensor $\sigma^{\mu\nu}$ has components $\sigma^{\mu\nu} \sim (\boldsymbol{\alpha}, i\boldsymbol{\Sigma})$ with

$$\boldsymbol{\Sigma} = \begin{pmatrix} \boldsymbol{\sigma} & 0 \\ 0 & \boldsymbol{\sigma} \end{pmatrix} \tag{6.18}$$

and we have

$$\frac{1}{2}\sigma_{\mu\nu}F^{\mu\nu} = \boldsymbol{\alpha}\cdot\mathbf{E} - i\boldsymbol{\Sigma}\cdot\mathbf{B} \tag{6.19}$$

with $\boldsymbol{\alpha} = \gamma_0\boldsymbol{\gamma}$.

Neither the Pauli matrices nor the $\gamma$ matrices are unique. Any set of Pauli matrices related by an arbitrary unitary transformation

$$\boldsymbol{\sigma} \to U\boldsymbol{\sigma}U^\dagger \tag{6.20}$$

with $U$ a unitary ($U^\dagger = U^{-1}$) two by two matrix will obey the algebraic conditions Eq. 1.69 and similarly any set of $\gamma$ matrices obtained by

$$\gamma^\mu \to U\gamma^\mu U^\dagger \tag{6.21}$$

with $U$ a unitary by $4 \times 4$ matrix will satisfy the constraints imposed on the Dirac matrices. We will find the standard form of the $\gamma$ matrices most useful for nonrelativistic problems and a different Weyl form more useful for relativistic problems.

## 6.2 ■ EQUATIONS OF SPINOR ELECTRODYNAMICS

An equation describing a spinor field interacting with an electrodynamic field is obtained as in Schrödinger theory and in the relativistic electrodynamics of charged bosons. We simply replace the ordinary derivative with the gauge covariant derivative $\partial^\mu \to D^\mu = \partial^\mu + ieA^\mu$ in the free particle Dirac equation and find

$$(i\gamma_\mu\partial^\mu - e\gamma_\mu A^\mu - m)\psi = 0. \tag{6.22}$$

This equation is covariant under a simultaneous gauge transformation of the electromagnetic potential and of the spinor field

$$A^\mu(x) \to A^\mu(x) + \partial^\mu \chi(x)\ ;\ \psi(x) \to e^{-ie\chi(x)} \tag{6.23}$$

where $\chi(x)$ is an arbitrary function of the coordinates $x$.

In what follows, we will need the Dirac conjugate wave function defined as

$$\bar{\psi} \equiv \psi^\dagger \gamma_0 \tag{6.24}$$

and the current density defined as

$$j^\mu = (\rho, \mathbf{j}) \equiv \bar{\psi}\gamma^\mu\psi = \left(\psi^\dagger\psi, \psi^\dagger \boldsymbol{\alpha}\psi\right). \tag{6.25}$$

The Hermitian conjugate of the Dirac equation is

$$\psi^\dagger\left(-i\gamma_\mu^\dagger \partial_\mu - e\gamma_\mu^\dagger A^\mu - m\right) = 0 \tag{6.26}$$

where the 4-derivative is understood to act to the left and for plane waves will be replaced by the 4-momentum. If we multiply by $\gamma_0$ on the left and right and use $\gamma_0^2 = 1$ and $\gamma_\mu^\dagger = \gamma_0\gamma_\mu\gamma_0$, the equation for $\bar{\psi}$ is

$$\bar{\psi}\left(i\gamma_\mu\partial^\mu + e\gamma_\mu A^\mu + m\right) = 0. \tag{6.27}$$

The Dirac equation with electromagnetic interaction and its conjugate imply (see exercise) the current conservation equation

$$\partial_\mu j^\mu = 0 \tag{6.28}$$

which is why $H$ was required to be Hermitian. This current density is easily seen to be gauge invariant. It represents the flow of electric charge in the complete theory of spinor electrodynamics described by the equations

$$\left(i\partial_\mu\gamma^\mu - m\right)\psi = eA^\mu\gamma_\mu\psi \tag{6.29}$$

$$\partial_\nu F^{\nu\mu} = e\bar{\psi}\gamma^\mu\psi. \tag{6.30}$$

It is left as an exercise to show that these equations follow from the gauge invariant Lagrangian

$$L = i\bar{\psi}\gamma_\mu D^\mu\psi - m\bar{\psi}\psi - \frac{1}{4}F_{\mu\nu}F^{\mu\nu}. \tag{6.31}$$

## 6.3 ■ NON-RELATIVISTIC SPIN 1/2 PARTICLES

The nonrelativistic limit $v << 1$ of the Dirac theory may be derived as follows. Isolate a rapid oscillation associated with the rest energy by writing positive energy solutions in the form

$$\psi(x) = e^{-imt} \begin{pmatrix} \eta(x) \\ \chi(x) \end{pmatrix}. \tag{6.32}$$

The 2-component spinors $\eta$ and $\chi$ then satisfy

$$\left(E - eA^0\right) \eta = \boldsymbol{\sigma} \cdot (\mathbf{p} - e\mathbf{A})\chi \tag{6.33}$$

and

$$\left(E - eA^0 + 2m\right) \chi = \boldsymbol{\sigma} \cdot (\mathbf{p} - e\mathbf{A})\eta. \tag{6.34}$$

For nonrelativistic motion in a weak potential, we may neglect the kinetic and potential energy terms relative to the rest mass on the left-hand side of the second equation and solve for

$$\chi = \frac{1}{2m} \boldsymbol{\sigma} \cdot (\mathbf{p} - e\mathbf{A})\eta \tag{6.35}$$

which indicates that $\chi$ is proportional to the space derivatives of $\eta$ and of order $v$ times $\eta$. (The 4-component Dirac spinor in the standard representation is loosely speaking a 2-component spinor $\eta$ together with a 2-component spinor $\chi$ associated with its momentum and hence no derivatives appear in the current density $\bar{\psi}\gamma\psi$.) Substitution into the first equation yields the Pauli equation:

$$E\eta = \frac{[\sigma \cdot (\mathbf{p} - e\mathbf{A})]^2}{2m} \eta + eA^0\eta. \tag{6.36}$$

Using the Pauli matrix identity

$$\mathbf{a} \cdot \boldsymbol{\sigma} \mathbf{b} \cdot \boldsymbol{\sigma} = \mathbf{a} \cdot \mathbf{b} + i\mathbf{a} \times \mathbf{b} \cdot \boldsymbol{\sigma} \tag{6.37}$$

and the corollary

$$(\mathbf{a} \cdot \boldsymbol{\sigma})^2 = \mathbf{a}^2, \tag{6.38}$$

the Pauli equation can be written as

$$i\partial_t \eta = H\eta = \left[ \frac{(-i\nabla - e\mathbf{A})^2}{2m} - \frac{g}{2} \frac{e}{2m} \boldsymbol{\sigma} \cdot \mathbf{B} + eA^0 \right] \eta \tag{6.39}$$

where $\mathbf{B} = \nabla \times \mathbf{A}$ is the magnetic field and $g = 2$.

The Dirac theory makes a definite prediction for the value of the magnetic moment which is born out by experiment. Let's pause and consider this fact in a little more detail. Starting with Pauli's equation in the expanded form Equation 6.39, we might consistently assume any value for the magnetic moment. The magnetic moments of the electron and muon, deduced from spectra of normal and muonic atoms, have the Dirac values with small calculable relativistic corrections.

The magnetic moment $<\boldsymbol{\mu}>= ge\hbar/2m <\sigma>$ divided by the spin angular momentum $(\hbar/2) <\boldsymbol{\sigma}>$ is $ge/m$ or $g$ times the value one expects for any classical distribution of moving charge. The deduction of quark magnetic moments from the properties of hadrons is more problematic, but the success of the quark model in the description of the mass spectrum of hadrons and scattering experiments described in more detail later leave little doubt quarks are best described as Dirac fermions. The proton and neutron are spin 1/2 fermions with magnetic moments $\mu_p = 2.792846e/(2m_p)$ and $\mu_n = -1.91315e/(2m_p)$ that differ from the Dirac values. This fact suggests these hadrons have structure and we will see in Chapter 10 that the values follow from a simple minded treatment of the correlated spin states of the quarks.

A gauge invariant modified Dirac equation accommodating an "anomalous" magnetic moment can be constructed:

$$\left[i\gamma_\mu\partial^\mu - e\gamma_\mu A^\mu - m + \frac{\kappa e}{4m} i\sigma_{\mu\nu}F^{\mu\nu}\right]\psi = 0. \tag{6.40}$$

The modified Dirac equation can be written as

$$\left[i\gamma_\mu\partial^\mu - e\gamma_\mu A^\mu - m + \frac{\kappa e}{2m} i\boldsymbol{\alpha}\cdot\mathbf{E} + \frac{\kappa e}{2m}\boldsymbol{\Sigma}\cdot\mathbf{B}\right]\psi = 0. \tag{6.41}$$

In finding the nonrelativistic limit with these terms present, we would neglect them both in solving for $\chi$. The last term then contributes to a magnetic interaction of the form $\boldsymbol{\sigma}\cdot\mathbf{B}$ and the net magnetic moment in Pauli's equation turns out to be

$$\mu = (1+\kappa)\,\frac{e}{2m}. \tag{6.42}$$

(The other term can be associated with the interaction of the moving magnetic moment with the magnetic field in its rest frame.) As discussed in detail Chapter 9, this theory as it stands with $\kappa$ a free parameter is not renormalizable so can only be interpreted as an effective low energy model. However, a term of this form appears in multi-particle quantum electrodynamics with $\kappa$ a calculable quantity and, in fact, the calculated and observed values for the electron and muon magnetic moments are in stunning agreement and provide a profound test of quantum electrodynamics.

## 6.4 ■ RELATIVISTIC SPINOR QUANTUM MECHANICS

In order to understand relativistic aspects of exotic atoms and quark composites, it is useful to be familiar with the corrections to the Pauli equation implied by the Dirac theory beyond the leading order magnetic term in the Pauli equation. These corrections in the context of atomic physics provided the first stringent tests of ordinary quantum mechanics based on Dirac theory and of the subsequently developed full quantum field theory. The general problem of relativistic bound states is complex. Even in formulating a classical mechanics of interacting relativistic charged spin 0 point particles, one must consider the fact that electromagnetic interactions are not instantaneous. The field at the location of one particle due to another depends on the location and acceleration of the second particle at an earlier time. Moreover, electromagnetic energy may be radiated and absorbed. An approximate description of a collection of interacting charged particles may found when radiation is neglected and terms in the energy of order $(v/c)^4$ are retained. A corresponding corrected Pauli theory of a fermion may be formulated. Beyond this order, multiparticle effects must be considered. It is these radiative corrections that will be discussed in Chapter 9.

We can find the next-to-leading approximation to the Dirac equation for an external field using a binomial/Taylor expansion

$$\chi = \left(E - eA^0 + 2m\right)^{-1} \boldsymbol{\sigma} \cdot (\mathbf{p} - e\mathbf{A})\eta \simeq \frac{1}{2m}\left(1 - \frac{E - eA^0}{2m}\right)\boldsymbol{\sigma} \cdot (\mathbf{p} - e\mathbf{A})\eta \tag{6.43}$$

which gives

$$\left(E - eA^0\right)\eta = \boldsymbol{\sigma} \cdot (\mathbf{p} - e\mathbf{A})\frac{1}{2m}\left(1 - \frac{E - eA^0}{2m}\right)\boldsymbol{\sigma} \cdot (\mathbf{p} - e\mathbf{A})\eta. \tag{6.44}$$

We are looking to arrive at a Pauli-like wave equation describing a single particle with a conserved probability density. In the Pauli equation, the probability density is $\rho = \eta^\dagger\eta$. If we have a solution with the integral of this quantity over all space equal to one at some time, the integral will remain equal to one. As we go beyond the leading approximation, the conserved quantity is

$$\rho = |\eta|^2 + |\chi|^2 \simeq |\eta|^2 + \frac{1}{4m^2}|\boldsymbol{\sigma} \cdot \mathbf{p}\eta|^2 \tag{6.45}$$

and differs from $\eta^\dagger\eta$ so in this approximation $\eta$ may no longer be regarded as a wave function with a conserved probability current. However, since

$$\int d\mathbf{x}\, \nabla\eta^\dagger \cdot \boldsymbol{\sigma}\boldsymbol{\sigma} \cdot \nabla\eta = \int d\mathbf{x}\, \eta^\dagger \mathbf{p}^2 \eta \tag{6.46}$$

by parts integration, the 2-component spinor

$$\phi = \left(1 + \frac{\mathbf{p}^2}{8m^2}\right)\eta \tag{6.47}$$

has a conserved density. We can substitute

$$\eta = \left(1 - \frac{\mathbf{p}^2}{8m^2}\right)\phi \tag{6.48}$$

into the first order equation to find a 2-component corrected Pauli equation

$$i\partial_t\phi = H\phi + \Delta H\phi \tag{6.49}$$

with $H$ the uncorrected Pauli Hamiltonian and

$$\Delta H = -\frac{\mathbf{p}^4}{8m^2} - \frac{e}{4m^2}\boldsymbol{\sigma}\cdot\mathbf{E}\times\mathbf{p} - \frac{e}{8m^2}\nabla\cdot\mathbf{E}. \tag{6.50}$$

The first term in Equation 6.50 represents the second term in the relativistic kinetic energy

$$\begin{aligned} K = \sqrt{m^2+\mathbf{p}^2} - m = m\left[\left(1+\frac{\mathbf{p}^2}{m^2}\right)^{1/2} - 1\right] \\ = \frac{\mathbf{p}^2}{2m} - \frac{\mathbf{p}^4}{8m^2} + \cdots . \end{aligned} \tag{6.51}$$

For motion in a spherically symmetric electric field, we can write $\mathbf{E} = -(\partial U/\partial r)\hat{\mathbf{r}}$, where $U(r)$ is the electric potential. The second term can then be written in the form

$$H_{\text{spin-orbit}} = -g_{\text{Thomas}}\boldsymbol{\mu}\cdot\mathbf{B}' = -\frac{e}{2m^2 r}\frac{\partial U}{\partial r}\mathbf{L}\cdot\mathbf{s}. \tag{6.52}$$

Here $\mathbf{B}' = -\mathbf{v}\times\mathbf{E}$ is the magnetic field in the instantaneous rest frame of the electron. The Thomas precession factor $g_{\text{Thomas}} = 1/2$ can be understood as a relativistic spin precession associated with the acceleration. The second form represents the interaction as a coupling of the spin and orbital angular momentum.

The last term in Equation 6.50 is an effective contact interaction with the source of the electric field. It can be understood in terms of a relativistic jitter of each component of the position vector independently by a half a Compton wavelength $\delta x_i \simeq \hbar/2mc$. If $P(\mathbf{y})$ is a probability for jitter $\mathbf{y}$ with $\int d\mathbf{y}\ P(\mathbf{y}) = 1$, the effective smeared potential may be found using the expression

$$\begin{aligned} < V(\mathbf{x}) > &= \int d\mathbf{y}\ P(\mathbf{y})V(\mathbf{x}+\mathbf{y}) \\ &= \int d\mathbf{y}\ P(\mathbf{y})[V(\mathbf{x}) + \Sigma_i y_i(\partial_i V)|_{\mathbf{x}} \\ &+\frac{1}{2}\Sigma_{i,j} y_i y_j(\partial_i\partial_j V)|_{\mathbf{x}} + \dots . \end{aligned} \tag{6.53}$$

If $P(\mathbf{y})$ does not correlate $y_i$ and $y_j$ and $< y_i >= 0$ then, writing $e\mathbf{E} = -\nabla eA^0 = -\nabla V$, we find

$$< V(\mathbf{x}) >= V(\mathbf{x}) + \frac{1}{6} < \delta\mathbf{x}^2 > \nabla^2 V(\mathbf{x}) = V(\mathbf{x}) - \frac{e}{8}\left(\frac{\hbar}{mc}\right)^2 \nabla\cdot\mathbf{E}. \tag{6.54}$$

As shown in an exercise, all three of these corrections lead to energy level shifts of order $\alpha^2$ times the Rydberg energy in hydrogen. Interestingly, for $s$-wave states which have nonvanishing wave function at the origin, the singularity at $r = 0$ in the spin-orbit interaction does not apply since the orbital angular momentum vanishes. For nonvanishing orbital angular momentum, the spin-orbit correction generally decreases with $l$ for fixed $n$ since the radius is on average larger. While the spin-orbit interaction does not affect the $s$-wave states, the last so-called Darwin term only affects $s$-wave states and the energy shift associated with this term is proportional to the square of the wave function at the origin which decreases when the principal quantum $n$ increases.

In the case of a fixed Coulomb potential, exact stationary solutions to the exact Dirac equation can be found. The spin precession and orbital motion are correlated, conserving total angular momentum $\mathbf{j} = \mathbf{L} + \mathbf{s}$, and the energies are

$$E_{jn} = m\left[1 + \frac{e^4}{\left(n - k_j\right)^2}\right]^{-\frac{1}{2}} \simeq m + \frac{E_0}{n^2}\left[1 + \frac{\alpha^2}{n^2}\left(\frac{n}{j + \frac{1}{2}} - \frac{3}{4}\right)\right]. \tag{6.55}$$

Here $k_j = j + 1/2 - \sqrt{(j + 1/2)^2 - e^4}$. To order $\alpha^4$, this energy spectrum is just what is found using our approximate Hamiltonian and time independent perturbation theory, combining spin and orbital angular momentum states to form states of fixed total angular momentum. In heavy elements, the effective potential is not inversely proportional to radius and exact solutions do not exist. The speeds of the inner electrons are of order $v \sim Z\alpha$, the energies are $E \simeq m_e(Z\alpha)^2$, and the fractional relativistic energy shifts are of order $(Z\alpha)^2$, larger than in hydrogen.

In a 2-body atomic system in which neither mass may be considered fixed, the interaction of each particle with the magnetic field as well as the electric field of the other must be considered. Thus, in describing the electron in hydrogen, we should include the vector potential $\mathbf{A}$ associated with the motion of the proton as well as with the magnetic moment of the proton. The vector potential associated with the motion is proportional to the speed of the proton and to its charge: $|\mathbf{A}_p| \sim ev_p/r$ where $r \sim a_0$ is the distance from the proton. The electron's speed is about $v_e = \alpha$ and the proton's speed is about $v_p = (m_e/m_p)v_e$. Examining the interaction terms in the Pauli equation, one finds the magnetic interaction energy is of order

$$\Delta E_A \sim e\mathbf{v_e} \cdot \mathbf{A_p} \sim (ev_e)\left(e\frac{m_e}{m_p}v_e/r\right) \sim \alpha^3/a_0. \tag{6.56}$$

Since $a_0^{-1} = \alpha m_e$ and $E_0 = \alpha^2 m_e/2$, we find $\Delta E_A \sim (m_e/m_p)\alpha^2 E_0$. We can conclude that these magnetic effects are small compared to the relativistic effects in hydrogen but comparable in an atom like positronium which will be discussed in more detail below.

## 6.5 ■ TWO-COMPONENT SPINORS

We have seen how Dirac's 4-component spinor equation reduces in the nonrelativistic limit to a theory of a 2-component spin 1/2 quantum particle. In relativistic processes, the magnitudes of all four components of the Dirac spinor are comparable. The relativistic theory can be conveniently cast in terms of a transformed set of Dirac matrices and an alternate pair of 2-component spinors. We now turn our attention to this formulation of the theory and, along the way, demonstrate the relativistic covariance of the Dirac equation.

We start by introducing the Hermitian matrix representations of a 4-vector such as the 4-velocity $v^\mu = p^\mu/m$. Two such matrix representations are defined by

$$v_\pm = v^\mu \sigma_\mu^\pm = v^0 \pm \mathbf{v}\cdot\boldsymbol{\sigma} = \begin{pmatrix} v^0 \pm v_z & \pm v_x \mp i v_y \\ \pm v_x \pm i v_y & v^0 \mp v_z \end{pmatrix} \tag{6.57}$$

where we define

$$\sigma_\mu^\pm = (1, \pm\boldsymbol{\sigma}),\ \sigma^{\mu\pm} = (1, \mp\boldsymbol{\sigma}). \tag{6.58}$$

The Pauli matrix identity Equation 6.37 may be used to show that the product of the two matrix representations in either order is the Lorentz invariant square of $v^\mu$ times the unit matrix:

$$v_- v_+ = v_+ v_- = v_0^2 - (\boldsymbol{\sigma}\cdot\mathbf{v})^2 = v_0^2 - \mathbf{v}^2 = v^2. \tag{6.59}$$

In these two by two matrix equations, it is customary to suppress writing the unit matrix. The preceding expressions are valid for any 4-vector $A^\mu$ sewn up in matrix form $A_\pm$. We are about to show that how the matrices $v_\pm$ in particular can be used to represent Lorentz boost transformations of the components of a general 4-vector $A^\mu$ in a $2\times 2$ matrix transformation form, and also how these matrices represent Lorentz boost transformations of the two 2-component spinors in Dirac's theory.

First note that the identity Equation 6.38 allows us to generalize Euler's formula

$$e^{ix} = \cos x + i \sin x \tag{6.60}$$

and write the expressions

$$e^{i\mathbf{n}\cdot\boldsymbol{\sigma}} = 1 + i\mathbf{n}\cdot\boldsymbol{\sigma} + (i\mathbf{n}\cdot\boldsymbol{\sigma})^2/2 + ... = \cos|\mathbf{n}| + i\frac{\mathbf{n}}{|\mathbf{n}|}\cdot\boldsymbol{\sigma}\sin|\mathbf{n}|. \tag{6.61}$$

Here the exponential of the matrix operator is defined by the usual series expansion of an exponential function and alternate terms have been combined and identified using the series expressions for the sin and cos functions. There is a similar generalization of the identity

$$e^x = \cosh x + \sinh x, \tag{6.62}$$

namely

$$e^{\mathbf{n}\cdot\boldsymbol{\sigma}} = 1 + \mathbf{n}\cdot\boldsymbol{\sigma} + (\mathbf{n}\cdot\boldsymbol{\sigma})^2/2 + ... = \cosh|\mathbf{n}| + \frac{\mathbf{n}}{|\mathbf{n}|}\cdot\boldsymbol{\sigma}\sinh|\mathbf{n}|. \tag{6.63}$$

Now if we express the 4-velocity in terms of rapidity, $v_\mu = (\cosh\eta, \mathbf{n}\sinh\eta)$ where $\tanh\eta = v$, we can write

$$v_\pm = \cosh\eta \pm \mathbf{n}\cdot\sigma\sinh\eta = e^{\pm\eta\mathbf{n}\cdot\boldsymbol{\sigma}} = (v_\mp)^{-1} \tag{6.64}$$

and, since $\boldsymbol{\sigma}^\dagger = \boldsymbol{\sigma}$ and the 4-velocity vector is real, we have

$$v_\pm^\dagger = v_\pm. \tag{6.65}$$

Since $(e^{\pm\frac{\eta}{2}\mathbf{n}\cdot\boldsymbol{\sigma}})^2 = e^{\pm\eta\mathbf{n}\cdot\boldsymbol{\sigma}}$, we can make the definitions

$$\sqrt{v_\pm} = e^{\pm\frac{\eta}{2}\mathbf{n}\cdot\boldsymbol{\sigma}} \tag{6.66}$$

and introduce

$$U_\pm = \sqrt{v_\pm}U = e^{\pm\frac{\eta}{2}\mathbf{n}\cdot\boldsymbol{\sigma}}U = \sqrt{v_\pm}U \tag{6.67}$$

with $U = e^{-i\frac{\theta}{2}\mathbf{m}\cdot\boldsymbol{\sigma}}$ an arbitrary unitary matrix. Then since $U_\pm^\dagger = U^\dagger v_\pm$, we can write

$$v_\pm = U_\pm 1 U_\pm^\dagger. \tag{6.68}$$

Now we interpret this expression as two distinct $2 \times 2$ matrix transformation representations of Lorentz transformations of the velocity 4-vector. The general transformation rule is

$$A_\pm \to U_\pm A_\pm U_\pm^\dagger \equiv A'_\pm. \tag{6.69}$$

Our expression $v_\pm = U_\pm 1 U_\pm^\dagger$ is the particular case of a transformation from the rest frame where $v_\pm = 1$ to an arbitrary boosted frame. The quantity $A_\pm$ is a matrix containing the components $A_\mu$ and the corresponding elements of the matrix $U_\pm A_\pm U_\pm^\dagger$ are the transformed components $A'_\mu$. The factor $\sqrt{v_\pm}$ gives the boost. The unitary matrix $U$ which has been included in the general definition of the $U_\pm$ corresponds to an active rotation by angle $\theta$ about direction $\mathbf{m}$.

In fact, for small $\theta$, we can write

$$U_\pm \simeq 1 - i\frac{\theta}{2}\mathbf{m}\cdot\boldsymbol{\sigma} \tag{6.70}$$

and apply this to the two by two matrix representation of, for example, a momentum vector $p^\mu = mv^\mu$. Using Equation 6.37 to reduce a triple product of Paul matrices, we find the transformation rule for a rotation alone reads

$$U_\pm(E \pm \mathbf{p}\cdot\boldsymbol{\sigma})U_\pm^\dagger = E \pm (\mathbf{p} + \theta\mathbf{m}\times\mathbf{p})\cdot\boldsymbol{\sigma} \equiv E' \pm \mathbf{p}'\cdot\boldsymbol{\sigma} \tag{6.71}$$

and gives the expected rotated vector components

$$p' = \left(E', \mathbf{p}'\right) = (E, \mathbf{p} + \theta\mathbf{m}\times\mathbf{p}). \tag{6.72}$$

Similarly, a pure infinitesimal boost gives the expected transformed energy and momentum. The general Lorentz transformation can be constructed from many infinitesimal transformation and in general

$$v^2 = v_\pm v_\mp \to U_\pm v_\pm U_\pm^\dagger U_\mp v_\mp U_\mp^\dagger \tag{6.73}$$

is invariant since $U_\pm^\dagger = U_\mp^{-1}$. This demonstrates the correctness of the interpretation.

We now apply these two dimensional Lorentz transformations to 2-component spinors, focusing on boost transformations. Starting with any 2-component rest frame spinor $\phi$, two boosted spinors may be constructed:

$$\phi \to \phi_\pm = \sqrt{v_\mp}\phi. \tag{6.74}$$

Since the boosts $\sqrt{v_\pm}$ are the inverses of the boosts $\sqrt{v_\mp}$, the two different boosted spinors are related by a double boost

$$\phi_\pm = \sqrt{v_\mp}\phi = \sqrt{v_\pm}\sqrt{v_\pm}\phi_\mp = v_\pm\phi_\mp. \tag{6.75}$$

Multiplying this expression by a common mass associated with the spinors and using $p^\mu = mv^\mu$ yields the expressions

$$\begin{aligned}(E - \mathbf{p}\cdot\boldsymbol{\sigma})\phi_+ &= m\phi_- \\ (E + \mathbf{p}\cdot\boldsymbol{\sigma})\phi_- &= m\phi_+.\end{aligned} \tag{6.76}$$

With $p^\mu \to i\partial^\mu$, these equations are Dirac's equation in so-called Weyl or spinor form and are essentially representations of Lorentz boost transformations relating the pair of 2-component spinors. In the presence of an electromagnetic field, the pair of 2-component equations may be written as

$$[i\partial - eA]_\mp\phi_\pm(x) = m\phi_\mp(x), \tag{6.77}$$

where $\partial_\pm = \partial^\mu\sigma_\mu^\pm$ and $A_\pm = A^\mu\sigma_\mu^\pm$.

The pair of 2-component spinor equations may be written in terms of a 4-component (Dirac) spinor

$$\psi = \begin{pmatrix}\phi_+ \\ \phi_-\end{pmatrix} \tag{6.78}$$

in the general form previously derived

$$i\gamma_\mu\partial^\mu\psi = m\psi \tag{6.79}$$

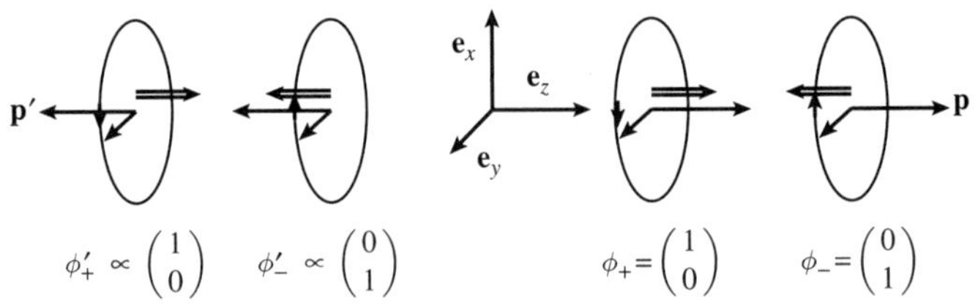


**FIGURE 6.2** Helicity states of an electron moving along a direction $\mathbf{e}_z$. The spin states $\phi_\pm$ represent spin angular momentum along the direction of momentum $\mathbf{s}\cdot\mathbf{p}/|\mathbf{p}| = \pm\hbar/2$. For a massive fermion such as the electron, a boost along the direction of motion can result in a new momentum $\mathbf{p}'$ of reversed sign and reversed helicity. Such a reversal is not possible for a massless particle. Dirac's equation may be interpreted as representing a boost from momentum $\mathbf{p}$ to $-\mathbf{p}$.

using Dirac matrices

$$\gamma^0 = \begin{pmatrix} 0 & 1 \\ 1 & 0 \end{pmatrix}, \ \boldsymbol{\gamma} = \begin{pmatrix} 0 & -\boldsymbol{\sigma} \\ \boldsymbol{\sigma} & 0 \end{pmatrix}, \ \boldsymbol{\alpha} = \begin{pmatrix} \boldsymbol{\sigma} & 0 \\ 0 & -\boldsymbol{\sigma} \end{pmatrix}. \tag{6.80}$$

In this representation of the Dirac matrices, we have

$$\gamma_5 = \begin{pmatrix} -1 & 0 \\ 0 & 1 \end{pmatrix}, \ \boldsymbol{\Sigma} = \begin{pmatrix} \boldsymbol{\sigma} & 0 \\ 0 & \boldsymbol{\sigma} \end{pmatrix}. \tag{6.81}$$

The two 2-component equations couple what is customarily called a left-handed 2-component spinor field $\phi_L \equiv \phi_-$ to what is called a right-handed field $\phi_R \equiv \phi_+$ provided $m \neq 0$. When the mass vanishes, no boost can convert one to the other. The equations for left- and right-handed components decouple and, since in this case $E = |\mathbf{p}|$, if we write $\mathbf{p} = E\hat{\mathbf{n}}$, we find the equations

$$\begin{aligned} (E - \mathbf{p}\cdot\boldsymbol{\sigma})\phi_+ &= 0 \\ (E + \mathbf{p}\cdot\boldsymbol{\sigma})\phi_- &= 0 \end{aligned} \tag{6.82}$$

reduce to

$$\begin{aligned} \hat{\mathbf{n}}\cdot\boldsymbol{\sigma}\phi_+ &= \phi_+ \\ \hat{\mathbf{n}}\cdot\boldsymbol{\sigma}\phi_- &= -\phi_- \end{aligned} \tag{6.83}$$

which indicate that the spin along the direction of motion is negative for the left-handed spinor and positive for the right-handed spinor.

Note that some authors define $\psi$ with the 2-component spinors interchanged and $\boldsymbol{\gamma} \to -\boldsymbol{\gamma}$. Our spinor and standard representation spinors and Dirac matrices are related by

$$\psi_{\text{Standard}} = \begin{pmatrix} \eta \\ \chi \end{pmatrix} = \frac{1}{\sqrt{2}} \begin{pmatrix} \phi_+ + \phi_- \\ \phi_+ - \phi_- \end{pmatrix} = U \begin{pmatrix} \phi_+ \\ \phi_- \end{pmatrix} \tag{6.84}$$

where the unitary transformation

$$U = U^\dagger = U^{-1} = \frac{1}{\sqrt{2}} \begin{pmatrix} 1 & 1 \\ 1 & -1 \end{pmatrix} \tag{6.85}$$

relates the standard and spinor forms: $\gamma^\mu_{\text{Standard}} = U\gamma^\mu_{\text{Spinor}}U^{-1}$.

## 6.6 ■ SPINOR PLANE WAVES

The representation of Lorentz transformations derived above permit us to construct solutions to the free particle Dirac equation of the form

$$\phi_\pm(x) = u^\pm e^{-ipx} \tag{6.86}$$

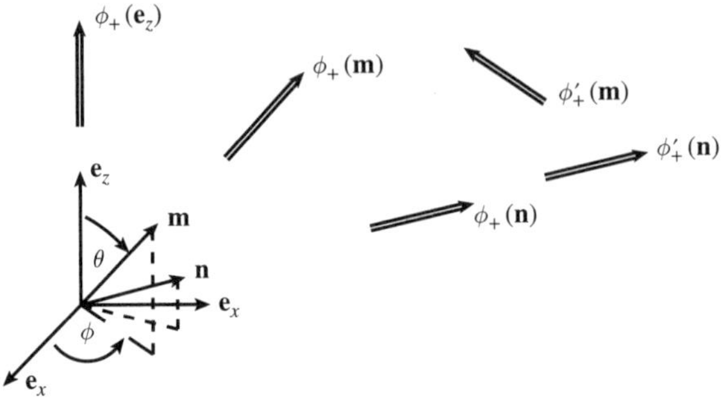


**FIGURE 6.3** The construction of plane wave solutions by boosting from the rest frame. A rotation applied to an eigenstate $\phi_+(\mathbf{e}_z)$ of $\boldsymbol{\sigma}\cdot\mathbf{e}_z$ in the rest frame produces an eigenstate $\phi_+(\mathbf{m})$ of $\boldsymbol{\sigma}\cdot\mathbf{m}$ along an arbitrary direction $\mathbf{m}$. A boost along direction $\mathbf{n}=\mathbf{m}$ produces a solution $\phi'_+(\mathbf{n})$.

where $u^{\pm}$ are two constant 2-component spinor amplitudes. We do this by boosting from the rest frame where the Dirac equation reduces to $\phi_+=\phi_-$ and the solution is any constant spinor. The construction is illustrated in Figure 6.3. In the rest frame, there are two linearly independent solutions we can denote by

$$\phi_{s=+1}(\hat{\mathbf{z}})=\begin{pmatrix}1\\0\end{pmatrix} \tag{6.87}$$

representing an eigenstate of $\sigma_z$ with eigenvalue $s=+1$ and

$$\phi_{s=-1}(\hat{\mathbf{z}})=\begin{pmatrix}0\\1\end{pmatrix} \tag{6.88}$$

representing an eigenstate of $\sigma_z$ with eigenvalue $s=-1$. These represent quantum mechanical states of the particle with spin $\pm 1/2$ along the $z$-direction. More generally, any two linearly independent combinations of these two basis states span the space of rest frame wave functions of the free spin 1/2 particle and such combinations have the quantum mechanical interpretation as representing the spin being aligned along some other direction.

For sake of generality, let's start with two component eigenstates of spin along some general direction $\mathbf{m}$ with eigenvalues $s=\pm 1$ (twice the spin), defined as solutions to

$$\mathbf{m}\cdot\boldsymbol{\sigma}\phi_s(\mathbf{m})=s\phi_s(\mathbf{m}). \tag{6.89}$$

These solutions may be constructed by rotation of the eigenstates of $\sigma_z$:

$$\phi_+(\mathbf{m})=e^{-i\frac{\phi}{2}\sigma_z}e^{-i\frac{\theta}{2}\sigma_y}\begin{pmatrix}1\\0\end{pmatrix}=\begin{pmatrix}e^{-i\phi/2}\cos\frac{\theta}{2}\\ e^{+i\phi/2}\sin\frac{\theta}{2}\end{pmatrix} \tag{6.90}$$

$$\phi_-(\mathbf{m})=e^{-i\frac{\phi}{2}\sigma_z}e^{-i\frac{\theta}{2}\sigma_y}\begin{pmatrix}0\\1\end{pmatrix}=\begin{pmatrix}-e^{-i\phi/2}\sin\frac{\theta}{2}\\ e^{+i\phi/2}\cos\frac{\theta}{2}\end{pmatrix} \tag{6.91}$$

where $\theta$ and $\phi$ are the spherical coordinate angles of the direction of the vector $\mathbf{m}$. Note that since

$$e^{\pm i\frac{\pi}{2}} = \pm i, \quad \sin\left(\frac{\pi}{2} - u\right) = \cos(u), \quad \cos\left(\frac{\pi}{2} - u\right) = \sin(u), \tag{6.92}$$

eigenstates of spin along opposite directions are related by

$$\phi_{\pm}(-\mathbf{m}) = \phi_{\pm}(\pi - \theta, \phi + \pi) = i\phi_{\mp}(\mathbf{m}). \tag{6.93}$$

These relationships will be useful when considering pairs of oppositely directed particles.

Suppose $\mathbf{n}$ is the direction of the momentum along which we wish to boost such a rest frame solution. The energy and momentum of the boosted solution may be expressed in terms of the rapidity $\eta$ of the boosted frame as $E = m\cosh(\eta)$ and $\mathbf{p} = m\sinh(\eta)\mathbf{n}$. The boosted spinor amplitudes given by the Lorentz transformation are

$$u_s^{\pm} = \sqrt{m}e^{\pm\frac{\eta}{2}\mathbf{n}\cdot\boldsymbol{\sigma}}\phi_s(\mathbf{m}). \tag{6.94}$$

The convenient normalization factor $\sqrt{m}$ with $m$ the mass is inserted here to give the waves an interpretation as representing $\sqrt{2EV}$ particles per unit volume as described below. (This $m$ does not denote $|\mathbf{m}|$ which has the value unity.)

In general, after a boost, an eigenstate of spin along direction $\mathbf{m}$ is no longer an eigenstate of spin along that direction - a boost changes the orientation of the spin direction. However, if we choose $\mathbf{m}$ as the direction $\mathbf{n}$ of the momentum, meaning we boost rest frame states which were already eigenstates of spin along the boost direction, we obtain states which remain eigenstates of spin along the momentum (helicity eigenstates) meaning

$$\mathbf{m}\cdot\boldsymbol{\sigma}u_s^{\pm}(\mathbf{m}) = su_s^{\pm}(\mathbf{m}). \tag{6.95}$$

This is easily seen by inserting the expression for the boosted spinor in case $\mathbf{n} = \mathbf{m}$:

$$\begin{aligned}
\mathbf{m}\cdot\boldsymbol{\sigma}u_s^{\pm}(\mathbf{m}) &= \mathbf{m}\cdot\boldsymbol{\sigma}\sqrt{m}e^{\pm\frac{\eta}{2}\mathbf{n}\cdot\boldsymbol{\sigma}}\phi_s(\mathbf{m}) \\
&= \sqrt{m}e^{\pm\frac{\eta}{2}\mathbf{n}\cdot\boldsymbol{\sigma}}\mathbf{m}\cdot\boldsymbol{\sigma}\phi_s(\mathbf{m}) = \sqrt{m}e^{\pm\frac{\eta}{2}\mathbf{n}\cdot\boldsymbol{\sigma}}s\phi_s(\mathbf{m}) \\
&= su_s^{\pm}(\mathbf{m}).
\end{aligned} \tag{6.96}$$

Using the identities

$$\sqrt{m}e^{\pm\eta/2} = \sqrt{E \pm p}, \tag{6.97}$$

we find that the Lorentz boost along the direction of momentum of a helicity state amounts to a multiplicative factor

$$u_s^{\pm} = \sqrt{m}e^{\pm s\frac{\eta}{2}}\phi_s(\mathbf{n}) = \sqrt{E \pm sp}\;\phi_s(\mathbf{n}) \tag{6.98}$$

and consequently we have a very convenient representation of the general solution of the Dirac equation for a plane wave of four momentum $p^\mu$ in terms of the eigenstates of spin along the direction of motion:

$$\psi_s = \begin{pmatrix} \sqrt{E + sp}\phi_s(\mathbf{n}) \\ \sqrt{E - sp}\phi_s(\mathbf{n}) \end{pmatrix} e^{-i(Et - p\mathbf{n}\cdot\mathbf{x})}. \tag{6.99}$$

While a general helicity eigenstate contains both left and right-handed components, in the ultra-relativistic limit in which $E = p$, one sees that only the upper or lower components of the 4-spinor are nonvanishing. The matter wave amplitudes approach

$$u_t^r \rightarrow \sqrt{2E}\phi_\pm(\mathbf{n})\delta_{rt} \tag{6.100}$$

where the Kronecker $\delta$-function expresses the fact that the amplitude vanishes unless the handedness $(r)$ coincides with helicity $(t)$. In this limit, the left and right-handed solutions are independent and a given helicity state corresponds to just one of the pair of 2-component spinors being non-vanishing. It is this fact that makes the Weyl or spinor representation especially simple in the ultra-relativistic case. In general however, a boost can reverse the helicity and, correspondingly, the left and right-handed 2-component spinors are both non-vanishing though related. Even so, the helicity representation of spin states is especially simple since helicity is independent of the orientation of the coordinate system. Hence we will generally specify the basis of spinor spin states in terms of helicity rather than as eigenstates of spin along some coordinate direction and use the spinor representation of the Dirac matrices in calculations of relativistic scattering processes.

## 6.7 ■ CHARGE CONJUGATION AND ANTI-MATTER

Negative energy solutions of the form

$$\phi_\pm = v^\pm e^{+ipx} \tag{6.101}$$

may be found by a construction similar to that just used. In the rest frame, the Dirac equation for this kind of wave reduces to $\phi_+ = -\phi_-$ and solutions may be found by boosting rest frame solutions, including the minus sign between $\phi^+$ and $\phi^-$ in the rest frame. We follow custom and choose, however, to define a special basis of negative energy solutions that correspond to the antiparticles of our solutions for particles. These are defined by consideration of symmetry of electrodynamics under charge conjugation. This choice is made in the context of interactions which connect the otherwise independent states we call matter and antimatter.

Unlike solutions to the Klein-Gordon equation for a scalar field, the complex conjugate of a solution to the Dirac (or Schrödinger) equation is not a solution and some care must be taken in its interpretation. Suppose $\phi_\pm(x)$ is a solution to Dirac's equation expressed in 2-component form as

$$[i\partial - eA]_\mp \phi_\pm(x) = m\phi_\mp(x). \tag{6.102}$$

The complex conjugate satisfies

$$[i\partial - eA]^*_\mp \phi^*_\pm = m\phi^*_\mp. \tag{6.103}$$

The complex conjugate of the Pauli matrices can be undone by a unitary transformation such as $U = i\sigma_y$ such that

$$U\boldsymbol{\sigma}^* U^\dagger = -\boldsymbol{\sigma},\ \boldsymbol{\sigma}^* = -U^\dagger \boldsymbol{\sigma} U. \tag{6.104}$$

Given $U$, we find we can write

$$\sigma^{\pm *}_\mu = (1, \pm\boldsymbol{\sigma}^*) = U^\dagger(1, \mp\boldsymbol{\sigma})U = U\sigma_\mp U^\dagger. \tag{6.105}$$

Since $(i\partial_\mu)^* = -i\partial_\mu$ and $A^*_\mu = A_\mu$, we have

$$[p - eA]^*_\pm = (p - eA)^{\mu *}\sigma^{\pm *}_\mu = (p^* - eA^*)^\mu U^\dagger \sigma^\mp_\mu U = -U^\dagger[p + eA]_\mp U \tag{6.106}$$

and substitution gives

$$-U^\dagger[p + eA]_\pm U\phi^*_\pm = m\phi^*_\mp. \tag{6.107}$$

Multiplying both sides of this equation by $-U$ gives

$$[i\partial + eA]_\pm U\phi^*_\pm = -mU\phi^*_\mp. \tag{6.108}$$

The charge conjugate spinors are defined by

$$C\phi_\pm \equiv \phi^c_\pm = \eta_\pm U\phi^*_\mp \tag{6.109}$$

where the two arbitrary phase factors satisfy $\eta_+\eta_- = -1$. Multiplying Equation 6.108 by $\eta_\mp$, we have

$$(i\partial + eA)_\pm \phi^c_\mp = -m\eta_\mp U\phi^*_\mp = +m\eta_\pm U\phi^*_\mp = m\phi^c_\pm. \tag{6.110}$$

We see that the charge conjugate pair of 2-component spinors are together a solution to Dirac's equation with reversed charge appropriate to an antiparticle for they satisfy the equation

$$[i\partial + eA]_\pm \phi^c_\mp = +m\phi^c_\pm. \tag{6.111}$$

Equivalently, $\psi \to \psi^c$ together with $A_\mu \to -A_\mu$ is a symmetry. We choose $\eta_\pm = \pm 1$, but note there is still an arbitrary overall phase. (Another choice $\eta_+ = \eta_- = i$ is used by some authors.) Then if we pick $U = i\sigma_y$, we have

$$C\phi_\pm = \pm \begin{pmatrix} 0 & 1 \\ -1 & 0 \end{pmatrix} \phi^*_\mp. \tag{6.112}$$

We now write the two independent solutions for energy $E$ and momentum $\mathbf{p}$ corresponding to antimatter as

$$\psi_s = \begin{pmatrix} v_s^+ \\ v_s^- \end{pmatrix} e^{+i(E-p\mathbf{n}\cdot\mathbf{x})} \tag{6.113}$$

where the two 2-component amplitudes are denoted by $v_s^\pm$ rather than $u_s^\pm$. Applied to our matter solutions, the charge conjugation operator produces amplitudes defined as

$$\begin{aligned} v_s^\pm &= \left[u_s^\pm\right]^c = \eta_\pm U\left(u_s^\mp\right)^* = \pm\sqrt{m}Ue^{\mp\frac{\eta}{2}\mathbf{n}\cdot\boldsymbol{\sigma}^*}U^\dagger U\phi_s^*(\mathbf{m}) \\ &= \pm\sqrt{m}e^{\pm\frac{\eta}{2}\mathbf{n}\cdot\boldsymbol{\sigma}}U\phi_s^*(\mathbf{m}). \end{aligned} \tag{6.114}$$

From the forms for the spinors $\phi_s(\mathbf{m})$ above, inspection shows that

$$i\sigma_y\phi_s^*(\mathbf{m}) = -s\phi_{-s}(\mathbf{m}) \tag{6.115}$$

and it follows that

$$v_s^\pm = \mp s\sqrt{m}e^{\pm\frac{\eta}{2}\mathbf{n}\cdot\boldsymbol{\sigma}}\phi_{-s}(\mathbf{m}). \tag{6.116}$$

For helicity states, these amplitudes are

$$v_s^\pm = \mp s\sqrt{m}e^{\mp s\frac{\eta}{2}}\phi_{-s}(\mathbf{n}) = \mp s\sqrt{E \mp sp}\phi_{-s}(\mathbf{n}) \tag{6.117}$$

or $v_{-s}^\pm = \pm s\sqrt{E \pm sp}\phi_s(\mathbf{n})$. In the rest frame of a positron, we have $p = 0$ and $E = m$ so the solutions reduce to

$$v_s^\pm = \mp s\sqrt{m}\phi_{-s}(\mathbf{n}) \tag{6.118}$$

so a spin +1/2 solution has right-handed component $v_+^+ = (+1)\phi_-$ while a spin -1/2 solution has right-handed component $v_-^+ = (-1)\phi_+$, the prefactors $(\pm 1)$ arising from the phase choice $\eta_\pm = \pm 1$ in defining charge conjugation. The left-handed components are the negative of the right-handed components as required by the Dirac equation in the rest frame.

The amplitudes $v_s$ have reversed spin projection and accompany a wave of reversed 4-momentum. The absorption of such a wave corresponds to the creation of a particle with the unreversed quantities. Incidentally, to within a phase factor, $v_s(E, \mathbf{p}) = u_{-s}(-E, -\mathbf{p})$. As we will see, the amplitude for a process in which a initial state $e^-$ is crossed to a final state $e^+$ is obtained through the replacement $u_s(E, \mathbf{p}) \to v_s(E, \mathbf{p})$ and the amplitudes $M_{xe^-\to y}$ and $M_{x\to e^+y}$ are, to within a phase, related by $s \to -s$ and $p \to -p$.

## 6.8 ■ PARITY

The simultaneous transformations

$$\mathbf{p} \to -\mathbf{p}\ ;\ \phi_\pm \to \eta_P\phi_\mp\ ;\ \mathbf{A} \to -\mathbf{A} \tag{6.119}$$

leave Dirac's equation

$$\left[\left(p^0 - eA^0\right) \mp (\mathbf{p} - e\mathbf{A}) \cdot \boldsymbol{\sigma}\right] \phi_{\pm} = m\phi_{\mp} \tag{6.120}$$

invariant. The invariance represents symmetry under space inversion called a parity transformation $P$. More precisely, suppose $\psi(x)$ or equivalently the pair $(\phi_+(x), \phi_-(x))$ is a solution for a given electromagnetic field $A(x)$. Define the transformed fields

$$\psi^P(x) = \eta_P \begin{pmatrix} \phi_-(x^P) \\ \phi_+(x^P) \end{pmatrix} \;;\; A^P_\mu(x) = \left[A^0\left(x^P\right), -\mathbf{A}\left(x^P\right)\right] \tag{6.121}$$

where $x^\mu = (t, \mathbf{x})$ and $x^{\mu\ P} = (t, -\mathbf{x})$. Then $\psi^P$ is a solution with $A^P$. Here $\eta_P$ is an arbitrary phase factor we take to be unity if we require $P^2 = 1$.

Under a parity transformation, $\phi_\pm \to \eta_P \phi_\mp$ and

$$\phi^c_\pm = \eta_\pm U \phi^*_\mp \to \eta^*_P \eta_\pm U \phi^*_\pm = -\eta^*_P \phi^c_\mp \tag{6.122}$$

so, for $\eta_P = \eta^*_P$, the anti-matter solutions are transformed with a minus sign relative to the matter solutions. Fermion parity eigenstates consequently have negative intrinsic parity relative to the corresponding anti-fermion parity eigenstates. This is a remarkable prediction of the Dirac equation with many consequences for electromagnetic and color processes which respect parity symmetry. In contrast, the relative parity of a boson and its antiparticle is positive.

The magnitude of a 2-component spinor of one type is not invariant under Lorentz transformation. However, the inner product of two spinors of different types is invariant:

$$\phi^\dagger_\pm \phi_\mp \to \phi^\dagger_\pm U^\dagger_\pm U_\mp \phi_\mp = \phi^\dagger_\pm \phi_\mp. \tag{6.123}$$

Under a parity transformation $\phi^\dagger_\mp \phi_\pm \to \phi^\dagger_\pm \phi_\mp$ so the quantities

$$\bar{\psi}\psi = \phi^\dagger_- \phi_+ + \phi^\dagger_+ \phi_- \;;\; \bar{\psi}\gamma^5\psi = \phi^\dagger_- \phi_+ - \phi^\dagger_+ \phi_- \tag{6.124}$$

are a Lorentz scalar and pseudoscalar (odd under parity). If we multiply the Dirac equation by $\bar{\psi}$, we have

$$\bar{\psi} p_\mu \gamma^\mu \psi = m\bar{\psi}\psi \tag{6.125}$$

and, since the right-hand side is a scalar and $p^\mu$ transforms as a 4-vector, the current density

$$j^\mu = \bar{\psi}\gamma^\mu\psi = \psi^\dagger(1, \alpha)\psi = j^{+,\mu} + j^{-,\mu} \tag{6.126}$$

transforms as a 4-vector. Here we define the right-handed and left-handed currents

$$j^{\alpha,\mu} = \phi^\dagger_\alpha \sigma^{-\alpha,\mu} \phi_\alpha = (\phi^\dagger_\alpha \phi_\alpha, \alpha\phi^\dagger_\alpha \boldsymbol{\sigma} \phi_\alpha) \tag{6.127}$$

with $\alpha = +1$ for right-handed and $\alpha = -1$ for left-handed. The quantity $\bar{\psi}\gamma^5\gamma^\mu\psi$ transforms as a pseudovector (space components odd under parity).

## 6.9 ■ ELECTROMAGNETIC CURRENTS

To recap, the 4-spinor plane wave solution to the free particle Dirac equation may be written as

$$\psi_p = u_s(p)e^{-ipx} = \begin{pmatrix} u_s^+ \\ u_s^- \end{pmatrix} e^{-ipx} \text{ (particle)}$$

$$\psi_{-p} = v_s(p)e^{ipx} = \begin{pmatrix} v_s^+ \\ v_s^- \end{pmatrix} e^{ipx} \text{ (antiparticle)}$$

where the subscript on $\psi$ distinguishes matter and antimatter solutions associated with the same 4-momentum. For helicity states in the spinor representation, we have

$$u_s(p) = \begin{pmatrix} \sqrt{E+sp}\phi_s(\mathbf{n}) \\ \sqrt{E-sp}\phi_s(\mathbf{n}) \end{pmatrix} \; ; \; v_s(p) = \begin{pmatrix} -s\sqrt{E-sp}\phi_{-s}(\mathbf{n}) \\ +s\sqrt{E+sp}\phi_{-s}(\mathbf{n}) \end{pmatrix}. \qquad (6.128)$$

The normalization of the 2-component spinors was fixed when we defined them as boosted from the rest frame and it is easily shown from the explicit form that

$$u^{\mp\dagger}u^{\pm} = -v^{\pm\dagger}v^{\mp} = m. \qquad (6.129)$$

It follows that

$$\bar{\psi}_p\psi_p = 2m \; , \; \bar{\psi}_{-p}\psi_{-p} = -2m. \qquad (6.130)$$

The quantity $\bar{\psi}\psi$ is a Lorentz invariant and we see that the particle solutions have a Lorentz invariant normalization equal to twice the proper (rest) energy. The antiparticle solutions have a negative normalization. Multiplying the Dirac equation by $\bar{\psi}$ we have

$$-i\bar{\psi}\partial_\mu\gamma^\mu\psi = m\bar{\psi}\psi \qquad (6.131)$$

and, inserting particle and antiparticle solutions, we find

$$p^\mu\bar{\psi}_p\gamma_\mu\psi_p = m\bar{\psi}_p\psi_p = 2m^2 \to \bar{\psi}_p\gamma_\mu\psi_p = 2p_\mu$$
$$-p^\mu\bar{\psi}_{-p}\gamma_\mu\psi_{-p} = m\bar{\psi}_{-p}\psi_{-p} = -2m^2 \to \bar{\psi}_p\gamma_\mu\psi_p = 2p_\mu. \qquad (6.132)$$

So, for both positive and negative energy solutions $\psi_{\pm p}$, the current is

$$j^\mu = \bar{\psi}\gamma^\mu\psi = 2p^\mu \qquad (6.133)$$

as can be confirmed by calculation with the plane waves above. Antiparticle solutions associated with the motion of charge opposite to that of particle solutions do not have current opposite to the corresponding particle solutions. In the Feynman rules motivated below, this oddity is compensated by interchange of initial and final states in transition currents and by the negative invariant normalization for antiparticle solutions.

The fundamental process of spinor electrodynamics is a photon interacting with a transition current which is the sum of the right-handed and left-handed transition currents

$$j_{fi} = \bar{\psi}_f \gamma \psi_i = j_{fi}^+ + j_{fi}^-. \tag{6.134}$$

Consider a transition between helicity eigenstates for an electron, $\psi_i = u_t e^{-ipx} \rightarrow \psi_f = u_r e^{-ip'x}$, in a frame where $\psi_i$ and $\psi_f$ have the same energy. The right-handed and left-handed currents are

$$j_{rt}^{\alpha} = \sqrt{E + \alpha rp}\sqrt{E + \alpha tp}\left[\phi_r^{\dagger}\left(\mathbf{n}'\right)\phi_t(\mathbf{n}), \alpha\phi_r^{\dagger}\left(\mathbf{n}'\right)\sigma\phi_t(\mathbf{n})\right] \tag{6.135}$$

where $\alpha = +1$ for the right-handed current and $\alpha = -1$ for the left-handed current and we have omitted the space dependent factor $\exp(-i(p - p')x)$. The total current $j_{rt}$ is the 4-vector

$$j_{rt} = j_{rt}^+ + j_{rt}^- = \left[A_{rt}(E)\phi_r^{\dagger}\left(\mathbf{n}'\right)\phi_t(\mathbf{n}), B(E)\phi_r^{\dagger}\left(\mathbf{n}'\right)\sigma\phi_t(\mathbf{n})\right] \tag{6.136}$$

where

$$A_{rt}(E) = \left(\sqrt{(E + rp)(E + tp)} + \sqrt{(E - rp)\,(E - tp)}\right). \tag{6.137}$$

and

$$B_{rt}(E) = \sqrt{(E + rp)(E + tp)} - \sqrt{(E - rp)(E - tp)}. \tag{6.138}$$

For equal helicity, $r = t$, the energy dependent factors are

$$\begin{aligned} A_{tt}(E) &= 2E \\ B_{tt}(E) &= 2tp \end{aligned} \tag{6.139}$$

while, for unequal helicity, $r \neq t$, the energy dependent factors are

$$\begin{aligned} A_{rt}(E) &= 2m \\ B_{rt}(E) &= 0. \end{aligned} \tag{6.140}$$

For transitions involving antiparticles, notice that

$$v_{-r}^{\alpha} = \alpha r\sqrt{E + \alpha rp}\phi_r \tag{6.141}$$

so one replacement $u_r \rightarrow v_{-r}$ multiplies the total current by a factor $r$ and reverses the left-handed current giving $j_{rt} \rightarrow r(j^+ - j^-)$ and therefore the coefficients of the time and space components of the total current are multiplied by $r$ and interchanged. Two replacements restores the form of the current. Hence, for $r = t$, the transition currents can be written as

$$\bar{u}_r'\gamma^{\mu}u_t = \bar{v}_{-r}'\gamma^{\mu}v_{-t} = \left[2E\phi_t^{\dagger}(\mathbf{n}')\phi_t(\mathbf{n}), 2tp\phi_t^{\dagger}(\mathbf{n}')\sigma\phi_t(\mathbf{n})\right]$$

$$\bar{u}_r'\gamma^{\mu}v_{-t} = \bar{v}_{-r}'\gamma^{\mu}u_t = \left[2p\phi_t^{\dagger}(\mathbf{n}')\phi_t(\mathbf{n}), 2tE\phi_t^{\dagger}(\mathbf{n}')\sigma\phi_t(\mathbf{n})\right] \tag{6.142}$$

and, for $r \neq t$, the transition currents can be written as

$$\bar{u}'_r \gamma^\mu u_t = -\bar{v}'_{-r} \gamma^\mu v_{-t} = \left[2m\phi_r^\dagger\left(\mathbf{n}'\right)\phi_t(\mathbf{n}), 0\right]$$
$$t\bar{u}'_r \gamma^\mu v_{-t} = r\bar{v}'_{-r} \gamma^\mu u_t = \left[0, 2m\phi_r^\dagger\left(\mathbf{n}'\right)\sigma\phi_t(\mathbf{n})\right]. \tag{6.143}$$

In the relativistic limit ($m = 0$), only transitions preserving helicity ($r = t$) have non-zero amplitude and the transition currents assume the simple form

$$\bar{u}'_t \gamma^\mu u_t = \bar{v}'_{-t} \gamma^\mu v_{-t} = \bar{u}'_t \gamma^\mu v_{-t} = \bar{v}'_{-t} \gamma^\mu u_t$$
$$= 2E\left[\phi_t^\dagger\left(\mathbf{n}'\right)\phi_t(\mathbf{n}), t\phi_t^\dagger\left(\mathbf{n}'\right)\sigma\phi_t(\mathbf{n})\right]. \tag{6.144}$$

In the nonrelativistic limit ($p \to 0$), the currents for $r = t$ reduce to

$$\bar{u}'_r \gamma^\mu u_t = \bar{v}'_{-r} \gamma^\mu v_{-t} = \left[2m\phi_t^\dagger\left(\mathbf{n}'\right)\phi_t(\mathbf{n}), 0, 0, 0\right] \tag{6.145}$$

$$\bar{u}'_r \gamma^\mu v_{-t} = \bar{v}'_{-r} \gamma^\mu u_t = \left[0, 2tm\phi_t^\dagger\left(\mathbf{n}'\right)\sigma\phi_t(\mathbf{n})\right] \tag{6.146}$$

while the currents for $r \neq t$ retain their form. The factor $2m$ comes from the normalization of the plane waves while the factor $\phi_t^\dagger(\mathbf{n}')\phi_t(\mathbf{n})$ in the time component of $\bar{u}'_r \gamma^\mu u_r$ is the projection of the final spin state onto the initial spin state and, as we will see, implies conservation of spin in nonrelativistic electron elastic scattering. The preceding expressions are used in evaluation of transition rates in what follows.

## 6.10 ■ PERTURBATIVE SPINOR ELECTRODYNAMICS

The equations of spinor electrodynamics in Lorenz gauge are Dirac's equation

$$\left[i\gamma_\mu \partial^\mu - m\right]\psi = e\gamma_\mu A^\mu \psi \tag{6.147}$$

and Maxwell's equations with a spinor field current density

$$\partial_\nu \partial^\nu A^\mu = e\bar{\psi}\gamma^\mu \psi. \tag{6.148}$$

The perturbative solution follows the now familiar pattern. Consider an initial electron wave $\psi_i$ interacting with an electromagnetic field $A^\mu$. Use Dirac's equation with the initial electron wave fixed in the interaction term

$$\left[i\gamma_\mu \partial^\mu - m\right]\psi = eA^\mu \gamma_\mu \psi_i. \tag{6.149}$$

To find the amplitude for a final electron described by $\psi_f = u_f e^{-ip_f x}$ for this first order equation, appeal to the Hamiltonian form

$$i\partial_t \psi = H_0 \psi + V\psi \; ; \; V = e\gamma_0 \gamma_\mu A^\mu \tag{6.150}$$

and the result of nonrelativistic scattering theory (see Equation 4.56)

$$S_{fi} = -i \int dt \int d\mathbf{x}\, \psi_f^\dagger V \psi_i = -i \int dx\, e\bar{\psi}_f \gamma_\mu \psi_i A^\mu = -i \int dx\, j_{fi} A. \tag{6.151}$$

Thus the amplitude of $\psi_f$ produced by the interaction term on the right-hand side of Dirac's equation in Lorentz covariant form is the projection of the interaction term onto $\bar{\psi}_f$ and with a phase $-i$. Note that in contrast, with the scalar wave equation which is second order in time, we project onto $\phi_f^\dagger$ with phase $+i$.

In the context of the Hamiltonian Dirac equation considered as a single particle theory, we can interpret "negative energy" antimatter solutions as representing matter moving backwards in time. Consequently in transition matrix elements, the amplitudes representing initial and final antimatter waves are interchanged. Thus an initial electron with 4-momentum $p_-$ and wave function $\psi_{p_-}$ which enters an interaction and emerges going backwards in time corresponds to a collision of a electron and positron. If the initial positron momentum is $p_+$ and its wave function is $\psi_{-p_+}$, then the transition current is

$$j_{fi} = -e\bar{\psi}_{-p^+} \gamma^\mu \psi_{p^-}. \tag{6.152}$$

The additional minus sign is required when projecting on to antiparticle solutions as may be surmised by reversing time in the Hamiltonian equation. Another way to see this is to start with a general interaction term $s(x)$ in Dirac's equation and multiply by $\gamma p + m$ to get a second order equation

$$[\gamma p - m]\psi = s(x) \rightarrow \left(\partial_\mu \partial^\mu + m^2\right)\psi = -(\gamma p + m)s. \tag{6.153}$$

Our waves have the invariant normalization $\bar{\psi}\psi = \pm 2m$ for positive and negative energy. So if we take the resonant contribution (factor $i$) and make a Lorentz invariant projection with $\bar{\psi}$ and divide by the invariant normalization factor $\bar{\psi}\psi = \pm 2m$, we find

$$S = i(-1) \int dx \left(\frac{\pm 1}{2m}\right) \bar{\psi}(\gamma p + m)s = \mp i \int dx\, \bar{\psi} s \tag{6.154}$$

where, since $\bar{\psi}(\gamma p + m) = 0$, when $p$ acts to the left, integration by parts gives the final equality. Thus, the amplitude for positron muon scattering is that for electron muon scattering with the electron spinor amplitudes replaced by positron spinor amplitudes and in opposite order, with the addition of a minus sign for the final projection onto the negative energy electron. The change in sign corresponds to the change in sign of the interaction energy between like charged particles and oppositely charged particles.

Consider now the amplitude for a transition between plane wave states

$$\psi_i = u_i e^{-ip_i x} \rightarrow \psi_f = u_f e^{-ip_f x} \tag{6.155}$$

where the 4-momentum vectors are $p_i$ and $p_f$, with absorption of a photon with 4-momentum $k$ represented by the wave

$$A^\mu = \epsilon^\mu e^{-ikx}. \tag{6.156}$$

Using the general expression Equation 6.151, we find

$$S_{fi} = iM(2\pi)^4 \delta(k + p_i - p_f) \tag{6.157}$$

where

$$iM = \bar{u}_f(-ie\gamma_\mu)u_i \epsilon^\mu. \tag{6.158}$$

The transition amplitude for plane wave states of an antiparticle represented by

$$\psi_i = v_i e^{ip_i x} \rightarrow \psi_f = v_f e^{ip_f x} \tag{6.159}$$

with absorption of a similar photon is found by the replacements $u_i \rightarrow v_i$ and $u_f \rightarrow v_f$, interchanging the initial and final state wave functions and adding a minus sign. The same expression for 4-momentum conservation results and

$$S_{fi} = iM(2\pi)^4 \delta\left(k + p_i - p_f\right) \tag{6.160}$$

where now

$$iM = -\bar{v}_i(-ie\gamma_\mu)v_f \epsilon^\mu. \tag{6.161}$$

As shown in Figure 6.4, the general amplitude can be represented by a Feynman diagram and, for plane wave states, we can associate factors in the matrix element with the parts of the diagram: a factor $-ie\gamma_\mu$ for the vertex sandwiched between spinor amplitudes $u_i$ or $v_f$ on the right and $\bar{u}_f$ or $v_i$ on the left, and a vector field amplitude $\epsilon^\mu$ for the photon with the vector field amplitude Lorentz index contracted with the vertex factor Lorentz index. The order of the spinor amplitudes follows the direction of an arrow following the electric charge. The minus sign for the antiparticle case must be added by hand.

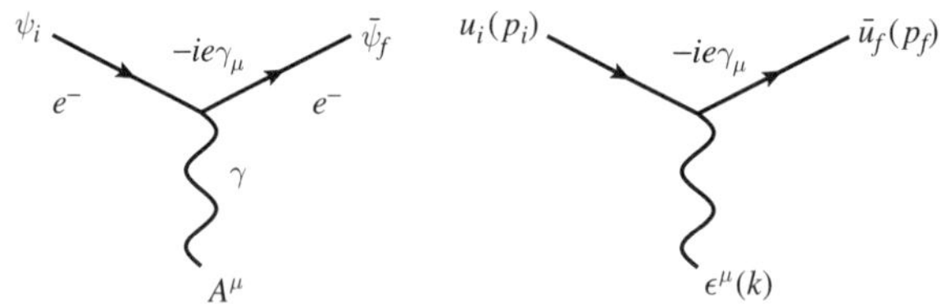


**FIGURE 6.4** Electromagnetic interaction of the electron. The general diagram at left represents for example the amplitude for an electron scattering in an electromagnetic field. The diagram at right represents the amplitude for an electron of momentum $p_i$ to absorb a photon of momentum $k$ and emerge with momentum $p_f$. The factors next to the external lines and the vertex appear in the amplitude.

The same rules follow from Maxwell's equations. Suppose a transition current fixed in the interaction term:

$$\partial_\nu \partial^\nu A^\mu = e\bar{\psi}_f \gamma^\mu \psi_i. \tag{6.162}$$

The amplitude for a photon $A = \epsilon e^{-ikx}$ to be radiated is the projection using $A^\dagger$ and a phase $i$ (with a minus sign since $\epsilon^\dagger \epsilon = -1$):

$$S = -i \int dx\ A^\dagger_\mu e\bar{\psi}_f \gamma^\mu \psi_i = iM(2\pi)^4 \delta(p_i - k - p_f) \tag{6.163}$$

where the emission amplitude is given by

$$iM = \epsilon_\mu{}^\dagger \bar{u}_f(-ie\gamma^\mu)u_i. \tag{6.164}$$

Here we see that, for emission, the photon polarization vector appears conjugated. The amplitude for absorption of this same photon is described by projecting on to an "antiphoton" wave $A^\dagger$ reversing $k$ in the 4-momentum conservation condition and replacing $\epsilon^\dagger_\mu$ with $\epsilon_\mu$. The invariant normalization is the same for photon and antiphoton so no additional minus sign is required.

We turn now to second order processes in spinor electrodynamics. Elastic scattering of an electron and muon

$$e_1^- \mu_2^- \to e_3^- \mu_4^- \tag{6.165}$$

is described by the diagram in Figure 6.5. We start with the equation for the electromagnetic field and use the electron transition current on the right-hand side with plane wave functions for the electron $\psi_1 = u_1 e^{-ip_1 x}$ and $\psi_3 = u_3 e^{-ip_3 x}$:

$$\partial_\nu \partial^\nu A^\mu = j^\mu_{31} \equiv e\bar{\psi}_3 \gamma^\mu \psi_1 = e\bar{u}_3 \gamma^\mu u_1 e^{i(p_3 - p_1)x}. \tag{6.166}$$

Next we find the mediating electromagnetic field ($q = p_3 - p_1$) as we did in scalar electron muon scattering:

$$\partial_\nu \partial^\nu A^\mu = e\bar{u}_3 \gamma^\mu u_1 e^{iqx} \to A^\mu = -\frac{1}{q^2} j^\mu_{31}. \tag{6.167}$$

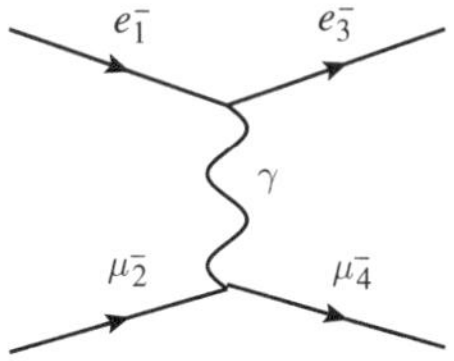


**FIGURE 6.5** Feynman diagram for electron and muon elastic scattering $e^-\mu^- \to e^-\mu^-$ by single photon exchange. The diagram is similar to that for scattering of distinguishable scalar charged particles.

The first approximation muon wave $\psi_2 = u_2 e^{-ip_2 x}$ coupled to this electromagnetic field in Dirac's equation gives

$$\left[i\gamma_\mu \partial^\mu - m_\mu\right]\psi = eA^\mu \gamma_\mu \psi_2. \tag{6.168}$$

To find the amplitude of $\psi_4 = u_4 e^{-ip_4 x}$ for this first order equation, project using $\bar{\psi}$ with phase $-i$. Therefore, the amplitude for $e^-\mu^-$ scattering is

$$S = -i\int dx\; e\bar{\psi}_4\gamma_\mu\psi_2 A^\mu = -i\int dx\; j_{42}A = iM(2\pi)^4\delta(p_3 + p_4 - p_1 - p_2) \tag{6.169}$$

where the invariant amplitude is

$$iM_{e^-\mu^-\to e^-\mu^-} = \left[\bar{u}_4\left(-ie\gamma^\mu\right)u_2\right]\left[\frac{-ig_{\mu\nu}}{q^2}\right]\left[\bar{u}_3\left(-ie\gamma^\nu\right)u_1\right]. \tag{6.170}$$

The same result can be found beginning with the muon transition current, finding the electromagnetic field, and then using Dirac's equation for the electron.

This expression for $M_{e^-\mu^-\to e^-\mu^-}$ exemplifies most of the Feynman rules of spinor quantum electrodynamics. The polarization states of an external initial fermion are represented by a spinor amplitude $u$. The polarization state of a final state fermion is represented by a factor $\bar{u}$ rather than $u^\dagger$. The same photon propagator

$$G_\gamma^{\mu\nu} = \frac{-ig^{\mu\nu}}{q^2} \tag{6.171}$$

that we used in scalar electrodynamics is associated with an internal photon line and, as we just found by considering emission or absorption, each vertex implies a factor $-ie\gamma_\mu$. The vertex factor contains the dimensionless coupling $e$ and combines two spinor polarization vectors to represent the corresponding transition current. Considering just the equation for the electromagnetic field, the amplitude for a final photon of polarization vector $\epsilon$ associated with a transition current $j_{fi} = e\bar{\psi}_f\gamma\psi_i$ is the projection

$$iM = -i\epsilon_\mu^\dagger j_{fi}^\mu \tag{6.172}$$

and, as in scalar electrodynamics, an outgoing photon corresponds to a factor $\epsilon^\dagger$ and incoming photon to a factor $\epsilon$.

We next consider a process involving antimatter. The Feynman diagram for positron-muon scattering

$$e_1^+\mu_2^- \to e_3^+\mu_4^- \tag{6.173}$$

is shown in Figure 6.6. The amplitude for this process is found by associating a wave $v_3 e^{ip_3 x}$ with the final positron and a wave $v_1 e^{ip_1 x}$ with the initial positron and

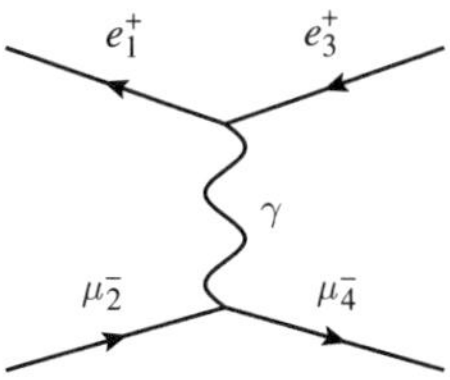


**FIGURE 6.6** Feynman diagram for positron and muon elastic scattering $e^+\mu^- \rightarrow e^+\mu^-$ by single photon exchange. The diagram is similar to that for $e^-\mu^- \rightarrow e^-\mu^-$ but an initial antiparticle appears as a final particle and a final antiparticle appears as a initial particle.

the current $j_{31} = \bar{u}_3\gamma^\mu u_1 \rightarrow -\bar{v}_1\gamma^\mu v_3$. We can then proceed as in $e^-\mu^- \rightarrow e^-\mu^-$ to find the electromagnetic field

$$\partial_\nu\partial^\nu A^\mu = e\bar{v}_1\gamma^\mu v_3 e^{i(p_3-p_1)x} \rightarrow A^\mu = -\frac{1}{q^2} j_{31}^\mu \tag{6.174}$$

and use this in the Dirac equation for the muon to find the scattering amplitude

$$iM_{e^+\mu^-\rightarrow e^+\mu^-} = -\left[\bar{u}_4(-ie\gamma^\mu)u_2\right]\left[\frac{-ig_{\mu\nu}}{q^2}\right]\left[\bar{v}_1(-ie\gamma^\nu)v_3\right]. \tag{6.175}$$

Now we turn to the process of muon pair production

$$e_1^- e_2^+ \rightarrow \mu_3^- \mu_4^+ \tag{6.176}$$

shown in Figure 6.7. The amplitude follows from our Feynman rules. A wave function factor $v_4 e^{ip_4x}$ is associated with creation of a $\mu^+$ of momentum $p_4$, a factor $\bar{v}_2 e^{-ip_2x}$ represents the $e^+$ destruction, and we have

$$iM_{e^+e^-\rightarrow\mu^+\mu^-} = (-1)^2\left[\bar{u}_3(-ie\gamma^\mu)v_4\right]\left[\frac{-ig_{\mu\nu}}{p^2}\right]\left[\bar{v}_2(-ie\gamma^\nu)u_1\right] \tag{6.177}$$

where two factors of $(-1)$ were included and cancel each other. The importance of the relative signs of these amplitudes becomes apparent in processes in which amplitudes interfere. The amplitude for $e_1^- e_2^+ \rightarrow e_3^- e_4^+$ may be found by combining our results for $e^-\mu^+ \rightarrow e^-\mu^+$ and $e^-e^+ \rightarrow \mu^-\mu^+$ and exemplifies the addition of amplitudes. A photon exchange amplitude and an annihilation amplitude add coherently and the total amplitude is

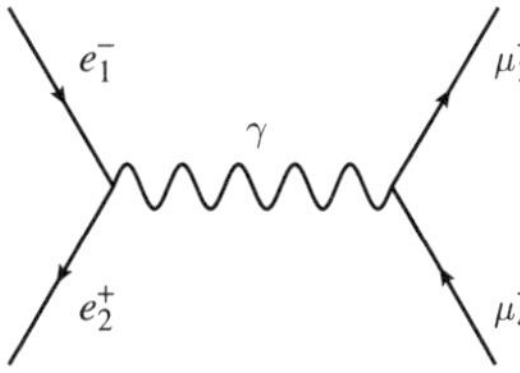


**FIGURE 6.7** Diagram describing $e^-e^+ \rightarrow \mu^-\mu^+$ in electrodynamics.

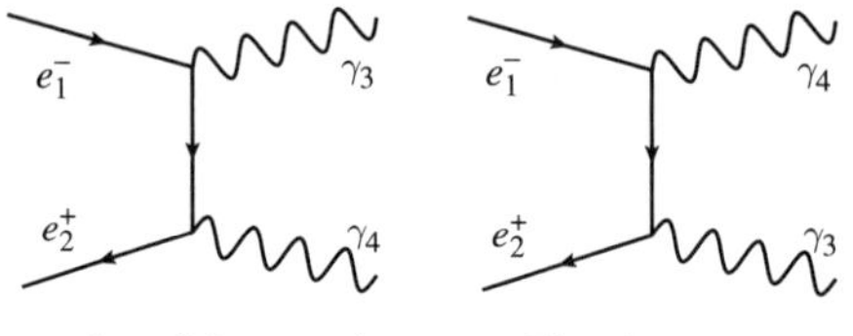


**FIGURE 6.8** Diagrams describing $e^- e^+ \to \gamma\gamma$. The electron appears in the intermediate state and a corresponding electron propagator in the amplitude. Two amplitudes related by interchange of the two indistinguishable photons must be superposed.

$$iM_{e^+e^-\to e^+e^-} = -\left[\bar{v}_2\,(-ie\gamma^\mu)\,v_4 \frac{-ig_{\mu\nu}}{q^2} \bar{u}_3\,(-ie\gamma^\nu)\,u_1\right]$$
$$+\left[\bar{u}_3\,(-ie\gamma^\mu)\,v_4 \frac{-ig_{\mu\nu}}{p^2} \bar{v}_2\,(-ie\gamma^\nu)\,u_1\right]. \qquad (6.178)$$

The squared amplitude which governs the probability distribution exhibits interference and the first amplitude includes one factor of $(-1)$ added by hand while the second amplitude had two such factors.

An intermediate state electron appears in the two photon annihilation process

$$e_1^- e_2^+ \to \gamma_3\gamma_4 \qquad (6.179)$$

described by the diagrams in Figure 6.8. To derive the propagator and the amplitude, start with an electron wave $\psi_1 = u_1 e^{-ip_1 x}$ and a photon wave $A_3 = \epsilon_3 e^{-ik_3 x}$ in the Dirac equation

$$[i\gamma_\mu \partial^\mu - m]\psi = e\gamma_\mu A_3^{\mu\dagger}\psi_1 = e\gamma_\mu \epsilon_3^{\mu\dagger} u_1 e^{-ipx} \qquad (6.180)$$

with $p = p_1 - k_3$. The corresponding intermediate fermion field has the formal representation

$$\psi = \frac{1}{\gamma_\nu p^\nu - m} e\gamma_\mu \epsilon_3^{\mu^\dagger} u_1 e^{-ipx}. \qquad (6.181)$$

The positron wave $\psi_2 = v_2 e^{ip_2 x}$ appears in a transition current in Maxwell's equations

$$\partial_\nu \partial^\nu A^\mu = e\bar{\psi}_2 \gamma^\mu \psi \qquad (6.182)$$

and the amplitude for a photon described by $A_4 = \epsilon_4 e^{-ik_4 x}$ is (project with factor $-i$ because of the normalization $\epsilon^\dagger \epsilon = -1$)

$$S = -i \int dx\; A_4^{\mu\dagger} e\bar{\psi}_2 \gamma_\mu \frac{1}{\gamma p - m} e\gamma^\mu A_3^{\mu\dagger}\psi_1 = iM(2\pi)^4 \delta(k_3 + k_4 - p_1 - p_2) \qquad (6.183)$$

where (adding an amplitude with photons exchanged)

$$iM_{e^+e^-\to\gamma\gamma} = \epsilon_4^{\mu\dagger}\bar{v}_2(-ie\gamma_\mu)\frac{i}{\gamma p - m}\epsilon_3^{\nu\dagger}(-ie\gamma_\nu)u_1 + (3 \Leftrightarrow 4). \qquad (6.184)$$

| Diagram element | Factor in amplitude |
|---|---|
| Ingoing fermion line | $u$ |
| Outgoing fermion line | $\bar{u}$ |
| Ingoing anti-fermion line | $v$ |
| Outgoing anti-fermion line | $-\bar{v}$ |
| Ingoing photon line | $\epsilon_\mu$ |
| Outgoing photon line | $\epsilon_\mu^\dagger$ |
| Vertex | $-ie\gamma$ |
| Internal photon line | $-ig_{\mu\nu}/q^2$ |
| Internal fermion line | $i/(\gamma p - m)$ |

**TABLE 6.1** Feynman rules for spinor electrodynamics.

Here we encounter the propagator for a spin 1/2 particle

$$G(p) = \frac{i}{\gamma p - m} = \frac{i}{\gamma p - m}\frac{\gamma p + m}{\gamma p + m} = i\frac{\gamma p + m}{p^2 - m^2}. \tag{6.185}$$

It is left as a problem to show that the numerator is the sum over spin states of the virtual electron.

We also encounter here an additional rule governing identical particles which can only be added by hand in our single particle approach to quantum field theory. Amplitudes for processes in which identical bosons appear interchanged must be added. Correspondingly, amplitudes for processes in which identical fermions or antifermions appear interchanged must be added with a minus sign for each interchange to be consistent with the exclusion principle. For example, the amplitude for $e_1^- e_2^- \to e_3^- e_4^-$ is

$$\begin{aligned} iM_{e^-e^-\to e^-e^-} &= [\bar{u}_4\,(-ie\gamma^\mu)\,u_2]\left[\frac{-ig_{\mu\nu}}{q^2}\right][\bar{u}_3\,(-ie\gamma^\nu)\,u_1] \\ &\quad - [\bar{u}_3\,(-ie\gamma^\mu)\,u_2]\left[\frac{-ig_{\mu\nu}}{q^2}\right][\bar{u}_4\,(-ie\gamma^\nu)\,u_1]. \end{aligned} \tag{6.186}$$

The Feynman rules for spinor QED are summarized in Table 6.1, in what follows we will apply the rules and previously developed results to some exemplary electromagnetic processes.

## 6.11 ■ ELASTIC SCATTERING

We will next see how to analyze the polarization dependence of the transition amplitudes and derive differential cross sections beginning with the simplest process, elastic scattering of distinguishable fermions. Consider $e^-\mu^-$ elastic scattering in the center of mass. The geometry of the collision is illustrated in Figure 6.9.

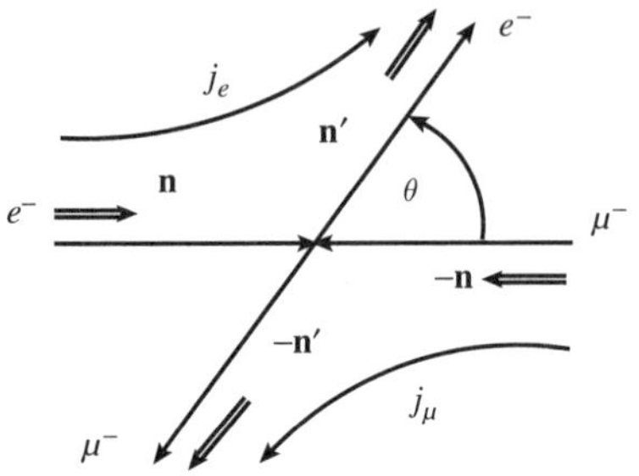


**FIGURE 6.9** Electron muon scattering in the center of mass frame. The electron enters at left along direction $\mathbf{n}$ and exits along direction $\mathbf{n}'$, scattering through angle $\theta$. The muon has opposite momentum and direction. The amplitude is proportional to the product of the helicity-dependent transition 4-currents $j_e$ and $j_\mu$. The small arrows indicate one of eight possible choices of the helicity states. In the ultra-relativistic case, helicity is conserved.

Denote the initial electron helicity and final electron helicity by $r$ and $r'$, and the initial electron direction by $\mathbf{n}$ and its final direction by $\mathbf{n}$'. Denote the initial and final muon helicities by $t$ and $t'$. The direction of the muon is opposite to that of the electron. For simplicity, we specialize to the ultrarelativistic case for which helicity is conserved. The transition current for the electron as given by Equation 6.142 is

$$j_r^e = 2E\left[\phi_r^\dagger\left(\mathbf{n}'\right)\phi_r\left(\mathbf{n}\right)\ ,\ r\phi_r^\dagger\left(\mathbf{n}'\right)\sigma\phi_r\left(\mathbf{n}\right)\right] \tag{6.187}$$

while the transition current for the muon is

$$j_t^\mu = 2E\left[\phi_t^\dagger\left(-\mathbf{n}'\right)\phi_t(-\mathbf{n})\ ,\ t\phi_t^\dagger\left(-\mathbf{n}'\right)\sigma\phi_t(-\mathbf{n})\right]. \tag{6.188}$$

Suppose the incident electron direction $\mathbf{n}$ is the $z$-direction and $\theta$ is the scattering angle in the $x-z$ plane. As we have seen, the explicit form of the electron spinors $\phi_r(\mathbf{n}')$ may be obtained by rotation of the eigenstates $\phi_r(\mathbf{n})$ of $\sigma_z$. Inserting expressions for these into the electron transition current, we find for helicity $r=+1$ the 4-vector

$$\begin{aligned} j_+^e &= 2E\left(\cos\frac{\theta}{2}, \sin\frac{\theta}{2}\right)\left[\begin{pmatrix}1 & 0\\ 0 & 1\end{pmatrix}, \begin{pmatrix}0 & 1\\ 1 & 0\end{pmatrix}, \begin{pmatrix}0 & -i\\ i & 0\end{pmatrix}, \begin{pmatrix}1 & 0\\ 0 & -1\end{pmatrix}\right]\begin{pmatrix}1\\ 0\end{pmatrix} \\ &= 2E\left(\cos\frac{\theta}{2}, \sin\frac{\theta}{2}\right)\left[\begin{pmatrix}1\\ 0\end{pmatrix}, \begin{pmatrix}0\\ 1\end{pmatrix}, \begin{pmatrix}0\\ i\end{pmatrix}, \begin{pmatrix}1\\ 0\end{pmatrix}\right] \\ &= 2E\left(\cos\frac{\theta}{2}, \sin\frac{\theta}{2}, i\sin\frac{\theta}{2}, \cos\frac{\theta}{2}\right). \end{aligned} \tag{6.189}$$

Combined with a similar expression for $r=-1$, the electron current can be written as

$$j_r^e = 2E\left(\cos\frac{\theta}{2}, \sin\frac{\theta}{2}, ir\sin\frac{\theta}{2}, \cos\frac{\theta}{2}\right). \tag{6.190}$$

The muon current is identical in form but rotated by $\pi$ around the $y$-axis so by simply rotating the electron current vector and changing the helicity label to that for the muon we find

$$j_t^\mu = 2E\left(\cos\frac{\theta}{2}, -\sin\frac{\theta}{2}, it\sin\frac{\theta}{2}, -\cos\frac{\theta}{2}\right). \tag{6.191}$$

The amplitude $M_{rt}$ for electron helicity $r$ and muon helicity $t$ is proportional to the inner product of these two 4-vectors and is

$$M_{rt} = \frac{e^2}{q^2} j_r^e j_t^\mu = \frac{e^2}{q^2}(2E)^2\left[2\cos^2\frac{\theta}{2} + \sin^2\frac{\theta}{2}(1+rt)\right]. \tag{6.192}$$

The following trigonometric identities are useful:

$$\begin{aligned}
\cos^2\left(\frac{\theta}{2}\right) &= \frac{1}{2}(1+\cos\theta) \\
\sin^2\left(\frac{\theta}{2}\right) &= \frac{1}{2}(1-\cos\theta) \\
\sin(A \pm B) &= \sin A\cos B \pm \cos A\sin B \\
\cos(A \pm B) &= \cos A\cos B \mp \sin A\sin B.
\end{aligned} \tag{6.193}$$

If $r = t$, the factor in brackets in $M_{rt}$ equals two, while, if $r = -t$, the factor is $2\cos^2(\theta/2) = 1 + \cos\theta$. There are two possible combinations for each so the spin averaged squared amplitude is

$$\overline{|M|^2} = \frac{1}{4}\sum_{r,t}|M_{rt}|^2 = \left(\frac{e^2}{q^2}\right)^2 2s^2\left(1+\cos^4\frac{\theta}{2}\right) \tag{6.194}$$

with $s = (2E)^2$. This expression governs the angular dependence of the scattering probability when the spin states are uncontrolled.

Given the squared matrix element, the scattering cross section may be found to be

$$\begin{aligned}
\frac{d\sigma_{e\mu\to e\mu}}{d\Omega} &= \frac{1}{64\pi^2 s}\overline{|M|^2} = \frac{1}{64\pi^2 s}\left[\frac{e^2}{s\sin^2(\theta/2)}\right]^2\left[2s^2\left(1+\cos^4\frac{\theta}{2}\right)\right] \\
&= \frac{\alpha^2}{2s}\frac{1+\cos^4\frac{\theta}{2}}{\sin^4\frac{\theta}{2}}
\end{aligned} \tag{6.195}$$

where we used the relativistic approximation

$$q^2 = t = -p'p = -2E^2(1-\cos\theta) = -s\sin^2\frac{\theta}{2}. \tag{6.196}$$

Since $\sin^2\frac{\theta}{2} = -t/s$, we can write $dt = \frac{s}{2}d\cos\theta$. Also,

$$\cos^4\frac{\theta}{2} = \left[1-\sin^2\frac{\theta}{2}\right]^2 = \left[1+\frac{t}{s}\right]^2 = \frac{u^2}{s^2} \tag{6.197}$$

since $s + t + u = 0$. Hence we can cast our result in invariant form

$$\frac{d\sigma_{e\mu\to e\mu}}{dt} = \frac{2}{s}\frac{d\sigma}{d\cos\theta} = 2\pi\frac{\alpha^2}{s^2}\frac{s^2+u^2}{t^2}. \tag{6.198}$$

This expression is confirmed by many experimental results as are the more detailed spin dependent cross section formulae. It differs from the expression for the elastic scattering cross section in scalar electrodynamics as given by Equation 5.142, reflecting the more complex structure of the electromagnetic current associated with spin 1/2. It is straightforward but lengthy to extend the calculation without the relativistic approximations $m_e = 0$ and $m_\mu = 0$ by using the exact expressions for the transition currents. At the end of this chapter we illustrate how such calculations may be facilitated using matrix techniques sometimes referred to as "trace technology."

Let us next connect the scattering amplitude to expectations in the nonrelativistic limit. The nonrelativistic Coulomb interaction is represented by the interaction density $j_0 A^0$, and conserves spin - the magnetic interaction is negligible. For $e^-\mu^-$ scattering, using the transition currents in the limit $E \to m$, the amplitude we compute in spinor electrodynamics is readily found to reduce to

$$M_{e^-\mu^-\to e^-\mu^-} = 2m_e 2m_\mu U(\mathbf{q}) \tag{6.199}$$

where

$$U(\mathbf{q}) = \frac{e^2}{\mathbf{q}^2} \tag{6.200}$$

is the Fourier transform of the repulsive interaction potential $e^2/(4\pi r)$ and the factors $2m_e 2m_\mu$ result from the relativistic normalization. Writing

$$q^2 = -2\mathbf{p}^2(1-\cos\theta) = -4\mathbf{p}^2\sin^2\left(\frac{\theta}{2}\right), \tag{6.201}$$

the Coulomb scattering differential cross section in the center of mass as a function of angle is

$$\frac{d\sigma_{e\mu\to e\mu}}{d\Omega} = \frac{1}{4\left(m_e+m_\mu\right)^2}\left[2m_e 2m_\mu\frac{e^2}{4\pi\mathbf{q}^2}\right]^2 = \frac{m^2\alpha^2}{16p^4}\frac{1}{\sin^4\left(\frac{\theta}{2}\right)} \tag{6.202}$$

where $m$ is the reduced mass. Aside from the normalization, our result for the amplitude and cross section are just those found with the Schrödinger equation. The minus sign in projecting anti-fermions reverses the sign of the potential for $e^-\mu^+$, the result one would anticipate by reversing the charge in Schrödinger theory.

In the scattering of electron and positron, an additional amplitude is present representing annihilation and formation of the pair via a virtual photon. The reader may find it instructive to show that this additional amplitude is non-vanishing for pairs of opposite spin and derives from the interaction $-\mathbf{j}\cdot\mathbf{A}$. The amplitude is important in relativistic scattering. In the nonrelativistic limit, the annihilation amplitude is negligible compared to the photon exchange amplitude. It corresponds to an effective potential energy which contributes a small energy shift in positronium comparable to other relativistic corrections.

## 6.12 ■ PAIR PRODUCTION

Electron positron colliders are used to produce and study heavy fermion anti-fermion pairs. We consider next the exemplary process

$$e_1^- e_2^+ \rightarrow \mu_3^- \mu_4^+ \tag{6.203}$$

proceeding through a state with an intermediate virtual photon. Since the final fermion type differs from the initial fermion type, only one (second order) amplitude is present. Using the Feynman rules, we find the amplitude has the form

$$M_{e^-e^+\rightarrow\mu^-\mu^+} = \frac{e^2}{p^2}\bar{u}_3\gamma_\nu v_4 \bar{v}_2\gamma^\nu u_1 \tag{6.204}$$

where the electron current is $j_e = \bar{v}_2\gamma u_1$, the muon current is $j_\mu = \bar{u}_3\gamma_\nu v_4$, and $p^2 = s$ is the center of mass energy squared.

As illustrated in Figure 6.10, in the center of mass, for electron direction $\mathbf{n} = \mathbf{z}$ and positron direction **-n**, the electron-positron transition current for electron helicity $t$ and positron helicity $-r$ follows from Equation 6.142 and Equation 6.143. In case $r = t$, we can calculate

$$\begin{aligned}
\bar{v}_{-r}\gamma^\mu u_t &= \phi_t^\dagger(-\mathbf{z})\,[2p, 2Et\boldsymbol{\sigma}]\,\phi_t(\mathbf{z})] = it\phi_{-t}^\dagger(\mathbf{z})\,[2pt, 2E\boldsymbol{\sigma}]\,\phi_t(\mathbf{z})] \\
&= it\phi_{-t}^\dagger(\mathbf{z})\,[2pt\phi_t(\mathbf{z}), 2E(\phi_{-t}(\mathbf{z}), it\phi_{-t}(\mathbf{z}), t\phi_t(\mathbf{z}))] \\
&= it2E[0, 1, it, 0]
\end{aligned} \tag{6.205}$$

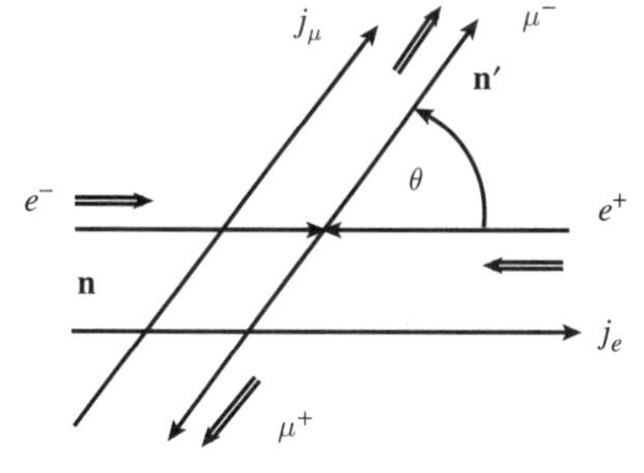


**FIGURE 6.10** The process $e^-e^+ \rightarrow \mu^-\mu^+$ in the center of mass frame. The electron enters at left along direction **n**. The $\mu^-$ exits along direction **n**′. The matrix element is proportional to the product of the transition 4-currents $j_e$ and $j_\mu$ which depend on the helicity states. One of eight possible choices of the helicity states is indicated by the small arrows.

where use was made of Equation 6.93 and of the expression

$$\boldsymbol{\sigma}\phi_t(\mathbf{z}) = (\phi_{-t}(\mathbf{z}), it\phi_{-t}(\mathbf{z})) \tag{6.206}$$

which is readily verified. In case $r \neq t$, we have

$$\begin{aligned}
\bar{v}_{-r}\gamma^{\mu}u_t &= \phi_r^{\dagger}(-\mathbf{z})[0, 2m\boldsymbol{\sigma}]\phi_t(\mathbf{z})] = i\phi_{-r}^{\dagger}(\mathbf{z})[0, 2m\boldsymbol{\sigma}]\phi_t(\mathbf{z}) \\
&= i\phi_{-r}^{\dagger}(\mathbf{z})[0, \phi_{-t}(\mathbf{z}), it\phi_{-t}(\mathbf{z}), t\phi_t(\mathbf{z})] \\
&= -it2m[0, 0, 0, 1].
\end{aligned} \tag{6.207}$$

Notice that opposite/same helicity states form a transverse/longitudinal current proportional to $[0, \epsilon_{\pm}]$ / $[0, \epsilon_0]$ as might be expected in combining two spin 1/2 states to form spin 1. The magnitude of longitudinal current is smaller than the magnitude of the transverse current by a factor $m/E$.

The muon current $\bar{u}_{r'}\gamma v_{-t'}$ has the same structure as the electron current with $r \to r'$ and $t \to t'$ but is rotated about the $y$-axis by the muon polar angle $\theta$. Dropping a common phase factor $it$, for $r' = t'$, we can rotate the electron current and find

$$\bar{u}_{r'}\gamma^{\mu}v_{-t'} = 2E[0, \cos\theta, it', -\sin\theta]. \tag{6.208}$$

Similarly, for $r' \neq t'$, we have

$$\bar{u}_{r'}\gamma^{\mu}v_{-t'} = -2m[0, \sin\theta, 0, \cos\theta]. \tag{6.209}$$

The products of the electron and muon currents for various helicity combinations are:

$$\begin{aligned}
j^{e}_{r=t}j^{\mu}_{r'=t'} &= -(2E)^2\left(\cos\theta - tt'\right) \\
j^{e}_{r=t}j^{\mu}_{r'\neq t'} &= (2E)\left(2m_{\mu}\right)\sin\theta \\
j^{e}_{r\neq t}j^{\mu}_{r'=t'} &= (2m_e)(2E)\sin\theta \\
j^{e}_{r\neq t}j^{\mu}_{r'\neq t'} &= -(2m_e)(2m_{\mu})\cos\theta.
\end{aligned} \tag{6.210}$$

Summed over the four possible $\mu^{\pm}$ polarizations and averaged over the $e^{\pm}$ polarizations, the squared amplitude may be calculated from these products and is

$$\begin{aligned}
\overline{|M_{e^+e^-\to\mu^+\mu^-}|^2} &= \frac{1}{4}\sum_{r,t,r',t'}|M_{e^+e^-\to\mu^+\mu^-}|^2 \\
&= e^4\left(1 + \cos^2\theta\left[1 + \left(\frac{m_e m_{\mu}}{E^2}\right)^2\right]\right. \\
&\quad \left. + \sin^2\theta\left[\left(\frac{m_e}{E}\right)^2 + \left(\frac{m_{\mu}}{E}\right)^2\right]\right).
\end{aligned} \tag{6.211}$$

The spin averaged differential cross section and its ultra-relativistic limit ($s = (2E)^2$) are

$$\frac{d\sigma_{e^+e^-\to\mu^+\mu^-}}{d\cos\theta} = \frac{p_\mu}{32\pi p_e s}\overline{|M|^2} \to \frac{\pi\alpha^2}{2s}\left(1+\cos^2\theta\right) \tag{6.212}$$

where $p_e = \sqrt{s-4m_e^2}/(2\sqrt{s}) \to \sqrt{s}/2$ and $p_\mu = \sqrt{s-4m_\mu^2}/(2\sqrt{s}) \to \sqrt{s}/2$. The total cross section and the ultra-relativistic limit are

$$\sigma_{e^+e^-\to\mu^+\mu^-} = \frac{4\pi}{3}\frac{\alpha^2}{s}\sqrt{\frac{s-4m_\mu^2}{s-4m_e^2}}\left(1+2\frac{m_e^2}{s}\right)\left(1+2\frac{m_\mu^2}{s}\right) \to \frac{4\pi}{3}\frac{\alpha^2}{s}. \tag{6.213}$$

This formula corresponds well with experimental results with electron positron colliding beam machnes. It dictates the electromagnetic production rate of pairs for any fermion of charge $q/e$ with the modification $\alpha^2 \to (q/e)^2\alpha^2$. The cross section for $e^+e^- \to \tau^+\tau^-$, which is identical to that for $e^+e^- \to \mu^+\mu^-$ except for the threshold energy, is illustrated in Figure 6.11. An example event appears in Figure 6.12. At center of mass energy $s \simeq m_Z^2$, the $Z$ boson appears as a resonance in $e^+e^- \to \tau^+\tau^-$. In quark pair production, the production of quark-antiquark bound states complicates the behavior near threshold.

The experimental cross section for $e^+e^-$ annihilation to quark-antiquark pairs which materialize as hadrons is shown as a function of center of mass energy in Figure 6.13. The $1/s$ behavior of the continuum production cross section that we have calculated is modified by intermediate resonant bound states. As the threshold is crossed for each quark flavor, the total cross section is increased in proportion to the square of the quark electric charge, times three since each quark flavor appears in three distinct colors. The ratio of cross sections for hadron production to muon pair production shows these steps more clearly and establishes that the number of colors states of each quark flavor is three. Although the additional QCD interactions of produced quark-antiquark pairs modify the cross sections near threshold, the angular distributions of the hadrons and the total rate above threshold indicate the quarks are fermions with the standard model electric charges.

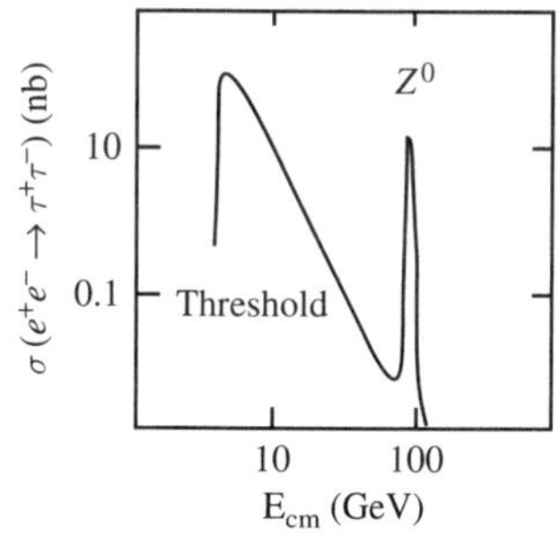


**FIGURE 6.11** Cross section for $e^+e^- \to \tau^+\tau^-$ as a function of center of mass energy $E_{cm}$. The threshold is at $2m_\tau$. The $Z$ boson contributes for $E_{cm}$ in excess of 90 GeV.

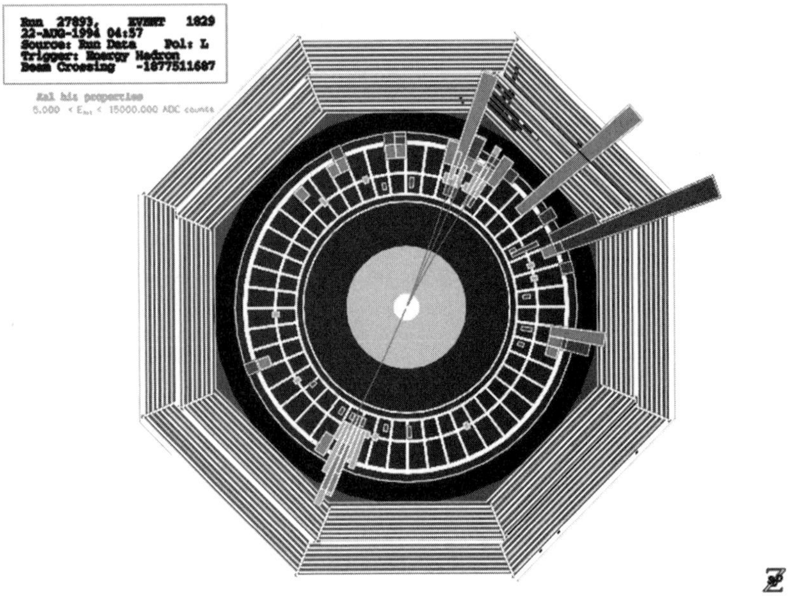


**FIGURE 6.12** Observation of $e^+e^- \to \tau^+\tau^-$ in the SLAC SLD detector. One decays to three pions and an invisible $\nu_\tau$ moving upwards, the other to an electron and two neutrinos moving downwards. [SLD collaboration]

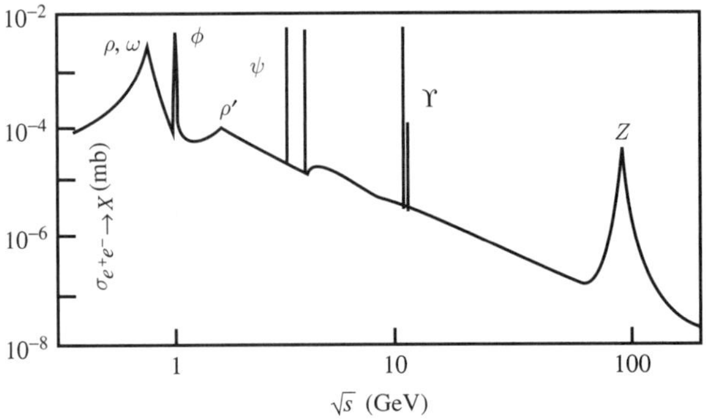


**FIGURE 6.13** Cross section for $e^+e^- \to$ hadrons. Outside regions of resonant enhancement and below threshold for $Z$ boson production, the cross section equals $\sigma_{e^+e^-\to\mu^+\mu^-} = 4\pi\alpha^2/(3s)$ times the number of flavors of $q\bar{q}$ pairs which may be produced, each contributing in proportion to the square of the quark electric charge. [after [C. Amsler et al. (Particle Data Group), Phys. Lett. B667, 1 (2008) and 2009 partial update for the 2010 edition] ]

The process $e^+e^- \to \mu^+\mu^-$ through an intermediate state photon averaged over spins can be described in terms of the resonant production and decay of a virtual spin 1 photon. The matrix element for decay of a photon (Figure 6.14) to $\mu^-\mu^+$, is

$$M_{A\to\mu^+\mu^-} = -eA_\mu j^\mu \tag{6.214}$$

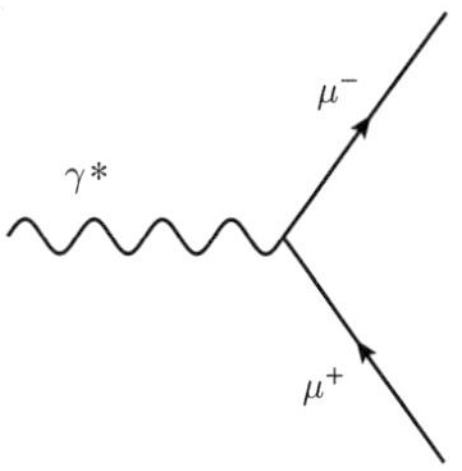


**FIGURE 6.14** Diagram describing the decay of a virtual photon $\gamma^* \to \mu^-\mu^+$.

where $j^\mu$ the fermion current and $A_\mu$ is the photon polarization vector. For fermion helicity state $r$ and antifermion helicity state $t$, in the rest frame of the photon, the current density is

$$j^\mu_{rt} = 2E[0, \cos\theta, it, -\sin\theta]\delta_{rt} + 2m[0, \sin\theta, 0, \cos\theta]\delta_{r,-t}. \tag{6.215}$$

For photon polarization vector $A = [0, 1, i, 0]/\sqrt{2}$, the matrix element is then

$$M_{e^+e^-\to\gamma} = -e\sqrt{2}E[\cos\theta - t]\delta_{rt} - e\sqrt{2}m\sin\theta\delta_{r,-t}. \tag{6.216}$$

Summed over spins, the square of the matrix element is

$$\begin{aligned}\Sigma|M_{e^+e^-\to\gamma}|^2 &= 2e^2\left[E^2(\cos\theta - 1)^2 + E^2(\cos\theta + 1)^2 + 2m^2\sin^2\theta\right]\\ &= 4e^2\left[E^2 + m^2 + \left(E^2 - m^2\right)\cos^2\theta\right].\end{aligned} \tag{6.217}$$

The decay rate is governed by the same matrix element

$$\begin{aligned}d\Gamma_{\gamma\to e^+e^-} &= \frac{\Sigma|M_{e^+e^-\to\gamma}|^2}{2\sqrt{s}}\left[dPS = \frac{d\Omega}{16\pi^2}\frac{p}{\sqrt{s}}\right]\\ &= \frac{e^2}{4\pi}\frac{p}{s}\left[E^2 + m^2 + (E^2 - m^2)\cos^2\theta\right]d\cos\theta.\end{aligned} \tag{6.218}$$

Integration over $\cos\theta$ and the substitutions $E = \sqrt{s}/2$ and $p = (s - 4m^2)^{\frac{1}{2}}/2$ gives an expression for the decay rate:

$$\sqrt{s}\Gamma_{\gamma\to e^+e^-} = \frac{1}{3}\frac{\alpha}{\sqrt{s}}\left(s - 4m^2\right)^{\frac{1}{2}}\left(s + 2m^2\right). \tag{6.219}$$

We can now use the resonance cross section formula Equation 4.169 to find

$$\begin{aligned}\sigma_{e^-e^+\to\mu^-\mu^+} &= \frac{4\pi(2j+1)}{(2s_e+1)(2s_e+1)p_e{}^2}\frac{\sqrt{s}\Gamma_{\gamma\to e^+e^-}\sqrt{s}\Gamma_{\gamma\to\mu^-\mu^+}}{\left(s - m_\gamma^2\right)^2 + m_\gamma^2\Gamma_\gamma^2}\\ &= \frac{12\pi}{s^2\left(s - 4m_e^2\right)}\frac{1}{3}\frac{\alpha}{\sqrt{s}}\left(s - 4m_e^2\right)^{\frac{1}{2}}\left(s + 2m_e^2\right)\end{aligned}$$

$$\times \frac{1}{3} \frac{\alpha}{\sqrt{s}} \left(s - 4m_\mu^2\right)^{\frac{1}{2}} \left(s + 2m_\mu^2\right)$$

$$= \frac{4\pi}{3} \frac{\alpha^2}{s} \sqrt{\frac{s - 4m_\mu^2}{s - 4m_e^2}} \left(1 + 2\frac{m_e^2}{s}\right)\left(1 + 2\frac{m_\mu^2}{s}\right). \qquad (6.220)$$

## 6.13 ■ ANNIHILATION AND POSITRONIUM

Pair annihilation to two photons $e_1^- e_2^+ \to \gamma_3 \gamma_4$ is especially interesting in the nonrelativistic limit. We have already written down the invariant amplitude corresponding to the diagrams in Figure 6.8:

$$M_{e^+e^- \to \gamma\gamma} = e^2 \epsilon_4^{\mu\dagger} \bar{v}_2 \gamma_\mu \frac{\gamma q + m}{q^2 - m^2} \epsilon_3^{\nu\dagger} \gamma_\nu u_1 + (3 \Leftrightarrow 4). \qquad (6.221)$$

In the center of mass, the electron and positron 4-momenta in the nonrelativistic limit are simply

$$p_1 = p_2 = m(1, 0). \qquad (6.222)$$

Each photon carries the energy equivalent of one electron mass and the photon 4-momenta may be expressed in terms of the direction of one of the photons as

$$k_3 = m(1, \mathbf{n})\ ; \ k_4 = m(1, -\mathbf{n}). \qquad (6.223)$$

The momentum transfer corresponding to the momentum of the intermediate state electron is $q = m(0, \mathbf{n})$ and $q^2 = -m^2$. The reduction of the amplitude is most simply carried out in the standard representation in which the Dirac equation reduces to the 2-component Pauli equation and only one of the 2-component spinors for a fermion is non-zero. The 4-spinor amplitude for the electron may be written as $u = \sqrt{2m}\begin{pmatrix} \eta \\ 0 \end{pmatrix}$ and that for the positron is $v = \sqrt{2m}\begin{pmatrix} 0 \\ \chi \end{pmatrix}$ where $\chi_s = (i\sigma_y)\eta_s^\dagger = -s\eta_{-s}$. The calculation requires that we consider only transverse final state photons corresponding to polarization vectors $e_\mu = (0, \mathbf{e})$, where $\mathbf{e}$ is a unit vector transverse to the photon momentum. The transition amplitude corresponding to the first of the two Feynman diagrams reduces to

$$M = e^2 \chi^\dagger \left[\mathbf{e}_4^\dagger \cdot \boldsymbol{\sigma}\mathbf{n} \cdot \boldsymbol{\sigma}\mathbf{e}_3^\dagger \cdot \boldsymbol{\sigma}\right] \eta \qquad (6.224)$$

and may be simplified using $\boldsymbol{\sigma} \cdot \mathbf{a}\boldsymbol{\sigma} \cdot \mathbf{b} = \mathbf{a} \cdot \mathbf{b} + i\boldsymbol{\sigma} \cdot \mathbf{a} \times \mathbf{b}$ and the conditions $\mathbf{e_3} \cdot \mathbf{n} = 0 = \mathbf{e_4} \cdot \mathbf{n}$ with the result

$$M^{n.r.}_{e^+e^- \to \gamma\gamma} = e^2 \chi^\dagger \left[i\mathbf{n} \cdot \mathbf{e}_3^\dagger \times \mathbf{e}_4^\dagger - \boldsymbol{\sigma} \cdot \mathbf{n}\mathbf{e}_3^\dagger \cdot \mathbf{e}_4^\dagger\right] \eta. \qquad (6.225)$$

When the amplitude with exchanged photons ($3 \Leftrightarrow 4$ and $\mathbf{n} \to -\mathbf{n}$) is added, only the term symmetric under the exchange survives and

$$M^{n.r.}_{e^+e^-\to\gamma\gamma} = 2e^2\chi^\dagger\eta\left[i\mathbf{n}\cdot\mathbf{e}_3^\dagger \times \mathbf{e}_4^\dagger\right]. \tag{6.226}$$

Since $\chi_s = -s\eta_{-s}$, it follows that only electron-positron pairs of opposite spin annihilate to two photons and that the amplitude changes sign under interchange of the electron and positron spin states. Moreover, since both polarization vectors are orthogonal to $\mathbf{n}$, the electric field vectors of the photons are orthogonal. For such states, the cross section is

$$\frac{d\sigma^{n.r.}_{e^+e^-\to\gamma\gamma}}{d\Omega} = \frac{1}{64\pi^2}|M|^2\frac{p_f}{p_i s} = \frac{1}{4(4\pi)^2}\left[2e^2\right]^2\frac{m}{(mv_{rel}/2)(2m)^2} = \frac{\alpha^2}{2m^2 v_{rel}}. \tag{6.227}$$

For given $\mathbf{n}$, there are two photon polarization pairs with nonvanishing amplitude and two of four possible initial spin state combinations can annihilate so this expression is the spin averaged cross section. The total cross section is obtained by integration over $2\pi$ so as to not double count indistinguishable final photon states and is

$$\sigma^{n.r.}_{e^+e^-\to\gamma\gamma} = \frac{\pi\alpha^2}{m^2 v_{rel}}. \tag{6.228}$$

This formula and the spin selection rules are important in understanding the annihilation of fermion anti-fermion bound states as described below.

A more lengthy calculation gives the center of mass spin averaged cross section

$$\frac{d\sigma_{e^+e^-\to\gamma\gamma}}{d\Omega} = \frac{\alpha^2}{sv}\left[\frac{1+v^2\left(1+\sin^2\theta\right)}{1-v^2\cos^2\theta} - \frac{2v^2\sin^4\theta}{\left(1-v^2\cos^2\theta\right)^2}\right] \tag{6.229}$$

where $s = 4E^2$ and $\mathbf{v} = \mathbf{p}/E$ is the velocity of the electron. While the nonrelativistic cross section is isotropic, the relativistic cross section is strongly peaked in the forward and backward directions. While in the nonrelativistic case, annihilation is possible only if the electron and positron spins are opposite so their helicities equal, in the ultra-relativistic case, helicity conservation applies and annihilation is possible only if the electron and positron spins are parallel and the helicity values opposite. The ultra-relativistic cross section turns out to be

$$\sigma^{UR}_{e^+e^-\to\gamma\gamma} = \frac{2\pi\alpha^2}{s}\left(\ln\left(\frac{s}{m^2}\right) - 1\right) \tag{6.230}$$

and decreases roughly in proportion to the inverse square of center of mass energy like the cross section for pair production.

A positronium atom is a bound state of electron and positron. The spin singlet ground state called parapositronium has a binding energy just 0.82 meV below that of the spin triplet state called orthopositronium. Both can decay to photons but do so with quite different decay rates. To find the decay rate of a bound system, one

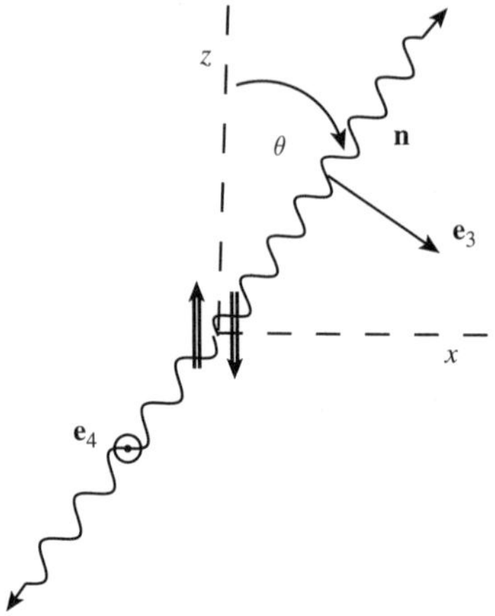


**FIGURE 6.15** Geometry of two photon annihilation $e_1^- e_2^+ \to \gamma_3\gamma_4$ in the center of mass of positronium. The two photons are emitted back to back. Annihilation takes place from $s$-wave states such as the ground state for which the net fermion spin vanishes. The matrix element favors orthogonal polarization vectors $\mathbf{e}_3$ and $\mathbf{e}_4$.

must in general use the the bound state wave function for the initial state in the matrix element. By expanding the initial state wave function in plane waves, the decay amplitude may be related to that for free particle interactions. The amplitude for annihilation in the nonrelativistic limit is momentum independent and the effective interaction potential $V(\mathbf{x}_{e^+} - \mathbf{x}_{e^-})$ obtained by Fourier transform is a Dirac $\delta$- function. Hence we may relate the two photon annihilation cross section to the positronium decay rate through a simple argument. If $v_{rel}\sigma = \Gamma$ is the reaction rate for one electron and one positron per unit volume, then for densities $\rho_\pm$, the reaction rate density is $\rho_+\rho_-\Gamma$. If $\rho_{\text{atom}}$ is the density of bound states (atoms) and if $\rho_+\rho_- = \rho_{\text{atom}}|\psi(0)|^2$ for an effectively point-like interaction, the rate per atom is obtained by multiplying the cross section by the relative flux $v|\psi(0)|^2$.

For $s$-wave positronium, the space wave function is $\psi(r) = e^{-r/a}/\sqrt{\pi a^3}$ with $a = 2\alpha^{-1}/m_e$, twice the Bohr radius of hydrogen due to the reduced mass. The annihilation amplitude for a spin singlet state $[\uparrow\downarrow - \downarrow\uparrow]/\sqrt{2}$ is $\sqrt{2}$ larger than the amplitude we computed for $\uparrow\downarrow$ as each component in the wave function contributes equally- the minus sign in the spin wave function is cancelled by the relative minus sign in the amplitude for exchanged spins. The minus sign also implies the radiation amplitudes cancel for the spin triplet $[\uparrow\downarrow + \downarrow\uparrow]/\sqrt{2}$, so neither this state nor the other symmetric spin states $\uparrow\uparrow$ and $\downarrow\downarrow$ can decay to two photons. Thus only the spin singlet can decay to two photons and in finding the singlet positronium lifetime we need four times the spin averaged annihilation cross section in computing the decay rate

$$\Gamma_{{}^1S_0(e^+e^-)\to\gamma\gamma} = 4v\sigma|\psi(0)|^2 = 4\pi\frac{\alpha^2}{m^2}|\psi(0)|^2 = m_e\alpha^5/2 = 8.03 \times 10^9\ \text{s}^{-1}. \tag{6.231}$$

The prediction is consistent with the experimental result $7.994 \pm 0.011\ \text{ns}^{-1}$.

Since the spin triplet states cannot decay to two photons and since single photon electric dipole radiative transitions between $s$-wave states are not permitted while magnetic dipole radiative transitions between orthopositronium and parapositronium are suppressed by phase space, the triplet states are rather stable. Three photon decay of spin triplet positronium is possible and is calculated to be

$$\Gamma_{^3S_1(e^+e^-)\to\gamma\gamma\gamma} = \frac{16}{9\pi}\left(\pi^2-9\right)\frac{\alpha^3}{m^2} \mid \psi(0)\mid^2 = \frac{2(\pi^2-9)}{9\pi} m_e\alpha^6 = 7.21\ (\mu s)^{-1} \tag{6.232}$$

which, when refined, agrees with experiment as well.

The prohibition of two photon decays of the triplet states does not depend on the nonrelativistic approximation. There is no total angular momentum 1 state of two photons consistent with Bose statistics (see Equation 10.27) so decay of the $l = 0$, $s = 1$, $j = 1$ positronium state to two photons would violate Bose symmetry or conservation of angular momentum. Moreover, the singlet and triplet states are eigenstates of charge conjugation symmetry. The singlet decay to two ($C = -1$) photons ($C = (-1)(-1) = +1$) implies the singlet is $C = +1$ while the triplet is $C = -1$. Conservation of $C$ parity implies the spin singlet (triplet) must decay to an even (odd) number number of photons.

A Schrödinger equation with relativistic corrections for a system such as positronium in which the motion of both particles is important may be obtained by finding a potential which reproduces the relativistic Born scattering amplitude. The result for positronium in which both particles have similar mass is somewhat more complex than the relativistic energy corrections we have calculated for hydrogen. For positronium, the annihilation diagram corresponds to a spin dependent contact interaction

$$U^{annihilation} = 2\pi\mu_B^2(3 + \boldsymbol{\sigma}_+\cdot\boldsymbol{\sigma}_-)\delta(\mathbf{r}) \tag{6.233}$$

and must be added to the single photon exchange potential and results in the Breit-Fermi Hamiltonian

$$\begin{aligned} H = {} & \frac{\mathbf{p}^2}{m} - \frac{\mathbf{p}^4}{4m^3} - \frac{\alpha}{4\pi r} + 4\pi\mu_B^2\delta(\mathbf{r}) \\ & - \frac{2\mu_B^2}{r}\left[\mathbf{p}^2 + \frac{\mathbf{r}\cdot(\mathbf{r}\cdot\mathbf{p})\mathbf{p}}{r^2}\right] + \frac{6\mu_B^2}{r^3}\left[\mathbf{r}\times\mathbf{p}\cdot\mathbf{s} + \frac{(\mathbf{s}\cdot\mathbf{r})^2}{r^2} - \frac{\mathbf{s}^2}{3}\right] \\ & + 4\pi\mu_B^2\left(\frac{7}{3}\mathbf{s}^2 - 2\right)\delta(\mathbf{r}) \end{aligned} \tag{6.234}$$

with $\mathbf{s} = (\boldsymbol{\sigma}_+ + \boldsymbol{\sigma}_-)/2$. The last term combines a spin-spin contact term with the annihilation potential. Only the last term differentiates the $l = 0$ orthopositronium ($s = 1$) and parapositronium ($s = 0$) energy levels. The triplet is higher in energy by $E_o - E_p = \frac{7}{12}\alpha^4 m$.

## 6.14 ■ NEUTRAL PION DECAY

The neutral pion decays almost exlusively to two photons. The positronium formula applied to a color singlet $(r\bar{r} + b\bar{b} + g\bar{g})/\sqrt{3}\, q\bar{q}$ meson is

$$\Gamma_{{}^1S_0(q\bar{q})\to\gamma\gamma} = 16\pi\alpha^2 e_q^4 N_c \frac{|\,\psi(0)\,|^2}{(2m_q)^2} \tag{6.235}$$

where $2m_q$ is the meson mass, $N_c = 3$ is the number of colors, and $e_q^2 = (e_u^2 - e_d^2)/\sqrt{2}$ for $(u\bar{u} - d\bar{d})/\sqrt{2}$. Taking $|\,\psi(0)\,|^2 \simeq 1/(\pi a^3)$ with $a \simeq 0.6$ fm leads to a decay rate

$$\frac{\Gamma_{\pi^0\to\gamma\gamma}}{m_\pi} = 48\frac{1}{137^2}\left(\frac{1}{18}\right)\left[\frac{0.2\text{ GeV fm}}{0.135\text{ GeV }0.6\text{ fm}}\right]^3 \simeq 2\times 10^{-3} \tag{6.236}$$

compared to the experimental result $5.8 \times 10^{-8}$. On the other hand, the prediction for $\eta_c = c\bar{c}$ is close to observation:

$$\frac{\Gamma_{\eta_c\to\gamma\gamma}}{m_{\eta_c}} = 48\frac{1}{137^2}\frac{16}{81}\left[\frac{0.2\text{ GeV fm}}{2.98\text{ GeV }0.6\text{ fm}}\right]^3 \simeq 7\times 10^{-7} \tag{6.237}$$

while the experimental result is $\sim 13 \times 10^{-7}$. The nonrelativistic model as applied to light hadrons has its limitations. A plausible explanation for the small $\pi^0$ decay rate in the naive quark model concerns the very small intrinsic $u$-quark and $d$-quark masses. In the limit of vanishing fermion mass, helicity is conserved by vector interactions and annihilation only occurs for total spin 1. The cross section is reduced by a factor $(m/E)^2$. For $E \simeq p \simeq 1/(0.6$ fm) and $m \simeq 2$ MeV, we can anticipate a suppression factor of $3 \times 10^{-4}$.

The rate for $\pi^0 \to \gamma\gamma$ is related to the empirical decay constant $f_\pi \simeq 93.3$ MeV determining the rate for $\pi^- \to \mu^+\nu_\mu$ (See Chapter 10) by chiral symmetry breaking field theory ideas according to which the pion is an excitation in a color interaction condensate. This theory predicts

$$\Gamma_{\pi^0\to\gamma\gamma} = \frac{\alpha^2}{64\pi^2}\frac{m_\pi^3}{f_\pi^2} = 7.3\text{ eV} \tag{6.238}$$

which agrees nicely with experiment. How $f_\pi$ appears in the numerator of the leptonic decay and the denominator of the electromagnetic decay is a fascinating story beyond the scope of perturbation theory and of this text. To convey a sense of phenomenological analysis, let's try to understand the so-called Dalitz decays $\pi^0 \to \gamma e^+e^-$ and $\pi^0 \to e^+e^-e^+e^-$ illustrated in Figure 6.16. We start with a calculation of the decay $\pi^0 \to \gamma\gamma$ using a relativistic matrix element constrained by symmetry to describe the conversion of the pion to photons. We then treat the external photons with spinor electrodynamics to describe their conversion to $e^+e^-$ pairs.

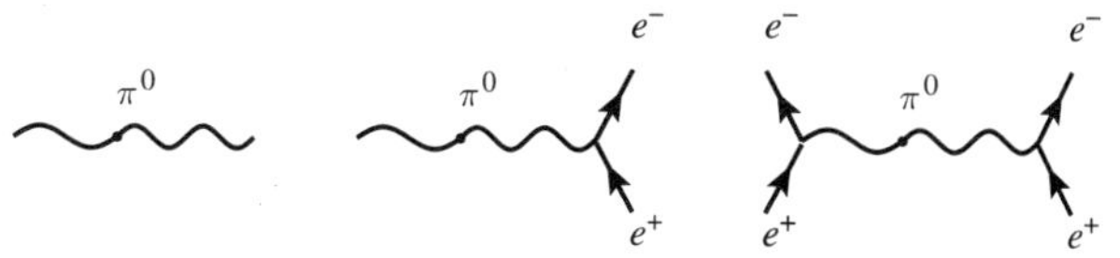


**FIGURE 6.16** Two photon and Dalitz decays of the $\pi^0$. The rates for the Dalitz decays $\pi^0 \to \gamma e^+e^-$ and $\pi^0 \to e^+e^-e^+e^-$ are suppressed by roughly a factor of $\alpha$ and $\alpha^2$ respectively.

The general form of the matrix element for $\pi^0$ decay is

$$M_{\pi^0\to\gamma\gamma} = A_{\mu\nu}\epsilon_1^{\mu\dagger}\epsilon_2^{\nu\dagger} \tag{6.239}$$

where the tensor $A(k_1, k_2)$ satisfies conditions of gauge invariance $A \cdot k_1 = A \cdot k_2 = 0$. Since the pion is a pseudoscalar, $A$ is a pseudotensor which suggests the form

$$M_{\pi^0\to\gamma\gamma} = g\epsilon_{\mu\nu\rho\sigma}\epsilon_1^{\mu\dagger}\epsilon_2^{\nu\dagger}k_1^\rho k_2^\sigma. \tag{6.240}$$

The scalar form factor $g$ contains information about the structure of the pion. It may depend upon the Lorentz invariants $k_1^2$, $k_2^2$, and $k_1k_2$. In $\pi^0 \to \gamma\gamma$, these Lorentz invariants have fixed values so we can treat $g$ as a constant. In Dalitz decays, the photons will be only slightly virtual and we will assume the same value of the form factor applies. Other processes such as photon-photon scattering $\gamma\gamma \to \phi^0 \to \gamma\gamma$ with one of the initial photons highly virtual and (in principle) elastic scattering $\gamma\pi^0 \to \gamma\pi^0$ with high momentum transfer can be used to explore the dependence of the form factor. We will not consider such processes.

We first use this matrix element to express the $\pi^0$ decay rate in terms of $g$. We can substitute $p = k_1 + k_2$ for $k_1$. In the rest frame, $p = (m_\pi, 0)$ and for polarization vectors $\mathbf{e}_1$ and $\mathbf{e}_2$, the matrix element is

$$M = -m_\pi g\left(\mathbf{e}_1^\dagger \times \mathbf{e}_2^\dagger\right) \cdot \mathbf{k}_1. \tag{6.241}$$

Only photon states transverse to $\mathbf{k} = \mathbf{k}_1 = -\mathbf{k}_2$ contribute, whether virtual or not. The Cartesian polarization vectors must be orthogonal and helicity values must be opposite. For such states , the matrix element is simply

$$M^2 = g^2m_\pi^2k^2 \tag{6.242}$$

where $k = m_\pi/2$. Including both photon helicity possibilities and dividing the phase space by two so as to not double count states of identical particles, we find the decay rate

$$\Gamma_{\pi^0\to\gamma\gamma} = \frac{1}{2m_\pi}2\left(g^2m_\pi^2k^2\right)\frac{1}{2}\frac{k}{4\pi m_\pi} = \frac{1}{8\pi}g^2k^3. \tag{6.243}$$

Notice that our $g$ has dimensions of inverse squared energy.

Assuming the Dalitz pair process $\pi^0 \to \gamma\gamma \to (e^+e^-)\gamma$ occurs dominantly by the external conversion of a photon to a $e^+e^-$ pair, and that the coupling $g$ of the source to virtual photon is approximately independent of its mass, we can relate the Dalitz pair rate to the two photon rate. To do this, we first derive a general result.

Consider a general decay proceeding through an intermediate state

$$a \to bV \to b(cd) \tag{6.244}$$

where $V$ is a virtual particle that decays to two particles denoted by $c$ and $d$. The matrix element has the form

$$M = \sum_i M_{a\to bV_j}\phi(s)M_{V_j\to cd} \tag{6.245}$$

where $\phi(s) = [s - m_V^2 + im_V\Gamma_V]^{-1}$ is a propagator factor, $j$ denotes the spin state of the virtual $V$, and $s$ is its invariant mass. The decay rate summed over final state spins is

$$d\Gamma = \frac{1}{2m_a}\Sigma_{b,c,d,i,j} M_{a,V_ib}M^*_{a,V_jb}M_{V_i,cd}M^*_{V_j,cd}|\phi|^2 dPS. \tag{6.246}$$

The Lorentz invariant phase space for a 3-body decay $p \to p_1 + p_2 + p_3$ can be expressed as a product of the phase space for the decay $p \to p_{12} + p_3$ where $p_{12} = p_1 + p_2$ times the phase space for decay $p_{12} \to p_1 + p_2$. Inserting a factor $d^4p_{12}\delta(p_{12} - p_1 - p_2) = 1$, and using the notation $d\mathbf{p}^L = d^3\mathbf{p}/[(2\pi)^3 2E]$, $p_{12} = p_1 + p_2$, $p_{12}^2 = s_{12}$, we have

$$\begin{aligned} dPS(p; p_1, p_2, p_3) &= (2\pi)^4\delta(p - p_{12} - p_3)d\mathbf{p}_1^L d\mathbf{p}_2^L d\mathbf{p}_3^L d^4p_{12}\delta(p_{12} - p_1 - p_2) \\ &= \delta^4(p - p_{12} - p_3)d\mathbf{p}_3^L d^4p_{12}dPS(p_{12}; p_1, p_2). \end{aligned} \tag{6.247}$$

We now make use of a property of the $\delta$-function that follows from its definition

$$\delta(f(x)) = (df/dx)^{-1}\delta(x) \tag{6.248}$$

to write

$$\delta\left(p_{12}^2 - s_{12}\right) = \delta\left(E_{12}^2 - \mathbf{p}_{12}^2 - s_{12}\right) = \delta\left(E_{12} - \sqrt{s_{12} + p_{12}^2}\right)(2E_{12})^{-1}. \tag{6.249}$$

Then we can write

$$\frac{1}{(2\pi)^4}d^4p_{12} = \frac{1}{(2\pi)^4}d^4p_{12}\delta\left(p_{12}^2 - s_{12}\right)ds_{12} = d\mathbf{p}_{12}^L\frac{ds_{12}}{2\pi} \tag{6.250}$$

and express the 3-body phase space as

$$dPS(p; p_1, p_2, p_3) = dPS(p; p_{12}, p_3)\frac{ds_{12}}{2\pi}dPS(p_{12}; p_1, p_2). \tag{6.251}$$

Hence we can separate matrix element and phase space factors and write the three body decay rate as

$$d\Gamma = \sum_{i,j} \frac{1}{2m_a} \sum_b M_{a,V_i b} M^*_{a,V_j b} dPS(a; bV) |\phi(s)|^2 \frac{ds}{2\pi}$$
$$\times \sum_{cd} M_{V_i,cd} M^*_{V_j,cd} dPS(V; cd). \tag{6.252}$$

The matrix elements $M_{V_i,cd}$ are orthonormal polynomials in the decay angles so integration over the phase space for the decay of the virtual $V$ gives

$$\sum_{cd} M_{V_i,cd} M^*_{V_j,cd} dPS(V; cd) = \delta_{ij} \Sigma_{cd} |M_{V_i,cd}|^2 dPS(V; cd) = \delta_{ij} 2\sqrt{s}\Gamma_{V\to cd} \tag{6.253}$$

and, using a similar expression for the decay of the $V$ to the initial state particles, we find the desired expression for the spin averaged sequential decay rate:

$$d\Gamma = \Gamma_{a\to bV} \frac{\sqrt{s}ds/\pi}{(s - m_V^2) + m_V^2 \Gamma_V^2} \Gamma_{V\to cd}. \tag{6.254}$$

We now return to the case of Dalitz decay of the $\pi^0$. In case one photon decays to $e^+e^-$ with invariant mass squared $s$, the decay momentum is $k = (m_\pi^2 - s)/(2m_\pi)$. Including a factor two since the decay is to different particles, the differential decay rate is

$$\begin{aligned} d\Gamma_{\pi^0\to\gamma e^+e^-} &= 2\Gamma_{\pi^0\to\gamma\gamma_s} \frac{1}{\pi} \frac{\sqrt{s}ds}{s^2} \Gamma_{\gamma_s\to e^+e^-} \\ &= \frac{g^2}{4\pi} \left[\frac{m_\pi^2 - s}{2m_\pi}\right]^3 \frac{ds}{s^{3/2}\pi} \left[\frac{\alpha}{3s}\left(s + 2m_e^2\right)\left(s - 4m_e^2\right)^{\frac{1}{2}}\right. \\ &= \frac{\alpha g^2}{96\pi^2} m_\pi^3 dx [1 - x]^3 x^{-\frac{5}{2}} [x + y][x - 2y]^{\frac{1}{2}} \end{aligned} \tag{6.255}$$

where $x = s/m_\pi^2$, and $y = 2m_e^2/m_\pi^2$. The pair mass spectrum is peaked just above the threshold $2y$ and falls roughly as $1/s$ as shown in Figure 6.17. Assuming the form factor $g$ is constant, integration gives for small $y$

$$\begin{aligned} I &= \int_{2y}^1 dx\, [1 - x]^3 x^{-\frac{5}{2}} [x + y][x - 2y]^{\frac{1}{2}} \\ &= \frac{1}{2}\sqrt{1 - 2y}\left(-7 + 13y + 2y^2\right) + \left(2 - 9y^2 + 2y^3\right) \ln\left[\frac{1 + \sqrt{1 - 2y} -}{\sqrt{2y}}\right] \\ &\simeq 2\left[\ln\left(\frac{m_\pi}{m_e}\right) - \frac{7}{4}\right]. \end{aligned} \tag{6.256}$$

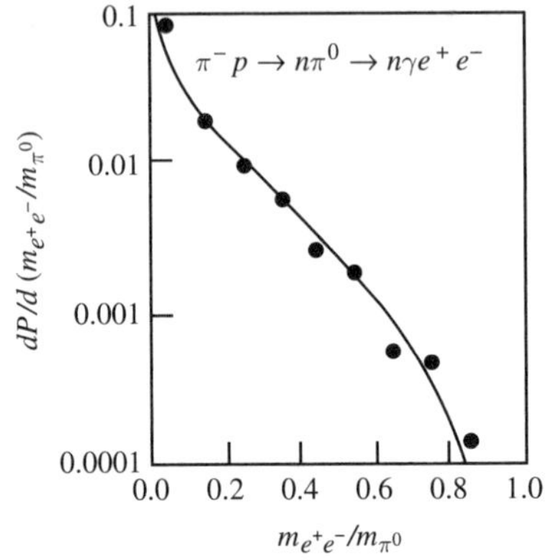


**FIGURE 6.17** Distribution of the invariant mass of the electron and positron in $\pi^0 \to e^+e^-$ compared to expectation. [After N. P. Samios, Phys. Rev. **121**, 275 (1961)]

The predicted branching ratio is just as observed:

$$\frac{\Gamma_{\pi^0\to\gamma e^+e^-}}{\Gamma_{\pi^0\to\gamma\gamma}} = \frac{4\alpha}{3\pi}\left[\ln\left(\frac{m_\pi}{m_e}\right) - \frac{7}{4}\right] = 1.19\% \tag{6.257}$$

It is possible for both photons to convert to pairs $e_1^- e_2^+$ and $e_3^- e_4^+$ - double Dalitz decay. The total amplitude is the difference between amplitudes with $1 \Leftrightarrow 3$, $M = M_{1234} - M_{3214}$. The individual photon decay amplitudes are large when pair masses are small and if $m_{12}$ and $m_{34}$ are small then $m_{32}$ and $m_{14}$ are large. Hence the interference term is small and $|M|^2 \simeq 2|M_{1234}|^2$ while the phase space must be divided by four as there are two pairs of identical particles. The single Dalitz decay formula generalizes to

$$d\Gamma_{\pi^0\to e^+e^-e^+e^-} = 2\Gamma_{\pi\to\gamma_r\gamma_s}\frac{1}{\pi}\frac{\sqrt{s}ds}{s^2}\Gamma_{\gamma_s\to e^+e^-}\frac{1}{\pi}\frac{\sqrt{r}dr}{r^2}\Gamma_{\gamma_r\to e^+e^-} \tag{6.258}$$

where now $k = \left[m_\pi^2 - \left(\sqrt{s}+\sqrt{r}\right)^2\right]^{\frac{1}{2}}\left[m_\pi^2 - \left(\sqrt{s}-\sqrt{r}\right)^2\right]^{\frac{1}{2}} \Big/ (2m_\pi)$. The branching fraction is

$$\frac{\Gamma_{\pi^0\to e^+e^-e^+e^-}}{\Gamma_{\pi^0\to\gamma\gamma}} = \left(\frac{\alpha}{3\pi}\right)^2 \int_{2y}^{1-2y} dx \int_{2y}^{(1-\sqrt{x})^2} dz\,(xz)^{-\frac{5}{2}}\left[1-\left(\sqrt{x}-\sqrt{z}\right)^{\frac{1}{2}}\right]^{\frac{3}{2}}$$

$$\left[1-\left(\sqrt{x}+\sqrt{z}\right)^{\frac{1}{2}}\right]^{\frac{3}{2}} \tag{6.259}$$

where $z = r/m_\pi^2$. The value of the integral is 57.65 and the expected branching ratio is $3.5 \times 10^{-5}$, close to the value $3.14 \pm 0.30 \times 10^{-5}$ observed.

The plane of each $e^+e^-$ pair is correlated with the polarization of the converted photon. In the two photon decay of a pseudoscalar meson like the $\pi^0$,

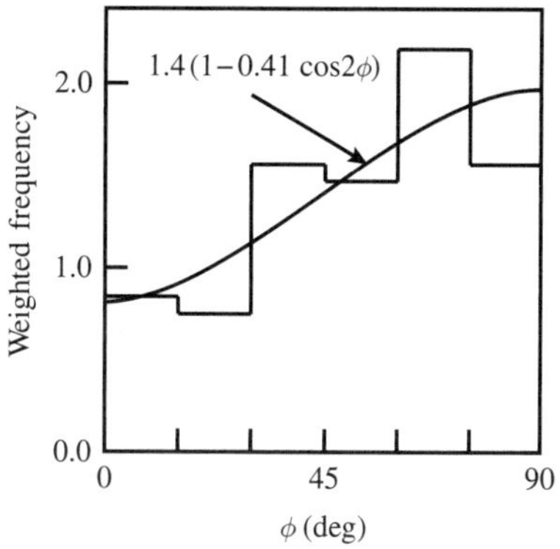


**FIGURE 6.18** Weighted frequency distribution of the angle between the planes of $e^+e^-$ pairs in $\pi^0 \to e^+e^-e^+e^-$. The fit to the data is shown. [After N. P. Samios *et al*, Phys. Rev. **126**, 1844 (1962)]

the photon polarizations are orthogonal while for a scalar meson they are parallel. Figure 6.18 shows the frequency distribution of the angle between the planes weighted depending on the pair masses and energy sharing for double Dalitz decay events observed in a bubble chamber with a stopping pion beam. The fitted correlation amplitude of $\alpha_{\text{exp}} = -0.41 \pm 0.24$ compares well to an expectation $\alpha_{\text{th}}^{\text{ps}} = -0.47$ for a pseudoscalar meson decay and is inconsistent with the expectation $\alpha_{\text{th}}^{\text{s}} = +0.47$ for a scalar meson decay. In this way, the parity of the $\pi^0$ was shown to be negative.

## 6.15 ■ LEPTONIC WIDTHS OF QUARKONIA

As a final illustration of spinor electrodynamics, we consider the annihilation process

$$q\bar{q} \to e^+e^- \tag{6.260}$$

of a nonrelativistic heavy quark and antiquark pair producing $e^+e^-$, assuming $\text{m}_\text{q} >> \text{m}_\text{e}$. We then apply the result to $q\bar{q}$ bound states called quarkonium as described by the diagram in Figure 6.19. The result would also apply to true muonium $(\mu^-\mu^+)_{\text{atom}}$. Figure 6.20 shows an example of event of this sort.

The matrix element in the center of mass has the form

$$M = ee_q \frac{j_q j_e}{(p_q + p_{\bar{q}})^2} = ee_q \frac{j_q j_e}{(2m_q)^2} \tag{6.261}$$

where we have approximated the center of mass energy by twice the heavy fermion mass, neglecting any binding energy. The nonrelativistic current associated with the heavy fermion for fermion helicity $t$ and anti-fermion helicity $s$ is given by Equation 6.146 as

$$j_q = \bar{v}_s \gamma u_t = -s \left[0, 2m_q \phi_{-s}^{\dagger} \sigma \phi_t\right] \tag{6.262}$$

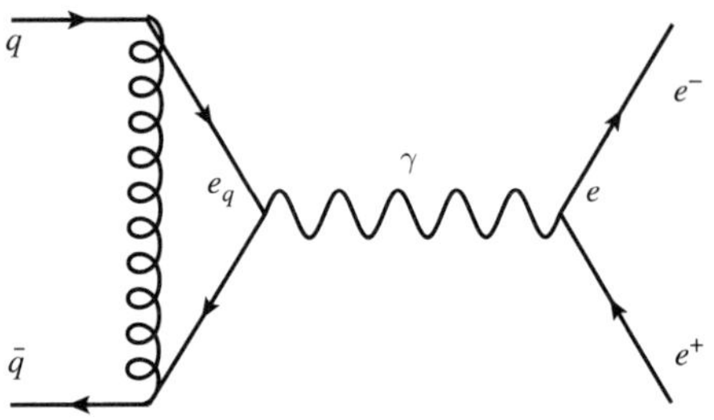


**FIGURE 6.19** Diagram describing the leptonic decay of a $q\bar{q}$ meson. The amplitude is proportional to the quark charge $e_q$.

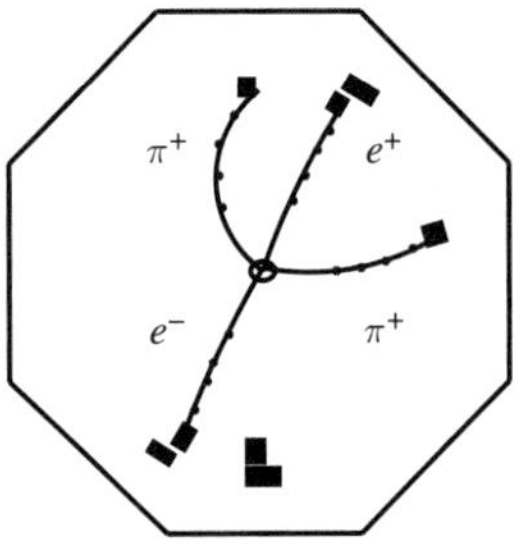


**FIGURE 6.20** Observation in the Mark I detector of $\psi'$ ($c\bar{c}$) production in $e^+e^-$ collisions at the SLAC SPEAR collider. The $\psi'$ decays to $J/\psi\pi^+\pi^-$. Annihilation of $c\bar{c}$ in the ground state results in $e^+e^-$. [SLAC photo archive, POW 991110]

and, using Equation 6.144, the ultra-relativistic ($E_e = m_q$) electron current for (conserved) electron helicity $r$ is

$$j_e = \bar{u}_r \gamma v_{-r} = 2m_q \left[\phi_r^\dagger \phi_r, r\phi_r^\dagger \boldsymbol{\sigma} \phi_r\right]. \tag{6.263}$$

Combining these expressions, we find

$$M_{r,st} = \frac{rsee_q}{3m_q^2} \left(\phi_{-s}^\dagger \boldsymbol{\sigma} \phi_t\right) \cdot \left(\phi_r^\dagger \boldsymbol{\sigma} \phi_r\right). \tag{6.264}$$

Referring to Figure 6.21, the electron and positron directions are opposite in the center of mass and we can write $\phi_r(\mathbf{n}_{e^+}) = \phi_{-r}(-\mathbf{n}_{e^-})$ and $\phi_{-s}(\mathbf{n}_{f^+}) = -s\phi_s(\mathbf{n}_{f^-})$ and choose $\mathbf{n}_f$ along the z axis and $\mathbf{n}_{e^-}$ along some direction $\mathbf{n}$ to write

$$M_{r,st} = -ee_f \left(\phi_s^\dagger (\mathbf{z})\, \boldsymbol{\sigma} \phi_t (\mathbf{z})\right) \cdot \left(\phi_r^\dagger (\mathbf{n})\, \boldsymbol{\sigma} \phi_{-r} (\mathbf{n})\right). \tag{6.265}$$

We have $\boldsymbol{\sigma}\phi_\pm = (\phi_\mp, \pm i\phi_\mp, \pm\phi_\pm)$ so we find, for the four possible spin states of the $q\bar{q}$ pair, the current density three vectors

$$\phi_+ \boldsymbol{\sigma} \phi_+ = (0, 0, 1)\ ;\ \phi_- \boldsymbol{\sigma} \phi_- = (0, 0, -1) \tag{6.266}$$

and

$$\phi_- \boldsymbol{\sigma} \phi_+ = (1, i, 0)\ ;\ \phi_+ \boldsymbol{\sigma} \phi_- = (1, -i, 0). \tag{6.267}$$

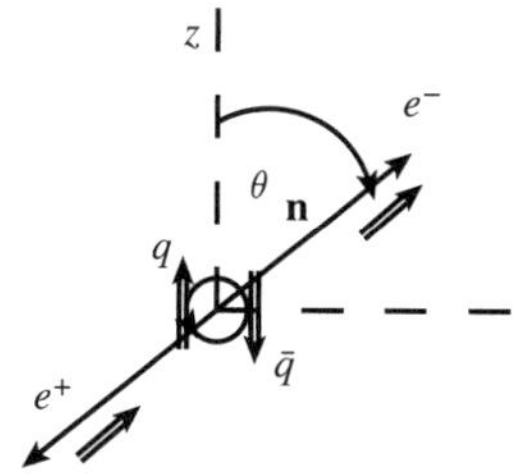


**FIGURE 6.21** Geometry of $q\bar{q}$ annihilation to $e^+e^-$. The spin quantization axis of the nonrelativistic $q$ and $\bar{q}$ is taken along the $z$-direction. The direction of the relativistic $e^-$ is $\mathbf{n}$, opposite to the direction of the $e^+$.

Corresponding electron transition current densities for electron angle $\theta$ relative the z-axis may be obtained by a rotation about the $y$ axis of the quark transition current densities. Helicity conservation for the relativistic electron and positron implies there are only two possible currents:

$$\mathbf{j}_+ = R_\theta \phi_+ \boldsymbol{\sigma} \phi_+ = (\cos\theta, -i, -\sin\theta) \tag{6.268}$$

and

$$\mathbf{j}_- = R_\theta \phi_- \boldsymbol{\sigma} \phi_- = (\cos\theta, +i, -\sin\theta). \tag{6.269}$$

The products of the electron current density and quark current density vectors are the angular factors given in Table 6.2. Here, $t$ is the $q$ spin along the $z$ axis and $s$ is the $\bar{q}$ helicity so its spin along the $-z$ axis. Using $\phi_s(\mathbf{z}) = s\phi_{-s}(-\mathbf{z})$ again, the initial vector states $\uparrow\uparrow$, $(\uparrow\downarrow + \downarrow\uparrow)/\sqrt{2}$, and $\downarrow\downarrow$ correspond to currents $-j_{-+}$, $(-j_{--} + j_{++})/\sqrt{2}$, and $j_{+-}$ while the scalar spin state $(\uparrow\downarrow - \downarrow\uparrow)/\sqrt{2}$ corresponds to $(-j_{--} - j_{++})/\sqrt{2} = 0$.

Summed over electron polarizations, the squared matrix element for a vector state is

$$\begin{aligned}
\overline{|M_{q\bar{q}\to e^+e^-}|^2} &= |M_{+,\uparrow\uparrow}|^2 + |M_{-,\uparrow\uparrow}|^2 \\
&= \frac{e^2 e_q^2}{(2m_q)^4}\left[(1+\cos\theta)^2 + (1-\cos\theta)^2\right] \\
&= \frac{e^2 e_q^2}{(2m_q)^4} 2\left(1+\cos^2\theta\right).
\end{aligned} \tag{6.270}$$

| $j_e/j_{q\bar{q}}$ | $j_{++}$ | $j_{-+}$ | $j_{+-}$ | $j_{--}$ |
|---|---|---|---|---|
| $j_+$ | $-\sin\theta$ | $\cos\theta + 1$ | $\cos\theta - 1$ | $\sin\theta$ |
| $j_-$ | $-\sin\theta$ | $\cos\theta - 1$ | $\cos\theta + 1$ | $\sin\theta$ |

**TABLE 6.2** Matrix elements for leptonic decays.

The differential cross section is

$$d\sigma = \frac{1}{64\pi^2}\overline{|M_{q\bar{q}\to e^+e^-}|^2}\frac{p'}{ps}d\Omega_e \tag{6.271}$$

and the angular integral is readily found to be

$$\int d\Omega_e \left(1+\cos^2\theta\right) = 2\pi\int_{-1}^{+1} dx \left(1+x^2\right) = \frac{16\pi}{3}. \tag{6.272}$$

The total cross section is therefore

$$\sigma_{q\bar{q}\to e^+e^-} = \frac{1}{2}\left(e_q/e\right)^2\left(\frac{e^2}{4\pi}\right)^2 \frac{m_q}{\left(m_q v_{cm}/2\right)\left(2m_q\right)^2}\frac{16\pi}{3} = \frac{4}{3}\frac{\left(e_q/e\right)^2\alpha^2}{m_q^2 v_{cm}}. \tag{6.273}$$

The decay rate for a vector bound state may be calculated following the example of positronium decay and is

$$\Gamma_{V\to e^+e^-} = v_{cm}\sigma|\psi(0)|^2 = \frac{4\pi}{3}(e_q/e)^2\alpha^2\frac{|\psi(0)|^2}{m_q^2}. \tag{6.274}$$

For a vector meson of mass $M = 2m_f$, the amplitude from the color singlet state $(r\bar{r}+b\bar{b}+g\bar{g})/\sqrt{3}$ is $3/\sqrt{3}$ larger so the decay rate is

$$\Gamma_{V_{\text{color singlet}}\to e^+e^-} = 16\pi\alpha^2 e_q^2\frac{|\psi(0)|^2}{M^2}. \tag{6.275}$$

For positronium, the probability density at the origin is

$$|\psi(0)|^2 = \frac{1}{\pi a^3} = \frac{\alpha^3 m_f^3}{8\pi}. \tag{6.276}$$

The color singlet analog requires the replacement $\alpha \to 4\alpha_s/3$ so

$$\Gamma_{V_{\text{color singlet}}\to e^+e^-} = \frac{16}{27}(e_q/e)^2\alpha^2\alpha_s^3 M. \tag{6.277}$$

We can turn this around and from the measured rates infer effective values of $\alpha_s$ as shown in Table 6.3. A cleaner result follows from comparison of electromagnetic decays such as

$$\Gamma_{\eta_c\to\gamma\gamma} = \frac{4\alpha^2 e_q^4 N_c}{m_q^2}|\psi(0)|^2 \tag{6.278}$$

and

$$\Gamma_{\psi\to e^+e^-} = \frac{4\alpha^2 e_q^2 N_c}{3m_q^2}|\psi(0)|^2 \tag{6.279}$$

| | M (GeV) | $\Gamma_{ee}$(KeV) | $e_q$ | $\alpha_s$ |
|---|---|---|---|---|
| $\Upsilon$ | 9.460 | 1.32 | $-\frac{1}{3}$ | 0.34 |
| $J/\psi$ | 3.097 | 5.26 | $+\frac{2}{3}$ | 0.49 |
| $\phi$ | 1.020 | 1.37 | $-\frac{1}{3}$ | 0.73 |
| $\omega$ | 0.782 | 0.62 | $\frac{1}{\sqrt{2}}(\frac{2}{3}-\frac{1}{3})$ | 0.48 |
| $\rho^0$ | 0.770 | 6.77 | $\frac{1}{\sqrt{2}}(\frac{2}{3}+\frac{1}{3})$ | 0.82 |

**TABLE 6.3** Leptonic widths of quarkonium states. The corresponding color fine structure constant $\alpha_s$ derived from the width is given.

where $N_c = 3$ and $e_q = 2/3$ and $m_q \simeq 1.5$ GeV. For a heavy quark system, the nonrelativistic model has a chance of approximating reality. Assuming the $\eta_c$ and $J/\psi$ are pure $c\bar{c}$ with the same space wave function, one expects

$$\frac{\Gamma_{\eta_c \to \gamma\gamma}}{\Gamma_{\psi \to e^+e^-}} = 3e_q^2 = \frac{4}{3}. \tag{6.280}$$

The observed rates $\Gamma_{\eta_c \to \gamma\gamma} = 4 \pm 1$ keV and $\Gamma_{\psi \to e^+e^-} = 5.2$ keV are consistent with the expectation.

## 6.16 ■ FERMION SPIN AVERAGING

This section provides a quick introduction to a common technique for computing spin averaged transition rates. A general scattering amplitude with some initial fermion polarization factor $u$ and final fermion factor $u'$ has the form $M = \bar{u}' A u$ where $A$ is some matrix. The conjugate is $(u'^{\dagger}\gamma_0 A u)^{\dagger} = u^{\dagger} A^{\dagger} \gamma_0^{\dagger} u' = \bar{u}\bar{A}u'$ where $\bar{A} = \gamma_0 A^{\dagger} \gamma_0^{\dagger}$ so the absolute square of the matrix element has the form

$$| M |^2 = \sum_{i,j,k,l} u'_i \bar{u}'_j A_{jk} u_k \bar{u}_l \bar{A}_{li} \tag{6.281}$$

where the indices denote the four components of the spinors. A sum over polarizations of such an expression may be accomplished using the expressions

$$\sum_{\alpha} u_{k,\alpha} \bar{u}_{l,\alpha} = \left(\hat{p} + m\right)_{kl} \equiv \rho_{kl} \; ; \; \sum_{\alpha} v_{k,\alpha} \bar{v}_{l,\alpha} = \left(\hat{p} - m\right)_{kl} \equiv \rho'_{kl}. \tag{6.282}$$

For our example, the squared matrix element when averaged over initial polarization states and summed over final polarization states can be expressed as a trace

$$\overline{| M |^2} = \frac{1}{2} \mathrm{tr}\left[\rho' A \rho \bar{A}\right]. \tag{6.283}$$

For photons, a similar result applies with $\rho_{kl} = -g_{kl}$.

The factor $A$ will include products of $\gamma$ matrices. From the relations $\gamma_\mu\gamma_\nu + \gamma_\nu\gamma_\mu = 2g_{\mu\nu}$ follow reduction formulae such as

$$\gamma_\mu\gamma^\mu = -4\ ;\ \gamma_\mu\gamma_\nu\gamma^\mu = -2\gamma_\nu\ ;\ \gamma_\lambda\gamma_\mu\gamma_\nu\gamma^\lambda = 4g_{\mu\nu}. \tag{6.284}$$

When $\overline{\mid M\mid^2}$ is expanded in factors and reduced, the final trace is a sum of the traces of the various terms. Since the $\gamma$ matrices have the same form in any reference frame, so must the trace of a product $\gamma_\mu\gamma_\nu...\gamma_\lambda$ so it is expressible in terms of products of "numerical tensors" $g_{\mu\nu}$, $g_{\mu\lambda}$,...$g_{\nu\lambda}$. Since these must have even rank, the trace of an odd number of $\gamma$ matrices vanishes. The trace is also unchanged under cyclic permutations of the factors. From this fact and the anti-commutation relations, one finds for example

$$\mathrm{tr}\left[\gamma_\mu\gamma_\nu\right] = 4g_{\mu\nu}\ ;\ \mathrm{tr}\left[\gamma_\mu\gamma_\nu\gamma_\alpha\gamma_\beta\right] = 16\left(g_{\mu\nu}g_{\alpha\beta} - g_{\mu\alpha}g_{\nu\beta} + g_{\mu\beta}g_{\nu\alpha}\right). \tag{6.285}$$

By way of illustration, consider $e\mu$ elastic scattering. The unpolarized cross section is

$$\frac{d\sigma}{dt} = \frac{\pi\alpha^2}{\left(p_e p_\mu\right)^2 - \left(m_e m_\mu\right)^2} f \tag{6.286}$$

where

$$f = \frac{1}{16t^2}\mathrm{tr}\left[\left(\gamma\hat{p}'_\mu + m_\mu\right)\gamma^\lambda\left(\gamma p_\mu + m_\mu\right)\gamma^\nu\right]\mathrm{tr}\left[\left(\gamma p'_e + m_e\right)\gamma_\lambda\left(\gamma p_e + m_e\right)\gamma_\nu\right] \tag{6.287}$$

From the reduction and trace formulae, one can establish that

$$\frac{1}{4}\mathrm{tr}\left[\left(\gamma p'_\mu + m\right)\gamma_\mu\left(\gamma p_\mu + m\right)\gamma_\nu\right] = g_{\mu\nu}\left(m^2 - p'p\right) + p'_\mu p_\nu + p'_\nu p_\mu \tag{6.288}$$

and then calculate

$$\mathrm{tr}\left[\left(\gamma p'_\mu + m\right)\gamma_\mu\left(\gamma p_\mu + m\right)\gamma_\mu\right) = 16m^2 - 8p'p. \tag{6.289}$$

One then arrives at the expressions

$$\begin{aligned} f &= \frac{2}{t^2}\left[\left(p_e p_\mu\right)^2 + \left(p_e p'_\mu\right)^2 + \frac{1}{2}\left(m_e^2 + m_\mu^2\right)\left(p_e - p'_e\right)^2\right] \\ &= \frac{1}{t^2}\left[\frac{1}{2}\left(s^2 + u^2\right) + \left(m_e^2 + m_\mu^2\right)\left(2t - m_e^2 - m_\mu^2\right)\right] \end{aligned} \tag{6.290}$$

where the invariants are $s = (p_e + p_\mu)^2$, $t = (p'_e - p_e)^2$, and $u = (p_e - p_\mu)^2$. In a frame in which the electron is at rest, the cross section can be cast in the form

$$\frac{d\sigma}{dT} = 2\pi\frac{\alpha^2}{m_e^2}\frac{1}{v_\mu^2 T^2}\left[1 - v_\mu^2\frac{T}{T^{max}} + \frac{1}{2}\left(\frac{m_e T}{E_\mu}\right)^2\right] \tag{6.291}$$

where $T = (E'_e - m_e)/m_e$ is the energy transfer to the electron in units of electron mass, and the maximum value is $T^{max} = 2\mathbf{p}_\mu^2/(m_e^2 + m_\mu^2 + 2m_e E_\mu)$. This recoil cross section describes the rate of production of hard knock-on electrons also called $\delta$-rays produced by a fast muon passing through matter. The soft portion of the spectrum is sensitive to atomic structure.

## 6.17 ■ FURTHER READING

The 1933 Nobel Prize in Physics was awarded to Erwin Schrödinger and Paul Dirac "for the discovery of new productive forms of atomic theory." Dirac introduced his equation in the seminal paper "The quantum theory of the electron," P. Dirac, Proceedings of the Royal Society of London, Series A, 6124 (1928).

The 1965 Nobel Prize in physics was awarded to Sin-Itiro Tomonaga, Julian Schwinger, Richard P. Feynman "for their fundamental work in quantum electrodynamics, with deep-ploughing consequences for the physics of elementary particles."

Cross section measurements for $e^+e^- \to \mu^+\mu^-, \tau^+\tau^-$ at high energy are found for example in Jade Collaboration, Phys. Lett. **B161**, 188 (1985) and Cello Collaboration, Phys. Lett. **B191**, 209 (1987). The angular distribution at $\sqrt{s} = 43$ GeV clearly shows the asymmetry due to the presence of the $Z$ boson.

A wealth of QED results including the Breit-Fermi Hamiltonian are derived in V. B. Berestetskii, E. M. Lifshitz and L. P. Pitaevskii, *Relativistic Quantum Theory*, Pergamon (1971). Complications due to anomalous moments and other form factors are described in H. M. Pilkuhn, *Relativistic Particle Physics*, Springer-Verlag (1979).

More detail on the 2-body problem in quantum mechanics with relativistic corrections may be found in Willis Lamb, Jr., Phys. Rev. **85**, 259 (1952).

The polarization dependence of $2\gamma$ annihilation is described in L. A. Page, Phys. Rev. **106**, 394 (1957) and W. H. McMaster, Rev. of Mod. Phys. **33**, 8 (1961). The theory of the correlation between the planes of $e^+e^-$ pairs in double Dalitz decay is given in N. Kroll and W. Wada, Phys. Rev. **98**, 1355 (1955).

An experimental result for the 2-photon decay lifetime of singlet state of positronium is D. W. Gidley et. al., Phys. Rev. Lett. **49**, 525 (1982). For 3-photon decay results, see G.S. Adkins, R.N. Fell, and J. Sapirstein, *Phys. Rev. Lett.* **84** (2000) and R.S. Vallery, P.W. Zitzewitz, and D.W. Gidley, *Phys. Rev. Lett.* **90**, 203402 (2003).

## 6.18 ■ PROBLEMS

---

**Problem 6.1. Pauli equation**

---

Use the relations $\sigma_i\sigma_j = \delta_{ij} + i\epsilon_{ijk}\sigma_k$ to show that

$$\frac{1}{2m}[(-i\nabla - e\mathbf{A})\cdot\sigma]^2\psi = \frac{1}{2m}(-i\nabla - e\mathbf{A})^2\psi + \frac{e}{2m}\mathbf{B}\cdot\sigma\psi.$$

Be careful with the gradient operator so as to not obtain an extra term proportional to $\boldsymbol{\sigma} \cdot \mathbf{A} \times \nabla$.

### Problem 6.2. Spinor rotations

Show that 2-component eigenstates $\phi_s(\mathbf{m})$ of spin along some direction $\mathbf{m}$ with eigenvalues $s = \pm 1$ (twice the spin) satisfying

$$\mathbf{m} \cdot \boldsymbol{\sigma} \phi_s(\mathbf{m}) = s \phi_s(\mathbf{m})$$

are rotated versions of the eigenstates of $\sigma_z$

$$\phi_+(\mathbf{m}) = e^{-i\frac{\phi}{2}\sigma_z} e^{-i\frac{\theta}{2}\sigma_y} \begin{pmatrix} 1 \\ 0 \end{pmatrix} = \begin{pmatrix} e^{-i\phi/2} \cos\frac{\theta}{2} \\ e^{+i\phi/2} \sin\frac{\theta}{2} \end{pmatrix}$$

$$\phi_-(\mathbf{m}) = e^{-i\frac{\phi}{2}\sigma_z} e^{-i\frac{\theta}{2}\sigma_y} \begin{pmatrix} 0 \\ 1 \end{pmatrix} = \begin{pmatrix} -e^{-i\phi/2} \sin\frac{\theta}{2} \\ e^{+i\phi/2} \cos\frac{\theta}{2} \end{pmatrix}.$$

### Problem 6.3. Spinor boosts

Using $E = m\cosh\eta$ and $p = \sqrt{E^2 - m^2}$, prove the following identities

$$\cosh\frac{\eta}{2} = \left[\frac{1}{2}(\cosh\eta + 1)\right]^{\frac{1}{2}} = \sqrt{\frac{E+m}{2m}} = \frac{E+m}{\sqrt{2m(E+m)}},$$

$$\sinh\frac{\eta}{2} = \left[\frac{1}{2}(\cosh\eta - 1)\right]^{\frac{1}{2}} = \sqrt{\frac{E-m}{2m}} = \frac{p}{\sqrt{2m(E+m)}}$$

and then show that the spinor boost transformations between a frame in which a particle of mass $m$ is at rest and a frame in which it has energy $E$ can be written as

$$e^{\pm\frac{\eta}{2}\mathbf{n}\cdot\boldsymbol{\sigma}} = \frac{E + m \pm \mathbf{p}\cdot\boldsymbol{\sigma}}{[2m(E+m)]^{\frac{1}{2}}}.$$

These may be used to provide an alternate representation of the upper and lower components of a 4-spinor in terms of a rest frame 2-component spinor.

### Problem 6.4. Spinor polarization tensor

Polarization tensors for fermions are defined by the outer products of four spinors

$$\rho(p) \equiv \sum_{s=1,2} u^s \bar{u^s}, \quad \bar{\rho}(p) = \sum_{s=1,2} v^s \bar{v^s}.$$

In the spinor representation of the $\gamma$ matrices, we can express the two 2-component parts of a 4-component spinor plane wave amplitude in terms of a rotated (factor $U$) then boosted (factor $\sqrt{p_\pm}$) rest frame 2-component spinor $\phi_s$ as $u^\pm = \sqrt{p_\pm}U\phi_s$ and $v_s^\pm = \pm\sqrt{p_\pm}U\phi_{-s}$. Use $\sqrt{p_\pm p_\mp} = m$ and $\sum_s \phi_s\phi_s^\dagger = 1$ to show that

$$\rho = \sum_s \begin{pmatrix} \sqrt{p_+}U\phi_s \\ \sqrt{p_-}U\phi_s \end{pmatrix} (\phi_s^\dagger U^\dagger \sqrt{p_-}, \phi_s^\dagger U^\dagger \sqrt{p_+}) = \begin{pmatrix} m & p_- \\ p_+ & m \end{pmatrix} = \gamma p + m$$

and similarly that $\bar{\rho} = \gamma p - m$ and then show that the numerator in the fermion propagator is

$$\gamma p + m = \frac{1}{2}[\rho(p) - \bar{\rho}(-p)].$$

The antifermion contribution $\bar{\rho}$ to the sum over intermediate states appears with the minus sign as might have been anticipated from the Feynman rules and the factor of (1/2) reflects the normalization of the amplitudes to $\sqrt{2m}$ in the rest frame.

---

**Problem 6.5. Spinor Electrodynamics Lagrangian**

---

a) Derive the equations of spinor electrodynamics from

$$L = i\bar{\psi}\gamma_\mu D^\mu \psi - m\bar{\psi}\psi - \frac{1}{4}F_{\mu\nu}F^{\mu\nu}$$

with $D^\mu = \partial^\mu + ieA^\mu$ by treating $\psi$ and $\bar{\psi}$ as independent fields and b) show the equations are covariant (retain their form) under the gauge transformation $A^\mu \to A^\mu + \partial^\mu \chi$ and $\psi \to U\psi = e^{-ie\chi}\psi$ while the Lagrangian is invariant.

---

**Problem 6.6. Current conservation in spinor electrodynamics**

---

Verify that solutions to the Dirac equation $\gamma_\mu(i\partial^\mu - eA^\mu)\psi = m\psi$ (describing a fermion field interacting with an electromagnetic field) satisfy the current conservation condition $\partial_\mu j^\mu = 0$ where $j^\mu = \bar{\psi}\gamma^\mu\psi$. Start by operating on $j^\mu$ with $\partial_\mu$ and then use the Dirac equation and its Dirac conjugate to replace the derivatives of the spinor field.

---

**Problem 6.7. Relativistic corrections to atomic energy levels**

---

The second approximation corrections to the nonrelativistic Hamiltonian derived from the Dirac equation are

$$V = -\frac{\mathbf{p}^4}{8m^3} - \frac{e}{4m^2}\boldsymbol{\sigma}\cdot\mathbf{E}\times\mathbf{p} - \frac{e}{8m^2}\nabla\cdot\mathbf{E}.$$

One can estimate the shifts in the energy levels of hydrogen using the nonrelativistic wave functions and the perturbation theory formula

$$\Delta E = \int d\mathbf{x}\, \psi^{\dagger}_{nlm}(\mathbf{x}) V \psi_{nlm}(\mathbf{x}).$$

Use $< nlm|r^m|nlm >\sim (n^2 a_0)^m$, $< nlm|p^m|nlm >\sim (l/ < r >)^m \simeq (na_0)^{-m}$ for $l \simeq n$ and $|\psi_{n00}(0)|^2 \sim (\pi(na_0)^3)^{-1/2}$ to show that each of these corrections is proportional to $\alpha^2 E_0$ where $E_0 = m_e\alpha^2/2$ but that the second one (spin-orbit correction) vanishes for $s$-wave states ($l$=0) while the third vanishes for all but $s$-wave states.

---

## Problem 6.8. Majorana fermion

A Majorana particle is a fermion that is identical to its antiparticle. a) Show that the definition of charge conjugate 2-component spinors imply the 2-component Majorana particle spinors satisfy

$$\phi_+ = +i\sigma_y\phi^*_- = +g\phi^*_- \ ; \ \phi_- = -i\sigma_y\phi^*_+ = -g\phi^*_+$$

where $g = i\sigma_y$ and the asterix denotes the complex conjugate. b) Show the equation of motion is

$$[E - \mathbf{p}\cdot\boldsymbol{\sigma}]\phi_+ = +mg\phi^*_+.$$

c) Show that the mass term in the Lagrangian for a Majorana fermion is

$$L_{\text{Majorana}} = -\frac{m}{2}\bar{\psi}\psi = +\frac{m}{2}\left[\phi^{\dagger}_+ g\phi^*_+ - \phi^t_+ g\phi_+\right] \tag{6.292}$$

where the superscript $t$ denotes the transpose. The first term destroys an antiparticle and creates an antiparticle while the second term destroys a particle and creates an antiparticle and does not preserve the number of Majorana particles minus antiparticles. ["Introduction to Majorana Masses," Philip D. Mannheim, *International Journal of Theoretical Physics*, Vol. 23, No. 7, 643-674 (1984)]

---

## Problem 6.9. Mott scattering

Consider elastic scattering of an electron of momentum $\mathbf{p}$ by a fixed Coulomb potential associated with a nucleus of charge $Ze$. Starting with

$$S = -i\int d^4x\, ej_{fi}A,$$

derive the spin dependent cross section in the frame of the nucleus

$$\frac{d\sigma}{d\Omega} = \frac{\alpha^2 Z^2}{4v^2\mathbf{p}^2}\frac{|\bar{u}_f\gamma_0 u_i|^2}{\sin^4\frac{\theta}{2}}.$$

Next calculate explicitly the transition current time component for transitions between helicity states and show that, averaged over the initial spin state and summed over the final spin state, the cross section is

$$\frac{d\sigma}{d\Omega} = Z^2\alpha^2 \frac{\cos^2\frac{\theta}{2} + m_e^2/\mathbf{p}^2}{4\mathbf{p}^2 \sin^4\frac{\theta}{2}}.$$

Show that, in the nonrelativistic limit, this differential cross section reproduces the Rutherford formula. For a relativistic fermion, the cross section has the relativistic Rutherford form with an additional factor $\cos^2(\theta/2)$ due essentially to helicity conservation and the projection of the final spin state onto the initial spin state. Notice this factor appears without a magnetic interaction and without a recoil correction.

---

**Problem 6.10. Selectron-electron scattering**

---

An electron scatters from a charged scalar particle by single photon exchange: $e_1\tilde{e}_2 \to e_3\tilde{e}_4$. a) Draw the Feynman diagram and, by combining the Feynman rules of scalar and spinor electrodynamics, justify the amplitude

$$iM \sim \bar{u}(3)\gamma_\mu(-ie)u(1)\left(-i\frac{g_{\mu\nu}}{q^2}\right)(-ie)(p_2 + p_4)$$

where $q^2 = (p_3 - p_1)^2 = (p_2 - p_4)^2$. The momenta are $p_1, p_2, p_3, p_4$ and the initial and final electron spinor amplitudes suppressing helicity labels are $u(1)$ and $u(3)$. b) Leaving the expression for the spinor transition current as it is, compare the matrix element in the rest frame of the pair of momenta $p_2$ and $p_4$ to the matrix element for scattering of an electron from a fixed scalar particle. c) Use the expression for the transition current in the center of mass of $p_1$ and $p_2$ for equal helicities

$$\bar{u}'_t\gamma^\mu u_t = 2E(\phi^\dagger(\mathbf{n}')\phi(\mathbf{n}), t\beta\phi^\dagger(\mathbf{n}')\sigma\phi(\mathbf{n})$$

following the example of $e - \mu$ elastic scattering to show that, in the center of mass, the amplitude for scattering without helicity flip is

$$M = -\frac{e^2}{\sin^2\frac{\theta}{2}}\left[\left(\beta_1^{-1}\beta_2^{-1} + \frac{3}{2}\right)\cos\frac{\theta}{2} - \frac{1}{2}\cos\frac{3}{2}\theta\right].$$

---

**Problem 6.11. Fermion pair to scalar pair**

---

Electron-positron annihilation into a virtual photon could produce a pair of charged Higgs bosons or supersymmetric scalar electrons. Calculate the cross section for $e^+e^- \to \tilde{e}^+\tilde{e}^-$ as follows. Start with an electron-positron annihilation current $j_e$ in the center of mass as in the example of $e^+e^- \to \mu^+\mu^-$ and neglect $m_e$. Combine this current with a charged scalar current $j_{\tilde{e}} = -ie(p_- - p_+) = -ie(0, 2\mathbf{p})$. The matrix element from combined spinor and scalar Feynman rules is $M \propto j_f j_h/s$. Average the cross section $d\sigma/d\cos\theta = |M|^2/(32\pi s)$ to over initial spins to derive the center of mass differential cross section

$$\frac{d\sigma_{e^+e^-\to\tilde{e}^+\tilde{e}^-}}{d\cos\theta} = \pi\frac{\alpha^2}{4s}\left(1 - \frac{4m_{\tilde{e}}^2}{s}\right)\sin^2\theta.$$

Compare the threshold behavior and angular distribution to that for fermion pair production such as $e^+e^- \to \mu^+\mu^-$.

---

**Problem 6.12. Electron-photon scattering**

---

Compton scattering of photons by electrons $e_1^-\gamma_2 \to e_3^-\gamma_4$ is described by the two second order diagrams in Figure 6.22. Note that the 4-particle vertex that contributes to $\tilde{e}\gamma \to \tilde{e}\gamma$ does not appear. a) Use the spinor electrodynamics Feynman rules to justify the expression for the total amplitude

$$M_{e\gamma\to e\gamma} = -e^2\epsilon_\mu^\dagger(4)\bar{u}(3)\gamma^\mu\frac{(\hat{p}_1 + \hat{p}_2 + m)}{s - m^2}\gamma^\nu u(1)\epsilon_\nu(2)$$
$$- e^2\epsilon_\nu^\dagger(4)\bar{u}(3)\gamma^\mu\frac{(\hat{p}_1 - \hat{p}_4 + m)}{u - m^2}\gamma^\nu u(1)\epsilon_\mu(2)$$

where $\hat{p} = p^\mu\gamma_\mu$, $m$ is the electron mass, and

$$s = (p_1 + p_2)^2,\ t = (p_1 - p_3)^2,\ u = (p_1 - p_4)^2.$$

b) Trace technology may be used to find the invariant cross section

$$\frac{d\sigma_{e\gamma\to e\gamma}}{dt} = \frac{8\pi r_e^2}{m^2 s'^2}\left[\left(\frac{1}{s'} + \frac{1}{u'}\right)^2 + \left(\frac{1}{s'} + \frac{1}{u'}\right) - \frac{1}{4}\frac{s'^2 + u'^2}{s'u'}\right]$$

where $s' = s/m^2 - 1$ and $u' = u/m^2 - 1$. Show that in the rest frame of the initial electron,

$$s' = 2\omega/m,\ ,\ u' = -2\omega'/m,\ t = -2\omega\omega'(1 - \cos\theta),\ \omega'^{-1} - \omega^{-1} = m^{-1}(1 - \cos\theta)$$

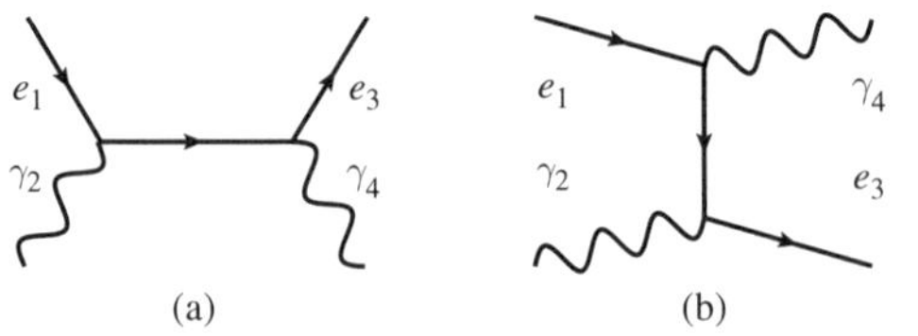


**FIGURE 6.22** Diagrams describing Compton scattering $e^-\gamma \to e^-\gamma$.

where $\omega$ and $\omega'$ are the initial and final photon energies and $\theta$ is the photon scattering angle, that

$$\frac{d\sigma_{e\gamma\to e\gamma}}{d\Omega} = \frac{1}{2}r_e^2\left(\frac{\omega'}{\omega}\right)^2\left(\frac{\omega'}{\omega} + \frac{\omega}{\omega'} - \sin^2\theta\right)$$

and that, for $\omega/m << 1$, the cross section approaches

$$\frac{d\sigma_{e\gamma\to e\gamma}}{d\Omega} \simeq \frac{1}{2}r_e^2(1+\cos^2\theta)$$

while, for $\omega/m << 1$, the cross section has a narrow peak in the *forward* direction.
c) Show that in the center of mass

$$s' = \frac{2}{m^2}\omega(\omega + E),\; u' = -\frac{2}{m^2}\omega(E + \omega\cos\theta),\; t = -2\omega^2(1-\cos\theta)$$

where $E = \sqrt{\omega^2 + m^2}$ and that the cross section has a peak in the *backward* direction.

---

**Problem 6.13. Fermion annihilation to photons**

---

The spin averaged differential cross section in the center of mass for $e_2^+e_1^- \to \gamma_3\gamma_4$ at high energy is

$$\frac{d\sigma_{e^+e^-\to\gamma\gamma}}{d\cos\theta} = \frac{2\pi\alpha^2}{s}\frac{1+\cos^2\theta}{\sin^2\theta}$$

where $\theta$ is the angle of a photon relative to the electron direction.
a) Start to derive this expression by using the Feynman rules to derive the matrix element with $m_e = 0$ in the form

$$M_{\alpha\beta\to ij} = e^2\left[\bar{v}\hat{e}_4\frac{\hat{p}_{31}}{p_{31}^2}\hat{e}_3 u + 3 \Leftrightarrow 4\right]$$

where $u$ and $v$ are 4-component spinors for the electron and positron, $e_3$ and $e_4$ are polarization 4-vectors for the photons, and $\hat{a} = a^0\gamma^0 - \mathbf{a}\cdot\gamma$. Let $\alpha$, $\beta$, $i$, and $j$ denote the spin states for the $e^-$, $e^+$ and photons.
b) Evaluate in the center of mass frame and take the $z$-axis along $\mathbf{p}$ of the $e^-$ and the $y$-axis orthogonal to the scattering plane so $\mathbf{p}_3 = (\sin\theta, 0, \cos\theta)$. Write the 4-component spinors in terms of two 2-component spinors as

$$u_+ = \sqrt{2E}\begin{pmatrix}\phi_+(\mathbf{z})\\ 0\end{pmatrix} \quad u_- = \sqrt{2E}\begin{pmatrix}0\\ \phi_-(\mathbf{z})\end{pmatrix}$$

$$\bar{v}_+ = \sqrt{2E}(\phi_+^\dagger(\mathbf{z}), 0) \quad \bar{v}_- = \sqrt{2E}(0, \phi_-^\dagger(\mathbf{z}))$$

and the gamma matrices in terms of Pauli matrices and use photon polarization vectors $e = (0, \mathbf{e})$ with $\mathbf{e}(1) = \hat{\mathbf{y}}$ and $\mathbf{e}(2) = (\cos\theta, 0, -\sin\theta)$. Multiply the $4 \times 4$ matrices considered as $2 \times 2$ matrices of Pauli matrices and show that the amplitude vanishes unless the electron and positron helicities are opposite so their spins parallel, and for these initial states find the following reduced formula expressed in terms of two by two matrices

$$M_{-\alpha\alpha ij} = \sqrt{s}e^2 \left[ \phi^{\dagger}_{-\alpha} \mathbf{e}_4 \cdot \boldsymbol{\sigma} \frac{\mathbf{p}_{31} \cdot \boldsymbol{\sigma}}{\mathbf{p}_{31}^2} \mathbf{e}_3 \cdot \boldsymbol{\sigma} \phi_\alpha + 3 \Leftrightarrow 4 \right].$$

This form of the amplitude is similar to the original. In the high energy limit, the two helicity states decouple and this amplitude could be derived from Feynman rules for 2-component spinors with $\gamma^\mu \to \sigma^\mu$ and $u \to u_\pm$.
c) Now use $\mathbf{a} \cdot \boldsymbol{\sigma} \mathbf{b} \cdot \boldsymbol{\sigma} = \mathbf{a} \cdot \mathbf{b} + i\boldsymbol{\sigma} \cdot \mathbf{a} \times \mathbf{b}$ to reduce the product of Pauli matrices and the properties of the two spinors to find (up to a phase)

$$M_{-\alpha\alpha ij} = \frac{\sqrt{s}e^2}{\mathbf{p}_{31}^2} (-\mathbf{p}_1 \cdot \mathbf{e}_3 \mathbf{e}_4 - \mathbf{e}_3 \cdot \mathbf{e}_4 \mathbf{p}_{31} + \mathbf{e}_4 \cdot \mathbf{p}_{31} \mathbf{e}_3) \cdot [\mathbf{e}_x + i\alpha \mathbf{e}_y] + 3 \Leftrightarrow 4.$$

Use the spinors and polarization vectors to find the amplitudes

$$M_{-\alpha\alpha 11} = -M_{-\alpha\alpha 22} = -2e^2 \cot\theta, \; M_{-\alpha\alpha 12} = M_{-\alpha\alpha 21} = 2e^2 i\alpha \frac{1}{\sin\theta}.$$

Finally, sum over photon spins and average over fermion spins to derive the differential cross section. The squared amplitudes and differential cross section are proportional to $\sin^{-2}\theta$. The photons appear along the fermion directions.

---

**Problem 6.14. Orthopositronium lifetime**

---

The magnetic dipole transition rate from orthopositronium to parapositronium is

$$\Gamma_{^3S_1 \to ^1S_0 \gamma} = \frac{2}{3} k^3 \mu^2$$

where $\mu = \mu_e^- - \mu_{e^+} = e/m_e$ and $k$ is the photon energy. Given the energy difference

$$E_{^3S_1} - E_{^1S_0} = \frac{7}{12} \alpha^4 m_e,$$

compute the radiative lifetime in seconds and compare to lifetime for othopositronium to annihilate to three photons.

## Problem 6.15. True muonium $(\mu^+\mu^-)_{\text{atom}}$

The name muonium is given to $(\mu^\pm e^\mp)_{\text{atom}}$. True muonium, $(\mu^+\mu^-)_{\text{atom}}$, analogous to positronium has not been observed. The spin singlet states of true muonium should decay to two photons or another even number. The triplet states can only decay to an odd number of photons and, unavailable to positronium triplet states due to energy conservation, a single photon intermediate state leading to $e^+e^-$ is possible. The lifetime for $s$-states (l = 0) of principal quantum number n, with spin zero ($^1S_0$) and one ($^3S_1$), are calculated to be

$$\tau_{n^3S_1(\mu^+\mu^-)\to e^+e^-} = \frac{6\hbar n^3}{\alpha^5 mc^2}\ ;\ \tau_{n^1S_0(\mu^+\mu^-)\to\gamma\gamma} = \frac{2\hbar n^3}{\alpha^5 mc^2}.$$

The $3^3S_1$ is linked to the $1^1S_0$ via radiative transitions through $2P$ states with lifetimes

$$\tau_{2P\to 1S\gamma} = \left(\frac{3}{8}\right)^8 \frac{2\hbar}{\alpha^5 mc^2}\ ;\ \tau_{3S\to 2P\gamma} = \left(\frac{5}{2}\right)^9 \frac{4\hbar}{3\alpha^5 mc^2}.$$

a) Verify the lifetimes $\tau_{1^1S_0\to\gamma\gamma} = 0.602$ ps, $\tau_{2^3S_1\to e^+e^-} = 1.81$ ps, $\tau_{2^1S_0\to\gamma\gamma} = 4.81$ ps, $\tau_{2^3S_1\to e^+e^-} = 14.5$ ps, $\tau_{3^1S_0\to\gamma\gamma} = 16.3$ ps, $\tau_{3^3S_1\to e^+e^-} = 48.8$ ps, $\tau_{3^3S_1\to 2P+\gamma} = 1.53$ ns, and $\tau_{2P\to 1^1S_0+\gamma} = 15.4$ ps.

b) The lifetimes of these true muonium states are short compared to the lifetime for free muon decay $\tau_\mu = 2.2\times 10^{-7}$ s. Show that for tau-onium, $(\tau^+\tau^-)_{\text{atom}}$, the lifetimes should be

$$\tau_{^1S_0(\tau^+\tau^-)\to\gamma\gamma} = 35.8\ \text{fs}$$

and

$$\tau_{^1S_1(\tau^+\tau^-)\to e^+e^-,\mu^+\mu^-,q\bar{q}} = 26.9\ \text{fs}$$

when account is taken of the quark charges in scaling the decay rate. The weak decay lifetime of the $\tau^\pm$ pair is $\tau_\tau/2$ where $\tau_\tau = 291$ fs and is proportional to $m_\tau^{-5}$. For what $\tau$ mass would the weak and electromagnetic decay rates be comparable? [Ref. Stanley J. Brodsky and Richard F. Lebed, "Production of the Smallest QED Atom: True Muonium $(\mu^+\mu^-)$," Phys. Rev. Lett. **102**, 213401 (2009)]

# CHAPTER 7

# Gauge Theory and Chromodynamics

Quantum chromodynamics (QCD) describes strong interactions of quarks with eight massless vector gluon fields. We develop the fundamental equations from a gauge principle and derive the Feynman rules. Many results in quantum electrodynamics carry over to QCD appropriately modified by color related factors.

## Contents



## 7.1 ■ GAUGE SYMMETRY AND GAUGE FIELDS

The theory of quantum electrodynamics derives from the requirement that the wave equations be invariant under the gauge symmetry transformations

$$q \to U(x)q, \; A^\mu \to A^\mu - \partial^\mu a(x) \tag{7.1}$$

where the complex field $q = (q_1 + iq_2)/\sqrt{2}$ describes a particle of charge $e$, $A$ is a neutral vector field, and $a$ an arbitrary function of position and

$$U(x) = e^{iea(x)}. \tag{7.2}$$

The unitary transformation $U(x)$ rotates $q_1(x)$ and $q_2(x)$ by angle $ea$. Symmetry under global transformations (constant $a$) expresses the physical equivalence of

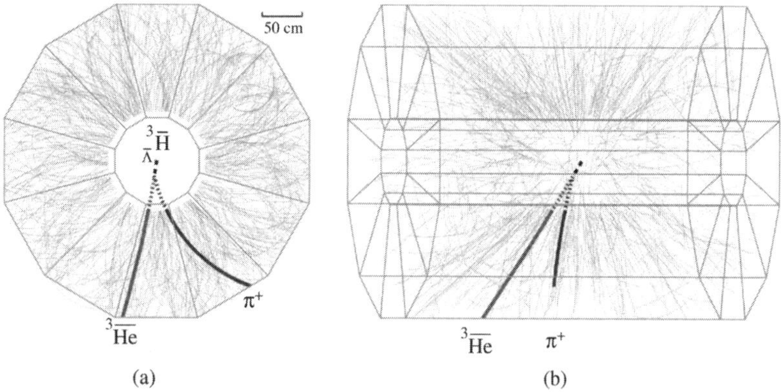


**FIGURE 7.1** Anti-hypertritium in heavy ion collisions. In high energy heavy ion collisions, chromodynamics converts energy into splashes of hadrons. This event display shows a candidate anti-hypertritium nucleus produced in an Au-Au collision with nucleon-nucleon center of mass energy 200 GeV at the Brookhaven National Laboratory Relativistic Heavy Ion Collider (RHIC). The $^{2}_{\bar{\Lambda}}\bar{H}$ nucleus is comprised of $\bar{p}$, $\bar{n}$, and $\bar{\Lambda}$ baryons. The decay $\bar{\Lambda} \to \bar{p}\pi^{+}$ leads to an anti-helium $^{3}_{2}\overline{He}$ and a charged pion. [From The STAR Collaboration, *Science Express*, 4 March 2010, DOI 10.1126/science.1183980]

$q_1$ and $q_2$. Gauge theory follows by demanding this symmetry be local (position dependent). The field equations are found simply by making the replacement

$$\partial^{\mu} \to D^{\mu} = \partial^{\mu} + ieA^{\mu} \tag{7.3}$$

in the free field kinetic energy Lagrangian for $q$ and adding a kinetic term $-F_{\mu\nu}F^{\mu\nu}/4$ for $A$.

This idea may be generalized. Let $q$ stand for a column vector of $N$ physically equivalent fields to be "gauged." Demand symmetry under a matrix transformation

$$q \to U(x)q \tag{7.4}$$

where $U$ is a unitary matrix which, by removing a phase factor $e^{i\theta(x)}$, we can suppose has determinant $|U| = 1$. Such a special unitary matrix is an element of a matrix group called $SU(N)$ and can be written in the form

$$U = e^{i\mathbf{a}(x)\cdot\mathbf{T}} \tag{7.5}$$

where $\mathbf{a}(x)$ is a vector of $N^2 - 1$ generalized rotation angles and $\mathbf{T}$ is a vector of Hermitian generator matrices of dimension $N \times N$. This expression introduces boldface notation for vectors of length $N^2 - 1$ of group elements. The inner product of such vectors is defined as

$$\mathbf{A} \cdot \mathbf{B} = \sum_{a=1}^{N^2-1} A_a B_a \equiv A_a B_a. \tag{7.6}$$

An index $a$ may appear upstairs or downstairs and we will use the convention that repeated indices are summed. The reuse of the boldface notation for vectors of group related quantities can be a little disorienting. Remember that we will be dealing principally with Lorentz covariant 4-vectors with their components labelled by a Lorentz index when needed and rarely with their 3-space (normal) vector components which are also normally expressed through boldface notation.

For $SU(2)$, the generators are the Pauli matrices

$$\mathbf{T} = \frac{\boldsymbol{\sigma}}{2}. \tag{7.7}$$

For $SU(3)$, the eight generators are denoted by

$$\mathbf{T} = \frac{\boldsymbol{\lambda}}{2} \tag{7.8}$$

where $\boldsymbol{\lambda}$ denotes eight so-called Gell-Mann matrices. The generators are conventionally normalized and satisfy the relations

$$T_a = T_a^\dagger, \quad \mathrm{tr}(T_a) = 0, \quad \mathrm{tr}(T_a T_b) = \frac{1}{2}\delta_{ab}, \quad \mathbf{T}^2 = \frac{N^2-1}{2N}. \tag{7.9}$$

The generators also satisfy commutation relations that we can write as

$$T_a T_b - T_b T_a = i f_{abc} T_c \tag{7.10}$$

where the structure constant tensor $f_{abc}$ can be expressed as

$$f_{abc} = -i\,\mathrm{tr}([T_a, T_b]_- T_c)/\mathrm{tr}(T_c^2) \tag{7.11}$$

and is completely antisymmetric: $f_{abc} = -f_{bac} = -f_{cba} = -f_{acb}$. It is convenient to define the cross product of vectors as

$$[\mathbf{A} \times \mathbf{B}]_a = f_{abc} A_b B_c. \tag{7.12}$$

For $SU(2)$, $f_{abc}$ is the antisymmetric tensor $\epsilon_{abc}$ and the cross product is the familiar outer product encountered in vector analysis in three dimensional Cartesian space.

We will write the generalized gauge covariant derivative as

$$D^\mu = \partial^\mu + M^\mu \tag{7.13}$$

where $\partial^\mu$ should be understood to be multiplying a unit matrix not expressed, and $M$ is matrix of fields. We assume $M$ to be anti-Hermitian as in electrodynamics and write it as

$$M^\mu = ig\mathbf{T} \cdot \mathbf{A}^\mu \tag{7.14}$$

where $\mathbf{A}^\mu$ is a vector of real fields called gauge fields. We can think of $igA_a^\mu$ as the projection of $M^\mu$ on to $T_a$. The projection on the $a$-direction is obtained by using the trace as an inner product for Hermitian matrices. In fact, using Equation 7.9, we have

$$\text{tr } T_a M^\mu = \text{tr } T_a i g A_b^\mu T_b = i g A_b^\mu \text{tr } T_a T_b = i g A_b^\mu 2 \frac{1}{2} \delta_{ab} = \frac{1}{2} i g A_a^\mu. \tag{7.15}$$

As an example, the gauge covariant equations for a collection $q$ of fermions are

$$[i D_\mu \gamma^\mu - m] q = 0 \tag{7.16}$$

where the mass is common to all members of $q$. Like other linear wave equations, these are covariant provided

$$D^\mu \phi \to U D^\mu q = U D^\mu (U^{-1} U) q = (U D^\mu U^{-1})(U q) \tag{7.17}$$

where we inserted a factor $U^{-1}U = 1$. (The operator $D^\mu$ contains the derivative operator $\partial^\mu$ but since $\partial^\mu (U U^{-1}) = 0$, this insertion is justified.) We therefore require

$$D^\mu \to U D^\mu U^{-1}. \tag{7.18}$$

To satisfy this condition, a transformation of the gauge fields $\mathbf{A}$ must accompany the unitary transformation of the gauged fields $q$ and it may be expressed in matrix form as

$$M^\mu \to M'^\mu = U M^\mu U^{-1} + U(\partial^\mu U^{-1}). \tag{7.19}$$

Let's write this rule in terms of the gauge fields explicitly as

$$\mathbf{T} \cdot \mathbf{A}'_\mu = U \mathbf{T} \cdot \mathbf{A}_\mu U^{-1} + i g^{-1} (\partial_\mu U) U^{-1}. \tag{7.20}$$

When $U$ is an element of the group $U(1)$, this rule reproduces the gauge transformation familiar from the theory of electromagnetism. A gauge transformation of the field $A$ simply adds to it the 4-gradient of a scalar function. When $N > 1$ implying elements of the transformation group do not generally commute, the term $U \mathbf{T} \cdot \mathbf{A}_\mu U^{-1}$ implies a general gauge transformation additionally mixes the different gauge fields.

For an infinitesimal transformation, $U \simeq 1 + i g \mathbf{T} \cdot \mathbf{a}$, we can exhibit an explicit form for the transformed gauge fields. The commutation relations of the generators $\mathbf{T}$ allow us to simplify the right-hand side of Equation 7.20 and then project out the rule for the fields $\mathbf{A}$. We arrive at the explicit form

$$\mathbf{A}'_\mu = \mathbf{A}_\mu - \mathbf{a} \times \mathbf{A}_\mu - \partial_\mu \mathbf{a}. \tag{7.21}$$

One can not write a simple expression for the general finite gauge transformation. However, under a global ($\partial_\mu \mathbf{a} = 0$) transformation, the inhomogeneous second term in Equation 7.1 vanishes and repeated infinitesimal transformations leads to the form

$$\mathbf{A}' = \lim_{n \to \infty} \left(1 - \frac{\mathbf{a}}{n} \times \right)^n \mathbf{A} = \exp[i \mathbf{a} \cdot \mathbf{T}] \mathbf{A} \tag{7.22}$$

where $\mathbf{T}$ in the argument of the exponential represents the vector of $N^2 - 1 \times N^2 - 1$ matrices defined by

$$\mathbf{T}_{a,bc} = -i f_{abc}. \tag{7.23}$$

It is left as an exercise to show that these matrices actually satisfy the same algebra as the $N \times N$ generators and so we overload the symbol $\mathbf{T}$ to denote both the $N \times N$ and $N^2 - 1 \times N^2 - 1$ matrices. The larger generator matrices generate $N^2 - 1 \times N^2 - 1$ transformation matrices $U = e^{-\mathbf{a}\cdot\mathbf{T}}$ appropriate for vectors. These transformations are said to form a vector representation of $SU(N)$.

We now define the field strength tensor matrix, a generalization of the gauge covariant field strength tensor of electrodynamics, by

$$F^{\mu\nu} = D^\mu D^\nu - D^\nu D^\mu = \partial^\mu M^\nu - \partial^\nu M^\mu + M^\mu M^\nu - M^\nu M^\mu. \tag{7.24}$$

It follows from Equation 7.18 that $F^{\mu\nu}$ transforms in a simple way, namely as

$$F^{\mu\nu} \to U F^{\mu\nu} U^{-1}. \tag{7.25}$$

We can write $F^{\mu\nu}$ in the form

$$F^{\mu\nu} = ig\mathbf{T} \cdot \mathbf{F}^{\mu\nu} \tag{7.26}$$

and project out a vector of field strength tensors

$$\mathbf{F}^{\mu\nu} = \partial^\mu \mathbf{A}^\nu - \partial^\nu \mathbf{A}^\mu - g\mathbf{A}^\mu \times \mathbf{A}^\nu. \tag{7.27}$$

Corresponding electric and magnetic components can be defined by

$$\partial^\mu A^\nu - \partial^\nu A^\mu - g\mathbf{A}^\mu \times \mathbf{A}^\nu = \begin{pmatrix} 0 & -\mathbf{E}_x & -\mathbf{E}_y & -\mathbf{E}_z \\ \mathbf{E}_x & 0 & -\mathbf{B}_z & \mathbf{B}_y \\ \mathbf{E}_y & \mathbf{B}_z & 0 & -\mathbf{B}_x \\ \mathbf{E}_z & -\mathbf{B}_y & \mathbf{B}_x & 0 \end{pmatrix}. \tag{7.28}$$

Still following the example of electrodynamics, we define the free field Lagrangian for the (gauge) vector fields $\mathbf{A}$ as

$$L_A \equiv \frac{1}{2g^2}\text{tr}\left(F_{\mu\nu}F^{\mu\nu}\right) = -\frac{1}{2}\text{tr}\left(T_a T_b\right) F^a_{\mu\nu} F^{b,\mu\nu} = -\frac{1}{4}\mathbf{F}_{\mu\nu} \cdot \mathbf{F}^{\mu\nu} \equiv -\frac{1}{4}\mathbf{F}^2 \tag{7.29}$$

where we used $\text{tr}(T_a T_b) = \delta_{ab}/2$. Since the trace of a product of matrices is invariant under cyclic permutations, the left-hand side of Equation 7.1 is easily found to be invariant under the transformation $F^{\mu\nu} \to U F^{\mu\nu} U^{-1}$ and, therefore, so is the right-hand side. The sum of the squares of the field strength tensors is invariant. Note that since each field strength tensor includes a cross product of gauge fields, only the whole package can be gauge invariant, not the square of any one field strength tensor.

Since $D_\mu D_\nu$ transforms covariantly under local symmetry transformations, under *local* transformations

$$\mathbf{T}\cdot\mathbf{F}' = U\mathbf{T}\cdot\mathbf{F}U^{-1}. \tag{7.30}$$

An infinitesimal transformation $U$ may be used to show this transformation rule is equivalent to

$$\mathbf{F}' = U\mathbf{F} \tag{7.31}$$

where the special unitary matrix $U$ is generated by the $N^2 - 1 \times N^2 - 1$ generators. In contrast, such a transformation rule applies to $\mathbf{A}$ (see Equation 7.22) only for global transformations. It follows from Equation 7.31 that the gauge covariant derivative $D_\mu$ constructed with the $N^2 - 1 \times N^2 - 1$ generators and acting on $\mathbf{F}$ is covariant under local transformations:

$$D_\mu \mathbf{F}^{\nu\rho} \to U D_\mu \mathbf{F}^{\nu\rho}. \tag{7.32}$$

One might therefore guess that a gauge covariant generalization of Maxwell's equations $\partial_\mu F^{\mu\nu} = j^\nu$ would have the structure

$$D_\mu \mathbf{F}^{\nu\rho} = \mathbf{j}^\rho \tag{7.33}$$

with $\mathbf{j}^\rho$ a similarly covariant vector of currents and this proves to be the case.

## 7.2 ■ GAUGE THEORY WITH FERMIONS

A gauge invariant Lagrangian for fermion fields $q$ together with gauge fields $\mathbf{A}$ may now be constructed:

$$L = \bar{q}\left(i\gamma_\mu D^\mu - m\right)q - \frac{1}{4}\mathbf{F}^2 \tag{7.34}$$

Several general features of gauge theory should be pointed out here. There are $N^2 - 1$ gauge fields, three for $N = 2$, eight for $N = 3$, fifteen for $N = 4$, and so on. A mass term of the general form $L_{AA} = \mathbf{A}^\dagger S\mathbf{A}$ with $S$ a constant matrix is not gauge invariant and precluded by gauge symmetry. Gauge fields are naturally massless. Also, gauge fields are 4-vector fields like the electromagnetic field and have spin 1. A fermion mass term $L_{\bar{q}q} = \bar{q}mq$ with $m$ a constant matrix is gauge invariant only if $m$ is proportional to the unit matrix. More precisely, left-handed and right-handed spinor fields appear independently in the spinor kinetic energy and interaction terms but are linked by the fermion mass term so $L$ or $R$ fields may be gauged alone ($g_L = g$, $g_R = 0$ or $g_L = 0$, $g_R = g$) only if $m = 0$. Since the coupling constant $g$ appears in $\mathbf{F}$, both $L$ and $R$ must have the same coupling if both coupling constants are non-vanishing. More generally, gauge symmetry requires a universal coupling. The gauge field Lagrangian alone is invariant under the parity transformation

$$(A_a^0, \mathbf{A}_a) \rightarrow (A_a^0, -\mathbf{A}_a) \tag{7.35}$$

applied simultaneously to all fields. In the theory of the three equivalent colors of a given quark flavor based on $SU(3)$, the quark has mass, its left-handed and right-handed components have equal color coupling, and parity is conserved. In the case of $SU(2)_L$ flavor gauge theory we will encounter, two quark flavors such as the $u$-quark and the $d$-quark are regarded as equivalent and their left-handed and right-handed components have different couplings. The quarks are naturally massless in the bare gauge theory. Their masses will appear through the special mechanism called spontaneous symmetry breaking. Finally, we point out that the dimension of $D$ ($[D]$) and hence of $gA$ is that of mass. That $[D] = [gA] = [m]$ follows by inspection of the terms in the Lagrangian and similarly $[F^2] = [\partial A] = [gA^2]$. It follows that the gauge boson fields have dimension $[A] = [m] = [L^{-1}]$, the fermion fields have dimension $[q] = [m]^{3/2} = [L]^{-3/2}$ as in Schrödinger's equation, while the gauge coupling constant is dimensionless: $[g] = 1$. The appearance of dimensionless coupling constants is an important prediction of gauge theory.

Euler's equation for $q$ is a matrix version of the fundamental equation of spinor electrodynamics

$$[i\gamma_\mu D^\mu - m]q = 0. \tag{7.36}$$

With interaction terms on the right-hand side, it can be written as

$$(i\gamma_\mu \partial^\mu - m)q = g\mathbf{T} \cdot \mathbf{A}_\mu \gamma^\mu q. \tag{7.37}$$

The matrix $M = ig\mathbf{T} \cdot \mathbf{A}$ here appears also in the interaction Lagrangian

$$L_{int} = \bar{q}(ig\mathbf{T} \cdot \mathbf{A})q \tag{7.38}$$

and links the different kinds of fermions. Experience with deriving the Feynman rules for electrodynamics immediately suggests the interpretation of such terms. As illustrated in Figure 7.2, a fermion of one kind may radiate a gauge boson and emerge as a different kind of fermion. The coupling constant for a transition $q_i A^a \rightarrow q_j$ may be read from the interaction term in the matrix Dirac equation or from the interaction Lagrangian and is

$$g_{q_i A_a q_j} = gT_{ji}^a. \tag{7.39}$$

For $SU(2)$, we have

$$M = i\frac{g}{2}\boldsymbol{\sigma} \cdot \mathbf{A} = i\frac{g}{2}\begin{pmatrix} A_3 & A_1 - iA_2 \\ A_1 + iA_2 & -A_3 \end{pmatrix} \tag{7.40}$$

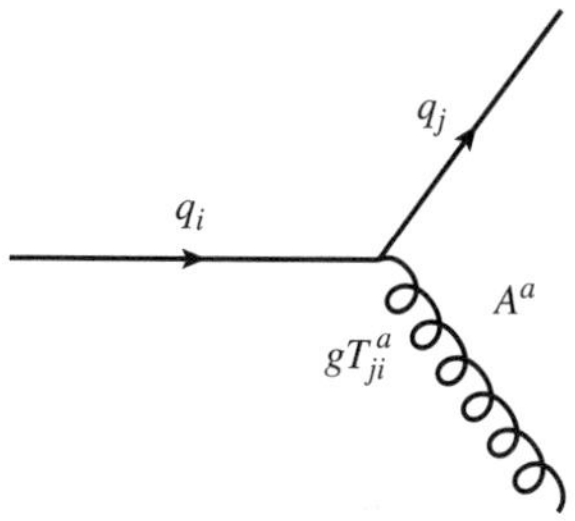


**FIGURE 7.2** Feynman diagram in a generic gauge theory. A gauged field of index $i$ is coupled to a gauged field of index $j$ through several gauge fields $A^a$. The vertex factor is proportional to a universal coupling constant $g$ and to the corresponding element of the generator matrix $T^a$.

For $SU(3)$, we have

$$M = i\frac{g}{2}\boldsymbol{\lambda} \cdot \mathbf{A} = i\frac{g}{2}\begin{pmatrix} A_3 + \frac{1}{\sqrt{3}}A_8 & A_1 - iA_2 & A_4 - iA_5 \\ A_1 + iA_2 & -A_3 + \frac{1}{\sqrt{3}}A_8 & A_6 - iA_7 \\ A_4 + iA_5 & A_6 + iA_7 & -\frac{2}{\sqrt{3}}A_8 \end{pmatrix}. \tag{7.41}$$

(The generator $T_i$ may be abstracted from these expressions by setting $A_j = \delta_{ij}$.) These matrices show precisely which fields or field superpositions are radiated by each possible fermion transition.

The Euler-Lagrange equations for the gauge fields are

$$-\partial_\mu \frac{\partial L}{\partial(\partial_\mu \mathbf{A}_\nu)} + \frac{\partial L}{\partial \mathbf{A}_\nu} = 0. \tag{7.42}$$

The gauge fields appear in the field strengths as well as in the interaction with the spinor fields and

$$\frac{\partial L}{\partial(\partial_\mu \mathbf{A}_\nu)} = -\mathbf{F}^{\mu\nu}\ ; \quad \frac{\partial L}{\partial \mathbf{A}_\nu} = -g\bar{q}\gamma^\nu \mathbf{T} q + g\mathbf{A}_\mu \times \mathbf{F}^{\mu\nu}. \tag{7.43}$$

Therefore we find

$$\partial_\mu \mathbf{F}^{\mu\nu} - g\bar{q}\gamma^\nu \mathbf{T} q + g\mathbf{A}_\mu \times \mathbf{F}^{\mu\nu} = 0. \tag{7.44}$$

We can define $N$ transition current densities associated with the fermions

$$\mathbf{j}_q^\nu \equiv -g\bar{q}\gamma^\nu \mathbf{T} q \tag{7.45}$$

and write the generalized Maxwell equations as

$$\partial_\mu \mathbf{F}^{\mu\nu} - g\mathbf{A}_\mu \times \mathbf{F}^{\mu\nu} = \mathbf{j}_q^\nu. \tag{7.46}$$

The gauge field equations may be written more succinctly and in a gauge covariant form

$$D_\mu \mathbf{F}^{\mu\nu} = \mathbf{j}_q^\nu \tag{7.47}$$

by defining the $(N^2 - 1) \times (N^2 - 1)$ covariant derivative appropriate for vectors

$$D_\nu = \partial_\nu - g\mathbf{A}_\nu \times . \tag{7.48}$$

This derivative matrix like the covariant derivative matrix in the spinor field equations can be written as $D^\mu = \partial^\mu + ig\mathbf{T} \cdot \mathbf{A}^\mu$ where $\mathbf{T}$ are the $N^2 - 1 \times N^2 - 1$ generator matrices of the vector representation of $SU(N)$. When operating on vectors such as the field strength tensors, the derivative is

$$\begin{aligned}\left[D_\mu \mathbf{F}^{\mu\nu}\right]_c &= \left[\left[\partial_\mu + ig\mathbf{T} \cdot \mathbf{A}_\mu\right] \mathbf{F}^{\mu\nu}\right]_c = \partial_\mu \mathbf{F}_c^{\mu\nu} + gf_{acb}A_{a,\mu}\mathbf{F}_b^{\mu\nu} \\ &= \left[\partial_\mu \mathbf{F}^{\mu\nu} - g\mathbf{A}_\mu \times \mathbf{F}^{\mu\nu}\right]_c .\end{aligned} \tag{7.49}$$

The gauge field equations may be written in terms of gauge fields $\mathbf{A}$ as

$$D_\nu \mathbf{F}^{\nu\mu} = (\partial_\nu - g\mathbf{A}_\nu \times)(\partial^\nu \mathbf{A}^\mu - \partial^\mu \mathbf{A}^\nu - g\mathbf{A}^\nu \times \mathbf{A}^\mu) = \mathbf{j}_q^\mu \tag{7.50}$$

Choosing the Lorentz condition $\partial_\mu \mathbf{A}^\mu = 0$, and moving interaction terms to the right-hand side, we arrive at

$$\begin{aligned}\partial_\nu \partial^\nu \mathbf{A}^\mu = \mathbf{j}_q^\mu &+ g\partial_\nu(\mathbf{A}^\nu \times \mathbf{A}^\mu) + g\mathbf{A}_\nu \times (\partial^\nu \mathbf{A}^\mu - \partial^\mu \mathbf{A}^\nu) \\ &- g^2 \mathbf{A}_\nu \times (\mathbf{A}^\nu \times \mathbf{A}^\mu).\end{aligned}$$

There is no mass term $m^2\mathbf{A}$ on the left-hand side—the free gauge fields correspond to massless particles. Even a diagonal mass term in the Lagrangian of the form $m\mathbf{A}_\mu \cdot \mathbf{A}^\mu$, while compatible with global $SU(N)$ symmetry, is not compatible with gauge symmetry. The purely gauge field terms on the right-hand side may be identified as current densities carried by the gauge fields.

A group with generators which do not commute is called non-abelian. For $N > 1$, the group $SU(N)$ is non-abelian. A general feature of a non-abelian gauge symmetry is the appearance of the nonlinear terms on the right-hand side of the gauge field equations and of contributions to conserved currents carried by the gauge fields themselves. Moreover, in the case of a non-abelian symmetry, covariance of the field equations and invariance of the Lagrangian under gauge transformations demand that the coupling $g$ be common to all members of a representation such a $q$, and in fact be universal. In familiar terms, the reason is that conserved charges flow between gauged fields such as our fermions $q$ through the gauge fields themselves. In contrast, an abelian $U(1)$ gauge symmetry with a single neutral gauge field permits different gauged fields to carry different values of charge. The electromagnetic charge of two fermions is never exchanged through the electromagnetic field. Hence the $U(1)$ gauge symmetry associated with electromagnetism is compatible with a charge +2/3 quark and a charge −1 lepton, in fact with any values of the charges.

## 7.3 ■ QUANTUM CHROMODYNAMICS

The success of the quark model in explaining the static properties of light hadrons motivated the introduction of color. In the quark model, each quark flavor appears in three equivalent so-called color states, the color degree of freedom being required to make hadron wave functions consistent with the exclusion principle. For example, the lowest mass spin 3/2 baryons containing three quarks of the same flavor can be described by a spatial and spin wave function that is symmetric under interchange, representing $s$-wave states of three quarks of identical flavor and spin projection, and but in an antisymmetric state of color.

The minimal number of colors required can be understood as follows. We first suppose a global symmetry $SU(N)$ with $N_c$ colors and attempt to explain the absence of $qq$ states and the existence of $q\bar{q}$, $qqq$, and $\bar{q}\bar{q}\bar{q}$ states in terms of color combinations. In particular, we suppose the observed combinations are color singlets. For $SU(2)$, the multiplicities of degenerate multiplets of $qq$, $q\bar{q}$, and $qqq$ color states are exactly those obtained by the vector addition of angular momentum for three spin 1/2 objects. Denote a multiplet by its multiplicity with a subscript $A$ for antisymmetric, $S$ for symmetric, and $M$, $A$ and $M_S$ for mixed symmetry referring to multiplets of states which are antisymmetric or symmetric under interchange of only a subset of of the constituents. Then we can summarize the possible $SU(2)$ multiplets as follows:

$$\begin{aligned} \mathbf{2}\otimes\mathbf{2} &= \mathbf{3}_S + \mathbf{1}_A \\ \mathbf{2}\otimes\bar{\mathbf{2}} &= \mathbf{3} + \mathbf{1} \\ \mathbf{2}\otimes\mathbf{2}\otimes\mathbf{2} &= \mathbf{2}_{M,A} + \mathbf{2}_{M,S} + \mathbf{4}_S. \end{aligned} \tag{7.51}$$

No antisymmetric color state $\mathbf{1}_A$ for three quarks exists as a multiplet of $SU(2)$. For $SU(3)$, the analogous combinations are

$$\begin{aligned} \mathbf{3}\otimes\mathbf{3} &= \mathbf{3}_A + \mathbf{6}_S \\ \mathbf{3}\otimes\bar{\mathbf{3}} &= \mathbf{8} + \mathbf{1} \\ \mathbf{3}\otimes\mathbf{3}\otimes\mathbf{3} &= \mathbf{10}_S + \mathbf{8}_{M,S} + \mathbf{8}_{M,S} + \mathbf{1}_A. \end{aligned} \tag{7.52}$$

Here, the necessary antisymmetric state exists so the minimal number of colors is three.

If we denote the three color states by $r$, $b$, and $g$, the antisymmetric color wave function common to all baryons is

$$\psi^{qqq}_{color} = \frac{1}{\sqrt{6}}\{|\,rgb > - \,|\,rbg > + \,|\,gbr > - \,|\,grb > + \,|\,brg > - \,|\,bgr >\}. \tag{7.53}$$

For mesons, treating $q$ and $\bar{q}$ as distinct, the color wave function is simply

$$\psi^{q\bar{q}}_{color} = \frac{1}{\sqrt{3}}\left(|\,r\bar{r} > + \,|\,g\bar{g} > + \,|\,b\bar{b} >\right). \tag{7.54}$$

Direct evidence that $N_c = 3$ is found in processes sensitive to the total number of equivalent quarks. The process of virtual photon decay is one example. The cross section for $e^+e^- \to$ hadrons increases with $\sqrt{s}$ as the threshold for each quark flavor is crossed. Normalizing to $\mu^+\mu^-$, excluding $\tau^+\tau^- \to \nu_\tau\bar{\nu}_\tau + X$, and assuming quark pairs fragment into hadrons with unit probability, we predict the ratio

$$R = \frac{\sigma_{e^+e^-\to\text{hadrons}}}{\sigma_{e^+e^-\to\mu^+\mu^-}} = \Sigma_i N_c e_i^2 \left(1 + \frac{\alpha_s(s)}{\pi}\right) \tag{7.55}$$

where the sum runs over flavors with pair masses less than the available center of mass energy. This ratio should be a staircase function of $\sqrt{s}$. Here, the charge of the $i$-th flavor $e_i$ is in units of the proton charge, and the final factor is a correction due to gluon radiation. Hadronic resonances and eventually the $Z$ boson complicate the picture but, for $\sqrt{s} > 12$ GeV, as shown in Figure 7.3, the ratio is consistent with

$$\frac{R(\sqrt{s} > 12\ \text{GeV})}{1 + \frac{\alpha_s}{\pi}} = 3\left(e_u^2 + e_d^2 + e_c^2 + e_s^2 + e_b^2\right) = \frac{11}{3} \tag{7.56}$$

for $\alpha_s \simeq 0.2$. In the absence of color, $R$ would be a factor three lower. The angular distribution distinguishes scalar boson and fermion pair production and confirms quarks are spin 1/2 and the validity of the leading order description.

Additional evidence appears in the decays of the $\tau^-$ through a virtual $W$ boson

$$\tau^- \to \nu_\tau W^- \Rightarrow \nu_\tau(e^-\bar{\nu}_e, \mu^-\bar{\nu}_\mu, d\bar{u}, s\bar{u}). \tag{7.57}$$

The coupling constant for the virtual $W$ boson decay to $e^+\nu_e$ and $\mu^+\nu_\mu$ is $g_W$. The $W$ boson couples to $d\bar{u}$ in each of $N_c$ colors with coupling $g_W V_{ud}$, and to $u\bar{s}$ in each of $N_c$ colors with coupling $g_W V_{us}$. Since $|V_{ud}|^2 + |V_{sd}|^2 \simeq 1$, if we treat the quarks as free particles and neglect phase space factors associated with their

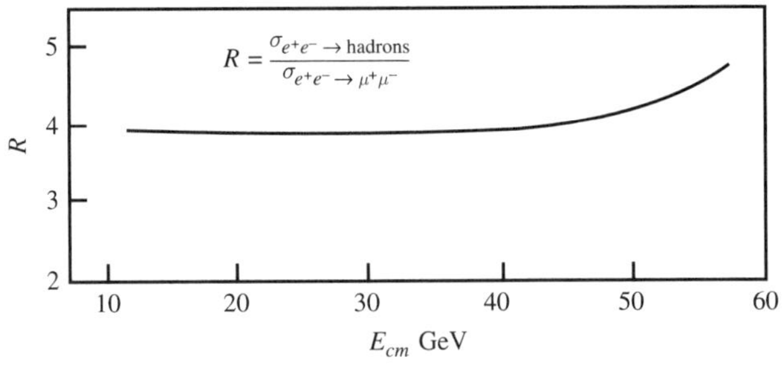


**FIGURE 7.3** Fit to the cross section ration $R = \sigma_{e^+e^-\to\text{hadrons}}/\sigma_{e^+e^-\to\mu^+\mu^-}$ as a function of center of mass energy. The hadronic cross section counts the number of color states of each quark flavor produced in proportion to the square of the quark electric charge and indicates that the number of colors is $N = 3$. [C. Caso et al., The European Physical Journal **C3**, 1 (1998).]

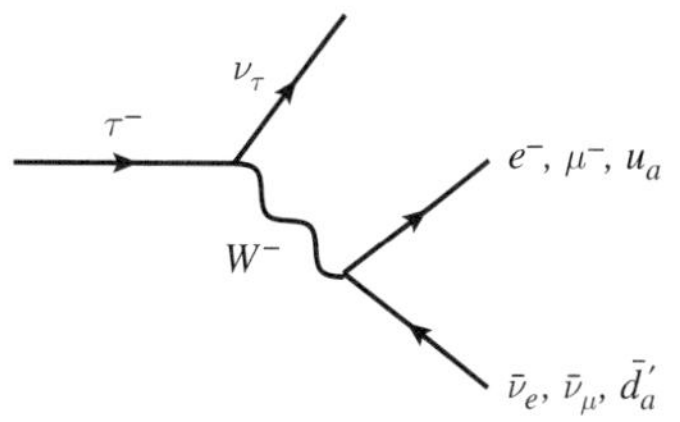


**FIGURE 7.4** Feynman diagram describing the decays of the $\tau^-$. The contribution of the quark decay modes to the total indicates that the number of colors is $N = 3$.

binding into hadrons, we expect the the decay rates for the two lepton pair modes and for the $N_c$ effective quark pair modes to be all equal and hence, for $N_c = 3$, predict a branching fraction $B_{\tau\to e\nu\nu} = B_{\tau\to\mu\nu\nu} = 1/5$. The observed branching fractions

$$B_{\tau^+\to\mu^+\nu_\mu\bar{\nu}_\tau} \simeq B_{\tau^+\to e^+\nu_e\bar{\nu}_\tau} = 17.5\% \tag{7.58}$$

are close to this value. Without color, we would expect 33%. Direct measurements of the decays of real W and Z bosons confirm the picture.

Quantum chromodynamics is a gauge theory of colored quarks based on the group of color transformations we will denote by $SU(3)_C$. Write the three color states of the $u$-quark for example as

$$q = \begin{pmatrix} u_1 \\ u_2 \\ u_3 \end{pmatrix} = \begin{pmatrix} u_r \\ u_b \\ u_g \end{pmatrix}. \tag{7.59}$$

Both numbers and the primary color labels $r$, $g$, and $b$ are in use. The primary color labels are a reminder that a uniform antisymmetric mix of color behaves as if color neutral. The Lagrangian and the field equations for a single quark flavor are precisely those introduced above with $N = 3$ and $\mathbf{T} = \boldsymbol{\lambda}/2$, and for multiple flavors we have

$$L = \Sigma_q \bar{q}(i\gamma_\mu D^\mu - m)q - \frac{1}{4}\mathbf{F}^2 \tag{7.60}$$

where the summation is over different flavors of quark, each flavor appearing with the same coupling $g$. The field equations are

$$\begin{aligned}(i\gamma_\mu\partial^\mu - m)q &= g\frac{\boldsymbol{\lambda}}{2}\cdot\mathbf{A}_\mu\gamma^\mu q \\ D_\mu\mathbf{F}^{\mu\nu} &= \Sigma_q\mathbf{j}_q^\nu.\end{aligned} \tag{7.61}$$

The eight gauge fields $\mathbf{A}$ are referred to as gluon fields and the corresponding spin 1 massless bosons are called gluons. The color fine structure constant is defined as

$$\alpha_s = \frac{g^2}{4\pi}. \tag{7.62}$$

The gauged fermion equations in color component form are

$$[\gamma_\mu p^\mu - m]q_a = \frac{g}{2}\lambda_{k,ab}\gamma_\mu A_k^\mu q_b. \tag{7.63}$$

By comparison with the corresponding equation for electrodynamics, we may identify the vertex factor for the transition $q_b A_k^\mu \to q_a$ as

$$-i\frac{g}{2}\lambda_{k,ab}\gamma_\mu. \tag{7.64}$$

The propagator for each gluon may be assumed to be identical in form to that for a photon in Lorenz gauge

$$G_k^{\mu\nu}(p) = \frac{-ig^{\mu\nu}}{p^2} \tag{7.65}$$

and the propagator for each quark is identical to that for a spin 1/2 charged particle in electrodynamics. We now turn to the study of these equations and to the derivation of the Feynman rules of the quantum theory.

Consider the elastic scattering of unlike flavors such as $u_a d_b \to u_c d_d$ by exchange of a single gluon of type $m$ as illustrated in Figure 7.5. The matrix element may be written down by analogy to the photon exchange amplitude and is

$$iM = \left[\bar{u}_c\left(-i\frac{g}{2}\lambda_{m,ca}\gamma_\mu\right)u_a\right]\left[\frac{-ig^{\mu\nu}}{q^2}\right]\left[\bar{d}_d\left(-i\frac{g}{2}\lambda_{m,db}\gamma_\nu\right)d_b\right] \sim \frac{g^2}{4}\lambda_{m,ca}\lambda_{m,db}. \tag{7.66}$$

The kinematic factors are identical to those in quantum electrodynamics. The amplitudes for the exchange of different gluon fields are coherent so we must implicitly sum over $m$. The relative coupling constants of each gluon type are determined by matrix elements of the generators of $SU(3)_C$ and this gives rise to the sum over products of Gell-mann matrix elements.

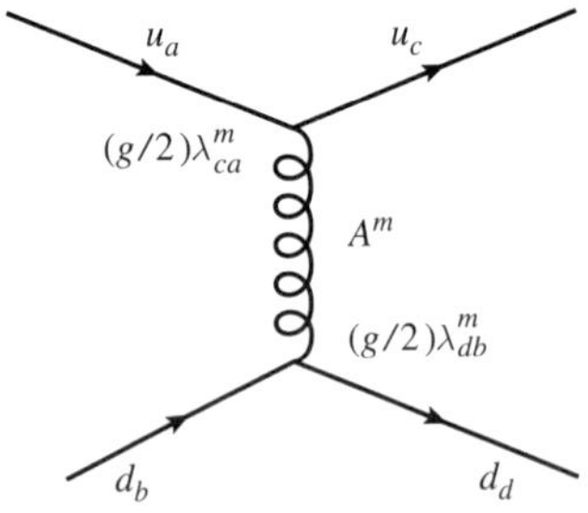


**FIGURE 7.5** Quark scattering by gluon exchange. The amplitude for the elastic scattering of a $u$-quark of color $a$ and a $d$-quark of color $b$ by exchange of a gluon of type $m$ yielding a $u$-quark of color $b$ and a $d$-quark of color $d$ includes the vertex factors $(g/2)T^m_{ba}$ and $(g/2)T^m_{dc}$.

The gauge field self-interactions derive from the terms

$$L_{int} = -\frac{g}{2}\left(\partial_\mu A^i_\nu - \partial_\nu A^i_\mu\right) f_{ijk} A^\mu_j A^\nu_k + \frac{g^2}{4} f_{ijk} f_{ilm} A^\mu_j A^\nu_k A^l_\mu A^m_\nu \tag{7.67}$$

and correspond to three-boson and four-boson diagrams as illustrated in Figure 7.6. Given that gluons carry color charge, the existence of such terms might have been anticipated based on experience with vector field electrodynamics. The amplitude for a triple-boson transition

$$A^a_1 + A^b_2 \rightarrow A^c_3 \tag{7.68}$$

may be found by inserting plane waves $\epsilon_1 e^{-ip_1 x}$ and $\epsilon_2 e^{-ip_2 x}$ in the O(g) terms on the right-hand side of the gauge field equation, projecting onto $\epsilon_3 e^{-ip_3 x}$, and erasing the coordinate dependence which produces momentum conservation. The projection is accompanied by a factor -i as in electrodynamics and each derivative gives -i times the momentum. The term $g\mathbf{A}_\mu \times (\partial^\mu \mathbf{A}^\nu - \partial^\nu \mathbf{A}^\mu)$ gives, treating each field independently,

$$iM_1 = (-i)^2 g \epsilon^{3\dagger}_\mu f_{bac}\left[\epsilon^\nu_2\left(p^\nu_1 \epsilon^\mu_1 - p^\mu_1 \epsilon^\nu_1\right) - \epsilon^\nu_1\left(p^\nu_2 \epsilon^\mu_2 - p^\mu_2 \epsilon^\nu_2\right)\right] \tag{7.69}$$

while the term $g\partial_\nu(\mathbf{A}^\nu \times \mathbf{A}^\mu)$ gives

$$iM_2 = (-i)^2 g \epsilon^{3\dagger}_\nu f_{bac}\left(p^1_\mu + p^2_\mu\right)\left[\epsilon^\mu_2 \epsilon^\nu_1 - \epsilon^\mu_1 \epsilon^\nu_2\right]. \tag{7.70}$$

With the Lorentz conditions $\epsilon p = 0$, the amplitude sum is

$$iM_{abc} = g f_{abc}\left[\epsilon_1 \cdot \epsilon_2 \epsilon^\dagger_3 \cdot (p_1 - p_2) + \epsilon_2 \cdot \epsilon^\dagger_3 \epsilon_1 \cdot (2p_2) - \epsilon^\dagger_3 \cdot \epsilon_1 \epsilon_2 \cdot (2p_1)\right]. \tag{7.71}$$

If all three bosons point into the vertex, $p_1 + p_2 + p_3 = 0$, the amplitude has the symmetrical form

$$iM = g f_{abc}[\epsilon_1 \cdot \epsilon_2 \epsilon_3 \cdot (p_1 - p_2) + \epsilon_2 \cdot \epsilon_3 \epsilon_1 \cdot (p_2 - p_3) + \epsilon_3 \cdot \epsilon_1 \epsilon_2 \cdot (p_3 - p_1)]. \tag{7.72}$$

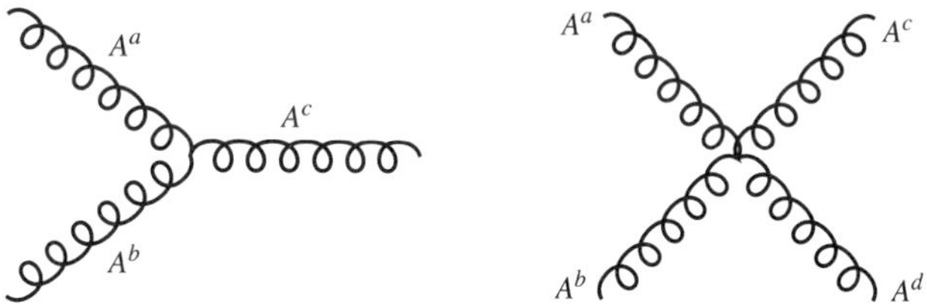


**FIGURE 7.6** Three-boson and four-boson interactions in chromodynamics. The three-boson vertex depends is a Lorentz tensor constructed from the gluon momenta and is proportional to $g$. The four-boson vertex is proportional to $g^2$.

A similar calculation produces the four-boson amplitude for the symmetrical process

$$A_1^a + A_2^b \to A_3^c + A_4^d \tag{7.73}$$

and the symmetrical form is

$$\begin{aligned} iM_{abcd} = &- ig^2[f_{abe}f_{cde}(\epsilon_1 \cdot \epsilon_3\epsilon_2 \cdot \epsilon_4 - \epsilon_1 \cdot \epsilon_4\epsilon_2 \cdot \epsilon_3) \\ &+ f_{ade}f_{bce}(\epsilon_1 \cdot \epsilon_2\epsilon_3 \cdot \epsilon_4 - \epsilon_1 \cdot \epsilon_3\epsilon_2 \cdot \epsilon_4) \\ &+ f_{ace}f_{dbe}(\epsilon_1 \cdot \epsilon_4\epsilon_2 \cdot \epsilon_3 - \epsilon_1 \cdot \epsilon_2\epsilon_3 \cdot \epsilon_4)]. \end{aligned} \tag{7.74}$$

These chromodynamic vertex factors may also be obtained directly from the corresponding terms in the interaction Lagrangian by insertion of plane waves.

## 7.4 ■ HIGH ENERGY CHROMODYNAMICS

Calculations of fundamental processes in quantum chromodynamics proceed as in spinor quantum electrodynamics and the results are often a transcription of quantum electrodynamics results with an appropriate so-called color factor. For example, consider elastic scattering of two quarks of unlike flavor by single gluon exchange:

$$q_i^a q_j^c \to q_i^b q_j^d \tag{7.75}$$

where $i$ and $j$ denote the flavor and $a$, $b$, $c$, and $d$ denote color. The amplitude is identical to that for

$$e^-\mu^- \to e^-\mu^- \tag{7.76}$$

except for a replacement

$$\alpha \to \alpha_s \mathbf{T}_{cd} \cdot \mathbf{T}_{ab} \tag{7.77}$$

where the inner product represent the summation over all possible gluons.

In general, color states are not determined and the experimental cross section is a sum over color states of final colored objects and an average over color states of initial colored objects. Matrix methods facilitate determining the color factor. Summation of squared amplitudes over all color labels with $\mathbf{T}^*_{ab} = \mathbf{T}^\dagger_{ba} = \mathbf{T}_{ba}$ gives

$$\sum_{abcd} |M|^2 = \sum_{ij} \sum_{ab} T^i_{ab} T^j_{ba} \sum_{cd} T^i_{cd} T^j_{dc} = 2 \tag{7.78}$$

since $\sum_{ab} T^i_{ab} T^j_{ba} = \mathrm{tr}[T^i T^j] = \delta^{ij}/2$ according to Equation 7.9. A factor 1/9 converts to an average over initial color states. With the transcription $\alpha \to \alpha_s$ and the

| Process | $(\pi\alpha_s^2)^{-1}\, d\sigma/dt$ | Process | $(\pi\alpha_s^2)^{-1}\, d\sigma/dt$ |
|---|---|---|---|
| $ud \to ud$ | $\frac{4}{9}\frac{s^2+u^2}{t^2}$ | $gg \to gg$ | $\frac{9}{2}\left(3 - \frac{tu}{s^2} - \frac{su}{t^2} - \frac{st}{u^2}\right)$ |
| $u\bar{d} \to u\bar{d}$ | $\frac{4}{9}\frac{s^2+u^2}{t^2}$ | $uu \to uu$ | $\frac{4}{9}\left(\frac{s^2+u^2}{t^2} + \frac{u^2+t^2}{s^2} - \frac{8}{27}\frac{s^2}{ut}\right)$ |
| $u\bar{u} \to d\bar{d}$ | $\frac{4}{9}\frac{t^2+u^2}{s^2}$ | $u\bar{u} \to u\bar{u}$ | $\frac{4}{9}\left(\frac{s^2+u^2}{t^2} + \frac{s^2+t^2}{u^2} - \frac{8}{27}\frac{u^2}{st}\right)$ |
| $u\bar{u} \to gg$ | $\frac{32}{27}\frac{t^2+u^2}{tu} - \frac{8}{3}\frac{t^2+u^2}{s^2}$ | $gg \to u\bar{u}$ | $\frac{1}{6}\frac{t^2+u^2}{tu} - \frac{3}{8}\frac{t^2+u^2}{s^2}$ |
| $gu \to gu$ | $\frac{u^2+s^2}{t^2} - \frac{4}{9}\frac{s^2+u^2}{su}$ | | |

**TABLE 7.1** Cross sections for QCD processes. In each case, spin and and color are averaged in the initial state and summed in the final state. [B. L. Combridge, J. Kripfganz, and J. Ranft, Phys. Lett. **B70**, 234 (1977).]

color factor 2/9, we find the relativistic cross section for elastic scattering by gluon exchange is

$$\frac{d\sigma_{q_iq_j\to q_iq_j,i\neq j}}{dt} = \frac{4}{9}\frac{\pi\alpha_s^2}{s^2}\frac{s^2+u^2}{t^2}. \tag{7.79}$$

Color averaged differential cross sections quark and gluon scattering in the ultra-relativistic limit are listed in Table 7.1.

Some amplitudes in quantum chromodynamics lack an electrodynamic counterpart. As illustrated in Figure 7.7, the processes $q\bar{q} \to gg$ and $gg \to qq$ include an amplitude with a three boson vertex. The process $gg \to gg$ described by the diagrams in Figure 7.8 is also without an electromagnetic analog. In computing the cross section for scattering with identical gluons, care must be taken to symmetrize the amplitude.

The first evidence for the existence of the gluon appeared in planar 3-jet events $e^+e^- \to q\bar{q}g$ ascribed to wide angle gluon radiation from one or the other of the final state quarks. The leading order diagram is shown in Figure 7.9 and the

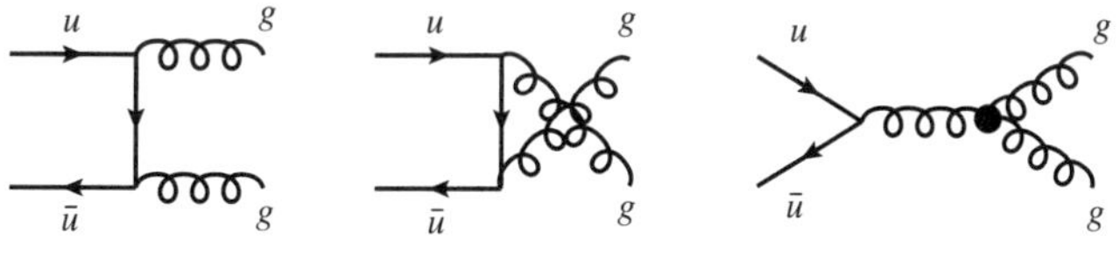


**FIGURE 7.7** Feynman diagrams for the process $u\bar{u} \to gg$. Diagrams (a) and (b) are similar to those for $u\bar{u} \to \gamma\gamma$. Diagram (c) represents the $ggg$ interaction. The amplitudes for color combinations must be summed.

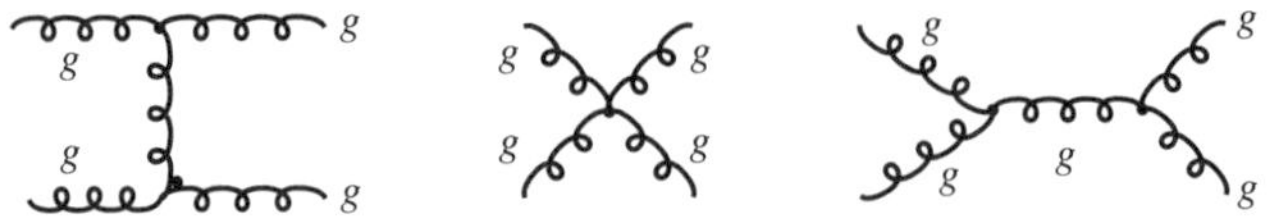


**FIGURE 7.8** Feynman diagrams for the process $gg \to gg$. Gluon scattering at this order in the coupling is without an electromagnetic analog.

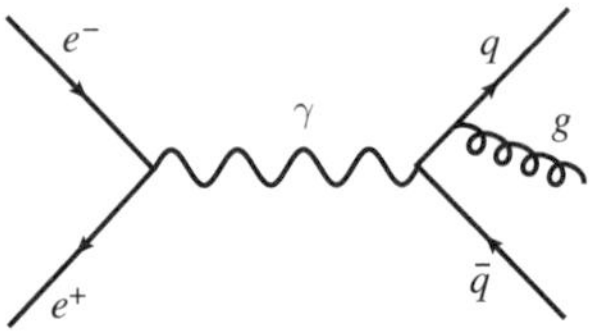


**FIGURE 7.9** Feynman diagrams for the process $e^+e^- \to q\bar{q}g$.

amplitude is proportional to $\alpha\sqrt{\alpha_s}T^c_{ab}$ where $a$, $b$, and $c$ are the color labels of the quark and antiquark and radiated gluon. The color factor is 4. Initial state gluon bremsstrahlung can be identified in $q\bar{q}$ annihilation to an electroweak boson plus a gluon. Another process with a relatively simple interpretation is so-called direct photon production which can be identified in $p\bar{p}$ and $p\pi$ collisions with differential cross section

$$\frac{d\sigma_{q\bar{q}\to\gamma g}}{dt} = \frac{8}{9}\pi\alpha\alpha_s\frac{t^2+u^2}{s^2tu}. \tag{7.80}$$

This process is analogous to $q\bar{q} \to \gamma\gamma$ with one photon replaced by a gluon. The process $q\bar{q} \to gg$ results from radiation of the final gluons from each quark leg also like $q\bar{q} \to \gamma\gamma$ but an additional amplitude contributes and corresponds to a diagram describing annihilation into a virtual gluon which decays via a three-boson coupling to the final gluons. The distribution of jet rapidity and transverse momentum in $p\bar{p}$ collisions resulting from $q\bar{q}$, $qg$, and $gg$ collisions is shown in Figure 7.10. Detailed studies of such processes are in good agreement with QCD predictions.

## 7.5 ■ COLOR COULOMB INTERACTION

In Chapter 6, we found the relationship between the single photon exchange amplitude and the Coulomb interaction potential. Leading order quantum chromodynamics predicts an analog color Coulomb potential of the form

$$V_{cdab} = \frac{\alpha_s}{r}\mathbf{T}_{cd}\cdot\mathbf{T}_{ab} \tag{7.81}$$

which depends on the color states. Color exchange leads to both attractive and repulsive interactions and we now show that the lowest energy bound states of quarks correspond to the color combinations of the naive quark model.

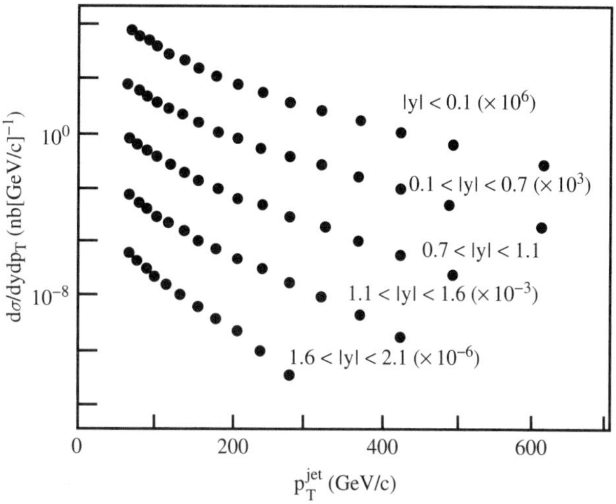


**FIGURE 7.10** Differential cross section for $p\bar{p} \to \text{j} + X$ as a function of jet rapidity $y$ and transverse momentum $p_T$ at $p\bar{p}$ center of mass energy 1.8 TeV. The data points shown are indistinguishable from next-to-leading-order QCD calculations. [T. Altonen *et al.* (CDF Collaboration) Phys. Rev. **D78**, 052006 (2008)]

The matrix representing the coupling of the octet of gluon fields to various color transition currents is

$$\mathbf{A} \cdot \mathbf{T} = \frac{1}{2} \begin{pmatrix} A_3 + \frac{1}{\sqrt{3}} A_8 & A_1 - i A_2 & A_4 - i A_5 \\ A_1 + i A_2 & -A_3 + \frac{1}{\sqrt{3}} A_8 & A_6 - i A_7 \\ A_4 + i A_5 & A_6 + i A_7 & -\frac{2}{\sqrt{3}} A_8 \end{pmatrix}. \tag{7.82}$$

It may also be expressed terms of eight fields $g_\alpha$ (three complex fields plus their complex conjugates plus two real fields) defined by equating this matrix to the alternate expression

$$\mathbf{A} \cdot \mathbf{T} = \frac{1}{\sqrt{2}} \begin{pmatrix} \frac{1}{\sqrt{2}} g_8 + \frac{1}{\sqrt{6}} g_7 & g_1 = g_{r\bar{b}} & g_2 = g_{r\bar{g}} \\ g_4 = g_{b\bar{r}} & -\frac{1}{\sqrt{2}} g_8 + \frac{1}{\sqrt{6}} g_7 & g_3 = g_{b\bar{g}} \\ g_6 = g_{g\bar{r}} & g_5 = g_{g\bar{b}} & -\sqrt{\frac{2}{3}} g_7 \end{pmatrix}. \tag{7.83}$$

As indicated in this second matrix, the charged fields $g_{1-6}$ may alternatively be labeled by the color and anticolor they bear and the neutral fields are in this notation

$$g_7 = \frac{1}{\sqrt{6}} g_{(r\bar{r}+g\bar{g}-2b\bar{b})}, \quad g_8 = \frac{1}{\sqrt{2}} g_{(r\bar{r}-g\bar{g})}. \tag{7.84}$$

The color indices in the two-index labeling scheme $g_{\alpha\bar{\beta}}$ represent the conservation of each of the three color charges. The gluons corresponding to complex fields are effectively bicolored objects. A diagram illustrating this idea is shown in

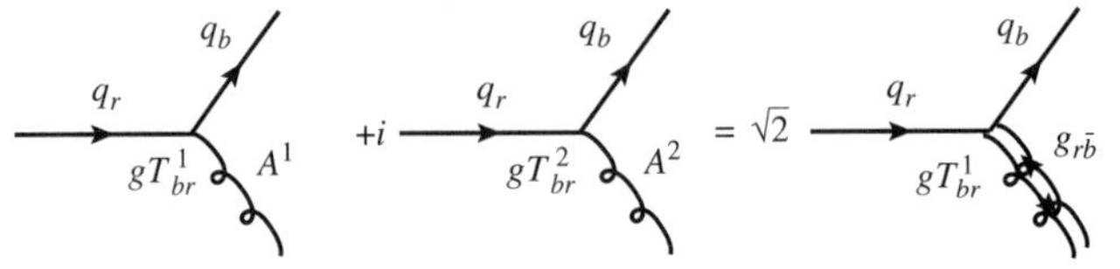


**FIGURE 7.11** Gluon emission amplitude. The quark transition $q_1 = q_r \to q_2 = q_b$ occurs by radiation of a combination of fields $A^1$ and $A^2$ corresponding to the field $g_1 = g_{r\bar{b}} = (A^1 + iA^2)/\sqrt{2}$.

Figure 7.11. Since the gauge fields collectively and coherently participate in any color interaction, there is nothing sacred about the using the fields $A_i$ as a basis in calculations. In fact, a gauge transformation will change the basis. So the $g_\alpha$ basis is an equally valid choice.

Gluon exchange between two quarks implies energy and momentum exchange and, as in photon exchange, a Coulomb interaction in the nonrelativistic limit. To understand the implications of color exchange, we evaluate the gluon exchange interaction energy for various $qq$ and $q\bar{q}$ states and examine individually the color exchange factors in the color Coulomb potential.

The matrix element $[\mathbf{T} \cdot \mathbf{A}]_{ab}$ specifies the linear combination of gluon fields coupled to a transition current $b \to a$. For example, a $q_r \to q_r$ transition produces $A_3$ with amplitude $1/2$ and $A_8$ with amplitude $1/(2\sqrt{3})$. Consider now the process $u_r d_r \to u_r d_r$ illustrated in Figure 7.12. The amplitude for exchange of $A_3$ with two $r \to r$ quark transitions is proportional to the emission vertex factor $1/2$ times the absorption vertex factor $1/2$. Exchange of $A_8$ adds to the interaction energy so the total amplitude is proportional to

$$M_{rr\to rr} = \left(\frac{1}{2}\right)^2 + \left(\frac{1}{2\sqrt{3}}\right)^2 = \frac{1}{3}. \tag{7.85}$$

In terms of the $g_i$ fields, two $r$ quarks exchange $g_7$ and $g_8$ and the coupling constants are $\frac{1}{\sqrt{6}}$ and $\frac{1}{\sqrt{2}}$ (just the amount of $r\bar{r}$ in the fields) times the factor of $\frac{1}{\sqrt{2}}$ outside the coupling matrix.

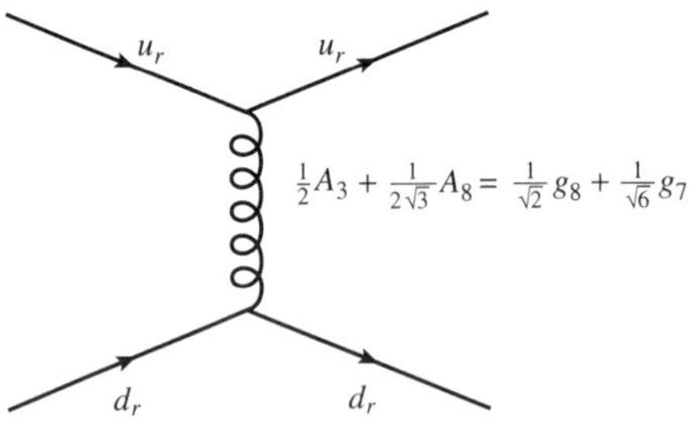


**FIGURE 7.12** Single gluon exchange with no color exchange. In the process described by this Feynman diagram, single gluon exchange occurs between quarks both in a red state and both quarks maintaining their color state. The gluon field that is exchanged is expressed explicitly in two forms.

Similarly, the $b \to b$ interaction is mediated by $A_3$ and $A_8$ and the relative strength is

$$M_{bb\to bb} = \left(-\frac{1}{2}\right)^2 + \left(\frac{1}{2\sqrt{3}}\right)^2 = \frac{1}{3} \tag{7.86}$$

while the $g \to g$ interaction is mediated by $A_8$ and

$$M_{gg\to gg} = \left(\frac{1}{\sqrt{3}}\right)^2 = \frac{1}{3}. \tag{7.87}$$

That the interaction amplitudes $M_{aa\to aa}$ are independent of color $a$ is a consequence of global color symmetry. The positive sign implies a repulsive interaction between like colors as in electrodynamics.

The interaction of two quarks of different color is determined by off-diagonal elements. Thus $A_1$ and $A_2$ in the combination $(A_1 + iA_2)/2$ mediate an $r \to b$ transition and in the combination $(A_1 - iA_2)/2$ mediate a $b \to r$ transition. For a color exchange interaction such as $u_r d_b \to u_b d_r$, the total amplitude is

$$M_{rb\to br} = \left(\frac{1}{2}\right)^2 + \left(\frac{-i}{2}\right)\left(\frac{i}{2}\right) = \frac{1}{2} \tag{7.88}$$

so the color exchange interaction between quarks is repulsive. The $A_3$ and $A_8$ contribute without color exchange to the interaction of quarks of different color and the amplitude is proportional to

$$M_{rb\to rb} = \left(\frac{1}{2}\right)\left(-\frac{1}{2}\right) + \left(\frac{1}{2\sqrt{3}}\right)^2 = -\frac{1}{6} \tag{7.89}$$

corresponding to an attractive force. As in electrodynamics, the amplitudes for gluon exchange between $q$ and $\bar{q}$ have an extra minus sign for the outgoing antifermion, reversing attractive and repulsive diagonal interactions. The order is also reversed in the transition current implying transposition of the coupling matrix so

$$M_{r\bar{r}\to r\bar{r}} = -\frac{1}{3} \ : \ M_{r\bar{r}\to b\bar{b}} = -\frac{1}{2} \ : \ M_{r\bar{b}\to r\bar{b}} = +\frac{1}{6}. \tag{7.90}$$

We see that the color conserving interaction between a quark and its antiquark is attractive as in electrodynamics but the color changing interaction is more attractive while color exchange is repulsive. We do not include the amplitude corresponding to the annihilation diagram $q\bar{q} \to g \to q\bar{q}$ here because, in the nonrelativistic limit, this amplitude is much smaller than that corresponding to gluon exchange.

To make sense of the various ways the color might be arranged to minimize the energy, we note that the nine color states of $q\bar{q}$ form a color singlet $(r\bar{r} + b\bar{b} + g\bar{g})/\sqrt{3}$ and a color octet, each multiplet transforming under global $SU(3)_C$ transformations into a linear combination of states from the same multiplet. Since the Hamiltonian of chromodynamics is invariant under such global transformations, the states within a multiplet have the same energy and it suffices to compute the energy for one state from each multiplet to discover the entire spectrum.

The amplitude for gluon exchange for the singlet state is a sum of nine terms

$$M_{\frac{(r\bar{r}+b\bar{b}+g\bar{g})}{\sqrt{3}} \rightarrow \frac{(r\bar{r}+b\bar{b}+g\bar{g})}{\sqrt{3}}} = \frac{1}{3}\left[3M_{r\bar{r}\rightarrow r\bar{r}} + 6M_{r\bar{r}\rightarrow b\bar{b}}\right] = -\frac{4}{3} \tag{7.91}$$

and corresponds to an attractive interaction. Gluon exchange for $q\bar{q}$ in a color octet state is exemplified by $M_{r\bar{b}\rightarrow r\bar{b}} = +1/6$ which is repulsive. The nine color states of $qq$ form a triplet transforming like antiquarks (3*)

$$\frac{1}{\sqrt{2}}(rb - br)\ ;\ \frac{1}{\sqrt{2}}(bg - gb)\ ;\ \frac{1}{\sqrt{2}}(gr - rg) \tag{7.92}$$

with an attractive interaction exemplified by

$$M_{\frac{1}{\sqrt{2}}(rb-br)\rightarrow\frac{1}{\sqrt{2}}(rb-br)} = M_{(rb\rightarrow rb)} - M_{(rb\rightarrow br)} = -\frac{2}{3}, \tag{7.93}$$

and a sextet

$$rr,\ bb,\ gg,\ \frac{1}{\sqrt{2}}(rb + br)\ ;\ \frac{1}{\sqrt{2}}(bg + gb)\ ;\ \frac{1}{\sqrt{2}}(gr + rg) \tag{7.94}$$

with a repulsive interaction exemplified by $M_{rr\rightarrow rr} = 1/3$. The triplet has an attractive interaction that is weaker than the color singlet $q\bar{q}$. Thus, of all two quark states, the color singlet $q\bar{q}$ is most tightly bound. This corresponds exactly to expectation in the quark model. There is, for example, only one $\pi^+_{ud}$ corresponding to the the color singlet. The higher mass triplet and sextet color states of a $ud$ pair presumably merge with multi-particle states and are not observable.

The color singlet state of three quarks $qqq$ can be written as a product of a $qq$ anti-triplet with a $q$ triplet

$$\begin{aligned} qqq_{singlet} &= \frac{1}{\sqrt{6}}\left[(rb - br)g + (bg - gb)r + (gr - rg)b\right] \\ &= \frac{1}{\sqrt{3}}\left[\bar{G}g + \bar{R}r + \bar{B}b\right]. \end{aligned} \tag{7.95}$$

Each $qq$ triplet interacting with itself contributes $(1/3)(-2/3) = -2/9$ (the first factor is the relative fraction of the triplet) to the interaction energy. There are three triplets so the di-quark energy is $-2/3$. The interaction of the di-quark with the remaining quark triplet contributes $-4/3$ so the total $qq$ interaction energy is $-2$. The color octet and color decuplet combinations of three quarks have interaction energy $-1/2$ and $+1$ so the color singlet is again the most tightly bound.

These calculations ignore relativistic corrections so are extremely naive for light quarks. They also ignore 3-body interactions generated by a three-gluon vertex and other higher order effects. But they do suggest that the observed color singlet mesons and baryons correspond to minimum energy configurations. The static color interaction potential found numerically for various color representations combining to form a singlet is shown in Figure 7.13. For all the color multiplets illustrated, the Coulomb form appears at small distances. At large distances the potential increases linearly with separation corresponding to a constant force. This large distance behavior is a manifestation of color confinement as described below.

Charmonium and $\Upsilon$ systems are nonrelativistic and considerable effort has gone into understanding the spectrum of $c\bar{c}$ and $b\bar{b}$ bound states. The color Coulomb potential alone should result in a mass spectrum similar to that of the hydrogen atom. The level structures are better described with a potential of the Cornell form

$$V = -\frac{4}{3}\frac{\alpha_s}{r} + kr \qquad (7.96)$$

plus relativistic corrections related to spin. In Figure 7.14, the observed spectrum of $b\bar{b}$ bound states is compared to expectations based on numerical calculations based on quantum chromodynamics which include corrections of order velocity squared. The average squared velocities in units of light speed are found to be 0.27-0.52 and 0.075-0.112 in the charmonium and $\Upsilon$ systems respectively. The average radii are in the range 0.43-1.47 fm and 0.24-0.93 fm respectively. The effective potential is linear above about 0.3 fm. Such studies have paved the way for calculations of the effective force between nucleons and hyperons which reproduce the repulsive core plus attractive light meson exchange model of nuclear physics and which will improve predictions of the many phases of nuclear and hyper-nuclear matter at high density.

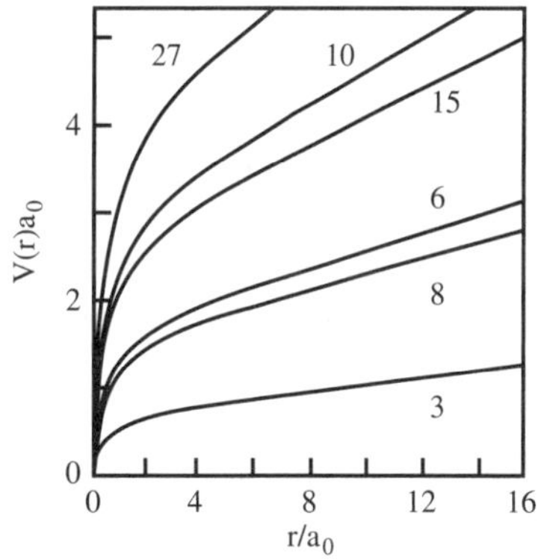


**FIGURE 7.13** Chromostatic potentials for various representations. The static potential energy for two colored objects in various representations combined as a color singlet state can be computed as a function of separation in lattice QCD. The results of one such computation is shown, fit to a Coulomb plus linear form. The separation $r$ is expressed in terms of a lattice spacing $a_\sigma = 0.085$ fm. The curve labeled by representation "3" corresponds to the fundamental quark plus antiquark potential. The representation "8" corresponds to gluon pairs. [Gunnar S. Bali, Nucl. Phys. Proc. Suppl. **83**, 422 (2000)]

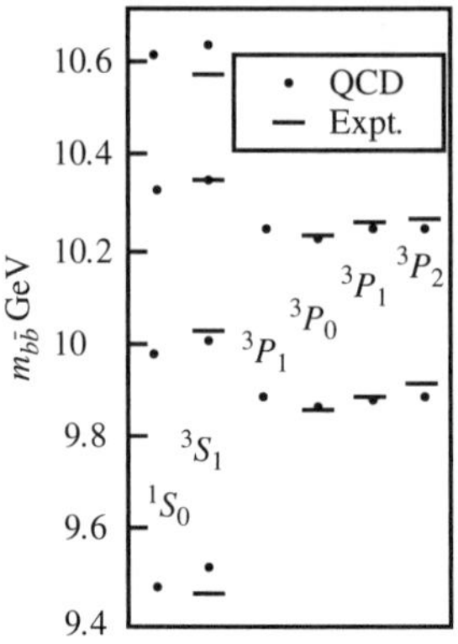


**FIGURE 7.14** Experimental spectrum of $b\bar{b}$ meson masses compared to lattice calculations for lattice spacing 1.5 fm. [Gunnar S. Bali, Klaus Schilling, Armin Wachter, Phys. Rev. **D56**, 2566 (1997), arXiv:hep-lat/9703019v2]

## 7.6 ■ CHARGE CONJUGATION IN CHROMODYNAMICS

Because gluons carry color, some care must be taken in considering the $C$ eigenvalues of multi-gluon states. Consider a hypothetical Lorentz charged scalar color triplet $\phi$ satisfying

$$D_\mu D^\mu \phi + m^2 \phi = 0 \tag{7.97}$$

with $D_\mu = \partial_\mu + ieA_\mu + ig\mathbf{T} \cdot \mathbf{A}_\mu$. The complex conjugate equation results from the charge conjugation operation $C$ defined by

$$\phi \to \phi^* \; ; \; A^\mu \to -A^\mu \; ; \; \mathbf{T} \cdot \mathbf{A}_\mu \to -\mathbf{T}^* \cdot \mathbf{A}_\mu = -\mathbf{T}^t \cdot \mathbf{A}_\mu \tag{7.98}$$

where we used $\mathbf{T}^\dagger = \mathbf{T} \to \mathbf{T}^* = \mathbf{T}^t$. In terms of real fields, $CA_i = \pm A_i$ where the plus sign applies for $i = 2, 5, 7$ and otherwise the minus sign applies. The real fields are charge conjugation eigenstates with a mix of eigenvalues.

Now consider the $C$ symmetry of multi-gluon states. The 64 color states of two gluons form various color multiplets with multiplicities that can be expressed as

$$\mathbf{8} \otimes \mathbf{8} = \mathbf{1} + \mathbf{8} + \mathbf{8} + \mathbf{10} + \overline{\mathbf{10}} + \mathbf{27}. \tag{7.99}$$

We will be interested in the color singlet combination. The matrix representation $\mathbf{T} \cdot \mathbf{A}$ of the wavefunction of one gluon is useful in this context. Under global $SU(3)_C$ transformations, $\mathbf{T} \cdot \mathbf{A} \to U\mathbf{T} \cdot \mathbf{A}U^{-1}$. The fact that the trace of a matrix product is invariant under cyclic permutations implies the trace of a product of $N$ factors of $\mathbf{T} \cdot \mathbf{A}$ is invariant so must be proportional to a combination of products of fields constituting a color singlet state of $N$ gluons. The color singlet two gluon state in particular is proportional to

$$\mathrm{tr}(\mathbf{T} \cdot \mathbf{A}_1 \mathbf{T} \cdot \mathbf{A}_2) = A_1^a A_2^b \mathrm{tr}(T^a T^b) = \frac{1}{2} \mathbf{A}_1 \cdot \mathbf{A}_2. \tag{7.100}$$

Using $(T^aT^b)^t = T^{bt}T^{at}$ and $\mathrm{tr}(M^t) = \mathrm{tr}(M)$, the two gluon state may be seen to transform under $C$ into itself,

$$\mathrm{tr}(\mathbf{T}\cdot\mathbf{A}_1\mathbf{T}\cdot\mathbf{A}_2) \to \mathrm{tr}(\mathbf{T}^t\cdot\mathbf{A}_1\mathbf{T}^t\cdot\mathbf{A}_2) = \mathrm{tr}(\mathbf{T}\cdot\mathbf{A}_1\mathbf{T}\cdot\mathbf{A}_2), \tag{7.101}$$

which shows the color singlet state of two gluons is always $C$-even. It follows that a $C$-odd color singlet state of quarks and gluons may not decay to two gluons. A $C$-even state may decay to two real gluons, unless $J^P = 1^-$. The decay of a vector particle to two massless vector particles is precluded by angular momentum conservation. A state with $J^{PC} = 1^{-+}, 1^{++}$ may decay to two photons or to two gluons if one or both vector particles is virtual and this possibility may not be neglected.

It is left as an exercise to show that the symmetric combination

$$\mathrm{tr}(\mathbf{T}\cdot\mathbf{A}_1\mathbf{T}\cdot\mathbf{A}_2\mathbf{T}\cdot\mathbf{A}_3) + \mathrm{tr}(\mathbf{T}\cdot\mathbf{A}_2\mathbf{T}\cdot\mathbf{A}_1\mathbf{T}\cdot\mathbf{A}_3)$$

represents a $C$-odd color singlet state of three gluons. It does not follow that a $C$-even state may not decay to three gluons - two gluons antisymmetric in color form an octet equivalent in color to one gluon so three gluons in a state $\mathbf{A}_1\cdot\mathbf{A}_2\times\mathbf{A}_3 = f^{abc}A_1^aA_2^bA_3^c$ are C-even! In this respect, QCD does not support a naive generalization of the QED result.

## 7.7 ■ QUARKONIUM DECAYS

We can apply what we have learned of chromodynamics to decays of quarkonia. Compare the two-photon process

$$q_a\bar{q}_a \to \gamma\gamma \tag{7.102}$$

with the two-gluon process

$$q_a\bar{q}_a \to g_cg_d. \tag{7.103}$$

Let's examine the differences between electrodynamic and chromodynamic decay rate calculations. Two gluons may be produced by $q\bar{q}$ annihilation analogous to the two-photon process, and also by $q\bar{q}$ annihilation into a single gluon which splits into two. For an initial color singlet state, $(q_r\bar{q}_r + q_b\bar{q}_b + q_g\bar{q}_g)/\sqrt{3}$, the splitting diagram does not contribute as the intermediate gluon is a member of a color octet. In the two-photon process, the intermediate state quark color must be that of the initial state while, in the two-gluon process, the intermediate state quark is a superposition of colors. The two-photon amplitude contains a factor three from the sum over initial state color components while the two-gluon amplitude, summed over the color components of the initial state and summed over intermediate state quark colors, is

$$M_{q_a\bar{q}_a\to g_cg_d} \sim \sum_{ae} T^c_{ae}T^d_{ea} = \mathrm{tr}T^cT^d = \frac{\delta^{cd}}{2}. \tag{7.104}$$

The amplitude for $A_1^a A_2^a$ is independent of $a$ and the two gluons are produced in a color singlet state. For each color, the two gluons are identical as in the two photon case so the phase space is that in electrodynamics. The ratio of amplitudes is therefore

$$\frac{M_{gg}}{M_{\gamma\gamma}} = \frac{\alpha_s}{6\alpha e_q^2}. \tag{7.105}$$

After including the eight possible choices for gluon color, the ratio of decays rates is predicted to be

$$\frac{\Gamma_{gg}}{\Gamma_{\gamma\gamma}} = \frac{2\alpha_s^2}{9e_q^4\alpha^2}. \tag{7.106}$$

The diagrams describing annihilation of $q\bar{q}$ into three gluons analogous to diagrams for annihilation to three photons must be augmented by diagrams describing decay to two gluons with one splitting into two. The latter diagrams are forbidden for $C$-odd initial states for which the chromodynamic result again follows from the electrodynamic result by analysis of the color factor and the ratio of decay rates is predicted to be

$$\frac{\Gamma_{ggg}}{\Gamma_{\gamma\gamma\gamma}} = \frac{5\alpha_s^3}{54e_q^6\alpha^3}. \tag{7.107}$$

The decay formulae for positronium are

$$\begin{aligned}\Gamma_{^1S_0(e^+e^-)\to\gamma\gamma} &= \frac{4\pi\alpha^2}{m_e^2} \mid \psi(0) \mid^2 \\ \Gamma_{^3S_1(e^+e^-)\to\gamma\gamma\gamma} &= \frac{16(\pi^2-9)\alpha^3}{9m_e^2} \mid \psi(0) \mid^2 .\end{aligned} \tag{7.108}$$

With color factors for decays of $s$-wave color singlets included, we have

$$\begin{aligned}\Gamma_{^1S_0(q\bar{q})\to\gamma\gamma} &= \frac{48\pi\alpha^2 e_q^4}{(2m)^2} \mid \psi(0) \mid^2 \\ \Gamma_{^3S_0(q\bar{q})\to e^+e^-} &= \frac{16\pi\alpha^2 e_q^2}{(2m)^2} \mid \psi(0) \mid^2 \\ \Gamma_{^1S_0(q\bar{q})\to gg} &= \frac{32\pi\alpha_s^2}{3(2m)^2} \mid \psi(0) \mid^2 \\ \Gamma_{^3S_1(q\bar{q})\to ggg} &= \frac{160(\pi^2-9)\alpha_s^3}{81(2m)^2} \mid \psi(0) \mid^2 .\end{aligned} \tag{7.109}$$

As an example application of these formulae, if we identify the rate for decay to gluons with the hadronic decay rate not associated with a virtual photon, the preceding formulae predict

$$\frac{\Gamma_{\psi \to e^+e^-}}{\Gamma_{\psi \to \text{hadrons}}} = \frac{18\pi\alpha^2}{5(\pi^2 - 9)\alpha_s^3} = \frac{5.94 \text{ KeV}}{74.2 \text{ KeV}} \tag{7.110}$$

from which a value $\alpha_s(m_\psi) = 0.2$ may be extracted.

## 7.8 ■ COLOR CONFINEMENT

The color potential derived from the spectra of heavy quark bound states increases linearly with separation for distances larger than a fundamental length we will call $R_s \sim 1$ fm implying an interaction energy that increases without bound. This behavior can be represented by introducing a running coupling constant for a scattering process

$$\alpha_s \to \alpha_s(|q^2|) \tag{7.111}$$

that is a decreasing function of 4-momentum transfer. We will see in Chapter 9 how this effective coupling constant is derived in high order calculations.

That $\alpha_s$ decreases with $|q^2|$ is a seminal feature of gauge theory and implies there is a scale at which $\alpha_s \simeq 1$. This scale turns out to be the hadronic scale $\Lambda_{QCD} = R_s^{-1} \sim 200$ MeV. Interestingly, there is no dimensional quantity in the fundamental gauge theory so the scale must be determined from experiment. For $q^2 >> \Lambda_{QCD}^2$, we may safely assume $\alpha_s << 1$ and this justifies the success of perturbative calculations of high energy processes. For $|q^2| < \Lambda_{QCD}$, perturbation theory is untenable. The nonperturbative interaction between quarks and gluons for $r > R_s$ is presumably the reason free quarks and gluons are never observed as free particles - as colored quarks separate from a color neutral system, energetics favors production of $q\bar{q}$ pairs which attach to and neutralize the color charge. Quarks and gluons are confined.

The increase in interaction energy with separation is associated with virtual gluons and turns out to be a general feature of non-abelian gauge theories. While virtual fermion pairs tend to hide bare charge corresponding to a vacuum permittivity $\epsilon > 1$, virtual gluons have a larger anti-screening effect corresponding to $\epsilon < 1$. It seems that $\epsilon \to 0$ at a distance $R_s$ from a bare color charge field energy becomes infinite. A dipole confined to a bubble of radius $R_s$ appears to have a finite energy and color neutral bound states are favored. As the two charges of a dipole separate, the field may be forced into a tube of constant color flux - a string. The stretching of this flux tube is plausibly the origin of the linear term in the Cornell potential.

Thus we arrive at a hybrid view of hadrons. The ground state of heavy quark pair such as the $b\bar{b}$ has a Bohr radius smaller than $\Lambda_{QCD}^{-1}$ and, consistent with nonrelativistic behavior, $\alpha_s$ is small at this scale. Light quarks such as the $u$-quark and $d$-quark which would have a Bohr radius much larger than $R_s$ are confined to a radius $R_s$. For light quark systems, $\alpha_s$ can not be considered small and correspondingly the light quarks are highly relativistic. Such systems are perhaps more aptly described by a bag model which supposes free quarks confined inside an effective spherical shell of radius of approximately $R_s$.

The effective potential at large separations is probed by the masses of high angular momentum hadronic states. Consider the sequences of hadronic excitations listed in Table 7.2 and joined by lines in Figure 7.15. The square of the mass is very nearly proportional to the angular momentum, $J = \alpha M^2$, with $\alpha_\rho = 0.84$, $\alpha_K = 0.86$ and $\alpha_N = 0.99$, $\alpha_\Delta = 0.92$, $\alpha_\Lambda = 0.94$ in $\mathrm{GeV}^{-2}$. These mass sequences can be described in a simple color flux tube model as follows. Suppose a massless $q\bar{q}$ pair connected by a string or flux tube of length $2L$ with rest energy per unit length $k$, as illustrated in Figure 7.16. If the string rotates such that the ends move at light speed, the speed at distance $r$ from the center is $v(r) = r/L$ and the energy is

$$M = 2\int_0^L \frac{kdr}{\sqrt{1 - v(r)^2}} = k\pi L. \tag{7.112}$$

| | | | |
|---|---|---|---|
| $\rho(770, 1^-)$ | $A_2(1318, 2^+)$ | $\rho_2(1690, 3^-)$ | $a_4(2050, 4^+)$ |
| $K(890, 1^-)$ | $K(1430, 2^+)$ | $K(1780, 3^-)$ | $K(2045, 4^+)$ |
| $N(940, \frac{1}{2}^+)$ | $N(1680, \frac{5}{2}^+)$ | $N(2220, \frac{9}{2}^+)$ | |
| $\Delta(1232, \frac{3}{2}^+)$ | $\Delta(1900, \frac{7}{2}^+)$ | $\Delta(2420, \frac{11}{2}^+)$ | |
| $\Lambda(1116, \frac{1}{2}^+)$ | $\Lambda(1820, \frac{5}{2}^+)$ | $\Lambda(2350, \frac{9}{2}^+)$ | |
| $\Lambda(1520, \frac{3}{2}^-)$ | $\Lambda(2100, \frac{7}{2}^-)$ | | |

**TABLE 7.2** Angular momentum and mass of light quark hadrons. Mass in MeV and $J^P$ are given in parentheses.

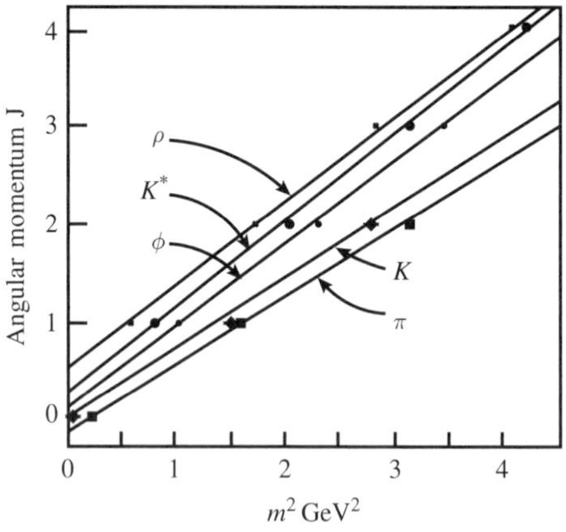


**FIGURE 7.15** Angular momentum versus mass squared of mesons. The linear dependence of sequences (Regge trajectories) is expected in a flux tube model of the QCD interaction at large distances. [Gunnar S. Bali, Phys. Rept. **343**, 1 (2001)]

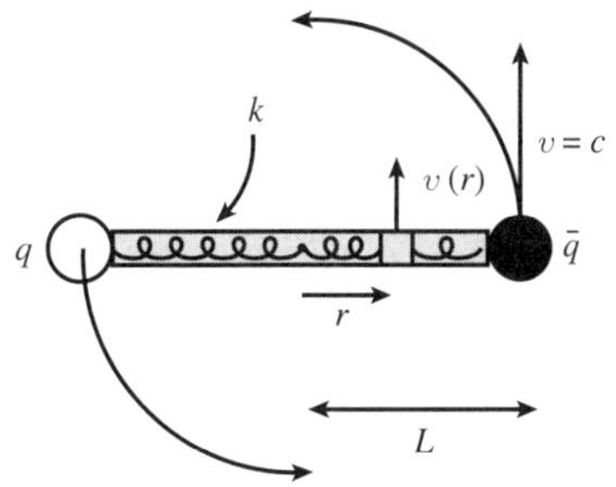


**FIGURE 7.16** String or flux tube model of a high orbital angular momentum meson. A tube of color electric flux with constant energy per unit length joins the quark and antiquark which orbit each other at light speed.

The orbital angular momentum is

$$J = 2\int_0^L \frac{kv(r)rdr}{\sqrt{1-v(r)^2}} = k\pi\frac{L^2}{2}. \tag{7.113}$$

Then $J = M^2/(2\pi k)$ and the experimental value $\alpha = 0.9$ GeV$^{-2}$ implies an energy density

$$k = 0.18 \text{ GeV}^2 \sim 1 \text{ GeV fm}^{-1}. \tag{7.114}$$

Constant linear energy density implies a linear potential energy $V = kr$. From Gauss's Law, the total flux from a charge $q = \sqrt{4\pi\alpha_s}$ is $\sqrt{\frac{4}{3}}q$ so, if the flux tube diameter is the confinement scale 1 fm, the expected energy density is for $\alpha_s \sim 1/2$

$$k = \frac{1}{2}\mathbf{E}^2 A = \frac{2}{3}\frac{q^2}{A} = \frac{32}{3d^2}\alpha_s \simeq 0.2 \text{ GeV}^2 \tag{7.115}$$

consistent with the experimental value. In chromodynamics, two quarks in a color triplet state attract with half the Coulomb binding energy of a singlet $q\bar{q}$ so a high angular momentum baryon may be a color triplet di-quark plus color triplet quark configuration with the same string constant as a meson.

The effect of color confinement on perturbative calculations bears some discussion. Consider for example the process $e^+e^- \to q\bar{q}$. The second order electromagnetic amplitude for production of a free $q$ and $\bar{q}$ may be calculated with quantum electrodynamics. Color is however confined - as the colored quarks separate, they dress themselves into mesons. Near threshold, the pair may bind resonantly into a single vector meson, a process not described by a perturbative calculation. Well above threshold however, intuition suggests the relatively soft process of energy fragmentation into hadrons is independent of the primary production process and this factorization is supported by higher order QCD calculations. Hence the total cross section for $e^+e^- \to$ hadrons is approximately that calculated treating the quarks as if they were free particles.

## 7.9 ■ COLOR DECONFINEMENT

While confinement of colored quarks and gluons is observed in experiments with hadronic matter at low energy density, lattice calculations suggest that, at a temperature equivalent to 150-200 MeV per quark, hadronic matter may form a deconfined state analogous to an electromagnetic plasma. Such a state should have been present in the early stages of the universe.

Collisions of gold nuclei with an energy up to 130 GeV per nucleon have been studied at the Brookhaven National Laboratory Relativistic Heavy Ion Collider (RHIC) to explore the phase diagram of hadronic matter and new results are emerging from heavy ion collisions at the LHC. In a single collision, thousands of particles may be produced from the available energy. Such collisions cannot be described in terms of the uncorrelated interactions of individual nucleons or quarks. In a simple picture, the Lorentz contracted colliding nuclei produce an elongated fireball of many interacting quarks and gluons which exhibit collective behavior.

Experiments examine the spectrum of produced particles, the propagation through the fireball of probe photons, leptons, quarks, and identifiable bound states such as vector mesons. The flow of energy as a function of impact parameter is also measured. The number and transverse momentum distributions of stable hadrons provide information on the temperature and density in the fireball. The spectrum of $e^+e^-$ pairs and photons provide a different measure of collision density and temperature. Correlations are used to tag probe particles. For example, a jet observed emerging from the periphery of the collision produced by a $qq \to qq$ process should be balanced by a jet corresponding to a quark which passes through and interacts with the fireball. It is observed that this probe quark does not produce a normal jet but instead a diffuse spray of low energy particles.

The collective properties of the hadronic matter are manifest in the azimuthal energy flow. When two nuclei collide at nonzero impact parameter, the overlap region has an almond shape. Leading order local QCD interactions within the overlap region will produce particles with a uniform azimuthal distribution. Collective behavior implies in effect a shorter particle wavelength along the short axis of the almond compared to the long axis and an azimuthal dependence to the transverse momentum distribution. Experimentalists fit the spectrum of particles for each event to the form

$$\frac{dN}{p_t dp_t dy d\phi} = \frac{1}{2\pi}\frac{dN}{p_t dp_t dy}\left[1 + \Sigma_{i=1}^{\infty} 2v_i \cos(i\phi - \phi_0)\right] \tag{7.116}$$

where $p_t$ is the magnitude of the transverse momentum, $y$ the rapidity, and $\phi$ the azimuthal angle in the transverse plane. The phase $\phi_0$ and the coefficient $v_1$ contain information on the impact parameter and orientation of the event. The coefficient $v_2$ is sensitive to collective properties.

The pressure and density gradient in a plasma-like state is expected to produce particles with an angular distribution distinct from that of an incompressible fluid. Plasma and fluid states are modeled with relativistic thermodynamics, relativistic transport equations, and relativistic hydrodynamics, the deviations from ideal hydrodynamics characterized by a shear viscosity parameter $\eta$. Present observations of energy flow suggest that at RHIC energies an unanticipated liquid state of matter is produced rather than a plasma. Bulk excitation modes and wakes associated with the passage of probe particles may further elucidate the picture.

The nuclear collisions at RHIC and at the LHC reproduce conditions in the universe a few microseconds after the big bang singularity. As the soup of quarks and antiquarks cools, coalescence into hadrons and anti-hadrons results. The tracks of a candidate hyper-anti-tritium nucleus produced in heavy ion collisions and decaying to anti-helium are shown in Figure 7.1. High statistics observations of the production of such nuclei will help test models of primordial nucleosynthesis which predict the primordial abundance of light isotopes.

## 7.10 ■ FURTHER READING

The 2004 Nobel Prize in Physics was awarded to David J. Gross, H. David Politzer, and Frank Wilczek for the development of QCD and "the discovery of asymptotic freedom in the theory of the strong interaction."

NonAbelian gauge theory is credited to C. N. Yang and R. L. Mills, Phys. Rev. **96**, 191 (1954). Seminal papers on QCD are H. Fritzsch, M. Gell-Mann, H. Leutwyler, Phys. Lett. **B 47**, 365 (1973), D. J. Gross, F. Wilczek, Phys. Rev. Lett. **30**, 1343 (1973), and H. D. Politzer, Phys. Rev. Lett. **30**, 1346 (1973).

A monograph on quantum chromodynamics is R. K. Ellis, W. J. Stirling, and B. R. Weber, *QCD and Collider Physics*, Cambridge, 1996. Some quirks of QCD are discussed in the short text F. J. Yndurain, *Quantum Chromodynamics*, Springer-Verlag, New York (1983). Aspects of QCD as a classical theory are discussed in K. Gottfried and V. F. Weisskopf, *Concepts of Particle Physics*, Vol. I and II, Oxford (1986). The classical chromodynamics of two sources is described in Roger A. Freedman, Lawrence Wilets, Stephen D. Ellis, and Earnest M Henley, Phys. Rev **D22**, 3128 (1980).

Visualizations of quantum chromodynamics by D. Leinweber are available at `http://www.physics.adelaide.edu.au/theory/staff/leinweber/VisualQCD/Nobel`.

For results of numerical computations (lattice QCD) of the effective potential between two quarks in hadronic bound states, see for example Y. Nakagawa, A. Nakamura, T. Saito, H. Toki, and D. Zwanziger, Properties of color-Coulomb string tension, Phys. Rev. **D73**, 094504 (2006) and "QCD forces and heavy quark bound states," and Gunnar S. Bali, Phys. Rept. **343**, 1 (2001), arXiv:hep-ph/0001312v2. An application to nuclear forces is "Hyperon-Nucleon Force from Lattice QCD," Hidekatsu Nemura, Noriyoshi Ishii, Sinya Aoki and Tetsuo Hatsuda, J. Phys. Lett **B2**, 3 (2009), arXiv:0806.1094v4 [nucl-th]. An overview of lattice QCD is Andreas S. Kronfeld, Science **322**, 1198 (2008) DOI: 10.1126/science.1166844.

The bag model was introduced in R Jaffe, K Johnson, C Thorn, V Weisskopf, Phys. REv. **d9**, 3471 (1974).

Annihilation and radiation of charmonium is reviewed in M. B. Voloshin, Progress in Particle and Nuclear Physics **61**, 455 (2008) and E. Eichten, S. Godfrey, H. Mahlke, J. L. Rosner, Report CLNS-07-1988, EFI-06-15, FERMILAB-PUB-07-006-T, Jan 2007. arXiv:hep-ph/ 0701208.

The applications to quarkonia of decay rate formulae derived in this chapter along with relativistic and higher order corrections are reviewed in W. Kwong, J. L. Rosner, and C Quigg. Ann. Rev. Nucl. Part. Sci. **37**, 325 (1987) and W. Kwong, P. B. Mackenzie, R. Rosenfeld, and J. L. Rosner, Phys. Rev. D37, 3210 (1988).

A comprehensive guide to the literature is Andreas S. Kronfeld, Chris Quigg, "Resource Letter: Quantum Chromodynamics,'' FERMILAB-PUB-10/040-T, arXiv:1002.5032v2 [hep-ph].

The four major experiments at RHIC are described in I. Arsene et al. [BRAHMS Collaboration], Nucl. Phys. **A757**, 1 (2005), K. Adcox et al. [PHENIX Collaboration], Nucl. Phys. **A757**, 184 (2005), B. B. Back et al. [PHOBOS Collaboration], Nucl. Phys. **A757**, 28 (2005), and J. Adams et al. [STAR Collaboration], Nucl. Phys. **A757**, 102 (2005).

Relativistic hydrodynamical models of high energy nuclear collisions are described in Kevin Dusling, Guy Moore, Derek Teaney, 'Radiative energy loss and v2 spectra for viscous hydrodynamics,' arXiv:0909.0754v2 [nucl-th] 8 Feb 2010. The coalescence model for the production of composite particles in high energy collisions is described in H. Sato, K. Yazaki, Phys. Lett. **B98**, 153 (1981).

Bound states of possible supersymmetric particles carrying color include squarkonium composed of a color triplet scalar quark plus color triplet anti-scalar quark forming color singlet bound states with a color Coulomb interaction similar to $q\bar{q}$ bound states but a distinct different parity and spin correlation and spectrum, and gluinonium composed of two color octet massive spin 1/2 particles. An introduction is E. Chikovani, V. Kartvelishvili, R. Shanidze, G. Shaw, Phys. Rev. **D53**, 6653 (1996).

## 7.11 ■ PROBLEMS

### Problem 7.1. Unitary matrix groups

A unitary matrix satisfies $U^\dagger = U^{-1}$. a) Use the matrix rule $(AB)^\dagger = B^\dagger A^\dagger$ to show that the product of two unitary transformations $U_1$ and $U_2$ is unitary. Since the unit matrix is unitary, the set of unitary matrices of a given dimension, with the usual matrix product, form what in mathematics is called a group.

b) A unitary matrix may be written as $U = \exp(i\mathbf{a}\cdot\mathbf{T})$ where $\mathbf{a}$ is a vector of real parameters and $\mathbf{T}$ are generators. Write

$$U = 1 + i\mathbf{a}\cdot\mathbf{T} + ...$$

and use a $UU^\dagger = 1$ to show that $\mathbf{T}^\dagger = \mathbf{T}$. Since for invertible square matrices, det $\exp(A) = \exp(\text{tr } A)$, we have tr $\mathbf{T} = 0$ for elements of the special unitary group $SU(N)$ of unitary matrices with unit determinant. c) Products of unitary matrices like $U(\mathbf{a})U(\mathbf{b})$ are unitary. Show that to second order

$$U(\mathbf{a})^{-1}U(\mathbf{b})^{-1}U(\mathbf{a})U(\mathbf{b}) = 1 + \mathbf{b}\cdot\mathbf{Ta}\cdot\mathbf{T} - \mathbf{a}\cdot\mathbf{Tb}\cdot\mathbf{T}.$$

Since this matrix is unitary, the commutator of two generators must be expressible in terms of generators as

$$T_aT_b - T_bT_a = if_{abc}T_c.$$

Take the trace of this expression and use tr $AB =$ tr $BA$ to verify tr $\mathbf{T} = 0$. d) A traceless Hermitian matrix is described by $N^2 - 1$ parameters. The linear space of traceless Hermitian matrices with the tr as an inner product can correspondingly be spanned by a basis of $N^2 - 1$ generator matrices (Figure 7.17) which may be chosen to satisfy the orthonormality conditions tr $T_aT_b = \delta_{ab}/2$ where the factor 1/2 is conventional. The non-vanishing $SU(3)$ structure constants for the generators we use are $f_{123} = 1$, $f_{147} = f_{246} = f_{257} = f_{345} = f_{516} = f_{637} = 1/2$ and $f_{458} = f_{678} = \sqrt{3}/2$.
The anti-commutator of two generators is Hermitian and therefore expressible in terms of the unit matrix and the generators:

$$T_aT_b + T_aT_b = (1/N)\delta_{ab} + d_{abc}T_c.$$

Use tr $\mathbf{T} = 0$ and tr$(T_aT_b) = \delta_{ab}/2$ to verify the coefficient of the unit matrix in this expression. Multiply by $T_c$ to show that

$$d_{abc} = 2\text{tr}([T_a, T_b]_+T_c]) \equiv 2\text{tr}([T_aT_b + T_bT_a]T_c]).$$

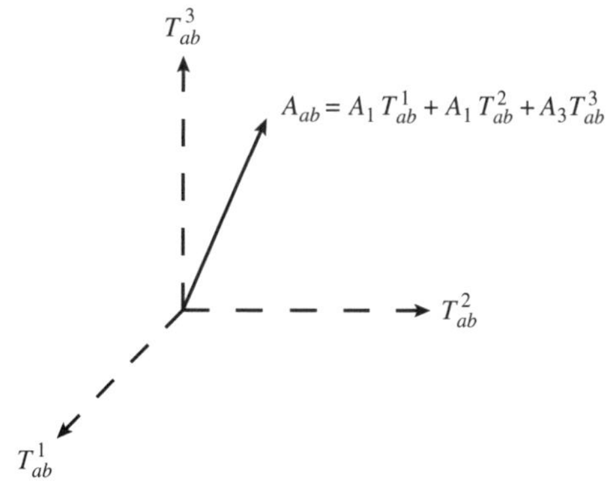


**FIGURE 7.17** Representation of the space generated by the Hermitian generator matrices $\mathbf{T}_{ab}$ of $SU(2)$. A sum of generator matrices with coefficients $A_1$, $A_2$ and $A_3$ corresponds to a 3-vector $\mathbf{A}$.

Use tr $ABC = \text{tr } CAB = \text{tr } BCA$ to deduce that $d_{abc}$ is completely symmetrical in its indices. Multiply the commutation relations by $T_c$ and take the trace to derive the expression

$$f_{abc} = -i\text{tr}([T_a, T_b]T_c)/\text{tr}(T_c^2).$$

---

**Problem 7.2. Vector representation of $SU(N)$**

---

Prove the Jacobi identity

$$[[A, B], C] + [[B, C], A] + [[C, A], B] = 0$$

holds for any matrices $A$, $B$, and $C$, where $[A, B] \equiv AB - BA$. Apply the identity to the generators of $SU(N)$ with $A = T_a$, $B = T_b$, $C = T_c$ and use the commutation relations $[T_a, T_b] = if^{abc}T_c$ to derive a quadratic relation between the structure constants. Show that this may be interpreted as the commutation relations for $N^2 - 1$ matrices $\mathbf{T}_V$ of dimension $N^2 - 1 \times N^2 - 1$ defined as

$$(\mathbf{T}_V^a)^{bc} = -if^{abc}.$$

These generate a group of $(N^2 - 1) \times (N^2 - 1)$ dimensional matrices which can transform $N^2 - 1$ dimensional vectors in correspondence with the $SU(N)$ matrices generated by $\mathbf{T}$.

---

**Problem 7.3. Isospin symmetry**

---

Consider a complex scalar (nucleon-like) $SU(2)$ doublet $N$ and real scalar (pion-like) triplet

$$N = \begin{pmatrix} p \\ n \end{pmatrix}, \quad \boldsymbol{\pi} = \begin{pmatrix} \pi_1 \\ \pi_2 \\ \pi_3 \end{pmatrix}$$

with masses $m_N$ and $m_\pi$ and the Lagrangian

$$L = \frac{1}{2}\partial_\mu \boldsymbol{\pi} \cdot \partial^\mu \boldsymbol{\pi} - \frac{1}{2}m_\pi^2 \boldsymbol{\pi}^2 + \partial_\mu N^\dagger \partial^\mu N - m_N^2 \boldsymbol{\pi}^2 - N^\dagger (g) \boldsymbol{\pi} \cdot \boldsymbol{\sigma} N.$$

For a 4-spinor nucleon doublet and a parity conserving interaction with a pseudoscalar triplet assumed, the free field Lagrangian would be that for spinor fields and coupling constant $g$ would be multiplied by $\gamma^5$. a) The interaction term is similar to the magnetic energy of a spin 1/2 charged fermion $U = \phi^\dagger \mathbf{B} \cdot \boldsymbol{\mu}\phi$. Using the analogy with rotations in space of 2-component spinors, show that this interaction term and the free field Lagrangian are invariant under global $SU(2)$ (isospin) transformations of the form

$$N \to UN = \exp\left(-i\frac{\theta}{2}\hat{\mathbf{n}} \cdot \boldsymbol{\sigma}\right) N, \quad \boldsymbol{\pi} \to R(\theta, \hat{\mathbf{n}})\boldsymbol{\pi}$$

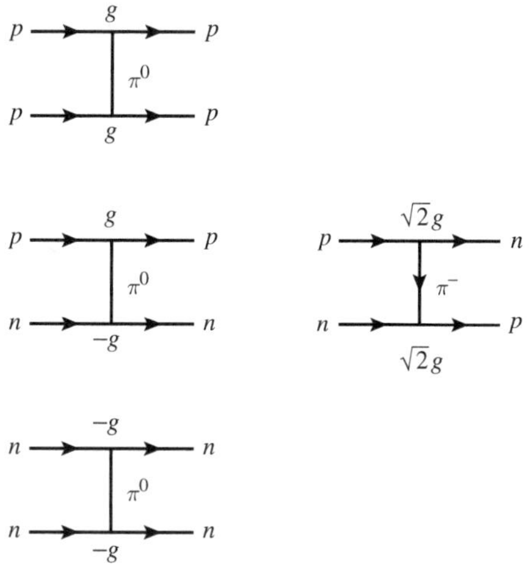


**FIGURE 7.18** Pion exchange diagrams.

where $R(\theta, \hat{\mathbf{n}})$ is an orthogonal (rotation) matrix by showing that show that

$$U\boldsymbol{\pi}\cdot\boldsymbol{\sigma}U^{\dagger} = (R\boldsymbol{\pi})\cdot\boldsymbol{\sigma}$$

is valid for an infinitesimal transformation with $R\boldsymbol{\pi} = \boldsymbol{\pi} - \theta\hat{\mathbf{n}}\times\boldsymbol{\pi}$. b) Using the fields defined by

$$\pi^{\mp} \equiv \frac{1}{\sqrt{2}}(\pi_1 \pm i\pi_2),\ \pi^0 \equiv \pi_3,$$

Show that the interaction term is

$$L_{\text{int}} = -g\left(p^{\dagger}\pi^0 p + \sqrt{2}p^{\dagger}\pi^{+}n + \sqrt{2}n^{\dagger}\pi^{-}p - n^{\dagger}\pi^0 n\right)$$

and implies Feynman diagrams for pion exchange shown in Figure 7.18.

c) Show that the scattering matrix elements for single pion exchange satisfy

$$M_{pp\to pp} = M_{\frac{1}{\sqrt{2}}(pn+np)\to\frac{1}{\sqrt{2}}(pn+np)} = M_{nn\to nn}.$$

---

**Problem 7.4. Field strength tensor**

---

The product of covariant derivatives in gauge theory is

$$\begin{aligned}D_\mu D_\nu\psi &= (\partial_\mu + ig\mathbf{T}\cdot\mathbf{A}_\mu)(\partial_\nu + ig\mathbf{T}\cdot\mathbf{A}_\nu)\psi \\ &= \partial_\mu\partial_\nu\psi + ig\mathbf{T}\cdot(\partial_\mu\mathbf{A}_\nu)\psi + ig\mathbf{T}\cdot\mathbf{A}_\nu\partial_\mu\psi + ig\mathbf{T}\cdot\mathbf{A}_\mu\partial_\nu\psi \\ &\quad -g^2\mathbf{T}\cdot\mathbf{A}_\mu\mathbf{T}\cdot\mathbf{A}_\nu\psi.\end{aligned}$$

Use $[T_a, T_b] = i f_{abc} T_c$ to show that the antisymmetric combination is

$$(D_\mu D_\nu - D_\nu D_\mu)\psi = ig\mathbf{T} \cdot \mathbf{F}_{\mu\nu}\psi$$

where $\mathbf{F}^{\mu\nu} = \partial^\mu \mathbf{A}^\nu - \partial^\mu \mathbf{A}^\nu - g\mathbf{A}^\mu \times \mathbf{A}^\nu$.

### Problem 7.5. Squark chromodynamics

Suppose a triplet of complex scalar fields (one squark flavor in a supersymmetric model) $\phi_i (i = 1, 2, 3)$ with color $SU(3)$ transformation rule

$$\phi = \begin{pmatrix} \phi_1 \\ \phi_2 \\ \phi_3 \end{pmatrix} \to U(x)\phi.$$

a) Give the form of the gauge covariant derivative $D^\mu$ acting on $\phi$ in terms of the coupling $g$, the $SU(3)$ generator matrices $\mathbf{T}$, and the eight gauge boson fields $\mathbf{A}^\mu$.
b) Give the gauge covariant Lagrangian in terms of $D_\mu$, $\phi$ and the field strength tensor $\mathbf{F}_{\mu\nu}$, assuming all three components of $\phi$ have mass $m$. Is the mass term for $\phi$ consistent with $SU(3)$ gauge invariance? If another flavor of squark is added to the theory, does gauge symmetry require its coupling $g$ be identical?

### Problem 7.6. Squark interactions

Suppose a triplet of complex scalar boson fields $\phi_i (i = 1, 2, 3)$ with color gauge interactions as in the previous problem. Following the example of scalar electrodynamics, show that the Euler-Lagrange equation for $\phi$ interacting with external gluon fields is

$$\partial_\mu \frac{\partial L}{\partial \partial_\mu \phi^\dagger} - \frac{\partial L}{\partial \phi^\dagger} = \partial_\mu (D^\mu \phi) - \left[-ig\mathbf{T} \cdot \mathbf{A}_\mu D^\mu \phi - m^2 \phi\right] = 0$$

or $D_\mu D^\mu \phi + m^2 \phi = 0$. Expand the Euler Lagrange equation above moving interaction terms to the right-hand side. For each of the interaction terms, draw a Feynman diagram and give the vertex factor for a general set of initial and final color indices.

### Problem 7.7. Two-jet cross sections

For a two-to-two process for massless particles, the Mandelstam variables satisfy $s + t + u = 0$ and $u$ and $t$ may be expressed in terms of $s$ and the scattering angle $\theta$ as

$$t = -s \sin^2 \frac{\theta}{2} \;;\; u = -s \cos^2 \frac{\theta}{2}.$$

Evaluate the cross section factors $(\pi\alpha_s^2)^{-1}d\sigma/dt$ given in Table 7.1 for $\theta = \pi/2$. In $p\bar{p}$ collisions, jet pairs with masses close to the kinematic limit are dominantly produced by valence quark scattering $u\bar{u}$, $u\bar{d}$, $d\bar{u}$, and $d\bar{d}$ because a gluon tends to carry a small fraction of the hadron energy. Is some final state quark flavor combination dominant at high masses at 90 degrees in the two-jet center of mass?

---

**Problem 7.8. Four gluon vertex factor**

---

Derive the 4-gluon vertex factor from the gluon field equations.

---

**Problem 7.9. Three gluon states**

---

Quantum chromodynamics is symmetric under the charge conjugation transformation $\mathbf{T}\cdot\mathbf{A} \to -\mathbf{T}^t\cdot\mathbf{A}$ and $C$ parity conservation applies therefore to hadron decays to multi-gluon states. Use the commutation and anti-commutation relations and $\mathrm{tr}\mathbf{T} = 0$ to show that

$$\mathrm{tr}[\mathbf{T}\cdot\mathbf{A}_1\mathbf{T}\cdot\mathbf{A}_2\mathbf{T}\cdot\mathbf{A}_3] \to -\mathrm{tr}[\mathbf{T}\cdot\mathbf{A}_1\mathbf{T}\cdot\mathbf{A}_2\mathbf{T}\cdot\mathbf{A}_3] - 4i\mathbf{A}_1\cdot\mathbf{A}_2\times\mathbf{A}_3$$

so $\mathrm{tr}[\mathbf{T}\cdot\mathbf{A}_1\mathbf{T}\cdot\mathbf{A}_2\mathbf{T}\cdot\mathbf{A}_3] + 1 \Leftrightarrow 2$ is $C$-odd. Using the explicit structure constants and $CA_i = +A_i$ for $i = 2, 5$, and 7, and $CA_i = -A_i$ for $i = 1, 3, 6$, and 8, show that $f^{abc}A_1^aA_2^bA_3^c$ is $C$-even.

---

**Problem 7.10. Leptonic decays of vector mesons and $\alpha_s$**

---

Apply the decay formulae for $\Gamma_{^3S_0(q\bar{q})\to e^+e^-}$ and $\Gamma_{^3S_1(q\bar{q})\to ggg}$ to the $\Upsilon$ to deduce $\alpha_s(m_\Upsilon)$. Take the value from the Particle Data Group for $\Gamma_{\Upsilon_{1s}\to e^+e^-}$ and use PDG values to calculate

$$\Gamma_{\Upsilon\to ggg} = \Gamma_\Upsilon - \Gamma_{\Upsilon\to e^+e^-} - \Gamma_{\Upsilon\to\mu^+\mu^-} - \Gamma_{\Upsilon\to\tau^+\tau^-} - \Gamma_{\Upsilon\to q\bar{q}}$$

assuming a calculated theoretical value $\Gamma_{\Upsilon\to q\bar{q}}/\Gamma_\Upsilon = 10\,\%$.

---

**Problem 7.11. Radiative decay to gluons**

---

The primary decay modes of the $\Upsilon(1S)$ are to three gluons, to a virtual photon, and to two gluons plus a photon. The decay widths should be proportional to $\alpha_s^3$, $\alpha^2$, and $\alpha_s^2\alpha$. The two gluon plus photon decay rate of an s-wave state is

$$\Gamma_{n^3S_1\to gg\gamma} = \frac{128(\pi^2-9)e_q^2\alpha\alpha_s^2}{9(2m)^2} \mid \psi_n(0)\mid^2 .$$

Derive this result from the electromagnetic 3-photon width $\Gamma_{n^3S_1\to\gamma\gamma\gamma}$ paying attention to the number of distinguishable final states (a factor 3!/2! is required), and apply it to $\Upsilon_{1s}$ to compare the ratio of $gg\gamma$ to $ggg$ final states and predict the fraction of decays to single photons. The observed fraction is $2.75 \pm 0.04$ (stat) $\pm 0.15$ (syst)%. B. Nemati *et al.*, Phys. Rev. **D55**, 5273 (1997).

---

### Problem 7.12. Toponium

---

Suppose a color interaction potential energy $V = -\frac{4}{3}\alpha_s/r$ for $t\bar{t}$ with $\alpha_s \simeq 0.1$ and $m_t = 171$ GeV. a) What is the "Bohr radius" in fm of toponium? b) Find the corresponding classical period in seconds for $t$ and $\bar{t}$ in a circular orbit about their center of mass at this radius. c) Estimate the top quark lifetime and the number of classical orbits before weak decay of the $t$ or $\bar{t}$.

CHAPTER

# 8 Electroweak Standard Model

The interactions mediated by W and Z bosons are called weak interactions and appear feeble at energies much smaller than the masses of these bosons. In the standard model, weak and electromagnetic interactions derive from gauge symmetries and in a sense are unified. The masses of the bosons and also of fermions are ascribed to spontaneous breaking of the gauge symmetry through the Higgs mechanism. In this chapter, this theory of electroweak interactions is introduced and applied.

## Contents



## 8.1 ■ ORIGINS OF WEAK INTERACTION THEORY

In this chapter, we will study the standard model of the weak interaction, the most complex of established theories in particle physics. We begin with a brief overview of select weak interaction phenomena that motivate the model and then turn to the gauge theory of weak interactions and unification with electromagnetic phenomena through the Higgs mechanism.

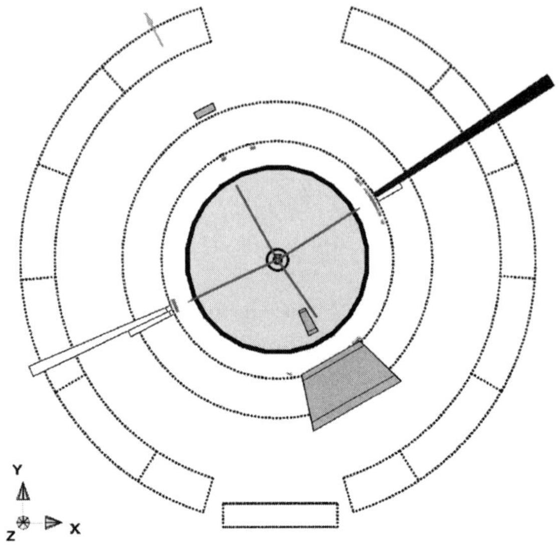


**FIGURE 8.1** Double $Z$ boson production event. An example of the electroweak process $e^+e^- \to ZZ$, this event was observed in the OPAL detector at LEP. One Z decays to $e^+e^-$ at upper right and lower left. The other decays to $\tau^+\tau^-$ with one $\tau$ decaying to muon and neutrinos (upper left) and the other $\tau$ decaying to a pion and neutrino lower right. [CERN photo archive, POW 971119]

Neutron $\beta$-decay $n \to pe^-\bar{\nu}_e$ and muon decay $\mu^- \to \nu_\mu e^-\bar{\nu}_e$ are evidence for a fundamental flavor changing interaction common to quarks and leptons. Neutron $\beta$-decay involves a quark transition $d \to u$ and the creation of a lepton anti-lepton pair $e^-\bar{\nu}_e$. The muon transition $\mu^- \to \nu_\mu$ is accompanied by the same lepton anti-lepton pair creation process. That the lifetimes for such flavor changing decay processes are long compared to electromagnetic and strong decay lifetimes also suggests a common origin in an independent so-called weak interaction.

The first field theory of the weak interaction is ascribed to Fermi who proposed in effect a 4-fermion contact interaction Lagrangian for nuclear $\beta$-decay, incorporating Dirac spinor fields for the proton, the newly discovered neutron, the electron, and the hypothetical neutrino. One candidate effective interaction Lagrangian is

$$L = G\bar{\psi}_p\gamma_\mu\psi_n\bar{\psi}_e\gamma^\mu\psi_\nu = Gj_\mu(h)j^\mu(l) \tag{8.1}$$

where $G$ is a constant and the flavor transition currents defined by

$$j_\mu(h) = \bar{\psi}_p\gamma_\mu\psi_n \text{ and } j_\mu(l) = \bar{\psi}_e\gamma^\mu\psi_\nu \tag{8.2}$$

are similar to the electromagnetic current $j = \psi_p\gamma\psi_p$. Such a Lagrangian can represent the low energy effective interaction generated by the exchange of a heavy particle between fermions. Though prescient, this form is not quite correct.

The development of models of the nucleus and radioactivity led to the discovery of spin altering weak interactions ($\Delta s = 1$, $\Delta L = 0$) and suggested model interactions with more general Lorentz structure. A critical extension to parity violating forms was motivated by parity violation in the weak decays of $K$ mesons. The observations of parity violation are described below. Suffice it to say that kaons have negative intrinsic parity and spin 0 yet decay to pairs of pions which each have negative parity and total parity positive. Parity violation in the weak interaction was confirmed in the correlation of electron momentum $\mathbf{p}$ with nuclear spin $\mathbf{s}$ in the decays of spin polarized nuclei and muons. If parity is conserved, mirror processes with $\mathbf{p} \to -\mathbf{p}$ and $\mathbf{s} \to \mathbf{s}$ should occur with equal probability while the angle between these vectors for electron velocity $v$ is consistent with $I(\theta) = 1 - v\cos\theta$, favoring emission anti-parallel to the nuclear spin.

From these origins emerged a refined theory of weak interactions in which weak flavor changing processes result from parity violating coupling of fermions to a charged vector boson, the $W^{\pm}$. The weak charge is the same for all lepton pairs and also for quark pairs with a certain unitary flavor twist. The apparent weakness of the interaction at energies small compared to about 100 GeV is a consequence of the large $W$ boson mass. Weak interactions distinguish right-handed and left-handed fermion field components defined by the Lorentz covariant expressions

$$\begin{aligned} \psi &= \frac{1}{2}(1+\gamma_5)\psi + \frac{1}{2}(1-\gamma_5)\psi = \psi_R + \psi_L \\ \psi_R &\equiv \frac{1}{2}(1+\gamma_5)\psi \\ \psi_L &\equiv \frac{1}{2}(1-\gamma_5)\psi. \end{aligned} \tag{8.3}$$

Only left-handed components couple to the $W$ boson. Put another way, the left and right-handed charges are maximally different, and parity is maximally violated. These observations are summarized by an interaction Lagrangian

$$L = -\frac{g}{\sqrt{2}} j^{\mu} W^{+}_{\mu} + h.c. \tag{8.4}$$

where $W^+$ is a complex 4-vector field representing the $W$ boson. The charged current for leptons is

$$j_{\mu} = \bar{\nu}_e O_{\mu} e + \bar{\nu}_{\mu} O_{\mu} \mu + \bar{\nu}_{\tau} O_{\mu} \tau \tag{8.5}$$

where the particle names are used to denote the corresponding 4-spinor fields and the Dirac operator $O_{\mu} = \gamma_{\mu}(1-\gamma_5)/2$ projects out the left-handed components.

To understand the Lorentz structure of the current, notice that $\gamma_5^{\dagger} = \gamma_5$ and $\gamma^{\mu}\gamma_5 = -\gamma_5\gamma^{\mu}$ allow us to write

$$\begin{aligned} \bar{\psi}_L &= \left(\frac{1-\gamma_5}{2}\psi\right)^{\dagger}\gamma^0 = \psi^{\dagger}\left(\frac{1-\gamma_5^{\dagger}}{2}\right)\gamma^0 \\ &= \bar{\psi}\frac{1}{2}\left(1-\gamma^0\gamma_5^{\dagger}\gamma^0\right) = \bar{\psi}\frac{1}{2}(1+\gamma_5). \end{aligned} \tag{8.6}$$

The Dirac matrix commutation relations imply

$$\left(1+\gamma^5\right)\gamma^\mu\left(1-\gamma^5\right)=\gamma^\mu\left(1-\gamma^5\right)^2=1-2\gamma^5+\gamma_5^2=2\left(1-\gamma^5\right). \quad (8.7)$$

Hence a transition current between purely left-handed fields may be expressed as

$$\bar{\psi}_L\gamma\psi_L=\bar{\psi}\frac{1}{4}\left(1+\gamma^5\right)\gamma^\mu\left(1-\gamma^5\right)\psi=\bar{\psi}\frac{1}{2}\gamma^\mu\left(1-\gamma^5\right)\psi. \quad (8.8)$$

This peculiar current is also often expressed as a difference between a vector current $\bar{\psi}\gamma^\mu\psi$ and a pseudovector current $\bar{\psi}\gamma^\mu\gamma^5\psi$ called an axial current as

$$\bar{\psi}_L\gamma\psi_L=\frac{1}{2}\bar{\psi}\gamma^\mu\psi-\frac{1}{2}\bar{\psi}\gamma^\mu\gamma^5\psi=\frac{1}{2}j_V^\mu-\frac{1}{2}j_A^\mu. \quad (8.9)$$

The so-called $V-A$ form of the fundamental interaction emerged from the analysis of angular distributions assuming scalar, pseudoscalar, vector, axial vector, and tensor interaction terms. The transition current may be most simply expressed in terms of the lower 2-component spinors in the spinor representation as

$$\bar{\psi}_L\gamma^\mu\psi_L=j_-^\mu=\phi_-^\dagger\sigma^{+,\mu}\phi_-=\phi_-^\dagger\left[1,-\boldsymbol{\sigma}\right]\phi_-. \quad (8.10)$$

The Feynman rules for the charged current weak interactions of leptons may be derived following the example of electrodynamics. The $W\nu e$ vertex factor can be read from the Lagrangian or field equations and is

$$-i\left(g/\sqrt{2}\right)\gamma_\mu\left(1-\gamma^5\right)/2. \quad (8.11)$$

Sandwiched between spinors in a transition matrix element for a process like $\mu^-\to\nu_\mu W^-$, this vertex factor projects out the left-handed components of the spinor fields. As derived in Chapter 5, the propagator for the vector boson is

$$G_{\mu\nu}^W(p)=\frac{-i\left(g_{\mu\nu}-p_\mu p_\nu/m_W^2\right)}{p^2-m_W^2} \quad (8.12)$$

When $p^2 << m_W^2$, the effective propagator is

$$G_{\mu\nu}^W\simeq\frac{ig_{\mu\nu}}{m_W^2} \quad (8.13)$$

and the effective Lagrangian for a second order process involving a virtual $W$ boson is a product of left-handed currents. For example, muon decay may be calculated from the effective Lagrangian

$$L=\frac{G_F}{\sqrt{2}}\bar{e}\gamma_\alpha\left(1-\gamma^5\right)\nu_e\bar{\nu}_\mu\gamma^\alpha\left(1-\gamma^5\right)\mu. \quad (8.14)$$

Here the fermion names denote corresponding 4-spinor fields and we have introduced the Fermi constant

$$G_F=\frac{g^2}{4\sqrt{2}m_W^2}. \quad (8.15)$$

This effective Lagrangian is used below to calculate angular and energy distribution of particles produced in muon decay and is in good agreement with experiment. The total muon decay rate is found to be

$$\Gamma_\mu = \frac{1}{192\pi^3} G_F^2 m_\mu^5 = \frac{1}{\tau_\mu}. \tag{8.16}$$

From the measured values of the lifetime $\tau_\mu$ and the mass $m_\mu$, the value of the Fermi constant

$$G_F \simeq 1.166 \times 10^{-5}\ \text{GeV}^{-2} \tag{8.17}$$

is deduced. An identical value is derived from other charged current processes involving leptons such as $\nu_\mu e^- \to \mu^- \nu_e$ which results from $W$ boson exchange.

With the development of the quark model, the weak interactions of hadrons came to be understood in terms of weak interactions of the constituent quarks and a complication emerged. The weak transitions $e^- \to \nu_e$ and $\mu^- \to \nu_\mu$ preserve lepton family and have identical weak coupling constant and Lorentz structure. The quark transitions exhibit the same Lorentz structure as the lepton transitions but are governed by an array of weak coupling constants. More precisely, the transitions $d \to u$, $s \to c$, and $b \to t$ between quarks within a family have slightly smaller weak coupling constants then the leptonic transitions $\nu_e \to e^-$, $\nu_\mu \to \mu^-$ and $\nu_\tau \to \tau^-$. Interfamily transitions including $s \to u$ and $d \to c$ have significantly smaller coupling constants.

Order was brought to the theory of weak interactions of the first two generations of quarks with the introduction of a unitary mixing model by Cabibbo. In the Cabibbo model, the Lagrangian for quarks is identical to that for leptons with the modified current density

$$\begin{aligned} j_{quark} &= \bar{u}_L \gamma\,(d_L \cos\theta_c + s_L \sin\theta_c) + \bar{c}_L \gamma\,(-d_L \sin\theta_c + s_L \cos\theta_c) \\ &= (\,\bar{u}_L \quad \bar{c}_L\,)\,\gamma \begin{pmatrix} \cos\theta_C & \sin\theta_C \\ -\sin\theta_C & \cos\theta_C \end{pmatrix} \begin{pmatrix} d_L \\ s_L \end{pmatrix} \end{aligned} \tag{8.18}$$

where the Cabibbo angle is $\theta_C = 14.04°$. The Cabibbo model was extended to three generations to encompass all flavor changing quark transitions by replacing the $d$-like fields in the interaction with those obtained by a 3 unitary matrix

$$\begin{pmatrix} d_L \\ s_L \\ b_L \end{pmatrix} \to V_{\text{CKM}} \begin{pmatrix} d_L \\ s_L \\ b_L \end{pmatrix} \tag{8.19}$$

where $V_{\text{CKM}}$ is called the Cabibbo-Kobayashi-Maskawa matrix. This model is successful in unifying the array of quark charged current interactions with those of leptons. All connect only left-handed fields and all share a common weak charge $g$.

A further complication first appeared in the study of neutrino interactions. In addition to such flavor changing charged current weak interactions as

$$\nu_\mu e^- \to \mu^- \nu_e \tag{8.20}$$

which derive from $W$ boson exchange, a neutral current interaction with strength comparable to the charged current weak interaction is observed in elastic scattering processes like

$$\nu_\mu e^- \to \nu_\mu e^- . \tag{8.21}$$

These processes may be interpreted as mediated by a neutral heavy vector boson called the $Z$. The presence of virtual $Z$ boson exchange is also seen in processes such as $\nu_e e^- \to \nu_e e^-$ and $\bar{\nu}_e e^- \to \bar{\nu}_e e^-$ in which both $W$ boson and $Z$ boson exchange contribute coherently. Like the electromagnetic interaction, the neutral current weak interaction does not alter fermion flavor. Like the photon, the $Z$ boson is found to couple to both left and right-handed electrons but with unequal strength.

## 8.2 ■ *P* AND *C* VIOLATION

Parity violation is a unique signature of weak interactions with a profound and still mysterious origin. The possibility of parity violation was first considered seriously after the discovery of the charged and neutral $K$ mesons and analysis of their decays. The weak decays of the $K^+$ to pions were at first ascribed to two different particles called $\theta^+$ and $\tau^+$ (not to be confused with the lepton that presently bears that name) of similar mass and lifetime with the decay modes:

$$\theta^+ \to \pi^+ \pi^0 \tag{8.22}$$

and

$$\tau^+ \to \pi^+ \pi^0 \pi^0, \pi^+ \pi^+ \pi^- . \tag{8.23}$$

The spin and parity of the pions are established through parity conserving electromagnetic and strong interactions. Some details are provided in Chapter 10. Given that the pions have odd parity and spin 0, while reflection of a wave function for two pions with angular momentum $L$ about their center of mass amounts to a factor $(-1)^L$ times the square of the intrinsic pion parity, parity conservation would imply the spin and parity of the $\theta^+$ is $J^P = 0^+, 1^-, ...$ independent of the of pion intrinsic parity . Assuming parity conservation, Dalitz was able to identify the spin and parity of the $\tau^+$ as $0^-$. To understand his conclusion, let $\mathbf{L}_1$ be the relative orbital angular momentum of the two identical pions and $\mathbf{L}_2$ be the angular momentum of the charged pion about the center of mass of the first two. The total angular momentum is $\mathbf{J} = \mathbf{L}_1 + \mathbf{L}_2$ and the parity of the three pion system is $P_{3\pi} = (-)^{L_1+L_2+1}$ when the intrinsic parity of the three pions is included. The two identical pions must be in a symmetric state to be consistent with exchange symmetry for bosons so $L_1$ is even. If $J = 0$, then $L_1 = L_2$ and $P = -1$. To establish that the spin $J$ of the $\tau^+$ was 0, Dalitz examined the energy sharing between the pions. For spin 0, no one pion is at all favored or disfavored, as observed. For nonzero angular momentum, one expects modifications to the smooth phase space distribution. For example, the condition in which the charged pion is at rest in the

$\tau^+$ rest frame is impossible for $J = 1$ since it could contribute only zero angular momentum while, for exchange symmetry, the two identical pions must have even angular momentum.

So, if the simplest choice $0^+$ for the $\theta^+$ was assumed based on parity conservation, parity conservation implies the two pion and three pion decays must originate from spin 0 mesons mesons of different parity but the same mass. The equality of the masses posed one puzzle. The equality of the lifetimes posed another - phase space would favor a shorter lifetime for the $\theta^+$. This quandary prompted C. N. Lee and T-D. Yang to propose tests of parity symmetry in the weak interaction. Parity violation in nuclear $\beta$ decay was first observed the following year by C. S. Wu *et al.* and was subsequently found to be a general feature of weak interactions. The $\theta^+$ and $\tau^+$ are the same particle, the $K^+$, with spin-parity $0^-$ like the pion. Parity is conserved in strong and electromagnetic interactions including those of the $K^+$ of the $K^+$ but the weak decays of the $\bar{s}$-quark in the $K^+$ and in other strangeness bearing hadrons violates parity.

The Wu experiment observed the decay of spin polarized cobalt nuclei

$$^{60}\mathrm{Co} \rightarrow {}^{60}\mathrm{Ni} + \mathrm{e}^- + \bar{\nu}_\mathrm{e} \tag{8.24}$$

and measured not just the electron energy spectrum, but the angular distribution of the electrons relative to the nuclear spin. (The polarization was produced by a technique called adiabatic demagnetization.) Symmetry under a transformation such as reflection requires the probability for a transition from one state to another to be equal to the probability for the transition between the transformed states under transformed conditions. In the initial state in the Wu experiment, the spin and magnetic moment of the nucleus are aligned parallel to a magnetic field $\mathbf{B}$. Under inversion, the magnetic field and magnetic moment are unchanged while the electron momentum $\mathbf{p}$ is reversed so transitions to states of momentum $\mathbf{p}$ and momentum $-\mathbf{p}$ should be equally probable. By rotational symmetry, the decay rate depends only on the polar angle $\theta$ between $\mathbf{p}$ and $\mathbf{B}$ so the rate for angle $\theta$ must be the same as the rate for $\pi - \theta$ and the average $< \mathbf{p} \cdot \mathbf{s} >$ should vanish. In other words, the angular distribution of the electron must be symmetric about the plane perpendicular to $\mathbf{B}$. Yet, the electron was observed preferentially emitted with momentum opposite to the spin direction!

Precisely the same effect is observed in the decay $\mu^- \rightarrow e^- \nu_\mu \bar{\nu}_e$. For spin polarized muons, the electron is emitted preferentially opposite to the direction of the spin vector of the $\mu^-$. Interestingly, for $\mu^+ \rightarrow e^+ \bar{\nu}_\mu \nu_e$, the positron is emitted preferentially along the spin direction of the $\mu^+$. Again parity is violated. These two processes are however related by $C$ as illustrated in Figure 8.2. The decay of a $\mu^+$ to a positron with angle $\theta$ relative to the $\mu^+$ spin direction is transformed by $C$ to the decay of a $\mu^-$ to an electron with angle $\theta$ relative to the $\mu^-$ spin direction. If $C$ is a symmetry, these two processes must have the same probability, contrary to observation. Hence $C$ is violated. It is now established that only the left-handed component of the Dirac spinors for all fermions and the right-handed component for anti-fermions interact with $W$ bosons while the left and right couple unequally to $Z$ bosons so $C$-violation is universal in weak interactions in addition to $P$-violation.

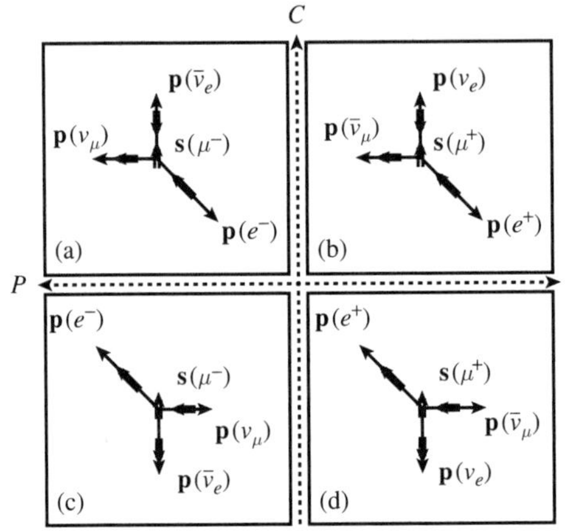


**FIGURE 8.2** The operations $C$ and $P$ applied to muon decay. a) An instance of the decay $\mu^- \to e^- \bar{\nu}_e \nu_\mu$. The muon is at rest. The spins of the decay products are shown superposed on their momenta. (b) The configuration obtained from (a) by charge conjugation $C$. c) The configuration obtained from (a) by inversion $P$. d) The configuration obtained from (a) by $CP$. Both $C$ and $P$ preserve the spin. $P$ reverses momenta. If $C/P$ is a symmetry then (b)/(c) are as likely as (a). In fact, both $P$ and $C$ symmetry is violated but $CP$ symmetry is respected in this decay.

The combined operation of $CP$ transforms the decay of a $\mu^+$ to positron with angle $\theta$ to the decay of a $\mu^-$ to electron with angle $\pi - \theta$ and the decay rates for these two transitions are equal. Thus $CP$ appears to be a a symmetry of the weak interaction. However, in fact $CP$ is violated in the charged current weak interactions of quarks (characterized by a phase angle $\theta$ in the CKM matrix) as was discovered in studies of neutral kaons.

## 8.3 ■ THEORY OF ELECTROWEAK INTERACTIONS

The theory of the electroweak interaction begins with the postulate that the left-handed fermions of a flavor pair such $e^-$ and $\nu_e$ are physically equivalent from the viewpoint of weak interactions. The physical symmetry is represented by unitary transformations of so-called weak isospin doublets

$$\begin{pmatrix} \nu_L \\ e_L \end{pmatrix} \to U \begin{pmatrix} \nu_L \\ e_L \end{pmatrix}. \tag{8.25}$$

Here $\nu_L$ is the left-handed electron neutrino, $e_L$ is the left-handed electron, and $U$ is an element of the group of matrices denoted by $SU(2)_L$ where the subscript $L$ is a reminder that the matrices act only on the left-handed components of the fermions. The elements of $SU(2)_L$ have the general conventional representation

$$U = \exp\left[i\mathbf{a} \cdot \mathbf{T}\right] = \exp\left[i\frac{1}{2}\theta\mathbf{n} \cdot \sigma\right] \tag{8.26}$$

where $\mathbf{a} = \theta\mathbf{n}$ is a vector of three real parameters and

$$\mathbf{T} = \frac{1}{2}\boldsymbol{\sigma} \tag{8.27}$$

denotes three generator matrices proportional to the Pauli matrices.

The standard model extends global weak isospin symmetry to a local gauge symmetry through the introduction of three real gauge fields denoted by $\mathbf{W}^\mu$ corresponding to the three generators of $SU(2)_L$. The Lagrangian for one $SU(2)_L$ doublet of fermions is obtained by the introduction of the $SU(2)_L$ gauge covariant derivative

$$\partial^\mu \rightarrow D^\mu = \partial^\mu + ig\mathbf{T} \cdot \mathbf{W}^\mu \tag{8.28}$$

acting on the left-handed flavor doublets. The gauged fields satisfy the matrix of equations

$$i\gamma^\mu \partial_\mu \begin{pmatrix} \nu_L \\ e_L \end{pmatrix} - m_e \begin{pmatrix} 0 \\ e_R \end{pmatrix} = ig\frac{\boldsymbol{\sigma}}{2} \cdot \mathbf{W}_\mu \gamma^\mu \begin{pmatrix} \nu_L \\ e_L \end{pmatrix}. \tag{8.29}$$

In this expression, the lepton names denote the left-handed 4-component spinor fields acted on by the Dirac matrices $\gamma_\mu$ and $\boldsymbol{\sigma}$ denotes $2 \times 2$ matrices operating on lepton flavor. Neglecting the term proportional to the electron mass $m_e$, and using 2-component spinors, we find the equations

$$\left(i\frac{\partial}{\partial t} - i\boldsymbol{\sigma} \cdot \nabla\right) \begin{pmatrix} \nu_L \\ e_L \end{pmatrix} = \frac{g}{2} \begin{pmatrix} W_3 & W_1 - iW_2 \\ W_1 + iW_2 & -W_3 \end{pmatrix} \begin{pmatrix} \nu_L \\ e_L \end{pmatrix} \tag{8.30}$$

where $W_i = W_i^0 + \boldsymbol{\sigma} \cdot \mathbf{W}_i$. Here the particle names denote the 2-component spinor fields acted on by the Pauli matrices $\boldsymbol{\sigma}$. We can identify the fields

$$W_\mu^\mp = \frac{1}{\sqrt{2}}(W_1 \pm iW_2)_\mu \tag{8.31}$$

with intermediate vector bosons $W^\mp$ mediating weak interactions. The right-handed fields $e_R$ and $\nu_R$ do not participate in charged current weak interactions and so must be assumed to be invariant (singlets) under $SU(2)_L$ transformations. Therefore they have no gauge interactions with any of the $W$ boson fields. In particular, the $W_3$ does not interact with the right-handed fields so can not be identified with the $Z$ boson or with the photon. This seeming stumbling block leads to the next crucial ingredient of the standard model theory - electroweak unification.

The $Z$ boson and simultaneously the photon may be accommodated through the introduction of an additional gauge symmetry denoted by $U(1)_Y$ associated with a new so-called hypercharge $Y$. (The subscript $Y$ is a reminder that the $U(1)$ hypercharge symmetry is not the gauge symmetry associated with electromagnetism.) The hypercharge gauge field is denoted by $B^\mu$. We will assume the field $B^\mu$ couples to both $R$ and $L$ gauged fields. Then a linear combination of $B$ and $W_3$ may be identified with the photon and the orthogonal combination with the $Z$ boson.

The construction of this $SU(2)_L \times U(1)_Y$ gauge theory requires extending the covariant derivative to

$$\partial^\mu \rightarrow D^\mu = \partial^\mu + ig' y B^\mu + ig\mathbf{T} \cdot \mathbf{W}^\mu. \tag{8.32}$$

Here $g$ is the coupling constant for $SU(2)_L$. The $SU(2)_L$ gauge transformations are

$$\begin{pmatrix} \nu_L \\ e_L \end{pmatrix} \rightarrow U \begin{pmatrix} \nu_L \\ e_L \end{pmatrix}, \quad e_R \rightarrow e_R$$
$$ig\mathbf{T} \cdot \mathbf{W} \rightarrow Uig\mathbf{T} \cdot \mathbf{W} U^{-1} + U\left(\partial^\mu U^{-1}\right). \tag{8.33}$$

Following convention, the constant $g'$ is a $U(1)_Y$ hypercharge scale factor and $y_R$ and $y_L$ are hypercharge values for the right-handed isosinglet and left-handed isodoublet which are unconstrained by gauge symmetry. The hypercharge of both members of the isodoublet must be identical or a $U(1)_Y$ transformation would not commute with an $SU(2)_L$ transformation of the isodoublet. The $U(1)_Y$ gauge transformations are therefore

$$\begin{pmatrix} \nu_L \\ e_L \end{pmatrix} \rightarrow e^{iy_L g' \chi} \begin{pmatrix} \nu_L \\ e_L \end{pmatrix}, \quad e_R \rightarrow e^{iy_R g' \chi} e_R$$
$$B^\mu \rightarrow B^\mu - \partial^\mu \chi. \tag{8.34}$$

One of the hypercharge values may be absorbed into the normalization of $g'$ and $y_L = -1/2$ is conventionally chosen for the $e_L$,$\nu_L$ doublet.

The $SU(2)_L \times U(1)_Y$ gauge invariant Lagrangian for the first generation leptons neglecting mass terms can be constructed following the example of quantum chromodynamics. Starting with the free field Lagrangian for fermions written in a way to make the global $SU(2)_L$ and $U(1)_Y$ symmetries apparent, substituting the gauge covariant derivative for the normal derivative, and adding kinetic Lagrangians for the gauge fields, we find

$$L = -\frac{1}{4}{F_B}^2 - \frac{1}{4}\mathbf{F}^2 + L_L + L_R. \tag{8.35}$$

Here the $U(1)_Y$ field strength tensor is given by

$$F_B^{\mu\nu} = \partial^\nu B^\mu - \partial^\nu B^\mu \tag{8.36}$$

while the $SU(2)_L$ gauge field strength tensor $\mathbf{F}$ has three components

$$F^i_{\mu\nu} = \partial_\mu W^i_\nu - \partial_\nu W^i_\mu - g\epsilon_{ijk} W^j_\mu W^k_\nu. \tag{8.37}$$

The right and left-handed gauged field Lagrangians for the fermions expressed as 4-component spinor fields are

$$L_R = \bar{e}_R i\gamma_\mu \left(\partial^\mu + ig' y_R B^\mu\right) e_R$$
$$L_L = \left(\bar{\nu}_L \quad \bar{e}_L\right) i\gamma_\mu \left(\partial^\mu + ig' y_L B^\mu + ig\mathbf{T} \cdot \mathbf{W}^\mu\right) \begin{pmatrix} \nu_L \\ e_L \end{pmatrix}. \tag{8.38}$$

Let us now focus on the interactions of fermions with the gauge fields. The gauge interactions of the left-handed components are contained in the matrix expression

$$L_{llV}^{L} = -\frac{1}{2}(\bar{\nu}_L \bar{e}_L)\gamma_\mu \begin{pmatrix} 2y_L g' B^\mu + g W_3^\mu & g\sqrt{2}W^+ \\ g\sqrt{2}W^- & 2y_L g' B^\mu - g W_3^\mu \end{pmatrix} \begin{pmatrix} \nu_L \\ e_L \end{pmatrix}. \quad (8.39)$$

The $\bar{\nu}_L e_L$ and $\bar{e}_L \nu_L$ matrix elements involving $W_1$ and $W_2$ have been written in terms of the (normalized) charged vector boson fields defined in Equation 8.31. Consider now the $\bar{\nu}_L \nu_L$ matrix element. With the conventional choice $y_L^e = -1/2$, the normalized combination

$$Z^\mu = \frac{1}{\sqrt{g^2 + g'^2}} \left(-g' B^\mu + g W_3^\mu\right) \quad (8.40)$$

couples the electrically neutral $\nu$ to $\bar{\nu}$ and is identified as the Z boson field. The orthogonal combination must therefore be identified with the electromagnetic field

$$A^\mu = \frac{1}{\sqrt{g^2 + g'^2}} \left(g B^\mu + g' W_3^\mu\right). \quad (8.41)$$

The $\bar{e}_L e_L$ matrix element then contains interactions of $e_L$ with both the $Z$ boson and the photon which may be made explicit by expressing $B^\mu$ and $W_3$ in terms of $Z^\mu$ and $A^\mu$. The mixing is conventionally expressed as

$$\begin{pmatrix} W_3 \\ B \end{pmatrix} = \begin{pmatrix} \cos\theta_W & \sin\theta_W \\ -\sin\theta_W & \cos\theta_W \end{pmatrix} \begin{pmatrix} Z \\ A \end{pmatrix} \quad (8.42)$$

where the Weinberg angle $\theta_W$ is defined by

$$\sin\theta_W = \frac{g'}{\sqrt{g^2 + g'^2}}\,; \quad \cos\theta_W = \frac{g}{\sqrt{g^2 + g'^2}}. \quad (8.43)$$

We have identified the gauge fields which correspond to the mass eigenstate vector fields observed in experiments. In terms of the $A$, $W$ boson, and $Z$ boson fields, and including now in addition the right-handed fermion fields, the interaction Lagrangian is

$$\begin{aligned} L_{llV} = &-\frac{g}{\sqrt{2}}\left(W^+ \bar{\nu}_L \gamma e_L + W^- \bar{e}_L \gamma \nu_L\right) - \frac{\sqrt{g^2 + g'^2}}{2} Z \bar{\nu}_L \gamma \nu_L \\ &-\frac{1}{2}\frac{g'^2 - g^2}{\sqrt{g^2 + g'^2}} Z \bar{e}_L \gamma e_L + \frac{y_R g'^2}{\sqrt{g^2 + g'^2}} Z \bar{e}_R \gamma e_R \\ &+\frac{g g'}{\sqrt{g^2 + g'^2}} A\left(\bar{e}_L \gamma e_L - y_R \bar{e}_R \gamma e_R\right). \end{aligned} \quad (8.44)$$

The last line in this expression represents the electromagnetic coupling of the electron. We now require the electromagnetic interaction of the electron field have the form

$$L_{eeA} = e\left(\bar{e}_L \gamma_\mu e_L + \bar{e}_R \gamma_\mu e_R\right) A^\mu \tag{8.45}$$

with $e = |e|$ and equal electromagnetic charge for $e_L$ and $e_R$. It follows that the hypercharge values of the leptons are

$$y_L = -1/2,\ y_R = -1 \tag{8.46}$$

and the electric charge $e$ of the electron is related to the universal weak charged current coupling $g$ by

$$e = gg'/\sqrt{g^2 + g'^2} = g \sin\theta_W. \tag{8.47}$$

Reviewing the entire interaction Lagrangian, we see the $W$ boson couples only to left-handed leptons, the $Z$ boson couples to left-handed neutrinos and to left-handed and right-handed electrons unequally, while the photon equally couples to equally to left-handed and right-handed electrons but not to neutrinos. The coupling constants are determined by two parameters, $g$ and $g'$, or $e$ and $\sin\theta_W$. The electroweak interactions of second and third generation leptons are identical to those of the first generation. To include these in the Lagrangian, make the replacements

$$\nu_L \to \nu_L^i,\ e_L \to e_L^i,\ e_R \to e_R^i \tag{8.48}$$

where the generation index is $i = e, \mu, \tau$ or $i = 1, 2, 3$.

The electroweak charges of quarks differ from the electroweak charges of leptons. The charged current interaction of the $u$-quark and $d$-quark is identical to the charged current interaction of $\nu_e$ and $e$ but $u$-like quarks are electrically charged, unlike the neutrinos, and their right-handed components must be included in addition to the right-handed components of the $d$-like quarks. We next show that quark hypercharge assignments distinct from lepton hypercharge assignments may be made that ensure that both right-handed $u$-like and $d$-like quarks emerge with a $L$-$R$ symmetric electromagnetic interaction.

Ignoring the generalization to flavor mixtures described below, the first generation quark electroweak interactions correspond to the interaction Lagrangian

$$\begin{aligned} L_{qqV} &= -g' y_R^u \bar{u}_R \gamma_\mu u_R B^\mu - g' y_R^d \bar{d}_R \gamma_\mu d_R B^\mu \\ &\quad - \frac{1}{2}\left(\bar{u}_L, \bar{d}_L\right)\gamma_\mu \begin{pmatrix} 2y_L g' B^\mu + g W_3^\mu & \sqrt{2} g W^+ \\ g\sqrt{2} W^- & 2y_L g' B^\mu - g W_3^\mu \end{pmatrix} \begin{pmatrix} u_L \\ d_L \end{pmatrix} \end{aligned} \tag{8.49}$$

where the $W_1$ and $W_2$ have been expressed in terms of $W^\pm$ fields and, since quarks couple to the same electroweak bosons as leptons, local symmetry requires the same $SU(2)_L$ coupling constant $g$. The hypercharge of the $SU(2)_L$ quark doublet is $y_L$. The hypercharge values of the $SU(2)_L$ singlets are denoted by $y_R^u$ and $y_R^d$.

The $\bar{u}_L d_L$ and $\bar{d}_L u_L$ matrix elements describe charged current interactions of the left-handed quarks identical to those of left-handed leptons. Using the expressions

$$2y_L g' B \pm g W_3 = \frac{-2y_L g'^2 \pm g^2}{\sqrt{g'^2 + g^2}} Z + \frac{(2y_L \pm 1)\, g' g}{\sqrt{g'^2 + g^2}} A, \tag{8.50}$$

we find the neutral current interaction Lagrangian for quarks is

$$\begin{aligned} L_{qqV^0} = &-\bar{u}_L \gamma u_L \left[ \frac{-2y_L g'^2 + g^2}{2\sqrt{g'^2 + g^2}} Z + \frac{(2y_L + 1)\, g' g}{2\sqrt{g'^2 + g^2}} A \right] \\ &-\bar{u}_R \gamma u_R \left[ \frac{-y_R^u g'^2}{\sqrt{g'^2 + g^2}} Z + \frac{y_R^u g' g}{\sqrt{g'^2 + g^2}} A \right] \\ &-\bar{d}_L \gamma d_L \left[ \frac{-2y_L g'^2 - g^2}{2\sqrt{g'^2 + g^2}} Z + \frac{(2y_L - 1)\, g' g}{2\sqrt{g'^2 + g^2}} A \right] \\ &-\bar{d}_R \gamma d_R \left[ \frac{-y_R^d g'^2}{\sqrt{g'^2 + g^2}} Z + \frac{y_R^d g' g}{\sqrt{g'^2 + g^2}} A \right]. \end{aligned} \tag{8.51}$$

The electromagnetic interaction Lagrangian for the two quarks should have the $L$-$R$ symmetric form

$$\begin{aligned} L_{qqA} = &-Q_u |e| (\bar{u}_L \gamma u_L + \bar{u}_R \gamma u_R) A + \\ &-Q_d |e| (\bar{d}_L \gamma d_L + \bar{d}_R \gamma d_R) A \end{aligned} \tag{8.52}$$

with $Q_u = +2/3$ and $Q_d = -1/3$. This follows if we choose the hypercharge assignments

$$y_R^u = Q_u = +2/3\,, \quad y_R^d = Q_d = -1/3 \text{ and } y_L = +1/6. \tag{8.53}$$

With $T_3 = \pm\frac{1}{2}$ for the left-handed flavor doublet members and $T_3 = 0$ for the right-handed singlets, the hypercharge values for right and left-handed quarks and leptons are related to their electric charges and isospin by the formula

$$Y = Q - T_3. \tag{8.54}$$

The expressions above describe all electroweak interactions in terms of the parameters $g$ and $g'$. A somewhat more condensed form is possible with suitable notation. If we introduce a 4-spinor doublet

$$\psi = \begin{pmatrix} \nu_e \\ e \end{pmatrix} \tag{8.55}$$

for leptons and

$$\psi = \begin{pmatrix} u \\ d \end{pmatrix} \tag{8.56}$$

for quarks and the weak isospin operators

$$T^+ = T_1 + iT_2 = \begin{pmatrix} 0 & 1 \\ 0 & 0 \end{pmatrix} \; ; \; T^- = T_1 - iT_2 = \begin{pmatrix} 0 & 0 \\ 1 & 0 \end{pmatrix}, \tag{8.57}$$

then the Lagrangian describing weak interactions of the first generation of either the quark or lepton pair is

$$L_{ff(W,Z)} = -\frac{g}{2\sqrt{2}}\bar{\psi}\gamma^\mu\left(1-\gamma^5\right)\left(T^+W_\mu^+ + T^-W_\mu^-\right)\psi - \bar{\psi}\gamma^\mu\left(g_v - g_a\gamma^5\right)\psi Z_\mu \tag{8.58}$$

where

$$g_v = (1/2)\,(g/\cos\theta_W)\left(T_3 - 2Q\sin^2\theta_W\right) \tag{8.59}$$

and

$$g_a = (1/2)\,(g/\cos\theta_W)\,T_3. \tag{8.60}$$

The axial and vector coupling constants are related to the left and right-handed coupling constants by

$$\begin{aligned} g_L &= g_v + g_a \\ g_R &= g_v - g_a. \end{aligned} \tag{8.61}$$

Table 8.1 lists all of the electroweak coupling constants in terms of $e$ and $\theta_W$.

Gauge symmetry constrains not just the interactions of fermions with gauge bosons but the boson self-interactions. When the fields $W_1$, $W_2$, $W_3$, and $B$ are expressed in terms of the fields $A$, $Z$, and $W^\pm$, the pure gauge portion of the Lagrangian

$$L_{\text{gauge}} = -\frac{1}{4}F_B^2 - \frac{1}{4}\mathbf{F}^2, \tag{8.62}$$

| | L | | | | | R | | | | |
|---|---|---|---|---|---|---|---|---|---|---|
| | $g_A$ | $t_3$ | $y$ | $g_W$ | $g_Z$ | $g_A$ | $t_3$ | $y$ | $g_W$ | $g_Z$ |
| $\nu$ | $0$ | $+\frac{1}{2}$ | $-\frac{1}{2}$ | $\frac{e}{2s_w}$ | $\frac{e}{s_wc_w}(-\frac{1}{2})$ | $0$ | $0$ | $0$ | $0$ | $0$ |
| $e$ | $-e$ | $-\frac{1}{2}$ | $+\frac{1}{2}$ | $\frac{e}{2s_w}$ | $\frac{e}{s_wc_w}(-s_w+\frac{1}{2})$ | $-e$ | $0$ | $-1$ | $0$ | $-\frac{e}{c_w}$ |
| $u$ | $+\frac{2}{3}e$ | $+\frac{1}{2}$ | $-\frac{1}{6}$ | $\frac{e}{2s_w}$ | $\frac{e}{s_wc_w}(+\frac{2}{3}s_w-\frac{1}{2})$ | $+\frac{2}{3}e$ | $0$ | $+\frac{2}{3}$ | $0$ | $\frac{2}{3}\frac{e}{c_w}$ |
| $d$ | $-\frac{1}{3}e$ | $-\frac{1}{2}$ | $+\frac{1}{6}$ | $\frac{e}{2s_w}$ | $\frac{e}{s_wc_w}(-\frac{1}{3}s_w+\frac{1}{2})$ | $-\frac{1}{3}e$ | $0$ | $-\frac{1}{3}$ | $0$ | $-\frac{1}{3}\frac{e}{c_w}$ |

**TABLE 8.1** Weak coupling constants $g_A$, $g_W$, and $g_Z$ of left and right-handed quarks and leptons to $\gamma$, $W$ boson, and $Z$ bosons. The hypercharge is given by $y = q - t_3$ with $q = g_A/e$. The coupling constants are written in terms of the charge $e$ of the positron and $s_w = \sin\theta_W$ and $c_w = \cos\theta_W$ where $\theta_W$ is the Weinberg angle. The left-handed coupling to the $Z$ is $g_Z^L = gc_w^{-1}(qe^{-1}s_w - t_3)$ and the right-handed coupling is $g_Z^R = gs_wc_w^{-1}qe^{-1}$ with $e = gs_w$.

can be expressed in terms of the physical fields. The resulting Lagrangian contains very specific terms corresponding to three-boson and four-boson interactions. The pure electroweak gauge field Lagrangian is

$$\begin{aligned} L_{\text{gauge}} = L_0 + L_{WWZ} + L_{WWA} \\ + L_{WWZZ} + L_{WWAA} + L_{WWZA} + L_{WWWW}. \end{aligned} \tag{8.63}$$

The noninteracting part of the gauge field Lagrangian expressed in terms of physical fields is

$$L_0 = -\frac{1}{4}F_A^2 - \frac{1}{4}F_Z^2 - \frac{1}{2}F_W^\dagger F_W \tag{8.64}$$

with $F_A^{\mu\nu} = \partial^\mu A^\nu - \partial^\nu A^\mu$, $F_Z^{\mu\nu} = \partial^\mu Z^\nu - \partial^\nu Z^\mu$ and $F_W^{\mu\nu} = \partial^\mu W^\nu - \partial^\nu W^\nu$ and $W_\mu \equiv W_\mu^-$. The interactions are expressed by the terms

$$\begin{aligned} L_{WWZ} &= -ig\cos\theta_W \left[ \left( W_\mu^\dagger W_\nu - W_\mu W_\nu^\dagger \right) \partial^\mu Z^\nu + F_W^{\mu\nu} Z_\mu W_\nu^\dagger - F_W^{\dagger\mu\nu} Z_\mu W_\nu \right] \\ L_{WWA} &= -ie \left[ \left( W_\mu^\dagger W_\nu - W_\mu W_\nu^\dagger \right) \partial^\mu A^\nu + F_W^{\mu\nu} A_\mu W_\nu^\dagger - F_W^{\dagger\mu\nu} A_\mu W_\nu \right] \\ L_{WWZZ} &= g^2\cos^2\theta_W \left[ W_\mu W_\nu^\dagger Z^\mu Z^\nu - W^\mu W_\mu^\dagger Z^\nu Z_\nu \right] \\ L_{WWAA} &= e^2 \left[ W_\mu W_\nu^\dagger A^\mu A^\nu - W^\mu W_\mu^\dagger A^\nu A_\nu \right] \\ L_{WWZA} &= eg\cos\theta_W \left[ W_\mu W_\nu^\dagger \left( Z^\mu A^\nu + Z^\nu A^\mu \right) - 2W^\mu W_\mu^\dagger Z^\nu A_\nu \right] \\ L_{WWWW} &= \frac{g^2}{2} W_\mu^\dagger W_\nu \left[ W^{\dagger\mu} W^\nu - W^\mu W^{\dagger\nu} \right]. \end{aligned} \tag{8.65}$$

The experimental observation of such interactions with the predicted coupling strength has been an important verification of the standard model. Note that no triple or quartic coupling of purely neutral bosons appears. Also, as discussed in Chapter 5, the $WWA$ electromagnetic interaction differs from that in a simple $U(1)_{EM}$ gauge theory of charged vector bosons. To date the triple boson interactions have been found to be characterized by the coupling constants predicted by the standard model. The four-boson interactions will be studied at the LHC.

In sum, the specification of an $\mathrm{SU(2)_L \times U(1)_Y}$ symmetry and knowledge of the electromagnetic coupling constant leads inexorably to a theory which predicts the existence of a charged and a neutral weak vector boson and prescribes universal flavor changing and flavor conserving spin dependent weak interactions for both quarks and leptons. Additionally, the theory predicts an elaborate collection of boson interactions. Astonishing as it may appear, this model works. However, several facts related to particle mass and the mixing of quark flavor pairs by weak interactions remain to be explained. To these issues we now turn our attention.

## 8.4 ■ HIGGS MECHANISM

The introduction of mass into the theory of electroweak interactions is problematic. Consider first that the gauge symmetry $SU(2)_L$ prohibits a Lagrangian fermion mass term $m\bar{e}e$ because a product of left and right spinors is not gauge invariant - left-handed factors are transformed while right-handed factors are invariant. Gauge symmetry also prohibits a gauge boson mass term such as $m_w^2 W_\mu W^\mu$. These problems do not arise in $SU(3)_C$ color gauge theory in which left and right-handed fermions on an equal footing while the vector bosons are massless.

The standard model solves the problem by postulating additional scalar fields. In brief, left and right fermion components are coupled to scalar fields with $SU(2)_L \times U(1)Y$ invariant interaction terms like $\phi\bar{\psi}_L\psi_R$ and the scalar fields have non-vanishing static vacuum (expectation) values. Effective fermion masses appear with symmetry violation ascribed to the properties of the vacuum. Moreover, these scalar fields have gauge interactions, so that terms analogous to $L_{int} = e^2|\phi|^2 A^2$ in scalar electrodynamics with $|\phi|$ constant generate effective masses for the weak vector bosons. This Higgs mechanism leads to a consistent theory of the unified electroweak interaction with massive $W$ and $Z$ bosons and a massless photon, massive fermions, and, as an added benefit, an explanation of weak interactions linking quarks of different generation.

The minimal addition is an $SU(2)_L$ doublet of complex scalar fields

$$\phi = \begin{pmatrix} \phi^+ \\ \phi^0 \end{pmatrix}. \tag{8.66}$$

To begin constructing the theory, let us consider the first generation of leptons and define

$$\psi_L \equiv \begin{pmatrix} \nu_L \\ e_L \end{pmatrix}. \tag{8.67}$$

We suppose a so-called Yukawa interaction Lagrangian

$$L_{ll\phi} = -G_e \left[\bar{e}_R \phi^\dagger \psi_L + \bar{\psi}_L \phi e_R\right] \tag{8.68}$$

where $G_e$ is called the Yukawa coupling constant. This Lagrangian is invariant under $SU(2)_L$ transformations

$$\phi' = U\phi,\ \phi'^\dagger = \phi U^\dagger,\ \psi_L' = U\psi_L,\ \bar{\psi}_L' = \bar{\psi}_L U^\dagger,\ e_R' = e_R,\ \bar{e}_R' = e_R \tag{8.69}$$

and, under $U(1)_Y$ hypercharge transformations, if each term in the Lagrangian is hypercharge neutral. Hypercharge neutrality requires

$$y_\phi = y_L - y_R = 1/2. \tag{8.70}$$

The gauge and self interactions of the scalar doublet fields are represented by the Lagrangian

$$L_\phi = (D^\mu\phi)^\dagger D_\mu\phi - V\left(\phi^\dagger\phi\right) \tag{8.71}$$

with $V(\phi^\dagger\phi)$ a Higgs potential energy function and

$$D_\mu = \partial_\mu + ig' y_\phi B_\mu + ig\mathbf{T}\cdot\mathbf{W}_\mu. \tag{8.72}$$

Now suppose the Higgs potential has the form

$$V = -\mu^2|\phi|^2 + \lambda^2|\phi|^4 \tag{8.73}$$

with $\mu^2 > 0$ and $\lambda^2 > 0$. As illustrated in Figure 8.3, this potential function has a local maximum at $|\phi| = 0$ and a minimum on the surface defined by

$$|\phi|^2 = |\phi^+|^2 + \phi^0|^2 = \frac{\mu^2}{2\lambda^2} \equiv \frac{v^2}{2}. \tag{8.74}$$

It is assumed that the field in the vacuum takes a static value on this surface. Expanded about the minimum of this potential, the scalar doublet fields may be cleverly represented by the form

$$\phi = e^{i\mathbf{a}\cdot\mathbf{T}}\begin{pmatrix} 0 \\ (v + H(x))/\sqrt{2} \end{pmatrix} \tag{8.75}$$

where the parameter

$$v = \mu/\lambda \tag{8.76}$$

is the minimum so-called vacuum expectation value of $|\phi|$, $H$ is a real scalar field called the standard model Higgs boson field representing the radial component of deviations from the vacuum expectation value, and the three real fields **a** describe the direction of these excursions. The variations associated with the parameters **a** have the form of an $SU(2)_L$ transformation and can be absorbed into a gauge transformation. The vacuum expectation value of the scalar doublet assumed to be

$$\phi_0 = \begin{pmatrix} 0 \\ v/\sqrt{2} \end{pmatrix} \tag{8.77}$$

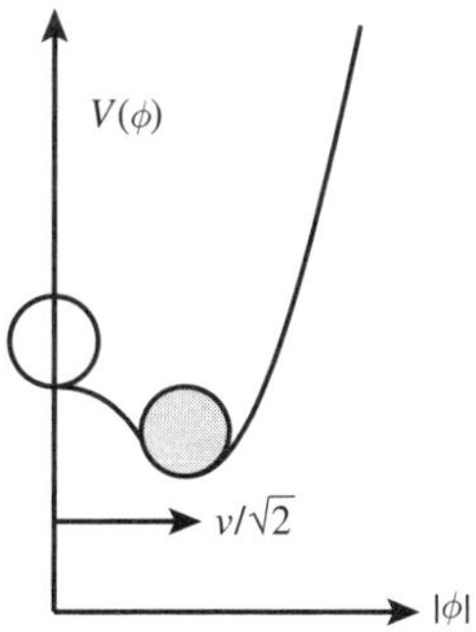


**FIGURE 8.3** Potential energy as a function of the magnitude of the Higgs doublet field. The standard model assumes a potential with a local maximum at $|\phi| = 0$ and a minimum at $|\phi| = v/\sqrt{2}$.

corresponds to isospin -1/2 and is invariant under an electromagnetic gauge transformation

$$\phi_0 \to \phi_0 \to e^{i\chi(x)(T_3+y_\phi)}\phi_0 = \phi_0 \tag{8.78}$$

but not under hypercharge or $SU(2)_L$ transformations. Although the underlying theory is fully symmetric, the postulated vacuum expectation value breaks $SU(2)_L$ and $U(1)_Y$ symmetry while leaving $U(1)_{EM}$ symmetry intact. This is an example of so-called spontaneous symmetry breaking.

With the substitution of the preceding spherical representation of the scalar fields, the Yukawa interaction reads

$$L_{ll\phi} = -m_e\bar{e}e - m_e H(x)\bar{e}e \tag{8.79}$$

where the electron mass in the first term is identified as

$$m_e = G_e v/\sqrt{2} \tag{8.80}$$

while the second term represents a coupling of the $H$ to the electron proportional to $m_e$. Since

$$\bar{e}e = e_L^\dagger e_R + e_R^\dagger e_L, \tag{8.81}$$

the interaction with the vacuum expectation value generates continuously transitions between $L$ and $R$ states as illustrated in Figure 8.4. Mass generation for the second and third generation leptons requires only different Yukawa coupling constants $G_\mu$ and $G_\tau$ to the common Higgs field.

The gauge interactions of $\phi$ expressed in terms of $v$ and $H$ imply mass terms for the $W$ boson and $Z$ boson and interactions between $H$ and the $W$ boson and $Z$ boson. The masses and interactions are found by expressing the Lagrangian $L_\phi$ in terms of $H$, $W$ and $Z$:

$$L_\phi = \frac{1}{2}\partial_\mu H \partial^\mu H - \frac{1}{2}m_H^2 H^2\left[1 + \frac{H}{v} + \left(\frac{H}{2v}\right)^2\right] \tag{8.82}$$

$$+ m_W^2 W_\mu^+ W^{-\mu}\left(1 + \frac{H}{v}\right)^2 + \frac{1}{2}m_Z^2 Z_\mu Z^\mu\left(1 + \frac{H}{v}\right)^2. \tag{8.83}$$

Here the mass of the Higgs field is identified as

$$m_H = \sqrt{2}\lambda v = \sqrt{2}\mu \tag{8.84}$$

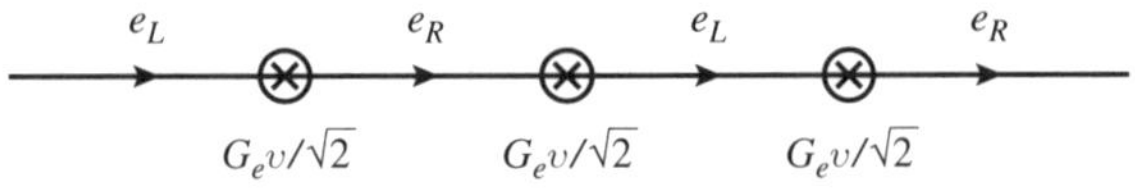


**FIGURE 8.4** Illustration of the interaction of an electron with the vacuum value of the Higgs field. The interaction flips the handedness of fermions.

and the gauge boson masses are identified as

$$m_W = \frac{gv}{2} = \frac{ev}{2\sin\theta_W} \tag{8.85}$$

and

$$m_Z = \frac{1}{2}v\sqrt{g^2 + g'^2} = \frac{ev}{\sin\theta_W\cos\theta_W} = \frac{m_w}{\cos\theta_W}. \tag{8.86}$$

From the expression $m_W = gv/2$ and measurements of $m_W$ and the weak coupling constant $g$, we find

$$v \simeq 246 \text{ GeV}. \tag{8.87}$$

Historically, low energy measurements ($E << m_W$) of rates for processes involving virtual W bosons, including the muon lifetime and charged current neutrino scattering cross sections, determined the ratio $g/m_W$ hence $v$. Measurements of neutral current process cross sections involving virtual $Z$ exchange are sensitive to $g'/m_Z$ and determined $\sin\theta_W$. Given the electron charge $e$ observed in electromagnetic processes, and the relationship between $m_W$ and $m_Z$, the unified theory permitted calculation of $g'$ and prediction of the $W$ boson and $Z$ boson masses. The discovery of the weak bosons at the predicted masses with all of the expected properties was a triumph of the model.

In the standard model, mass results from interaction of naturally massless fields with a background component in the vacuum and is intimately linked to other dynamics. In prosaic terms, the background field reduces the speed of particles from light speed, rather like glass reduces the effective speed of light. Other than the relationship between $m_W$ and $m_Z$, the values of the indices of refraction or masses are not determined in the model but we can say that heavier particles are more strongly coupled to the scalar fields. In particular, the coupling of fermions to the Higgs field is proportional to fermion mass for the minimal Higgs model. Notice also that the mass of the Higgs field is not specified by the model. It is determined by the parameter $\mu$ in the scalar potential. If we think of the mass $m_H$ as a consequence of the self interaction of the scalar fields, the value is a feature of the dynamics of the Higgs sector alone and not connected with matter. It must be determined from fundamental physics beyond the standard model.

## 8.5 ■ ELECTROWEAK INTERACTIONS OF QUARKS

The extension of this spontaneously broken gauge theory to quarks entails some complication. Right-handed $u$-quark, $c$-quark, and $t$-quark states most certainly exist, unlike right-handed neutrinos (in the simple model, neutrino mass may be neglected) and require suitably generalized Yukawa interactions. As we will see, analysis of a minimal model for quark-Higgs field interactions relates the matrix of intergenerational charged current weak interactions to the Higgs mechanism.

For three quark generations, the portion of the electroweak Lagrangian involving the quark fields has the form

$$L_{\text{quark}} = L_0 + L_{qq\phi} \tag{8.88}$$

with kinetic and gauge interaction terms

$$L_0 = i\bar{u}_a^R \gamma D_{2/3} u_a^R + i\bar{d}_a^R \gamma D_{-1/3} d_a^R + i\bar{q}_a^L \gamma D_{1/6} q_a^L \tag{8.89}$$

and the most general $SU(2)_L$ and $U(1)_Y$-invariant Yukawa interaction is a collection of mass-like terms

$$L_{qq\phi} = -\frac{\sqrt{2}}{v}\left(\bar{u}_a^R m_{ab} \phi_c^\dagger q_b^L + \bar{d}_a^R M_{ab} \phi^\dagger q_b^L\right) + h.c.\,. \tag{8.90}$$

In this expression, the indices $a$ and $b$ run over generations and we use notation distinguishing hypercharge states

$$u_a^R\,(y = 2/3) = (u_R, c_R, t_R)\,;\, d_a^R\,(y = -1/3) = (d_R, s_R, b_R) \tag{8.91}$$

and

$$q_a^L\,(y = 1/6) = \left(\begin{pmatrix} u_L \\ d_L \end{pmatrix}, \begin{pmatrix} c_L \\ s_L \end{pmatrix}, \begin{pmatrix} t_L \\ b_L \end{pmatrix}\right) \equiv \begin{pmatrix} u^L \\ d^L \end{pmatrix}. \tag{8.92}$$

The gauge covariant derivatives are $D_Y = \partial + iYg'a + ig\vec{T}\cdot\vec{b}$ where the $SU(2)_L$ term $ig\vec{T}\cdot\vec{b}$ applies only to $D_{1/6}$. In writing the most general $SU(2)$-invariant Yukawa interaction of the scalar doublet to fermions, we have reused the notation for the scalar doublet and its vacuum value

$$\phi\,(y = +1/2) = \begin{pmatrix} \phi^+ \\ \phi^0 \end{pmatrix} \to \begin{pmatrix} 0 \\ (v + H)/\sqrt{2} \end{pmatrix} \tag{8.93}$$

and we take advantage of the magic of the Pauli matrices which implies the conjugate of the scalar doublet defined by

$$\phi_c\,(y = -1/2) = i\sigma_2\phi^* = \begin{pmatrix} \phi^{0\dagger} \\ -\phi^{+\dagger} \end{pmatrix} \to \begin{pmatrix} (v + H)/\sqrt{2} \\ 0 \end{pmatrix} \tag{8.94}$$

transforms like $\phi$ under an $SU(2)$ transformation:

$$\phi \to U\phi;\, \phi_c \to U\phi_c. \tag{8.95}$$

The Yukawa coupling matrices $m_{ab}$ and $M_{ab}$ for $\phi$ constant play the role of mass matrices.

Consider unitary transformations mixing flavors of the same hypercharge

$$u^R \to Uu^R;\, d^R \to Dd^R;\, q^L \to Qq^L \tag{8.96}$$

with unitary matrices $U$, $D$, and $Q$. The massless Lagrangian $L_0$ is invariant, while the Yukawa interaction undergoes a matrix transformation

$$m \to U^\dagger m Q;\ M \to D^\dagger M Q. \tag{8.97}$$

Now any complex matrix $m$ can be diagonalized with left and right unitary matrices. In fact, $mm^\dagger$ is Hermitian and can be diagonalized with a unitary transformation $U$ to positive diagonal form

$$mm^\dagger \to U^\dagger mm^\dagger U = \begin{pmatrix} m_u^2 & 0 & 0 \\ 0 & m_c^2 & 0 \\ 0 & 0 & m_t^2 \end{pmatrix} \tag{8.98}$$

which implies we can write $m$ as a diagonal mass matrix times a unitary matrix $W$:

$$m = \begin{pmatrix} m_u & 0 & 0 \\ 0 & m_c & 0 \\ 0 & 0 & m_t \end{pmatrix} W \equiv m_{diag} W. \tag{8.99}$$

So picking $U$ to diagonalize $mm^\dagger$ and then $Q = W^\dagger$, we can make $m$ diagonal. Similarly, by a choice of $D$, we can bring $(MQ)(MQ)^\dagger$ to diagonal form and hence $MQ$ to the form $M_{diag}V^\dagger$ with

$$M = \begin{pmatrix} m_d & 0 & 0 \\ 0 & m_s & 0 \\ 0 & 0 & m_b \end{pmatrix} \equiv M_{diag} W. \tag{8.100}$$

Another transformation with $D = V$ gives the canonical form

$$m \to m_{diag};\ M \to V_{\text{CKM}} M_{diag} V_{\text{CKM}}^\dagger \tag{8.101}$$

where $V_{\text{CKM}}$ is a unitary matrix called the Cabibbo-Kobayashi-Maskawa (CKM) matrix.

With these transformations, the Yukawa interaction portion of the Lagrangian becomes

$$L_{\text{Yukawa,quarks}} = -\frac{\sqrt{2}}{v}\left(\bar{u}_a^R m_{diag,ab} \phi_c^\dagger q_b^L + \bar{d}_a^R V M_{diag,ab} V_{\text{CKM}}^\dagger \phi^\dagger q_b^L\right) + h.c. \tag{8.102}$$

or, expressed in terms of the Higgs field $H$ and static value $v$,

$$\begin{aligned} L_{\text{Yukawa,quarks}} = & -\bar{u}_a^R m_{diag,ab}\left(1 + \frac{H}{v}\right) u_b^L \\ & -\overline{\left(V_{\text{CKM}}^\dagger d\right)}_a^R M_{diag,ab}\left(1 + \frac{H}{v}\right)\left(V_{\text{CKM}}^\dagger d_b^L\right) + h.c.. \end{aligned} \tag{8.103}$$

The transformations define a basis in which the $u$-like quarks with charge +2/3, along with a unitary mixture $V_{\text{CKM}}^{\dagger} d$ of charge -1/3 $d$-like quarks, are mass eigenstates. We can change our basis for $d$-like quarks to the mass eigenstates by the replacement $d^R \to V_{\text{CKM}} d^R$ (i.e. $D = V_{\text{CKM}}$), which does not affect the Lagrangian, and also $d^L \to V_{\text{CKM}} d^L$. Under the latter transformation, the portion of the Lagrangian which is not completely invariant is

$$L_0^q = i\bar{q}_a^L \gamma D_{1/6} q_a^L. \tag{8.104}$$

In this Lagrangian, the neutral terms of the form $\bar{q}^L \gamma q^L$ are invariant but the charged weak interaction is not. The final result of the transformation to the $d$-like quark mass eigenstate basis is a modification of the charged current with $d$-like quark fields replaced with the unitary mixture

$$\begin{pmatrix} d_L \\ s_L \\ b_L \end{pmatrix} \to V_{\text{CKM}} \begin{pmatrix} d_L \\ s_L \\ b_L \end{pmatrix} = \begin{pmatrix} V_{ud} & V_{us} & V_{ub} \\ V_{cd} & V_{cs} & V_{cb} \\ V_{td} & V_{ts} & V_{td} \end{pmatrix} \begin{pmatrix} d_L \\ s_L \\ b_L \end{pmatrix}. \tag{8.105}$$

The off-diagonal elements determine the relative strength of $d$-like quark to $u$-like quark transitions which change not only quark flavor but generation. The amplitude for a transition with a $d$-like quark in the initial state or $\bar{d}$-like quark in the final state is proportional to the corresponding CKM matrix element and the amplitude for a $u$-like quark in the initial state or $\bar{u}$-like quark in the final state is proportional to the conjugate of the CKM matrix element. For example, the amplitude for $s \to uW^-$ is proportional to $gV_{us}/\sqrt{2}$ and the amplitude for $u \to sW^+$ is proportional to $gV_{us}^*/\sqrt{2}$.

An $N \times N$ unitary matrix is described by $N^2$ real parameters. By redefining the phases of the $2N$ fermion fields, leaving the mass matrix diagonal, $2N - 1$ parameters may be defined away as $V_{\text{CKM}}$ is unchanged by a phase common to all fields. The CKM matrix may be expressed in terms of four parameters, e.g.

$$V_{\text{CKM}} = \begin{pmatrix} 1 & 0 & 0 \\ 0 & c_{23} & s_{23} \\ 0 & -s_{23} & c_{23} \end{pmatrix} \begin{pmatrix} c_{13} & 0 & s_{13}e^{-i\delta} \\ 0 & 1 & 0 \\ -s_{13}e^{i\delta} & 0 & c_{13} \end{pmatrix} \begin{pmatrix} c_{12} & s_{12} & 0 \\ -s_{12} & c_{12} & 0 \\ 0 & 0 & 1 \end{pmatrix} \tag{8.106}$$

with angles $\theta_{23}$, $\theta_{13}$, $\theta_{12}$ and $\delta$ and the notation $c_{ij} = \cos\theta_{ij}$ and $s_{ij} = \sin\theta_{ij}$. The matrix elements are determined or constrained by comparing hadronic weak processes to muon decay. The magnitude of $V_{ud}$ is derived from from nuclear $\beta$ decay, the magnitude of $V_{us}$ from $K \to \pi e \nu$, the magnitude of $V_{cs}$ from $D$-meson decay, and so on. Some of these measurements will be considered in Chapter 10. The magnitudes of the matrix elements are approximately

$$|V_{\text{CKM}}| = \begin{pmatrix} 0.97 & 0.22 & 0.003 \\ 0.22 & 0.97 & 0.04 \\ 0.01 & 0.04 & 1 \end{pmatrix}. \tag{8.107}$$

In the approximation $c_{23} = c_{13} = 1$ and $s_{23}s_{13} = 0$,

$$V_{\rm CKM} = \begin{pmatrix} \cos\theta_c & \sin\theta_c & s_{13}e^{i\delta} \\ -\sin\theta_c & \cos\theta_c & s_{23} \\ s_{12}s_{23} - s_{13}e^{i\delta} & -s_{23} & 1 \end{pmatrix} \tag{8.108}$$

where $\theta_c = \theta_{12}$ is the Cabibbo angle. The phase $\delta$ is associated with $CP$-violation and is discussed in Chapter 10.

We could repeat this analysis for leptons replacing $u$-like flavors with neutrinos and $d$-flavors with charged leptons. We would then expect analogous mixing in the leptonic charged currents described by a replacement of the vector of $e$-like leptons by a unitary mixture $e_L \to V e_L$. However, if the neutrino mass matrix were diagonal or all neutrino masses negligibly small, a redefinition $\nu_R \to V\nu_R$ together with $\nu_L \to V\nu_L$ would render the charged current diagonal, leave the neutral current diagonal, without affecting the mass matrix. Neutrino mass will be considered in more detail later.

Having formulated the electroweak standard model, we turn now to the analysis of a few exemplary processes. The 2-body decays of $W$ and $Z$ bosons are the simplest weak processes to calculate and we start with these. The $\beta$-decay of the muon which proceeds through a virtual $W$ boson is representative of weak decays of all unstable fermions excluding the $t$-quark. The calculation is somewhat involved but is considered next. A variety of neutrino interactions with fixed target electrons and nuclei exemplify weak scattering. Neutrino interactions in particular have been important in probing the structure of nucleons and in understanding newly discovered properties of neutrinos themselves. The production and decay of single and multiple electroweak bosons and the properties of the $t$-quark provide striking verification of the theory. We will also take a brief look at the properties of the Higgs boson. Applications to hadronic decays and mixing phenomena which verify the CKM flavor mixing model are found in Chapter 10.

## 8.6 ■ *W* AND *Z* BOSON DECAYS

The $W$ and $Z$ bosons are unstable and decay to fermion pairs. The decay rates are of fundamental interest and calculation of these rates is perhaps the simplest application of the standard model of electroweak interaction. The prototype 2-body decay is

$$W^- \to e^-\bar{\nu}_e. \tag{8.109}$$

The matrix element for the decay is a simple transcription of the interaction Lagrangian

$$L_{Wee} = -\frac{g}{\sqrt{2}} W^\mu \bar{e}_L \gamma_\mu e_L \tag{8.110}$$

with the boson field replaced by its polarization vector and the fermion fields replaced by spinors

$$M_{W^- \to e^- \bar{\nu}_e} = -\frac{g}{\sqrt{2}} \epsilon^\mu \bar{u}_e^L \gamma_\mu v_\nu^L . \tag{8.111}$$

In the $W$ boson rest frame, the $W$ boson polarization vector has the form $\epsilon = (0, \boldsymbol{\epsilon})$. In 2-component form the left-handed current is

$$\bar{u}_e^L \gamma_\mu v_\nu^L = u_e^L \sigma_\mu^+ v_\nu^L \tag{8.112}$$

with left-handed 2-component electron spinor

$$u_e^L = \sqrt{E_e - sp}\phi_s (\mathbf{n}_e) \tag{8.113}$$

where $\mathbf{n}_e$ is the electron direction and $s$ its helicity. For the anti-fermion, we need the lower two components of the 4-component spinor, namely

$$v_s^- = s\sqrt{E + sp}\phi_{-s} (\mathbf{n}_\nu) \tag{8.114}$$

which implies positive helicity in the massless limit - the negative helicity fermion and positive helicity anti-fermion amplitudes dominate in charged weak interactions. For a massless antineutrino, neglecting phase factors, we have

$$v_\nu^L = \sqrt{2E_\nu}\phi_{-1} (\mathbf{n}_\nu) = \sqrt{2E_\nu}\phi_{+1} (\mathbf{n}_e) . \tag{8.115}$$

The transition current obtained with these 2-component spinors is

$$j_s^L = \left[ u_e^\dagger v_\nu, -u_e^\dagger \boldsymbol{\sigma} v_\nu \right] = \sqrt{2E_\nu}\sqrt{E_e - sp} \left[ \phi_s^\dagger (\mathbf{n}_e) \phi_{+1} (\mathbf{n}_e) , -\phi_s^\dagger (\mathbf{n}_e) \boldsymbol{\sigma} \phi_{+1} (\mathbf{n}_e) \right] . \tag{8.116}$$

Notice that the amplitude for a positive helicity electron is suppressed relative to that for negative helicity by a factor $\sqrt{(1 - p/E)/2} = \sqrt{(1 - v)/2}$. If we choose the electron direction $\mathbf{n}_e$ to be along the $z$-axis, then we can calculate the three-vector

$$\begin{aligned} \phi_s^\dagger (\hat{\mathbf{z}}) \boldsymbol{\sigma} \phi_+ &= \phi_s^\dagger (\hat{\mathbf{z}}) \left( \begin{pmatrix} 0 & 1 \\ 1 & 0 \end{pmatrix}, \begin{pmatrix} 0 & -i \\ i & 0 \end{pmatrix}, \begin{pmatrix} 1 & 0 \\ 0 & -1 \end{pmatrix} \right) \begin{pmatrix} 1 \\ 0 \end{pmatrix} \\ &= \phi_s^\dagger (\hat{\mathbf{z}}) (\phi_-, i\phi_-, \phi_+) \end{aligned} \tag{8.117}$$

and find the current 4-vectors

$$j_{+1}^L = \sqrt{2E_\nu (E_e - p)} [1, 0, 0, -1] \ ; \ j_{-1}^L = \sqrt{2E_\nu (E_e + p)} [0, -1, -i, 0] . \tag{8.118}$$

More generally, suppose $\mathbf{n}_e$ is in the $x - z$ plane with angle $\theta$ relative to $z$-axis. Rotation of the currents just found around the $y$-axis gives

$$\begin{aligned} j_{+1}^L &= \sqrt{2E_\nu (E_e - p)} [1, -\sin\theta, 0, -\cos\theta] \\ j_{-1}^L &= \sqrt{2E_\nu (E_e + p)} [0, -\cos\theta, -i, \sin\theta] . \end{aligned} \tag{8.119}$$

These currents are convenient for we may now choose polarization vectors $\epsilon_\pm = (1, \pm i, 0)/\sqrt{2}$ and $\epsilon_0 = (0, 0, 1)$ for the $W$ boson and find the matrix elements for various spin states given in Table 8.2. The squared matrix element summed over lepton helicity for decay from state $\epsilon_0$, for example, is

$$\sum |M_{W^- \to e^- \bar{\nu}_e}|^2 = g^2 E_\nu \left[(E_e - p)\cos^2\theta + (E_e + p)\sin^2\theta\right] \tag{8.120}$$

where $E_e = (m_W^2 + m_e^2)/(2m_W)$ and $p_e = (m_W^2 - m_e^2)/(2m_W)$. Writing the 2-body phase space as

$$dPS = \frac{1}{16\pi^2}\frac{p_e}{m_W}d\Omega_e, \tag{8.121}$$

we can now find the differential decay rate

$$d\Gamma_{W^- \to e\bar{\nu}_e} = \frac{|M_{W^- \to e\bar{\nu}_e}|^2}{2m_W}dPS = \frac{g^2}{32\pi^2 m_W^2}E_\nu p_e\left[(E_e + p_e) - 2p_e\cos^2\theta\right]d\Omega_e. \tag{8.122}$$

Integration over decay angles give the total decay rate

$$\Gamma_{W^- \to e^- \bar{\nu}_e} = \frac{g^2 m_w}{48\pi}(1 - x)^2\left(1 + \frac{1}{2}x\right) \tag{8.123}$$

where $x = (m_e/m_W)^2$. One may verify that the decay rate is independent of $W$ boson polarization. Neglecting $m_e$ compared to $m_W$ and using the Fermi constant $G_F = g^2/(4\sqrt{2}m_W^2)$, we arrive at the alternate expressions

$$\Gamma_{W \to e\nu} = \frac{G_F m_W^3}{6\pi\sqrt{2}} = \frac{\alpha m_W}{12\sin^2\theta_W}. \tag{8.124}$$

These expressions apply also to the $\mu\nu_\mu$ and $\tau\nu_\tau$ decay modes when the lepton masses are neglected. Including a factor of three for the sum over color states and a CKM matrix element, we have:

$$\Gamma_{W^- \to \bar{u}d} = \Gamma_{W^- \to \bar{c}s} = 3\cos^2\theta_c \Gamma_{W^- \to l\nu} \tag{8.125}$$

and

$$\Gamma_{W^- \to \bar{s}d} = \Gamma_{W^- \to \bar{c}d} = 3\sin^2\theta_c \Gamma_{W^- \to l\nu}. \tag{8.126}$$

| | Factor | $\epsilon_+$ | $\epsilon_0$ | $\epsilon_-$ |
|---|---|---|---|---|
| $j_+$ | $\sqrt{2E_\nu(E_e - p)}$ | $\frac{\sin\theta}{\sqrt{2}}$ | $\cos\theta$ | $\frac{\sin\theta}{\sqrt{2}}$ |
| $j_-$ | $\sqrt{2E_\nu(E_e + p)}$ | $-\frac{1-\cos\theta}{\sqrt{2}}$ | $-\sin\theta$ | $\frac{1+\cos\theta}{\sqrt{2}}$ |

**TABLE 8.2** Matrix elements for $W$ boson decay.

Note that $W^- \to \bar{t}b$ is forbidden by energy conservation, so the total rate, neglecting $\bar{c}b$, is $3+6=9$ times the rate to $e\nu_e$ and the branching fraction for $W \to e\nu_e$ is 1/9.

The calculation of the rate for $W \to e\nu$ may be reinterpreted to predict decays rates for the $Z$ boson to light fermion pairs. The decay amplitude for $Z \to \bar{\nu}\nu$ is related to that for $W \to e\nu$ in the limit of negligible electron mass by the coupling ratio

$$\frac{\sqrt{g'^2+g^2}}{g\sqrt{2}} = \frac{1}{\sqrt{2}\cos\theta_W} = \frac{1}{\sqrt{2}}\frac{m_Z}{m_W} \tag{8.127}$$

so we can deduce from the $W$ boson decay rate the partial rate

$$\Gamma_{Z\to\bar{\nu}\nu} = \frac{G_F m_Z^3}{12\pi\sqrt{2}} = \frac{g^2}{96\pi}\frac{m_Z^3}{m_W^2} = \frac{\alpha m_Z}{24\sin^2\theta_W\cos^2\theta_W}. \tag{8.128}$$

For decays to charged leptons, we have a left-handed amplitude of strength relative to $W \to e\nu$ given by the ratio

$$\frac{g'^2-g^2}{g\sqrt{2}\sqrt{g^2+g'^2}} = \frac{\left(2\sin^2\theta_W-1\right)}{\sqrt{2}\cos\theta_W} \tag{8.129}$$

and a right-handed amplitude of relative strength

$$\frac{g'^2\sqrt{2}}{g\sqrt{g^2+g'^2}} = \frac{2\sin\theta_W}{\sqrt{2}\cos\theta_W} \tag{8.130}$$

for which the space components of the transition currents have reversed sign. The sign change does not affect the square of the amplitude so left and right together give

$$\Gamma_{Z\to e^+e^-} = \left[4\sin^4\theta_W + \left(2\sin^2\theta_W - 1\right)^2\right]\Gamma_{Z\to\bar{\nu}\nu}. \tag{8.131}$$

To derive the decay rates to quark pairs, compare the left-handed and right-handed quark coupling constants

$$g_L = \frac{1}{2\cos\theta_W}\left(2T_3 - 2Q\sin^2\theta_W\right);\; g_R = \frac{1}{2\cos\theta_W}\left(-2Q\sin^2\theta_W\right) \tag{8.132}$$

to the neutrino coupling $g_{Z\nu\nu} = g/(2\cos\theta_W)$. The decay rate to $q\bar{q}$ counting three colors is

$$\Gamma_{Z\to q\bar{q}} = 3\left(R_q^2 + L_q^2\right)\Gamma_{Z\to\nu\bar{\nu}} \tag{8.133}$$

where $R_q = -2Q\sin^2\theta_W$ and $L_q = 2T_3 - 2Q\sin^2\theta_W$. Summing over permitted $u$-like quarks, we have

$$\Gamma_{Z\to u\bar{u}+c\bar{c}} = 3\left[1 + \left(1 - \frac{8}{3}\sin^2\theta_W\right)^2\right]\Gamma_{Z\to\nu\bar{\nu}} \tag{8.134}$$

while the sum over $d$-like quarks is

$$\Gamma_{Z\to d\bar{d}+s\bar{s}+b\bar{b}} = \frac{9}{2}\left[1+\left(1-\frac{4}{3}\sin^2\theta_W\right)^2\right]\Gamma_{Z\to\nu\bar{\nu}}. \tag{8.135}$$

The various partial widths have been measured and the agreement with the predicted partial widths and with the total width deduced from the resonant line shape is excellent. In fact, the measurements are sufficiently precise that higher order radiative corrections and QCD effects must be included in the theoretical prediction and experimental analysis.

## 8.7 ■ TOP QUARK DECAY

The decay of the top quark is unique among heavy flavor decays in that, since $m_t > m_W$, the $W$ boson produced is real. The matrix element for $t \to W^+ b$ can be read from the standard model Lagrangian and is

$$M_{t\to W^+b} = -V_{tb}\frac{g}{\sqrt{2}}\bar{b}_L\gamma t_L\epsilon^\dagger. \tag{8.136}$$

Neglecting $m_b$ and phase factors, we can use the two dimensional spinor form of the current given in Equation 8.10. Let $\mathbf{n}_t$, $\mathbf{n_b}$, and $\mathbf{n}_W$ denote the particle directions, and $s$ the top quark helicity. In the $t$-quark rest frame,

$$t_L = \sqrt{2m_t}\phi_s(\mathbf{n}_t)\ ;\ b_L = \sqrt{2E_b}\phi_{-1}(\mathbf{n}_b) = \sqrt{2E_b}\phi_{+1}(\mathbf{n}_W). \tag{8.137}$$

Take $\mathbf{n}_W$ as the $z$-direction and $\theta$ to be the angle between $\mathbf{n}_W$ and $\mathbf{n}_t$. For $s = +1$, the fermion current (Equation 8.10) is proportional to

$$j^\mu = \phi_{+1}^\dagger(1, -\boldsymbol{\sigma})\begin{pmatrix} C \\ S \end{pmatrix} = (C, -S, iS, -C) \tag{8.138}$$

where $C = \cos\frac{\theta}{2}$ and $S = \sin\frac{\theta}{2}$. The $W$ boson polarization vectors may be found by boosting from the $W$ boson rest frame and written as

$$\epsilon_0^\dagger = \left(\frac{p_W}{m_W}, \frac{E_W}{m_W}\epsilon_0\right);\ \epsilon_\pm^\dagger = (0, \epsilon_\mp) \tag{8.139}$$

where $\epsilon_0 = (0, 0, 1)$ and $\epsilon_\pm = (1, \pm i, 0)/\sqrt{2}$. Combining these expression yields

$$j\epsilon_0^\dagger = C\frac{E_W + p_W}{m_W}\ ;\ j\epsilon_+^\dagger = 0\ ;\ j\epsilon_-^\dagger = S\sqrt{2}. \tag{8.140}$$

Kinematics determines the energy and momentum of the $W$ boson and $b$-quark to be

$$p_W = p_b = E_b = \frac{m_t^2 - m_W^2}{2m_t}\ ;\ E_W = \frac{m_t^2 + m_W^2}{2m_t} \tag{8.141}$$

so $(E_W + p_W)/m_W = m_t/m_W$. The total decay rate is found by summing the squared amplitudes and is

$$\begin{aligned}
\Gamma_{t\to W^+b} &= \frac{1}{2m_t}\left[|M_0|^2 + |M_-|^2\right] dPS \\
&= \frac{1}{2m_t} 2m_t\,(2E_b)\,\frac{g^2}{2}\left[C^2\left(\frac{m_t}{m_W}\right)^2 + 2S^2\right]\frac{p_b d\Omega_b}{16\pi^2 m_t} \\
&= \frac{g^2}{8\pi}\frac{E_b^2}{m_t}\left[\frac{1}{2}(1+\cos\theta)\left(\frac{m_t}{m_W}\right)^2 + 2\frac{1}{2}(1-\cos\theta)\right] d\cos\theta \\
&= \frac{g^2}{64\pi}\frac{m_t^3}{m_W^2}\left(1 - \frac{m_W^2}{m_t^2}\right)^2\left(1 + 2\frac{m_W^2}{m_t^2}\right) \qquad (8.142)
\end{aligned}$$

where we have set $V_{tb} = 1$. Note that this expression is not well behaved as $m_W \to 0$. For $m_t >> m_W$, the top is strongly coupled to the longitudinal component of the $W$ boson which derives from the Higgs field.

The angular distributions based on reconstruction of $t\bar{t} \to W^+bW^-\bar{b}$ events in $p\bar{p}$ collisions are consistent with this calculation. With electroweak and QCD corrections and $m_t = 172$ GeV, the value $\Gamma_{t\to Wb} \simeq 1.4$ GeV is estimated, consistent with upper limits of about 10 GeV inferred from the width of reconstructed $t$-quark mass distributions at the Tevatron.

## 8.8 ■ THREE-BODY DECAYS OF HEAVY FERMIONS

In the standard model, heavy quarks and heavy charged leptons decay to lighter flavors through a virtual $W$ boson intermediate state. Muon decay is the prototype for all such decays. Understanding its pure charged current weak decay was a benchmark in the development of the theory of the weak interaction. The 3-body final state entails some complexity in the calculation of the decay distributions, so take a deep breath.

The matrix element for the decay $\mu_1^- \to \nu_2 e_3^- \bar{\nu}_4$ that follows from the Feynman rules is

$$M_{\mu^-\to\nu_\mu e^-\bar{\nu}_e} = \frac{g^2}{2m_W^2} u_2^\dagger \sigma_\mu^+ u_1 u_3^\dagger \sigma^{+\mu} v_4 \qquad (8.143)$$

where the $W$ boson propagator was approximated by its low energy form $ig^{\mu\nu}/m_W^2$. In the muon rest frame, take the $z$-direction along the electron and the muon spin along some direction $\mathbf{n}_1$ in the $x - z$ plane. Neglecting all lepton masses, the unnormalized spinor amplitudes are

$$u_1 = \phi_+(\mathbf{n}_1) = \begin{pmatrix} E_1^* C_1 \\ E_1 S_1 \end{pmatrix}$$

$$u_2 = \phi_-(\mathbf{n}_2) = \begin{pmatrix} -E_2^* S_2 \\ E_2 C_2 \end{pmatrix}$$

$$u_3 = \phi_- (\mathbf{n}_3) = \begin{pmatrix} -E_3^* S_3 \\ E_3 C_3 \end{pmatrix} = \begin{pmatrix} 0 \\ 1 \end{pmatrix}$$

$$v_4 = \phi_- (\mathbf{n}_4) = \begin{pmatrix} -E_4^* S_4 \\ E_4 C_4 \end{pmatrix} = i \begin{pmatrix} E_2^* S_4 \\ E_2 C_4 \end{pmatrix} \tag{8.144}$$

where $C_i = \cos\theta_i$, $S_i = \sin\theta_i$, and $E_i = e^{-\frac{i\phi_i}{2}}$ with $\phi_1 = 0$ and, by momentum conservation, $\phi_4 = \phi_2 + \pi$. The electron current may then be expressed as

$$u_3^\dagger \sigma_\mu v_4 = \left[E_2 C_4, -E_2^* S_4, -i E_2^* S_4, E_2 C_4\right]. \tag{8.145}$$

This should be multiplied by the product of relativistic spinor normalizations $\sqrt{m 2E_\nu 2E_{\bar{\nu}} 2E_e}$. The muon current may also be similarly expressed and the product of the currents is

$$j_\mu j_e = u_2^\dagger \sigma_\mu u_1 u_3^\dagger \sigma^\mu v_4 \propto 2S_2 \left[S_1 S_4 - \left(E_2 E_1^*\right)^2 C_1 C_4\right]. \tag{8.146}$$

Using $2S^2 = (1 - \cos\theta)$, $2C^2 = (1 + \cos\theta)$, and $2SC = \sin\theta$ to simplify the squared matrix element, including the spinor normalization factors, and combining with a similar calculation for muon spin down along $\mathbf{n}_1$, one finds

$$\begin{aligned} |M_{\mu^- \to \nu_\mu e^- \bar{\nu}_e}|^2 = \left(\frac{g^2}{2m_W^2}\right)^2 & 8m E_\nu E_{\bar{\nu}} E_e (1 - \cos\theta_\nu) \\ & \times \left[1 + \lambda \cos\theta_{\bar{\nu}} \cos\theta_e - \lambda \sin\theta_{\bar{\nu}} \sin\theta_e \cos(\phi_{\bar{\nu}} - \phi_\mu)\right] \end{aligned} \tag{8.147}$$

where $\lambda = \pm 1$ for muon spin parallel or anti-parallel to $\mathbf{n}_1$. Here $\phi_{\bar{\nu}} - \phi_\mu$ is the azimuthal angle of the neutrino pair relative to the electron and $\theta_e \equiv \theta_\mu$ is the angle between the electron direction and the muon quantization axis.

With the definition $E_f = E_\nu - E_{\bar{\nu}} - E_e$, the decay rate can be written as

$$\begin{aligned} d\Gamma_{\mu^- \to \nu_\mu e^- \bar{\nu}_e} &= \frac{|M_{\mu^- \to \nu_\mu e^- \bar{\nu}_e}|^2}{(2\pi)^5 \, 2m 2E_\nu 2E_{\bar{\nu}} 2E_e} \delta\left(m - E_f\right) \delta\left(\mathbf{p}_2 + \mathbf{p}_\nu + \mathbf{p}_{\bar{\nu}}\right) d\mathbf{p}_2 d\mathbf{p}_\nu d\mathbf{p}_{\bar{\nu}} \\ &= \frac{|M_{\mu^- \to \nu_\mu e^- \bar{\nu}_e}|^2}{(2\pi)^5 \, 2m 2E_\nu 2E_{\bar{\nu}} 2E_e} \delta\left(m - E_f\right) p_e^2 dp_e d\Omega_e p_\nu^2 dp_\nu d\Omega_\nu \end{aligned} \tag{8.148}$$

where the second expression results after integration over $d\mathbf{p}_{\bar{\nu}}$ and the antineutrino energy is

$$E_{\bar{\nu}} = \sqrt{\mathbf{p}_{\bar{\nu}}^2} = \left[\mathbf{p}_e^2 + \mathbf{p}_\nu^2 + 2|\mathbf{p}_e||\mathbf{p}_\nu| \cos\theta_{\nu e}\right]^{1/2}. \tag{8.149}$$

Reckoning the neutrino angles relative to the electron direction and writing $d\Omega_\nu = d\phi_\nu d\cos\theta_\nu$, integration over $d\phi_\nu$ eliminates the $\cos\phi_\nu$ term in the squared matrix element and gives a factor $2\pi$. Changing variables with $dE_\nu/d\cos\theta_\nu = |\mathbf{p}_e||\mathbf{p}_\nu|/E_{\bar{\nu}}$ gives

$$\begin{aligned}
d\Gamma_{\mu^-\to\nu_\mu e^-\bar{\nu}_e} &= \left(\frac{g^2}{2m_W^2}\right)^2 \frac{1}{2}\left(1-\cos\theta_\nu\right)\left[1+\lambda\cos\theta_{\bar{\nu}}\cos\theta_e\right](2\pi)^{-4} \\
&\quad \delta\left(m-E_f\right) E_e E_\nu E_{\bar{\nu}} dE_{\bar{\nu}} d\Omega_e \\
&= \left(\frac{g^2}{2m_W^2}\right)^2 \frac{1}{2}\left(1-\cos\theta_\nu\right)\left[1+\lambda\cos\theta_{\bar{\nu}}\cos\theta_e\right](2\pi)^{-3} \\
&\quad E_e E_\nu E_{\bar{\nu}} dE_{\bar{\nu}} d\cos\theta_e
\end{aligned} \tag{8.150}$$

where, after integration over $dE_{\bar{\nu}}$, the antineutrino energy $E_{\bar{\nu}} = m - E_e - E_\nu$ is determined by the electron and neutrino energies, and integration over the azimuth of the electron was performed.

Momentum conservation gives $\mathbf{p}_{\bar{\nu}}^2 = (\mathbf{p}_2 + \mathbf{p}_\nu)^2$ or

$$\cos\theta_\nu = \frac{E_{\bar{\nu}}^2 - E_e^2 - E_\nu^2}{2E_e E_\nu} \tag{8.151}$$

and momentum conservation along the electron direction implies

$$E_\nu\cos\theta_\nu + E_{\bar{\nu}}\cos\theta_{\bar{\nu}} = -E_e. \tag{8.152}$$

and hence we have the relationship

$$\cos\theta_{\bar{\nu}} = -\frac{E_e + E_\nu\cos\theta_\nu}{E_{\bar{\nu}}}. \tag{8.153}$$

Using this relationship, the decay rate may be expressed in terms of $E_e$, $E_\nu$ and $\cos\theta_e$. For fixed $E_e$, the neutrino energy ranges from $m/2 - E_e$ to $m/2$. Integration over $E_\nu$ gives the observable electron spectrum. Setting $(g^2/2m_W^2)^2 = 8G_F^2$, the spectrum is given by

$$\frac{d\Gamma_{\mu^-\to\nu_\mu e^-\bar{\nu}_e}}{dE_e d\cos\theta_e} = \frac{G_F^2 m^2 E_e^2}{24\pi^3}\left(3 - 4\frac{E_e}{m} + \lambda\left[1 - \frac{4E_e}{m}\right]\cos\theta_e\right). \tag{8.154}$$

This expression shows how the electron angular distribution for fixed electron energy is correlated with the muon spin direction.

The maximum electron energy is $E_{\max}$= 52.8 MeV. For muon spin down, we can write the probability distribution for normalized electron energy $y = E_e/E_{\max}$ as

$$\frac{dP}{dy d\Omega_e} = \frac{n(y)}{2\pi}\left[1 - \alpha(y)\cos\theta_e\right] \tag{8.155}$$

where we define the number density and asymmetry functions

$$n(y) = y^2(3 - 2y)$$
$$\alpha(y) = \frac{2y-1}{3-2y}. \tag{8.156}$$

These functions are shown in Figure 8.5. The electron tends to emerge with a large fraction $y$ of the available energy. For $y > 1/2$, the asymmetry is positive and the electron tends to be emitted opposite to the muon spin. For $y < 1/2$, it is negative and the electron tends to be emitted along the muon spin direction.

The energy spectrum that results from integration over angles,

$$\frac{d\Gamma_{\mu^-\to\nu_\mu e^-\bar{\nu}_e}}{dE_e} = \int_{-1}^{+1} d\cos\theta_e \frac{d\Gamma}{dE_e d\cos\theta_e} = \frac{G_F^2 m^2 E_e^2}{12\pi^3}\left(3 - 4\frac{E_e}{m}\right), \tag{8.157}$$

is independent of muon spin. One more integration leads to the total decay rate:

$$\Gamma_{\mu^-\to e^-\nu_\mu\bar{\nu}_e} = \int_0^{\frac{m}{2}} dE_e \frac{d\Gamma}{dE_e} = \frac{G_F^2 m^5}{192\pi^3}. \tag{8.158}$$

In the preceding calculation we neglected all lepton masses for simplicity. In a general 3-body weak decay such as $b \to cs\bar{c}$, the final state fermion masses may not be negligible. The calculation must consider all combinations of the spin states of the initial and final particles and becomes quite complex. We illustrate by extending the calculation of muon decay including the electron mass, but neglecting the mass of the neutrino and the mass of the antineutrino. The result is directly applicable to a process like $b \to ce^-\bar{\nu}_e$.

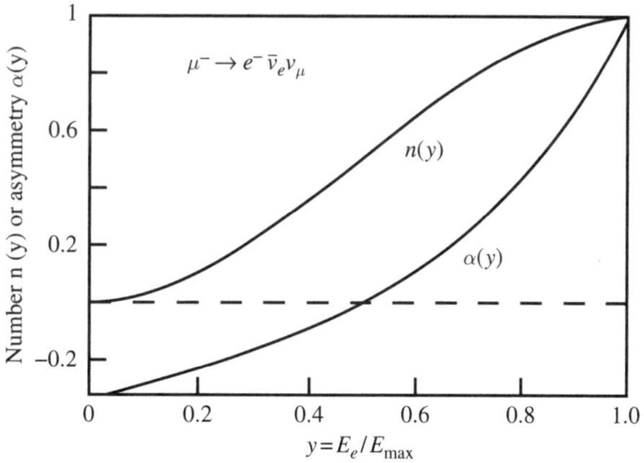


**FIGURE 8.5** Number density and asymmetry functions describing the electron energy and angle distribution in the decay $\mu^- \to e^-\bar{\nu}_e\nu_\mu$. The electron tends to emerge with a large fraction $y$ of the available energy. For $y > 1/2$, the asymmetry is positive. For $y < 1/2$, it is negative.

For variety, we consider $\mu^+ \to \bar{\nu}_\mu \nu_e e^+$. The Lorentz invariant matrix element is

$$M_{\mu^+\to\bar{\nu}_\mu\nu_e e^+} = \frac{g^2}{2m_W^2} j_e \cdot j_\mu = \frac{g^2}{2m_W^2} \bar{u}^L(\nu_e)\gamma v^L(e^+)\cdot \bar{v}^L(\mu^+)\gamma v^L(\nu_\mu) \tag{8.159}$$

where

$$u^L(\nu_e) = \sqrt{2E_\nu}\phi_{-1}(\mathbf{n}_\nu)\ ;\ v^L(\bar{\nu}_\mu) = \sqrt{2\bar{E}}\phi_{-1}(\mathbf{n}_{\bar{\nu}}) \tag{8.160}$$

and

$$v^L(e^+) = s\sqrt{E_e + sp_e}\phi_{-s}(\mathbf{n}_e)\ ;\ v^L(\mu^+) = \sqrt{E_\mu + rp_\mu}\phi_{-r}(\mathbf{n}_\mu). \tag{8.161}$$

It is easiest to evaluate $M$ in the center of mass of the two neutrinos where we can orient our coordinate system so that $\mathbf{p}_\mu = \mathbf{p}_e$ is along the $z$-axis and $\mathbf{p}_{\bar{\nu}} = -\mathbf{p}_\nu$ has polar angle $\theta$ relative to the $z$-axis and lies in the $x - z$ plane. The electron current for position helicity $s$ is proportional to

$$\begin{aligned} j_e^s &= \left[\phi_-^\dagger(\mathbf{n}_\nu)\phi_{-s}(\mathbf{n}_e), -\phi_-^\dagger(\mathbf{n}_\nu)\boldsymbol{\sigma}\phi_{-s}(\mathbf{n}_e)\right] \\ &= \left[\phi_+^\dagger(\mathbf{n}_{\bar{\nu}})\phi_{-s}(\mathbf{n}_e), -\phi_+^\dagger(\mathbf{n}_{\bar{\nu}})\boldsymbol{\sigma}\phi_{-s}(\mathbf{n}_e)\right]. \end{aligned} \tag{8.162}$$

Using the notation

$$\phi_+ = \begin{pmatrix}1\\0\end{pmatrix}\ ;\ \phi_- = \begin{pmatrix}0\\1\end{pmatrix}, \tag{8.163}$$

we can find $\boldsymbol{\sigma}\phi_- = (\phi_+, i\phi_+, -\phi_-)$ and, with the abbreviations $S = \sin\frac{\theta}{2}$, $C = \cos\frac{\theta}{2}$, for s=1, the electron current is

$$j_e^+ = [C, S][\phi_-, -\phi_+, i\phi_+, -\phi_-] = [S, -C, iC, S]. \tag{8.164}$$

For $s = -1$, we have $\boldsymbol{\sigma}\phi_+ = (\phi_-, i\phi_-, \phi_+)$ and the electron current is

$$j_e^- = [C, S][\phi_+, -\phi_-, -i\phi_-, -\phi_+] = [C, -S, -iS, -C]. \tag{8.165}$$

The muon current for muon helicity $r$ is

$$j_\mu^r = \left[\phi_{-r}^\dagger(\mathbf{n}_\mu)\phi_-(\mathbf{n}_{\bar{\nu}}), -\phi_{-r}^\dagger(\mathbf{n}_\mu)\boldsymbol{\sigma}\phi_-(\mathbf{n}_{\bar{\nu}})\right] = \left[\phi_{-r}^\dagger\phi_-(\mathbf{n}_{\bar{\nu}}), -\phi_{-r}^\dagger\boldsymbol{\sigma}\phi_-(\mathbf{n}_{\bar{\nu}})\right]. \tag{8.166}$$

Inserting the two component spinor for the muon neutrino gives

$$j_\mu = \phi_{-r}^\dagger\left[\begin{pmatrix}-S\\C\end{pmatrix}, \begin{pmatrix}-C\\S\end{pmatrix}, \begin{pmatrix}iC\\iS\end{pmatrix}, \begin{pmatrix}S\\C\end{pmatrix}\right] \tag{8.167}$$

from which the muon currents for the two muon helicity states follow:

$$j_\mu^+ = (C, S, iS, C) \ ; \ j_\mu^- = (-S, -C, iC, S)\,. \qquad (8.168)$$

The products of electron 4-current times muon 4-current for the various helicity combinations are

$$\begin{aligned} j_e^+ j_\mu^+ &= SC - (-SC - SC + SC) = 2\sin\frac{\theta}{2}\cos\frac{\theta}{2} = \sin\theta \\ j_e^+ j_\mu^- &= -2\sin^2\frac{\theta}{2} = \cos\theta - 1 \\ j_e^- j_\mu^+ &= 2\cos^2\frac{\theta}{2} = \cos\theta + 1 \\ j_e^- j_\mu^- &= -2\sin\frac{\theta}{2}\cos\frac{\theta}{2} = -\sin\theta. \end{aligned} \qquad (8.169)$$

The squared matrix element summed over initial and final polarization may be computed from these and is given by

$$\overline{|M_{\mu^+\to\bar\nu_\mu\nu_e e^+}|^2} = \frac{g^4}{4m_W^2} 2E_\nu 2E_{\bar\nu}\left[a\sin^2\theta + b\,(1-\cos\theta)^2 + c\,(1+\cos\theta)^2\right] \qquad (8.170)$$

where $a = (E_\mu + p_\mu)(E_e + p_e) + (E_\mu - p_\mu)(E_e - p_e)$, $b = (E_\mu - p_\mu)(E_e + p_e)$ and $c = (E_\mu + p_\mu)(E_e - p_e)$. The integration over over neutrino angles may be performed using

$$\int d\Omega\ (1\pm\cos\theta)^2 = \frac{16\pi}{3}\ ;\ \int d\Omega\ \sin^2\theta = \frac{8\pi}{3} \qquad (8.171)$$

and, if we divide by two and by $4\pi$, we obtain the squared matrix element summed over final polarization and averaged over the initial polarization and over neutrino angles for fixed neutrino energy:

$$\overline{|M_{\mu^+\to\bar\nu_\mu\nu_e e^+}|^2} = 2\frac{g^4}{m_W^4}E_\nu^2\left[E_\mu E_e - p_\mu^2/3\right]. \qquad (8.172)$$

The differential decay rate in the muon rest frame is given by the general formula

$$d\Gamma_{\mu^+\to\bar\nu_\mu\nu_e e^+} = \frac{1}{2m_\mu}\overline{|M_{\mu^+\to\bar\nu_\mu\nu_e e^+}|^2}dPS. \qquad (8.173)$$

The Lorentz invariant phase space for a 3-body decay can be expressed as (see Equation 6.251)

$$dPS(p\to p_1+p_2+p_3) = dPS(p\to p_{12}+p_3)\frac{ds_{12}}{2\pi}dPS(p_{12}\to p_1+p_2)\,. \qquad (8.174)$$

The 2-body phase space for the two neutrinos becomes after angular integration

$$dPS(p_{12} \to p_1 + p_2) = \frac{1}{16\pi^2} \frac{p}{\sqrt{s_{12}}} d\Omega \to \frac{1}{8\pi} \tag{8.175}$$

and so, for the muon decay to electron plus aneutrino pair of invariant mass $\sqrt{s_{12}}$, we have

$$dPS\left(\mu \to e + \sqrt{s_{12}}\right) = \frac{1}{4\pi} \frac{p_e}{m_\mu}. \tag{8.176}$$

Hence the differential decay rate in the muon rest frame reduces to

$$d\Gamma = \frac{g^4}{m_W^4} \frac{s_{12}}{m_\mu^2} \left[ E_e E_\mu - \frac{p^3}{3} \right] p_e d s_{12} \tag{8.177}$$

where $p_e$ refers to the electron momentum in the muon rest frame while $E_e$, $E_\mu$, and $p$ refer to the rest frame of the neutrinos. To evaluate these energies in terms of the electron energy in the muon rest frame, use

$$m_\mu^2 = (p_e + p_{12})^2 = m_e^2 + s_{12} + 2E_e\sqrt{s_{12}} \tag{8.178}$$

and

$$m_e^2 = \left(p_\mu - p_{12}\right)^2 = m_\mu^2 + s_{12} - 2E_\mu\sqrt{s_{12}} \tag{8.179}$$

to find

$$E_e E_\mu = \frac{1}{4s_{12}} \left[ \left(m_\mu^2 - m_e^2\right)^2 - s_{12}^2 \right]. \tag{8.180}$$

The quantity $p$ is the momentum of the muon or electron in the neutrino rest frame. The momentum of the neutrino pair or electron in the muon rest frame is $p_e$ so the relative 4-velocity is

$$u = \left[ \sqrt{s_{12} + p_e^2}, p_e \right] / \sqrt{s_{12}} \tag{8.181}$$

and the 4-momentum of the muon in the neutrino pair rest frame is $m_\mu u$ and hence

$$p = p_e \frac{m_\mu}{\sqrt{s_{12}}}. \tag{8.182}$$

Finally, the electron energy in the muon frame is

$$E = \frac{1}{2m_\mu} \left(m_\mu^2 + m_e^2 - s_{12}\right) \tag{8.183}$$

so we can write

$$s_{12} = m_\mu^2 + m_e^2 - 2m_\mu E. \tag{8.184}$$

Collecting results, we write the decay rate as

$$d\Gamma_{\mu^+\to\bar{\nu}_\mu\nu_e e^+} = \frac{1}{4^5\pi^3}\frac{g^4}{m_W^4 m_\mu^2} p_e ds_{12}\left[\left(m_\mu^2 - m_e^2\right)^2 - s_{12}^2 - \frac{4}{3}p_e^2 m_\mu^2\right]$$
$$= \frac{1}{6(4\pi)^3}\frac{g^4}{m_W^4} m_\mu \sqrt{E^2 - m_e^2}dE$$
$$\times\left[-2m_e^2 + 3m_\mu^{-1}\left(m_\mu^2 + m_e^2\right)E - 4E^2\right] \tag{8.185}$$

where $p_e = \sqrt{E^2 - m_e^2}$ and $s_{12} = m_\mu^2 + m_e^2 - 2m_\mu E$ were used to arrive at the final expression for the electron energy spectrum. Excepting the region where the electron is nonrelativistic, the shape of the spectrum is well described by the simpler form

$$\frac{d\Gamma_{\mu^+\to\bar{\nu}_\mu\nu_e e^+}}{dE} \sim E^2\left(3m_\mu - 4E\right). \tag{8.186}$$

The total decay rate is obtaining by integration over electron energy and neglecting the electron mass is

$$\Gamma_{\mu^+\to\bar{\nu}_\mu\nu_e e^+} = \frac{1}{192\pi^3}G_F^2 m_\mu^5 f\left[1 - \frac{\alpha\left(m_\mu\right)}{\pi}\left(\pi^2 - \frac{25}{4}\right)\right] \tag{8.187}$$

where the dependence on $m_e$ is encapsulated in the factor

$$f(x) = 1 - 8x + 8x^3 - x^4 - 12x^2\ln x \simeq 1, \tag{8.188}$$

with $x = (m_e/m_\mu)^2$. As an indication of the limits of our leading order calculation, we have included the correction factor from a higher order calculation [A. Sirlin, Rev. Mod. Phys. **50**, 573 (1978)] proportional to $\alpha$ where

$$\alpha\left(m_\mu\right)^{-1} = \alpha^{-1} - \frac{2}{3\pi}\log\left(\frac{m_\mu}{m_e}\right) + \frac{1}{6\pi} \simeq 136. \tag{8.189}$$

The lifetime of the muon is about two microseconds and the exponential time dependence can be measured directly to infer the decay rate. This fundamental measurement determines $G_F$. The calculation is applicable to the decay of the $\tau^\pm$ to electron or muon plus neutrinos. The $\tau$ lifetime is a fraction of one ps and inferred from the decay length distribution of $\tau$ leptons of known energy. Using measured rates for $\tau$ and $\mu$ decay, we can compute the ratio

$$\left(\frac{m_\mu}{m_\tau}\right)^5\frac{B_{\tau\to e\nu}\Gamma_\tau^{tot}}{\Gamma_{\mu\to e\nu}} = \left(\frac{0.1056}{1.777}\right)^5\frac{(0.174)2.20\times 10^{-6}\mathrm{s}}{0.291\times 10^{-12}\ \mathrm{s}} = 0.97 \tag{8.190}$$

Including $f_\mu = 0.962$, the ratio is consistent with a universal coupling. (A more precise comparison requires inclusion of radiative corrections.)

The complexity of this muon decay calculation illustrates how much information is contained in the leading order matrix element. It predicts the correlations between the momenta and helicity states of the four particles participating in the process. This kind of analysis can be made in the more general case of the decay of a fermion to three massive fermions considering all three final state masses and combinations of six final helicity states. A more general matrix element form can be considered and limits placed on non-standard model terms. It should be clear that such an analysis is quite lengthy. Such calculations and comparisons with experiment may be found in technical literature.

## 8.9 ■ NEUTRINO SCATTERING

Scattering of neutrinos from fixed targets illustrates the exchange of intermediate vector bosons. Consider the process

$$\nu_\mu + e^- \to \mu^- + \nu_e \tag{8.191}$$

shown in Figure 8.6. For center of mass energy much less than $m_W$, the matrix element is

$$\begin{aligned} M_{\nu_\mu e^- \to \mu^- \nu_e} &= -i\frac{4G_F}{\sqrt{2}}\bar{u}(\mu)\gamma_\mu \frac{1}{2}(1-\gamma_5)\,u\left(\nu_\mu\right)\bar{u}\left(\nu_e\right)\gamma_\mu \frac{1}{2}(1-\gamma_5)\,u(e) \\ &= -i\frac{4G_F}{\sqrt{2}} j(\mu) j(e). \end{aligned} \tag{8.192}$$

The scattering cross section may be obtained following the example of muon electron elastic scattering in quantum electrodynamics. If fermion masses are neglected, the left-handed muonic current in the center of mass frame is (compare to Equation 6.187)

$$j(\mu) = \left[2E\phi_t^\dagger\left(\mathbf{n}'\right)\phi_t\left(\mathbf{n}\right), 2tp\phi_t^\dagger\left(\mathbf{n}'\right)\sigma\phi_t\left(\mathbf{n}\right)\right] \tag{8.193}$$

where $t = -1$, $p = E$ is the momentum of each fermion, $\mathbf{n}$ is the muon direction, and $\mathbf{n}$' is the direction of the muon neutrino. Taking $\mathbf{n}$ along the $z$-direction with scattering angle $\theta$ in the $x - z$ plane, one finds the muon current

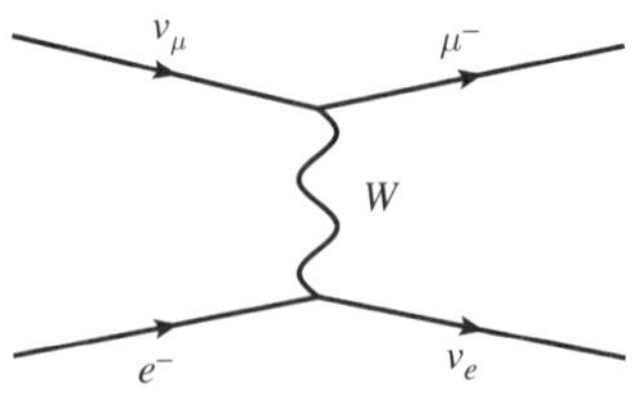


**FIGURE 8.6** Feynman diagram for the charged current scattering process $\nu_\mu e^- \to \mu^- \nu_e$.

$$j(\mu) = 2E\left[\cos\left(\frac{\theta}{2}\right), -\sin\left(\frac{\theta}{2}\right), -i\sin\left(\frac{\theta}{2}\right), -\cos\left(\frac{\theta}{2}\right)\right] \tag{8.194}$$

and the electron current follows by 180 degree rotation of the muon current about the $y$-axis. The product of these currents is

$$j(\mu)j(e)=(2E)^2\left[\cos^2\left(\frac{\theta}{2}\right) - \left(-\sin^2\left(\frac{\theta}{2}\right) - \sin^2\left(\frac{\theta}{2}\right) - \cos^2\left(\frac{\theta}{2}\right)\right)\right]=2s \tag{8.195}$$

where $s = (2E)^2$ is the center of mass energy squared. The differential scattering cross section is therefore

$$\frac{d\sigma}{d\Omega} = \frac{1}{64\pi^2 s}\overline{|M|^2} = \frac{1}{64\pi^2 s}\frac{1}{2}\left(\frac{4G_F}{\sqrt{2}}2s\right)^2 = \frac{G_F^2}{4\pi^2}s \tag{8.196}$$

where a factor of 1/2 arises from averaging over target electron polarization. The differential cross section is isotropic in the center of mass and the total cross section which follows by integration over angles is

$$\sigma_{\nu_\mu e^- \to \mu^- \nu_e} = \frac{G_F^2}{\pi}s. \tag{8.197}$$

Notice that the cross section is proportional to neutrino beam energy ($s \simeq 2m_e E$) in scattering from stationary electrons. The observed cross section provides a measurement of the Fermi constant consistent with that deduced from muon decay.

Elastic scattering of neutrinos is sensitive to $Z$ boson exchange. The processes $\nu_\mu e^- \to \nu_\mu e^-$ and $\bar{\nu}_\mu e^- \to \bar{\nu}_\mu e^-$ shown in Figure 8.7 are pure $Z$ boson exchange while a process such as $\nu_e e^- \to \nu_e e^-$ results from the coherent amplitudes for $Z$ boson exchange and $W$ boson exchange. Since the $Z$ boson couples only to left-handed neutrinos like the $W$ boson, but to both left and right-handed electrons, the pure $Z$ boson exchange cross sections are proportional to $G_F^2 s$ times a quadratic combination of the electron left and right-handed coupling constants or, equivalently, the axial and vector coupling constants. For the kinematic assumptions used above, one predicts the ratio

$$\frac{\sigma_{\nu_\mu e^- \to \nu_\mu e^-}}{\sigma_{\nu_\mu e^- \to \mu^- \nu_e}} = \left(\frac{g_v + g_a}{2}\right)^2 + \frac{1}{3}\left(\frac{g_v - g_a}{2}\right)^2 \tag{8.198}$$

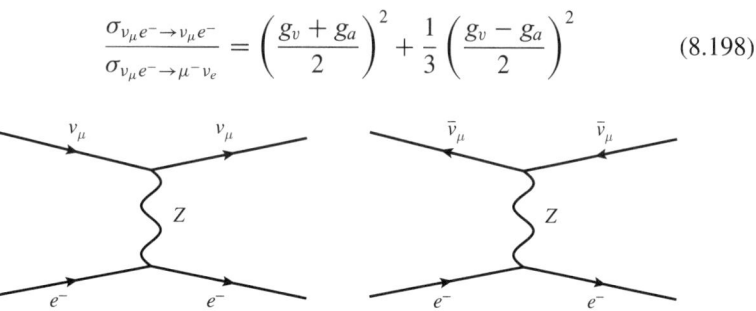

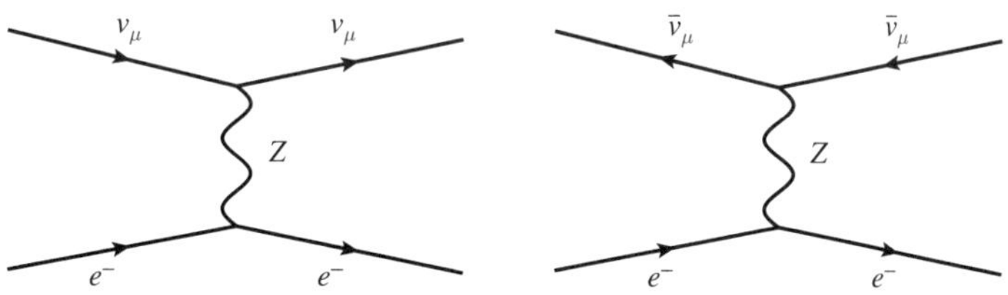


**FIGURE 8.7** Feynman diagrams for the purely neutral current scattering process $\nu_\mu e^- \to \nu_\mu e^-$ and $\bar{\nu}_\mu e^- \to \bar{\nu}_\mu e^-$. The cross sections for these processes are easily related to that for that charged current process $\nu_\mu e^- \to \mu^- \nu_e$.

and also the ratio

$$\frac{\sigma_{\bar{\nu}_\mu e^- \to \bar{\nu}_\mu e^-}}{\sigma_{\nu_\mu e^- \to \mu^- \nu_e}} = \left(\frac{g_v - g_a}{2}\right)^2 + \frac{1}{3}\left(\frac{g_v + g_a}{2}\right)^2 \tag{8.199}$$

where the vector and axial vector coupling constants for the electron that appear in these expressions are $g_v = -1/2 + 2\sin^2\theta_W$ and $g_a = -1/2$. These cross sections are of order $10^{-42}\text{cm}^2$ for $E$=1 GeV.

From the neutrino elastic scattering cross sections, the value of $\sin^2\theta_W \simeq 0.22$ is determined. As described previously, this value was crucial in predicting $m_W$ and $m_Z$. In scattering neutrinos from target electrons, the recoil electron appears typically at small laboratory angles of order $\sqrt{m_e/E_\nu}$. An accessible variable is the energy fraction $y = E_e/E_\nu$ ranging from 0 up to close to unity correlated with the scattering angle in the center of mass. The differential cross sections are quadratic functions of $y$ with coefficients functions of $g_v$ and $g_a$ and the detailed distributions also agree with calculation.

Further confirmation of the electroweak model is found in scattering of neutrinos from nucleons. A typical experiment uses a multi-GeV neutrino beam and observes an inelastic process like

$$\nu_\mu + N \to \mu^- + X \tag{8.200}$$

resulting from $W$ boson exchange where $X$ denotes hadrons. A $W^+$ can be absorbed by a $d$-quark and only a few GeV energy deposit is sufficient to create a multi-hadron state $X$. Suppose, in the center of mass of the collision, the $d$-quark carries a fraction $x$ of the nucleon momentum with probability density $d(x)$. The cross section follows from that for $\nu_\mu + e^- \to \mu^-\nu_e$ by averaging over the center of mass energy of the neutrino and quark:

$$\frac{\sigma_{\nu_\mu N \to \mu^- + X}}{\sigma_{\nu_\mu e^- \to \mu^- \nu_e}} = \frac{M_N}{m_e}\int_0^1 dx\, x d(x). \tag{8.201}$$

For a heavy nucleus, the distribution function $d(x)$ should be replaced by its average for protons and neutrons, and this average should be the same for $u$-quarks and for $d$-quarks. Consequently, the scattering cross sections for both charged and neutral current scattering from heavy nuclei are all proportional to

$$\bar{x} = \frac{1}{2}\Sigma_{p,n}\int_0^1 dx\, x < d(x) > \tag{8.202}$$

and the cross sections for a process including $Z$ boson exchange divided by the cross section for a charged current process depends only on weak interaction couplings. In particular, we have the predictions

$$\frac{\sigma_{\nu_\mu N \to \nu_\mu X}}{\sigma_{\nu_\mu N \to \mu^- X}} = \frac{1}{2} - \sin^2\theta_W + \frac{20}{27}\sin^4\theta_W \tag{8.203}$$

$$\frac{\sigma_{\bar{\nu}_\mu N \to \bar{\nu}_\mu X}}{\sigma_{\nu_\mu N \to \mu^- X}} = \frac{1}{2} - \sin^2\theta_W + \frac{20}{9}\sin^4\theta_W. \qquad (8.204)$$

Measurements of these ratios provide a determination of $\sin^2\theta_W$ complementary to that from neutrino electron scattering and the detailed kinematic distributions provide a test of the interactions of first generation quarks with the $Z$ boson.

For reference, we list the cross sections for various neutrino scattering processes in engineering units. The standard model cross sections for electron and muon (or tau) neutrinos scattering elastically from stationary electrons are

$$\begin{aligned}
\sigma_{\nu_e e^- \to \nu_e e^-} &= \tfrac{G_F^2 s}{\pi}\left[\left(\tfrac{1}{2} + s_W^2\right)^2 + \tfrac{1}{3}s_W^4\right] \simeq 9.5 \times 10^{-49}\ \mathrm{m}^2 \left(\frac{\mathrm{E}_\nu}{1\ \mathrm{MeV}}\right) \\
\sigma_{\bar{\nu}_e e^- \to \bar{\nu}_e e^-} &= \tfrac{G_F^2 s}{\pi}\left[\tfrac{1}{3}\left(\tfrac{1}{2} + s_W^2\right)^2 + s_W^4\right] \simeq 4.0 \times 10^{-49}\ \mathrm{m}^2 \left(\frac{\mathrm{E}_\nu}{1\ \mathrm{MeV}}\right) \\
\sigma_{\nu_\mu e^- \to \nu_\mu e^-} &= \tfrac{G_F^2 s}{\pi}\left[\left(\tfrac{1}{2} - s_W^2\right)^2 + \tfrac{1}{3}s_W^4\right] \simeq 1.6 \times 10^{-49}\ \mathrm{m}^2 \left(\frac{\mathrm{E}_\nu}{1\ \mathrm{MeV}}\right) \\
\sigma_{\bar{\nu}_\mu e^- \to \bar{\nu}_\mu e^-} &= \tfrac{G_F^2 s}{\pi}\left[\tfrac{1}{3}\left(\tfrac{1}{2} - s_W^2\right)^2 + \tfrac{1}{3}s_W^4\right] \simeq 1.3 \times 10^{-49}\ \mathrm{m}^2 \left(\frac{\mathrm{E}_\nu}{1\ \mathrm{MeV}}\right)
\end{aligned} \qquad (8.205)$$

where $s_W^2 = \sin^2\theta_W \simeq 0.23$. Neutrino nucleon scattering at low energy may be characterized by a vector coupling constant $g_V$ and an axial vector coupling constant $g_A$ (see Equation 10.216). (In general $g_V$ and $g_A$ for nucleons should be replaced by weak interaction structure functions. For discussions of these structure functions, the reader should consult the additional reading.) The neutrino nucleon charged current cross sections for energies less than 1 GeV are independent of neutrino flavor and given by

$$\begin{aligned}
\sigma_{\nu_e n \to e^- p} &= \frac{G_F^2 E_\nu^2}{\pi}\left(g_V^2 + 3g_A^2\right)\left(1 + \frac{Q}{E_\nu}\right)\left[1 + 2\frac{Q}{E_\nu} + \frac{Q^2 - m_e^2}{E_\nu^2}\right]^{1/2} \\
\sigma_{\bar{\nu}_e n \to e^- + p} &= \frac{G_F^2 E_\nu^2}{\pi}\left(g_V^2 + 3g_A^2\right)\left(1 - \frac{Q}{E_\nu}\right)\left[1 - 2\frac{Q}{E_\nu} + \frac{Q^2 - m_e^2}{E_\nu^2}\right]^{1/2}
\end{aligned} \qquad (8.206)$$

where the vector and axial vector coupling constants are $g_V = 1$ and $g_A = 1.23$, and $Q = 1.3$ MeV is the nucleon mass difference. In the second reaction, $E$ must exceed $Q$. The approximate inelastic cross section for neutrino energy above a few MeV is

$$\sigma_{inelastic} = \frac{G_F^2 E_\nu^2}{\pi}\left(g_V^2 + 3g_A^2\right) = 9.3 \times 10^{-48}\ \mathrm{m}^2 \left(\frac{E_\nu}{1\ \mathrm{MeV}}\right)^2. \qquad (8.207)$$

The neutral current neutrino nucleon elastic cross sections are

$$\sigma_{\nu n \to \nu n} = \frac{G_F^2 E_\nu^2}{\pi}\left(1+3g_A^2\right) \simeq 9.3 \times 10^{-48}\ \mathrm{m}^2 \left(\frac{\mathrm{E}_\nu}{1\ \mathrm{MeV}}\right)^2$$

$$\sigma_{\nu p \to \nu p} = \frac{G_F^2 s}{\pi}\left(1+3g_A^2\right)\left[\left(\frac{1}{4}-2s_W^2\right)^2+4s_W^4\right] \simeq 6.0 \times 10^{-50}\ \mathrm{m}^2 \left(\frac{\mathrm{E}_\nu}{1\ \mathrm{MeV}}\right)^2. \tag{8.208}$$

These may be compared with the high energy inclusive cross sections above several GeV valid for all neutrino flavors

$$\sigma_{\nu N \to l^- X} \simeq 6.7 \times 10^{-43}\left(\frac{E_\nu}{1\ \mathrm{GeV}}\right)\mathrm{m}^2 \tag{8.209}$$

and for antineutrino flavors

$$\sigma_{\bar{\nu} N \to l^+ X} \simeq 3.4 \times 10^{-43}\left(\frac{E_\nu}{1\ \mathrm{GeV}}\right)\mathrm{m}^2. \tag{8.210}$$

## 8.10 ■ *Z* AND *W* BOSON PRODUCTION AT COLLIDERS

It is a formidable task to produce a real weak boson in the laboratory. The production of a $W$ boson by colliding a meson containing an antiquark against a fixed nucleon requires a minimum beam energy $E_m > m_W^2/2m_p \sim 3$ TeV. Colliding a $\bar{\nu}_e$ against an atomic electron requires a neutrino energy a factor $m_p/m_e \sim 200$ larger still. Electroweak bosons were first observed at the CERN Super Proton Synchrotron (SPS) $p\bar{p}$ collider, created through the processes

$$u\bar{d} \to W^+ \to e^+\nu_e,\ \mu^+\nu_\mu \tag{8.211}$$

$$d\bar{u} \to W^- \to e^-\bar{\nu}_e,\ \mu^-\bar{\nu}_\mu \tag{8.212}$$

$$u\bar{u} \to Z \to e^+e^-,\ \mu^+\mu^- \tag{8.213}$$

and

$$d\bar{d} \to Z \to e^+e^-,\ \mu^+\mu^-. \tag{8.214}$$

The bosons have since been studied at the LEP and Stanford $e^+e^-$ colliders, at the Fermilab Tevatron, and at the LHC. In this section, we examine these important so-called Drell-Yan processes.

The process of electron positron annihilation to form a charged fermion $f$ and its antiparticle $\bar{f}$

$$e^+e^- \rightarrow f\bar{f} \tag{8.215}$$

occurs through $s$-channel coherent formation of both $\gamma$ and $Z$ as shown in Figure 8.8. We ignore the small $H$ boson $s$-channel and $t$-channel elastic scattering amplitudes. The photon propagator has a pole at $s = 0$ while the propagator for the $Z$ boson has a pole at $s = m_Z^2$. The total cross is proportional to $1/s$ at low energy. As the energy increases, the interference of the electromagnetic amplitude with the weak amplitude becomes appreciable and at the $Z$ boson pole the weak amplitude dominates. The interference of the two coherent amplitudes for center of mass energy less than $m_Z$ provided early evidence for the $Z$ boson. Parity violation in the weak decay of the $Z$ boson is observed in the charge asymmetry. Note that the second amplitude applies to the invisible decay $Z \rightarrow \nu\bar{\nu}$. Such a decay may be tagged by initial state radiation.

For electron helicity $r$ and final fermion helicity $t$, with the inclusion of the heavy boson propagator and relevant coupling constants, the amplitude that follows from the electroweak Feynman rules has the form

$$M_{e^+e^-\rightarrow f\bar{f}} = j^t_{f\mu}\left[eq\frac{-g^{\mu\nu}}{s} + g_r g'_t \frac{g^{\mu\nu} - p^\mu p^\nu/m_Z^2}{s - m_Z^2 - im_Z\Gamma_Z}\right] j^r_{e\nu}. \tag{8.216}$$

Here $g_r$ and $g'_t$ are the helicity dependent neutral current weak coupling constants for the initial and final fermions and $q$ is the charge of $f$. In the center of mass with $z$-axis along the electron direction, with the electron mass neglected, the electron current for helicity $r$ is

$$j^r_e = 2E(0, -r, i, 0) \tag{8.217}$$

so $j_e p = 0$ and the second term in the numerator of the $Z$ boson propagator may be dropped. The form of current $j^t_f$ for outgoing fermion helicity $t$ may be obtained by rotating the form $j^t_e$ by the scattering angle and the product of currents is

$$j^r_e j^t_f = s\,(1 - rt\cos\theta) \tag{8.218}$$

where $\theta$ is the angle of $f$ relative to the electron direction. Therefore the amplitude reduces to

$$M_{e^+e^-\rightarrow f\bar{f}} = s\,(1 - rt\cos\theta)\left[\frac{eq}{s} + \frac{g_r g'_t}{s - m_Z^2 - im_Z\Gamma_Z}\right]. \tag{8.219}$$

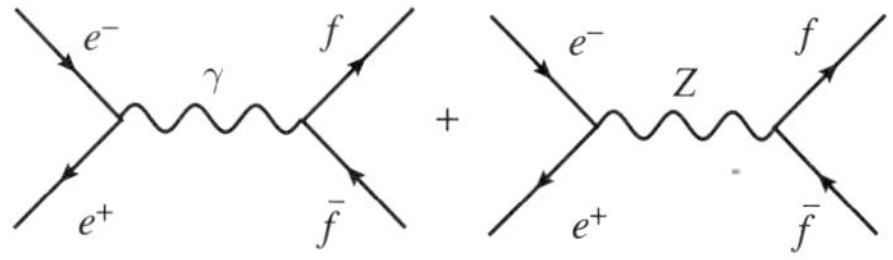


**FIGURE 8.8** Feynman diagrams for $e^+e^- \rightarrow f\bar{f}$ where $f$ is a charged fermion.

The squared amplitude contains an electromagnetic term, a weak term, and an energy dependent interference term:

$$|M_{e^+e^-\to f\bar{f}}|^2 = s^2(1 - rt\cos\theta)^2 \left[\frac{(eq)^2}{s^2} + \frac{\left(g_r g_t'\right)^2}{\left(s-m_Z^2\right)^2 + m_Z^2\Gamma_Z^2} + \frac{eq\,g_r g_t}{s}\frac{2\left(s-m_Z^2\right)}{\left(s-m_Z^2\right)^2 + m_Z^2\Gamma_Z^2}\right]. \tag{8.220}$$

The differential cross section has the form

$$\frac{d\sigma_{e^+e^-\to f\bar{f}}}{d\cos\theta} = \frac{1}{32\pi s}\frac{1}{4}\Sigma_{r,t}|M_{e^+e^-\to f\bar{f}}|^2 = \frac{1}{128\pi\Gamma^2}\left[A(s)\left(1+\cos^2\theta\right) + 2B(s)\cos\theta\right] \tag{8.221}$$

where, at the $Z$ boson pole, neglecting the amplitude for an intermediate photon, we have

$$A(m_Z) = \left(g_+g_+'\right)^2 + \left(g_-g_-'\right)^2 + \left(g_+g_-'\right)^2 + \left(g_-g_+'\right)^2 \tag{8.222}$$

and

$$B\left(m_Z\right) = \left(g_+g_+'\right)^2 + \left(g_-g_-'\right)^2 - \left(g_+g_-'\right)^2 - \left(g_-g_+'\right)^2. \tag{8.223}$$

The parity violating weak interaction leads to an asymmetry used to measure the Weinberg angle. In terms of forward and backward cross sections

$$\sigma_F = \int_0^1 d\cos\theta\frac{d\sigma_{e^+e^-\to f\bar{f}}}{d\cos\theta}\ ;\ \sigma_B = \int_{-1}^0 d\cos\theta\frac{d\sigma_{e^+e^-\to f\bar{f}}}{d\cos\theta}, \tag{8.224}$$

the asymmetry at a given energy is defined as

$$\alpha(s) = \frac{\sigma_F - \sigma_B}{\sigma_F + \sigma_B} = \frac{3}{4}\frac{B}{A}. \tag{8.225}$$

For $e^+e^- \to \mu^+\mu^-$, we have $g_\pm' = g_\pm$ and

$$g_+ = g_R = \frac{g}{2\cos\theta_W}\left[2x_w\right]\ ;\ g_- = g_L = \frac{g}{2\cos\theta_W}\left[2x_w - 1\right] \tag{8.226}$$

with $x_w = \sin^2\theta_W$ so

$$A\left(m_Z\right) = g_+^4 + g_-^4 + 2g_+^2g_-^2 = \left(g_+^2 + g_-^2\right)^2 = \left(g_v^2 + g_a^2\right)^2 \tag{8.227}$$

and

$$B\left(m_Z\right) = \left(g_+^2 - g_-^2\right)^2 = \left(g_+^2 - g_-^2\right)\left(g_+^2 + g_-^2\right) = g_v^2g_a^2. \tag{8.228}$$

We can express the asymmetry in terms of the Weinberg angle as

$$\alpha = \frac{3}{4}\frac{g_a^2 g_v^2}{g_v^2 + g_a^2} = \frac{3}{4}\left(\frac{1 - 4x_W}{1 - 4x_W + 8x_W^2}\right)^2 . \tag{8.229}$$

In general the asymmetry is a function of energy and reflects the interference between the photon and $Z$ boson contributions to the amplitude and for $q\bar{q}$ production the coupling constants differ from those for charged leptons. The asymmetry is negative and increasing linearly with $s$ at low energy. Already at $\sqrt{s} = 35$ GeV, the value $\alpha_{\mu^+\mu^-}(s) \simeq 0.1$. At $\sqrt{s} = m_Z$ since $x_W \simeq 0.25$, the asymmetry is relatively small and above the pole becomes positive.

After the discovery of resonant $Z$ boson production in $\bar{p}p$ collisions, two $e^+e^-$ machines (SLC and LEP I) operated with $\sqrt{s} \simeq m_Z$ and verified the preceding formulae in detail. The total width of the $Z$ boson is measured by observation of the production rate variation as the center of mass energy is scanned through the region of the $Z$ boson mass. In Figure 8.9, experimental results are compared with expectations assuming the $Z$ boson can decay to $e^+e^-$, $\mu^+\mu^-$, $\tau^+\tau^-$, plus $u\bar{u}$, $d\bar{d}$, $c\bar{c}$, $s\bar{s}$, $b\bar{b}$ in three colors and plus $\nu\bar{\nu}$ pairs in two through four massless flavors, in an attempt to see evidence for new particles coupled to the $Z$ boson. For example, if a fourth generation of quarks and leptons existed with all of the charged particles heavier than $m_Z/2$ but with a light neutrino, the $Z$ boson could decay invisibly to the fourth generation neutrino and its width would be increased. The data clearly

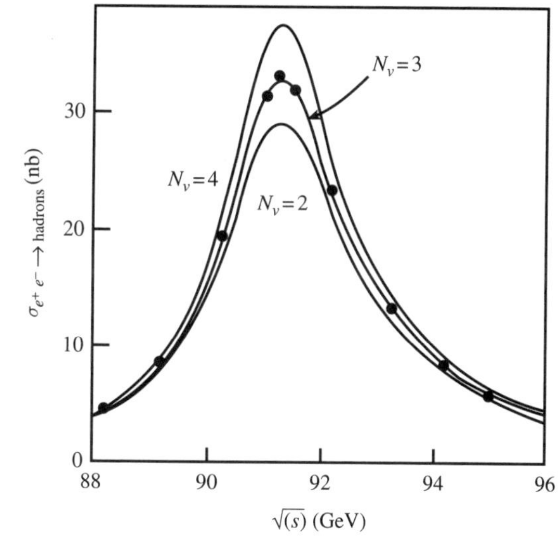


**FIGURE 8.9** Cross section for $e^+e^- \to$ hadrons as a function of center of mass energy in the region of the $Z$ boson resonance as measured by the ALEPH experiment at LEP. The fits are based on standard model expectations with 2-4 generations of essentially massless neutrinos. The observed width of the resonance indicates the number of generations of light neutrinos coupling to the $Z$ boson is three. [CERN Report CERN-DI-9008004 ]

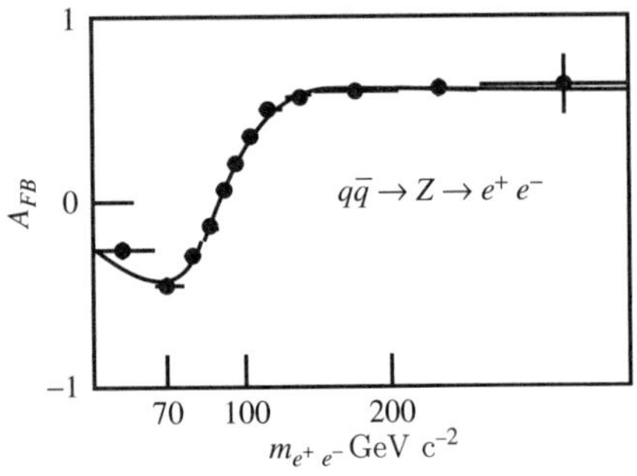


**FIGURE 8.10** Charge asymmetry $A_{FB}$ in $q\bar{q} \to Z, \gamma \to e^+e^-$ events as a function of mass at the Tevatron. Preliminary data are compared to standard model expectations. (CDF Electroweak Group)

indicate that if there is a fourth generation of quarks and leptons, then the fourth generation neutrino must have a rest mass larger than about $m_Z/2$.

The $W$ boson and $Z$ boson were discovered in $p\bar{p}$ collisions, produced through $q\bar{q}$ fusion processes analogous to $e^+e^- \to Z$. Colliding quarks have an uncontrolled distribution of invariant mass and uncontrolled total momentum along the direction of the colliding beams. The quark fusion processes produce bosons with a distribution of momentum along the beam direction determined by the momentum distributions of the two quarks participating in the collision. The leptonic decay modes are distinctive experimental signatures. These are shown for the $W$ boson in Figure 8.11. Both electron and muon are stable and readily identified. The presence of the neutrino from $W$ boson decay is inferred from observation of significant unbalanced momentum transverse to the direction of the beams. In $Z$ boson decay, the invariant mass of the lepton pairs peaks at $m_Z$. In $W$ boson decay, the momentum component of the neutrino along the beam direction is unknown but the transverse momentum of the lepton ranges up to $m_W/2$ which provides a signature. Boson decays to quark pairs resulting in two jet events are buried in QCD processes.

The calculation of vector boson production in $p\bar{p}$ collisions is similar to the calculation of $Z$ boson production in $e^+e^-$ collisions. The cross section for $Z$ boson production for given $q\bar{q}$ center of mass energy is in fact identical to that for $e^+e^-$ annihilation and must be convolved with the quark energy distributions to correspond to observations. The process of $W$ boson production and decay exhibits

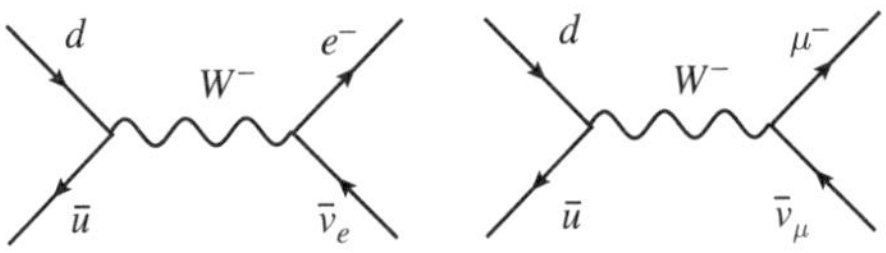


**FIGURE 8.11** Processes used to discover the $W$ boson in $p\bar{p}$ collisions. A single $W$ boson can be produced in a $p\bar{p}$ collider through $q\bar{q}$ fusion.

more strongly the parity violation characteristic of weak interactions. We consider this process next, with the polarization of the intermediate boson explicit.

The matrix element for

$$\bar{u}d \rightarrow W^- \rightarrow e^-\bar{\nu}_e \tag{8.230}$$

neglecting quark and lepton masses may be written as

$$M_{\bar{u}d\rightarrow W^-\rightarrow e^-\bar{\nu}_e} = \frac{g^2}{2} V_{ud}\, j_{l\mu} \frac{\Sigma_r \epsilon_r^{\mu} \epsilon_r^{\dagger\nu}}{s - m_W^2 - i m_W \Gamma} j_{q\nu} \tag{8.231}$$

where the quark and lepton currents are

$$j_q = \bar{v}_u^L \gamma u_d^L \text{ and } j_l = \bar{u}_e^L \gamma v_\nu^L \tag{8.232}$$

and the $W$ boson polarization vector for helicity $r$ is $\epsilon_r$. Write the quark spinor amplitude as

$$u_d^L = \sqrt{2E_d}\phi_{-1}(\mathbf{n}_d) \tag{8.233}$$

where $\mathbf{n}_d$ is the quark direction, and write antiquark spinor amplitude as

$$v_u^L = \sqrt{2E_u}\phi_{-1}(\mathbf{n}_{\bar{u}}) = \sqrt{2E_u}\phi_{+1}(\mathbf{n}_d)\,. \tag{8.234}$$

In the $W$ boson rest frame with $\mathbf{n}_d = \mathbf{z}$, the quark current is

$$j_q = \sqrt{s}\,[\phi_{+1}\phi_{-1}, -\phi_{+1}\sigma\phi_{-1}] = \sqrt{s}\,[0, -1, +i, 0] \tag{8.235}$$

and, using $W$ boson polarization vectors $\epsilon_\pm = [0, 1, \pm i, 0]/\sqrt{2}$ and $\epsilon_0 = (0, 0, 0, 1)$, one finds

$$\epsilon_r^\dagger \cdot j_q = \sqrt{s}\sqrt{2}\delta_{r,-1} \tag{8.236}$$

which implies the $W^-$ is produced with spin $\mathbf{s}_W$ anti-parallel to the direction of the proton momentum $\mathbf{p}_p$. The quantity $\mathbf{s}_W \cdot \mathbf{p}_p$ is $P$-odd so parity is violated in the production. Similarly, as a consequence of the left-handed weak coupling, a $W^+$ produced by $u\bar{d}$ fusion is polarized anti-parallel to the proton direction.

The lepton current can be written as

$$j_l = \sqrt{s}\,[0, -\cos\theta, -i, \sin\theta] \tag{8.237}$$

and the matrix element factor

$$j_l \cdot \epsilon_- = \sqrt{s}\,(1 + \cos\theta)\,/\sqrt{2} \tag{8.238}$$

favors $e^-$ emission along the proton direction, opposite to the spin of the $W^-$. Combining these factors, the matrix element may be written as

$$M_{\bar{u}d\rightarrow W^-\rightarrow e^-\bar{\nu}_e} = V_{ud}\frac{g^2}{2} s \frac{1 + \cos\theta}{s - m_W^2 - i m_w \Gamma}. \tag{8.239}$$

The cross section averaged over the spins and quark colors $\left(\text{factor } 3 \times \frac{1}{3} \times \frac{1}{3}\right)$ may then be found to be

$$d\sigma = \frac{1}{32\pi s}(4n_C)^{-1}|M_{\bar{u}d\to W^-\to e^-\bar{\nu}_e}|^2 d\cos\theta \tag{8.240}$$

where $n_C = 3$ is the number of colors and only quarks of opposite color may annihilate. Integration over polar angle and use of the formulae

$$\Gamma_f = \Gamma_{W\to e\nu} = \frac{g^2 m_W}{48\pi}\ ;\ \Gamma_i = \Gamma_{W\to\bar{u}d} = \frac{g^2 m_W}{48\pi} n_C |V_{ud}|^2 \tag{8.241}$$

gives the resonance form for the total quark level cross section

$$\sigma_{\bar{u}d\to W^-\to e^-\bar{\nu}_e} = \frac{4\pi}{3}\frac{\Gamma_i\Gamma_f}{\left(s - m_W^2\right)^2 + m_W^2\Gamma^2} \simeq \frac{4\pi^2}{3}\frac{\Gamma_{d\bar{u}}\Gamma_{e\nu}}{\Gamma m_W}\delta\left(s - m_W^2\right) \tag{8.242}$$

where the last expression approximates the resonance with a $\delta$-function.

The approximation of the resonance as a $\delta$-function may be used to write a relatively simple expression for the production cross section in $p\bar{p}$ collisions. Write the quark momenta as

$$p_{\bar{u}} = x_u p_{\bar{p}} \text{ and } p_d = x_d p_p \tag{8.243}$$

where $x_u$ and $x_d$ are the fractions of the hadron longitudinal momenta carried by the quarks which interact. Define the center of mass energy of the $p\bar{p}$ system as $s = (p_p + p_{\bar{p}})^2$ and the center of mass energy of the $q\bar{q}$ collision as

$$\hat{s} = (p_{\bar{u}} + p_d)^2 \simeq x_u x_d s. \tag{8.244}$$

Then the $\delta$-function may be written as

$$\delta\left(\hat{s} - m_w^2\right) = \frac{1}{s}\delta\left(x_u x_d - \tau\right) \tag{8.245}$$

where $\tau = m_W^2/s$. The production cross section is obtained by integrating over the probability distribution functions $u(x_u)$ for a $u$-quark to carry momentum fraction $x_u$ and $d(x_d)$ a $d$-quark to carry momentum fraction $x_d$ and is found to be

$$\sigma_{\bar{p}p\to W\to e\nu} = \frac{4\pi^2\Gamma}{3m_W^2} B_{u\bar{d}} B_{e\nu} \int dx_u dx_d\ u\left(x_u\right) d\left(x_d\right) \delta\left(x_u x_d - \tau\right). \tag{8.246}$$

If naively we estimate the momentum fraction of a quark as 1/3, the center of mass energy available in a typical $q\bar{q}$ collision is $\hat{s} = s/9$ so to produce a mass $m_W$= 80 GeV requires a collider of center of mass energy 720 GeV. The $W$ boson and $Z$ boson were discovered in 1983 at the CERN $Sp\bar{p}S$ collider operating at a center of mass energy of 540 GeV where the total cross sections for $W^+$, $W^-$,

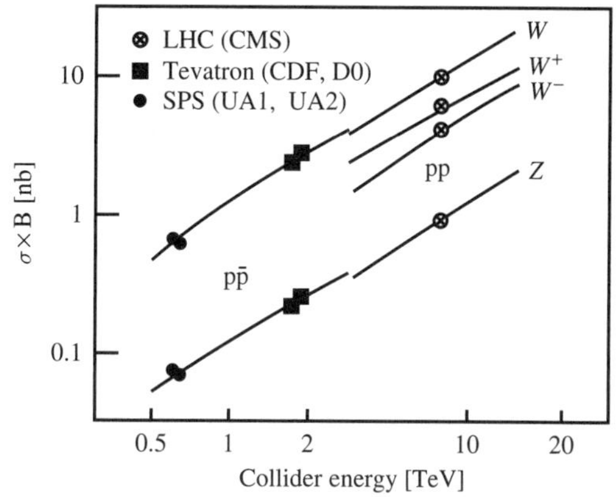


**FIGURE 8.12** Measurements of inclusive cross sections times branching ratio as a function of collider center of mass energy for production of $W$ and $Z$ bosons in $p\bar{p}$ and $pp$ colliders. The points are measurements. The lines are the predictions of next to next to leading order calculations. ["Measurements of Inclusive $W$ and $Z$ Cross Sections in $pp$ Collisions at $\sqrt{s} = 7$ TeV," CMS collaboration, (2010), arXiv:1012.2466v2 [hep-ex]]

and $Z$ bosons are of order 1 nb. The branching fraction for $W^+ \to e\nu$ is about 9% while the branching fraction for $Z \to e^+e^-$ is about 3% so $W$ boson decays to leptons in $p\bar{p}$ collision are observed roughly six times as often as $Z$ boson decays to leptons. At the LHC $pp$ collider, the Drell-Yan processes proceed through fusion of combinations of so-called valence quarks and antiquark members of quark-antiquark pairs deriving from the gluon content of the proton. The cross sections for single weak boson production in hadron colliders are shown in Figure 8.12.

## 8.11 ■ MULTIPLE WEAK BOSON PRODUCTION

The LEP 2 electron-positron collider reached a center of mass energy in excess of 200 GeV in 1999, sufficient to produce pairs of $W$ bosons. At about the same time, sufficient integrated luminosity was achieved to observe weak boson pair production ($W^+W^-$ and $W^\pm Z$) in Tevatron experiments.

As shown in Figure 8.13, $W$ boson pair production at LEP results from $e^+e^-$ annihilation into a photon or $Z$ boson ($s$-channel) or through a neutrino exchange with $W$ boson radiation from each initial fermion. The three boson interaction terms contribute to the $s$-channel amplitude. The $t$-channel amplitude, which requires only that the $W$ boson couple to leptons, actually dominates near threshold. The contribution of the $H$ is negligible because it couples extremely weakly to the nearly massless electron. A complete analysis of standard model predictions for $W^+W^-$ pair production entails calculation of the sum of the three amplitudes for four choices of lepton spin and nine choices of boson spin as functions of the center of mass energy and $W^+$ direction. Of special interest is a yet more complex analysis in which the three boson vertex is permitted to have a general form consistent with

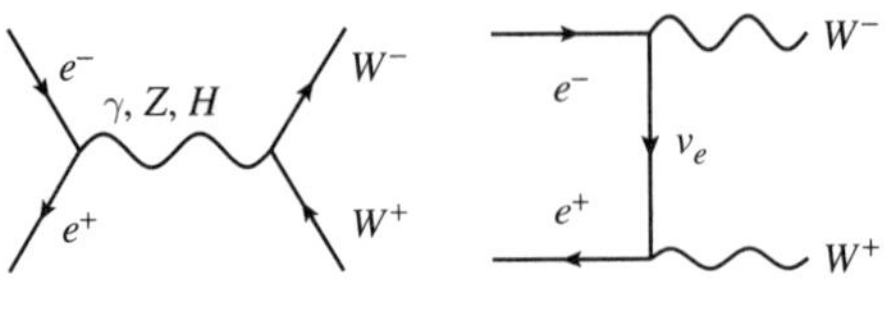


**FIGURE 8.13** Feynman diagrams for production of $W^+W^-$ in $e^+e^-$ collisions. In the standard model, this process occurs through fusion into $\gamma$ or $Z$ boson and also by $\nu$ exchange. The interference of these amplitudes in the standard model leads to suppression of the cross section that would calculated based on neutrino exchange alone.

Lorentz covariance, the anomalous terms representing structure in the electroweak boson vertex outside the standard model. Such form factor terms could represent traditional compositeness in the weak bosons or the effects of vertex corrections (see Chapter 9) due to virtual particles outside the standard model. We content ourselves with the comment that the $W^+$ tends to emerge close to the positron direction and transversely polarized bosons are favored, and with a brief look at the total cross section.

The tree level spin averaged standard model cross section is calculated to be

$$\begin{aligned}\sigma_{e^+e^-\to W^+W^-} = \frac{\pi\alpha^2\beta}{2s\sin^4\theta_W}\Bigg\{&\left[1+2x+2x^2\right]\frac{1}{\beta}\ln\left(\frac{1+\beta}{1-\beta}\right)-\frac{5}{4}\\ &+\frac{m_Z^2\left(1-2\sin^2\theta_W\right)}{s-m_Z^2}\left[2\left(2x+x^2\right)\frac{1}{\beta}\ln\left(\frac{1+\beta}{1-\beta}\right)-\frac{x}{12}-\frac{5}{3}-x\right]\\ &+\frac{m_Z^4\left(8\sin^4\theta_W-4\sin^2\theta_W+1\right)\beta^2}{48\left(s-m_Z^2\right)^2}\left[x^2+20x+12\right]\Bigg\}\end{aligned} \tag{8.247}$$

where $s = (2E_b)^2$ is the center of mass energy squared, $x = m_W^2/s$, and $\beta = (1-4x)^{1/2}$ is the $W$ boson velocity in the center of mass. The cross section rises rapidly near threshold and in the absence of photon and $Z$ boson contributions would increase linearly with $s$. In the standard model, the interference of amplitudes leads to a peak cross section of about 20 pb and, for large $s$, the cross section decreases

$$\sigma_{e^+e^-\to W^+W^-}\left(s >> m_W^2\right) \simeq \frac{\pi\alpha^2}{2s\sin^4\theta_W}\ln\frac{s}{m_W^2}. \tag{8.248}$$

In Figure 8.14, the observed cross section is shown to agree with standard model expectation. A precise comparison requires careful consideration of many higher order diagrams and from such comparisons the mass and width of the $W$ boson are extracted. In addition, limits are placed on the influence of anomalous values of the coupling constants in the standard model and on addition non-standard contributing effective Lagrangian terms.

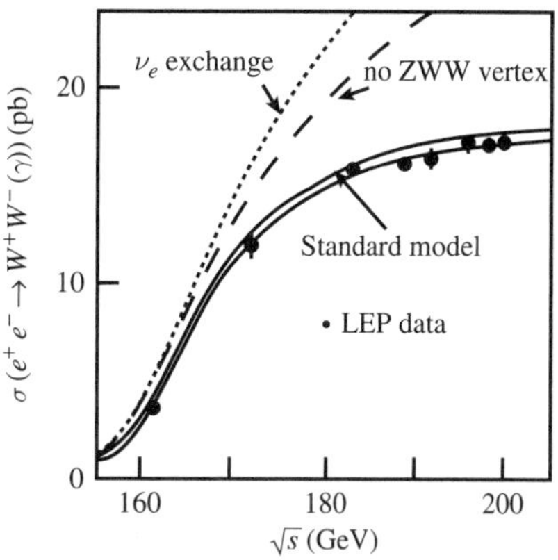


**FIGURE 8.14** Cross section for $e^+e^- \to W^+W^-$ near threshold. The fit to the data is based on a model including substantial corrections for initial and final state radiation and other higher order effects. [LEP Electroweak Working Group, reported in A.Denner, S.Dittmaier, M.Roth, D.Wackeroth, arXiv:hep-ph/0005074v1]

The production of $W^+W^-$ at a $p\bar{p}$ collider occurs through the $u\bar{u}$ and $d\bar{d}$ analogs of the $e^+e^-$ process. The diagrams for the case of $u\bar{u}$ collisions are shown in Figure 8.15. Again the $s$-channel amplitudes ($q\bar{q}$ fusion into a virtual photon or $Z$ boson which couples to a $W^+W^-$ pair) are sensitive to the three boson interaction terms in the standard model and again the contribution of the $H$ is negligible. Quark exchange rather than neutrino exchange contributes a $t$-channel tree level amplitude and the contributions of three generations of intermediate state quarks are coherent. In the $p\bar{p}$ observations of $W^+W^-$ production, the production of $t\bar{t}$ pairs with $t \to W^+b$ and $\bar{t} \to W^-\bar{b}$ is another source of pairs of real $W$ bosons. Such events are distinguished from the continuum production process by the presence of the jets of particles associated with the $b$-quarks. The production of $W^\pm Z$ in a $p\bar{p}$ collider results from $q\bar{q}$ fusion to a virtual $W$ boson conjoined with radiation of a $Z$ boson from the virtual $W$ boson or initial quarks. Pairs of $Z$ bosons may also be produced. These processes have been studied by the CDF and D0 collaborations at the Fermilab Tevatron and by the ATLAS and CMS experiments at the LHC.

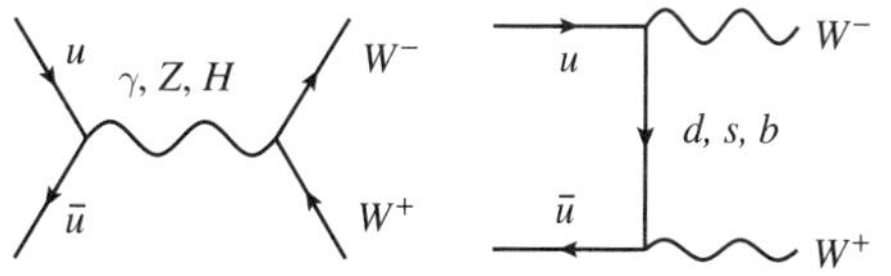


**FIGURE 8.15** Diagrams describing $W^+W^-$ production at a $p\bar{p}$ collider. The case of $u\bar{u}$ annihilation is shown. The quark exchange plays the role of $\nu$ exchange in $e^+e^- \to W^+W^-$.

## 8.12 ■ THE HIGGS BOSON

The symmetry breaking mechanism of the standard model requires one scalar Higgs boson which has been the object of extensive searches but which still eludes discovery. The minimal standard model predicts that the coupling constant for the interaction of the $H$ with a particle of mass $m$ is proportional to $m$. The mass of the $H$ is a feature of the dynamics of the Higgs field related to the shape of the Higgs potential and a free parameter. We conclude our presentation of the electroweak standard model with calculations of some of the properties of the minimal standard model Higgs particle. It should be said that the electroweak symmetry breaking mechanism could be more complex and entail multiple Higgs particles, both charged and neutral. Such is the case in the minimal supersymmetric standard model described in Chapter 11.

The interaction of the minimal standard model Higgs boson with a fermion is described by the interaction Lagrangian

$$L_{Hff} = -\frac{m_f}{v}\bar{\psi}\psi H = -\frac{m_f}{v}\left(\phi_-^\dagger\phi_+ + \phi_+^\dagger\phi_-\right)H. \tag{8.249}$$

The coupling constants for the light fermions $e^+e^-$, $u\bar{u}$, and $d\bar{d}$ available in colliders are of order

$$g_{Hff} = \frac{m_f}{v} \sim \frac{1\ \text{MeV}}{250\ \text{GeV}} = 4 \times 10^{-6}. \tag{8.250}$$

So, even though the width of the $H$ resonance is narrow for low $m_H$, s-channel $H$ production directly through annihilation of light quarks and electrons in colliders is impractical. In $e^+e^-$, $pp$, and $p\bar{p}$ colliders, the $H$ may be produced via a higher order process in which heavy particles such as a virtual $W^+$ boson and $W^-$ boson or a $t$-quark and $\bar{t}$-quark fuse. It is also possible to produce a Higgs boson through (Higgsstrahlung) radiation from a heavy particle such as a $W$ boson, $Z$ boson, or $t$-quark. Such processes are illustrated in Figure 8.16.

For example, in $p\bar{p}$ collisions, the $H$ can be produced in association with a $W$ boson through the reaction

$$u\bar{d} \to W^{+*} \to W^+H \tag{8.251}$$

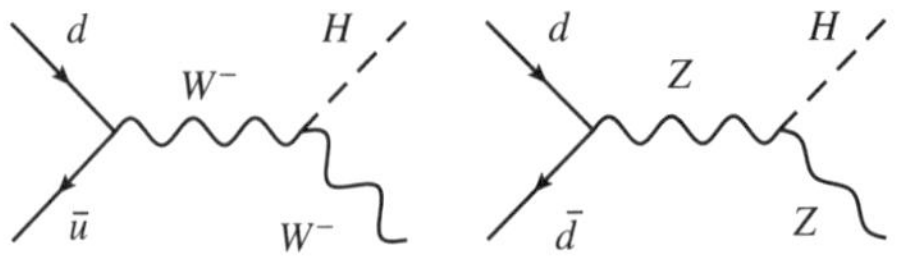


**FIGURE 8.16** Examples of Higgsstrahlung processes in which a Higgs boson $H$ is radiated by a heavy particle.

where the asterix is conventionally used simply to emphasize that the intermediate state $W$ boson is virtual. The relevant terms in the Lagrangian are

$$L = L_{udW} + L_{HWW} = \frac{gV_{ud}}{\sqrt{2}} W^{\dagger} \bar{d} \gamma \left(\frac{1-\gamma_5}{2}\right) u + \frac{m_W^2}{2v} H W^{\dagger} W. \tag{8.252}$$

The amplitude that follows from the Feynman rules is

$$M_{u\bar{d}\to W^+H} = -\frac{\alpha m_W V_{ud}}{\sqrt{2}\sin^2\theta_W} \epsilon_\mu^* \left(\frac{-g^{\mu\nu} + p^\mu p^\nu / m_W^2}{p^2 - m_W^2}\right) \bar{d} \gamma_\nu \left(\frac{1-\gamma_5}{2}\right) u \tag{8.253}$$

where $\epsilon$ is the polarization vector of the final $W^+$ and $p = p_u + p_{\bar{d}}$ is the momentum of the virtual $W^+$. The cross section corresponding to this single amplitude is

$$\sigma_{u\bar{d}\to W^+H} = \frac{\pi\alpha^2 |V_{ud}|^2}{18\sin^4\theta_W} \frac{p_H\left(p_H^2 + 3m_W^3\right)}{\sqrt{s}\left(s - m_W^2\right)^2} \tag{8.254}$$

where $\sqrt{s} = \sqrt{p^2}$ is the center of mass energy and $p_H$ is the Higgs boson momentum in the $u\bar{d}$ center of mass.

The cross section for $H$ produced in association with a $Z$ boson

$$q\bar{q} \to W^{+*} \to W^+ H \tag{8.255}$$

can be similarly computed and is given by

$$\sigma_{q\bar{q}\to ZH} = \frac{\pi\alpha^2\left(l^2 + r^2\right)}{36\sin^4\theta_W \cos^4\theta_W} \frac{p_H\left(p_H^2 + 3m_W^3\right)}{\sqrt{s}\left(s - m_W^2\right)^2} \tag{8.256}$$

with $l = 2(t_3 - Q\sin^2\theta_W)$ and $r = -2Q\sin^2\theta_W$ denoting the left and right neutral current coupling constants of the $u$-quark or $d$-quark. These cross sections must be convolved with quark momentum distribution functions to derive the associated production cross sections for the processes

$$p\bar{p} \to (W, Z)H + X \tag{8.257}$$

used to search for the $H$ at Tevatron experiments.

Experiments at LEP searched for the process

$$e^+e^- \to ZH \tag{8.258}$$

and placed a lower bound of 114 GeV on the mass of the $H$. The cross section for Higgstrahlung at LEP is (Ref: B. W. Lee, C. Quigg, and H. B. Thacker, Phys. Rev. **D16**, 1519 (1977))

$$\sigma_{e^+e^-\to HZ} = \frac{\pi\alpha^2\left[1 + \left(1 - \sin^2\theta_W\right)^2\right]}{24\sin^4\theta_W\cos^4\theta_W} \frac{3m_Z^2 + k^2}{\sqrt{s}\left(s - m_Z^2\right)^2} \tag{8.259}$$

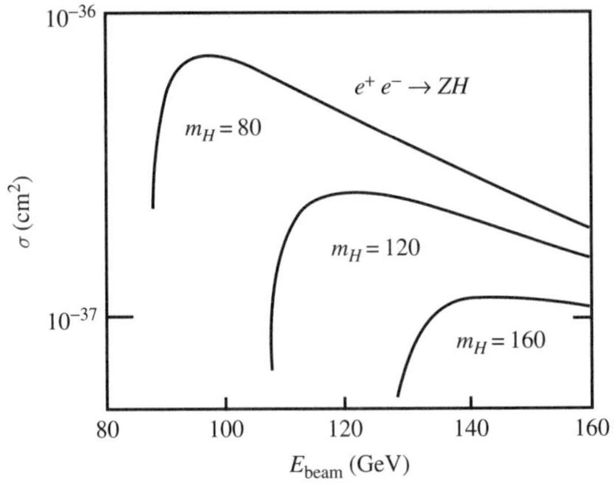


**FIGURE 8.17** Cross section for production of the Higgs boson in association with a $Z$boson in $e^+e^-$ collisions as a function of beam energy.

where $\sqrt{s}$ is the center of mass energy and $k$ is the momentum of the $Z$ boson. The Higgstrahlung cross section for various Higgs boson mass values is shown in Figure 8.17.

Multiple vector boson production can have the appearance of Higgstrahling when searching for $H$ decay to $b\bar{b}$ for a $Z$ boson also has this signature decay. The cross section for $ZZ$ production at LEP is [ I. Hinchliffe in R. Donaldson, R. Gustafson, and F. Paige, *Elementary Particle Physics and Future Facilities*, Snowmass 1982)]

$$\sigma_{e^+e^-\to ZZ} = \frac{\pi\alpha^2\left[1+6\left(1-4\sin^2\theta_W\right)^2+\left(1-4\sin^2\theta_W\right)^4\right]}{64s\sin^4\theta_W\cos^4\theta_W} \cdot \left\{\frac{4m_Z^4+s^s}{s\left(s-2m_Z^2\right)}\ln\frac{(1+\beta)}{(1-\beta)}-\beta\right\} \tag{8.260}$$

where $\beta = (1-4m_Z^2/s)^{1/2}$ is the velocity of the $Z$ boson. This expression peaks at about 5 pb while the Higgstrahlung production cross section peaks at about 0.25 pb for $m_H \simeq 120$ GeV. Precise measurement of the mass of one $Z$ boson observed through its decay to $e^+e^-$ or $\mu^+\mu^-$ permits inference of the mass of the recoiling $b$-quark plus $\bar{b}$-quark pair which are observed as particle jets with poorly measured energy. In this way, a signal for $H$ events with a narrow recoil mass distribution can be observed atop a slowly varying background from $Z \to b\bar{b}$.

A minimal standard model Higgs boson of mass $m_H < 2m_W$ decays predominantly to the heaviest fermion anti-fermion pair $f\bar{f}$ consistent with the kinematic constraint $m_H > 2m_f$. It is established that $m_H > 2m_b$ and global fits to electroweak parameters suggest $m_H \sim 120$ GeV $< 2m_t$ so the dominant decay should be to $b\bar{b}$. We next consider in more detail the decays of a Higgs boson, beginning with the decay to fermion pairs.

The matrix element for the decay $H \rightarrow f\bar{f}$ that follows from the Lagrangian $L_{Hff}$ is

$$M_{H\rightarrow f\bar{f}} = -\frac{m_f}{v}\left(v_t^{-\dagger}(\bar{\mathbf{n}})\, u_s^{+}(\mathbf{n}) + v_t^{+\dagger}(\bar{\mathbf{n}})\, u_s^{-}(\mathbf{n})\right) \tag{8.261}$$

where $s$ is the helicity of the fermion and $t$ the helicity of the anti-fermion. Inserting the spinor amplitudes, we find

$$M_{H\rightarrow f\bar{f}} = -\frac{m_f}{v}\left(\sqrt{E+sp}\sqrt{E+tp} - \sqrt{E-sp}\sqrt{E-tp}\right)\phi_{-t}^{\dagger}(\bar{\mathbf{n}})\,\phi_s(\mathbf{n})\Big]. \tag{8.262}$$

For $s = t$, $M_{H\rightarrow f\bar{f}}$ is non-zero and proportional to $2pm_f/v$. Hence summed over final state polarizations, and in the case of a $q\bar{q}$ final state summed over colors, we have

$$\overline{|M_{H\rightarrow f\bar{f}}|^2} = N_c 8p^2 m_f^2/v^2 \tag{8.263}$$

where $N_c$ is the number of colors. From this expression, the decay rate may be computed to be

$$\Gamma_{H\rightarrow f\bar{f}} = \frac{\overline{|M|^2}}{2m_H} PS = \frac{1}{2m_H}\left[N_c 8p^2\left(\frac{m_f}{v}\right)^2\right]\frac{1}{4\pi}\frac{p}{m_H} = \frac{N_C}{\pi}\left(\frac{m_f}{m_H v}\right)^2 p^3. \tag{8.264}$$

Using the two relations

$$p = \frac{1}{2}\sqrt{m_H^2 - 4m_f^2};\; v^2 = \left(\frac{2m_W}{g}\right)^2 = \frac{1}{\sqrt{2}G_F}, \tag{8.265}$$

the decay rate may be cast in the alternate form

$$\Gamma_{H\rightarrow f\bar{f}} = N_c\frac{\sqrt{2}G_F}{8\pi}m_f^2 m_H\left[1 - 4\left(\frac{m_f}{m_H}\right)^2\right]^{\frac{3}{2}}. \tag{8.266}$$

Taking $G_F = 1.166 \times 10^{-5}$ GeV$^2$ and $m_b$= 5 GeV, the fractional width has the numerical value

$$\frac{\Gamma_{H\rightarrow b\bar{b}}}{m_H} = 3\frac{\sqrt{2}}{8\pi}G_F m_f^2 = 1.5 \times 10^{-6}. \tag{8.267}$$

While reconstruction of bottom quark pair masses with this level of precision is beyond the capability of experiments, it is significant that the fractional width is small. This result motivates searches for the rare decay $H \rightarrow \gamma\gamma$ which proceeds through a quark loop and for which the improved experimental mass resolution could compensate for the reduced event rate.

The decay $H \rightarrow \gamma\gamma$ results from amplitudes described by diagrams in which the $H$ couples to a charged fermion or $W$ boson loop as illustrated in Figure 8.18. The fermions radiate the two photons at separate vertices. The amplitude for an internal

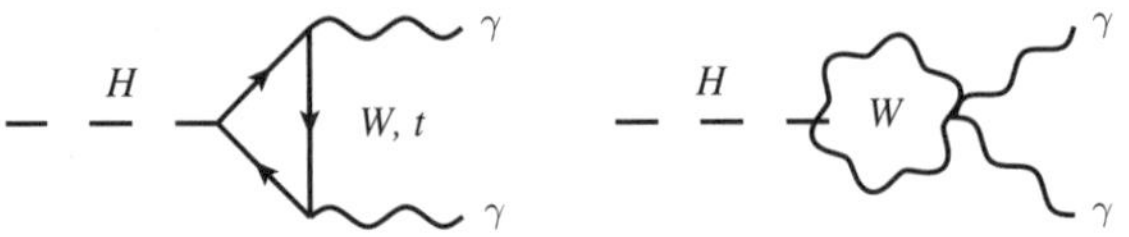


**FIGURE 8.18** Contributions to the rare decay $H \to \gamma\gamma$ from the $W$ boson and top quark. As the energy of a photon may be measured more precisely than the energy of a jet originating from a $b$-quark, the rare $\gamma\gamma$ decay may be used instead of the more likely decay $H \to b\bar{b}$ to search for the $H$.

$W$ boson includes an additional four-boson contribution to photon pair production. The calculation of the loop requires renormalization techniques discussed in the next chapter and the result has the form

$$\Gamma_{H\to\gamma\gamma} = \frac{\alpha^2 g^2 m_H^3}{1024\pi^3 m_w^2} |\Sigma_i N_c e_i^2 F_i|^2 \tag{8.268}$$

where $e_i$ is the charge of the particle in the loop, and $N_c$ is the number of colors (3 for quarks). The factor $F_i$ is a dimensionless function of the ratio $m_i/m_H$ of the mass $m_i$ of the particle in the loop to $m_H$ and it depends on the spin of the particle in the loop. The calculation finds not unexpectedly that only the heavy charged $t$-quark and $W$ boson are significant. The contribution from the $W$ boson with $F_W \simeq 6$ is of opposite sign to and dominates the contribution of the $t$-quark with $F_t \simeq -1$. (A fourth generation heavier quark would change this conclusion.) This expression predicts a branching fraction for the two photon channel below 0.3%. The Higgs

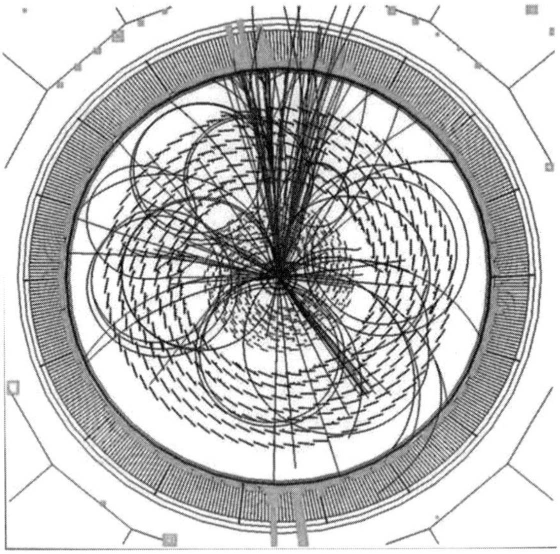

**FIGURE 8.19** Simulated $H \to ZZ \to (e^+e^-)(q\bar{q})$ event in the CMS detector. [CERN Photo archive, POW 990707]

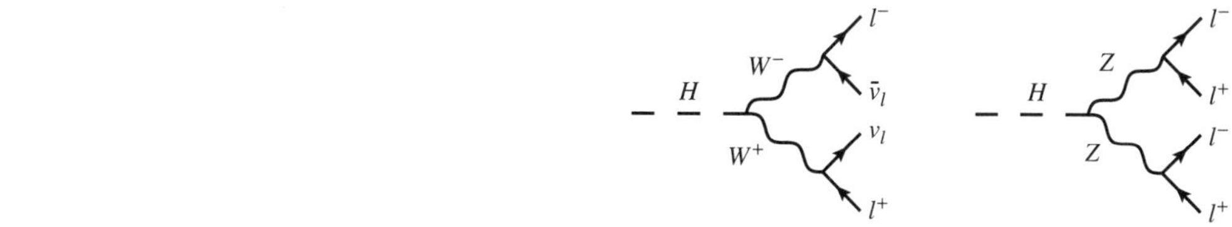


**FIGURE 8.20** Principal decay modes of a heavy standard model Higgs boson. If $m_H$ exceeds twice the rest mass of a weak boson, it decays principally to a pair of such bosons. The leptonic decays of the weak bosons shown are the most easily distinguished in experiments.

boson can decay to weak boson pairs if $m_H$ is sufficiently high. The leptonic signatures of such decays are illustrated in Figure 8.20. If $m_H > 2m_W = 162$ GeV, then the decay rate for $H \to W^+W^-$ will dominate the decay rate for $H \to b\bar{b}$. The interaction term responsible for $H \to W^+W^-$ is

$$L_{HWW} = \frac{2m_w^2}{v} HW^+ \cdot W^- \tag{8.269}$$

and the matrix element for decay to $W^-$ with polarization $t$ and direction $\mathbf{n}$ plus $W^+$ with polarization $s$ in the center of mass is

$$M^{st}_{H\to W^+W^-} = \frac{2}{v} m_w^2 \epsilon_t^\dagger(\mathbf{n}) \cdot \epsilon_s(-\mathbf{n}) . \tag{8.270}$$

Polarization vectors for the $W^-$ are

$$\epsilon_0 = \left( \frac{p}{m_W}, \frac{E}{m_W}\epsilon_0 \right) \; ; \; \epsilon_\pm = (0, \epsilon_\pm) \tag{8.271}$$

and for the $W^+$

$$\epsilon_0^\dagger = \left( -\frac{p}{m_W}, \frac{E}{m_W}\epsilon_0 \right) \; ; \; \epsilon_\pm^\dagger = (0, \epsilon_\mp) \tag{8.272}$$

where the decay momentum is $p = \sqrt{m_H^2 - 4m_W^2}/2$ and $E = m_H/2$. Examining the inner product of polarization vector combinations, we find

$$M^{00}_{H\to W^+W^-} \propto \left(p^2 + E^2\right)/m_W^2 \text{ and } M^{+-}_{H\to W^+W^-} = M^{-+}_{H\to W^+W^-} \propto 1 \tag{8.273}$$

which implies that decays to pairs of longitudinally polarized $W$ bosons dominates for large $m_H$. Summing the squared matrix element over final state polarizations, using $1/v = g/(2m_W)$, the total decay rate is found to be

$$\begin{aligned} \Gamma_{H\to W^+W^-} &= \frac{1}{2m_H} \frac{4}{v^2} m_w^4 \left[ \left( \frac{p^2 + E^2}{m_w^2} \right)^2 + 2 \right] \frac{1}{4\pi} \frac{p}{m_H} \\ &= \frac{g^2}{64\pi m_W^2} m_H^3 (1 - 4x)^{\frac{1}{2}} \left(1 - 4x^2 + 12x^4\right) \end{aligned} \tag{8.274}$$

with $x = m_W/m_H$.

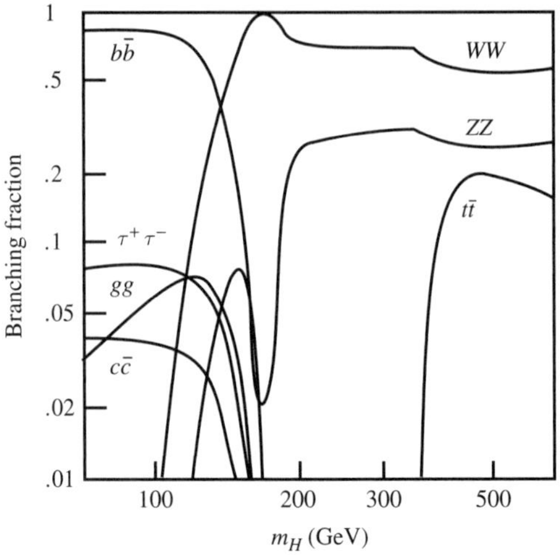


**FIGURE 8.21** Branching fractions for various decay modes of the standard model Higgs boson as a function of $m_H$. For $m_H < 2m_W$, the decay $H \to b\bar{b}$ has the highest branching fraction. For $m_H > 2m_W$, the decay $h \to W^+W^-$ dominates. [A. Djouadi, J. Kalinowski, M. Spira, Comput. Phys. Commun. **108**, 56-74 (1998), arXiv:hep-ph/9704448]

The decay $H \to ZZ$ becomes important in case $m_H > 2m_Z$. The interaction Lagrangian governing $H \to ZZ$

$$L_{HZZ} = m_Z^2 Z \cdot Z \frac{h}{v} \tag{8.275}$$

is similar to that for the $H \to WW$ and we can transcribe our result above. Treating the two $Z$ field factors as independent requires a factor two relative to the $H \to WW$ amplitude but the phase space must be divided by two so as to not double count indistinguishable states. Then, replacing $m_W \to m_Z$ where appropriate, we find

$$\Gamma_{H\to ZZ} = \frac{g^2}{128\pi m_Z^2} m_H^3 (1-4x)^{\frac{1}{2}} \left(1 - 4x^2 + 12x^4\right) \tag{8.276}$$

with $x = m_Z/m_H$. For larger $m_H$, the decay to $t\bar{t}$ icontributes and described by Equation 8.266. For such a large mass, the fractional width of the Higgs boson resonance is no longer small. The branching fractions of the $H$ are shown as functions of $m_H$ in Figure 8.21.

## 8.13 ■ FURTHER READING

The 1957 Nobel Prize in Physics was awarded jointly to Chen Ning Yang and Tsung-Dao Lee "for their penetrating investigation of the so-called parity laws which has led to important discoveries regarding the elementary particles."

The 1979 Nobel Prize in Physics was awarded to Sheldon Glashow, Abdus Salam, and Steven Weinberg for the development of the $SU(2)_L \times U(1)_Y$ model of the electroweak interaction. The gestation of the theory with many references may be found in their Nobel lectures. For a review of alternative early models, see B. W. Lee,"Perspectives on Theory of Weak Interactions," NAL-THY-92 (1972).

The 1984 Nobel Prize in physics was awarded to Carlo Rubbia, Simon van der Meer "for their decisive contributions to the large project, which led to the discovery of the field particles $W$ and $Z$, communicators of weak interaction."

The 2004 Nobel Prize in physics was awarded to Yoichiro Nambu "for the discovery of the mechanism of spontaneous broken symmetry in subatomic physics", the other half jointly to Makoto Kobayashi and Toshihide Maskawa "for the discovery of the origin of the broken symmetry which predicts the existence of at least three families of quarks in nature."

The idea that spontaneous symmetry breaking in gauge theory could lead to massive vector gauge bosons was introduced in P. W. Higgs, Phys. Rev. Lett. **13**, 508 (1964) and Phys. Rev. **145**, 1156 (1966).

Classic texts on weak interactions are Eugene Commins, *Weak Interactions*, McGraw-Hill (1973), and L. B. Okun, *Leptons and Quarks*, North-Holland Physics Publishing (1984).

Observations of electroweak physics are numerous. The reader is invited to consult the literature for results especially from the LHC experiments CMS, ATLAS, and ALICE, the Tevatron experiments D0 and CDF, the b-factory experiments BABAR and BELLE, the DESY experiments ZEUS and H1, the LEP experiments ALEPH, DELPHI, L3 and OPAL, and the SLC experiment SLD.

An introduction to electroweak physics at LEP dedicated to single and multiple weak boson production is John Ellis and Roberto Peccei (editors), "Physics at LEP, Vol. 1 and 2," CERN 86-01 and 86-02. A recent Tevatron multiple boson production observation is "Observation of WZ Production," A. Abulencia *et al.* (CDF Collaboration), Phys. Rev. Lett. **98**, 161801 (2007).

For an explication of non-standard structure in the electroweak three gauge boson vertex, see K. J. F. Gaemers and G. J. Gounaris, Z. Phys. **C1**, 259 (1979).

The Higgsstrahlung process was first proposed by S. L. Glashow, D. V. Nanopoulos, and A. Yildiz, Phys. Rev. D **18**, 1724 (1978). Limits on the production of a Higgs boson in association with a $Z$ boson in $e^+e^-$ collisions have been published by the LEP experiments. The Tevatron experiments continue to search for production of a Higgs boson in association with $W$ and $Z$ bosons. See for example V. M. Abazov, et al. (D Collaboration), Phys. Rev. Lett. **97**, 161803 (2006); A. Abulencia, et al. (CDF Collaboration), Phys. Rev. Lett. **97**, 081802 (2006); Daniela Bortoletto, J. Phys.: Conference Series 110, 042005 (2008). Theoretical predictions for Higgs production at the Tevatron are reviewed in Julien Baglio, Abdelhak Djouadi, arXiv:1003.4266v1 [hep-ph]. For predictions of production cross sections relevant to the LHC, see recent literature. The decays widths and branching fractions of the standard model and supersymmetric extension Higgs bosons are described in A. Djouadi, J. Kalinowski, M. Spira, Comput. Phys. Commun. **108**, 56-74 (1998), arXiv:hep-ph/9704448. A review of electroweak symmetry breaking and standard model Higgs production is Abdelhak Djouadi, Phys. Rep. **457**, 1-216 (2008).

## 8.14 ■ PROBLEMS

### Problem 8.1. Neutrino rocket

A cascade of parity violating leptonic decays of a polarized lepton is a crude rocket for charge, powered by neutrino emission. Suppose the muon in $\tau^- \to \nu_\tau \bar{\nu}_\mu \mu^-$ has the maximal energy in the $\tau^-$ rest frame and produces an electron with maximal energy in the same direction. Find the muon energy, momentum, and velocity in the $\tau^-$ rest frame and the electron energy and momentum in the muon rest frame and use a Lorentz transformation to find the electron energy and momentum in the $\tau^-$ rest frame. Show that the results are equivalent to a two body decay $\tau^- \to e^- X$ with $m_X = 0$ which implies the intermediate rocket stage in this case does not effect the energy given to the payload (the electron). The correlation between spin and momentum in parity violating processes could contribute to acceleration of polarized macroscopic objects such as a neutron star. See for example G. M. Fuller *et al*, "Pulsar kicks from a dark-matter sterile neutrino", Phys. Rev. **D68**, 103002 (2003).

### Problem 8.2. Single top production

At the Fermilab $p\bar{p}$ collider, top quarks are observed produced in pairs. Figure 8.22 shows an image of a candidate event for the single top production process

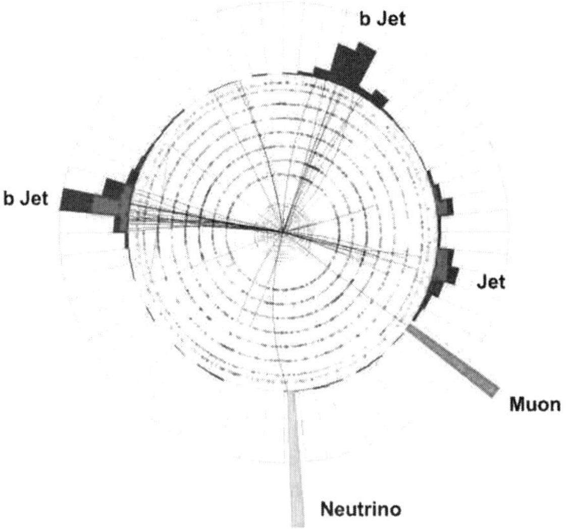


**FIGURE 8.22** Candidate event for production of a single top quark in $p\bar{p}$ collisions. The top quark decayed and produced a bottom quark jet ($b$-jet), a muon, and a neutrino. (D0 Collaboration, Fermilab Image FN0366)

$p\bar{p} \to t + X$ at the Fermilab Tevatron. What leading order diagrams can contribute to single top production? What sorts of leptons and jets result? ["Observation of Single Top-Quark Production", The D0 Collaboration, Phys. Rev. Lett. **103**, 092001 (2009)]

---

**Problem 8.3. Statistics in $Z$ boson parameter estimation**

---

Suppose a sample of $e^+e^-$ pairs with invariant mass values $\mu_i$ clustered around the mass of the $Z$ boson. Assume these derive from a Gaussian distribution with mean $m_Z$ estimated as $\bar{\mu} \equiv \Sigma\mu_i/n$ and variance $\sigma_Z^2$ estimated as $\sigma_\mu^2 = n(n-1)^{-1} < (\mu - \bar{\mu})^2 >$ where $n$ is the number of events. For Gaussian distributions, the variance on $\mu$ is $\sigma_Z^2/n$ and the variance on $\sigma_\mu^2$ is $2\sigma_Z^4/(n-1)$. The variance on $\sigma_\mu$ is $\sigma_Z^2/(2(n-1))$ and the error on the error is $\sigma_\sigma \simeq \sigma_Z/\sqrt{2n}$. The values of the mass and full width at half maximum derived from many measurements are $m_Z = 91.1876 \pm 0.0021$ GeV and $\Gamma = 2.4952 \pm 0.0023$ GeV. How many events would be required to achieve the quoted errors given statistical fluctuations?

---

**Problem 8.4. Electroweak boson gauge interactions**

---

The $SU(2)_L \times U(1)_Y$ gauge theory Lagrangian is

$$L = -\frac{1}{4}F_B^2 - \frac{1}{4}\mathbf{F}^2,$$

with physical fields $A$, $Z$, $W^+$, and $W^-$ defined by

$$W^1 = \frac{W^- + W^+}{\sqrt{2}}, \quad W^2 = \frac{W^- - W^+}{\sqrt{2}i}, \quad W^3 = cZ + sA, \quad B = -sZ + cA.$$

a) Starting with g=0, substitute the definitions to derive the noninteracting Lagrangian

$$L_0 = -\frac{1}{2}W^+_{\mu\nu}W^{-\mu\nu} - \frac{1}{4}A_{\mu\nu}A^{\mu\nu} - \frac{1}{4}Z_{\mu\nu}Z^{\mu\nu}$$

where we use the notation $V_{\mu\nu} = \partial_\mu V_\nu - \partial_\nu V_\mu$. b) Show that the terms linear in $g$ have the form

$$\begin{aligned} L_g &= g\Big[W^1{}_{\mu\nu}W^{2\mu}W^{3\nu} + W^2{}_{\mu\nu}W^{3\mu}W^{1\nu} + W^3{}_{\mu\nu}W^{1\mu}W^{2\nu} \\ &= \frac{g}{2i}\Big[\big(W^- + W^+\big)_{\mu\nu}\big(W^- - W^+\big)^\mu(cZ + sA)^\nu \\ &\quad + \big(W^- - W^+\big)_{\mu\nu}(cZ + sA)^\mu\big(W^- + W^+\big)^\nu \\ &\quad + (cZ + sA)_{\mu\nu}\big(W^- + W^+\big)^\mu\big(W^- - W^+\big)^\nu. \end{aligned}$$

c) Use $\Sigma_a \epsilon^{abc}\epsilon^{ajk} = \delta^{bj}\delta^{ck} + \delta^{bk}\delta^{cj}$ to show that the terms quadratic in $g$ are

$$\begin{aligned} L_{g^2} = \frac{g^2}{4}\big[&2W^-{}_\mu W^{+\mu} + (cZ+sA)_\mu (cZ+sA)^\mu \\ &+ \left(W^-{}_\mu W^{+\nu} + W^+{}_\mu W^{1\nu} + (cZ+sA)_\mu (cZ+sA)^\nu\right) \\ &\times \left(W^{-\mu} W^+{}_\nu + W^{+\mu} W^1{}_\nu + (cZ+sA)^\mu (cZ+sA)_\nu\right)\big]. \end{aligned}$$

---

**Problem 8.5. Higgs boson interactions**

---

The standard model Higgs Lagrangian is

$$L_\phi = \left(D_\mu \phi\right)^\dagger D^\mu \phi + \mu^2 |\phi|^2 - \lambda^2 |\phi|^4$$

with gauge covariant derivative

$$D_\mu \phi = \left[\partial_\mu + i\frac{g'}{2}B_\mu + ig\mathbf{T}\cdot\mathbf{W}_\mu\right]\begin{pmatrix} 0 \\ (v+H)/\sqrt{2} \end{pmatrix}.$$

Use the definitions

$$B = -sZ + cA;\ W^3 = (cZ + sA),\ W^\pm = \frac{1}{\sqrt{2}}\left(W^1 \mp iW^2\right)$$

and $m_W = gv/2$, $m_Z = (1/2)v\sqrt{g^2+g'^2}$, $m_H = \sqrt{2}\lambda v$ to write this as

$$\begin{aligned} L_\phi = &\frac{1}{2}\partial_\mu H \partial^\mu H + \frac{1}{2}m_H^2 H + \frac{1}{2}\left(\frac{m_Z^2}{v}\right)^2 Z_\mu Z^\mu + \frac{m_W^2}{v^2}W^+W^-\Big)(v+H)^2 \\ &+ \frac{\mu^2}{2}|v+H|^2 - \frac{\lambda^2}{4}|v+H|^4. \end{aligned}$$

---

**Problem 8.6. Prediction of weak boson masses**

---

The 4-fermion Lagrangian of Fermi theory is

$$L = \frac{G_F}{\sqrt{2}}\bar{e}\gamma_\alpha\left(1-\gamma^5\right)\nu_e \bar{\nu}_\mu \gamma^\alpha \left(1-\gamma^5\right)\mu.$$

Verify the standard model expressions

$$G_F = \sqrt{2}e^2/\left(8m_W^2 \sin^2\theta_W\right) \text{ and } v = 2^{-1/4}/\sqrt{G_F}.$$

Given $G_F = 1.16637(1) \times 10^{-5}\text{GeV}^{-2}$ from muon decay and $\sin^2\theta_W = 0.239$ from low energy neutral current neutrino scattering cross sections, calculate $m_W$,

$m_Z$, and $v$ in GeV. (Radiative corrections change these predictions.) The discovery of the $W$ and $Z$ bosons at close to these expected masses was a triumph of the standard model.

## Problem 8.7. Leptonic decay of the $W$ boson

The matrix element for the decay $W \to e\bar{\nu}_e$ is

$$M_{W^- \to e^- \bar{\nu}_e} = -\frac{g}{2}\epsilon^{\mu} j_{\mu}$$

where $\epsilon$ is the polarization 4-vector of the $W$ and the leptonic current for positive and negative electron helicity and electron polar angle $\theta$ in the $W$ rest frame is

$$j_+ = \sqrt{2E_\nu\,(E_e - p)}\,[1, -\sin\theta, 0, -\cos\theta]$$
$$j_- = \sqrt{2E_\nu\,(E_e + p)}\,[0, -\cos\theta, -i, \sin\theta]\,.$$

Verify by explicit calculation that the decay rate summed over lepton helicity and integrated over decay angle is independent of $W$ boson polarization as required by rotational symmetry.

## Problem 8.8. $W$ boson decay to heavy quarks

The matrix element for the decay of the $W$ boson to $q\bar{q}$ is

$$M_{W^- \to q\bar{q}} = -\frac{gV_{ab}}{\sqrt{2}}\epsilon_W^{\mu}\bar{u}\sigma_\mu^{+} v$$

with $V_{ab}$ the appropriate CKM matrix element. In the rest frame of the $W$ boson, the 2-component spinors are

$$u_s = \sqrt{E_q - sp}\phi_s(\mathbf{n}_q),\ \ v_r = r\sqrt{E_{\bar{q}} + rp}\phi_{-r}\left(\mathbf{n}_{\bar{q}} = ir\sqrt{E_{\bar{q}} + rp}\phi_r(\mathbf{n}_q\right).$$

Show the decay rate for $W$ boson decay $W \to cb$ for three colors retaining quark masses is

$$\Gamma_{W\to cb} = \frac{N_c g^2 |V_{cb}|^2}{8\pi m_W^2} p\left(E_b E_{\bar{c}} + \frac{1}{3}p^2\right)$$

and verify the alternate expression

$$\Gamma_{W\to cb} = \frac{g^2}{12\pi}|V_{cb}|^2 m_W\left(1 - 2\left(x_c + x_b + x_c x_b\right) + x_c^2 + x_b^2\right)^{\frac{1}{2}}$$
$$\times\left(1 - \frac{x_b + x_c}{2} - \frac{(x_b - x_c)^2}{2}\right)$$

where $x_q = (m_q/m_W)^2$. Similar expressions apply for other allowed quark pair combinations.

---

**Problem 8.9. Spectator model for charmed hadron decays**

---

Assume the decay rate of a hadron containing a $c$-quark plus light quarks is dominated by $c \to sW^+$ with $W^+ \to e^+\nu_e$, $\mu^+\nu_\mu$, and $u\bar{d}$, the latter in three colors. The decay rate scaled from $\Gamma_{\mu^- \to \nu_\mu \bar{\nu}_e}$ is therefore

$$\Gamma_{c \to x} = \frac{5}{192\pi^3} G_F^2 m_c^5 f\left(m_s^2/m_c^2\right)$$

with $f(x) = 1 - 8x + 8x^3 - x^4 - 12x^2 \ln x \simeq 1$. Take $m_s = m_\phi/2 = 0.510$ GeV and $m_c = m_{J/\psi}/2 = 1.55$ GeV. Show that $f = 0.45$ and that this model predicts the lifetime of a charmed hadron is 1.41 ps. Measured values $\tau_{\Lambda_c^+ = udc} = 0.206 \pm 0.012$ ps, $\tau_{D^+ = c\bar{d}} = 1.057 \pm 0.015$ ps, $\tau_{D^0 = c\bar{u}} = 0.415 \pm 0.004$ ps are smaller, presumably as a result of non-spectator amplitudes. For example, in $D^0$ decay, the $c \to sW^+$ can be followed by internal absorption $W^+\bar{u} \to d$.

---

**Problem 8.10. Invisible Z decay**

---

Show that the differential decay rate for $Z \to \nu\bar{\nu}$ as a function of the neutrino polar angle for $Z$ boson spin +1 along the $z$-axis is

$$\frac{d\Gamma_{Z \to \nu\bar{\nu}}}{d\cos\theta_\nu} = \frac{\left(g^2 + g'^2\right) m_Z \sin^2\theta}{128\pi}$$

and that $\Gamma_{Z \to \nu\bar{\nu}} = G_F\sqrt{2}m_Z^3/(24\pi)$. The two neutrino decay of the $Z$ boson is invisible but contributes to the width of the $Z$ boson resonance.

---

**Problem 8.11. Decays to a virtual $W$ boson**

---

Consider the decay such as $a \to bW \to bcd$. The spin averaged decay rate for virtual boson mass $\sqrt{s}$ can be written as

$$d\Gamma_{a \to bW \to bcd} = \Gamma_{a \to bW} \frac{\sqrt{s}/\pi}{\left(s - m_W^2\right)^2 + m_W^2\Gamma_W^2} ds \Gamma_{W \to cd}.$$

Apply this formula to a decay of a fermion of mass $m$ to massless fermions, e.g. $\mu^- \to e^-\bar{\nu}_e\nu_\mu$. Approximate the denominator by $m_W^4$ and transcribe the results for $\Gamma_{W \to e\nu}$ and $\Gamma_{t \to Wb}$ to find

$$d\Gamma_{\mu \to (e\nu)\nu} = \left[\frac{g^2}{64\pi}\frac{m^3}{s}\left(1 - \frac{s}{m^2}\right)^2\left(1 + 2\frac{s}{m^2}\right)\right]\frac{\sqrt{s}ds}{\pi m_W^4}\left[\frac{g^2}{48\pi}\sqrt{s}\right]$$

$$= \frac{g^4}{64 \cdot 48\pi^3}\frac{m^5}{m_W^4} dx(1 - x)^2(1 + 2x)$$

where $x = s/m^2$ ranges from 0 to 1. Integrate over $s$ to find the decay rate

$$\Gamma_{\mu\to e\nu\nu} = \frac{g^4}{6144\pi^3}\frac{m_\mu^5}{m_W^4} = \frac{G_F^2 m_\mu^5}{192\pi^3}.$$

---

**Problem 8.12. Higgs boson exchange**

---

The Lagrangian describing interactions of electrons and muons with the Higgs field $H(x)$ is

$$L_{int} = -\frac{m_e}{v}H\bar{\psi}_e\psi_e - \frac{m_\mu}{v}H\bar{\psi}_\mu\psi_\mu.$$

Calculate the spin averaged elastic scattering cross section $d\sigma_{e^-\mu^-\to e^-\mu^-}/dt$ for Higgs boson exchange.

---

**Problem 8.13. Muon collider**

---

The Higgs boson coupling to a fermion of mass $m_f$ is proportional to $m_f$ so direct $H$ production in $e^+e^-$ colliders has a low cross section relative to a $\mu^+\mu^-$ collider. Suppose a $\mu^+\mu^-$ collider with $L = 10^{33}$ cm$^2$ s$^{-1}$. The resonance cross section for $\mu\mu \to H \to X$ is

$$\sigma = \frac{4\pi\Gamma_{H\to\mu\mu}\Gamma_{H\to x}}{\left(s - m_H^2\right)^2 + m_H^2\Gamma_H^2}.$$

For $m_H \simeq= 120$ GeV, show that at the peak

$$\sigma \simeq \frac{4\pi}{3m_H^2}\left(\frac{m_\mu}{m_b}\right)^2$$

and estimate the inclusive event rate at the peak assuming the spread of beam energy is small compared to the width of the resonance.

---

**Problem 8.14. Higgs boson width**

---

What is the fractional width $\Gamma_{H\Rightarrow X}/m_h$ in percent if $m_H = 500$ GeV? Why is $H \to l^+l^-l^+l^-$ a “golden” discovery mode channel at the LHC and what background process produces similar events?

CHAPTER

# 9 Advanced Calculations

Calculations of high order processes in quantum field theory lead to important new considerations. Radiative corrections and renormalization are discussed along with applications.

## Contents



## 9.1 ■ INTRODUCTION TO RADIATIVE CORRECTIONS

Calculations of transition amplitudes in quantum field theory beyond lowest order involve both an increase in complexity and new considerations. In this chapter we look at this important subject beginning with the process of radiation associated with a collision. The problem of interpreting the result serves as an entry point to the general problem of interpretation of higher order amplitudes in field theories of point-like particles. The resolution of the issues through renormalization is essential to understanding the important higher order calculations in general use.

Let's start with the process of bremsstrahlung (braking radiation) where, in classical terms, the acceleration of a charge in scattering in a fixed potential leads to radiation. The diagram shown in Figure 9.2 shows photon emission by a final state electron in electron nucleus elastic scattering. In a general, all charged particles participating in any leading order process can contribute to a next-to-leading order process in which a final state photon appears and the amplitudes for final photon emission are coherent.

An amplitude associated with radiation is obtained from the amplitude without radiation by adding a propagator describing the intermediate state electron and a

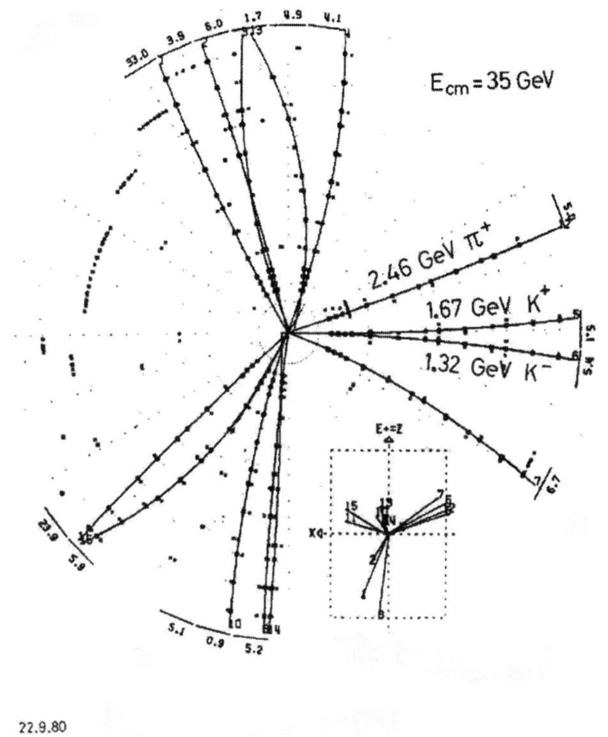


**FIGURE 9.1** Observation of $e^+e^- \to q\bar{q}g$ in the TASSO detector at the DESY PETRA collider. The quarks and the gluon appear as jets of particles. Gluon bremsstrahlung is an example of a radiative correction in quantum field theory. [DESY photo archive, POW 990901]

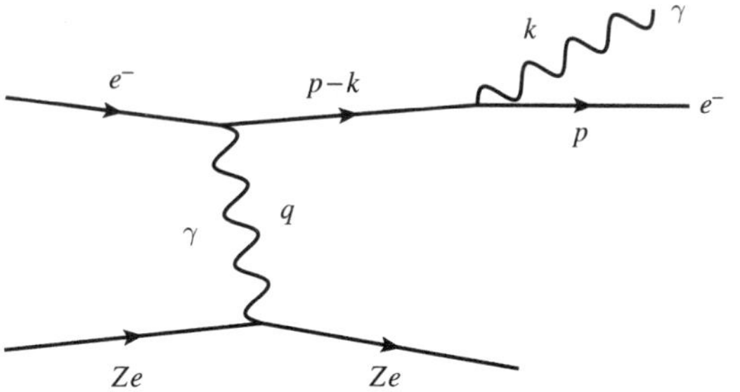


**FIGURE 9.2** Elastic scattering of an electron from a nucleus of charge $Ze$ with final state bremsstrahlung radiation. The amplitude for photon momentum $k$ and final electron momentum $p - k$ is related to that without the photon and final electron momentum $p$.

final photon wave function factor. These changes amount to a replacement of an initial state electron spinor for example as follows:

$$u_p \to -i\frac{(p-k)\gamma + m}{(p-k)^2 - m^2}\epsilon_k^\dagger(-ie)\gamma u_p. \tag{9.1}$$

Here $p$ denotes the 4-momentum of the electron and $k$ the 4-momentum of the photon. In case $|\mathbf{k}| << |\mathbf{p}|$, the momenta in the underlying process are little changed and the replacement may be approximated as

$$u_p \simeq -e\frac{p\gamma + m}{2pk}\epsilon^\dagger \gamma u_p. \tag{9.2}$$

If we now use the expression

$$(p\gamma)(\epsilon^\dagger\gamma) = 2\epsilon^\dagger p - (\epsilon^\dagger\gamma)(p\gamma) \tag{9.3}$$

and the fact that the free particle wave function obeys Dirac's equation

$$p\gamma u_p = m u_p, \tag{9.4}$$

we arrive at the approximate replacement

$$u_p \to -e\frac{p\epsilon^\dagger}{pk}u_p. \tag{9.5}$$

Similarly the amplitude for final state radiation can be obtained from the amplitude for the underlying process by the replacement

$$\bar{u}_p \to e\bar{u}_p\frac{p\epsilon^\dagger}{pk}. \tag{9.6}$$

Thus, for small $|\mathbf{k}|$, the amplitude $M$ for potential scattering of an electron with single photon emission factorizes as

$$M = M_0\left(\frac{p_f\epsilon^\dagger}{p_f k} - \frac{p_i\epsilon^\dagger}{pk}\right) \tag{9.7}$$

where $M_0$ is the amplitude without radiation.

In the case of scattering from a heavy particle with emission of a soft photon, if $d\sigma_{el}/d\Omega$ is the cross section for elastic scattering without photon emission, the cross section for scattering with single photon emission has the form

$$\frac{d\sigma}{d\Omega} = \frac{d\sigma_{el}}{d\Omega}\left(\frac{p_f\epsilon^\dagger}{p_f k} - \frac{p_i\epsilon^\dagger}{pk}\right)^2 \frac{e^2 d^3\mathbf{k}}{(2\pi)^3 2\omega} \equiv \frac{d\sigma_{el}}{d\Omega}f. \tag{9.8}$$

The polarization vector for a transverse photon has the form $\epsilon = (0, \mathbf{e})$ and, writing $p_f = E(1, \mathbf{v}_f, p_i = E(1, \mathbf{v}_i)$, and $k = \omega(1, \mathbf{n})$, the factor multiplying the elastic cross section is

$$f \equiv \frac{\alpha}{4\pi^2\omega}\left|\frac{\mathbf{v}_f \cdot \mathbf{e}}{1 - \mathbf{v}_f \cdot \mathbf{n}} - \frac{\mathbf{v}_i \cdot \mathbf{e}}{1 - \mathbf{v}_i \cdot \mathbf{n}}\right|^2 d\omega d\Omega_{\mathbf{k}}. \tag{9.9}$$

The sum over polarizations may be done with the formulae

$$\begin{aligned} \Sigma_{1,2} e_i e_j &= \delta_{ij} - n_i n_j \\ \Sigma(\mathbf{a} \cdot \mathbf{e})(\mathbf{b} \cdot \mathbf{e}) &= \mathbf{a} \cdot \mathbf{b} - (\mathbf{a} \cdot \mathbf{n})(\mathbf{b} \cdot \mathbf{n}) = (\mathbf{a} \times \mathbf{n})(\mathbf{b} \times \mathbf{n}) \end{aligned} \tag{9.10}$$

and results in the expression

$$d\sigma = d\sigma_{el}\alpha\left[\frac{\mathbf{v}_f \times \mathbf{n}}{1 - \mathbf{v}_f \cdot \mathbf{n}} - \frac{\mathbf{v}_i \times \mathbf{n}}{1 - \mathbf{v}_i \cdot \mathbf{n}}\right]^2 \frac{d\omega}{\omega}\frac{d\Omega}{4\pi^2}. \tag{9.11}$$

Given that the photon energy is small, single photon emission does not appreciably alter the initial or final electron velocity. The preceding expression shows that in this case the photon energy spectrum is inversely proportional to photon energy and the denominators favor emission along the initial and final directions. Integrating over photon energy, we find the cross section for elastic scattering with production of a photon within some frequency range is

$$\sigma \sim \alpha \ln \frac{\omega^{max}}{\omega^{min}} \sigma_{el}. \tag{9.12}$$

This result becomes infinite as $\omega^{min} \to 0$.

It is natural to demand a minimum frequency below which a photon may not be observed in a real experiment. For a typical energy resolution of 1% of the maximum energy, the ratio of the cross section for scattering with a resolved photon to the elastic cross section is roughly $\alpha \ln 100 \simeq 0.03$. The cross section for scattering with emission of a photon of energy less than the experimental resolution remains, however, formally infinite. This so-called infrared divergence involving soft radiation is one of several divergences found in higher order calculations in QED and in field theories in general. Amplitudes for loop diagrams in which a photon is emitted and reabsorbed also diverge as the photon momentum vanishes. In addition, so-called ultraviolet divergences appear in loop diagrams which involve arbitrarily large internal fermion momenta.

The divergence problems in spinor (and scalar but not vector) quantum electrodynamics may be cured by careful consideration of the physics as formalized in renormalization of parameters such as mass and charge. First, arbitrary low and high energy cutoffs are introduced to render divergent integrals finite. Then the results of higher order calculations are constrained to reproduce measured finite values of mass and charge for real particles with their accompanying fields. When the physical cross section for scattering with a photon below some resolution is calculated including not just the external radiation diagrams but loop diagrams which mix in amplitudes of comparable order in the coupling, the infrared divergences cancel leaving a finite result. The ultraviolet cutoff dependence of loop diagrams is absorbed into the difference between the measured and original bare parameter values leaving finite answers for real particle transition amplitudes that are independent of the cutoff. This is illustrated below. There remain collinear divergences involving hard but parallel massless particles corresponding to the indistinguishability of these states. Thus the full calculation of bremsstrahlung produces a cross section proportional to $\log(\sqrt{s}/m)$ which diverges as $m \to 0$.

The results of higher-order calculations in renormalizable theories are independent of the ultraviolet cutoff so independent of unknown physics at very high energies or small length scales. A non-renormalizable theory gives results which depend on the ultraviolet cutoff suggesting some new short distance physics must come to the rescue. If that physics is unknown, the analytic form of the modifications to low energy amplitudes is not prescribed and the theory lacks predictive power. In practice, renormalization is possible only for a restricted class of theories in which all ultraviolet divergences result from a few basic loop diagrams. Only in such cases may the divergences be absorbed into a finite number of basic parameters once and for all. Generally, such theories involve stabilizing symmetries and conservation laws.

The condition of renormalizability restricts field theories to those with coupling constants of mass dimension $\geq 0$ which strongly constrains the allowed types of interactions of simple form. The action $S = \int dx\ L$ is dimensionless so the Lagrangian density has natural dimension of mass to the fourth power: $[L] = m^4$. A derivative has dimension of mass $[\partial] = m$. Boson fields have dimension of mass $[\phi] = m$ and fermion fields have $[\psi] = m^{\frac{3}{2}}$. Hence a theory of scalar fields may only include self-interactions up to the fourth power:

$$L_\phi = (\partial\phi)^2 + m_1^2\phi^2 + m\phi^3 + \lambda\phi^4. \qquad (9.13)$$

For spinor fields interacting with scalar and vector fields, the only allowed terms are

$$L_\psi = \bar{\psi}\partial\psi + m\bar{\psi}\psi + e\bar{\psi}\gamma_\mu A^\mu\psi + f\bar{\psi}\gamma_\mu\gamma_5 A^\mu\psi + g\phi\bar{\psi}\psi \qquad (9.14)$$

where the coupling constants are dimensionless. The terms represent vector and axial vector currents coupled to the vector field and a Yukawa coupling to a scalar. Because it has mass dimension six, a four-fermion self interaction $(\bar{\psi}\psi)^2$ is not permitted. For vector fields, the allowed interaction terms have the form

$$L_A = \partial_\mu A^\nu \partial_\nu A^\mu + e(\partial_\mu A^\mu)^2 + f A^2 \partial_\mu A^\mu + g A^4. \qquad (9.15)$$

A fairly simple argument produces the restriction on the types of interactions in renormalizable theories. Consider a loop diagram such as the photon self-energy diagram in Figure 9.3. An internal boson line corresponds to a propagator which is, at high energy, proportional to $1/p^2$ while an internal fermion line

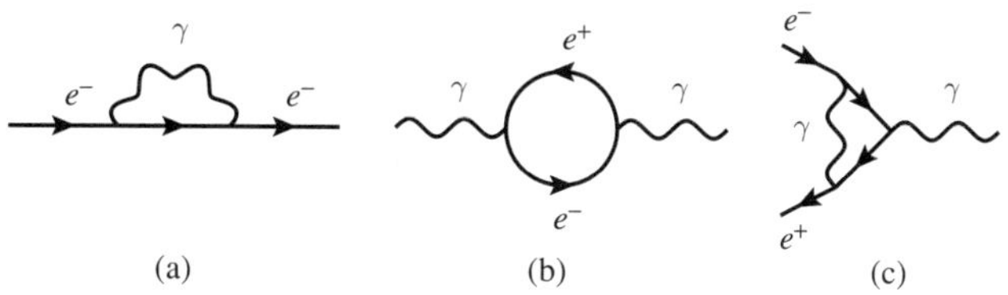


**FIGURE 9.3** Loop diagrams which correct propagators and vertices in electrodynamics.

corresponds to a propagator proportional to $\hat{p}/p^2 \sim 1/p$. The momenta within the loop are constrained by the external lines but each unconstrained momentum requires integration over $d^4p$. The superficial degree of divergence of the amplitude is defined as

$$D = 4I - 2B - F \tag{9.16}$$

with $I$ the number of integrations, $B$ the number of boson propagators, and $F$ the number of fermion propagators. For the QED photon self-energy loop, $I = 1$, $B = 0$, $F = 2$ and $D = 2$. The divergence will have to be absorbed by renormalization. Higher order calculations will entail, for example, a loop with a boson exchange within the loop and this internal line corresponds to $\Delta I = 1$, $\Delta B = 1$, $\Delta F = 2$ and $\Delta D = 0$. We see that the more complex diagram is not more divergent. On the other hand, a four-fermion interaction would imply a diagram with two fermions spanning the loop with $\Delta I = 2$ and $\Delta F = 4$ so $\Delta D = 4$. The higher order diagram is much more divergent indicating that this theory will get out of hand.

There is a flaw in the argument for massive vector fields. Although a mass term $L_m = m_A^2 A^2$ in the Lagrangian appears allowed on dimensional grounds, the second term in the propagator for a massive vector field

$$G^{\mu\nu} = \frac{-g^{\mu\nu} + p^\mu p^\nu/m^2}{p^2} = \frac{-g^{\mu\nu}}{p^2} + \frac{p^\mu p^\nu}{m^2 p^2} \tag{9.17}$$

approaches a constant at high energy rather than being proportional to $1/p^2$ implying diagrams with internal massive vector bosons diverge faster than those with massless vector bosons. Hence pure electrodynamics of a charged massive vector field is not renormalizable. It turns out massive vector particles are allowed if the mass results from spontaneous symmetry breaking of a gauge theory. Such is the case for the physical $W$ and $Z$ bosons. Since the massive vector gauge fields in the standard model are coupled only to conserved currents, the $p^\mu p^\nu/m^2$ term in the propagator is not effective. That the currents are conserved is connected to gauge symmetry. In case of a spontaneously broken symmetry, both the symmetry and renormalizability are subtly hidden.

Non-renormalizable interactions appear in effective low energy theories which are useful only in tree level calculations. An example is the four-fermion weak interaction coupling postulated by Fermi to describe weak interactions. Such nonrenormalizable terms produce amplitudes which when extrapolated to high energy violate unitarity. Unitarity can be saved by the appearance at high energy of new degrees of freedom which soften the interaction. Such new degrees of freedom are characterized by form factors in effective theories. The appearance of the $W$ boson softens Fermi's point fermion interaction ($G_F$ becomes a propagator), although not enough to save unitarity by itself. The $W$ boson mass is ultimately softened by gauge symmetry restoration at high energy.

## 9.2 ■ QED LOOPS AND RENORMALIZATION

Let us now take a closer look at some quantum electrodynamics amplitudes beyond tree level. In particular, consider diagrams corresponding to the replacement of electron and photon lines and vertex points in a tree level diagram by the diagram elements shown in Figure 9.3. Diagram $a$ in which a fermion emits and absorbs a photon represents a correction to electron self-energy. Diagram $b$ in which a photon converts to an intermediate electron-positron pair represents a correction to the photon self-energy. The vertex correction diagram $c$ conjoins the electron self-energy correction with an interaction.

The electron self-energy loop amounts to a replacement of the electron propagator

$$\frac{i}{p\gamma - m} \to \frac{i}{p\gamma - m} + \frac{i}{p\gamma - m}\frac{\Sigma(p)}{i}\frac{i}{p\gamma - m} \simeq \frac{i}{p\gamma - m - \Sigma(p)} + O\left(\alpha^2\right) \tag{9.18}$$

where

$$-i\Sigma(p) = (-ie)^2 \int \frac{dk}{(2\pi)^4}\frac{-ig_{\mu\nu}}{k^2}\gamma^\mu \frac{i}{(p-k)\gamma - m}\gamma^\nu \tag{9.19}$$

and we used the operator relation

$$\frac{1}{A-B} = \frac{1}{A} + \frac{1}{A}\frac{1}{B}\frac{1}{A} + \frac{1}{A}\frac{1}{B}\frac{1}{A}\frac{1}{B}\frac{1}{A}\cdots \tag{9.20}$$

and expanded in $\alpha$. The numerator and denominator in the integrand can be expressed as

$$\begin{aligned} N &= g_{\mu\nu}\left(\gamma^\mu\gamma(p-k) + m\right)\gamma^\nu = -2(p-k)\gamma + 4m \\ D^{-1} &= \left[k^2\left[(p-k)^2 - m^2\right]\right]^{-1} = \int_0^1 dx\,\left[k^2 - 2pkx + \left(p^2 - m^2\right)x\right]^{-2}. \end{aligned} \tag{9.21}$$

The integral representation for the denominator allows a shift in momentum integration from $k$ to $K \equiv k - xp$ and we can write

$$-i\Sigma(p) = -e^2 \int_0^1 dx\, [2(x-1)\gamma p + 4m]I \tag{9.22}$$

where

$$I = \int \frac{dK}{(2\pi)^4} \left[K^2 - M^2\right]^{-2}, \; M^2 = p^2x^2 + \left(m^2 - p^2\right)x. \tag{9.23}$$

In doing the integration, one must write $m^2 = m^2 + i\epsilon$ and take the limit $\epsilon \to 0$. Then

$$\frac{\partial I}{\partial M^2} = \frac{1}{i(4\pi)^2 M^2} \to I = \frac{-i}{(4\pi)^2} \ln\left(\frac{M^2}{\mu_2^2}\right) \tag{9.24}$$

with $\mu_2$ an integration constant that serves as a high energy cutoff of the integration. After integration over $x$, one finds

$$\Sigma = A + Bp\gamma \tag{9.25}$$

where

$$A = \frac{\alpha}{\pi} m \left[2 - \ln\frac{m^2}{\mu_2^2} + \frac{\left(m^2 - p^2\right)}{p^2} \ln\left(1 - \frac{p^2}{m^2}\right)\right]$$

$$B = \frac{\alpha}{4\pi}\left[\ln\frac{m^2}{\mu_2^2} - 3 - \frac{m^2 - p^2}{p^2}\left(1 + \frac{m^2 + p^2}{p^2} \ln\frac{m^2 - p^2)}{m^2}\right)\right]. \tag{9.26}$$

In terms of $A$ and $B$, the effect on the electron propagator can be expressed as

$$G = \frac{i}{p\gamma - m} \to \frac{iZ_2}{p\gamma - m_R} \tag{9.27}$$

where $Z_2 = (1 - B)^{-1}$ is called a wave function renormalization factor and the renormalized mass is

$$m_R = \frac{m + A}{1 - B} \simeq m + (A + mB). \tag{9.28}$$

We see that the loop, in effect, changes the mass of the electron from its bare value $m$ to a value that depends on $p^2$. This can be understood if we think of the bare electron as surrounded by a cloud of virtual particles (here photons) the availability of which depends on the energy in the virtual electron rest frame. Setting $m_R$ equal to the experimental mass $m_e = 0.511$ MeV defined by $p^2 = m^2$, we require

$$\delta m \simeq (A + mB)|_{p^2=m^2} \simeq \frac{\alpha m}{4\pi}\left(5 - 3\ln\frac{m^2}{\mu_2^2}\right) = 0 \tag{9.29}$$

which fixes $\mu_2$ and gives $Z_2 = 1 - \alpha/(3\pi)$.

We turn now to the vacuum polarization diagram. The $e^+e^-$ loop gives the photon propagator the form

$$\frac{-ig_{\mu\nu}}{q^2} \to \frac{-ig_{\mu\nu}}{q^2} + \frac{-ig_{\mu\alpha}}{q^2}\Pi^{\alpha\beta}(q)\frac{-ig_{\beta\nu}}{q^2} + \cdots. \tag{9.30}$$

The tensor $\Pi$ follows from the Feynman rules with two internal propagators (note the minus sign) and is

$$\Pi_{\mu\nu} = -(-ie)^2 \mathrm{tr} \int \frac{dk}{(2\pi)^4}\, \gamma_\mu \frac{1}{k\gamma - m}\gamma_\nu \frac{1}{(k-q)\gamma - m} \tag{9.31}$$

where the trace is over the internal gamma matrices and $k$ is the internal 4-momentum. Putting $K = k - xq$, this can be written as

$$\Pi_{\mu\nu} = -4e^2 \int_0^1 dx\, (I_1 + I_2) \tag{9.32}$$

where we define $M^2 = m^2 + q^2(x^2 - x)$ and

$$I_1 = \int \frac{dK}{(2\pi)^4} \frac{2K_\mu K_\nu - g_{\mu\nu}K^2}{\left(K^2 - M^2\right)^2}$$

$$I_2 = \int \frac{dK}{(2\pi)^4} \frac{\left(x^2 - x\right)\left(2q_\mu q_\nu - g_{\mu\nu}q^2\right) + m^2 g_{\mu\nu}}{\left(K^2 - M^2\right)^2}. \tag{9.33}$$

Integration yields

$$\Pi_{\mu\nu}(q) = \frac{ie^2}{2\pi^2}\left(q_\mu q_\nu - g_{\mu\nu}q^2\right) f(q, m, \mu_3) \tag{9.34}$$

with

$$f = \int_0^1 dx\, \left(x^2 - x\right) \ln \frac{m^2 + q^2\left(x^2 - x\right)}{\mu_3^2} \tag{9.35}$$

where $\mu_3$ is an integration constant serving to cut off the integration at high momentum. (The current conservation condition $q^\mu \Pi_{\mu\nu} = 0$ is imposed in arriving at these expressions.) The corrected photon propagator is

$$\begin{aligned} G_{\mu\nu} &= \frac{-ig_{\mu\nu}}{q^2} + \frac{-ig_{\mu\alpha}}{q^2}\Pi^{\alpha\beta}(q)\frac{-ig_{\beta\nu}}{q^2} \\ &= \frac{-ig_{\mu\nu}}{q^2} + \frac{-ig_{\mu\alpha}}{q^2}\left[\frac{ie^2}{2\pi^2}\left(q^\alpha q^\beta - g^{\alpha\beta}q^2\right) f\,(q, m, \mu_3)\right]\frac{-ig_{\beta\nu}}{q^2} \\ &= \frac{-ig_{\mu\nu}}{q^2}\left(1 + \frac{e^2}{2\pi^2}f\right) - i\frac{e^2}{2\pi^2}f\frac{q_\mu q_\nu}{q^4}. \end{aligned} \tag{9.36}$$

Consider now $e_1^- \mu_2^- \to e_3^- \mu_4^-$ scattering with the modified photon propagator. The matrix element has the form

$$M = j_{13}^{\mu} G_{\mu\nu} j_{24}^{\nu}. \tag{9.37}$$

The term in $G_{\mu\nu}$ proportional to $q_\mu q_\nu$ does not contribute because current conservation implies $qj_{13} = qj_{24} = 0$. The effect of including the loop amounts to multiplying the scattering amplitude by a factor

$$Z_3\left(q^2, m, \mu_3\right) = 1 + \frac{e^2}{2\pi^2} f \simeq 1 + \frac{\alpha}{3\pi}\left(\ln\frac{m^2}{\mu_3^2} - \frac{q^2}{5m^2} + \cdots\right) \tag{9.38}$$

where $f$ was expanded as a function of $q^2$. In fact, this conclusion applies wherever the propagator appears in electrodynamics and can be interpreted as a replacement of the charge of each particle by a renormalized charge defined by

$$e_R = \sqrt{Z_3} e. \tag{9.39}$$

Since $f$ is a function of $q^2$, this renormalized charge is a function of $q^2$ also.

The limit $q \to 0$ corresponds to Coulomb scattering and electrostatics by which the charge of the electron is measured. The term proportional to $q^2$ (Uehling potential) vanishes in this limit. Apparently we should identify the renormalized charge $e_R$ with the static charge of the electron measured in experiments and identify the charge $e$ in the field equations with an idealized bare charge. Hence we define a momentum transfer dependent so-called running coupling constant

$$\alpha(q^2, m, \mu_3) = \frac{e_R^2}{4\pi} \tag{9.40}$$

and identify the value

$$\alpha(q^2 = 0, m, \mu_3) = \frac{e_R^2}{4\pi} = \frac{Z_3(q^2 = 0, m, \mu_3)e^2}{4\pi} \simeq \frac{1}{137}. \tag{9.41}$$

We will see below how the vertex correction cancels the factor $Z_2$ from the electron self-energy leaving only the factor $Z_3$ to be interpreted as a charge renormalization and we then fix the parameter $\mu_3 = m$ so that $e = e_R(q^2 = 0, m, m)$, thereby absorbing this arbitrary parameter into the definition of the unobservable bare charge. Note that a loop on an external photon line may be neglected. Such a photon may be regarded as having been produced by a distant source and the loop has $q^2 = 0$ so only adds a factor $\sqrt{Z_3}$ to the charge at the source and within the system of interest.

The running of the coupling constant has a quasi-classical interpretation. As shown in Figure 9.4, we can think of a bare charge as surrounded by virtual $e^+e^-$ pairs with density decreasing with distance $r$ from the bare charge. The pairs will tend to be polarized—the positrons will be closer to the bare electron charge than the electrons. The cloud of virtual pairs then behaves as a dielectric medium with a

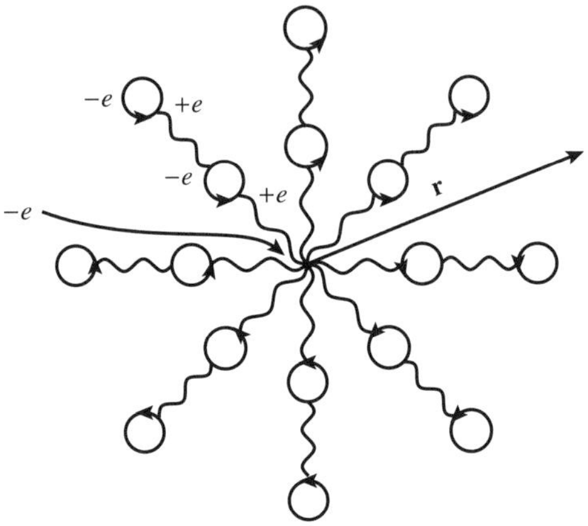


**FIGURE 9.4** Illustration of vacuum polarization due to virtual $e^+e^-$ pairs surrounding a bare charge. The polarization charge modifies the net charge within a radius $r$, screening the bare charge.

distance dependent dielectric constant $\epsilon(r)$. The charge within a radius $r$ is smaller than without polarization. The bare charge is screened and appears to grow as the wavelength of an external probe photon shrinks.

We turn now to the vertex correction which amounts to the replacement

$$-ie\gamma_\mu \to -ie\gamma_\mu - ie\Lambda_\mu \tag{9.42}$$

where the Feynman rules give for electron momentum $p$ and $p'$

$$\Lambda_\mu(p', p) = (-ie)^2 \int \frac{dk}{(2\pi)^4} \frac{-i}{k^2} \gamma_\nu \frac{i}{(p'-k)\gamma - m} \gamma_\mu \frac{i}{(p-k)\gamma - m} \gamma^\nu. \tag{9.43}$$

In case the initial and final electron are both real so $p^2 = m^2 = p'^2$, $\Lambda$ is a function only of $q = p' - p$ and is calculated to be

$$\Lambda_\mu = \gamma_\mu F_1\left(q^2\right) - \frac{\sigma_{\mu\nu} q^\nu}{2m} F_2\left(q^2\right) \tag{9.44}$$

where $\sigma_{\mu\nu} = [\gamma^\mu, \gamma^\nu]/2$ [see Equation 6.17]. The form factors are

$$F_1\left(q^2\right) = -\frac{\alpha}{\pi}\left\{\left[\ln\frac{m^2}{\mu_1^2} - \frac{3}{2} + \frac{1}{2\omega}F\right] - (4\ln\eta + 5)\frac{2\omega}{Q^2}Fm^2 - \left(4\ln\eta + \frac{7}{2}\right)\omega F\right\}$$

$$F_2\left(q^2\right) = \left(\frac{\alpha m^2 \omega}{\pi Q^2}\right) F \tag{9.45}$$

where $\mu_1$ is a high momentum cutoff, $\eta$ is a low momentum cutoff associated with infrared photons, and

$$F = \ln\frac{1+\omega}{1-\omega}, \; \omega = \left(1 + \frac{4m^2}{Q^2}\right)^{-1/2} \tag{9.46}$$

and $Q^2 = -q^2 > 0$ is introduced. For $Q^2 << m^2$,

$$F_1\left(q^2 \simeq 0\right) \simeq \frac{\alpha}{4\pi}\left(\frac{11}{2} - \ln\frac{m^2}{\mu_1^2} + 4\ln\eta\right) - \left(\frac{1}{6} + \frac{4}{3}\ln\eta\right)\frac{q^2}{m^2}$$
$$F_2\left(q^2\right) \simeq 0 \simeq \frac{\alpha}{2\pi}. \tag{9.47}$$

The amplitude for the interaction of an electron with an external electromagnetic field when the vertex diagram is included may be found via the replacement

$$e\gamma_\mu \to e\left(\gamma_\mu + \Lambda_\mu\right) = e\left(\left[1 + F_1\left(q^2\right)\right]\gamma_\mu - \frac{\sigma_{\mu\nu}q^\nu}{2m}F_2\left(q^2\right)\right). \tag{9.48}$$

The form factor $F_1$ that multiplies the $\gamma_\mu$ can be interpreted as implying an additional charge renormalization factor given by

$$Z_1^{-1} = 1 + \frac{\alpha}{4\pi}\left\{\left[\frac{3}{2} - \ln\frac{m^2}{\mu_1^2} - \frac{1}{2\omega}F(\omega)\right] + (4\ln\eta + 5)\frac{2\omega m^2}{Q^2}F(\omega)\right.$$
$$\left. + \left(2\ln\eta + \frac{7}{4}\right)2\omega F(\omega)\right\}. \tag{9.49}$$

The total charge renormalization from the vacuum polarization, electron self-energy, and vertex diagrams can be written as

$$e_R = \frac{Z_2}{Z_1}\sqrt{Z_3}e \tag{9.50}$$

but it can be shown that $Z_1 = Z_2$. In fact when $q = p' - p \to 0$, the vertex must be expressible in the form

$$\bar{u}(p)\Lambda_\mu u(p) = \left(Z_1^{-1} - 1\right)\bar{u}(p)\gamma_\mu(p, p)u(p) \tag{9.51}$$

where $Z_1$ is a constant since the only other vector $p$ is the same as $m\gamma$ when sandwiched between spinors. Comparison with the electron self-energy integral for $p' = p$ shows that

$$\Lambda_\mu(p, p) = -\frac{\partial\Sigma(p)}{\partial p^\mu} \tag{9.52}$$

since

$$\frac{\partial}{\partial p^\mu}\frac{1}{p\gamma - m} = -\frac{1}{p\gamma - m}\gamma_\mu\frac{1}{p\gamma - m}. \tag{9.53}$$

The derivative of the expression for $\Sigma$ shows that $Z_1 = Z_2$.

Given $Z_1 = Z_2$, setting $p^2 = m^2$ in $Z_2$ and $Q^2 = 0$ in $Z_1$ and $\mu_1 = \mu_2$, we require $\ln\eta = -\frac{5}{8}$. Using this one finds

$$e_R(Q) = e\left\{1 + \frac{\alpha}{\pi}\int_0^1 dx\left[\left(x - x^2\right)\ln\frac{Q^2\left(x - x^2\right) + m^2}{\mu_3^2}\right]\right\} \tag{9.54}$$

and, for $Q^2 << m^2$,

$$e_R(Q) \simeq e\left[1 + \frac{\alpha}{2\pi}\left(\frac{1}{3}\ln\frac{m^2}{\mu_3^2} + \frac{1}{15}\frac{Q^2}{m^2}\right)\right]. \tag{9.55}$$

Setting $e_R(Q^2 \to 0) = e = 1.602 \times 10^{-19}$ C implies

$$\mu_3 = m. \tag{9.56}$$

We arrive at an interpretation in which the cutoffs disappear and we are left with an electron with effective structure functions $F_1$ and $F_2$ and a running coupling constant that increases with $Q^2$. The running can be described by introducing the Beta function

$$\beta(\alpha, Q) \equiv \frac{\partial}{\partial \ln Q}\alpha_R(Q) = Q\frac{\partial}{\partial Q}\alpha_R(Q) \tag{9.57}$$

which allows one to compute the value of $\alpha_R$ at a different $Q^2$ from a known value. From the expression for $e_R$, one can find that for spinor electrodynamics,

$$\beta(\alpha, Q) = \frac{2\alpha^2}{3\pi} - \frac{4\alpha^2 m^2}{\pi Q^2}\left\{1 + \frac{2m^2}{x^2}\ln\frac{x^2 - Q^2}{x^2 + Q^2}\right\} \tag{9.58}$$

where $x = \sqrt{Q^4 + 4Q^2m^2}$. For $Q^2 >> 4m^2$, $\beta \simeq 2\alpha^2/(3\pi)$ is a constant and we have roughly

$$\alpha_R(Q) = \frac{\alpha_R(m)}{1 - \frac{2\alpha_R(m)}{3\pi}\ln\frac{Q}{m}}. \tag{9.59}$$

More generally, one must sum the contributions of all particles which may participate in loop diagrams with proper accounting for their mass. That the coupling grows with energy suggests there is an energy at which $\alpha \sim 1$ (although that seems impossible to prove through perturbation theory!). For electrodynamics, that scale is much larger than scales at which new physics is expected to complicate the picture so this problem can be ignored.

Richard Feynman remarked: "The shell game that we play...is technically called renormalization. But no matter how clever the word, it is still what I would call a dippy process!" However odd the reinterpretation of the theory, experiments bear out the results of the higher order renormalized calculations. The impact of radiative corrections was first observed in slight shifts in atomic energy levels. The nonrelativistic interaction between an electron and proton including just the running coupling constant correction corresponds to the potential

$$V(\mathbf{x}) = -\int \frac{d\mathbf{q}}{(2\pi)^3} e^{i\mathbf{q}\cdot\mathbf{x}} \frac{\alpha(\mathbf{q}^2)}{\mathbf{q}^2} \tag{9.60}$$

When $\alpha(\mathbf{q}^2)$ is expanded in $\mathbf{q}^2$ as

$$\alpha(\mathbf{q}^2) = \alpha(0) + \alpha'\mathbf{q}^2 + \cdots, \tag{9.61}$$

the first correction gives a constant integrand and one finds

$$V(\mathbf{x}) = -\frac{\alpha}{|\mathbf{x}|} - \frac{4\alpha^2}{15m^2}\delta(\mathbf{x}). \tag{9.62}$$

The energy shift for a hydrogen atom is

$$\Delta E = \int d\mathbf{x}\,|\psi(\mathbf{x})|^2 \left[-\frac{4\alpha^2}{15m^2}\delta(\mathbf{x})\right] = -\frac{4\alpha^2}{15m^2}|\psi(0)|^2. \tag{9.63}$$

Only $s$-wave states are affected and, for the 2$s$ state, the shift is $\Delta E = -0.1123 \times 10^{-6}$ eV.

The effect of vacuum polarization and of the other radiative corrections on atomic energy levels are well established. The energies associated with the exact solutions to the Dirac equation for a Coulomb potential for atomic number $Z$ are

$$\begin{aligned} E_{n,j} &= m_e \left[1 + \left(\frac{Z\alpha}{n - \left(j + \frac{1}{2} + \sqrt{\left(j+\frac{1}{2}\right)^2 - Z^2\alpha^2}\right)}\right)^2\right]^{-1/2} \\ &= m_e \left[1 - \frac{(Z\alpha)^2}{2n^2} - \frac{(Z\alpha)^4}{2n^4}\left(\frac{n}{j+\frac{1}{2}} - \frac{3}{4}\right) + \dots\right] \end{aligned} \tag{9.64}$$

where $n$ is the principal quantum number and $j$ is the quantum number associated with the total angular momentum $\mathbf{j} = \mathbf{L} + \mathbf{s}$. Notice that to order $(Z\alpha)^2$, this spectrum is that obtained with Klein-Gordon equation (Equation 5.9) with the replacement of $l$ by $j \pm 1/2$. Two $j$ states for a given $l$ give rise to fine structure. The single particle Dirac theory therefore predicts the $(n = 2, j = 1/2)$ $2S_{1/2}$ state should be degenerate with the $(n = 2, j = 1/2)2P_{1/2}$ state while, as illustrated in Figure 9.5 and discovered by Willis Lamb, they differ in energy by some $4 \times 10^{-6}$ eV. The energy difference corresponds to an electromagnetic emission or absorption frequency of 1058 MHz and resonance absorption of radio waves of this frequency may be used to induce transitions between the levels.

The splitting, called the Lamb shift, is accounted for by multiparticle radiative corrections. To order $\alpha^5$ in the energy, the radiative corrections include the vacuum polarization change in the potential corresponding to a loop in the single photon exchange diagram plus the vertex correction at the electron vertex and the electron self-energy. The radiative corrections are calculated to produce an additional energy shift

$$\begin{aligned} \Delta E(n, l = 0) &= \frac{4Z^4}{3\pi n^3} m_e \alpha^5 \left[\ln\left(\frac{m_e}{2\omega(n,0)}\right) + \frac{5}{6} - \frac{1}{5}\right], \\ \Delta E(n, l > 0) &= \frac{4Z^4}{3\pi n^3} m_e \alpha^5 \left[\ln\left(\frac{\alpha^2 m_e}{2\omega(n,l)}\right) + \frac{3}{8}\frac{c_{lj}}{2l+1}\right] \end{aligned} \tag{9.65}$$

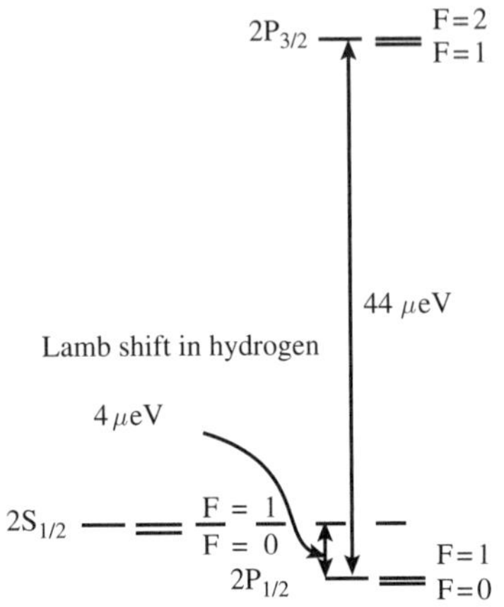


**FIGURE 9.5** The Lamb shift in the $n = 2$ energy levels of hydrogen. The $2S$ and $2P$ level degeneracy valid even in the exact solutions to the relativistic Dirac equation for a Coulomb potential is broken by radiative corrections.

where $c_{lj} = 1/(l+1)$ for $j = l + 1/2$ and $c_{lj} = -1/l$ for $j = l - 1/2$ and $\omega(n,l)$ is an average excitation energy. [See. H. Bethe, L. Brown, Phys. Rev. **77**, 370 (1949)] Included here is a spin-orbit term that derives from the anomalous magnetic moment associated with the vertex correction. The term proportional to -1/5 represent the vacuum polarization contribution. The logarithm derives from the difference in self-energy of a bound state electron from a free electron and $\omega_n$ is roughly an average of energy level differences $E_{nj} = E_n - E_j$ weighted by ln $E_{nj}/m_e$. In hydrogen $\omega(n,0) \simeq 8.3\alpha^2 m_e$ and $\omega(2,1) \simeq 0.5\alpha^2 m_e$. Averaged over the atomic wave functions, the average energy difference $E_{2S_{1/2}} - E_{2P_{1/2}}$ amounts to 1052.1 MHz of which the electron mass renormalization accounts for 1017 MHz, the anomalous magnetic moment contributes +68 MHz, and the vacuum polarization contribution is -27.1 MHz. Including fourth order radiative corrections and mass and recoil effects, the theoretical splitting of 1057.864 (14) MHz compares well with the experimental value 1057.862(20) MHz. The vacuum polarization effect is only $\sim$ 2.5% of the Lamb shift in hydrogen. In muonic and mesonic atoms in which the muon or meson orbits the nucleus more closely than in hydrogen, the Uehling potential corresponding to the vacuum polarization correction gives the dominant contribution to the Lamb shift.

The most stringent test of quantum electrodynamics is often said to be the corrections to the electron magnetic moment. We have seen how the vertex correction gives rise to a term proportional to $F_2\sigma_{\mu\nu}$. This term may be interpreted as representing an anomalous magnetic moment. (See the discussion of $\mu_p$ and $\mu_n$ in Chapter 6.) If we write the electron magnetic moment as

$$\mu = \frac{g}{2}\frac{e}{2m_e}\sigma, \tag{9.66}$$

then the vertex correction translates to

$$\frac{g}{2} = 1 + \frac{\alpha}{2\pi} - 0.328\frac{\alpha^2}{\pi^2} + \cdots \tag{9.67}$$

where higher order terms modify the structure functions $F_1$ and $F_2$ and the calculated $O(\alpha^2)$ correction is given. The term proportional to $\alpha^2$ corresponds the diagrams shown in Figure 9.7.

The anomalous magnetic moment factor $a_e = g - 2$ has been measured with high precision using a single electron caught in a Penning trap. The difference between the electron's cyclotron frequency and its spin precession frequency in a magnetic field is proportional to $g - 2$. An extremely high precision measurement of the quantized energies of the cyclotron orbits, or Landau levels, of the electron, compared to the quantized energies of the electron's two possible spin orientations, gives a value for the electron's spin g-factor:

$$\mathrm{a_e} = 1.159\ 652\ 180.73(28) \times 10^{-3}. \tag{9.68}$$

The anomalous magnetic moment of the muon is measured by observing the decays of spin polarized muons in a storage ring. Polarized muons are produced in pion decay as discussed in Chapter 10. As described in Chapter 8, electrons appear preferentially in a direction opposite to the muon spin. The projection of the muon

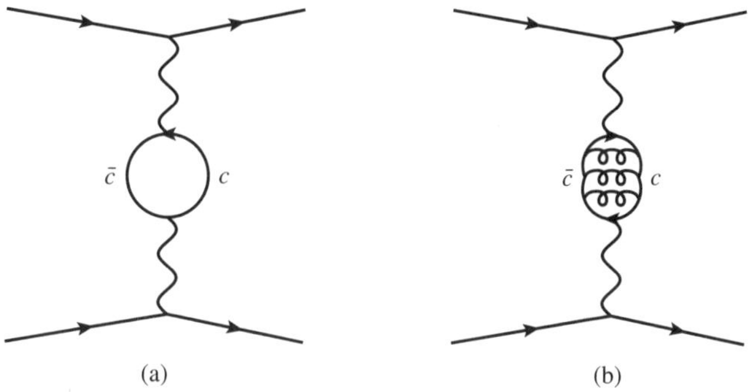


**FIGURE 9.6** The contribution of a heavy quark to vacuum polarization in electrodynamics.(a) The leading order loop contribution. (b) Higher order contribution of a bound state resonance. The coupling of a resonance to the photon is derived empirically from production and decay data.

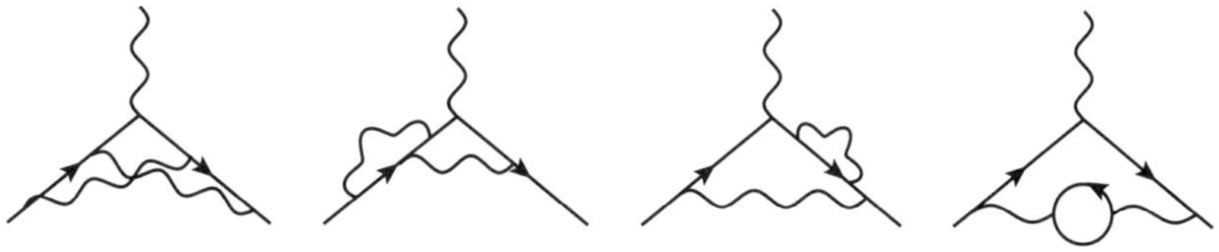

**FIGURE 9.7** Contributions of the order $\alpha^2$ to the electron magnetic moment.

spin onto the plane of the muon orbit rotates with anomaly-dependent frequency $\omega_a$ and the number of electrons above any energy threshold observed inside or outside the orbit is modulated at frequency $\omega_a$. Calculations within the standard model with a precision of $10^{-10}$ require not only the electrodynamic radiative correction up to the eighth order but also the hadronic contribution up to order $\alpha_s^3$ and the electroweak contribution up to the two-loop order. Such calculations are not for the faint of heart. As illustrated in Figure 9.6, quark bound state resonances such as the $J/\psi$ contribute and are not exactly calculable. However, their contribution may related to empirical results such as the decay $J/\psi \to e^+e^-$. The anomalous magnetic moment of the muon is measured to be

$$\mathrm{a}_\mu(\mathrm{Expt}) = 11659208.0(6.3)^{-10}. \tag{9.69}$$

The $\alpha^5$ order prediction of quantum electrodynamics is

$$\mathrm{a}_\mu(\mathrm{QED}) = 11\,658\,471.\,958(115) \times 10^{-10}. \tag{9.70}$$

Contributions $a_\mu(\mathrm{Weak}) = 15.4(0.2) \times 10^{-10}$ and $a_\mu(\mathrm{Hadronic}) = 694(6) \times 10^{-120}$ have been calculated. The experimental and theoretical errors on $a_\mu$ are at present about $5 \times 10^{-10}$. The magnetic moment is in principle sensitive to effects of virtual particles beyond the standard model. This fact motivates advancing the experimental precision with improved muon storage ring dedicated facilities.

## 9.3 ■ ASYMPTOTIC FREEDOM AND UNIFICATION

Calculations of higher order processes in gauge theories other than spinor electrodynamics employ similar techniques. For example, in scalar electrodynamics, the running coupling constant can be computed takes the form

$$\tilde{\alpha}\left(Q^2\right) = \frac{\tilde{\alpha}\left(m^2\right)}{1 - \frac{\tilde{\alpha}\left(m^2\right)}{12\pi}\ln\left(\frac{Q^2}{m^2}\right)} \tag{9.71}$$

The correction factor $\tilde{\alpha}/(12\pi)$ is one fourth that in spinor electrodynamics and indicates that the gauge interactions of charged particles of different spin have different Beta functions. As in spinor electrodynamics, the coupling grows with energy.

Qualitatively different results emerge in the renormalization of spinor quantum chromodynamics. Both fermion and boson loops correct the gluon propagator (Figure 9.8) and there is an additional three-gluon modification to the vertex (Figure 9.9). That the theory is renormalizable is subtle. After renormalization, the effect of next-to-leading order corrections can be expressed by the introduction of a running strong interaction coupling constant

$$\alpha_s \to \alpha_s\left(\mu^2\right)\left[1 + \frac{\alpha_s\left(\mu^2\right)}{4\pi}\ln\left[\frac{-q^2}{\mu^2}\right]\left(\frac{2n}{3} - \frac{13N_c}{6} - \frac{6N_c}{4}\right) + \cdots\right], \tag{9.72}$$

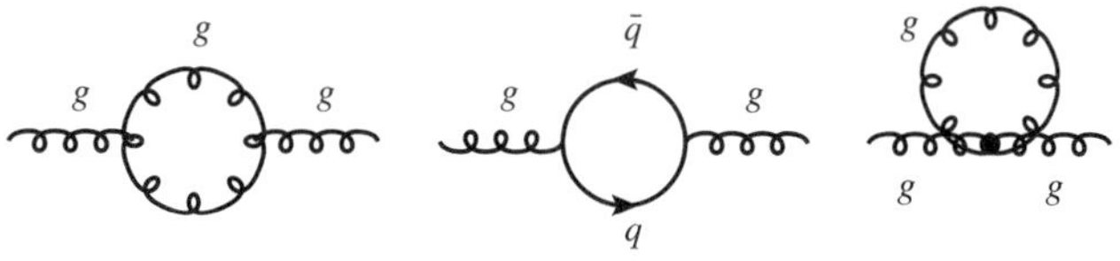


**FIGURE 9.8** Leading corrections to the gluon propagator. Both quarks and gluons appear in loops.

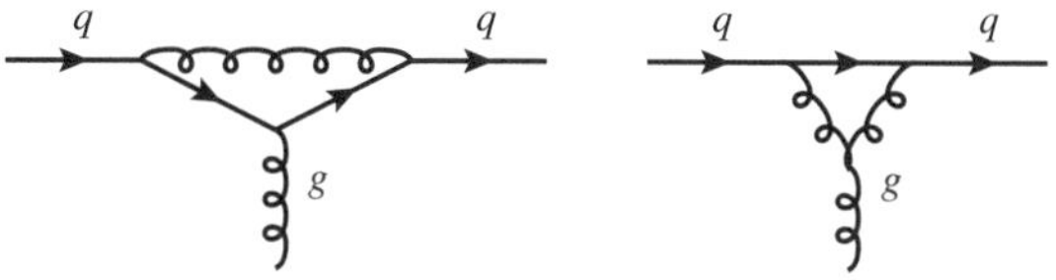


**FIGURE 9.9** Second order corrections to the quark-gluon vertex. Diagram (a) has an analog in electrodynamics. Diagram (b) does not.

where the three terms are the corrections due to quark and gluon loops in the propagator and to the vertex respectively. Here $n$ is the number of quark flavors and $N_c = 3$ the number of colors. Unlike electrodynamics where the coupling constant may be fixed by classical measurements at vanishing momentum transfer (long distances), in chromodynamics, color is confined so an arbitrary scale of momentum transfer squared $\mu^2$ must be introduced where $\alpha_s$ must be determined by experiment. The variation in the strong coupling constant with energy scale has been observed in a variety of measurements. For example, the ratio

$$R3\left(\sqrt{s}\right) = \frac{\sigma_{e^+e^-\to 3\text{ jets}}}{\sigma_{e^+e^-\to\text{hadrons}}} \tag{9.73}$$

is shown in Figure 9.10 as a function of center of mass energy $\sqrt{s} = E_{cm}$. To first approximation,

$$R3\left(\sqrt{s}\right) = \frac{\sigma_{e^+e^-\to q\bar{q}g}}{\sigma_{e^+e^-\to q\bar{q}}} \simeq N_c\alpha\left(\sqrt{s}\right) \tag{9.74}$$

where $N_c = 3$ is the number of colors. The remarkable fact is that the theoretical and experimental coupling decreases with energy implying it is small at high energy, despite the fact that quarks and hadrons interact strongly with effective coupling constants of order one at low energy. This behavior is called asymptotic freedom. The implication of course is that there is a scale at which the $\alpha_s \sim 1$ and this can explain color confinement. The specification of the choice of momentum scale when considering terms beyond leading order can be confusing. Consider the measurement of $\alpha_s$ from the ratio

$$R = \frac{\sigma_{e^+e^-\to\text{hadrons}}}{\sigma_{e^+e^-\to\mu^+\mu^-}} = 3\sum_q Q_q^2\left\{1 + \frac{\alpha_s}{\pi} + f(s)\left(\frac{\alpha_s}{\pi}\right)^2 \ldots\right\} \tag{9.75}$$

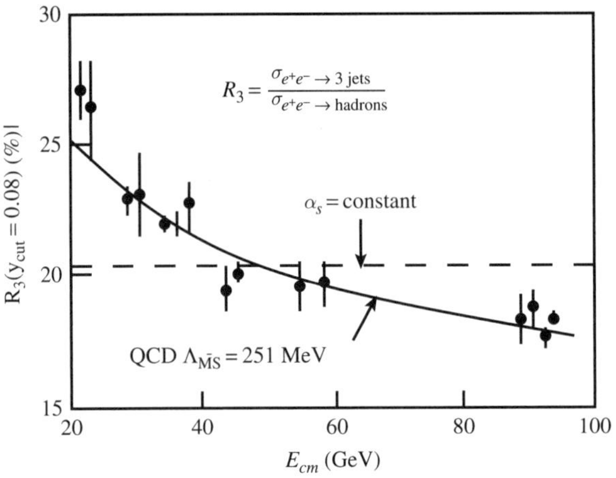


**FIGURE 9.10** The fraction of three jet events in hadronic events in $e^+e^-$ collisions. In so far as the total number of hadronic events is proportional to $\sigma_{e^+e^-\to q\bar{q}}$ and the number of three jet events is proportional to $\sigma_{e^+e^-\to q\bar{q}g}$, we expect $R3 = N_c\alpha_s$. The curve shows the predicted energy dependence. [S. Bethke, Prog. Part. Nucl. Phys. **58**, 351 (2007)]

When working to any finite order, the function $f$ and the $\alpha_s$ defined by the finite expansion depend (logarithmically) on another completely arbitrary scale introduced in renormalization so the $\alpha_s$ defined by, for example, the second order expansion depends upon the renormalization scale. This renormalization scale is not unique - it depends upon just how renormalization is carried out. It is customary to choose the renormalization scheme called the modified minimal subtraction scheme $\overline{MS}$ with its cut off parameter $\Lambda$ which minimizes the dependence on renormalization scheme of the $\alpha_s$ defined at a given order. Then $\mu^2$ is written as $\Lambda^2_{\overline{MS}}$. Having specified the renormalization scale, to compute $\alpha_s$ at some gluon $q^2$, we must choose a starting point such as $-q^2 = \sqrt{s} = m_Z$ to define $\alpha_s(M_z, \Lambda_{\overline{MS}})$. To leading order this $\Lambda_{\overline{MS}}$ is in fact the scale at which $\alpha_s = 1$. Determinations of $\alpha_s$ from a variety of experiments are shown in Figure 9.11.

The fermion loop vacuum polarization increases the coupling constant at large $q^2$ but the boson loop (and vertex) corrections have opposite sign and consequently $\alpha_s$ is predicted and observed to decrease at large momentum transfer. That the effective coupling constant is large in the static limit leading to color confinement, yet small at large momentum transfer permitting perturbative calculations was a major triumph of the gauge theory of color.

Quasi-classical arguments suggest an interpretation of the difference in the behavior of the chromodynamic and electrodynamic running coupling constants. As discussed in Chapter 7, virtual fermion pairs in either electrodynamics and chromodynamic screen the bare charge, behaving like a normal condensed matter medium with dielectric constant $\epsilon > 1$. In chromodynamics, the vector boson pair contribution is anti-screening and dominates, resulting in a dielectric constant $\epsilon < 1$. Relativistic invariance of the vacuum implies $\epsilon\mu = 1$ where the fermion

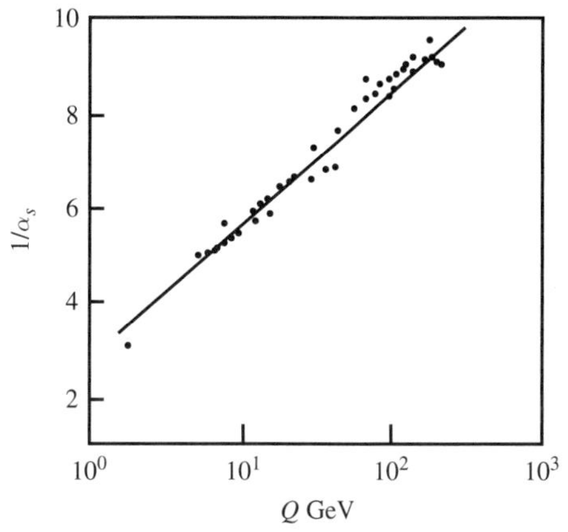


**FIGURE 9.11** Measurements of the strong coupling $\alpha_s^{-1}$(Q). The strong coupling constant is deduced from a variety of measurements including hadronic $\tau$-decay, quarkonium decay, and $e^+e^-$, $e^\pm p$ and $p\bar{p}$ collisions.The curve is the QCD prediction for the combined world average value at $Q = m_Z$. [Andreas S. Kronfeld, Chris Quigg, FERMILAB-PUB-10/040-T arXiv:1002.5032v2 [hep-ph] ]

pair contribution to the permeability is $\mu > 1$ while the vector boson contribution is $\mu < 1$. Calculation of the shift in the energy of the quantum vacuum confirms the results that the vacuum in spinor electrodynamics is diamagnetic while the spinor chromodynamic vacuum is paramagnetic. These calculations assume each fermion mode in a box has occupancy 1/2 and is subject to an external magnetic field (QED) or analog color potential (QCD). The gluons contribute color magnetic moments to the paramagnetism. But it is not just the vector versus spinor spin magnetism that leads to asymptotic freedom. Instead the result appears to derive principally from the exchange of conserved charges in nonabelian gauge interactions. Thus a colored quark can send its color off on a loop and be invisible to a probe of that color. When the nonlinear terms in the classical gauge field equations are interpreted as effective charge densities, the color cross product of a probe field with the electric field of a quark charge behaves an anti-screening charge density. Also, the end result is not specific to $SU(3)$ gauge theory. It is found that pure $SU(2)$ gauge theory is asymptotically free like $SU(3)_C$. For this reason, the $SU(2)_L$ coupling $g$ decreases with energy like the $SU(3)_C$ coupling. But unlike the case of $SU(3)_C$ found in nature, the scale at which the weak coupling $g^2/(4\pi)$ approaches unity is negative - the weak coupling is still weak at low energy.

The general expression for the running of a coupling constant $\alpha_i$ in standard model gauge theories is

$$\alpha_i(q^2) = \frac{\alpha_i(\mu^2)}{1 - \beta_i \ln \frac{q^2}{\mu^2}}. \tag{9.76}$$

The standard model parameter values at a scale equal to $m_Z$

$$\begin{aligned} m_Z &= 91.184 \pm 0.022 \\ \alpha_{\text{EM}}^{-1} &= 127.9 \pm 0.09 \\ \sin^2\theta_W &= 0.2315 \pm 0.0003 \\ \alpha_s &= 0.112 \pm 0.005 \end{aligned} \tag{9.77}$$

are related to the coupling constants for $U(1)_Y$, $SU(2)_L$, and $SU(3)_C$ by

$$\begin{aligned} \alpha_1 &= \frac{5}{3}\frac{\alpha_{\text{EM}}}{\cos^2\theta_W} \\ \alpha_2 &= \frac{\alpha_{\text{EM}}}{\sin^2\theta_W} \\ \alpha_3 &= \alpha_s. \end{aligned} \tag{9.78}$$

The Beta functions are usually expressed as $\beta_i = b_i/(2\pi)$ with

$$\begin{aligned} b_Y &= -\frac{20}{9}n_g = -\frac{5}{3}\cdot 4 \\ b_2 &= \frac{22}{3} - \frac{4}{3}n_g = \frac{13}{3} \simeq 4 \\ b_3 &= 11 - \frac{4}{3}n_g = 7 \end{aligned} \tag{9.79}$$

where $b_Y = (5/3)b_1$, $n_g$ is the number of generations of fermions. Here it is assumed the scale is much larger than any of the masses. For scales below the weak scale, only contributions from particles with masses $m < \mu$ should be included and Beta function should be expressed as a sum of a collection of step functions.

That $\alpha_{\text{EM}}$ increases while $\alpha_s$ decreases with energy implies there is an energy at which $\alpha = \alpha_s$, and it turns out this energy is not far on a logarithmic scale from the point where $\alpha_W = \alpha_s$. This energy turns out to be of order $10^{16-17}$ GeV. It is naive to suppose no particles or other physics cloud the extrapolation of the coupling constants from the weak scale to such an energy but the possibility of such coupling constant unification motivates so-called Grand Unified Theories (GUT) in which, at an extremely high energy, all particles and interactions are equivalent and the three coupling constants are related by symmetry. Presumably spontaneously symmetry breaking distinguishes subsets of interactions at low energy and, since new particles in such models can modify the evolution of the coupling constants, the coupling evolution from the GUT scale to the weak scale is not given by the standard model expectations. In Figure 9.12, the evolution of all three coupling constants is shown in a supersymmetric grand unified model with a particle spectrum tuned so that grand unification occurs, and in a model in which the number of particles grows exponentially at a much smaller scale due to excitations in extra dimensions.

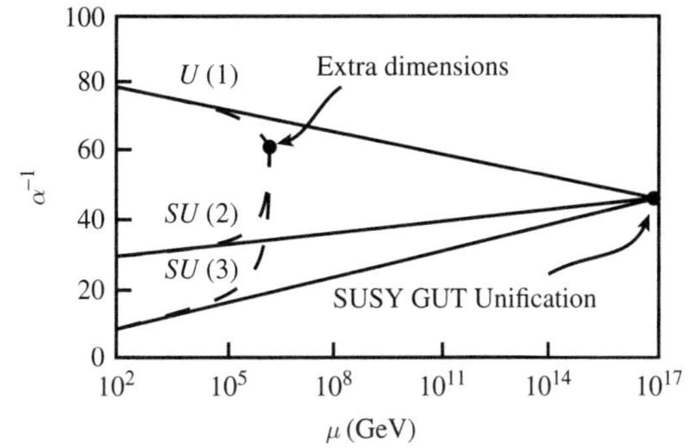


**FIGURE 9.12** Schematic evolution of coupling constants with energy. In a grand unified model, the observed gauge symmetries $U(1)_Y$, $SU(2)_L$ and $SU(3)_C$ emerge from the breakdown of an encompassing symmetry valid at a grand unification scale $\Lambda_{\rm GUT} \sim 10^{17}$ GeV. Also shown is the evolution in a scenario in which new degrees of freedom appear that significantly alter the evolution.

## 9.4 ■ APPLICATIONS OF HIGH ORDER CALCULATIONS

The computation of radiative corrections entails evaluation of the amplitudes associated with many Feynman diagrams. Such computations are highly technical and require the use of computer programs which can perform symbolic manipulations and reductions to core integrals. By way of illustration, Figure 9.13 shows the result of an analytic calculation including terms of order $\alpha_s^4$ of the static potential

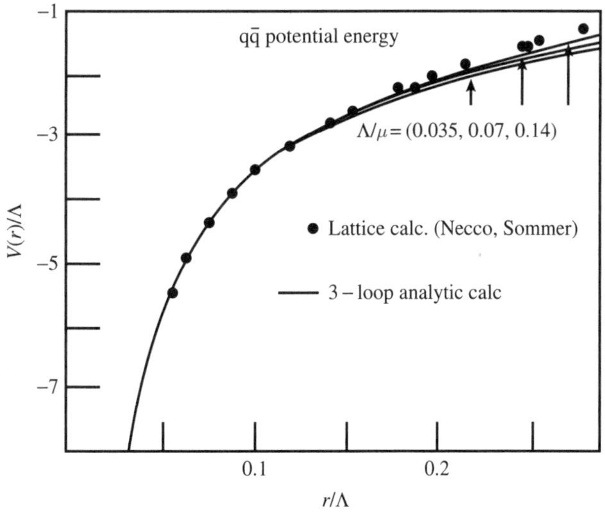


**FIGURE 9.13** Computed static color interaction potential at three-loop order compared with the lattice computations. The curves correspond to several choices of renormalization scale. The points represent the results of independent lattice QCD calculations. [C. Anzai, Y. Kiyo, and Y. Sumino, Phys. Rev. Lett. **104**, 112003 (2010)]

between a color and anti-color quark pair. The calculation compares well with that obtained by static numerical modeling of the color fields on a lattice. The potential is the Fourier transform of a momentum space amplitude corresponding to the summation of over 2000 diagrams.

Computations of QCD processes including radiative corrections are important in the interpretation of experimental data at present energies for which $\alpha_s \simeq 0.1 \sim 10\alpha_{\text{EM}}$. It is not unusual that the leading order cross section calculated for a production process qualitatively describes the shapes of particle distribution but radiative corrections alter the predicted overall rate by a so-called a K factor with a value between 1 and 2. In fact, the leading order cross section is sensitive to the choice of renormalization scale and higher order calculations are necessary to reduce the scale sensitivity. At the Large Hadron Collider, understanding and unraveling many important processes including Higgs production through gluon fusion require next-to-next-to-leading order (NNLO) calculations. Once an arcane art, computation of radiative correction is becoming an necessary industry. Figure 9.14 illustrates the difference between leading order (LO) and next-to-leading order calculations of a differential distributions of the photon-lepton separation and of jet transverse momentum in

$$pp \to W^{\pm}\gamma \text{ jet} + \text{X}$$

at the LHC. It is found that the NLO prediction is roughly a factor 1.4 larger than the LO prediction.

Perhaps the most striking application of higher order calculations has been to understanding the plethora of extremely precise measurements of electroweak processes. For example, a combined fit of one loop level calculations has been made of quantities related to weak boson exchange in neutrino scattering, the production and decay of a single $Z$ boson at LEP and the SLC, and of the production and decay of a $W$ boson at LEPII and the Tevatron. The rates, asymmetries, and polarizations at the one loop level are sensitive to the contributions of virtual Higgs bosons.

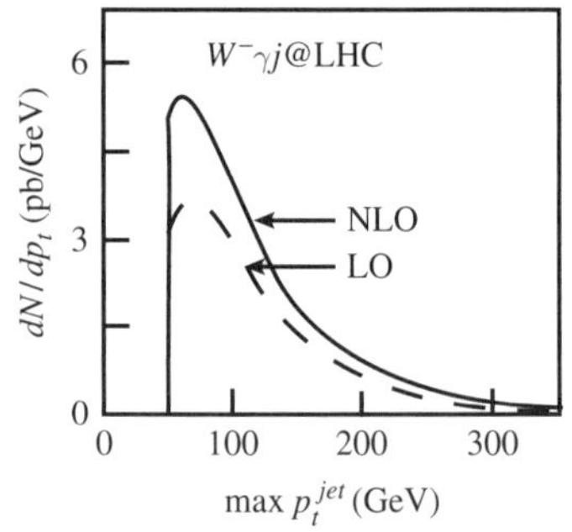


**FIGURE 9.14** Differential distribution of the photon-lepton separation in $pp \to W^{\pm}\gamma$ jet + X at the LHC in leading order (LO) and next-to-leading order (NLO) calculations. The two distributions differ roughly by a scale factor of 1.4. [D. Zeppenfeld *et al*, RADCOR 2009]

The output of the fit included the values $m_H = 139.1(+134.2\ -76.5)$ GeV and $\alpha_s = 0.1195 \pm 0.0030$, with $\chi^2$/nd.o.f. $= 15.1/14$. [V. A. Novikov, L. B. Okun, A. N. Rozanov and M. I. Vysotsky,"Theory of Z boson decays," Rep. Prog. Phys. **62**, 1275 (1999).] The value derived for the standard model Higgs mass is just above the limit from direct searches at LEP, albeit with an error permitting a mass in excess of 200 GeV.

As this result indicates, precision measurements and correspondingly precise high order calculations can sense the presence of virtual particles which may too massive to produce directly. Examples of historic proportion include the so-called GIM mechanism and the prediction of the existence of the $c$-quark based on calculation of the fourth order weak mixing amplitude in the neutral kaon system discussed in Chapter 10, and evidence for the $Z$ boson in the parity violating asymmetry in low energy $e^+e^-$ collisions discussed in the Chapter 8. An example is the attempt to deduce the mass of the standard model Higgs boson from the effect of virtual Higgs bosons on the ratio of the top quark mass to $W$ mass. As illustrated in Figure 9.15 from the ALEPH, CDF, D0, DELPHI, L3, OPAL, SLD Collaborations, the LEP Electroweak Working Group 1, the Tevatron Electroweak Working Group 2, and the SLD electroweak and heavy flavor groups, the standard model expectations for $m_W$ and $m_t$ depend on $m_H$. The 68% confidence level contour about the values of $m_t$ and $m_W$ measured directly at the Tevatron and LEPII favor a Higgs mass near the lower bound $m_H > 114$ GeV derived from direct searches at LEP and encourages experimenters to reduce the errors on $m_W$ and $m_t$.

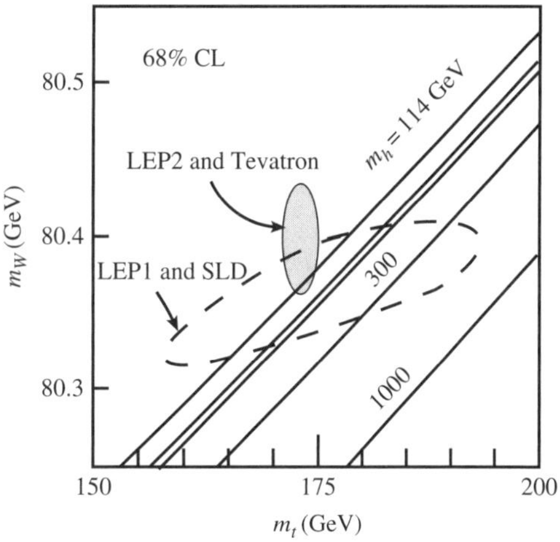


**FIGURE 9.15** Contour (68% confidence level) surrounding the mean values of $m_t$ and $m_W$ measured by LEP and Tevatron experiments. Also shown is the standard model relationship between $m_t$ and $m_W$ for various values of the Higgs mass. A search for associated production of $HZ$ at LEP places a lower bound $m_h > 114$ GeV. A search for $h \to W^+W^-$ at the Tevatron excludes a window near $m_H = 2m_W = 160$ GeV. Values derived from fits to older data are shown by the dashed contour. The measurements favor a value for $m_H$ near the lower limit from LEP. [arXiv:0911.2604v2 [hep-ex] ]

## 9.5 ■ FURTHER READING

The 1955 Nobel Prize in Physics 1955 was divided equally between Willis Eugene Lamb "for his discoveries concerning the fine structure of the hydrogen spectrum" and Polykarp Kusch "for his precision determination of the magnetic moment of the electron."

The 1999 Nobel Prize in physics was awarded to Gerardus 't Hooft, Martinus J.G. Veltman for their work establishing the renormalizability of electroweak gauge theory.

Renormalization is described in a number of field theory textbooks in which you will find references to original sources. Some of these are B. De Wit and J. Smith, Field *Theory in Particle Physics*, North-Holland (1986), F. Mandl and G. Shaw, *Quantum Field Theory*, John Wiley and Sons, (1986), T. Cheng and L. Li, *Gauge theory of elementary particle physics*, Clarendon Press (1984), T. D. Lee, *Particle Physics and Introduction to Field Theory*, Harwood Academic Publishers (1982), and Michio Kaku, *Quantum Field Theory*, Cambridge (1993). Additional field theory texts which discuss higher order effects include Thomas Banks, *Modern Quantum Field Theory: A Concise Introduction*, Cambridge University Press (2008). M. E. Peskin and D. V. Schroeder, *An Introduction to Quantum Field Theory*, Addison-Wesley (1995). Claude Itzykson and Jean-Bernard Zuber, *Quantum Field Theory*, McGraw-Hill (1980).

The presentation of renormalization here is based on Guang-jiong Ni, Guo-hong Yang, Rong-tang Fu, Haibin Wang, "Running Coupling Constants of Fermions with Masses in Quantum Electro Dynamics and Quantum Chromo Dynamics," arXiv:hep-ph/9906364v1.

Possible effects of extra dimensions on running coupling constants are described in Keith R. Dienes, Emilian Dudas, and Tony Gherghetta, Nucl. Phys. **B537**, 47 (1999), arXiv:hep-ph/9806292v2.

To imagine a universe in which the QCD confinement scale is astronomical, see H. Georgi, *Weak Interactions and Modern Particle Theory*, Benjamin/Cummings, Reading MA (1984), page 41.

Measurements of the electron and muon anomalous magnetic moments are B. Odom, D. Hanneke, B. D'Urso, and G. Gabrielse, Phys. Rev. Lett. **97**, 030801 (2006) and G. W. Bennett it et al., Phys. Rev. **D73**, 72003 (2006).

The leading-order vertex correction and the electron anomalous magnetic moment are described in C. M. Sommerfield, Phys. Rev. **107**, 327 (1957). The state of the art in calculating the anomalous magnetic moments of the electron and muon is "Tenth-order lepton g-2: Contribution from diagrams containing a sixth-order light-by-light-scattering subdiagram internally," T. Aoyama, K. Asano, M. Hayakawa, T. Kinoshita, M. Nio, N. Watanabe, RIKEN-TH 182, arXiv:1001.3704v2 [hep-ph].

Measurement of the Lamb shift is described in W. Lamb and C. Retherford, "Fine Structure of the Hydrogen Atom by a Microwave Method," Phys. Rev. **72**, 241 (1947). A review of relativistic corrections in atomic physics focused on muonic atoms is E. Borie and G. RInker, Re. Mod. Phys. **54**, 67 (1982).

Atomic physics including radiative, recoil, and nuclear corrections are reviewed in Peter J. Mohr, Barry N. Taylor, and David B. Newell, "CODATA recommended values of the fundamental physical constants," Rev. Mod. Phys. **80**, 633 (2008)]

Computing higher order so-called radiative corrections in the standard model is an enterprise entailing a deep understanding of renormalization and in some cases parameterizations of non-computable quantities from data. For state of the art results, consult conference proceedings such as RADCOR 2009 - 9th International Symposium on Radiative Corrections (Applications of Quantum Field Theory to Phenomenology), Proceedings of Science, `http://www.itp.uzh.ch/radcor2009/index.html`. For a review of the theory and applications of electroweak parameters in the context of higher order calculations, see for example V. A. Novikov, L. B. Okun, A. N. Rozanov and M. I. Vysotsky, "Theory of Z boson decays," Rep. Prog. Phys. **62**, 1275 (1999).

The computation of radiative corrections for a two particle to two particle process at the 1-loop level can entail the summation of amplitudes corresponding to 100 diagrams. Automation of the generation of diagrams and their symbolic reduction is achieved in comprehensive packages for perturbative calculations including FeynArts, CompHEP, and MadGraph. See the websites for these packages for additional information.

## 9.6 ■ PROBLEMS

### Problem 9.1. Corrected Coulomb potential

The radiatively corrected Coulomb potential is obtained from the Fourier transform to coordinate space of the radiatively corrected elastic scattering amplitude in momentum space. It has the form [see E. M. Lifshitz and L. P. Pitaevskii, *Relativistic Quantum Therory, Part 2*, Pergamon, 426 (1973)]

$$V(r) = \frac{e}{r}\left[1 + \frac{2\alpha}{3\pi}\int_1^\infty dx\, e^{-mrx}\left(1 + \frac{1}{2x^2}\frac{\sqrt{(x^2-1)}}{x^2}\right)\right].$$

The limiting forms are

$$V_{mr<<1} = \frac{e}{r}\left[1 + \frac{2\alpha}{3\pi}\left(\ln\frac{1}{mr} - C - \frac{5}{6}\right)\right]$$

$$V_{mr>>1} = \frac{e}{r}\left(1 + \frac{\alpha}{4\sqrt{\pi}}\frac{e^{-2mr}}{(mr)^{\frac{3}{2}}}\right)$$

where $C = 0.577\ldots$ is Euler's constant. a) What is the ratio $R$ of the effective charge at the Bohr radius of hydrogen $r = a_0$ to that for $r \to \infty$? b) For what value of $r$ is the effective charge doubled? What is the corresponding energy scale $E = \hbar c/r$ in GeV? How does this compare to the scale $10^{16-17}$ GeV associated with grand unification?

### Problem 9.2. Anomalous electron magnetic moment

Consider the amplitude for elastic scattering in an external field $e_1 A_{\text{ext}} \to e_2$ resulting from the vertex correction term proportional to $\sigma_{\mu\nu}$:

$$M_{fi}^{\mathbf{B}} = -\frac{e}{2m}\frac{\alpha}{2\pi}\bar{u}_2\sigma_{\mu\nu}u_1 k^\nu A_{\text{ext}}^\mu(k).$$

Assume a static magnetic field $A_{\text{ext}}(k) = (0, \mathbf{A}(k))$ and $k^\mu = (0, \mathbf{k})$ with $|\mathbf{k}| \to 0$ and show that

$$M_{fi}^{\mathbf{B}} = -\frac{e}{2m}\frac{\alpha}{2\pi}(\bar{u}_2\boldsymbol{\Sigma}u_1)\cdot\mathbf{B}$$

with $\mathbf{B} = i\mathbf{k}\times\mathbf{A}(k)$. Take the nonrelativistic limit with $u = \sqrt{2m}\begin{pmatrix}\eta\\0\end{pmatrix}$ and compare to the amplitude for scattering in a static electric field

$$M_{fi}^{\mathbf{E}} = -e\bar{u}_2\gamma_0 u_1 A^0$$

to deduce that the electron behaves as if it has a contribution to its magnetic moment given by

$$\delta\mu = \frac{e}{2m}\frac{\alpha}{2\pi}.$$

### Problem 9.3. Photon-photon scattering

The total cross section for light scattering by light as described in Figure 9.16 can be expressed in terms of the classical electron radius $r_2 = e^2/m_e$ for center of mass energy $\sqrt{s} << m_e$ as

$$\sigma_{\gamma\gamma\to\gamma\gamma}\left(\sqrt{s} << m_e\right) = \frac{973}{10125\pi}\alpha^2 r_e^2\left(\frac{\sqrt{s}}{2m_e}\right)^6.$$

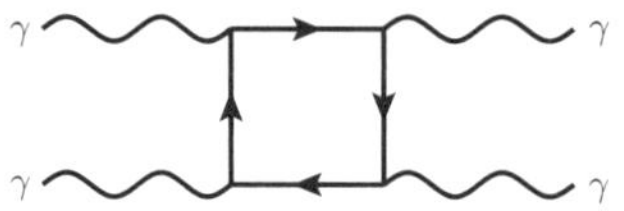


**FIGURE 9.16** Fourth order process responsible for photon-photon elastic scattering. Such a diagram corresponds to a nonlinearity in the classical field equations.

For $\sqrt{s} >> m_e$, the cross section is

$$\sigma_{\gamma\gamma\to\gamma\gamma}\left(\sqrt{s} >> m_e\right) = 18.8\frac{\alpha^4}{s}.$$

If two similar beams of *incoherent* photons of energy 1 eV and cross sectional area 1 $mm^2$ are collided over an interaction region one meter in length, what single beam power in watts corresponds to a scattering rate of 1 $s^{-1}$? This estimate neglects multiphoton processes. This process has not been observed but the related process of photon scattering by a Coulomb field via virtual electron-positron pairs (Delbruck scattering) is established. [Ref. R. Karplus and M. Neuman, Phys. Rev. **83**, 776 (1951), G. Jarlskog *et. al.*, Phys. Rev **D8**, 3813 (1973), A.I. Milstein 1, M. Schumacher Phys. Rep. **243**,183 (1994)].

---

**Problem 9.4. Lamb shift**

---

Compute the total energy shift $\Delta E =< \psi|V|\psi >$ for the 2s state of hydrogen in $\mu$eV using

$$\Delta V = \frac{4\alpha^2}{3m_e^2}\left(\ln\frac{m_e}{\omega(n,l)} + \frac{5}{6} - \frac{1}{5}\right)\delta(\mathbf{x}) + \frac{\alpha^2}{4\pi m_e^2 r^3}\boldsymbol{\sigma}\cdot\mathbf{L}$$

with $\omega_0 = 8.3\alpha^2 m_e$ and $|\psi(0)|^2 = 1/(\pi a_0^3 n^3)$. Compute the contribution in $\mu$eV of the vacuum polarization

$$\Delta V_{v.p.} = \frac{4\alpha^2}{3m_e^2}\left(-\frac{1}{5}\right)\delta(\mathbf{x}).$$

Compute the total and vacuum polarization shifts for muonic hydrogen assuming the same $\Delta V_{v.p.}$ resulting from $e^+e^-$ pairs, the other shifts and factors scaled to the muon mass.

---

**Problem 9.5. Electron self energy**

---

The electron self energy correction which dominates the Lamb shift in hydrogen may be estimated as resulting from a smearing of the electron position through interaction with fluctuations in the vacuum electromagnetic field. Translate the ground state energy of a field mode with frequency $\omega$ in a volume $V$ into a time averaged electric field strength

$$\frac{1}{V}\frac{1}{2}\omega = \frac{1}{2}\left(\overline{\mathbf{E}^2} + \overline{\mathbf{B}^2}\right) = \overline{\mathbf{E}^2}.$$

Show that the mean square displacement is

$$\overline{\mathbf{x}^2} = \frac{e^2}{m^2\omega^4}\overline{\mathbf{E}^2}.$$

Multiply by the density of states and integrate over frequency to find the incoherent mean square displacement

$$< \overline{\mathbf{x}^2} > = \frac{e^2}{2m^2\pi^2} \ln \frac{\omega_{max}}{\omega_{min}} = \frac{e^2}{2m^2\pi^2} \ln \frac{\omega_{max}}{\omega_{min}} \simeq \frac{2\alpha}{m^2\pi} \ln \frac{2}{Z^2\alpha^2}$$

where an integration cutoff $\omega_{min} \simeq \frac{1}{2}Z^2\alpha^2 m_e$ is chosen to reflect the fact that the perturbing fields are ineffective when $\omega$ is small compared to an excitation energy and a limit $\omega_{max} \simeq m$ results from the growth in the electron mass for relativistic motions. Compare $\sqrt{\overline{\mathbf{x}^2}}$ in hydrogen to the Bohr radius and to the proton radius. Now, write the potential energy as

$$U(\mathbf{r} + \mathbf{x}) = U(\mathbf{r}) + \mathbf{x} \cdot \nabla U(\mathbf{r}) + \frac{1}{2} x_i x_j \frac{\partial^2 U}{\partial r_i \partial r_j}$$

to find for a central potential

$$< \overline{\Delta U} > = \frac{1}{6} < \overline{\mathbf{x}^2} > \nabla^2 U.$$

Use the Coulomb potential with

$$\nabla^2 U = 4\pi Z\alpha\delta(\mathbf{r}), \ |\psi_{nl}(0)|^2 = Z^2/(\pi a_0^3 n^3)$$

to compute the average energy shift

$$\Delta E_{nl} = \frac{4}{3\pi n^3} Z^3\alpha^5 m \ln(\frac{2}{Z^2\alpha^2}).$$

---

**Muon $g - 2$ measurement**

---

The cyclotron frequency $\omega_c$ and the spin precession frequency $\omega_s$ for a muon moving in a circular orbit in the $x - y$ plane in a uniform magnetic field of strength $\mathbf{B} = B\hat{z}$ are

$$\omega_c = -\frac{eB}{m_\mu \gamma}, \ \omega_s = -g\frac{eB}{2m_\mu} + (\gamma - 1)\frac{eB}{\gamma m_\mu}$$

where $\gamma = (1 - v^2/c^2)^{-1/2}$. The rate of electrons from the spin orientation dependent decay $\mu^- \to \nu_\mu e^- \bar{\nu}_e$ that arrive at a fixed detector oscillates at the difference between these frequencies

$$\omega_a = \omega_s - \omega_c = -\left(\frac{g-2}{2}\right)\frac{eB}{m_\mu} = -a_\mu \frac{eB}{m_\mu}.$$

The difference frequency is used to measure $a_\mu$. If electrostatic quadrupoles are used to focus the beam in the $z$-direction, an electric field $\mathbf{E}$ perpendicular to $\mathbf{B}$ results in an altered frequency

$$\omega_a = -\frac{e}{m_\mu}\left[a_\mu \mathbf{B} - \left(a_\mu - \frac{1}{\gamma^2 - 1}\right)\frac{\mathbf{v}\times\mathbf{E}}{c^2}\right].$$

Show that a) the coefficient of the $\mathbf{v} \times \mathbf{E}$ term vanishes for $\gamma = 29.3$ corresponding to momentum $p = 3.094$ GeV c$^{-1}$, b) the time dilated muon lifetime for such a momentum is 64.4 $\mu$s, and c), for $B = 1.45$ T, the orbital radius for $B = 1.45$ T is 7.1 m and $\tau_a = 2\pi\,|\omega_a^{-1}| \simeq 4.365$ $\mu$s. (Ref: G.W. Bennett *et al*, Phys. Rev. **D73**, 072003 (2006), arXiv:hep-ex/0602035v1]

## Problem 9.6. Anti-screening in gauge theory

In $SU(N)$ gauge theory, the analog of Gauss's Law for the electric fields is

$$\nabla_i \mathbf{E}_i = g\mathbf{A}_i \times \mathbf{E}_i + g\boldsymbol{\rho}$$

where the index $i = 1 - 3$ refers to the spatial coordinate and is implicitly summed while boldface is used to denote the vectors of fields and charge densities. The first term can be considered an effective charge density. Consider $SU(2)$ with three types of charge and suppose a stationary point charge of type 1 at the origin interacting with a field $\mathbf{A}_i^3$ directed along the $z$-axis. Show that the effective charge

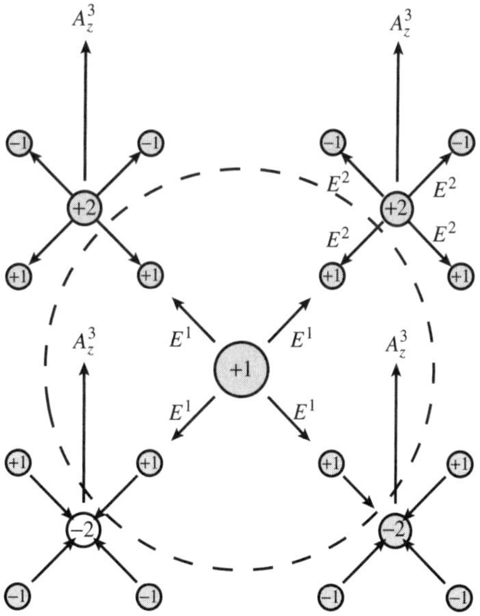


**FIGURE 9.17** Anti-screening in gauge theory.

density resulting from the combination of this field and the first approximation Coulomb field of the stationary charge is a charge distribution of type 2 and positive for positive $z$ and negative for negative $z$. Then show that the combination of the electric field of one of the positive type 2 charges with $\mathbf{A}_i^3$ yields a charge density of type 1 that is negative (due to the antisymmetry of the cross product) in the $+z$-direction from that charge and positive in the $-z$-direction. This results in an anti-screening of the original type 1 charge as show in Figure 9.17. [K. Gottfried and V. F. Weisskopf, *Concepts of Particle Physics*, Vol. II, Oxford (1986).]

---

**Problem 9.7. Unification in the MSSM**

---

The evolution of a coupling constant $\alpha_i \equiv g_i^2/4\pi$ is given by

$$\frac{d}{d\ln\mu}\alpha_i^{-1}(\mu) = -\frac{b_i}{2\pi}, \quad \alpha_i^{-1}(\mu) = \alpha_i^{-1}(m_Z) - \frac{b_i}{2\pi}\ln\frac{\mu}{m_Z}$$

for reference scale $m_Z$. In the MSSM model (Chapter 11) for $n_g$ generations of fermions and $n_H$ Higgs doublets,

$$(b_1, b_2, b_3) = \left(2n_g + \frac{3}{10}n_H, -6 + 2n_g + \frac{1}{2}n_H, -9 + 2N_g\right)$$

with $\alpha_1 = (5/3)\alpha_Y$, and $b_1 = (3/5)b_Y$. At $m_Z = 91.17$ GeV, experiment gives

$$\begin{aligned}\alpha_Y^{-1}(m_Z)|_{\bar{M}S} &= 98.29 \pm 0.13\\ \alpha_2^{-1}(m_Z)|_{\bar{M}S} &= 29.61 \pm 0.13\\ \alpha_3^{-1}(m_Z)|_{\bar{M}S} &= 8.5 \pm 0.5.\end{aligned}$$

Take $n_g = 3$ and $n_H = 2$ and show that

$$\alpha_1(m_{\rm GUT}) \simeq \alpha_2(m_{\rm GUT}) \simeq \alpha_3(m_{\rm GUT}) \simeq \frac{1}{24}$$

at $m_{\rm GUT} \simeq 2 \times 10^{16}$ GeV. [ Keith R. Dienes, Emilian Dudas, Tony Gherghetta, Nucl. Phys. **B537**, 47 (1999), arXiv:hep-ph/9806292v2, H. P. Nilles, Phys. Rep. **110**, 1 (1984)]

CHAPTER

# 10 Hadrons

Classification of hadrons by spin, parity, and flavor symmetry suggests the three color quark constituent model of $q\bar{q}$ mesons and $qqq$ baryons. QCD motivated models explain hadron properties and the scattering of their constituents. Weak interactions within hadrons account for curious flavor oscillations and symmetry violation effects. Much may be understood without detailed calculations.

## Contents



## 10.1 ■ OVERVIEW

This chapter examines the properties of hadrons in the context of the standard model. Hadrons are complex relativistic multiparticle systems. Despite their intrinsic complexity, much of their behavior can be understood through simple models and considerations of symmetry. We begin with a review of symmetries which govern the spectrum of light bound quark states and their decay modes. We then examine the naive quark model flavor wave functions for light hadrons and QCD motivated explanations of the spectrum of both light and heavy flavor mesons and baryons. This is followed by a description of the parton model used to describe the interactions of the constituents of hadrons at high energy. We next look at reflection and charge conjugation symmetry violation in weak interaction mixing of neutral hadrons which provide insight into the quark current mixing

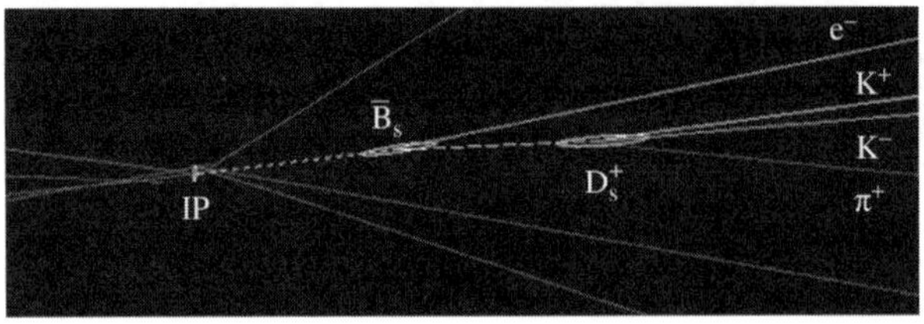


**FIGURE 10.1** Reconstruction of the decay of a $b\bar{s}$ meson $B_s$ produced in $e^+e^-$ collisions at LEP. The weak transition $b \to cW \to ce^+\nu_e$ after a 2 mm travel distance results in a $c\bar{s}$ meson $D_s^+$, which decays to $\pi^+K^+K^-$. [ALEPH collaboration, CERN photo archive, POW 980401]

described by the CKM matrix. The chapter ends with two phenomenological calculations which synthesize many of the ideas.

Symmetries and associated conservation laws are especially important in understanding the fundamental interactions of particle physics. The general connection between symmetries and conservation laws is described in Chapter 1. Quantum mechanics text books show in detail how invariance under translation in time and space and under rotation imply energy, linear momentum, and angular momentum conservation. The generators of the group of rotations of a scalar wave function are the angular momentum operators

$$\mathbf{L} = \mathbf{r} \times \mathbf{p} = -i\mathbf{r} \times \nabla \tag{10.1}$$

and eigenfunctions of $\mathbf{L}^2$ and $L_z$, known as the spherical harmonics $Y_{lm}$, naturally appear in the description of problems with spherical symmetry. The formal algebra of the quantum theory of rotations can be more generally applied to collections of component functions which describe particles with internal angular momentum.

Rotational symmetry implies the energy eigenstates of any closed system, no matter how complex, are multiplets of total angular momentum. The total internal angular momentum of a closed system in a energy eigenstate is called the spin. The spins of stable particles are often determined by their magnetic dipole interaction with macroscopic or atomic magnetic fields. Analysis of decays and other interactions must be used to determine the spin of unstable particles.

We will make extensive use of the classification of the space of products of wave functions which are eigenfunctions of angular momentum into subspaces of fixed total angular momentum, the so-called theory of the vector addition of angular momentum. The general theory shows that the states obtained from a direct product of states of angular momentum $j_1$ with states of angular momentum $j_2$ are states of angular momentum $j = |j_1 - j_2|$ through $j = j_1 + j_2$. For example, the four possible products of the basis states of two 2-component spinors can be combined linearly to form a rotational scalar of angular momentum 0 plus three linearly independent states which are transformed into one another under rotations and are a basis for states of angular momentum 1. (See Equation 1.91)

Reflection symmetry implies conservation of parity ($P$). Additional discrete symmetries in relativistic particle physics include interchange of particles and antiparticles called charge conjugation ($C$), symmetry under time reversal ($T$), and also flavor transformations. In the standard model, $P$, $C$ and $T$ are individually symmetries of electromagnetic and color interactions. Weak interactions violate all three but not the product $CPT$. Symmetry under interchange of two identical bosons and fermions requires their joint wave function be symmetric or antisymmetric.

## 10.2 ■ DISCRETE SYMMETRIES

Let us start our consideration of the consequences of symmetry under the time reversal transformation $T$. One important consequence of time reversal invariance is that a particle is prohibited from having an electric dipole moment **d**. In fact, since the spin operator **s** is the only vector associated with a particle, **d** must be proportional to **s**. But $T$ reverses **s** yet not an electric field **E** so an interaction

$$H = \mathbf{d} \cdot \mathbf{E} \tag{10.2}$$

is not $T$-invariant. Polar molecules invariably have degenerate ground states, induced not intrinsic dipole moments **d**, and $H$ is not $T$-odd (see problem). Sensitive searches for electric dipole moments of elementary particles have as yet to discover evidence for $T$-violation. The experimental limit on the electric dipole moment of the neutron $d_n < 10^{-25}$ $e$-cm corresponds to a separation of an electronic charge of about $10^{-12}$ of the neutron diameter.

Time reversal symmetry relates the magnitudes of transition amplitudes for forward and time reversed processes. We have

$$|M_{a\to b}| = |M_{-b\to -a}| \tag{10.3}$$

where $-a$ denotes the state obtained from a state $a$ by reversing all $T$-odd quantum numbers (momenta and spins). Symmetry under $PT$, which leaves momenta unchanged and reverses spins, implies equal transition rates for forward and reversed processes if all spins vanish or if spins are averaged in the initial and final states (detailed balance). The spin of the $\pi^+$ was first determined by comparing the reaction for proton fusion into a deuteron

$$pp \to \pi^+ d \tag{10.4}$$

to the reverse reaction at the same center-of-mass energy making use of detailed balance. The spin averaged cross section in the center of mass for two initial particles 1 and 2 leading to two final particles 3 and 4 is generally

$$\frac{d\sigma_{12\to 34}}{d\Omega} = \frac{2\pi}{(2s_1+1)(2s_2+1)} \sum_{s_1 s_2 s_3 s_4} |\, M_{s_1 s_2 s_3 s_4} \,|^2 \, \frac{p^2}{(2\pi)^3 v_i v_f} \tag{10.5}$$

for initial spin states denoted by $s_1$ and $s_2$ and final spins $s_3$ and $s_4$. For our forward reaction with two initial protons, $s_1 = s_2 = s_p = \frac{1}{2}$. For the reverse reaction with initial pion and deuteron, $s_1 = s_\pi$ and $s_2 = s_d = 1$, and the cross section must be divided by two to account for the fact that the two protons are indistinguishable. The ratio

$$d\sigma_{\pi^+ d\to pp}/d\sigma_{pp\to\pi^+ d} = \frac{3}{4}(2s_\pi + 1)(p_{\pi^+}/p_p)^2 \tag{10.6}$$

fits observation for charged pion spin zero.

It should also be noted that the resonance cross section $\sigma_{12\to c\to 34}$ for the production of a state $c$ in collisions of two particles 1 and 2 with decay to two particles 3 and 4 is proportional to $(2s_c + 1)$ and can be used to determine the spin $s_c$ of the intermediate state if the branching fractions are known. A classic example is the spin 3/2 nucleon excitation $\pi N \Leftrightarrow \Delta(1232)$. The spin of a particle may also be inferred from the angular distributions of its decay products. The directions of the decay products in $\pi^\pm \to l^\pm\nu$ and $\pi^0 \to \gamma\gamma$ always appear isotropic in the rest frame consistent with spin 0 and no preferred direction. Now consider the production of a resonance in a two particle to two particle reaction such as

$$\pi\pi \to \rho \to \pi\pi. \tag{10.7}$$

(A clashing pion beam machine has never been built but low momentum transfer reactions such as $\pi^- p \to \rho^- p \to \pi^-\pi^0 p$ have been successfully interpreted as collisions with virtual pions in a cloud about the nucleon.) The $\rho$ is a spin 1 vector object and the (scalar) production matrix element in the center of mass must be proportional to a vector amplitude (spin wave function) $\mathbf{e}_\alpha^\dagger$. The only other vector in the center of mass is the relative momentum of the pions

$$\mathbf{p} = \mathbf{p}_2 - \mathbf{p}_1 = 2\mathbf{p}_2 \tag{10.8}$$

so the matrix element for the production must the form

$$M = f\mathbf{e}_\alpha^\dagger \cdot \mathbf{p} \tag{10.9}$$

where $f$ is a constant. The decay matrix element has a similar form, so, summed over polarization states, the amplitude for the 2-to-2 reaction is

$$M \propto \sum_\alpha \mathbf{e}_\alpha^\dagger \cdot \mathbf{p}\mathbf{e}_\alpha \cdot \mathbf{p}' = \mathbf{p}\cdot\mathbf{p}' \propto \cos\theta \tag{10.10}$$

where $\theta$ is the scattering angle and we used the spin completeness relation

$$\sum_{\alpha=1,3} \mathbf{e}_\alpha^{i\dagger}\mathbf{e}_\alpha^j = \delta^{ij}. \tag{10.11}$$

In other words, the initial two pion state contains all values of $l$ but has no angular momentum along $\mathbf{p}$ so couples through a spin 1 state only to an outgoing wave $Y_{10} \propto \cos\theta$. The final state angular distribution is proportional to $\cos^2\theta$.

A general analysis of the consequences of rotational invariance for a reaction in which two particles scatter through an intermediate state is possible. The transition amplitude is expressed in the center of mass in terms of 2-body angular momentum eigenstates. The amplitudes and phases of resonant intermediate states may be extracted by fitting the final state angular distribution to the sum of resonant contributions as functions of center of mass energy.

Reflection symmetry implies energy eigenstates are also eigenstates of parity so particle states may be classified by parity. In particular, an energy eigenstate of a two particle closed system is also an angular momentum eigenstate so, in the simplest case, is described by an internal angular momentum wave function which is a spherical harmonic function $Y_{Lm}$ of the angles of the vector joining the two particles. Under reflection,

$$Y_{lm} \rightarrow (-)^l Y_{lm} \tag{10.12}$$

so the parity is correlated with the angular momentum.

Pions turn out to have negative intrinsic parity relative to nucleons. The experimental demonstration entails a clever application of quantum mechanics to pion nucleon interactions. A $\pi^-$ slowed by ionization energy loss in deuterium may be captured in an atomic state about a deuteron forming an unstable $\pi^- d$ atom. The strong process

$$\pi^- d \rightarrow nn \tag{10.13}$$

leads to the demise of the atom and formation of two neutrons. This process occurs on contact and is initiated only from a state with $l = 0$ with non-zero amplitude at zero separation. The total angular momentum before and after the reaction in this case must be $j = s_d = 1$. The final neutron pair spin state is symmetric or antisymmetric if its total spin $s$ is 1 or 0 respectively while the parity of the final neutron pair orbital wave function $Y_{lm}$ is $(-)^l$. The exclusion principle requires $(-1)^{l+s+1} = -1$. So $l + s$ is even and $j = 1$ permits $(l, s) = (0,1), (1,0), (1,1)$, or $(2,1)$. Of these possibilities, only $(l, s) = (1,1)$ has $l + s$ even, so the neutron pair appears only in a spin 1 $p$-wave state.

Parity conservation requires

$$P_\pi P_d = P_n^2(-1)^l = -1. \tag{10.14}$$

The deuteron is a spin 1 $s$-wave bound state of neutron plus proton and consequently has parity $P_d = P_p P_n$. The evidence for the structure of the deuteron wave function includes the spin and magnetic moment $\mu_d$ determined from the hyperfine structure in deuterium atoms. The magnetic moment is found to be $\mu_d \simeq \mu_p + \mu_n$ as expected for the spin 1 simple combination of the two constituent angular momenta. (An admixture of spin 1 $d$-wave is required to explain the electric quadrupole moment of the deuteron but this does not affect the parity assignment.) Given the deuteron parity, we deduce from parity conservation that

$$P_\pi P_n P_p = -1. \tag{10.15}$$

At least one of the parities must be odd. Assuming even parity for $p$ and $n$, $P_\pi$ is odd.

What about the neutral pion? It's lifetime ($\tau = 8.4 \times 10^{-17}$ s) is so short that it is almost unimaginable to produce and study subsequent strong interactions of the neutral pion. Negative parity of the $\pi^0$ was cleverly established by observation of anti-correlated photon polarizations in double Dalitz decay. While the $\pi^0$ decays most often to two photons, one or both may materialize as an electron-positron pair (Dalitz decay). As described in Chapter 6, due to the additional electromagnetic vertex in the diagram, the single Dalitz decay probability is suppressed relative to the two photon mode by a factor of roughly $\alpha$ - the branching fraction is 1.15%. Double Dalitz decay

$$\pi^0 \to \gamma\gamma \to (e^+e^-)(e^+e^-) \tag{10.16}$$

occurs with a branching ratio of $3.2 \times 10^{-5}$. The pair plane is correlated to the polarization of the (virtual) photon.

For initial pion amplitude $\psi_\pi$, photon polarization vectors $\mathbf{e}^1$ and $\mathbf{e}^2$, and center of mass relative photon momentum $\mathbf{k}$ with $\mathbf{e}^i \cdot \mathbf{k} = 0$, the decay amplitude symmetric in the two photons must be proportional to $\mathbf{e}^1 \cdot \mathbf{e}^2 \psi_\pi$ or $\mathbf{k} \cdot \mathbf{e}^1 \times \mathbf{e}^2 \psi_\pi$. Since $\mathbf{e}^1$, $\mathbf{e}^2$, and $\mathbf{k}$ are reversed under inversion, the first form favoring parallel polarization vectors applies to a scalar and the second to a pseudoscalar. Observations correspond to the pseudoscalar form and establish that the $\pi^0$ has negative parity. Perpendicular polarization in the analogous case of positronium was verified by Wu and Shaknov using the polarization dependence of Compton scattering to analyze the photon polarization correlation.

The quark model and parity conservation in quantum electrodynamics and quantum chromodynamics explain such results when a curious characteristic of fermions is recognized. Let's return to the process of the pion deuteron fusion process writing it as

$$\pi^-\tilde{d} \to nn \tag{10.17}$$

where $\tilde{d}$ denotes the deuteron. (Unfortunately, the standard symbols are D for the deuterium atom and $d$ for the deuteron which is the nucleus. Since $d$ refers to the $d$-quark in particle physics, in what follows we need a different symbol for the deuteron.) In the quark model, the $\pi^-$ is an $s$-wave spin 0 $\bar{u}d$ bound state while the proton and neutron contain $uud$ and $udd$ respectively with all quarks in $s$-wave relative to the center of mass. The strong interaction annihilation of the $\bar{u}$ in the $\pi^-$ with a $u$-quark in the proton (within the deuteron) permits the remaining remaining $d$-quark of the pion to attach to $u$-quark and $d$-quark forming a neutron. This neutron is then not bound to the original neutron in the deuteron and thus $\pi^-\tilde{d} \to nn$.

Now consider the parities of the players in this process. The quarks in the nucleons have no angular momentum so, if the intrinsic parities of the $u$-quark and $d$-quark are $P_u$ and $P_d$, we expect the nucleon, pion and deuteron parities to be

$$P_p = P_u^2 P_d \; ; \; P_n = P_d^2 P_u \; ; \; P_{\pi^-} = P_{\bar{u}} P_d \; ; \; P_{\tilde{d}} = P_p P_n = P_d P_u. \tag{10.18}$$

The condition for parity conservation can be expressed in terms of the quark parities as

$$P_{\pi^-} P_{\bar{d}} = P_{\bar{u}} P_d^2 P_u = P_{\bar{u}} P_u = -1 \tag{10.19}$$

and reduces to the constraint that the $u$-quark and $\bar{u}$-quark have opposite intrinsic parity. Remarkably, the Dirac equation requires opposite relative parity of spin 1/2 particle and antiparticle as discussed in Chapter 6 (see Equation 6.122). By contrast, the relative parity of a boson and its antiparticle is even. Parity conserving electromagnetic and color interactions conserve the number of quarks plus antiquarks $n_q + n_{\bar{q}}$ of each flavor so $P_q P_{\bar{q}}$ is measurable for any flavor only when $\Delta n_q$ for that flavor is odd. The relative parity of different flavors is not measurable.

By convention, parity $P = +1$ is assigned to $e^-$, $\mu^-$, $\tau^-$, $u$, $d$, $c$, $s$, $t$, and $b$ and P$= -1$ to their antiparticles. Ground state baryons have parity $P = +1$ while the corresponing anti-baryons have $P = -1$. A $q\bar{q}$ state with relative orbital angular momentum $l$ has parity

$$P_{\text{meson}} = P_q P_{\bar{q}} (-1)^l = (-1)^{l+1}. \tag{10.20}$$

The spin $\mathbf{j}$ of a $q\bar{q}$ meson is the vector sum of the total $q\bar{q}$ spin and relative orbital angular momentum so $q\bar{q}$ mesons follow the spin/parity sequence shown in Table 10.1, applicable also to positronium and protonium ($p\bar{p}$).

Charge conjugation ($C$) interchanges particles and antiparticles, reversing charge but not kinematic variables, and is important in relativistic process involving anti-matter. Charge conjugation symmetry requires equal probabilities for processes with particles replaced by antiparticles. If $C$ (or a product transformation such as $PT$) is a symmetry, a particle (system) is degenerate in mass with its charge conjugate, eg. $m_{\pi^-} = m_{\pi^+}$. A neutral system may be a $C$ eigenstate with $C = \pm 1$. Like $P$, $C$ is conserved in electromagnetic and color interactions but not weak interactions.

The electromagnetic field associated with sign reversed charges is reversed in sign so the photon is naturally defined as $C$-odd. Applied to a particle and antiparticle state of orbital angular momentum $l$, $C$ produces a charge parity factor $(-1)^l$ from the space wave function and from the spin wave function a factor $(-1)^{s+1}$ for spin 1/2 fermions (antisymmetric if $s = 0$, symmetric if $s = 1$) and $(-1)^s$ for bosons. Another factor $-1$ connected with Fermi statistics is required for fermions

| l | s | $j^P$ |
|---|---|---|
| 0 | 0 | $0^-$ |
| 0 | 1 | $1^-$ |
| 1 | 0 | $1^+$ |
| 1 | 1 | $0^+, 1^+, 2^+$ |

**TABLE 10.1** Spin and parity of fermion-antifermion bound states.

which may be motivated as follows. A particle-antiparticle pair may annihilate and reform so we may not follow the state of one constituent - the nonrelativistic description is somewhat inadequate. Instead, regard particle and antiparticle as different states of one particle distinguished by a charge wave function

$$\psi = \psi_{\text{space}} \psi_{\text{spin}} \psi_{\text{charge}}. \tag{10.21}$$

If we denote the different charge states by $+1$ and $-1$, the charge wave function of a pair may be expressed in terms of the symmetric and antisymmetric forms

$$\psi_{\text{charge}} = ((+1, -1) \pm (-1, +1))/\sqrt{2}. \tag{10.22}$$

If $C$ is a symmetry, then energy eigenstates are also eigenstates of $C$ and charge parity reflects the symmetry of the charge wave function

$$C\psi = \psi_{\text{space}} \psi_{\text{spin}} C\psi_{\text{charge}}. \tag{10.23}$$

Under interchange, require for bosons

$$P_{12}\psi = (-)^l \psi_{\text{space}} (-)^s C\psi_{\text{charge}} = \pm\psi \tag{10.24}$$

and for fermions

$$P_{12}\psi = (-)^l \psi_{\text{space}} (-)^{s+1} C\psi_{\text{charge}} = \pm\psi. \tag{10.25}$$

So the $C$ parity of a fermion plus anti-fermion bound state is

$$C f\bar{f} = (-1)^{l+s} f\bar{f}. \tag{10.26}$$

In quantum field theory, the relative charge parity of fermion and antifermion emerges from the anti-commutation relations between the corresponding creation and annihilation operators.

Charge conjugation symmetry together with parity conservation explains numerous facets of strong and electrodynamic phenomena which can not be calculated explicitly. For example, neutral pseudoscalar mesons are $C$-even while neutral vector mesons are $C$-odd. For $N$ photons, $C = (-)^N$, and $C = +1$ for any number of $\pi^0$ mesons. Consequently, the $C = +1$ states $\pi^0$ and $\eta$ may decay to two photons while the $C = -1$ states $\omega$, $\rho^0$, and $\phi$ may not. The minimum number of photons is three in neutral vector meson decay. Limits such as $B_{\pi^0 \to 3\gamma} < 3.1 \times 10^{-8}$ confirm $C$ conservation in electromagnetic interactions. Naively, this branching fraction would be expected to be or order $\alpha \sim 10^{-2}$.

As another example, consider a $\pi^+\pi^-$ pair with relative angular momentum $l$. The charge parity is $C = (-1)^l$ so the decay $\rho^0 \to \pi^+\pi^-$ is allowed since $l$ must be one to conserve angular momentum but the decay $\rho^0 \to \pi^0\pi^0$ is precluded by $C$ symmetry. The relative orbital angular momentum $l$ of two neutral pions must be even by Bose symmetry so the decay $\rho^0 \to \pi^0\pi^0$ is also precluded by interchange symmetry. Pseudoscalar mesons such as the $\eta$ and $\eta'$ could decay strongly to $\pi^+\pi^-$ with $l = 0$ but such decays would violate $P$ and decays to $\pi^0\pi^0$ with $l = 0$ are consistent with both $P$ and $C$ conservation but not with Bose symmetry. These particles decay electromagnetically (to two photons) and are long lived compared to naive expectation. Note that one may draw a Feynman diagram corresponding to a disallowed process such as $\pi^0 \to \gamma\gamma\gamma$. When symmetry precludes a decay, it implies complete cancellation in the coherent sum of amplitudes representing a transition from a particular initial state or a final state. That cancellation does not necessarily apply to momentum eigenstates.

We have seen that $C$ symmetry forbids the decay of a $C = -1$ vector particle such as a neutral vector meson from decaying to two photons. Two photon decays of vector particles are actually forbidden by other considerations. Two photon states with $j$ odd and $P = -1$ or $j = 1$ and $P = +1$ do not exist because the massless photon has only transverse helicity states. In fact, consider two photons with relative momentum $\mathbf{k}$ and polarization vectors $\mathbf{e}_1$ and $\mathbf{e}_2$ in their center of mass. A two photon wave function which transforms as a spin 1 object must be proportional to a combination of the following vectors:

$$\boldsymbol{\psi} \propto \mathbf{e}_1 \times \mathbf{e}_2,\ (\mathbf{e}_1 \cdot \mathbf{e}_2)\mathbf{k},\ (\mathbf{e}_1 \cdot \mathbf{k})\mathbf{e}_2,\ (\mathbf{e}_2 \cdot \mathbf{k})\mathbf{e}_1,\ \mathbf{k} \times (\mathbf{e}_1 \times \mathbf{e}_2). \tag{10.27}$$

The first two are antisymmetric under $\mathbf{e}_1 \Leftrightarrow \mathbf{e}_2$ and $\mathbf{k} \to -\mathbf{k}$ so inconsistent with Bose statistics while the others vanish by transversality: $\mathbf{e}_1 \cdot \mathbf{k} = \mathbf{e}_2 \cdot \mathbf{k} = 0$.

The quantum numbers $j^{PC} = 0^{-+}$ ($\pi^0, \eta...$ ) and $j^{PC} = 1^{--}$ ( $\rho^0, \omega...$ ) are well established and a profound test of the quark model. Some quantum numbers such as $j^{PC} = 1^{-+}$ correspond to no $q\bar{q}$ state and would indicate a non-standard form of matter, such as a hybrid meson composed of a normal meson bound together with a gluon. For details about the determinations of the quantum numbers of the many known hadrons and information about states which may not be classified as $q\bar{q}$ mesons and $qqq$ baryons, the reader is referred to the literature.

## 10.3 ■ STRONG INTERACTION ISOSPIN

The universality of the color interaction prescribed by gauge symmetry implies a global symmetry of strong interactions under transformation of quark flavors for fixed color, when mass differences and electroweak interactions are neglected. This global flavor symmetry may be represented by unitary transformations of a column vector $q$ of quark flavors of fixed color

$$q \to Uq\ ;\ \bar{q} \to U^*\bar{q} \tag{10.28}$$

with $U$ an element of the matrix group $SU(N)$ for $N$ flavors. This symmetry requires that hadronic matter appears in degenerate multiplets of states related by flavor transformation and it relates amplitudes for processes differing by flavor transformation. The mass generation mechanism and also the weak and electromagnetic interactions distinguish flavors and violate this symmetry. Only the color interaction portion of the total standard model Lagrangian is invariant. But, in so far as $m_u$ and $m_d$ are both negligible compared to the hadron mass scale and electroweak effects are often negligible, symmetry under mixing of up and down flavors, called $SU(2)_I$ strong isospin symmetry, is closely respected in strong interactions of hadrons. The strange quark mass is not negligible on the scale of the masses of light hadrons so up, down, and strange flavor symmetry $SU(3)_f$ is less well respected on this scale.

As we have seen in Chapter 7, a $SU(2)$ transformation may be represented as

$$U = e^{i\frac{\theta}{2}\mathbf{n}\cdot\boldsymbol{\sigma}} \tag{10.29}$$

where $\boldsymbol{\sigma}$ are the Pauli matrices. The fundamental multiplets are the $u$-quark and $d$-quark doublet and the $\bar{u}$-quark and $\bar{d}$-quark doublet which transform as

$$\begin{pmatrix} u \\ d \end{pmatrix} \rightarrow U \begin{pmatrix} u \\ d \end{pmatrix} \tag{10.30}$$

$$\begin{pmatrix} \bar{u} \\ \bar{d} \end{pmatrix} \rightarrow U^* \begin{pmatrix} \bar{u} \\ \bar{d} \end{pmatrix}. \tag{10.31}$$

Do not confuse this doublet and these transformations with the weak isospin doublets and the transformations $SU(2)_L$ which apply only to the left-handed components of the quark fields. The $SU(2)_I$ transformations we are considering apply only to the $u$-quark and $d$-quark flavors and equally to left-handed and right-handed field components.

Since $SU(2)$ transformations also represent spinor space rotations, $SU(2)_I$ isospin is formally similar to ordinary spin. Multiplets of isospin are analogous to angular momentum multiplets: the members of a multiplet are degenerate eigenstates of total isospin $\mathbf{I}^2$, distinguished by one component conventionally chosen to be $I_3$. Isospin quantum numbers and flavor wave functions of multi-quark states follow the usual rules for the vector addition of angular momentum.

To include the antiparticle doublet in constructing wave functions through vector addition, one can invoke the identity

$$i\sigma_2\boldsymbol{\sigma}^*(i\sigma_2)^{-1} = -\boldsymbol{\sigma} \tag{10.32}$$

which follows from the properties of the Pauli matrices to show that

$$i\sigma_2\begin{pmatrix} \bar{u} \\ \bar{d} \end{pmatrix} \rightarrow i\sigma_2 e^{+i\frac{\theta}{2}\mathbf{n}\cdot\boldsymbol{\sigma}^*}[i\sigma_2]^2\begin{pmatrix} \bar{u} \\ \bar{d} \end{pmatrix} = e^{+i\frac{\theta}{2}\mathbf{n}\cdot\boldsymbol{\sigma}} i\sigma_2\begin{pmatrix} \bar{u} \\ \bar{d} \end{pmatrix}. \tag{10.33}$$

Hence the quantity

$$i\sigma_2\begin{pmatrix} \bar{u} \\ \bar{d} \end{pmatrix} = \begin{pmatrix} 0 & 1 \\ -1 & 0 \end{pmatrix}\begin{pmatrix} \bar{u} \\ \bar{d} \end{pmatrix} = \begin{pmatrix} \bar{d} \\ -\bar{u} \end{pmatrix} \tag{10.34}$$

transforms like the iso-spinor $\begin{pmatrix} u \\ d \end{pmatrix}$ and may be combined with it in product representations. This convenience does not generalize to $SU(3)_I$ for which the transformations of the fundamental triplet are not mathematically equivalent to the transformations of its complex conjugate.

According to the rules of angular momentum addition, two spin 1/2 doublets combine to form a spin 1 triplet and spin 0 singlet. Thus, two quarks may form an isospin 1 triplet

$$uu, \frac{ud + du}{\sqrt{2}}, dd \tag{10.35}$$

and an isospin 0 singlet

$$\frac{ud - du}{\sqrt{2}} \tag{10.36}$$

In other words, the two quark wave function may be written as

$$\psi = \psi_{\text{space}} \psi_{\text{spin}} \psi_{flavor} \tag{10.37}$$

where, in the flavor wave function, the first flavor state in the product refers to particle 1 and the second to particle 2. Vector addition of a third quark doublet leads to two isospin 1/2 doublets and an isospin 3/2 quartet. These are the expected multiplets of three quark systems (baryons) made up of $u$-quarks and $d$-quarks. A quark plus antiquark form the triplet

$$u\bar{d}, \frac{d\bar{d} - u\bar{u}}{\sqrt{2}}, -d\bar{u} \tag{10.38}$$

and a singlet

$$\frac{1}{\sqrt{2}}(u\bar{u} + d\bar{d}). \tag{10.39}$$

These are the expected multiplets of quark plus antiquark systems (mesons) composed of $u$, $d$, $\bar{u}$, and $\bar{d}$.

In the quark model, the proton and neutron and their spin 1/2 excitations are identified with the iso-doublet combination of three quarks. The near equality of the proton and neutron masses is a consequence of the equivalence of $u$-quark and $d$-quark in the internal color dynamics. The three spin 0 pions $\pi^+, \pi^0, \pi^-$ ($m_\pi \simeq$ 135 MeV) and excitations such as the spin 1 $\rho^+, \rho^0, \rho^-$ ($m_\rho \simeq 770$ MeV) are $q\bar{q}$ flavor triplets. The corresponding singlets have different masses and behavior and may be (tentatively) identified with the pseudoscalar $\eta(547)$ and vector $\omega(782)$. Here symmetry offers an explanation for why one combination of $u\bar{u}$ and $d\bar{d}$, states which communicate through annihilation, remains degenerate with $u\bar{d}$ and $d\bar{u}$ states, while the other combination has a vastly different rest energy.

With the notation $(I, I_3)$ for isospin quantum numbers, we can write $p = (\frac{1}{2}, +\frac{1}{2})$, $n = (\frac{1}{2}, -\frac{1}{2})$ and $\pi^+ = (1, +1)$, $\pi^0 = (1, 0)$, $\pi^- = (1, -1)$. The $I_3$ assignment follows the empirical rule $I_3 = Q - B/2$ ($Q$ is the charge, $B$ the baryon number). This rule follows from the quark model with the assignment $(1/2, +1/2)$ to the $u$-quark and $(1/2, -1/2)$ to the $d$-quark.

Multi-nucleon states may also be classified by isospin and, in fact, the notion of isospin symmetry originated in its empirical application in nuclear physics. Two nucleon isospin states form a triplet and a singlet:

$$\begin{aligned}(1,+1) &= pp \\ (1,0) &= \frac{1}{\sqrt{2}}(pn+np) \\ (1,-1) &= nn \\ (0,0) &= \frac{1}{\sqrt{2}}(pn-np). \end{aligned} \qquad (10.40)$$

The flavor mixings are physical - $p$ and $n$ can change places by exchanging a quark or meson. This classification applies in particular to two nucleon bound states. The deuteron is an iso-singlet bound state with symmetric total spin $s = 1$ and symmetric orbital angular momentum $l = 0$ (and 2) orbital wave functions. The isospin wave function is antisymmetric in accordance with a generalized exclusion principle - in the isospin formalism, the nucleons are regarded as different flavor states of a single fundamental fermion. Iso-triplet states of two nucleons with $l = 0$ and $s = 0$ are consistent with the exclusion principle and predicted to be degenerate, but experimentally unbound - isospin symmetry does not require that states of different total isospin be similarly behaved. Tri-nucleon states have $I = 1/2$ or 3/2. There exist only two tri-nucleon bound states, $^3$H (tritium) and $^3$He, both with $J^P = 1/2^+$, and these constitute an iso-doublet. The $I = 3/2$ states are not bound. The only bound state of four nucleons is $^4$He, an iso-singlet. The degeneracy of nuclear levels related by isospin is described in texts on nuclear physics.

Isospin symmetry relates strong interaction amplitudes. Consider the proton deuteron fusion process

$$\mathrm{p} + \mathrm{d} \to {}^3\mathrm{He} + \pi^0 \qquad (10.41)$$

and the similar process

$$\mathrm{p} + \mathrm{d} \to {}^3\mathrm{H} + \pi^+. \qquad (10.42)$$

Since the deuteron isospin is $I_d = 0$, the initial state for either reaction has isospin quantum numbers $(I, I_3) = (\frac{1}{2}, +\frac{1}{2})$ and isospin conservation implies the final state must have the same quantum numbers. The results of the vector addition of normal angular momentum may be transcribed to express the final state as a superposition of isospin $I = 1$ and isospin $I = 0$ with appropriate Clebsch-Gordan coefficients as

$$\left(\frac{1}{2}, +\frac{1}{2}\right) = \sqrt{\frac{2}{3}}\left(1, -\frac{1}{2}\right) - \sqrt{\frac{1}{3}}\left(0, +\frac{1}{2}\right) = \sqrt{\frac{2}{3}}\pi^{+}\,{}^3\mathrm{H} - \sqrt{\frac{1}{3}}\pi^{0}\,{}^3\mathrm{He}. \qquad (10.43)$$

This expression implies that amplitude for the second reaction is $-\sqrt{2}$ times that for the first. At the same center of mass energy, the ratio of the cross sections is 1:2. The isospin is $(1, +1)$ in

$$p + p \Leftrightarrow d + \pi^+. \tag{10.44}$$

An isospin rotation of the flavor of both sides of this equation gives

$$\frac{1}{\sqrt{2}}(pn + np) \Leftrightarrow d + \pi^0 \tag{10.45}$$

so the ratio of the differential cross section for $pn \to d\pi^0$ to that for $pp \to d\pi^+$ at any angle is

$$\frac{d\sigma_{pn\to d\pi^0}}{d\sigma_{pp\to d\pi^+}} = \frac{1}{2}. \tag{10.46}$$

In $d + d \to {}^4\text{He} + \pi^0$, the initial state isospin is (0,0) as both deuterons have $I = 0$. The final state is composed of an iso-singlet and iso-triplet so has isospin (1,0). This reaction would violate isospin conservation and is not observed.

The isospin quantum numbers of light mesons were deduced from a variety of observations. The $\eta$ was discovered in

$$\pi^+ + d \to p + p + \eta \to p + p + (\pi^+\pi^-\pi^0) \tag{10.47}$$

as a peak near 550 MeV in the distribution of three pion mass $m_{\pi^+\pi^0\pi^-}$. Decays to neutral pions and to two photons were subsequently discovered. An analog charged $\eta$ is not observed implying that the isospin of the $\eta$ is $I_\eta = 0$. Isospin is respected in the strong production but may be violated in the electromagnetic decays. It is tempting to assign isospin $I = 3/2$ to the four spin 0 kaons all of which have similar mass. However, since the $\pi^-$ isospin is (1,-1), the proton isospin is $(1/2, 1/2)$, and the $\Lambda$ isospin is (0,0), the observed strong interaction

$$\pi^- + p \to K^0 + \Lambda \tag{10.48}$$

implies the $I_{K^0} = 3/2$ or 1/2 with $I_3 = -1/2$ while

$$\pi^- + n \to K^0 + K^- + n \tag{10.49}$$

implies the $K^-$ also has $I_3 = -1/2$. Hence the $K^0$ and $K^-$ can not be members of a quartet and must in fact belong to two strange iso-doublets with the quantum numbers:

$$K^+ = \left(\frac{1}{2}, +\frac{1}{2}\right);\ K^0 = \left(\frac{1}{2}, -\frac{1}{2}\right) \tag{10.50}$$

$$\bar{K}^0 = \left(\frac{1}{2}, +\frac{1}{2}\right);\ K^- = \left(\frac{1}{2}, -\frac{1}{2}\right). \tag{10.51}$$

The vector mesons $\rho^+$, $\rho^0$, $\rho^-$ and $\omega$ are all nearly degenerate and it is tempting to assign $I$ = 3/2 to these particles also. The $\rho^+$ is observed in

$$\pi^+ + p \to \rho^+ + p \to (\pi^+\pi^0) + p \tag{10.52}$$

as a peak in the distribution of the two pion invariant mass $m_{\pi^+\pi^0}$ at 780 MeV with a full width $\Gamma = 150$ MeV. The large width implies a strong interaction decay which is isospin conserving. For decays to two $I = 1$ pions, the possibilities are $I_\rho = 0, 1$, or 2. In

$$\pi^- + p \to \rho^- + p \to (\pi^-\pi^0) + p \tag{10.53}$$

a similar peak is observed. In

$$\pi^+ + p \to \pi^+ + \pi^+ + \pi^- + p \tag{10.54}$$

a peak is observed in $m_{\pi^+\pi^-}$ corresponding to the $\rho^0$ but no peak is observed in $m_{\pi^+\pi^+}$ - there is no $\rho^{++}$. Three distinct degenerate particles are found, the $\rho^+$, $\rho^0$, and $\rho^-$, and these are an iso-triplet like the pions. On the other hand, in the process

$$\pi^+ + p \to \pi^+ + p + (\pi^+\pi^-\pi^0), \tag{10.55}$$

a peak appears at $m_{\pi^+\pi^-\pi^0} = 783$ MeV with decay width $\Gamma$ only 9.8 MeV. The different mass, lifetime, and decay modes distinguish the iso-singlet $\omega$ from the $\rho$ states.

The $K^*(890)$ states are observed via their decays to $K\pi$. For example, in $K^-p \to K^-\pi^0 p$, there is a peak at $m_{K^-\pi^0} = 890$ MeV with $\Gamma = 50$ MeV. The strong decay implies $I_{K^*} = 3/2$ or 1/2. The iso-doublet pairs $K^{*+}$, $K^{*0}$ and $\bar{K}^{*0}$, $K^{*-}$ are excited pseudoscalar kaons. The Clebsch-Gordan decomposition

$$\left(\frac{1}{2}, +\frac{1}{2}\right) = \sqrt{\frac{2}{3}}\left(+1, -\frac{1}{2}\right) - \sqrt{\frac{1}{3}}\,(0, +1/2) = \sqrt{\frac{2}{3}}\pi^+K^+ - \sqrt{\frac{1}{3}}\pi^0\bar{K}^0 \tag{10.56}$$

implies the relative decay rates $B_{\bar{K}^{0*}\to\pi^+K^-}/B_{\bar{K}^{0*}\to\pi^0K^0} = 2$.

In the simplest cases, it is easy to understand isospin conservation in constituent terms. Consider strong decays of vector mesons. Suppose equal amplitude for formation of a $u\bar{u}$ or $d\bar{d}$ pair between $q$ and $\bar{q}$. Put $\pi^+ = u\bar{d}$, $\pi^0 = (-u\bar{u} + d\bar{d})/\sqrt{2}$, $\pi^- = -d\bar{u}$ and $\eta = (u\bar{u} + d\bar{d})/\sqrt{2}$ so $u\bar{u} = (\eta - \pi^0)/\sqrt{2}$ and $d\bar{d} = (\eta + \pi^0)/\sqrt{2}$, and put $K^+ = u\bar{s}$ and $K^0 = d\bar{s}$. Then the vector kaon decays are

$$K^+(890) = u\bar{s} \to u\bar{u}u\bar{s} + u\bar{d}d\bar{s} = \frac{1}{\sqrt{2}}\left(\eta - \pi^0\right)K^+ - \pi^+K^- \tag{10.57}$$

and

$$K^0(890) = d\bar{s} \to d\bar{u}u\bar{s} + d\bar{d}d\bar{s} = -\pi^-K^+ - \frac{1}{\sqrt{2}}\left(\eta + \pi^0\right)K^-. \tag{10.58}$$

The coefficients give the relative amplitudes. The decays of non-strange vector mesons may be similarly resolved. If the amplitudes for the transitions

$$u\bar{d} \to u\bar{u}u\bar{d} + u\bar{d}d\bar{d} \tag{10.59}$$

$$u\bar{u} \to u\bar{u}u\bar{u} + u\bar{d}d\bar{u} \tag{10.60}$$

and

$$d\bar{d} \to d\bar{u}u\bar{d} + d\bar{d}d\bar{d} \tag{10.61}$$

are equal, then

$$\begin{aligned} u\bar{d} &\to \frac{1}{\sqrt{2}}\left[\eta\pi^+ + \pi^+\eta + \pi^+\pi^0 - \pi^0\pi^+\right] \\ \frac{1}{\sqrt{2}}(-u\bar{u} + d\bar{d}) &\to \frac{1}{\sqrt{2}}\left[\eta\pi^0 + \pi^0\eta - \pi^+\pi^- - \pi^-\pi^+\right] \\ \frac{1}{\sqrt{2}}(u\bar{u} + d\bar{d}) &\to \frac{1}{\sqrt{2}}\left[\eta\eta + \pi^0\pi^0 + \pi^+\pi^- - \pi^-\pi^+\right]. \end{aligned} \tag{10.62}$$

If the states on the left are $\rho^+$, $\rho^0$, and $\omega$, then the $\rho^+$ decays to an equal superposition of two pion states and the following branching ratios are equal

$$B_{\rho^+ \to \pi^0\pi^+} = B_{\rho^0 \to \pi^+\pi^-} \tag{10.63}$$

while the amplitude for $\rho^0 \to \pi^0\pi^0$ vanishes. Bear in mind that the naive quark model of light hadrons is simplistic. But the symmetry applies generally to multi-particle states and processes so, in the absence of subtle cancellations, we can expect the naive model to give correct results.

## 10.4 ■ $SU(3)$ FLAVOR SYMMETRY

Were the masses of the $u$-quark, $d$-quark, and $s$-quark equal and electroweak interactions negligible, a three flavor symmetry $SU(3)_f$ would be valid in strong interactions. A multiplet structure associated with $SU(3)_f$ flavor symmetry was historically important in the development of the quark model of hadrons even though this symmetry is violated. The violation is now understood to result principally from the fact that $m_s$ is much larger than $m_u$ and $m_d$ and comparable to the mass scale of light hadrons.

Nine $q\bar{q}$ mesons may be formed with three flavors. The $SU(3)_f$ singlet is $\eta_0 = (u\bar{u} + d\bar{d} + s\bar{s})/\sqrt{3}$ and the remaining eight states are connected by $SU(3)_f$ transformations. These are an $SU(2)_I$ iso-singlet $\eta_8 = (u\bar{u} + d\bar{d} - 2s\bar{s})/\sqrt{6}$, an $SU(2)_I$ iso-triplet $(\pi^+, \pi^0, \pi^-)$, and a pair of $SU(2)_I$ iso-doublets, $(K^+ = u\bar{s}$, $K^0 = d\bar{s})$, and $(\bar{K}^0 = s\bar{d}$, $K^- = s\bar{u})$. It is expected that $SU(3)_f$-violating effects will not mix states of different total isospin but may still mix the iso-singlet states. Hence the flavor wave functions of physical candidates are written as

$$\eta(547) = \eta_8 \cos\theta - \eta_0 \sin\theta \tag{10.64}$$

$$\eta'(958) = \eta_8 \sin\theta + \eta_0 \cos\theta \tag{10.65}$$

with a similar expression for the vector mesons $\omega(782)$ and $\phi(1020)$.

The mixing angles are probed by the electromagnetic decays $\phi \to \eta\gamma$, $\eta' \to \rho^0\gamma$ and decays to two photons. In two photon decays of pseudoscalar mesons, each component of the flavor wave function should contribute an annihilation amplitude proportional to the square of the quark charge so:

$$\begin{aligned} e^2_{\pi^0\to\gamma\gamma} &= \frac{1}{\sqrt{3}}\left[e_u^2 - e_d^2\right] = \frac{1}{3\sqrt{2}} \\ e^2_{\eta_8\to\gamma\gamma} &= \frac{1}{\sqrt{6}}\left[e_u^2 + e_d^2 - 2e_s^2\right] = \frac{1}{3\sqrt{6}} \\ e^2_{\eta_0\to\gamma\gamma} &= \frac{1}{\sqrt{3}}\left[e_u^2 + e_d^2 + e_s^2\right] = \frac{2}{3\sqrt{3}}. \end{aligned} \tag{10.66}$$

If each decay width is proportional to the cube of the mass, $\Gamma_{P\to\gamma\gamma} \propto e^2 m_P^3$, the ratios of reduced rates are expected to be

$$\begin{aligned} 3\frac{\Gamma_{\eta\to\gamma\gamma}/m_\eta^3}{\Gamma_{\pi^0\to\gamma\gamma}/m_\pi^3} &= \left[\cos\theta - \sqrt{8}\sin\theta\right]^2 = 3.0 \pm 0.3 \\ \frac{3}{8}\frac{\Gamma_{\eta'\to\gamma\gamma}/m_{\eta'}^3}{\Gamma_{\pi^0\to\gamma\gamma}/m_\pi^3} &= \left[\cos\theta + \frac{1}{\sqrt{8}}\sin\theta\right]^2 = 0.61 \pm 0.5. \end{aligned} \tag{10.67}$$

The experimental results suggest $\theta \approx -18$ deg. The $m_P^3$ dependence of the width normally assumed follows from a Lorentz invariant interaction $L_{int} = ge^2 m_P \phi_P \mathbf{E}\cdot\mathbf{B}$ motivated by chiral field theory.

The vector mesons are usually assumed to be nearly

$$\phi = s\bar{s}\ ; \ \omega = \frac{1}{\sqrt{2}}(u\bar{u} + d\bar{d}) \tag{10.68}$$

since the natural width ratios

$$\Gamma_{\rho^0} : \Gamma_\omega : \Gamma_\phi = \left[\frac{1}{\sqrt{2}}(e_u - e_d)\right]^2 : \left[\frac{1}{\sqrt{2}}(e_u + e_d)\right]^2 : e_s^2 = 9:1:2 \tag{10.69}$$

are comparable to the experimental values $8.8 \pm 2.6 : 1 : 1.70 \pm 0.41$. The $\phi$ decomposes into $K^+K^-$ and $K^0\bar{K}^0$ with 88% probability by creation of a $u\bar{u}$ or $d\bar{d}$ which combine with the $s\bar{s}$. In the $\omega$, double pair creation and recombination leads to $\pi^+\pi^0\pi^-$ (88 %). The $q\bar{q}$ annihilation into pions presumably accounts for the remaining decays. The $\omega$ and $\rho^0$ are nearly degenerate so the mixing mechanism which splits the $\pi^0$ and $\eta$ is not operative. A plausible explanation is that two gluon color singlet intermediate states are possible for pseudoscalars but precluded for vectors by $C$-symmetry. (See Equation 7.100.) Decays of $\phi$ to pions must proceed through states containing a minimum of three gluons and such processes are suppressed relative to single gluon intermediate states by a factor of $\alpha_s^2$. This suppression is called the Zweig rule.

In writing meson flavor wave functions, we have assumed particle 1 is a quark and particle 2 is an antiquark. The possibility of annihilation suggests introduction of a charge wave function so

$$\psi = \psi_{\text{space}}\psi_{\text{spin}}\psi_{isospin}\psi_{\text{charge}} \tag{10.70}$$

where for a single particle $\psi_{\text{charge}} = \pm 1$. In this formalism, the iso-singlet state of two quarks is a "charge" +2 state with flavor wave function

$$\psi_{flavor} = \psi_{isospin}\psi_{\text{charge}} = \frac{1}{\sqrt{2}}(\uparrow\downarrow - \downarrow\uparrow)(+1, +1) = \frac{1}{\sqrt{2}}(ud - du). \tag{10.71}$$

Two iso-singlet states of $q\bar{q}$,

$$\begin{aligned}\psi_{flavor} &= \frac{1}{\sqrt{2}}(\uparrow\downarrow - \downarrow\uparrow)\frac{1}{\sqrt{2}}[(+1, -1) \pm (-1, +1)] \\ &= \frac{1}{2}(-u\bar{u} + d\bar{d}) \pm \frac{1}{2}(\bar{d}d - \bar{u}u),\end{aligned} \tag{10.72}$$

are distinguished by interchange symmetry. Similarly, states containing a $u$-quark and $\bar{d}$-quark may be formed in two ways

$$\psi_{flavor} = \frac{1}{\sqrt{2}}(u\bar{d} \pm \bar{d}u). \tag{10.73}$$

An $s$-wave, ordinary spin 0, isospin 1 state like a pion has a symmetric charge wave function. An $s$-wave, ordinary spin 0, isospin 0 state like the $\eta$ is charge antisymmetric. The charge conjugation operator $C$ applies to both the charge and isospin components of the flavor wave function, reversing both charge and $I_3$. Extended flavor wave functions of the light mesons are listed in the Table 10.2.

The combination of the isospin rotation $u \to d$, $d \to -u$ followed by $C$ is a strong interaction symmetry called $G$:

$$G : \begin{pmatrix} u \\ d \end{pmatrix} \to \begin{pmatrix} \bar{d} \\ -\bar{u} \end{pmatrix}. \tag{10.74}$$

Systems of equal numbers of $u$-quarks and $d$-quarks and antiquarks may be eigenstates with $G$-parity even or odd. The $\pi$, $\omega$, and $\phi$ are $G$-odd while the $\eta$, $\eta'$, and $\rho$ are $G$-even. Conservation of $G$ parity allows the decay $\rho \to 2\pi$ but forbids the decay $\rho \to \eta\pi$ and requires the $\omega$ to decay to a minimum of three pions. The $\omega$ decay to $3\pi^0$ or $\eta\pi^0$ is forbidden by $C$. The phase space suppression of the $\pi^+\pi^0\pi^-$ mode explains the narrow width of the $\omega$ relative to the $\rho$. The $\eta$ must decay to an even number of pions. However, two pion decays would violate $P$ while kinematics almost excludes four pion decay so the $\eta$ is stable with respect to strong interactions and this is why it is observed to decay by the relatively slow electromagnetic process to $\eta \to \gamma\gamma$ and to have a narrow width. The relatively

| | Wave function | $j^{PC}$ | $I$ | $I_3$ | $G$ |
|---|---|---|---|---|---|
| $\pi^+$ | $\frac{1}{\sqrt{2}}(u\bar{d} + \bar{d}u)$ | $0^{-+}$ | 1 | +1 | −1 |
| $\pi^0$ | $\frac{1}{2}(-u\bar{u} - \bar{u}u + d\bar{d} + \bar{d}d)$ | $0^{-+}$ | 1 | 0 | −1 |
| $\pi^-$ | $\frac{1}{\sqrt{2}}(-d\bar{u} - \bar{u}d)$ | $0^{-+}$ | 1 | −1 | −1 |
| $K^+$ | $\frac{1}{\sqrt{2}}(u\bar{s} + \bar{s}u)$ | $0^{-+}$ | 1/2 | +1/2 | |
| $K^0$ | $\frac{1}{\sqrt{2}}(d\bar{s} + \bar{s}d)$ | $0^{-+}$ | 1/2 | −1/2 | |
| $\bar{K}^0$ | $\frac{1}{\sqrt{2}}(s\bar{d} + \bar{d}s)$ | $0^{-+}$ | 1/2 | +1/2 | |
| $K^-$ | $-\frac{1}{\sqrt{2}}(s\bar{u} + \bar{u}s)$ | $0^{-+}$ | 1/2 | −1/2 | |
| $\eta_8$ | $\frac{1}{6}(-u\bar{u} - \bar{u}u - \bar{d}d - d\bar{d} + 2s\bar{s} + 2\bar{s}s)$ | $0^{-+}$ | 0 | 0 | +1 |
| $\eta_0$ | $\frac{1}{3}(-u\bar{u} - \bar{u}u - \bar{d}d - d\bar{d} - s\bar{s} - \bar{s}s)$ | $0^{-+}$ | 0 | 0 | +1 |
| $\rho^+$ | $\frac{1}{\sqrt{2}}(u\bar{d} - \bar{d}u)$ | $1^{-}$ | 1 | +1 | +1 |
| $\rho^0$ | $\frac{1}{2}(-u\bar{u} + \bar{u}u + d\bar{d} - \bar{d}d)$ | $1^{--}$ | 1 | 0 | +1 |
| $\rho^-$ | $\frac{1}{\sqrt{2}}(-d\bar{u} + \bar{u}d)$ | $1^{-}$ | 1 | −1 | +1 |
| $K^{+*}$ | $\frac{1}{\sqrt{2}}(u\bar{s} + \bar{s}u)$ | $1^{-}$ | 1/2 | +1/2 | |
| $K^{0*}$ | $\frac{1}{\sqrt{2}}(d\bar{s} + \bar{s}d)$ | $1^{--}$ | 1/2 | −1/2 | |
| $\bar{K}^{0*}$ | $\frac{1}{\sqrt{2}}(s\bar{d} + \bar{d}s)$ | $1^{--}$ | 1/2 | +1/2 | |
| $K^{-*}$ | $-\frac{1}{\sqrt{2}}(s\bar{u} + \bar{u}s)$ | $1^{-}$ | 1/2 | −1/2 | |
| $\omega$ | $\frac{1}{2}(-u\bar{u} + \bar{u}u + \bar{d}d - d\bar{d})$ | $1^{--}$ | 0 | 0 | −1 |
| $\phi$ | $\frac{1}{\sqrt{2}}(s\bar{s} + \bar{s}s)$ | $1^{--}$ | 0 | 0 | −1 |

**TABLE 10.2** Extended meson wave functions.

slow decays to $\pi^+\pi^-\pi^0$ and $3\pi^0$ are $C$-conserving but $G$-violating. Similarly, the $\omega$ decays relatively slowly to $\pi^+\pi^-$ ($B_{\omega\to\pi^+\pi^-} = 1.4\%$).

If we denote the three color states by $r$, $b$, and $g$, the antisymmetric color wave function common to all baryons is

$$\psi^{qqq}_{color} = \frac{1}{\sqrt{6}}\{|\, rgb > - \,|\, rbg > + \,|\, gbr > - \,|\, grb > + \,|\, brg > - \,|\, bgr >\}. \tag{10.75}$$

For mesons, the color wave function is

$$\psi^{q\bar{q}}_{color} = \frac{1}{\sqrt{6}}\left\{\left(|\, r\bar{r} > + \,|\, g\bar{g} > + \,|\, b\bar{b} >\right) \pm \left(|\, \bar{r}r > + \,|\, \bar{g}g > + \,|\, \bar{b}b >\right)\right\} \tag{10.76}$$

where the plus and minus sign is chosen so that the extended wave functions of the mesons are fully antisymmetric. Charge conjugation $C$ then applies to the entire internal wave function $\psi_{flavor}\psi_{color}\psi_{\text{charge}}$.

Eight spin 1/2 baryons and ten somewhat heavier spin 3/2 baryons composed of $u$-quarks, $d$-quarks, and $s$-quarks are observed and represent multiplets of $SU(3)_f$ striated in $SU(2)_I$ flavor multiplets. The masses indicate the number of strange quarks and the isospin and electric charge fix the $u$-quark and $d$-quark content. The baryon octet comprises the following isospin multiplets:

$$\Xi^-_{ssd}(1321) \quad \Xi^0_{ssu}(1315) \tag{10.77}$$

$$\Sigma^-_{sdd}(1197) \;\; \Sigma^0_{sud}(1192) \;\; \Sigma^+_{suu}(1189) \tag{10.78}$$

$$\Lambda_{sud}(1115) \tag{10.79}$$

$$n_{udd}(939) \quad p_{uud}(938). \tag{10.80}$$

The quark content is shown in the subscript, the mass in MeV is given in parenthesis, and isospin $I_3$ increases to the right. The baryon decuplet multiplets are

$$\Omega^-_{sss}(1672) \tag{10.81}$$

$$\Xi^-_{ssd}(1535) \quad \Xi^0_{ssu}(1532) \tag{10.82}$$

$$\Sigma^-_{sdd}(1385) \;\; \Sigma^0_{sud}(1384) \;\; \Sigma^+_{suu}(1383) \tag{10.83}$$

$$\Delta^-_{ddd}(1232) \;\; \Delta^0_{udd}(1232) \;\; \Delta^+_{uud}(1232) \;\; \Delta^{++}_{uuu}(1232). \tag{10.84}$$

The spin 1/2 baryons decay weakly with lifetimes of order $10^{-10}$ s. Principle decays include the pion emission processes in which one $s$-quark converts to $u$-quark: $\Xi^- \to \Lambda + \pi^-$, $\Xi^0 \to \Lambda + \pi^0$, $\Sigma^- \to n + \pi^-$, $\Sigma^+ \to n + \pi^+$, $p + \pi^0$, and $\Lambda \to p + \pi^-$. An exception is the electromagnetic decay $\Sigma^0 \to \Lambda + \gamma$ ($\tau = 7 \times 10^{-20}$ s) permitted by the fact that the $\Lambda$ and $\Sigma^0$ have common constituents. The spin 3/2 baryons decay strongly by pion emission to the octet states conserving strangeness. The resonance widths of these states are in the MeV range, indicating the strong origin of such processes. Examples are

$$\Xi(1530) \to \Xi(1315) + \pi \;\; (\Gamma = 10 \text{ MeV}) \tag{10.85}$$

$$\Sigma(1385) \to \Sigma(1190) + \pi, \Lambda + \pi \;\; (\Gamma \simeq 35 \text{ MeV}) \tag{10.86}$$

and

$$\Delta \to N + \pi \;\; (\Gamma = 120 \text{ MeV}). \tag{10.87}$$

The $\Omega^-$ however has no analog octet $sss$ state into which it can decay strongly by pion emission and strong decay through kaon emission to $\Xi^0(1315) + K^-(493)$ is excluded by energy conservation. Hence, the $\Omega^-$ decays weakly:

$$\Omega^- \to \Lambda + K^-, \Xi^0 + \pi^0 \;\; (\tau = 0.8 \times 10^{-10} \text{ s}). \tag{10.88}$$

The lowest mass tri-quark states ought to be spherically symmetric with vanishing relative angular momentum of each quark pair. The spins may add to total angular momentum $j = 1/2$ or 3/2. A decuplet baryon state with $j_z = +3/2$ is, for example,

$$\Omega^- = s^\uparrow s^\uparrow s^\uparrow. \tag{10.89}$$

The analogous flavor-spin wave function for the $\Delta^+$ is

$$\Delta^+ = u^\uparrow u^\uparrow d^\uparrow. \tag{10.90}$$

Lowering each spin in turn by application of the angular momentum lowering operator $j_1^- + j_2^- + j_3^-$ gives the $j_z = +1/2$ state

$$\begin{aligned}\Delta^+ &= \frac{1}{\sqrt{3}}\left(u^\downarrow u^\uparrow d^\uparrow + u^\uparrow u^\downarrow d^\uparrow + u^\uparrow u^\uparrow d^\downarrow\right)\\ &= \sqrt{\frac{2}{3}}\frac{\left[u^\downarrow u^\uparrow + u^\uparrow u^\downarrow\right]}{\sqrt{2}} d^\uparrow + \sqrt{\frac{1}{3}} u^\uparrow u^\uparrow d^\downarrow.\end{aligned} \tag{10.91}$$

The change $\sqrt{\frac{2}{3}} \to -\sqrt{\frac{1}{3}}$, $\sqrt{\frac{1}{3}} \to \sqrt{\frac{2}{3}}$ produces the orthogonal $j_z = \frac{1}{2}$ state

$$p = \frac{1}{\sqrt{6}}\left(2u^\uparrow u^\uparrow d^\downarrow - (u^\uparrow u^\downarrow + u^\downarrow u^\uparrow) d^\uparrow\right). \tag{10.92}$$

The spin-flavor wave functions of other baryons on the rim of the octet are similar, e.g.

$$\Sigma^- = \frac{1}{\sqrt{6}}\left(2d^\uparrow d^\uparrow s^\downarrow - (d^\uparrow d^\downarrow + d^\downarrow d^\uparrow) s^\uparrow\right). \tag{10.93}$$

The wave function of the $\Sigma^+$ follows from the replacement $dd \to uu$ and that of the $\Sigma^0$ from the replacement $dd \to ud$:

$$\Sigma^0 = \frac{1}{\sqrt{6}}\left(2u^\uparrow d^\uparrow s^\downarrow - (u^\uparrow d^\downarrow + u^\downarrow d^\uparrow) s^\uparrow\right). \tag{10.94}$$

The wave function of the $\Lambda$ is the iso-singlet orthogonal to the wave function of the $\Sigma^0$:

$$\Lambda = \frac{1}{\sqrt{2}}\left(u^\uparrow d^\downarrow - u^\downarrow d^\uparrow\right) s^\uparrow. \tag{10.95}$$

In baryons of the form $aab$, the two $a$-quarks are in a symmetric spin 1 state so the addition of the spin 1/2 quark $b$ can produce either spin 1/2 or spin 3/2. For three different flavors $uds$, the $u$-quark and $d$-quark pair may be spin 0 or spin 1. The spin 0 combination is an iso-singlet and the addition of the $s$ gives the spin

1/2 $\Lambda$. The spin 1 combination is an iso-triplet member and the addition of the $s$ gives spin 1/2 and spin 3/2 neutral iso-triplet members.

The preceding construction treats the quarks as distinguishable particles. To correct this oversight, start by noticing that the state $\Omega^- = s^\uparrow s^\uparrow s^\uparrow$ with all quarks in the same $s$-wave state about the center of mass is completely symmetric under interchange of any pair of quarks. To be consistent with the exclusion principle for spin 1/2 fermions, the wave function must be multiplied by another internal wave function completely antisymmetric in some other internal variable - color. If it is boldly assumed that all octet and decuplet baryons are antisymmetric in color, in a sense color neutral, and all have symmetric space wave functions, then the spin-flavor wave functions must all be completely symmetric when the quarks are treated as indistinguishable. It is precisely this assumption that appears justified by quantum chromodynamics. As described in the Chapter 7, color singlet states in a leading order calculation have minimal energy and running of the color coupling constant suggests that color is confined by non-perturbative chromodynamics.

To construct symmetric spin flavor wave functions for three quarks each in three flavors, we first consider the spin and flavor separately. Spin 3/2 wave functions for three quarks are obtained by applying the symmetrical lowering operator $j^- = j_1^- + j_2^- + j_3^-$ to the symmetric state $\uparrow\uparrow\uparrow$ and are completely symmetric under interchange of any pair. Symmetric spin 1/2 wave functions are not so simple. These may be separated into $SU(2)_{\text{spin}}$ multiplets of mixed symmetry which are antisymmetric and symmetric in two particles and result from the vector addition of two spin 1/2 to form spin 1 and spin 0 followed by the vector addition of the third spin to the spin 0 to form spin 1/2 or to the spin 1 to form spin 1/2 and spin 1. This algebra and the symmetry of such wave functions is summarized as follows:

$$\mathbf{2} \otimes \mathbf{2} \otimes \mathbf{2} = \mathbf{2}_{M,A} \oplus \mathbf{2}_{M,S} \oplus \mathbf{4}_S. \tag{10.96}$$

Possible flavor states of three quarks in three flavors may be decomposed into representations of $SU(3)_f$ flavor transformations as an antisymmetric singlet, two octets of mixed symmetry, and a decuplet of completely symmetric flavor states:

$$\mathbf{3} \otimes \mathbf{3} \otimes \mathbf{3} = \mathbf{1}_A \oplus \mathbf{8}_{M,S} \oplus \mathbf{8}_{M,A} \oplus \mathbf{10}_S. \tag{10.97}$$

These multiplets result from the analogous vector addition of two triplets to form an sextet plus an anti-triplet, followed by the vector addition of a third triplet to the sextet to form an octet plus a decuplet and the addition of the triplet to the anti-triplet to form a singlet and another octet. Exactly this algebra appears in our discussion of the transformation properties of multi-gluon states in the context of color symmetry. These representations may now be combined to discover completely symmetric spin-flavor states.

Of the $8 \times 27$ spin-flavor states, a fraction are symmetric and exactly what is required to outfit the ground state baryon wave functions. The decuplet is the symmetric $\mathbf{4}_S$ of spin and symmetric $\mathbf{10}_S$ of flavor. Octet wave functions which are similarly completely symmetric in spin-flavor turn out to be combinations of the form

$$\psi = \frac{1}{\sqrt{2}}[\mathbf{2}_{M,S}\mathbf{8}_{M,S} - \mathbf{2}_{M,A}\mathbf{8}_{M,A}]. \tag{10.98}$$

For example, the spin up state of a proton is

$$\begin{aligned} p^{\uparrow} &= \frac{u(du+ud)-2duu}{\sqrt{6}} \times \frac{\uparrow(\downarrow\uparrow+\uparrow\downarrow)-2\downarrow\uparrow\uparrow}{\sqrt{6}} \\ &\quad -\frac{u(ud-du)}{\sqrt{2}} \times \frac{\uparrow(\downarrow\uparrow-\uparrow\downarrow)}{\sqrt{2}} \\ &= \frac{1}{\sqrt{18}}\begin{pmatrix} udu(2\uparrow\downarrow\uparrow-\uparrow\uparrow\downarrow-\downarrow\uparrow\uparrow) \\ +uud(-\uparrow\downarrow\uparrow+2\uparrow\uparrow\downarrow-\downarrow\uparrow\uparrow) \\ +duu(-\uparrow\downarrow\uparrow-\uparrow\uparrow\downarrow+2\downarrow\uparrow\uparrow) \end{pmatrix}. \end{aligned} \tag{10.99}$$

The wave functions of other perimeter baryons such as the neutron follow from flavor interchange. Starting with

$$\Sigma^{+\uparrow} = \frac{1}{\sqrt{18}}\begin{pmatrix} usu(2\uparrow\downarrow\uparrow-\uparrow\uparrow\downarrow-\downarrow\uparrow\uparrow) \\ +uus(-\uparrow\downarrow\uparrow+2\uparrow\uparrow\downarrow-\downarrow\uparrow\uparrow) \\ +suu(-\uparrow\downarrow\uparrow-\uparrow\uparrow\downarrow+2\downarrow\uparrow\uparrow) \end{pmatrix}, \tag{10.100}$$

lowering $u \to d$ by application of the isospin lowering operator $I^- = I_1^- + I_2^- + I_3^-$ gives

$$\Sigma^{0\uparrow} = \frac{I^-}{\sqrt{2}}\Sigma^{+\uparrow} = \frac{1}{6}\begin{pmatrix} (dsu+usd)(2\uparrow\downarrow\uparrow-\uparrow\uparrow\downarrow-\downarrow\uparrow\uparrow) \\ +(dus+uds)(-\uparrow\downarrow\uparrow+2\uparrow\uparrow\downarrow-\downarrow\uparrow\uparrow) \\ +(sdu+sud)(-\uparrow\downarrow\uparrow-\uparrow\uparrow\downarrow+2\downarrow\uparrow\uparrow) \end{pmatrix}. \tag{10.101}$$

The $\Lambda$ is the orthogonal iso-singlet

$$\Lambda^{\uparrow} = \begin{pmatrix} dus(\uparrow\downarrow\uparrow-\downarrow\uparrow\uparrow) \\ +dsu(\uparrow\uparrow\downarrow-\downarrow\uparrow\uparrow) \\ +uds(\downarrow\uparrow\uparrow-\uparrow\downarrow\uparrow) \\ +usd(\downarrow\uparrow\uparrow-\uparrow\uparrow\downarrow) \\ +sud(\uparrow\downarrow\uparrow-\uparrow\uparrow\downarrow) \\ +sdu(\uparrow\uparrow\downarrow-\uparrow\downarrow\uparrow) \end{pmatrix}. \tag{10.102}$$

The light baryon wave functions of the quark model are validated by the static and transition baryon magnetic moments which follow from the superposition of the constituent magnetic moments. Orbital contributions to the magnetic moment vanish in the nonrelativistic model since the quarks are in $s$-wave states. We may write the magnetic moment operator for a baryon in the nonrelativistic form

$$\boldsymbol{\mu} = \Sigma_i \mu_i \boldsymbol{\sigma}_i \tag{10.103}$$

where $\mu_i$ is the magnitude of the magnetic moment of $i$-th quark. The proton magnetic moment is given by the expectation value

$$
\begin{aligned}
< p^{\uparrow}|\mu_z|p^{\uparrow} > &= \frac{1}{18}\begin{pmatrix} 4(2\mu_u - \mu_d) & +\mu_d & +\mu_d \\ +\mu_d & +4(2\mu_u - \mu_d) & +\mu_d \\ +\mu_d & +\mu_d & +4(2\mu_u - \mu_d) \end{pmatrix} \\
&= \frac{4}{3}\mu_u - \frac{1}{3}\mu_d \equiv \mu_p. \qquad (10.104)
\end{aligned}
$$

The magnetic moments of other octet baryons containing two quarks of the same flavor are similar, e.g.

$$\mu_n = \frac{4}{3}\mu_d - \frac{1}{3}\mu_u. \qquad (10.105)$$

The $u$-quark and $d$-quark are in an antisymmetric spin 0 state in the $\Lambda$ so do not contribute to its magnetic moment which equals the magnetic moment of the $s$-quark.

If the quarks behave as nonrelativistic Dirac particles inside the baryons, the magnetic moments are

$$\mu_u = +\frac{2}{3}\frac{e}{2m_u};\ \mu_d = -\frac{1}{3}\frac{e}{2m_d};\ \mu_d = -\frac{1}{3}\frac{e}{2m_s}. \qquad (10.106)$$

If the effective masses satisfy $m_u \simeq m_d$ then $\mu_p/\mu_n = -3/2$. The observed ratio is -1.45989781(125). A fair description of all octet baryon magnetic moments (Table 10.3) results from effective masses $m_u = m_d \simeq 300$ MeV and $m_s \simeq 500$ MeV.

The $u$-quark and $d$-quark have masses small compared to the color scale, $m_u \sim m_d << 0.1$ GeV. Values $m_u = 1 - 5$ MeV and $m_d = 3 - 9$ MeV are estimated from $m_\pi$ and $m_K$ by chiral symmetry model calculations. Such values imply light

| Baryon | Magnetic Moment (Expt.) | Model Value | Magnetic moment (Fit) |
|---|---|---|---|
| p | +2.79 | $\frac{4u-d}{3}$ | Input |
| n | -1.91 | $\frac{4d-u}{3}$ | Input |
| $\Lambda$ | -0.61 | $s$ | Input |
| $\Sigma^+$ | +2.42 | $\frac{4u-s}{3}$ | +2.67 |
| $\Sigma^-$ | -1.16 | $\frac{4d-s}{3}$ | -1.09 |
| $\Xi^0$ | -1.25 | $\frac{4s-u}{3}$ | -1.44 |
| $\Xi^-$ | -0.69 | $\frac{4s-d}{3}$ | -0.49 |

**TABLE 10.3** Comparison of baryon magnetic moment values in units of $e/2m_p$ to values in the naive quark model.

hadrons are filled with relativistic particles. To understand how $u$-quark and $d$-quark masses may be so small on the scale of hadrons despite the much higher effective values inferred from baryon magnetic moments, consider a light quark confined to a sphere (bag model) of radius $R \simeq 1$ fm as illustrated in Figure 10.2. Solution of the Dirac equation subject to the boundary condition that the normal component of the current density vanish at the radius $R$ gives the spin 1/2 ground state energy

$$E_{bag} = \sqrt{m_q^2 + p^2} \tag{10.107}$$

where $p$ is determined by the transcendental equation

$$\tan(pR) = \frac{pR}{1 - m_q R - R\sqrt{p^2 + m_q^2}}. \tag{10.108}$$

For a massless quark, $x \equiv pR = 2.06$ while the limit $m_q \to \infty$ corresponds to $x = \pi$. The kinetic energy of a confined massless quark $p \simeq x/1$ fm $= 400$ MeV is already comparable to the mass of light hadrons. The ground state magnetic moment turns out to be

$$\mu = e_q \frac{R}{12} \frac{4x - 3}{x(x-1)} = \frac{e_q}{2m_{\text{eff}}} \tag{10.109}$$

where the last equality defines an effective mass $m_{eff}$. For a massless quark, the effective mass is

$$m_{\text{eff}} = \frac{12}{5R} \simeq 500 \text{ MeV}. \tag{10.110}$$

This value is comparable to that required for $u$-quark and $d$-quark Pauli magnetic moments to explain baryon magnetic moments and vector meson radiative decays in the nonrelativistic quark model. In the bag model, a hadron is a collection of

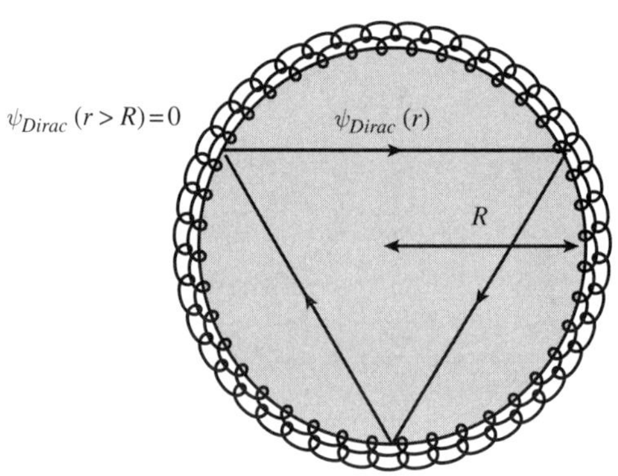


**FIGURE 10.2** Bag model of light quark hadrons. In this simple model, relativistic quarks are confined within a sphere of radius $R \sim \Lambda_{\text{QCD}}^{-1}$.

non-interacting relativistic quarks in a common bag and the nonrelativistic quark model magnetic moments follow from the spin-isospin assumptions used in the naive nonrelativistic quark model.

Within the octet, the average nucleon mass, the average of $\Sigma$ and $\Lambda$ masses, and the $\Xi$ masses are 939, 1154, and 1318 MeV, while within the decuplet, the $\Delta$, $\Sigma$, $\Xi$ and $\Omega$ masses are 1232, 1384, 1533, and 1672 MeV. The effective mass of the strange quark is evidently something like 150 MeV larger than the effective $u$-quark and $d$-quark masses. The details of the light hadron mass spectrum may be qualitatively understood in terms of spin dependent forces expected for a vector gluon interaction between fermions. In what follows we review the Hamiltonian including relativistic corrections in quantum electrodynamics to motivate a simple model which we apply to light hadrons.

In quantum electrodynamics, the Hamiltonian for two fermions of charge $+e$ and $-e$ including terms of order $(v/c)^2$ is

$$H = H_1^0 + H_2^0 + U \tag{10.111}$$

where for either particle $H^0 = m + \mathbf{p}^2/2m - \mathbf{p}^4/8m^3$ and

$$\begin{aligned} U = {} & \frac{\alpha}{r} - \frac{\pi\alpha}{2}\left(\frac{1}{m_1^2} + \frac{1}{m_2^2}\right)\delta(\mathbf{r}) \\ & - \frac{\alpha}{2m_1 m_2 r}\left[\mathbf{p}_1 \cdot \mathbf{p}_2 + \frac{\mathbf{r}\cdot(\mathbf{r}\cdot\mathbf{p}_1)\,\mathbf{p}_2}{r^2}\right] - \frac{\alpha}{4m_1^2 r^3}\mathbf{r}\times\mathbf{p}_1\cdot\boldsymbol{\sigma}_1 \\ & + \frac{\alpha}{4m_2^2 r^3}\mathbf{r}\times\mathbf{p}_2\cdot\boldsymbol{\sigma}_2 - \frac{\alpha}{2m_1 m_2 r^3}\left[\mathbf{r}\times\mathbf{p}_1\cdot\boldsymbol{\sigma}_2 - \mathbf{r}\times\mathbf{p}_2\cdot\boldsymbol{\sigma}_1\right] \\ & + \frac{\alpha}{4m_1 m_2}\left[\frac{\boldsymbol{\sigma}_1\cdot\boldsymbol{\sigma}_2}{r^3} - 3\frac{\boldsymbol{\sigma}_2\cdot\mathbf{r}\boldsymbol{\sigma}_2\cdot\mathbf{r}}{r^5} - \frac{8\pi}{3}\boldsymbol{\sigma}_1\cdot\boldsymbol{\sigma}_2\delta(\mathbf{r})\right] \end{aligned} \tag{10.112}$$

where $\mathbf{r}$ is the separation. The various terms are the Coulomb potential, Darwin terms which affect $s$-waves and correspond to smearing of point-like sources over a Compton wavelength, $\mathbf{L}\cdot\mathbf{L}$ orbital magnetic interactions, $\mathbf{L}\cdot\mathbf{S}$ spin-orbit magnetic interactions, and $\mathbf{S}\cdot\mathbf{S}$ spin-spin magnetic interactions. For $q\bar{q}$ in a color singlet state bound through the color interaction, we should make the replacement

$$\alpha \to -(4/3)\alpha_s. \tag{10.113}$$

(For $qq$ in a color triplet, $\alpha \to -(2/3)\alpha_s$.) For $s$-wave color singlet states, the spin dependent interaction

$$H_{hf} = -\frac{8\pi\alpha}{3}\mu_1\cdot\mu_2\delta(\mathbf{x}_1 - \mathbf{x}_2) \to -\frac{8\pi\alpha_s}{9}\frac{\boldsymbol{\sigma}_1\cdot\boldsymbol{\sigma}_2}{m_1 m_2}\delta(\mathbf{x}_1 - \mathbf{x}_2) \tag{10.114}$$

is the $q\bar{q}$ analog of the interaction of electron and nucleon magnetic moments resulting in hyperfine structure in atomic energy levels. Electromagnetic annihilation, possible for $q\bar{q}$ of the same flavor, adds an additional term

$$U^a = \frac{\pi\alpha}{2m^2}(3 + \boldsymbol{\sigma}_1\cdot\boldsymbol{\sigma}_2)\delta(\mathbf{r}). \tag{10.115}$$

Color singlet mesons may not annihilate into octet gluons so this term is absent in color interactions.

Only the hyperfine interaction distinguishes the different spin wave functions of $s$-wave states. As this interaction is inversely proportional to the two masses

$$H_{hf} \sim 1/m_i m_j, \tag{10.116}$$

the resulting energy level differences decrease with quark mass. Although kinetic energy and spin independent corrections also vary with quark mass, the simplistic formulae

$$\begin{aligned} M(q_1\bar{q}_2) &= m_1 + m_2 + 8a_m \frac{\mathbf{s_1} \cdot \mathbf{s_2}}{m_1 m_2} \\ M(q_1 q_2 q_3) &= m_1 + m_2 + m_3 + \sum_{i \neq j} 4a_b \frac{\mathbf{s}_i \cdot \mathbf{s}_j}{m_i m_j} \end{aligned} \tag{10.117}$$

with $a = 4\pi\alpha_s \mid \psi(0) \mid^2 /9$ provide a good representation of the spectrum of meson and baryons masses. For $q\bar{q}$ mesons, with $\mathbf{s} = \mathbf{s}_1 + \mathbf{s}_2$, we can write

$$\mathbf{s}_1 \cdot \mathbf{s}_2 = \left(\mathbf{s}^2 - \mathbf{s}_1^2 - \mathbf{s}_2^2\right)/2 = s(s-1) - \frac{3}{2} \tag{10.118}$$

which takes the value 1/4 for s = 1 and −3/4 for s = 0, so the vector mesons are higher in mass than the pseudoscalar mesons and we have the following expectations for light meson masses:

$$\begin{aligned} m_\pi &= 2m_u - 6a_m \frac{1}{m_u^2} \\ m_K &= m_u + m_s - 6a_m \frac{1}{m_u m_s} \\ m_\eta &= \frac{1}{3}\left(2m_u - 6a_m \frac{1}{m_u^2}\right) + \frac{2}{3}\left(2m_s - 6a_m \frac{1}{m_s^2}\right) \\ m_\rho = m_\omega &= 2m_u + 2a_m \frac{1}{m_u^2} \\ m_{K^*} &= m_u + m_s + 2a_m \frac{1}{m_u m_s} \\ m_\phi &= 2m_s + 2a_m \frac{1}{m_s^2}. \end{aligned} \tag{10.119}$$

For baryons containing equal mass quarks, the sum of pairwise hyperfine interactions is proportional to

$$\sum_{i \neq j} \mathbf{s}_i \cdot \mathbf{s}_j = \frac{1}{2}\left(\mathbf{s}^2 - \sum \mathbf{s}_i^2\right) \tag{10.120}$$

which has the value +3/4 for s = 3/2 and −3/4 for s = 1/2, explaining why the decuplet states are higher in mass than the octet states, and we have the predictions

$$\begin{aligned} m_N &= 3m_u - 3a_b\frac{1}{m_u^2} \\ m_\Delta &= 3m_u + 3a_b\frac{1}{m_u^2} \\ m_\Omega &= 3m_s + 3a_b\frac{1}{m_s^2}. \end{aligned} \tag{10.121}$$

For baryons containing two light quarks, 1 and 2, and a strange quark 3, write

$$M = 2m_u + m_s + 4K\left(\frac{\mathbf{s}_1 \cdot \mathbf{s}_2}{m_u^2} + \frac{(\mathbf{s}_1 + \mathbf{s}_2)\cdot \mathbf{s}_3}{m_u m_s}\right). \tag{10.122}$$

With $\mathbf{s}_{12} = \mathbf{s}_1 + \mathbf{s}_2$, we can use the expressions

$$\begin{aligned} \mathbf{s}_1 \cdot \mathbf{s}_2 &= \left(\mathbf{s}_{12}^2 - \mathbf{s}_1^2 - \mathbf{s}_2^2\right)/2 \\ \mathbf{s}_{12} \cdot \mathbf{s}_3 &= \left(\mathbf{s}^2 - \mathbf{s}_{12}^2 - \mathbf{s}_3^2\right)/2 \end{aligned} \tag{10.123}$$

to find the formula

$$\begin{aligned} M = 2m_u + m_s + 2a_b\,&[(s_{12}\,(s_{12}+1) - 3/2)\,\frac{1}{m_u^2} \\ &+ (s(s+1) - s_{12}(s_{12}+1) - 3/4)\frac{1}{m_u m_s}. \end{aligned} \tag{10.124}$$

The wave function of a baryon with $I = 0\,(1)$ is symmetric (antisymmetric) under interchange of light quark flavor so symmetric (antisymmetric) under exchange of light quark spin corresponding to $s_{12} = 1(0)$. Hence

$$\begin{aligned} m_\Lambda &= 2m_u + m_s - a_b\left[\frac{3}{m_u^2}\right] \\ m_\Sigma &= 2m_u + m_s + a_b\left[\frac{1}{m_u^2} - \frac{4}{m_u m_s}\right] \\ m_{\Sigma^*} &= 2m_u + m_s + a_b\left[\frac{1}{m_u^2} + \frac{2}{m_u m_s}\right]. \end{aligned} \tag{10.125}$$

The results of a fit of this model to light hadron masses are shown in Table 10.4. The fit gave each particle equal weight except the $\eta'$ which is problematic and was excluded, assumed $m_u = m_d$, and permitted the values of the effective masses and wave functions of the light quarks to depend on the environment. Excepting the $\eta'$, the observed masses are accounted for to within a few percent.

Thus, the mass difference relation $m_{K^*} - m_K < m_\rho - m_\pi$ is ascribed to the smaller effective magnetic moment of the strange quark. Similarly, as the

| Hadron | mass (expt.) | mass (fit) | hadron | mass (expt.) | mass (fit) |
|---|---|---|---|---|---|
| $\eta$' | 958. | 353. | $D$ | 1869. | 1910. |
| $\pi$ | 140. | 144. | $D_s$ | 1969. | 2005. |
| $K$ | 496. | 489. | $D^*$ | 2007. | 2034. |
| $\eta$ | 549. | 563. | $D_s^*$ | 2112. | 2111. |
| $\rho$ | 776. | 772. | $J/\psi$ | 3097. | 3009. |
| $K^*$ | 892. | 890. | $\eta_c$ | 2980. | 3049. |
| $\phi$ | 1020. | 1028. | $\Lambda_c$ | 2284. | 2318. |
| $n$ | 939. | 943. | $\Sigma_c$ | 2455. | 2455. |
| $\Lambda$ | 1116. | 1116. | $\omega_c$ | 2704. | 2704. |
| $\Sigma$ | 1193. | 1179. | $B$ | 5279. | 5245. |
| $\Xi$ | 1318. | 1326. | $B_s$ | 5369. | 5331. |
| $\Delta$ | 1232. | 1239. | $B^*$ | 5325. | 5285. |
| $\Sigma$ | 1384. | 1380. | $B_c$ | 6400. | 6301. |
| $\Xi$ | 1533. | 1526. | $\Upsilon$ | 9460. | 9583. |
| $\Omega$ | 1672. | 1678. | $\Lambda_b$ | 5624. | 5590. |

| Parameter | light meson | light baryon | heavy meson | heavy baryon |
|---|---|---|---|---|
| $m_u$ | 307.5 | 363.6 | 484.0 | 473.9 |
| $m_s$ | 481.9 | 536.6 | 565.0 | 561.5 |
| $a$ | 78.5 | 49.3 | 48.6 | 49.7 |

**TABLE 10.4** Hadron masses compared to fit with a simple spin dependent interaction model. The values of the parameters (in MeV) $m_u$, $m_s$, and $a$ are allowed to float according to context. The heavy quark masses are $m_c = 1519.4$ and $m_b = 4791.0$ MeV. The $\eta'$ is excluded from the fit.

strangeness increases in the baryons, the hyperfine effect decreases. A difference in $a$ values for the mesons and baryons is expected due to the size difference $|\psi(0)|^2 \sim 1/R^3$ with $R_{q\bar{q}}/R_{qqq} \simeq .6/.8$. The QCD color factor of two between $q\bar{q}$ and qq interactions was already included.

The $\eta$ and $\eta'$ masses are not well predicted by the simple formula for any value of mixing parameter although $m_\eta$ corresponds rather precisely to the prediction for $m_{\eta_8}$. It is plausible that the masses are shifted through the mixing interaction applicable to these states. Suppose amplitudes $\alpha$ for $u\bar{u} \leftrightarrow gg$ and $\beta$ for $s\bar{s} \leftrightarrow gg$ lead to mixing $u\bar{u}$, $d\bar{d}$, and $s\bar{s}$ components as described by the model equation

$$i\frac{\partial}{\partial t}\psi = H\psi = \begin{pmatrix} E_u + \alpha^2 & \alpha^2 & \alpha\beta \\ \alpha^2 & E_u + \alpha^2 & \alpha\beta \\ \alpha\beta & \alpha\beta & E_s + \beta^2 \end{pmatrix} \begin{pmatrix} \psi_{u\bar{u}} \\ \psi_{d\bar{d}} \\ \psi_{s\bar{s}} \end{pmatrix} \tag{10.126}$$

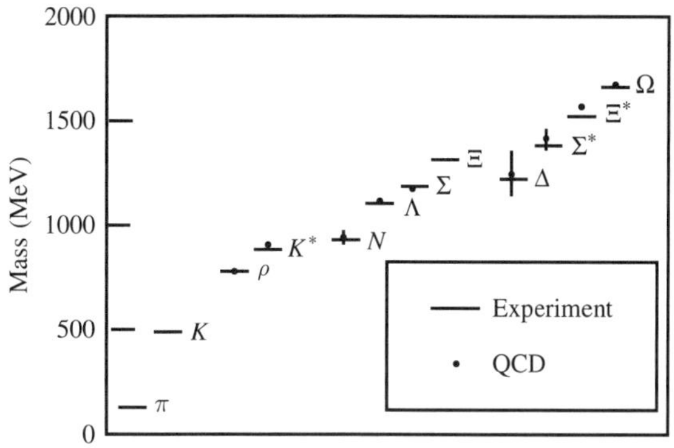


**FIGURE 10.3** Lattice quantum chromodynamics calculation of light hadron masses. Horizontal lines represent experimental masses. The $\pi$, $K$ and $\Sigma$ masses are used as to fix light quark and strange quark masses and the overall scale. The calculated values are shown with estimated numerical errors. [S. Drr *et. al*, Science **322**, 1224 (2008) DOI: 10.1126/science.1163233]

where $E_u = 2m_u - 6a/m_u^2$ and $E_s = 2m_s - 6a/m_s^2$. Then $\pi^0 = (u\bar{u} - d\bar{d})/\sqrt{2}$ is an eigenstate while the other two eigenstates are mixtures of $\eta_+ = (u\bar{u} + d\bar{d})/\sqrt{2}$ and $s\bar{s}$, and the model reduces to

$$i\frac{\partial}{\partial t}\psi = H\psi = \begin{pmatrix} E_u + \alpha^2 & \sqrt{2}\alpha\beta \\ \sqrt{2}\alpha\beta & E_s + \beta^2 \end{pmatrix} \begin{pmatrix} \psi_{\eta^+} \\ \psi_{s\bar{s}} \end{pmatrix}. \tag{10.127}$$

Given $E_u$ and $E_s$ from fits to the other light mesons, $m_\eta$ and $m_{\eta'}$ determine the mixing amplitudes $\alpha = 262$ MeV and $\beta = 66$ MeV and the mixtures

$$|\eta' > = 0.29|\eta_+ > \ +0.84|s\bar{s} > \quad |\eta > = 0.71|\eta_+ > \ -0.54|s\bar{s} > \tag{10.128}$$

which are similar to those deduced from the $\eta \to \gamma\gamma$ and $\eta' \to \gamma\gamma$ decay rate ratio. That $gg$ mixing causes a positive energy shift of about the right magnitude is supported by a bag model calculation which gives $< u\bar{u}|H|u\bar{u} > = 0.39\,\alpha_s^2/R$, $< u\bar{u}|H|s\bar{s} > = 0.29\,\alpha_s^2/R$, $< s\bar{s}|H|s\bar{s} > = 0.18\,\alpha_s^2/R$. See John F. Donaghue and Harold Gomm, Phys. Rev. **D28**, 2800 (1983).

Hadron masses have been calculated numerically on a space-time lattice based on quantum chromodynamics including the effects of the fluctuating vacuum - the appearance of light quark and antiquark pairs. In Figure 10.3, the results for the masses of light hadrons (based on extrapolating calculations for lattice spacing range of 0.065 fm to 0.125 fm to zero spacing) are shown to reproduce the observed values to within a few percent.

## 10.5 ■ PARTON MODEL OF HADRON INTERACTIONS

Leptons can be used to probe the content of nucleons in so-called deep inelastic scattering experiments. At energies in excess of the scale of hadronic binding, hadrons are successfully modeled as collections of free and independent so-called partons which include so-called valence quarks identified with the quarks of the naive quark model, a gluon gas corresponding to the color fields of the bag or string model, and additional clouds of quark antiquark pairs of all flavors called sea quarks. High energy interactions between leptons and nucleons in the parton picture are illustrated in Figure 10.4.

Given that quarks are confined, the success of this impulse approximation model was something of a surprise, yet helped buttress the naive quark model and was later justified by the decrease in coupling with momentum transfer in quantum chromodynamics. In simple terms, in deep inelastic scattering experiments, an incoming lepton scatters elastically from a quark to leading order. The scattered quark materializes as a shower of color neutral hadronic matter largely uncorrelated with untouched fragments of the target nucleon. The complexity of the nucleon fields at length scales of order 1 fm are largely irrelevant except in so far as they determine the distributions of parton momentum.

Only gross knowledge of the density in momentum space is gained of the quarks and gluons in these hard scattering processes. The correlations between partons are generally not measurable. To leading order, the gluon fields are transparent to leptons but higher order effects such as final state gluon radiation and coupling to the target gluon component are observed. The lepton scattering experiments have yielded parton momentum distribution functions which are borne out in hadron-hadron collisions modeled as parton-parton scattering.

The cross section for $e^-\mu^-$ elastic scattering by single photon exchange serves as a model for the analogous lepton quark scattering and can be written as

$$\frac{d\sigma_{e^-\mu^-\to e^-\mu^-}}{dy} = 2\pi\frac{\alpha^2}{q^4}s\left[1+(1-y)^2\right] \tag{10.129}$$

where $y = (\cos\theta_{cm} - 1)/2$, $\theta_{cm}$ is the scattering angle in the center of mass, and $s$ is the square of the energy in the center of mass. The first factor in square brackets represents scattering with equal helicity (LL+RR) and the second with opposite helicity. The cross section for scattering of an electron or muon from a quark has

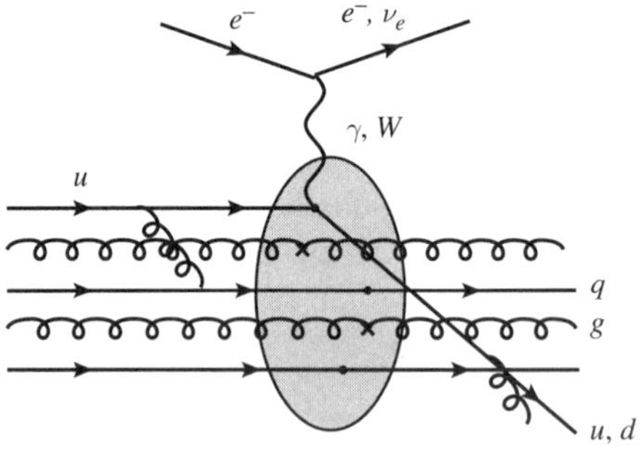


**FIGURE 10.4** Interaction of an electron with a nucleon in the parton model. The lepton exchanges an electroweak boson with a quark which is scattered out of the nucleon. The constituents of the nucleon are regarded as free particles in the first approximation.

the identical form but is multiplied by $e_q^2$ where $e_q = 2/3$ for $u$-like quarks and $q_q = -1/3$ for $d$-like quarks.

Suppose a quark carries fraction $x$ of the momentum of a proton. Neglecting the transverse momentum of the quark relative to the lepton initial direction, and neglecting lepton and quark masses, we may write the $e^- q$ center of mass energy squared as

$$\hat{s} = (p_e + p_q)^2 = 2p_e p_q = 2p_e p_p x = sx \tag{10.130}$$

with $s$ the lepton hadron center of mass energy. If $f_q(x)dx$ is the probability of momentum fraction in $dx$ about $x$ in a hadron $h$, the cross section for scattering of a lepton from this hadron is

$$\frac{d\sigma_{e^-h\to e^-h}}{dxdy} = 2\pi \frac{\alpha^2}{q^4} s\left[1 + (1-y)^2\right] x F_p \tag{10.131}$$

where $F_p = \Sigma_q e_q^2 f_q(x)$. The factorization into a point-like cross section appropriate for spin 1/2 quarks times a function of $x$ provided the initial evidence for the parton model. The functions which specify the probability that a quark or gluon carries a fraction $x$ of a hadron's momentum are called parton distribution functions and are derived by comparing this formula to observations.

Experiments measure the charged lepton initial and final energies $E_i$ and $E_f$ and the lepton scattering angle $\theta$. The majority of experiments have been performed with lepton beams scattering from stationary nucleons and these quantities are measured in the laboratory frame. (Experiments at the HERA electron proton collider were an exception.) The laboratory quantities may be related to the lepton parton center of mass variables as follows. The 4-momentum transfer is $q = p_f - p_i$ and in the laboratory frame

$$q^2 = -2p_f p_i = 2E_f E_i(\cos\theta_L - 1). \tag{10.132}$$

The lepton-proton invariant mass squared is $s = 2m_p E_i$. In the lepton quark center of mass, the scattering is elastic with

$$E_{i,cm} = p_i v = E_{f,cm} = p_f v \equiv E_{cm} \tag{10.133}$$

where $v$ is the 4-velocity of the center of mass. Since $qv = 0$, the speed of the center of mass is

$$v_{cm} = \frac{E_f - E_i}{E_f \cos\theta_L - E_i}. \tag{10.134}$$

The proton energy is $E_{p,cm} = p_p v$ and the quark energy is $E_{q,cm} = E_{cm} = p_i v$ so

$$\begin{aligned} x &= \frac{p_{q,cm}}{p_{p,cm}} = \frac{p_i v}{v_{cm} p_p v} = \frac{E_i(v_{cm}^{-1} - 1)}{m_p} \\ &= \frac{E_f E_i(\cos\theta_L - 1)}{m_p(E_f - E_i)} = \frac{-q^2}{2m_p v} \end{aligned} \tag{10.135}$$

where $\nu = E_i - E_f$ and we substituted the expressions for $v_{cm}$ and for $q^2$. The scattering angle in the lepton-quark center of mass follows from $q^2 = 2E_{cm}^2(\cos\theta_{cm} - 1) = \hat{s}y \to y = q^2/\hat{s}$.

In scattering from nucleons $N = (p, n)$, neglecting the possibility of any constituents heavier than the $s$-quark, and letting $q_N(x)$ denote the distribution function for quark $q$ in nucleon $N$, we have

$$F_N = \frac{4}{9}(u_N + \bar{u}_N) + \frac{1}{9}(d_N + \bar{d}_N + s_N + \bar{s}_N). \tag{10.136}$$

For a proton, the quantity $u_p$ is expected to have a valence quark part corresponding to the quarks of the naive quark model and a contribution from a sea of relatively soft $u\bar{u}$ pairs:$u_p = u_p^{val} + u_p^{sea}$. Remarkably, it is found that $u_p \neq d_p$ but isospin symmetry implies $u_n = d_p$ and $d_n = u_p$. With $F_n$ extracted by comparing scattering from deuterium with scattering from hydrogen, and assuming equal sea contributions in the neutron and proton, an interesting quantity is

$$F_p - F_n = \frac{1}{3}\left(u_p^{val} - d_p^{val}\right). \tag{10.137}$$

This function represents the valence quarks alone and it appears as a broad peak around $x = 1/3$, a not unexpected value, the broadening roughly as expected from Fermi motion. In contrast, the simple distributions of $u$-quark, $d$-quark, and $s$-quark momenta in nucleons as extracted from combined fits to electromagnetic and weak scattering all peak at $x = 0$ where a large number of soft sea quarks are to be found.

Integration of the experimental $F_p$ and $F_n$ over momentum neglecting $s$ and $\bar{s}$ gives two experimental constraints

$$\int_0^1 dx\, xF_p = \frac{4}{9}\epsilon_u + \frac{1}{9}\epsilon_d = 0.18$$
$$\int_0^1 dx\, xF_n = \frac{1}{9}\epsilon_u + \frac{4}{9}\epsilon_d = 0.12 \tag{10.138}$$

with $\epsilon_u = \int dx\, x(u + \bar{u})$ the fraction of momentum carried by $u$-quark and $\bar{u}$-quark and $\epsilon_d$ the $d$-quark and $\bar{d}$-quark momentum fraction. Solving for these quantities gives $\epsilon_u = 0.36$ and $\epsilon_d = 0.18$ leaving 46% of the momentum presumably carried by glue to which electroweak probes are blind. Processes like direct photon production $qg \to q\gamma$ and $gg \to q\bar{q}, gg$ probe the gluons.

Nucleons are also probed with neutrino beams. Charged current neutrino interactions such as

$$\nu_\mu d \to u\mu^- \; \nu_\mu\bar{u} \to \bar{d}\mu^-$$
$$\bar{\nu}_\mu u \to d\mu^+ \; \bar{\nu}_\mu\bar{d} \to \bar{u}\mu^+ \tag{10.139}$$

complement electromagnetic scattering in studies of parton distribution functions. The charged current weak interaction couples only left-handed fermions and right-handed anti-fermions so $\nu q$ and $\bar{\nu}\bar{q}$ couple with equal handedness while $\bar{\nu}q$ and

$\nu\bar{q}$ couple with opposite handedness. With the replacement $\alpha/q^2 \to g^2/(2m_W^2)$ in the electron-muon scattering cross section, the parts corresponding to these combinations of handedness are the neutrino-quark cross sections

$$\frac{d\sigma_{\nu d\to u\mu^-}}{dxdy}\left(\nu d \to u\mu^-\right) = \pi\left(\frac{g^2}{2m_W^2}\right)^2 sxf_d \tag{10.140}$$

$$\frac{d\sigma_{\bar{u}\nu d\to u\mu^+}}{dxdy}\left(\bar{u}\nu d \to u\mu^+\right) = \pi\left(\frac{g^2}{2m_W^2}\right)^2 s(1-y)^2xf_u. \tag{10.141}$$

That weak neutrino scattering results are consistent with electromagnetic charged lepton scattering results validates the parton model and provides a host of tests of the standard model of lepton and quark electroweak interactions.

The simple parton model of lepton scattering from nucleons provides a basis for understanding higher order processes in hadronic physics. For example, a virtual $W$ boson or $\gamma$ exchanged between a lepton and a hadron can interact with a valence quark which has already emitted a gluon and which therefore has a lower than normal momentum. The probability of such gluon emission is a function of $q^2$ so $F(x) \to F(x, q^2)$ and the factorization of the $x$ and $q^2$ dependences of scattering cross sections observed in the leading order scattering breaks down in this next to leading order process. The probe can also strike one or the other of a $q\bar{q}$ pair into which a gluon has just transformed. The gluon and quark distributions are coupled quantities when considered at next to leading order. Accounting for such complications, the $q^2$ dependence of the functions $F$ induced by such processes may be calculated with quantum chromodynamics and agrees well with experiment.

The HERA electron-proton collider provided extensive data with which to evaluate the structure of the proton. In Figure 10.5, the gluon distribution derived from the scattering cross section for $e^-p \to e^-X$ as a function of 4-momentum transfer $Q^2 = -q^2$ and parton momentum fraction $x$ are compared to calculations. Good fits to the data are found using the evolution with $q^2$ predicted by quantum chromdynamics. From such measurements combined with results from hadron hadron collisions, the parton distribution functions continue to be refined. The distribution functions for valance up and down quarks and for gluons and sea quarks are shown in Figure 10.6. Notice that the functions have been multiplied by $x$. For $x$ values close to one, the valence quarks dominate while, for small $x$ values, gluons and sea quarks dominate.

Hadron-hadron collisions at high energy may be described in the parton model by regarding both hadrons as collections of independent partons and combining parton-parton interactions. Denote the fraction of the hadron momentum carried by a constituent of type $i$ in a hadron of type $a$ of momentum $P_a$ by $x_i = p_i/P_a$ and let $f_i^a(x)$ be the probability distribution for finding in a hadron of type $a$ a constituent of type $i$ carrying momentum fraction $x$. Neglecting constituent and hadron masses compared to the available energies, the constituent center of mass energy squared is

$$\hat{s} = (p_i + p_j)^2 \simeq x_i x_j s = \tau s \tag{10.142}$$

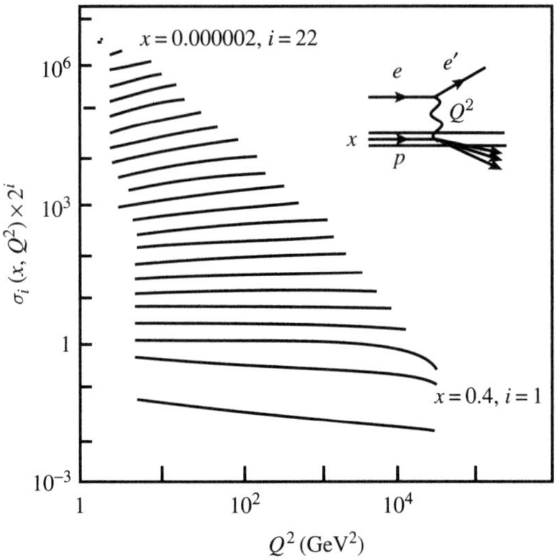


**FIGURE 10.5** Cross sections $\sigma(x, Q^2)$ for $ep \to e + X$ at the HERA $ep$ collider. The measured cross section is shown as a function for 4-momentum transfer $Q^2$ and parton momentum fraction $x$. In the parton model in first approximation, $\sigma(x, Q^2)$ is independent of $Q^2$. [Cristinel Diaconu, ICHEP08]

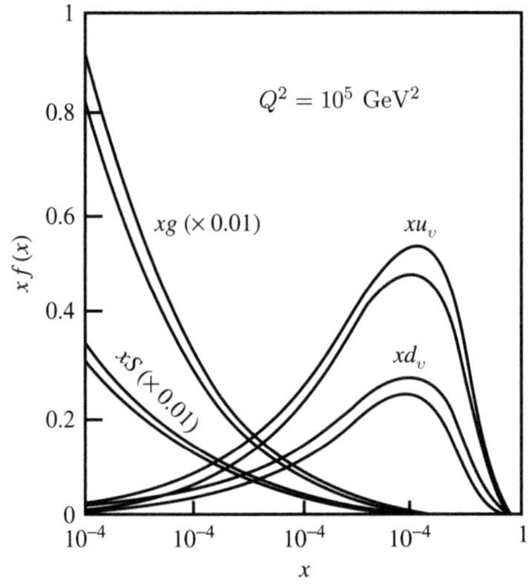


**FIGURE 10.6** Parton distribution functions deduced from $ep$ collisions. The distribution functions are compared with global fits to distributions in lepton-hadron and hadron-hadron scattering processes. The distribution functions for valence $u$-quarks $u_v(x, Q^2)$, valence $d$-quarks $d_v(x, Q^2)$, gluons $g(x, Q^2)$, and sea quarks $S(x, Q^2)$ are plotted multiplied by $x$ as functions of $x = E_{parton}/E_{proton}$ for 4-momentum transfer squared $Q^2 = 10^5$ GeV$^2$, a value appropriate for the LHC. The bands indicate the uncertainties. [Cristinel Diaconu, ICHEP08]

where $\tau = x_i x_j$ is the available fraction of the collision energy squared. If $d\sigma_{ij}(p_i, p_j \to f)$ is the differential cross section for a constituent process to produce a final state $f$, the cross section $d\sigma_{P_a,P_b \to f+X}$ for a collision of hadrons of types $a$ and $b$ to produce $f$ along with fragments $X$ is the incoherent sum

$$d\sigma_{P_a,P_b \to f+X} = \sum_{ij} dx_a dx_b f_i^a(x_a) f_j^b(x_b) d\sigma_{ij}(p_i, p_j \to f). \tag{10.143}$$

When the final state $f$ carries color, sum over colors assuming the color is neutralized by a dressing and fragmentation process with unit probability, and average over initial colors by dividing by $N_c = 3$ for a quark and $N_c = 8$ for a gluon.

If particle $a$ is not a hadron but a lepton, then $\sum_i f_i^a(x_a) = \delta(x_a - 1)$ and the expression above reduces to the simpler formula

$$d\sigma_{P_a,P_b \to f+X} = \sum_{j} dx_b f_j^b(x_b) d\sigma_{ij}(p_i, p_j \to f). \tag{10.144}$$

For example, in electromagnetic scattering of muons from nucleons, the constituent process in invariant form is

$$d\sigma_{\mu q \to \mu q} = \frac{\alpha^2 e_q^2}{8\pi} \frac{\hat{s}^2 + \hat{u}^2}{\hat{s}^2 \hat{t}^2} d\hat{t}. \tag{10.145}$$

In scattering of hadrons from hadrons, it is customary to introduce new variables $\tau$ and $x = x_a - x_b$ such that

$$x_{a,b} = \left[(x^2 + 4\tau)^2 \pm x\right] / \sqrt{2}. \tag{10.146}$$

The cross section for an incoherent sum of parton-parton collisions is then expressed as

$$d\sigma = \sum_{ij} \frac{dx d\tau}{(x^2 + 4\tau)^{\frac{1}{2}}} f_i^a(x_a) f_j^b(x_b). \tag{10.147}$$

The rapidity of a particle is defined as

$$y = \frac{1}{2} \ln \left[ \frac{E + p}{E - p} \right] \tag{10.148}$$

and the rapidity of the colliding pair is $y = \dfrac{1}{2} \ln[x_a / x_b]$. In terms of pair rapidity, we have $x_{a,b} = \sqrt{\tau} e^{\pm y}$ and

$$d\sigma = dy d\tau f_i^a(x_a) f_j^b(x_b) d\sigma_{ij}. \tag{10.149}$$

For production of a system with invariant mass squared $m^2$, writing $\delta(s_{ij} - m^2) = \delta(\tau - m^2/s)/s$, the cross section is

$$d\sigma = \sum_{ij} dx_a dx_b f_i^a(x_a) f_j^b(x_b) \sigma_{ij}(\hat{s}) \tag{10.150}$$

where the integral gives the effective flux. If several constituents contribute to the process, the total cross section is the sum over constituents.

As an example, let us consider lepton pair production through a virtual photon (Drell-Yan process) in $\bar{p}p$ collisions. The constituent cross section is

$$\sigma_{q\bar{q}\to l^+l^-}(\hat{s}) = \frac{4\pi}{3}\frac{\alpha^2}{\hat{s}}e_q^2 \tag{10.151}$$

and the differential cross section for production of a pair of invariant mass squared $m^2 = s\tau$ and pair rapidity $y$ is

$$\frac{d\sigma_{p\bar{p}\to l^+l^-}}{dy d\tau} = \frac{4\pi}{3}\frac{\tau\alpha^2}{m^2}\frac{1}{3}\sum_q e_q^2 f_q^p(x_p) f_{\bar{q}}^{\bar{p}}(x_{\bar{p}}) \tag{10.152}$$

where the color average pre-factor is 1/3 since for each color the probability of collision with the anti-color is 1/9 but all three colors contribute, and the sum is over quark flavors. This process is sensitive to a certain weighted sum of quark distributions.

In parton scattering, it must not be forgotten that experiments observe not the bare partons themselves but instead jets of color neutral hadrons. Generally, the process of conversion from parton to jet, called fragmentation and hadronization, largely factorizes from the parton level collision. The energy of a parton may materialize in many forms but for parton energy in excess of a few GeV, the energy and momentum in a cone about the initial parton direction closely matches that of the parton. The essence of the fragmentation process is an incoherent showering of final state radiation and pair conversion not unlike the development of an electromagnetic shower in matter. Quantum chromodynamics predicts the initial perturbative stages and the evolution with initial parton momentum transfer and non-perturbative models like the string model motivate statistical models of the ultimate distribution of hadronic energy fragments. For more information, the reader is referred to the literature on Monte Carlo event generators. Such models do not purport to predict details such as the probability that a $b$ appears ultimately in a meson or baryon. Such details are determined empirically. Not surprisingly, materializing as a baryon is less probable than materializing as a meson, appearing bound with a $u$-quark is equally likely as appearing bound with a $d$-quark, strange and heavy quarks appear in jet fragmentation more rarely than light quarks, and production of excited states like a $B^*$ is common.

## 10.6 ■ NEUTRAL KAONS

The two charge conjugate neutral strange pseudoscalar mesons $K^0 = \bar{s}d$ and $\bar{K}^0 = s\bar{d}$ are produced by strangeness conserving strong interactions such as $\pi^- p \to \Lambda K^0$. They may be identified by processes such as $K^0 p \to K^+ n$, $\bar{K}^0 p \to \Sigma^+ \pi^0$, and $\bar{K}^0 n \to K^- p$. Like charged kaons, neutral kaons decay weakly to two and three pions and to $\pi^{\pm}l^{\mp}\nu$. The decay $K^0 \to \pi^+\pi^-$ violates parity. In fact, $P_{K^0} = -1$ while $P_{\pi^+\pi^-} = (-)^l = +1$. The decay also violates charge symmetry since $C_{\pi^+\pi^-} = (-)^l = +1$ while the $K^0$ is not a $C$ eigenstate.

Transitions between $K^0$ and $\bar{K}^0$ proceed through second order weak interactions. Large distance mixing can be understood to be a consequence of processes such as $K^0 \Leftrightarrow \pi^+\pi^- \Leftrightarrow \bar{K}^0$. At short distances, mixing is described by box diagrams involving two $W$ bosons as shown in Figure 10.7. Flavor conservation precludes mixing by electromagnetic or strong interactions. Thus the $K^0$ and $\bar{K}^0$ may be regarded as degenerate eigenstates of the Hamiltonian $H_0 = H_{em} + H_{strong}$ but not of $H = H_0 + H_{weak}$. If $CP$ is conserved by the weak interaction, the true eigenstates are $CP$ eigenstates:

$$| K_1 > = \frac{| K^0 > +CP \mid K^0 >}{\sqrt{2}} \tag{10.153}$$

$$| K_2 > = \frac{| K^0 > -CP \mid K^0 >}{\sqrt{2}} \tag{10.154}$$

where $CP \mid K_1 > = + \mid K_1 >$ and $CP \mid K_2 > = - \mid K_2 >$. Two pion states with $l = 0$ have $CP = (-)^l = +1$ so the $K_1$ may decay to two pions while the $K_2$ decays only into three pions. ($CP$-odd $\pi^+\pi^-\pi^0$ states of total angular momentum 0 exist with $\pi^+\pi^-$ orbital angular momentum greater than 0 but are suppressed.) So the $K_1$ and $K_2$ may be expected to have different lifetimes and slightly different masses, rather like ortho-positronium and para-positronium. The phase space for the simpler two pion decay is larger and the $K_1$ has the shorter lifetime.

A $K^0$ produced by the strong interaction evolves by weak mixing into a superposition of $K^0$ and $\bar{K}^0$:

$$| \psi(t) > = \frac{1}{\sqrt{2}} \left[ | K_1(0) > e^{-iE_1t} + | K_2(0) > e^{-iE_2t} \right] \tag{10.155}$$

$$\begin{aligned} &= \frac{1}{2} \mid K^0 > \left[ e^{-iE_1t} + e^{-iE_2t} \right] \\ &+ \frac{1}{2} \mid \bar{K}^0 > \left[ e^{-iE_1t} - e^{-iE_2t} \right] \end{aligned} \tag{10.156}$$

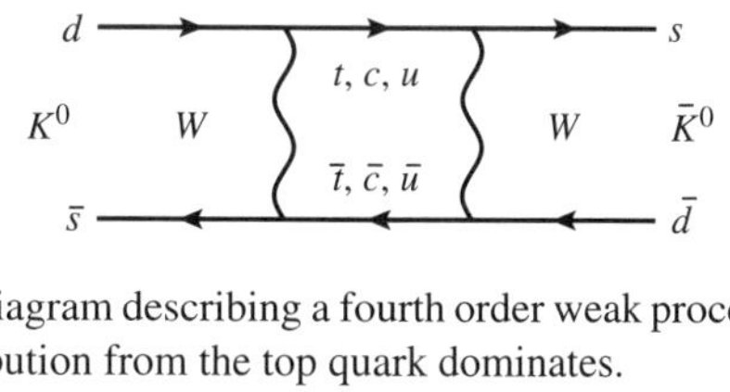


**FIGURE 10.7** Box diagram describing a fourth order weak process contributing to $K^0 - \bar{K}^0$ mixing. The contribution from the top quark dominates.

where, in the rest frame, $E_i = m_i - i\Gamma_i/2$ with $m_i$ the mass and $\Gamma_i$ the total decay width. The probability of finding a $K^0$ at proper time $t$ is

$$|< K^0 \mid \psi(t) >|^2 = \frac{1}{4}\left[e^{-\Gamma_1 t} + e^{-\Gamma_2 t} + 2e^{\frac{t}{2}(\Gamma_1+\Gamma_2)}\cos(\Delta m t)\right] \tag{10.157}$$

with $\Delta m = m_2 - m_1$. Long lived neutral kaons predicted by the weak mixing hypothesis were first observed in the 3-body decay modes $K \to \pi l \nu$ and $K \to \pi^+\pi^-\pi^0$. Oscillation in strangeness content is cleanly illustrated by events such as the following observed in a hydrogen bubble chamber. A $\pi^-(\bar{u}d)p \to \Lambda(sud)K^0(\bar{s}d)$ with $\Lambda \to p\pi^-$ was observed and $K^0$ production inferred by strangeness conservation in the strong interaction. A vertex along the $K^0$ direction was found in which a $\Sigma^+$ (uus) and $\pi^0$ were produced with the $\Sigma^+$ identified by its decay to $p\pi^0$ and the $\pi^0$'s reconstructed by observation of conversion to electron positron pairs. Evidently, the $\bar{s}$ in the $K^0$ had evolved into an $s$-quark. Strangeness oscillation is also observed through the semi-leptonic decay modes $K^0 \to \mu^+\nu_\mu\pi^-$ and $\bar{K}^0 \to \mu^-\bar{\nu}_\mu\pi^+$. These decays result from $\bar{s}$-quark or $s$-quark radiating a $W^+$ or $W^-$ boson. The virtual $W$ boson decays to leptons $l^+\nu_l$ or $l^-\bar{\nu}_l$. The lepton charge tags the strange quark flavor. By observations of oscillations, two neutral K meson states are identified called "kay-long" and "kay-short" with lifetimes $\tau_S = 0.892 \times 10^{-10}$ s and $\tau_L = 5.18 \times 10^{-8}$ s and mass difference $\Delta m \equiv m_L - m_S = 3.521 \times 10^{-6}$ eV.

A neutral beam made by colliding a proton beam with a fixed target followed by a sweeping magnet contains equal numbers of $K^0$ and $\bar{K}^0$ produced by strong interactions. (It will also contain $\gamma$, $n$, $\bar{n}$, $\Lambda$...) Weak evolution produces a double exponential distribution of vertices, $K_S \to \pi\pi$ and $K_L \to 3\pi$. After a few $\tau_S$, only $K_L$ survive. In 1964, Christenson, Cronin, Fitch and Turley observed well downstream of a production target extraordinary two pion vertices. The number, though small, was well in excess of expectation from the exponentially suppressed $K_S$ component of the beam and could only be interpreted as $CP$-violating decays $K_L \to \pi^+\pi^-$ with a branching fraction of about $2 \times 10^{-3}$. The existence of $CP$-violation was confirmed by the observation of $K_L \to \pi^0\pi^0$ and of a charge asymmetry in $K_L \to e^\pm\pi^\mp\nu$ and $K_L \to \mu^\pm\pi^\mp\nu$.

Unitarity, Lorentz covariance and the connection between spin and statistics in quantum field theory require $CPT$ symmetry so $CP$-violation implies $T$-violation. The observation of $CP$-violation in the neutral kaon system motivates searches for $T$, $CP$, and $CPT$ violation effects in a variety of processes. The standard model predicts $CP$-violation arises through a single complex phase in the CKM matrix and it predicts weak mixing and $CP$-violating effects in the evolution of bottom flavored mesons $B_d^0 = d\bar{b}$ and $B_s^0 = s\bar{b}$ and in charmed mesons as well as in strange mesons. Historically, $CP$-violation was discovered prior to the development of the standard model and spawned an exploration of kaon physics and a phenomenological analysis based on elementary principles of quantum mechanics to which we will devote a little attention. A similar approach is in use today modeling neutrino oscillation phenomena as described in Chapter 11.

The violation of $CP$ implies the long and short lived kaons are mixtures of $CP$ eigenstates $K_1$ and $K_2$ that can be expressed as

$$\begin{aligned} | K_L> &= \frac{| K_2> +\epsilon \mid K_1>}{\sqrt{(}1+ \mid \epsilon \mid^2)} \\ &= \frac{1}{\sqrt{2(1+ \mid \epsilon \mid^2)}}\left[(1+\epsilon) \mid K^0> -(1-\epsilon) \mid \bar{K}^0>\right] \\ | K_S> &= \frac{| K_1> +\epsilon \mid K_2>}{\sqrt{(}1+ \mid \epsilon \mid^2)} \\ &= \frac{1}{\sqrt{2(1+ \mid \epsilon \mid^2)}}\left[(1+\epsilon) \mid K^0> +(1-\epsilon) \mid \bar{K}^0>\right]. \end{aligned} \quad (10.158)$$

Here $\epsilon$ is a complex amplitude with, in this case, a small magnitude. The question arises as to the origin of the mixing. The observed lack of any energy dependence to the parameter $\epsilon$ suggests it is intrinsic to the kaon as a closed system. The size of the effect suggests it is a weak interaction phenomenon although some new interaction might play a role. From a phenomenological point of view, it remains then to determine if the $CP$-violation occurs in the process of mixing of $K_1$ and $K_2$ or directly in the decay processes.

To address this question, one can study other decay modes. Given that $|\epsilon|$ is small, to first order, one expects the semi-leptonic decays to tag strangeness and that $\Gamma_{K^0\to\pi^-l^+\nu} = \Gamma_{\bar{K}^0\to\pi^+l^-\bar{\nu}} \equiv \Gamma_{Kl3}$ and $\Gamma_{\bar{K}^0\to\pi^-l^+\nu} = \Gamma_{K^0\to\pi^+l^-\bar{\nu}} = 0$. So we expect the semi-leptonic decays of the $K_L$, for example, to be given by

$$\Gamma_{K_L\to\pi^-l^+\nu} = \frac{1}{2}\frac{\mid 1+\epsilon \mid^2}{1+ \mid \epsilon \mid^2}\Gamma_{Kl3} \text{ and } \Gamma_{K_L\to\pi^+l^-\bar{\nu}} = \frac{1}{2}\frac{\mid 1-\epsilon \mid^2}{1+ \mid \epsilon \mid^2}\Gamma_{Kl3} \quad (10.159)$$

so the lepton charge asymmetry defined by

$$\delta = \frac{\Gamma_{K_L\to\pi^-l^+\nu} - \Gamma_{K_L\to\pi^+l^-\bar{\nu}}}{\Gamma_{K_L\to\pi^-l^+\nu} + \Gamma_{K_L\to\pi^+l^-\bar{\nu}}} = \frac{2\text{Re}\epsilon}{1+ \mid \epsilon \mid^2} \simeq 2\text{Re}\epsilon \quad (10.160)$$

should measure the real part of the $CP$-violation parameter $\epsilon$. The experimental values are $\delta_\mu = (3.04 \pm 0.25) \times 10^{-3}$ and $\delta_e = (3.33 \pm 0.14) \times 10^{-3}$ and are consistent with this expectation.

The $CP$-violating mixing of $K_1$ and $K_2$ implies we must correct our previous expression for the evolution of a neutral kaon for which we assumed the $K_1$ and $K_2$ were the energy eigenstates. To first order in $\epsilon$, an initially pure $K^0$ state follows the evolution

$$| \psi(t)> = \frac{1-\epsilon}{\sqrt{2}}\left[| K_S> e^{-iE_St} + \mid K_L> e^{-iE_Lt}\right]. \quad (10.161)$$

This expression implies the squared decay amplitude for decay to a two pion final state by the evolving neutral kaon mixture divided by the squared decay amplitude for the same two pion decay mode of the eigenstate $K_S$ will have the rest frame time dependence given by

$$\frac{|<\pi\pi \mid M \mid \psi(t)>|^2}{|<\pi\pi \mid M \mid K_S>|^2} = e^{-\Gamma_S t} + \mid \eta \mid^2 e^{-\Gamma_L t} + 2e^{-(\Gamma_S+\Gamma_L)t/2} \mid \eta \mid \cos(\phi - \Delta m t) \quad (10.162)$$

and this is what is observed. Here a parameter $\eta = \mid \eta \mid e^{i\phi}$ is defined for charged and neutral pions by

$$\eta_{+-} = \frac{<\pi^+\pi^- \mid M \mid K_L>}{<\pi^+\pi^- \mid M \mid K_S>} ; \ \eta_{00} = \frac{<\pi^0\pi^0 \mid M \mid K_L>}{<\pi^0\pi^0 \mid M \mid K_S>}, \quad (10.163)$$

and we can write

$$\eta = \frac{<\pi\pi \mid M \mid K_2> + \epsilon <\pi\pi \mid M \mid K_1>}{<\pi\pi \mid M \mid K_1> + \epsilon <\pi\pi \mid M \mid K_2>} = \frac{\epsilon + x}{1 + \epsilon x} \simeq \epsilon + x \quad (10.164)$$

where $x = <\pi\pi \mid M \mid K_2> / <\pi\pi \mid M \mid K_1>$ measures the amount of direct $CP$-violation in the decay $K_2 \rightarrow \pi\pi$ and was assumed small, that is, most two pion decays result from the $CP$ conserving decay of the $K_2$. The experimental results $\mid \eta_{00} \mid = (2.22 \pm 0.01) \times 10^{-3}$ and $\mid \eta_{+-} \mid = (2.23 \pm 0.01 \times 10^{-3}$ with $\phi_{+-} = 43.51 \pm 0.05$ deg and $\phi_{00} = 43.52 \pm 0.05$ deg are consistent with $x = 0$ for both charged and neutral pions and the results are consistent with the semi-leptonic decay results.

The comparison of the neutral and charge pion decay modes can be rendered a bit more incisive by invoking isospin symmetry and other observations. In $K \rightarrow 2\pi$, the orbital angular momentum quantum number of the two pions must be $l = 0$ and the two final state bosons must have a symmetric isospin wave function. In case $I = 0$, the wave function is proportional to $\boldsymbol{\pi}_1 \cdot \boldsymbol{\pi}_2 = \pi_1^+\pi_2^- + \pi_1^-\pi_2^+ + \pi_1^0\pi_2^0$ while I $= 2$ is represented by the traceless tensor $\pi_1^i\pi_2^j + \pi_1^j\pi_2^i - (2/3)\boldsymbol{\pi}_1 \cdot \boldsymbol{\pi}_2$. The ratio of branching fractions $B_{K_1\rightarrow\pi^+\pi^-}/B_{K_1\rightarrow\pi^0\pi^0} \simeq 2$ suggests only the $I = 0$ state is produced as in charged kaon decays. In fact, in the decay $K^+ \rightarrow \pi^+\pi^0$, the two pions have $I_3 = +1$ so $I = 0$ is ruled out leaving the choice $I = +2$. Experiment finds this $\Delta I = 3/2$ transition is suppressed, $\Gamma_{K^+\rightarrow\pi^+\pi^0}/(\Gamma_{K_1\rightarrow\pi^+\pi^-} + \Gamma_{K_1\rightarrow\pi^0\pi^0}) << 1$. The suppression (called the $\Delta I = \frac{1}{2}$ rule) may result from cancelations in combined weak and chromodynamic amplitudes. Whatever its explanation, the suppression simplifies the phenomenological analysis of the $CP$-violating $2\pi$ decays. The direct $CP$-violation parameter for the $\pi^+\pi^-$ mode is

$$\epsilon' \equiv x_{+-} = \frac{<\pi^+\pi^- \mid M \mid K^0> - <\pi^+\pi^- \mid M \mid \bar{K}^0>}{<\pi^+\pi^- \mid M \mid K^0> + <\pi^+\pi^- \mid M \mid \bar{K}^0>}. \quad (10.165)$$

The decay amplitudes appearing here are written in terms of the two isospin states as

$$< \pi^+\pi^- \mid M \mid K^0 > = \frac{1}{\sqrt{3}} A_2 e^{i\delta_2} + \frac{\sqrt{2}}{3} A_0 e^{-i\delta_0}$$
$$< \pi^+\pi^- \mid M \mid \bar{K}^0 > = \frac{1}{\sqrt{3}} A_2^* e^{i\delta_2} + \frac{\sqrt{2}}{3} A_0^* e^{i\delta_0} \qquad (10.166)$$

when $CPT$ is assumed to relate the particle and antiparticle weak decay amplitudes. The phase factors represent the effect of pion strong interactions and are half of the $s$-wave phase shifts for $\pi\pi$ elastic scattering at $\sqrt{s} = M_K$. The connection emerges from partial wave unitarity with weak interactions treated as a perturbation. The phase of $A_0$ may be chosen as real and, assuming $\mid A_2/A_0 \mid << 1$, substitution gives

$$\epsilon' = i\frac{e^{i(\delta_2-\delta_0)}}{\sqrt{2}} \frac{\text{Im}A_2}{A_0}. \qquad (10.167)$$

With a similar resolution of the $\pi^0\pi^0$ decay amplitudes into isospin amplitudes, assuming these characterize both charged and neutral decays, one has

$$\eta_{+-} = \epsilon + \epsilon'$$
$$\eta_{00} = \epsilon - 2\epsilon' \qquad (10.168)$$

which relate $x_{+-} = \epsilon'$ to $x_{00} = -2\epsilon'$. The ratio of these quantities provides a sensitive probe of $\epsilon'$. The experimental average reported by the PDG is $Re(\epsilon'/\epsilon) \simeq \epsilon'/\epsilon = (1.65 \pm 0.26) \times 10^{-3}$. Thus direct $CP$-violation appears to be small but not zero. In the standard model, $CP$ violation is attributed to a complex phase in the CKM matrix. Calculations of the parameters $\epsilon$ and $\epsilon'$ appear to be consistent with experiment but rely on estimates of uncertain hadronic matrix elements. In any event, it appears the $CP$-violation in the neutral kaon system does not require an explanation outside the standard model. Perhaps more importantly, the values of $\epsilon$ and $\epsilon'$ constrain extensions of the standard model which include new $CP$-violating interactions.

That neutral kaons $K_0$ and $\bar{K}_0$ mix and that the mixtures have long lifetimes results in a remarkable coherent scattering phenomenon called regeneration. A similar phenomenon affects neutrino oscillations as described in Chapter 11. Consider a $K_L$ beam interacting with condensed matter. A $K_L \simeq (K_0 - \bar{K}^0)/\sqrt{2}$ emerges from an elastic collision with a nucleus or electron in matter in a superposition of states

$$\mid K_{out} > = \frac{f \mid K^0 > - \bar{f} \mid \bar{K}^0 >}{\sqrt{2}} = \frac{(f + \bar{f})}{2} \mid K_L > + \frac{(f - \bar{f})}{2} \mid \bar{K}_S > \qquad (10.169)$$

where $f(\theta)$ and $\bar{f}(\theta)$ are the $K^0$ and $\bar{K}^0$ elastic scattering amplitudes. Elastic scattering of neutral kaons in condensed matter is dominated by diffraction from nuclei. Heavy nuclei appear nearly but not quite equally black to $K^0$ and $\bar{K}^0$. The particle and antiparticle total cross sections and elastic scattering amplitudes must approach each other at high energy under fairly general assumptions (Pomeranchuk theorem). But at beam energies of order 10 GeV, the $K^0$ and $\bar{K}^0$ react differently with nuclei, $f \neq \bar{f}$, and a short lived $K_S$ component to the state is regenerated.

According to the general theory of waves, the coherent superposition of scattering amplitudes from many nuclei in the forward direction may be summed and the effect of the medium represented by a refractive index $n$ changing the wave in vacuum to a slower wave

$$e^{i(kz-\omega t)} \rightarrow e^{i((k/n)z-\omega t)}. \tag{10.170}$$

The refractive index is given by

$$n^{-1} = 1 + 2\pi N f(0)/k^2 \tag{10.171}$$

where $N$ is the number density of targets and, for de Broglie waves, the wave vector $k$ is the particle momentum. For a mixture of kaon and anti-kaon, when the refractive effect is small, neglecting decay, we can write an approximate expression

$$\begin{aligned} | K_{out} > &= \frac{e^{ikz}}{\sqrt(2)}\left[e^{i2\pi Nzf/k} \mid K^0 > - e^{i2\pi Nz\bar{f}/k} \mid \bar{K}^0 >\right] \\ &\simeq e^{ikz}\left[\left(1 + 2\pi i Nz\frac{f+\bar{f}}{2k}\right) \mid K_L > \right. \\ &\quad \left. + \left(2\pi i Nz\frac{f-\bar{f}}{2k}\right) \mid K_S > .\right. \end{aligned} \tag{10.172}$$

The regeneration amplitude $\rho \equiv 2\pi i Nz(f - \bar{f})/k$ is modified when decays are included.

If the energy transfer to the nucleus $\mathbf{q}^2/2m_N$ is comparable to a typical nuclear excitation energy, the nucleus will be left in many different final states and regeneration is incoherent. Coherence requires the phase differences of amplitudes arriving at the observation point be nearly zero. For near forward scattering, the contribution at position $z$ from nuclei at $z'$ is

$$A = e^{ip_L z'}\frac{f-\bar{f}}{2}e^{ip_S(z-z')} \tag{10.173}$$

and the phase difference between nuclei separated by $\Delta z$ is $(p_S - p_L)\Delta z - p_S \Delta z \theta^2/2$. The energy transfer to the nucleus is negligible, so

$$E_S = E_L = p_S^2 + m_S^2 = p_L^2 + m_L^2 \tag{10.174}$$

which implies

$$p_S - p_L \simeq \Delta m(m_L/p_L) \simeq \Delta m/\gamma \tag{10.175}$$

where $\Delta m = m_L - m_S$. Hence longitudinal coherence requires $\Delta z < \gamma/\Delta m \sim \gamma\tau_S$ - coherence is maintained only over a length comparable to the oscillation length. The maximum scattering angle for coherent regeneration is $\theta^2 \leq 2$ $(p_S - p_L)/p \rightarrow \theta \leq 10^{-7}$ rad.

The state which exits a weak regenerator is a coherent mixture of $K_L$ and $K_S$ given by

$$| \psi(t) > \simeq e^{-iE_L t} \mid K_L > + \rho e^{-iE_S t} \mid K_S > . \qquad (10.176)$$

The rate of downstream two pion decays is proportional to

$$I = \mid \rho \mid^2 e^{-\Gamma_S t} + \mid \eta \mid^2 e^{-\Gamma_L t} + \mid \rho \mid \mid \eta \mid e^{-\frac{t}{2}(\Gamma_S + \Gamma_L)} \cos(\phi). \qquad (10.177)$$

Here $\phi = \Delta m t + \phi_\rho - \phi_\eta$ and $\phi_\rho$ and $\phi_\eta$ are the phases of the regeneration amplitude and of $\eta$. By varying the regenerator strength, both the magnitude and phase of the $CP$-violating amplitude $\eta$ may be extracted.

In Figure 10.8 is shown the distribution of transverse momentum relative to the beam direction for $K \to \pi^+\pi^-$ events in two parallel $K_L$ beams, one of which passes through a regenerator. In the regenerator beam, high $p_t^2$ events arise from inelastic processes and, near the peak at $p_t^2 = 0$ representing coherent regeneration, a contribution from diffractively regenerated $K_S$ is evident. Radiative decays $K \to \pi^+\pi^-\gamma$, background from misidentified $\pi^\pm l^\pm \nu$ decays, and regeneration from miscellaneous material account for the events at $p_t^2 > 0$ in the vacuum beam. Coherent interference between $K_L$ and regenerated $K_S$ decaying to two pions is illustrated in Figure 10.9.

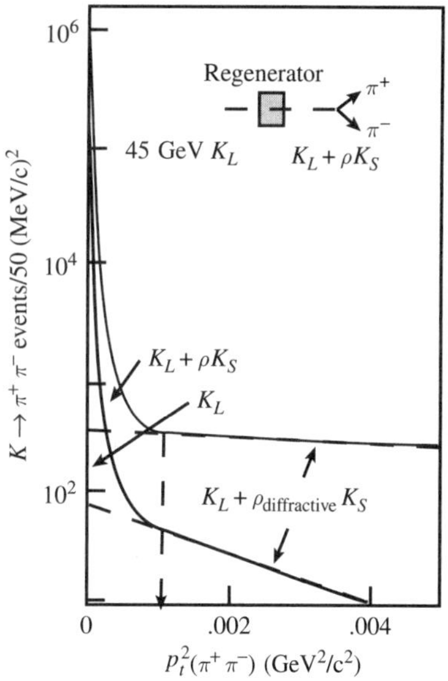


**FIGURE 10.8** Transverse momentum of $K \to \pi^+\pi^-$ in a 45 GeV $K_L$ beam with and without a regenerator. Coherent regeneration of $K_S$ yields a large enhancement in the forward direction. Diffractive regeneration from nuclei is also evident. [L. K. Gibbons *et al.* (The E731 Collaboration) Phys. Rev. **D55**, 6625 (1997).]

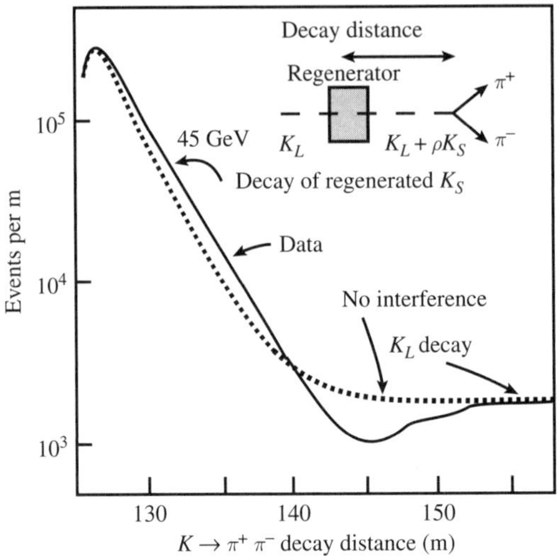


**FIGURE 10.9** Interference between $K_L$ and $K_S$ following a regenerator observed through $K \to \pi^+\pi^-$ decays. The short lived component represents principally the regenerated $K_S$. The $K_L$ decay length is much longer. The data are compared to the double exponential distribution which would be observed in the absence of coherence between $K_L$ and $K_S$. [A. Alavi-Harati *et al.*, Phys. Rev. **D67**, 012005 (2003)]

## 10.7 ■ WEAK INTERACTIONS IN $B^0$ AND $D^0$ SYSTEMS

Heavy quark neutral mesons analogous to the $K^0_{d\bar{s}}$ are the $D^0_{c\bar{u}}$, $B_d \equiv B^0_{d\bar{b}}$ and $B_s \equiv B^0_{s\bar{b}}$. The heavy mesons have many open decay channels, much shorter lifetimes than neutral kaons, and lack the decay mode constraints which produce the large lifetime difference between weakly mixed kaons. But weak mixing has been observed in these mesons and tremendous interest is attached to checking standard model predictions for mixing and $CP$-violation in these systems.

The general phenomenological analysis used to describe such systems is modeled after the analysis of the neutral kaon system and proceeds as follows. Assume $H = H^0 + H^w$ and states $a$ and $b = \bar{a}$ are eigenstates of the strong and electromagnetic interaction $H^0$. The two state Hamiltonian is written as

$$H = M - \frac{i}{2}\Gamma \tag{10.178}$$

where $M$ and $\Gamma$ are Hermitian. The time evolution of a two state system of particle and antiparticle is then governed by the equation

$$i\frac{d}{dt} \mid \psi >= H \mid \psi >= \begin{pmatrix} M_{aa} - \frac{i}{2}\Gamma_{aa} & M_{ab} - \frac{i}{2}\Gamma_{ab} \\ M_{ba} - \frac{i}{2}\Gamma_{ba} & M_{bb} - \frac{i}{2}\Gamma_{bb} \end{pmatrix} \mid \psi > . \tag{10.179}$$

Perturbation theory allows us to represent the matrix elements in terms of the weak Hamiltonian as follows:

$$M_{aa} = m_a + H^w_{aa} + \sum \frac{H^w_{an} H^w_{na}}{m_a - E_n} + \cdots$$
$$M_{bb} = m_b + H^w_{bb} + \sum \frac{H^w_{bn} H^w_{nb}}{m_b - E_n} + \cdots$$
$$M_{ba} = M^\dagger_{ab} = H^w_{ba} + \sum \frac{H^w_{bn} H^w_{na}}{m_a - E_n} + \cdots . \quad (10.180)$$

Suppose $CPT \mid a >=< b \mid$. Symmetry under $CPT$ implies

$$m_a =< a \mid H^0 \mid a >=< b \mid H^0 \mid b >= m_b. \quad (10.181)$$

The decay matrix in the rest frame is

$$\Gamma_{ba} = 2\pi \sum H^w_{bc} H^w_{ca} \delta(E_c - m_a). \quad (10.182)$$

Assuming $CPT$-invariance, we have $H^w_{ca} = H^w_{a'c'}$ where $c'$ is the $CP$ conjugate with reversed spins. As all spin states are included in calculating the decay rate, $\Gamma_{aa} = \Gamma_{bb}$ and $\Gamma_{ab} = \Gamma^*_{ba}$. Then $H_{aa} = H_{bb}$ and the eigenvalues resulting from characteristic equation

$$\mid H - E \mid = (H_{aa} - E)^2 - H_{ab} H_{ba} = 0 \quad (10.183)$$

are given by

$$E_\pm = m_\pm - \frac{i}{2}\Gamma_\pm = M_{aa} - \frac{i}{2}\Gamma_{aa} \pm pq. \quad (10.184)$$

The eigenstates may be written as

$$\mid a_\pm >= \frac{1}{|p^2| + |q^2|}(p \mid a > \pm q \mid b > \quad (10.185)$$

where the mixing parameters are

$$p = \sqrt{H_{ab}} \quad (10.186)$$

and

$$q = \sqrt{H_{ba}}. \quad (10.187)$$

The masses and decay rates of the eigenstates are

$$m_\pm = M_{aa} \pm \mathrm{Re}[pq] \quad (10.188)$$

and

$$\Gamma_\pm = \Gamma_{aa} \mp 2\mathrm{Im}[pq]. \quad (10.189)$$

The following definitions of parameters are useful

$$\Delta m = m_+ - m_- = 2\mathrm{Re}[pq]\;;\; \Delta\Gamma = \Gamma_+ - \Gamma_- = -4\mathrm{Im}[pq]$$
$$\bar{\Gamma} = \frac{\Gamma_{aa} + \Gamma_{bb}}{2} = \Gamma_{aa}\;;\; \frac{p}{q} = \frac{1+\epsilon}{1-\epsilon}\;;\; \epsilon = \frac{p-q}{p+q}. \tag{10.190}$$

Symmetry under $CP$ implies $p = q$ and $\epsilon = 0$. The $CP$-even and $CP$-odd eigenstates are

$$| a_1 > = \frac{1}{\sqrt{2}}(| a > + | b >)$$
$$| a_2 > = \frac{1}{\sqrt{2}}(| a > - | b >) \tag{10.191}$$

and the true eigenstates are

$$| a_+ > = \frac{1}{\sqrt{1+|\epsilon|^2}}(| a_2 > +\epsilon \mid a_1 >)$$
$$| a_- > = \frac{1}{\sqrt{1+|\epsilon|^2}}(| a_1 > +\epsilon \mid a_2 >). \tag{10.192}$$

The time evolution of a general state with initial amplitudes $A_\pm$ for the states $| a_\pm >$ can be written as

$$\begin{aligned} | \psi(t) > &= A_+ e^{-iE_+t} \mid a_+ > + A_- e^{-iE_-t} \mid a_- > \\ &= A_+ e^{-iE_+t}\left[\frac{p}{\sqrt{|p|^2+|q|^2}} \mid a > + \frac{q}{\sqrt{|p|^2+|q|^2}} \mid b >\right] \\ &\quad + A_- e^{-iE_-t}\left[\frac{p}{\sqrt{|p|^2+|q|^2}} \mid a > - \frac{q}{\sqrt{|p|^2+|q|^2}} \mid b >\right] \end{aligned} \tag{10.193}$$

In particular, a pure $| a >$ state at $t = 0$ has $A_+ = A_- = \sqrt{|p|^2 + |q|^2}/(2p)$ and evolves into the mixture

$$| \psi(t) > = \frac{1}{2}\left(e^{-iE_+t} + e^{-iE_-t}\right) | a > + \frac{1}{2}\left(e^{-iE_+t} - e^{-iE_-t}\right)\frac{q}{p} \mid b > . \tag{10.194}$$

The probability for state $| a >$ at time t is

$$| < a \mid \psi(t) > |^2 = \frac{1}{4}\left(e^{-\Gamma_+ t} + e^{-\Gamma_- t} + 2e^{-\bar{\Gamma}t}\cos\Delta m t\right) \tag{10.195}$$

while the probability for $| b >$ is

$$| < b \mid \psi(t) > |^2 = \frac{1}{4}\frac{|q|^2}{|p|^2}\left(e^{-\Gamma_+ t} + e^{-\Gamma_- t} - 2e^{-\bar{\Gamma}t}\cos\Delta m t\right). \tag{10.196}$$

The probability of observing $| a >$ in a time interval $dt$ is proportional to $dP = |< a \mid \psi(t) >|^2 \, dt$ so, starting with state $|a >$, the probability of observing state $|a >$ integrated over all time can be calculated and is given by

$$P_{aa} = \int_0^\infty dt \frac{dP}{dt} = \frac{\bar{\Gamma}}{2}\left[\frac{1}{\bar{\Gamma}^2 - \left(\frac{\Delta\Gamma}{2}\right)^2} + \frac{1}{\bar{\Gamma}^2 + (\Delta m)^2}\right]. \tag{10.197}$$

The probability for observing state $|b >$ is therefore

$$P_{ba} = 1 - P_{aa} = \int_0^\infty dP = \frac{|q|^2}{|p|^2}\frac{\bar{\Gamma}}{2}\left[\frac{1}{\bar{\Gamma}^2 - \left(\frac{\Delta\Gamma}{2}\right)^2} - \frac{1}{\bar{\Gamma}^2 + (\Delta m)^2}\right]. \tag{10.198}$$

The ratio of these two probabilities is

$$r = \frac{P_{aa}}{P_{ba}} = \frac{|q|^2}{|p|^2}\frac{x^2 + y^2}{2 + x^2 - y^2} \tag{10.199}$$

with $x = \Delta m/\bar{\Gamma}$ and $y = \Delta\Gamma/(2\bar{\Gamma})$. Starting with state $|b >$ leads to the ratio

$$\bar{r} = \frac{P_{ab}}{P_{bb}} = \frac{|p|^2}{|q|^2}\frac{x^2 + y^2}{2 + x^2 - y^2}. \tag{10.200}$$

If $a$ and $b = \bar{a}$ are pair produced and each evolves independently, event pairs of type $ab$, $aa$, $ba$ and $bb$ result and the ratio of mixed to unmixed pairs is

$$R = \frac{N_{aa} + N_{bb}}{N_{ab} + N_{ba}} = \frac{P_{ab}(1 - P_{ba}) + (1 - P_{ab})P_{ba}}{P_{ba}P_{ab} + (1 - P_{ba})(1 - P_{ab})}. \tag{10.201}$$

In the absence of $CP$-violation, these ratios are

$$r = \bar{r} = \frac{x^2 + y^2}{2 + x^2 - y^2}\ ; \ R = \frac{2r}{1 + r^2}. \tag{10.202}$$

In $p\bar{p}$ collisions, $B\bar{B}$ and $D\bar{D}$ meson pairs are generally created incoherently and may be distinguished in their semileptonic decays by the transverse momenta of the leptons relative to the meson direction. In $e^+e^- \to \Upsilon(4s) \to B_d\bar{B}_d$, the $B\bar{B}$ pairs produced through a resonance of fixed $CP$ are entangled and the evolution and decay fractions are somewhat different. In both cases, observation of like sign lepton pairs provides evidence for mixing within neutral $B$ mesons. The $B_d$ mixing parameters are found to be $\chi_d \equiv P_{ba} = 0.18$, $x_d = 0.73$. Much stronger mixing is ($x_s \simeq 18$) observed in the $B_s$ system.

Measurements of the time dependence of mixing requires knowledge of the initial flavor in addition to the flavor at decay. For $b\bar{b}$ quark pairs produced at a $p\bar{p}$ collider, the initial flavor is correlated to the charge of associated pions. For example, if a $\bar{b}$-quark combines with a $d$-quark to form a $B^0$, the $\bar{d}$-quark tends to form a $\pi^+$. The correlation implies a same-side-tagging principle can be used to determine on a statistical basis the initial flavor of neutral $B$ mesons, while the leptonic decays $B^0 \to l^+ D^- X$ and $B^0 \to l^+ D^{*-} X$ can be used to tag the flavor at decay. The secondary charmed mesons can for example be reconstructed in the decay modes $D^- \to K^+\pi^-\pi^-$ and $D^{*-} \to \bar{D}^0\pi^-$ with $\bar{D}^0 \to K^+\pi^-, K^+\pi^-\pi^+\pi^-$. Neglecting $CP$-violation and decay rate differences between the two neutral $B_d$ eigenstates, the rate of charged leptons is expected to have the form

$$\frac{dN_\pm}{dt} = \frac{1}{2} e^{-t/\tau}(1 \pm \cos \Delta m_d t). \tag{10.203}$$

Decays of $B^\pm$ mesons which may not oscillate provide a check on the procedure. The results shown in Figure 10.10 are used to measure the mass difference $\Delta m_d = 0.47$ ps$^{-1}$. Corresponding results for the $B_s$ system shown in Figure 10.11 give $\Delta m_s \simeq 18$ ps$^{-1}$.

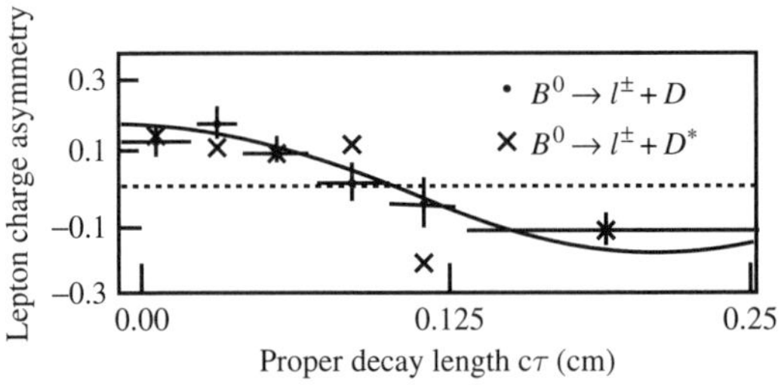


**FIGURE 10.10** $B^0 \Leftrightarrow \bar{B}^0$ oscillations. The time dependence of the quark flavor is observed by leptonic decays $B^0 \to l^+ D^{(*)-} X$ with initial flavor identified by charge of correlated pions. [The CDF Collaboration, Phys. Rev. Lett. **80**, 2057 (1998).]

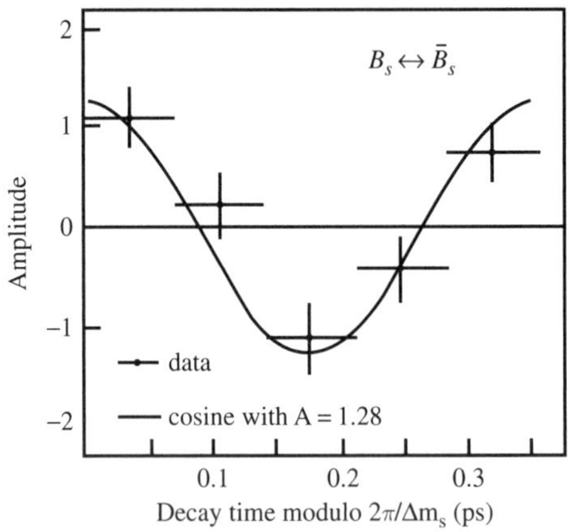


**FIGURE 10.11** $B_s \Leftrightarrow \bar{B}_s$ oscillations. The time dependence of the quark flavor is observed through $B_s \to D_s + X$ where $X$ denotes several hadronic states. The event rate is plotted as a function of decay time modulo the fitted oscillation period governed by $\Delta m_s$. [A. Abulencia *et al.*, The CDF Collaboration, Phys. Rev. Lett. **97**, 242003 (2006).]

In a spectator model in which the heavy quark decays weakly to three lighter quarks followed by recombination, $D$ meson mixing may result from the transitions

$$\begin{aligned} c\bar{u} &\to dW^+\bar{u} \to d\bar{d}u\bar{u} \\ c\bar{u} &\to sW^+\bar{u} \to s\bar{s}u\bar{u} \end{aligned} \tag{10.204}$$

followed by reverse transitions to form the antiparticle of the initial meson. The amplitudes for the transitions are proportional to the products of CKM matrix elements $V_{cd}^* V_{ud}$ and $V_{cs}^* V_{us}$ while the amplitude for the dominant decay $c\bar{u} \to s\bar{d}u\bar{u}$ is proportional to $V_{cs}^* V_{ud}$. The mixing rate is small compared to the decay rate and mixing difficult to observe: $x = \Delta m/2\bar{\Gamma} << 1, y = \Delta\Gamma/(2\bar{\Gamma}) << 1$. The amplitude for $B_d = d\bar{b} \to d\bar{c}c\bar{d}$ is proportional to $V_{bc}^* V_{cd}$ compared to the decay transition amplitude for $d\bar{b} \to d\bar{c}d\bar{u}$ proportional to $V_{bc}^* V_{ud}$ while $B_s$ mixing can occur through $s\bar{b} \to s\bar{c}\bar{s}c$ which is the dominant decay channel with amplitude proportional to $V_{bc}^* V_{sc}$. Thus one might expect large mixing in the $B_s$ system and small mixing in the $B_d$ system. However, the spectator model amplitude is not the only possibility. The amplitude for a "box diagram" process such as $d\bar{b} \to W^+W^-$ with $u$-quark, $c$-quark, and $t$-quark exchange followed by $W^+W^- \to \bar{d}b$, and the related processes shown in Figure 10.12 turns out to be proportional to the mass of the exchanged quark. Denoting the light quark in a $B$ meson by $q = s, d$, the dominant top quark contribution gives a mixing amplitude

$$M_{B_q\bar{B}_q} = f(V_{tb}V_{tq}^*)^2 m_t^2 \tag{10.205}$$

where the factor $f$ derives from the structure of the meson. This "short distance" effect appears to enhance $B$ mixing relative to $D$ mixing in which the highest mass flavor in the box diagrams is the $b$-quark. In terms of our general parameters, for $B^0\bar{B}^0$ mixing,

$$\frac{q}{p} = \frac{V_{tb}^* V_{td}}{V_{tb} V_{td}^*} \tag{10.206}$$

and, since in the standard phase convention $V_{tb} \simeq 1$ and $V_{tb} \sim e^{-i\delta}$, this ratio is directly related to the $CP$-violating phase $\delta : q/p = \exp(-2i\delta)$.

The decays of bottom flavored mesons have been studied by the CDF and D0 collaborations at the Tevatron, by the BABAR collaboration at the SLC, and by the BELLE collaboration at KEK. Arguably the most astounding result of these studies is the observation of $CP$-violation in the $B$ system. In the decays of a pseudoscalar $B$ meson to fully reconstructable $CP$ eigenstates such as

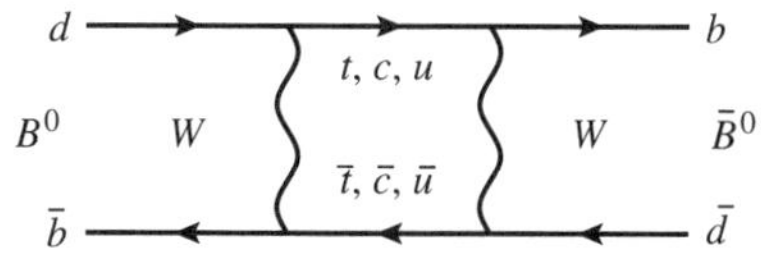


**FIGURE 10.12** Feynman diagram describing $B^0\bar{B}^0$ mixing. The contribution from the top quark dominates.

$$B^0 \to J/\psi K_{S,L} \text{ or } \psi(2s) K_{S,L}, \tag{10.207}$$

the final state is $CP = +1$ for $K_S$ and $CP = -1$ odd for $K_L$. The time distributions for the decay of $B^0$ to a state with $CP = \pm 1 \equiv \eta_f$ in the standard model have the form

$$\Gamma_\pm(t) = \frac{e^{-t/\tau}}{4\tau}\{1 \mp \mathrm{Im}[\lambda_f]\sin(\Delta m_B t)\} \tag{10.208}$$

where $\lambda_f = (q/p)\bar{A}_f/A_f$ with $A_f$ the amplitude for $B^0 \to f$ and $\bar{A}_f$ the amplitude for $\bar{B}^0 \to f$. In the spectator model, the decay amplitude for $\bar{b} \to \bar{c}W^+ \to \bar{c}c\bar{s}$ is proportional to $V^*_{cb}V_{cs}$ and since $K_S \simeq (K^0_{d\bar{s}} - \bar{K}^0_{d\bar{s}})/\sqrt{2}$, the ratio is

$$\begin{aligned}\frac{\bar{A}_f}{A_f} &= \eta_f \frac{V_{cb}V^*_{cs}}{V^*_{cb}V_{cs}}\left(\frac{q}{p}\right)_K \\ &= \eta_f \frac{V_{cb}V^*_{cs}V_{cs}V^*_{cd}}{V^*_{cb}V_{cs}V^*_{cs}V_{cd}} = \eta_f\end{aligned} \tag{10.209}$$

while the mixing factor is governed by the ratio of CKM matrix elements $(q/p)_B = \exp(-2i\delta)$ and consequently

$$\lambda_f = -\eta_f \sin(2\delta). \tag{10.210}$$

The time-dependent $CP$-violating asymmetry is defined by

$$A_{CP} = \frac{\Gamma_+(t) - \Gamma_-(t)}{\Gamma_+(t) + \Gamma_-(t)} = -\eta_f \sin(2\beta)\sin(\Delta m_B t) \tag{10.211}$$

where the customary Wolfenstein parameter name $\beta = \delta$ is used. The results from the $b$-factory experiments BABAR and BELLE in Figure 10.13 clearly show an asymmetry of the expected form. From a variety of such measurements, these experiments have extracted the lifetimes and mixing parameters from which the values of CKM matrix elements related to $b$-quarks may be derived.

It has become customary to describe the CKM matrix in terms of the Wolfenstein parameters $\lambda$, $A$, $\beta = \delta$, $\bar{\rho} \simeq \rho$, and $\bar{\eta} \simeq \eta$ where

$$s_{12} = \lambda,\ s_{23} = A\lambda^2,\ s_{13}e^{i\delta} = A\lambda^3(\rho + i\eta). \tag{10.212}$$

Unitarity of $V_{\mathrm{CKM}}$ implies the constraints

$$\Sigma_i V_{ij}V^*_{ik} = \delta_{jk}\ ,\ \Sigma_j V_{ij}V^*_{kj} = \delta_{ik}. \tag{10.213}$$

The products of matrix elements in these sums, divided by one of them, can be considered as phasors in the complex plane. Then the phasor sums which vanish can be represented as closed triangles, one side of which is of unit length. In the standard parametrization, $CP$-violation is characterized by the single complex phase and this phase implies these triangles have non-zero area. An example is shown in Figure 10.14 along with experimental constraints. It may be seen that

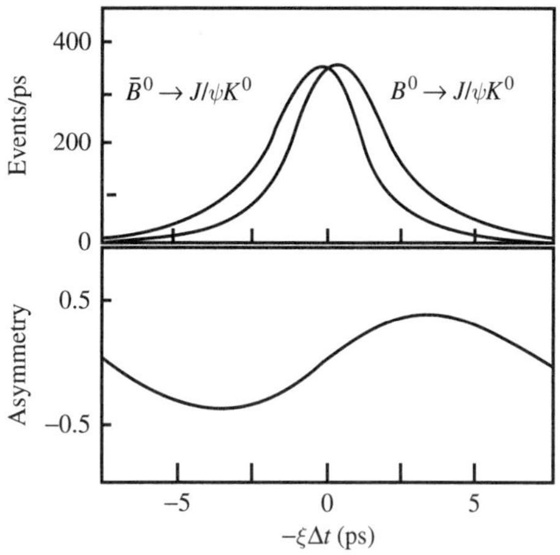


**FIGURE 10.13** $CP$-violation in $B^0$ decay. Distribution of decay position for tagged $B^0$ and $\bar{B}^0$ decays to $CP$ eigenstates such as $J/\psi K_S$ observed b-factory experiments BELLE and BABAR, and the derived $CP$-violating asymmetry. The asymmetry oscillation reflects $CP$-violation in the the weak interaction mixing $B^0 \leftrightarrow \bar{B}^0$.

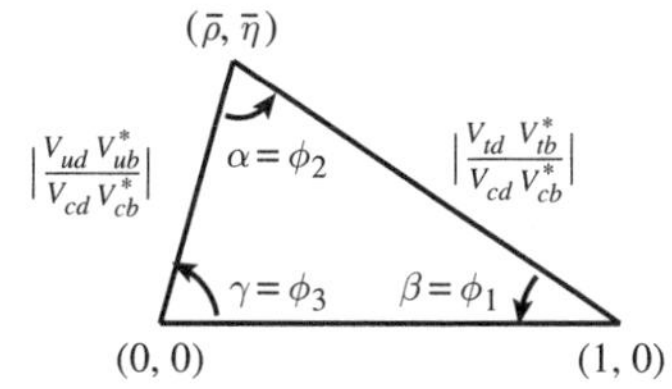


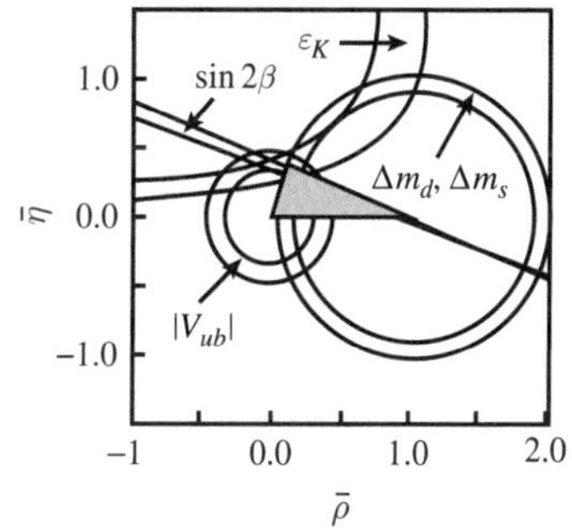


**FIGURE 10.14** CKM matrix unitary triangle. (a) Graphical representation of one unitarity condition for the CKM matrix. (b) Experimental constraints on the location of the uppermost vertex from neutral kaon decay ($\epsilon_k$), neutral $B_d$ and $B_s$ meson oscillations ($\Delta m_d$ and $\Delta m_s$), $B$ decay rates ($|V_{ub}|$) and $B$ direct $CP$-violation observations ($\sin 2\beta$) [Particle Data Group, C. Amsler et al., Phys. Lett. **B667**, 1 (2008)]

observations are consistent with unitarity and that the $CP$-violation observed in neutral strange quark and bottom quark systems are consistent with the single source embodied in the standard model. A global fit assuming unitarity by the Particle Data Group in 2008 derives the Wolfenstein parameter values

$$\lambda = 0.226 \pm 0.001, \quad A = 0.82 \pm 0.02, \quad \bar{\rho} = 0.135 \pm 0.03, \quad \bar{\eta} = 0.35 \pm 0.02 \tag{10.214}$$

and the errors indicate that the matrix elements in this context are known to within a few percent.

## 10.8 ■ PHENOMENOLOGICAL CALCULATIONS

The weak interactions of light quarks in hadrons to a large extent motivated the standard model electroweak theory. The analysis of such processes is limited by incomplete knowledge of hadronic wave functions but various approximations and parameterizations lead to interesting results. So-called semi-leptonic decays in which a quark transition produces a virtual $W$ boson which decays to lepton plus neutrino are the simplest to understand. Non-leptonic weak decays in which a virtual $W$ boson decays to light quarks leading to hadrons are more complicated. In this section, a few example processes are considered.

The original weak interaction, neutron $\beta$ decay, occurs via a weak transition $d \to uW^-$ with $W^- \to e^-\bar{\nu}_e$ leading to

$$n \to pe^-\bar{\nu}. \tag{10.215}$$

The matrix element has the form

$$M_{n\to pe^-\bar{\nu}} = \frac{G_F V_{ud}}{\sqrt{2}} \bar{u}_e\gamma_\mu(1-\gamma_5)v_\nu \bar{u}_p\gamma^\mu(g_V - g_A\gamma_5)u_n \tag{10.216}$$

where the hadronic current has been expressed in the most general form consistent with Lorentz invariance in the limit of vanishing decay momentum, and the CKM matrix element $V_{ud}$ appears explicitly. The form factors $g_V$ and $g_A$ are in principle functions of $q^2 = (p_p - p_n)^2 \sim 1\ \text{MeV}^2$ but, since $q^2$ is small compared to an inverse fm squared, the form factors may be assumed constant. The vector portion of the hadronic current is related to the normalized electromagnetic current $j_p = \bar{u}_p\gamma u_p$ by an isospin rotation so $g_V = 1$. No corresponding argument constrains the axial portion and experimentally $g_A \simeq 1.26$.

The decay rate calculated with this matrix element is found to be

$$\Gamma_{n\to pe^-\bar{\nu}} = \frac{G_F^2|V_{ud}|^2}{2\pi^3}(g_V^2 + 3g_A^2)m_e^5 f(E_{max}/m_e) \tag{10.217}$$

with $f(x) = (1/15)(2x^4 - 9x^2 - 8)\sqrt{x^2-1} + x\ln[x + \sqrt{x^2-1}]$. The analysis may be extended to nuclear decay. In particular, the decay of $^{14}O(0^+) \to {}^{14}N^*(0^+)$ involves only the vector current since the spins and parities are the same and leptons, which effectively originate at a point, carry away no angular momentum. The states are connected by an isospin transformation and the decay rates used to determine $V_{ud}$.

With the Particle Data Group value $V_{ud} = 0.974$ extracted from nuclear $\beta$ decay, using $E_{max}/m_e = 2.53$ for neutron decay along with $G_F = 1.116 \times 10^{-5}\ \text{GeV}^{-2}$, the calculated neutron lifetime is 965 s compared to the experimental and somewhat controversial result

$$\tau_n = 885.7 \pm 0.8\ \text{s}. \tag{10.218}$$

The differential decay rate prediction should be multiplied by a function $F_C(E_e)$ to correct for the final state Coulomb interaction between the electron and proton. This correction function has the approximate form $F_C = x/(1 - e^{-x})$ with $x = 2\pi\alpha/v_e$ and increases the predicted decay rate and lowers the predicted lifetime by about 3%. Analysis of the angular correlation between the proton and electron separately determines $g_A/g_V$. An electroweak radiative correction increases the predicted rate by about 4%. At present, consistency of the prediction and measurement of neutron decay requires a value $|V_{ud}| = 0.971$ slightly smaller than the value derived from nuclear beta decay.

That $g_V = 1$ derives from the hypothesis of conservation of the vector current (CVC) in strong and weak interactions. In the limit $m_u = m_d$, the Lagrangian of quantum chromodynamics is symmetric under $u$-$d$ flavor transformations. The fundamental currents are vectors and the vector isospin currents are conserved. Were $m_u = m_d = 0$, the Lagrangian of quantum chromodynamics would be actually symmetric under independent rotations of left and right spinors (chiral symmetry) and would conserve both the vector and axial vector isospin currents. The axial vector current seems to be only approximately conserved.

It is interesting to see what the quark model predicts for $g_V$ and $g_A$. In the nonrelativistic limit, the weak current connecting the $d$-quark and $u$-quark is

$$j = \frac{g}{\sqrt{2}}(u^\dagger d, -u^\dagger \boldsymbol{\sigma} d) \tag{10.219}$$

where $u$ and $d$ are 2-component spinors and nonrelativistic normalization has been assumed. The vector current contributes the time component and the axial vector current contributes the space components. The proton wave function is

$$\begin{aligned} p^\uparrow = \frac{1}{\sqrt{18}}&[2u^\uparrow d^\downarrow u^\uparrow + 2u^\uparrow u^\uparrow d^\downarrow + 2d^\downarrow u^\uparrow u^\uparrow \\ &-u^\uparrow u^\downarrow d^\uparrow - u^\uparrow d^\uparrow u^\downarrow - u^\downarrow d^\uparrow u^\uparrow \\ &-d^\uparrow u^\downarrow u^\uparrow - d^\uparrow u^\uparrow u^\downarrow - u^\downarrow u^\uparrow d^\uparrow] \end{aligned} \tag{10.220}$$

and the corresponding neutron wave function $n^\uparrow$ is obtained by interchanging $u$ and $d$ and adding a minus sign. The proton wave function may be formally obtained from the neutron wave function by application of the isospin raising operator $T^+ = T_1^+ + T_2^+ + T_3^+$ using with $T^+ d = u$ and $T^+ u = 0$.

Consider now the matrix element between neutron and proton for emission of a $W$ boson with, for example, polarization 4-vector $\epsilon^\mu = (1, 0, 0, 1)$. The vector current contributes

$$\epsilon^\dagger j_V = p^\uparrow \frac{g}{\sqrt{2}} T^+ n^\uparrow = \frac{g}{\sqrt{2}}. \tag{10.221}$$

This matrix element is just that expected for point-like nucleons - thus $g_V = 1$. Now consider the contribution of the axial vector current. When $T^+\sigma_z$ is applied to the neutron wave function resulting in successive replacements $d^\uparrow \to u^\uparrow$ and $d^\downarrow \to -u^\downarrow$, terms with $d^\downarrow$ pick up a minus sign and the result is the proton wave function with the replacement of coefficients $2/\sqrt{18} \to 2/\sqrt{18}$ and $-1/\sqrt{18} \to -3/\sqrt{18}$ so projecting onto the proton wave function yields

$$p^{\uparrow} T^{+} \sigma_z n^{\uparrow} = \frac{1}{18}[4+4+4+3+3+3+3+3+3] = \frac{5}{3}. \tag{10.222}$$

The axial matrix element is therefore given by

$$\epsilon^{\dagger} j_A = p^{\uparrow} \frac{g}{\sqrt{2}} T_{+} \sigma_z n^{\uparrow} = \frac{5}{3} \frac{g}{\sqrt{2}} \tag{10.223}$$

so the naive quark model implies $g_A/g_V = 5/3$. The experimental ratio $g_A/g_V$ is closer to 5/4.

The deviation of the experimental result from the nonrelativistic model prediction can be understood in a bag model to be a consequence of relativistic corrections. In the bag picture, nucleons are modeled by confined quark spherical waves with $j_z = \pm 1/2$ with the same isospin structure as in the nonrelativistic model. The ratio of coupling constants is

$$\frac{g_A}{g_V} = \frac{< p^{\uparrow}|T^{+}\sigma_z|n^{\uparrow} >}{< p^{\uparrow}|T^{+}|n^{\uparrow} >} = \frac{5}{3} < \sigma_z > . \tag{10.224}$$

The total angular momentum is $j_z = l_z + \sigma_z/2$ and the explicit solutions give

$$\frac{g_A}{g_V} = \frac{5}{3}\left[1 - \frac{2x-3}{3(x-1)}\right] \tag{10.225}$$

with $x = 2.04$ the quark momentum in units of inverse bag radius.

The results for neutron beta decay are displayed to indicate the level of success obtained from a mixture of quark level field theory with naive quark model and symmetry ideas. Flavor rotation may be used to relate the neutron decay to other baryon beta decays including hyperon decays. The $\beta$ decays of kaons may be similarly analyzed. Ever more sophisticated analyses of heavy quark weak decays are required to understand the plethora of weak decay modes accessible to charm and bottom flavors hadrons. For such analyses, the reader should consult the literature.

The pion weak decay $\pi^- \to \mu^- \bar{\nu}_\mu$ posed an interesting conundrum prior to the discovery of parity violation. The decay can be understood to result from $\bar{u}d$ fusion into a $W$ boson followed by leptonic decay of the $W$ boson

$$\pi^- \to \bar{u}d \to W^- \to \mu\bar{\nu}_\mu. \tag{10.226}$$

Since $m_\pi = 139$ MeV, $m_e = 0.5$ MeV, and $m_\mu = 105$ MeV, the available lepton kinetic energy and phase space favors decay to electron rather than muon. The apparent conundrum is why the pion rarely decays to $e^-\bar{\nu}_e$ when we know that the $W$ boson couples with equal strength to $e^-$ and $\mu^-$.

The explanation is illustrated in Figure 10.15. The matrix element has the low energy form

$$M_{\pi^- \to \mu^- \bar{\nu}_\mu} = \frac{g^2 V_{ud}}{2} j_\pi^\mu \frac{g_{\mu\nu}}{m_W^2} j_{l\nu} \simeq G_F \sqrt{2} V_{ud} 2 j_\pi \cdot j_{l\nu} \tag{10.227}$$

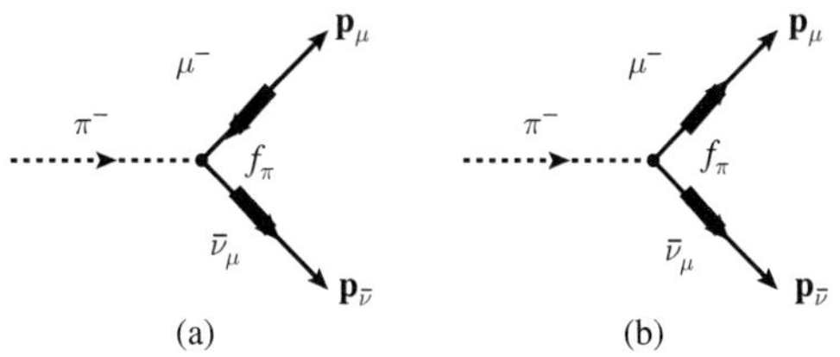


**FIGURE 10.15** The decay $\pi^- \to \mu^- \bar{\nu}_\mu$. Two possible final states are (a) a negative helicity $\mu^-$ and (b) a positive helicity $\mu^-$. The charged current weak interaction couples $\mu_L^+$ and $\bar{\nu}_R$. The amplitude for (a) is inconsistent with angular momentum conservation. The amplitude for (b) is proportional to $m_\mu$. The corresponding amplitude for $\pi^- \to e^- \bar{\nu}_e$ is proportional to $m_e$.

where $j_l = \bar{u}_l \gamma \frac{1}{2}(1-\gamma_5)v_\nu$ is the left-handed lepton transition current coupling to the virtual $W$ boson, $j_\pi$ represents the weak transition $\pi^- \to W^-$, and $p$ is the system 4-momentum satisfying $p^2 = m_\pi^2$. On grounds of Lorentz invariance, the hadronic current has the form

$$j_\pi = < 0|\bar{u}\gamma \frac{(1-\gamma_5)}{2} d|\pi^- > = \frac{1}{2} < 0|\bar{u}\gamma\gamma_5 d|\pi^- > = \frac{i}{2} f_\pi p \tag{10.228}$$

where only the axial current contributes since the $\pi^-$ wave function is odd under parity. In the pion rest frame, $j_\pi = (1/2)m_\pi f_\pi(1,0,0,0)$. The left-handed lepton current with nonvanishing time component $j_{+1}$ represents electron helicity opposite to the neutrino helicity conserving angular momentum along the decay direction. The matrix element is

$$M_{\pi^- \to \mu^- \bar{\nu}_\mu} = m_\pi f_\pi G_F \sqrt{2} V_{ud} \sqrt{2E_\nu(E_l - p)} \tag{10.229}$$

and then, since the decay is isotropic, the decay rate is easily found to be

$$\Gamma_{\pi^- \to \mu^- \bar{\nu}_\mu} = \frac{|M_{\pi^- \to \mu^- \bar{\nu}_\mu}|^2}{8\pi m_\pi^2} p_l. \tag{10.230}$$

The lepton energy is $E_l = (m_\pi^2 + m_l^2)/(2m_\pi)$ and the momentum is $p_l = p_\nu = E_\nu = (m_\pi^2 - m_l^2)/(2m_l)$. Hence

$$\begin{aligned}\Gamma_{\pi^- \to \mu^- \bar{\nu}_\mu} &= \frac{G_F^2 |V_{ud}|^2 f_\pi^2}{2\pi}\left[E_\nu = \frac{m_\pi^2 + m_\mu^2}{2m_\pi}\right]\left[E_e - p = \frac{m_\mu^2}{m_\pi}\right]\left[p = \frac{m_\pi^2 - m_\mu^2}{2m_\pi}\right]\\ &= \frac{G_F^2 |V_{ud}|^2 f_\pi^2 m_\pi}{8\pi}\left(1 - \frac{m_\mu^2}{m_\pi^2}\right)^2 m_\mu^2.\end{aligned} \tag{10.231}$$

The rate is proportional to $m_\mu^2$ as a consequence of the left-handed leptonic coupling. Decays to electrons are suppressed relative to muons by a factor of $(m_e/m_\mu)^2$

despite the fact that much more phase space is available to the electron channel. The helicity suppression explains the otherwise apparent violation of weak lepton universality. Given $G_F$ and $V_{ud}$, the pion decay constant is found to be $f_\pi = 130$ MeV. A similar calculation applies to $K \to l\nu$ with $f_\pi \to f_K, V_{ud} \to V_{us}, m_\pi \to m_K$.

Extensions of these ideas give successful descriptions of other weak processes involving hadrons which can be illustrated by

$$\tau^- \to W^-\nu_\tau \to \pi^-\nu_\tau. \tag{10.232}$$

Assume the amplitude for the conversion $W^- \Leftrightarrow \pi^-$ is determined by $\pi^- \to \mu^-\bar{\nu}$. The matrix element has the form

$$M_{\tau^-\to\pi^-\nu_\tau} = \sqrt{2}G_F V_{ud} f_\pi p_\pi \cdot j^L_{\tau\nu} \tag{10.233}$$

where the left-handed leptonic current is $j = \phi^\dagger_\nu(1, -\boldsymbol{\sigma})\phi_\tau$. In the $\tau$ rest frame, the neutrino spinor is

$$\phi_\nu = \sqrt{2E_\nu}\phi_{-1}(\mathbf{n}_\nu) = \sqrt{2E_\nu}\phi_{+1}(\mathbf{n}_\pi) = \sqrt{2E_\nu}\begin{pmatrix} C \\ S \end{pmatrix} \tag{10.234}$$

with $\mathbf{n}_\nu$ the neutrino direction and $\mathbf{n}_\pi = -\mathbf{n}_\nu$ the direction of the pion and we have chosen $\theta$ the polar angle of $\mathbf{n}_\pi$ and used the notation $S = \sin\frac{\theta}{2}$, $C = \cos\frac{\theta}{2}$. We can choose the $\tau$ spin along the $z$-axis and we have

$$\phi_\tau = \sqrt{m_\tau}\phi_s(\mathbf{z}) = \sqrt{m_\tau}\begin{pmatrix} 1 \\ 0 \end{pmatrix} \text{ or } \sqrt{m_\tau}\begin{pmatrix} 0 \\ 1 \end{pmatrix}. \tag{10.235}$$

The current for the $\tau$ spin state $s_z(\tau) = +1$ is

$$\begin{aligned} j(s_z = +1) &= \sqrt{2m_\tau E_\nu}[C, S][\phi_+, -\phi_-, -i\phi_-, -\phi_+] \\ &= \sqrt{2m_\tau E_\nu}[C, -S, -iS, -C] \end{aligned} \tag{10.236}$$

and the current for the $\tau$ spin state $s_z(\tau) = -1$

$$\begin{aligned} j(s_z = -1) &= \sqrt{2m_\tau E_\nu}[C, S][\phi_-, -\phi_+, i\phi_+, +\phi_-] \\ &= \sqrt{2m_\tau E}[S, -C, +iC, +S]. \end{aligned} \tag{10.237}$$

The pion momentum is $p_\pi = E_\pi(1, v\sin\theta, 0, v\cos\theta)$. The relations

$$\sin(\theta - \frac{\theta}{2}) = \sin\theta\cos\frac{\theta}{2} - \cos\theta\sin\frac{\theta}{2} \tag{10.238}$$

$$\cos(\theta - \frac{\theta}{2}) = \cos\theta\cos\frac{\theta}{2} + \sin\theta\sin\frac{\theta}{2} \tag{10.239}$$

can be used to write

$$j(s_z=+1)\cdot p=\sqrt{2m_\tau E_\nu}E_\pi(1+v)\cos\frac{\theta}{2}$$
$$j(s_z=-1)\cdot p=\sqrt{2m_\tau E_\nu}E_\pi(1+v)\sin\frac{\theta}{2}. \tag{10.240}$$

The pion energy and momentum are

$$E_\pi=\frac{m_\tau^2+m_\pi^2}{2m_\tau}\;;\;p_\pi=\frac{m_\tau^2-m_\pi^2}{2m_\tau} \tag{10.241}$$

so $E_\pi(1+v)=E_\pi+p_\pi=m_\pi$ and, since

$$\cos^2\frac{\theta}{2}=\frac{1}{2}(1+\cos\theta)\;;\;\sin^2\frac{\theta}{2}=\frac{1}{2}(1-\cos\theta), \tag{10.242}$$

the squared matrix element is

$$|M^\pm_{\tau^-\to\pi^-\nu_\tau}|^2=G_F^2|V_{ud}|^2f_\pi^2m_\tau^3E_\nu(1\pm\cos\theta). \tag{10.243}$$

The spin average is angle independent and the 2-body phase space is

$$PS=E_\nu/(4\pi m_\tau) \tag{10.244}$$

so the decay rate is

$$\Gamma_{\tau^-\to\pi^-\nu_\tau}=\frac{1}{2m_\tau}\frac{1}{2}\left[|M^+|^2+|M^-|^2\right]PS=\frac{1}{16\pi}G_F^2|V_{ud}|^2f_\pi^2m_\tau^3\left(1-\frac{m_\pi^2}{m_\tau^2}\right)^2. \tag{10.245}$$

Hence we arrive at a prediction of decay rate of the $\tau$ to a single pion in terms of the pion decay constant which may be expressed in terms of the $\tau^-\to e^-\bar{\nu}_e\nu_\tau$ decay rate as

$$\Gamma_{\tau\to\pi\nu}\simeq\frac{12\pi^2f_\pi^2|V_{ud}|^2}{m_\tau^2}\Gamma_{\tau\to e\nu\bar{\nu}} \tag{10.246}$$

and a similar line of reasoning leads to

$$\Gamma_{\tau\to K\nu}\simeq\frac{12\pi^2f_K^2|V_{us}|^2}{m_\tau^2}\Gamma_{\tau\to e\nu\bar{\nu}}\simeq 0.6\,\Gamma_{\tau\to e\nu\bar{\nu}}. \tag{10.247}$$

We can take this one step further. The $W$ boson and $\gamma$ interactions of the first generation quark isodoublet $q$ is expressed by the Lagrangian

$$L=-\frac{g}{2\sqrt{2}}\bar{q}\gamma_\mu(1-\gamma^5)T^+qW^+-\frac{g}{2\sqrt{2}}\bar{q}\gamma_\mu(1-\gamma^5)T^-qW^-$$
$$+e\left(\frac{1}{2}\bar{q}\gamma_\mu T_3q+\frac{1}{6}\bar{q}\gamma_\mu q\right)A^\mu \tag{10.248}$$

where $T^{\pm}$ are isospin raising and lowering operators and the electromagnetic current has been written as a sum of iso-vector and iso-scalar currents:

$$j_{\rm EM} = \frac{2}{3}\bar{u}\gamma u - \frac{1}{3}\bar{d}\gamma d = \frac{1}{2}\bar{q}\gamma T_3 q + \frac{1}{6}\bar{q}\gamma q. \tag{10.249}$$

The Lorentz vector current $\bar{u}\gamma_\mu d$ is a (spherical) component of an iso-vector which is $G$-even like the $\rho$ meson and is responsible for the leading order weak conversion $\rho^- \to W^-$. Iso-vector currents connecting isospin eigenstates are related by isospin symmetry in the strong interaction. For example, consider the decay

$$\tau^- \to \rho^- \nu_\tau. \tag{10.250}$$

The transition vector current describing the fusion of a $d$-quark and $\bar{u}$-quark in a $\rho^-$ to form a $W^-$ which is operative here can be related to the transition vector current describing the fusion of $u$-quark and $\bar{u}$-quark in a $\rho^0$ to form a virtual photon. So the decay $\rho \to \gamma \to e^+e^-$ can be used to estimate how the $\rho^-$ couples to the $W^-$ and, since the amplitude for $\tau^- \to W^-\nu_\tau$ may be calculated, the decay rate for $\tau^- \to \nu_\tau\rho^-$ may be predicted. (The Lorentz axial vector iso-vector $\bar{u}\gamma_\mu\gamma^5 d$ is $G$-odd like the $\pi$ and has no electromagnetic counterpart.)

The electromagnetic annihilation process $\rho^0 \to e^+e^-$ is described by the amplitude

$$M_{\rho^0\to e^+e^-} = \frac{1}{\sqrt{2}} f_\rho \epsilon \frac{4\pi\alpha}{q^2} \bar{u}_e \gamma u_e \tag{10.251}$$

where $f_\rho$ is a form factor which will assumed to be constant as in pion decay. The electromagnetic current appears and, since $\rho = (u\bar{u} - d\bar{d})/\sqrt{2}$, the factor $\frac{1}{\sqrt{2}}$ in the amplitude results from the sum over quark charges:

$$M_{\rho^0\to e^+e^-} \sim \left[\frac{1}{\sqrt{2}}[e_u - e_d]\right] = \left[\frac{1}{\sqrt{2}}\left(\frac{2}{3} - \left(\frac{-1}{3}\right)\right)\right] = \frac{1}{\sqrt{2}}. \tag{10.252}$$

From this amplitude, the leptonic decay rate is found to be

$$\Gamma_{\rho^0\to e^+e^-} = \frac{4}{3}\pi\alpha^2 \frac{f_\rho^2}{2}\frac{1}{m_\rho^3}. \tag{10.253}$$

The Lorentz vector iso-vector current in the weak decay leads to a matrix element of the form

$$M_{\tau^-\to\rho^-\nu_\tau} = \sqrt{2}G_F V_{ud} f_\rho \epsilon_\mu j^\mu_{\nu\tau} \tag{10.254}$$

where $\epsilon$ is the polarization vector of the $\rho^-$ and $\epsilon \cdot p_\rho = 0$. Using this form, the decay rate is

$$\Gamma_{\tau^-\to\rho^-\nu_\tau} = \frac{1}{16\pi^2}\frac{G_F^2 f_\rho^2}{m_\rho^2} m_\tau^3 |V_{ud}|^2 \left(1 - \frac{m_\rho^2}{m_\pi^2}\right)^2 \left(1 + 2\frac{m_\rho^2}{m_\tau^2}\right). \tag{10.255}$$

Comparison leads to the prediction

$$\frac{\Gamma_{\tau\to\rho\nu}}{\Gamma_{\tau\to e\nu_e\nu_\tau}} = 18\alpha^{-2}\frac{m_\rho}{m_\tau^2}|V_{ud}|^2\Gamma_{\rho\to e^+e^-}. \quad (10.256)$$

that the reader is invited to compare to measurements.

## 10.9 ■ FURTHER READING

The 1980 Nobel Prize in Physics was awarded jointly to James Watson Cronin and Val Logsdon Fitch "for the discovery of violations of fundamental symmetry principles in the decay of neutral K-mesons." Regeneration is described in M. L. Good, Phys. Rev. **106**, 591 (1957).

Much of the material of this chapter is discussed in introductory texts on particle physics with emphasis on phenomenology. See for example D. H. Perkins, *Introduction to High Energy Physics*, Addison-Wesley (2000).

Seminal papers on parity violation include T. D. Lee and C. N. Yang, Phys. Rev. **104**, 254 (1956), C. S. Wu et al., Phys. Rev. **105**, 1413 (1957), and C. N. Yang, Phys. Rev. **77**, 242 (1950). Selection rules for the dematerialization of a particle into two photons are found in R. Plano et. al., Phys. Rev. Lett. **9**, 114 (1962).

A recent example of helicity decomposition analysis to determine the spin and parity of a hadron is K. Abe *et al.* (Belle Collaboration),"Experimental constraints on the possible $J^{PC}$ quantum numbers of the X(3872)," hep-ex/0505038, and "Analysis of the Quantum Numbers $J^{PC}$ of the X(3872) Particle," A. Abulencia *et al.* (CDF Collaboration), Phys. Rev. Lett. **98**,132002 (2007).

Gluon mixing in neutral hadrons is described in T. DeGrand, R. L. Jaffe, K. Johnson, and J. Kiskis, Phys. Rev. **D12**, 2060 (1975). String and bag models are described in K. Johnson and C. B. Thorn, Phys. Rev. **D13**, 1934 (1976). QCD motivated fits to the mass spectrum are described in A. De Rujula, H. Georgi, and S. L. Glashow, Phys. Rev. **D12**, 147 (1975) and Phys. Rev. Lett. **37**, 398 (1976). The experimental status of glue-balls is reviewed in V. Crede, C. A. Meyer, Progress in Particle and Nuclear Physics **63**, 74 (2009).

The naive quark model gives an incomplete description of hadronic resonances. Outstanding mysterious states are reviewed in Eberhard Klempt, "Glueballs, Hybrids, Pentaquarks: Introduction to Hadron Spectroscopy and Review of Selected Topics," lectures at 18th Annual Hampton University Graduate Studies, Jefferson Lab, Newport News, Virginia, June 2-20, 2003, arXiv:hep-ph/0404270v1.

A review of the properties of the neutron including measurement of the neutron lifetime, $g_A$ and extraction of $|V_{ud}|$, and the gravitational properties of ultracold neutrons is Hartmut Abele, Progress in Particle and Nuclear Physics **60**, 1 (2008).

A paper describing the state of the art in calculating the nuclear force is "Helium Nuclei in Quenched Lattice QCD," T. Yamazaki, Y. Kuramashi, A. Ukawa, for the PACS-CS Collaboration, UTCCS-P-57, UTHEP-601, and arXiv:0912.1383v1 [hep-lat]. For baryon-baryon interactions, see Silas R. Beane *et. al.*, UNH-09-06, JLAB-THY-09-1116, NT@UW-09-26, IUHET-539, ATHENA-PUB-09-019, arXiv:0912.4243v1 [hep-lat].

Fermion flavor mixing model is described in N. Cabibbo, Phys. Rev. Lett. **10**, 531 (1963) and M. Kobayashi and T. Maskawa, Prog. Theor. Phys. **49**, 652 (1973), and L. Wolfenstein, Phys. Rev. Lett. **51**, 1945 (1983).

A plethora of measurements of weak decay and oscillation phenomena used to extract the elements of the CKM matrix are summarized at the website of the Heavy Flavor Averaging Group `http://www.slac.stanford.edu/xorg/hfag/`. See also C. Amsler et al., Phys. Lett. **B667**, 1 (2008) and more recent reviews from the Particle Data Group and the CKM Fitter group website `http://www.slac.stanford.edu/xorg/ckmfitter/ckm\_workinggroup.html`

While mixing of strange and bottom flavored mesons is established, evidence for mixing in charmed mesons remains to be confirmed. See M. Staric *et al.* (Belle Collaboration), Phys. Rev. Lett. **98**, 211803 (2007), B. Aubert *et al.* (BABAR Collaboration), Phys. Rev. Lett. **98**, 211802 (2007), and T. Aaltonen *et. al.* (CDF Collaboration ) Phys. Rev. Lett.**100**, 121802 (2008).

For one explication of parton model applications to high energy *pp* collisions, see E. Eichten, I. Hinchliffe, K. Lane, C. Quigg, Rev. Mod. Phys. **56**, 579 (1984). A Handbook of Perturbative QCD was published in Rev. Mod. Phys. **67**, 157 (1995). See also J. M. Campbell *et al*, Rep. Prog. Phys. **70**, 89 (2007).

For current information on parton distribution functions provided by two groups, see P. Nadolsky *et al*, "Implications of CTEQ global analysis for collider observables," asXiv:0802.0007 [hep-ph], the website of the Coordinated Theoretical-Experimental Project on QCD (CTEQ) `http://www.phys.psu.edu/~cteq/`, and A. D. Martin, W. J. Stirling, R. S. Thorne, and G. Watt, "Parton distributions for the LHC," Eur. Phys. J. **C63**, 189 (2009), arXiv:0901.0002v3 [hep-ph], and the website `http://projects.hepforge.org/mstwpdf/`.

## 10.10 ■ PROBLEMS

---

**Problem 10.1. Vector addition of angular momentum**

---

The total spin angular momentum of a pair of particles is represented by an operator

$$\mathbf{J} = \mathbf{s}_1 + \mathbf{s}_2.$$

Like all angular momentum operators, the components satisfy the commutation relations

$$J_a J_b - J_b J_a = i\epsilon_{abc} J_c$$

and the eigenstates of $\mathbf{J}^2$ and $J_z$ may be denoted by $|j, m>$. For the vector addition of two identical angular momenta, where $\mathbf{s}_1^2 = \mathbf{s}_2^2 = s(s+1)$, we have

$\mathbf{J}^2$ = j(j+1) where $j = j_{min}, j_{min}+1, ..., j_{max}, j_{min} =\mid s_1 - s_2 \mid= 0$, and $j_{max} = s_1 + s_2 = 2s$. The lowering operator is defined as $J^- = s_1^- + s_2^- = J_x - iJ_y$ and the commutation relations imply

$$J^- \mid j, m >= \sqrt{(j(j+1) - m(m-1))} \mid j, m-1 >$$

with similar expressions for $s_1^-$ and $s_2^-$. For two spin 1 particles, the total spin $j$ can be 2, 1, or 0. a) Starting with the maximal state

$$\mid j = 2, m = 2 >=\mid m_1 = +1 >\mid m_2 = +1 >,$$

use the lowering operator $J^- = s_1^- + s_2^-$ and orthogonality arguments to construct explicitly the spin wave functions for all possible eigenstates of $\mathbf{J}^2$ and $J_z$ . Check your results with a table of Clebsch-Gordan coefficients. b) Following the procedure you used for spin 1, show the eigenstates of total angular momentum alternate in symmetry character, with the states of highest $j$ being symmetric.

---

**Problem 10.2. Dipole moments and $CPT$**

---

A general theorem in field theory requires invariance under the combined symmetry operation of charge conjugation ($C$), reflection in space ($P$), and time reversal ($T$). Suppose a particle is in a combined static uniform electric and magnetic field and the interaction energy is

$$U = -\mu < \mathbf{s} \cdot \mathbf{B} > -d < \mathbf{s} \cdot \mathbf{E} >$$

where $\mu\mathbf{s}$ is the magnetic moment and $d\mathbf{s}$ is the electric dipole moment of the particle. Argue that $U$ changes sign under $CPT$ while particles and antiparticles are interchanged so $CPT$ symmetry requires $\mu$ and $d$ vanish for a neutral particle such as a Majorana neutrino. [Ref: Norman Ramsey, Rep. Prog. Phys. **45**, 95 (1982)]

---

**Problem 10.3. Induced electric dipole moment**

---

Consider a molecule with two nearly degenerate states of opposite parity, $H\psi_\pm = E_\pm\psi_\pm$, $P\psi_\pm = \pm\psi_\pm$ with $0 < (E_- - E_+)/E_+ << 1$. In an external electric field, the states will mix. Diagonalize the Hamiltonian

$$H = \begin{pmatrix} E_+ & e \\ e & E_- \end{pmatrix}$$

where $e = \mathbf{d} \cdot \mathbf{E} = \sum q_i \mathbf{x}_i \cdot \mathbf{E}$ and show that for $E_+ - E_- << e$, the new energy eigenvalues are $E'_\pm \simeq \dfrac{E_+ + E_-}{2} \pm \mid \mathbf{d} \cdot \mathbf{E} \mid$ while for $E_+ - E_- >> e$, $E'_\pm \simeq E_\mp \pm \dfrac{(\mathbf{d} \cdot \mathbf{E})^2}{E_- - E_+}$ so the induced energy shift is quadratic in $\mathbf{E}$. See P. Claver and

G. Jona-Lasino, Phys. Rev. **A33**, 2245 (1986) and Giovanni Jona-Lasinio, Carlo Presilla, and Cristina Toninelli Phys. Rev. Letts. **88**, 123001-1 (2002) for a discussion of the localization in pyramidal molecules like $NH_3$.

---

**Problem 10.4. Protonium decay to pions**

---

A stopping antiproton is captured into an s-wave state in a hydrogen target, forming a $(p\bar{p})_{atom}$ protonium atom of spin $s$ either 0 or 1 with $P_{p\bar{p}} = (-1)^{l+1}$, $C = (-)^{l+s}$, and $CP = (-)^{s+1}$. What two and three pion decays are permitted and which forbidden by $C$, $P$ and $G$? Consider combinations of neutral and charged pions. In analyzing the quantum numbers of three pions, consider one pion plus a two pion system.

---

**Problem 10.5. Isospin related decays**

---

The $\Delta^+(1236)$ is an $I_3 = 1/2$ member of an isospin 3/2 multiplet and its natural width indicates a strong, isospin conserving decay. The $\Sigma^+(1385)$ and $\Sigma^+(1193)$ are both I = 1, $I_3 = +1$ with strangeness -1. Predict the branching ratios

$$\frac{\Gamma(\Delta^+(1236) \to \pi^+ n)}{\Gamma(\Delta^+(1236) \to \pi^0 p)} \text{ and } \frac{\Gamma(\Sigma^+(1385) \to \Sigma^+\pi^0)}{\Gamma(\Sigma^+(1385) \to \Sigma^0\pi^+)}.$$

---

**Problem 10.6. Isospin related cross sections**

---

Use isospin conservation to predict the ratio of cross sections for the kaon-deuteron reactions $\sigma_{K^- d\to\pi^-\Lambda p}/\sigma_{K^- d\to\pi^0\Lambda n}$ at the same center of mass energy.

---

**Problem 10.7. Protonium decay to kaons**

---

What combinations of two or more $K_1$ and $K_2$ mesons produced by protonium $(\bar{p}p)_{atom}$ annihilation at rest in an $s$-wave, $J = 0$ or $J = 1$ state are consistent with Bose symmetry and conservation of $C$ and $P$ and strangeness? What combinations are allowed in decay of the $\phi(1020)$?

---

**Problem 10.8. Conservation of $CP$ in kaon decay**

---

Show that, although $C$ and $P$ may be separately violated in weak interactions, if the product of $C$ and $P$ is a symmetry, then $K_1 \to 2\pi^0$ and $K_1 \to \pi^+\pi^-$ are allowed, $K_2 \to 2\pi^0$ and $K_2 \to \pi^+\pi^-$ are forbidden, $K_1 \to 3\pi^0$ is forbidden, $K_2 \to 3\pi^0$ is allowed, while $K_1 \to \pi^+\pi^-\pi^0$ and $K_2 \to \pi^+\pi^-\pi^0$ are allowed.

**Problem 10.9. Rare neutral pion decays**

Consider double Dalitz decay $\pi^0 \to (e^+e^-)_{\text{atom}}(e^+e^-)_{\text{atom}}$ in which both $e^+e^-$ pairs appear bound as ground state $s$-wave positronium atoms in spin singlet (parapositronium $P_S$) or triplet (orthopositronium $P_L$) forms with rest frame lifetimes $\tau_{P_S \to \gamma\gamma} = 1.25 \times 10^{-10} s$ and $\tau_{P_L \to \gamma\gamma\gamma} = 1.37 \times 10^{-7} s$ respectively. Which combinations of final states are allowed by $P$-invariance, $C$-invariance, and Bose symmetry? Explain why $C$ conservation requires the decay $\pi^0 \to e^+e^-$ be fourth order. Is a naive estimate of the branching fraction for this decay mode consistent with observation?

**Problem 10.10. Optical model for refractive index**

Consider an amorphous medium with a number density $N$ scattering centers per unit volume each giving rise to elastic scattering amplitude $f(\theta)$. For a thin slab of thickness $dz$ orthogonal to an incident wave $e^{ikz}$, show that the scattered wave in cylindrical coordinates amounts to

$$\psi_s(\mathbf{x}) = \int d\mathbf{x}' N(\mathbf{x}') \frac{e^{ik|\mathbf{x}-\mathbf{x}'|}}{|\mathbf{x}-\mathbf{x}'|} f(\theta)$$
$$\simeq 2\pi N dz \int_0^\infty dR\, e^{ikR}[f(0) + f'(0)\theta(R)] \simeq -2\pi N dz \frac{f(0)}{ik} e^{ikz}$$

where we assumed $f'(0) = 0$ for a finite range interaction and neglected the oscillating term at $R = \infty$. Exponentiate the forward amplitude to show that a plane wave in this medium has the form $e^{ikz/n - N\sigma_T z/2}$ with refractive index $n^{-1} = 1 + 2\pi N \text{Re}[f(0)]/k^2$. Use the optical theorem (see any quantum mechanics text) which states that the total cross section for interaction is proportional to the imaginary part of the forward elastic scattering amplitude $\sigma_T = \dfrac{4\pi}{k}\text{Im}(f(0))$.

**Problem 10.11. Color hyperfine interaction splittings**

The mass $M_V$ of a vector meson is larger than the mass $M_P$ of the corresponding pseudoscalar meson. The color hyperfine interaction model predicts

$$M_V - M_P \propto |\psi(0)|^2/(m_q m_{\bar{q}}).$$

Take $m_u = m_d = 300$ MeV, $m_c = m_{J/\psi}/2$ and $m_b = m_\Upsilon/2$ and suppose $|\psi(0)|^2 \propto r^{-3}$ with $r = 0.6$ fm for light mesons and $r = 0.3$ fm for heavy mesons. Scale $m_\rho - m_\pi = 630$ MeV to predict $m_{J/\psi} - m_{\eta_c}$, $m_{D^*} - m_D$ and $m_{B^*} - m_B$ and compare to observations.

## Problem 10.12. B meson decay

By comparison to the decay $\pi^+ \to \mu + \nu_\mu$, justify the expression for the decay rate

$$\Gamma_{B^+ \to l^+ \nu_l} = \frac{G_F |V_{ub}|^2 f_B^2}{8\pi} m_b m_l^2 (1 - (m_l/m_B)^2)$$

and calculate the decay rate and branching fraction

$$\Gamma_{\mathrm{B}^+ \to \mathrm{l}^+ \nu_\mathrm{l}} / \Gamma_{\mathrm{B}^+ \to \text{ anything}}$$

for $l^+ = \tau^+$, $\mu^+$, and $e^+$, assuming $\mathrm{f_B} = 190$ MeV, $|\mathrm{V_{ub}}| = 3.94 \times 10^{-4}$ and $\Gamma_{\mathrm{B}^+}^{-1} = \tau_{\mathrm{B}^+} = 1.64 \times 10^{-12}$ s.

## Problem 10.13. Fragmentation functions

In the simplest independent fragmentation model of Field and Feynman, given a quark $q_0$, a pair $q_1\bar{q}_1$ is generated with light quark flavor ratios $u : d : s =$ 0.42; 0.42 : 0.14 and a $q_0\bar{q}_1$ meson formed with mean transverse momentum $< p_t > = 0.35$ GeV and carrying a fraction $z$ of the $q_0$ longitudinal momentum given by $E' + p_z' = z(E + p_z)$ with probability density

$$\frac{dP}{dz} = 1 - a + a(b+1)(1-z)^b, \ a = 0.96, \ b = 3$$

or, for a $c$-quark or $b$-quark pair,

$$\frac{dP}{dz} \propto z(1-z)^2/((1-z)^2 + \epsilon z)^2$$

with $\epsilon_c = 0.8\,\mathrm{GeV}^2/m_c^2$ and $\epsilon_b = 0.5\,\mathrm{GeV}^2/m_b^2$. (The parameters are the result of fits to PEP, PETRA, and LEP data with ISAJET.) Plot and compare the normalized light quark and $c$-quark distributions. [Ref: R. D. Field, and R. P. Feynman, Nucl. Phys. **B136**, 1 (1978), C. Peterson, D. Schlatter, I. Schmitt, and P. M. Zerwas, Phys. Rev. **D27**, 105 (1983)]

CHAPTER

# 11 Beyond the Standard Model

The standard model is successful but incomplete. This chapter provides some orientation to extensions of the standard model and to new physical phenomena which may be discovered at the Large Hadron Collider and future machines.

**Contents**



## 11.1 ■ GRAND UNIFICATION AND SUPERSYMMETRY

The standard model can be summarized as follows. Matter is comprised of fundamental spin 1/2 fermions and vector gauge bosons. Fermions appear as leptons and quarks in exactly three matching generations of pairs and participate in $SU(3)_C \times SU(2)_L \times U(1)_Y$ gauge interactions. The $SU(2)_L \times U(1)_Y$ symmetry is spontaneously broken at a scale $v \sim 0.25$ TeV leaving $SU(3)_C$ and $U(1)_{EM}$ symmetry. The low energy interactions are mediated by the $W$ boson, the $Z$ boson, the photon, and eight gluons. Non-Abelian gauge symmetry underlies the universality of the couplings of fermions with gauge fields. The condensation of quarks into hadrons is also a consequence of non-Abelian gauge symmetry. The model does not explain the number of fermion generations, the number and type of gauge interactions, unconstrained coupling constants, the source of $P$, $C$, and $T$ violation, and the origin of spontaneous symmetry breaking. Many explanations have been explored since the standard model was first formulated and some of these will now be described.

As a starting point, consider one oddity embedded in the standard model: parity violation. While a chirality preference at the molecular and macroscopic scale can

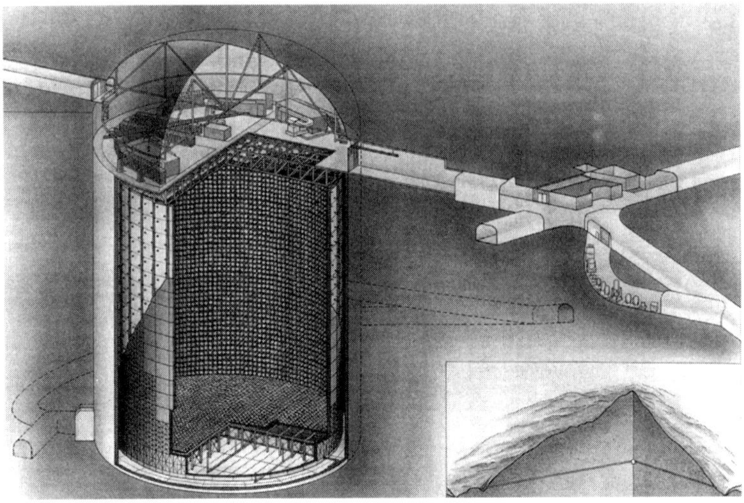

**FIGURE 11.1** The Super-Kamiokande water Cherenkov light detector in Hida, Japan, used in detecting neutrino oscillations and in searches for nucleon instability, both effects representing physics beyond the standard model. [SuperK website gallery.]

be explained by evolution, nothing fundamental in the subatomic world was known to distinguish fundamental physical processes from their mirror images until the discovery of parity violation in particle physics in the 1950s. In the standard model, parity violation appears exclusively in the weak interaction. It is natural to suppose parity is a spontaneously broken symmetry and that left right symmetry obtains at high energy.

A simple extension of the standard model postulates two gauge symmetries $SU(2)_L$ and $SU(2)_R$, each associated with three gauge bosons. As illustrated in Figure 11.2, both symmetries are assumed spontaneously broken but at different scales, the $SU(2)_L$ at the $v_L = v$ and the $SU(2)_R$ at a much larger scale $v_R$

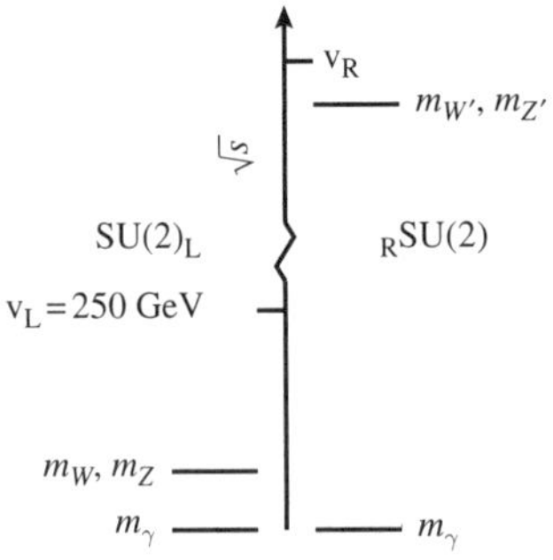


**FIGURE 11.2** Schematic of the symmetry breaking in a left-right symmetric model. The symmetry breaking scale $v_R$ for $SU(2)_R$ much exceeds the scale $v_L = v$ for $SU(2)_L$, and for center of mass energy $\sqrt{s} << v_R$, the interactions mediated by the gauge bosons of $SU(2)_R$ are suppressed.

implying much larger masses for the weak bosons coupled to right-handed fermions. The interactions associated with the gauge bosons of $SU(2)_R$ are suppressed at presently accessible energies. Such a left-right symmetric model motivates searches for heavy analogs of the $W$ and $Z$ bosons called $W'$ and $Z'$ at colliders and searches for virtual particle effects in precision studies of weak interactions. Tevatron experiments have placed lower bounds on the masses of such bosons of order 0.5 TeV (limits depend upon model dependent couplings) and LHC experiments can extend such limits by an order of magnitude.

Consider next the commonalities of quarks and leptons. The $SU(3)_C$ and $SU(2)_L$ gauge symmetries reflect the physical equivalence of the members of quark color triplets and of left-handed weak isospin doublets and require universal coupling constants. It is natural to imagine the physical equivalence of the fermions of a given generation or even the equivalence of all fermions and to postulate a larger symmetry such as $SU(5)$ that encompasses $SU(3)_C \times SU(2)_L \times U(1)_Y$. One must invoke a cascade of spontaneous symmetry breaking beginning at a high scale $m_{\text{GUT}}$ resulting in the lesser gauge symmetries seen at presently accessible scales. The merger of running color and electroweak coupling constants (see Figure 9.12) suggests $m_{\text{GUT}} = 10^{15-17}$ GeV.

Such grand unified theories (GUTs) imply new gauge interactions and Higgs bosons. The features of such models can be illustrated schematically as follows. The minimal $SU(5)$ model accommodates a lepton pair and one quark flavor in three colors in a fundamental ($\mathbf{5}^*$) representation written as

$$q_5^* = (d_R^c, d_B^c, d_G^c, e, -\nu_e)_L \tag{11.1}$$

where $c$ denotes the charge conjugate of a right-handed field. (The particular assignment of fields in this representation is chosen and arranged to ease identification of the color and weak isospin symmetries that ultimately emerge.) Just as a product of two fundamental 2-dimensional representations of $SU(2)$ symmetry includes a symmetric (spin 1) triplet, the product of two 5-dimensional representations of $SU(5)$ symmetry includes a symmetric representation which is 10-dimensional and accommodates the remaining states for one generation: three colors each of $u_L$, $u_R$, and $d_L$, plus the $e_R$.

The construction of the gauge theory follows the general prescription given in Chapter 7. The gauge covariant derivative is

$$D^\mu = \partial^\mu + ig_5 \mathbf{T} \cdot \mathbf{A}^\mu \tag{11.2}$$

where $\mathbf{T}$ is a vector of generator matrices for $SU(5)$ and $\mathbf{A}$ is a vector of $SU(5)$ gauge fields. Recall that $SU(N)$ has $N^2 - 1$ generators and each generator is associated with a gauge boson so $SU(5)$ gauge symmetry implies 24 related forces and vector quanta. Twelve of the bosons can be identified with the gluons and electroweak bosons, four of which correspond to diagonal $SU(5)$ generators. Each off-diagonal generator (analogous to $\sigma_x$ and $\sigma_y$ in $SU(2)$) corresponds to symmetry transformations of one fermion flavor into another and to a process in which the emission of the associated gauge boson transforms the one fermion flavor into the other fermion flavor. Hence there are twelve real gauge fields or equivalently six fractionally charged colored gauge fields (analogous to $W_1$ and $W_2$ in $SU(2)$) called leptoquarks dubbed $X_a^{-4/3}$ and $Y_a^{-1/3}$. These mediate transitions such as

$$d_R^c \to uY^{-1/3} \tag{11.3}$$

and lepton number and quark number violating processes such as those illustrated in Figure 11.3.

Breakdown of $SU(5)$ symmetry in a realistic model is arranged so that leptoquarks acquire masses of order the high GUT scale $v_{GUT}$ leaving room for a further breakdown at the familiar weak scale $v_{\text{weak}} \equiv v$. This is achieved through a generalization of the standard model Higgs mechanism with a **24** of scalar fields acquiring a vacuum expectation value $v_{24} = m_{\text{GUT}}$ and an additional **5** with a vacuum expectation value $v_5 = v_{\text{weak}}$. Yukawa interactions that are $SU(5)$-invariant are postulated such that all the scalar particles acquire mass of order $v_{24}$ except one standard model Higgs boson $H$, while none of the fermions acquire mass at the GUT scale. The standard model applies at the weak scale and below.

By virtue of leptons and quarks being related through $SU(5)$ symmetry, the fractional relationship between quark and lepton electric charges can be explained and it emerges that the ratio of $SU(2)_L$ and $U(1)_Y$ coupling constants $g'$ and $g$ are related by

$$g'^2 = \frac{3}{5}g^2 \tag{11.4}$$

which translates to the prediction for the Weinberg angle at the GUT scale

$$\sin^2\theta_W(m_{GUT}) = \frac{g'^2}{g^2 + g'^2} = \frac{3}{8}. \tag{11.5}$$

Quark and lepton number violation mediated by interactions with leptoquarks implies rare baryon number violating processes like

$$p \to e^+\pi^0,\ p \to \mu^+\pi^0. \tag{11.6}$$

The proton lifetime (and the lifetime of the neutrons bound in nuclei) is at least the age of the universe. Direct searches for nucleon decay monitor vast numbers of particles simultaneously and place much stronger bounds on visible decays. The

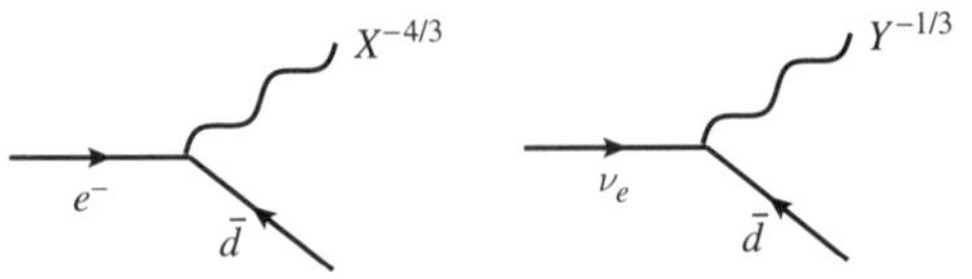


**FIGURE 11.3** Lepton and quark number violating transitions in the $SU(5)$ grand unified theory mediated by charged leptoquark gauge bosons.

Kamiokande collaboration has placed a limit on the mean lifetime for the reactions (11.6) of

$$\tau_{p\to l\pi^0} > 6.6 \times 10^{35}\ \text{yr}, \tag{11.7}$$

close to the estimates in supersymmetric grand unified models. [H. Nishino et al. (Kamiokande collaboration) Phys. Rev. Lett. **102**, 141801 (2009). doi:10.1103/PhysRevLett.102.141801]

The unitary symmetries in the standard model link only particles of identical spin. Supersymmetry (SUSY) combines transformations between the different spin states of fermions and bosons with Lorentz transformation symmetry. Invoking supersymmetry is a popular approach to generalizing the standard model but supersymmetric models are quite complex and we can provide only a sketch of the ideas.

A SUSY multiplet must contain equal numbers of fermion and boson degrees of freedom and the simplest multiplets are one helicity state of a massless self-conjugate fermion together with a scalar field, one helicity state of a charged fermion together with a complex scalar field, one of the two helicity states of a massless vector field with one of the two helicity states of a fermion, and so on. The standard so-called minimal supersymmetric standard model (MSSM) supposes all known particles are transformed by supersymmetry to new particles of lesser spin.

More precisely, in order to accommodate parity violation, each complex chiral state of a (non self-conjugate) fermion is paired with a complex scalar field. The super-partner of a fermion field is denoted by the fermion field name with a tilde. Thus the super-partner of a green left-handed up quark $u_L^G$ is the complex colored squark scalar field $\widetilde{u_L^G}$ or more succinctly

$$q = \begin{pmatrix} u_l \\ \tilde{u}_L \end{pmatrix}. \tag{11.8}$$

The partners of leptons are called sleptons and sneutrinos. The partners of gauge vector bosons are spin 1/2 fermions called gauginos. The partner of the left helicity massless real hypercharge gauge vector boson $B_L$ is a left-handed self-conjugate fermion $\tilde{B}_L$ and together the partners of the two helicity states of $B$ are a self conjugate bino $\tilde{B}$. Similarly the partner of the photon is the photino $\tilde{\gamma}$ and the partner of the $Z$ boson is the zino $\tilde{Z}$ while the partner of the massless $W^{\pm}$ helicity states are the two helicity states of a charged fermion $\tilde{W}$ called a wino.

Imposing the gauge symmetries of the standard model particles in the context of supersymmetry requires the partner members of a multiplet of supersymmetry have $SU(3)_C \times SU(2)_L \times U(1)_Y$ gauge interactions. Therefore the squark fields appear in three colors, for example $\tilde{u}_L^a$, and are subject to scalar chromodynamics. Similarly, squarks and sleptons are subject to scalar electroweak interactions. In particular, the left-handed squarks $\tilde{u_L}$ couple to the $W$ boson while the right-handed squarks $\tilde{u_R}$ do not. The minimal model of electroweak symmetry breaking requires two $SU(2)_L$ doublets of complex scalar fields rather than one and these are denoted by

$$H_u = \begin{pmatrix} H_u^+ \\ H_u^0 \end{pmatrix}, \quad H_d = \begin{pmatrix} H_d^0 \\ H_d^- \end{pmatrix}. \tag{11.9}$$

The doublet $H_u$ with $Y = +1/2$ gives mass to $u$-like quarks, and the doublet $H_d$ with $Y = -1/2$ gives mass to $d$-like quarks and leptons. These Higgs doublets have left-handed fermion partners called higginos. The standard model neutral scalar Higgs field $H$ is a linear combination of $H_u^0$ and $H_d^0$. After breaking of $SU(2)_L \times U(1)_Y$ symmetry, the $W$ boson and $Z$ boson emerge as massive and the known particles acquire their known masses.

The unbroken SUSY Lagrangian describes degenerate particles and sparticles with gauge interactions. SUSY breaking through the Higgs mechanism requires a matrix of Yukawa interactions between the Higgs scalar doublets and the usual fermions plus additional new quadratic through quartic interaction terms between scalar Higgs and SUSY scalar fields, constrained by the gauge symmetries. The minimal supersymmetric model (MSSM) adopts a particularly simple form for the many interaction terms that may be freely specified.

If supersymmetry were unbroken, the sparticles would all be degenerate with their partners but none are observed to date so sparticles must be supposed to develop masses large compared to particle masses. The SUSY breaking terms which give sparticles high masses may be chosen independently of spontaneous breaking of gauge symmetry because, for example, scalar field mass terms like $m^2|\phi|^2$ are gauge invariant, and hence there are many possibilities. The MSSM adopts minimal sparticle and gaugino mass terms, sparticle Yukawa interactions with the Higgs doublets, and corrections to the Higgs potential which are consistent with field theory consistency conditions in an explicit model of symmetry breaking. While minimal, there are approximately 100 free parameters in the model. Phenomenological constraints motivate simplified versions in which only a handful of new parameters are required beyond those in the standard model.

A discrete symmetry called $R$-parity is usually assumed in SUSY theories that enforces conservation of the net particle number and independently super-particle number. It outlaws terms which link quark and lepton SUSY multiplets that would imply baryon and lepton number violation in contradiction to experiment. (Baryon number and lepton number are conserved accidentally in the standard model.) Consequently the decays of superparticles ultimately lead to a lightest stable (super) particle (LSP), perhaps the photino, which is a candidate for the dark matter component of the universe. Such a particle would be weakly interacting and neutral and its production, like production of a neutrino, would lead to apparent missing energy.

Supersymmetry implies SUSY particle interactions with known gauge particles are similar to the interactions of normal matter. Rates for processes such as squark pair production illustrated in Figure 11.4 are calculable functions of the SUSY particle mass spectrum. Searches for SUSY particles at the Tevatron have placed lower bounds on the masses of SUSY particles in the range $m_{SUSY} \sim 200$ GeV.

Extensions to the standard model such as supersymmetry that entail many new particles have many unconstrained parameters and too many possible incarnations

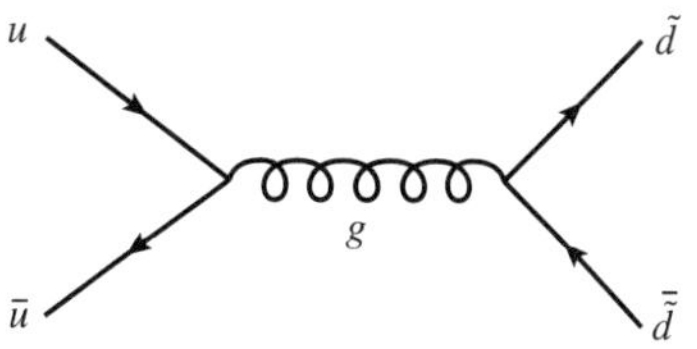


**FIGURE 11.4** Pair production of squarks through the color interaction. Such processes occur at rates comparable to the production of $q\bar{q}$ pairs of the same mass.

to review here. However, to get the flavor of investigations, let us consider a relatively straightforward generic extension of the Higgs sector to two Higgs doublets as appear in the MSSM. Suppose $\Phi_1$ and $\phi_2$ are two complex $SU(2)_L$ doublet scalar fields with $Y = 1$. A general gauge invariant potential takes the form (See Howard E. Haber and Maria Krawczyk, "The $CP$-violating two-Higgs doublet model", CERN 2006-009, 5(2006).)

$$\begin{aligned} V &= m_{11}^2\Phi_1^\dagger\Phi_1 + m_{22}^2\Phi_2^\dagger\Phi_2 - \left(m_{12}^2\Phi_1^\dagger\Phi_2 + \text{h.c.}\right) \\ &+ \frac{1}{2}\lambda_1\left(\Phi_1^\dagger\Phi_1\right)^2 + \frac{1}{2}\lambda_2\left(\Phi_2^\dagger\Phi_2\right)^2 + \frac{1}{2}\lambda_3\Phi_1^\dagger\Phi_1\Phi_2^\dagger\Phi_2 + \frac{1}{2}\lambda_4\Phi_1^\dagger\Phi_2\Phi_2^\dagger\Phi_1 \\ &+ \left\{\frac{1}{2}\lambda_5\left(\Phi_1\Phi_2\right)^2 + \left[\lambda_6\Phi_1^\dagger\Phi_1 + \lambda_7\Phi_2^\dagger\Phi_2\right)\right]\Phi_1^\dagger\Phi_2 + h.c.\right\} \end{aligned} \quad (11.10)$$

where $m_{12}^2, m_{22}^2$, and $\lambda_{1-4}$ are real parameters while $m_{12}^2$ and $\lambda_{5-7}$ are complex. The number of parameters may be reduced by requiring $CP$-invariance and requiring that the general interactions with fermions do not include tree-level flavor changing neutral interactions on which there are strong experimental constraints.

The condition for electroweak symmetry breaking is that the scalar potential have a minimum at nonzero values for the scalar fields for which

$$\frac{\partial V}{\Phi_1} = \frac{\partial V}{\Phi_2} = 0. \quad (11.11)$$

Then requiring $U(1)_{EM}$ gauge symmetry, the minimum in the potential occurs at values that can be written as

$$< \Phi_1 >= \frac{1}{\sqrt{2}}\begin{pmatrix} 0 \\ v_1 \end{pmatrix}, \quad < \Phi_2 >= \frac{1}{\sqrt{2}}\begin{pmatrix} 0 \\ v_2 \end{pmatrix} \quad (11.12)$$

where $v_1$ and $v_2$ are real vacuum expectation values. With this symmetry breaking in the $SU(2)_L \times U(1)_Y$ model, the effective vacuum expectation value parameter in the fermion weak interaction is given by

$$v^2 = v_1^2 + v_2^2 = \frac{4m_W^2}{g^2} = (246\text{ GeV})^2 \quad (11.13)$$

and, of the eight real scalar degrees of freedom in the bare theory, three appear as the longitudinal degrees of freedom of the $W^\pm$ and $Z$ bosons and there are five remaining. Two are $CP$-even neutral scalar fields, $h^0$ and $H^0$ with $m_{H^0} \leq m_{H^0}$, one is a $CP$-odd scalar $A^0$, and two form a charged scalar $H^{\pm}$. We see that the physical Higgs sector grows from one to five scalar particles.

It is customary in two Higgs doublet models to define an angle $\beta$ related to the vacuum expectation values by

$$t_\beta \equiv \tan\beta \equiv \frac{v_2}{v_1} \tag{11.14}$$

and to use the magnitude $v$ and direction $\beta$ along with the parameters in the Higgs potential above to characterize the Higgs sector. The physical neutral $CP$-even Higgs fields are no longer the radial deviations $\eta_1 = \sqrt{2}\mathrm{Re}\Phi_1 - v_1$ and $\eta_2 = \sqrt{2}\mathrm{Re}\Phi_2 - v_2$ but instead the combinations

$$\begin{aligned} H^0 &= \eta_1 \cos\alpha + \eta_2 \sin\alpha \\ h^0 &= -\eta_1 \sin\alpha_1 + \eta_2 \cos\alpha \end{aligned} \tag{11.15}$$

where the mixing angle $\alpha$ and the masses $m_{h^0}$ and $m_{H^0}$ may be expressed in terms of $\beta$ and the parameters of the Higgs potential.

The theory contains a variety of single Higgs interactions with multiple bosons including couplings of the following groups of particles:

$$H^0W^+W^-,\ H^0ZZ,\ ZA^0h^0,\ W^{\pm}H^{\mp}h^0,\ ZW^{\pm}H^{\mp}h^0,\ W^{\pm}H^{\mp}\gamma h^0. \tag{11.16}$$

The coupling constants for these interactions turn out to be proportional to $\cos(\beta - \alpha)$. Other single and double Higgs boson interactions with bosons represented by

$$h^0W^+W^-,\ h^0ZZ,\ ZA^0A^0,\ W^{\pm}H^{\mp}H^0,\ ZW^{\pm}H^{\mp}H^0,\ W^{\pm}H^{\mp}\gamma H^0 \tag{11.17}$$

have couplings proportional to $\sin(\beta - \alpha)$. If the pattern of interactions with fermions is that in the MSSM in which one of the Higgs doublets couples to $u$-type quarks and the other to $d$-type quarks, then $\alpha$ and $\beta$ characterize the interactions of the $CP$-even neutral Higgs boson to quarks with (for the third generation of quarks) Yukawa coupling constants

$$g_{h^0b\bar{b}} = -\frac{\sin\alpha}{\cos\beta},\quad g_{h^0t\bar{t}} = +\frac{\cos\alpha}{\sin\beta},\quad g_{H^0b\bar{b}} = +\frac{\cos\alpha}{\cos\beta},\quad g_{H^0t\bar{t}} = +\frac{\sin\alpha}{\sin\beta} \tag{11.18}$$

while the $CP$-odd Higgs boson interactions are $g_{A^0b\bar{b}} = g_{A^0t\bar{t}} = \tan\beta$ and the charged Higgs boson has the coupling

$$g_{H^-t\bar{b}} = \frac{g}{2\sqrt{2}m_W}\left[m_t \cot\beta(1+\gamma_5) + m_b \tan\beta(1-\gamma_5)\right]. \tag{11.19}$$

Such an extension of whatever origin already implies a rich range of particle production and decay signatures to unravel at the LHC provided the particle masses are within the accessible range.

Note that in the two Higgs doublet model, if $|\sin(\beta - \alpha)| \simeq 0$, some interactions like $h^0W^+W^-$ and $h^0ZZ$ are suppressed relative to the standard model while others are enhanced. Neither the $h^0$ nor the $H^0$ mimic the minimal standard model Higgs boson in this case. A decoupling limit exists which describes a case of $m_{H^\pm} >> v$ in which one of the neutral Higgs bosons has the properties of the minimal standard model Higgs boson while the other two have masses similar to $m_{H^\pm}$. In such a case, the minimal model applies at the weak scale while new scalars appear only at a higher mass scale. These two possibilities exemplify the kinds of phenomenology associated with viable extensions of the minimal standard model.

## 11.2 ■ PRESENT AND FUTURE ACCELERATOR BASED EXPERIMENTS

The signatures of the standard model and of extended models at the LHC have been extensively studied. Aspects of the standard model which may be illuminated at the LHC include triple electroweak boson production, virtual weak boson fusion, new charm and bottom hadron flavored hadrons, and chromodynamics phase transitions of quark matter. Physics beyond the standard model may be manifest in the production of novel vector and scalar bosons and supersymmetric particles. A focal point is the search for the standard model Higgs boson $H$.

Higgs boson production occurs rarely through leading order collisions of the valence constituents of the protons. The $u$-quark and $d$-quark are quite light and consequently weakly coupled to the $H$. The $H$ is produced predominantly through gluon fusion forming heavy particles

$$gg \to H \tag{11.20}$$

as illustrated by the loop diagram in Figure 11.5. In this process, a gluon from one proton collides with one from the other proton to form a heavy quark pair which fuses to form the $H$. The $t$-quark contribution dominates. The reverse process leads to the decay $H \to gg$. Processes in which a valence quark in each of the colliding protons radiates a $W$ or $Z$ boson and these bosons fuse to form a Higgs boson (vector boson fusion) have a smaller but significant cross section.

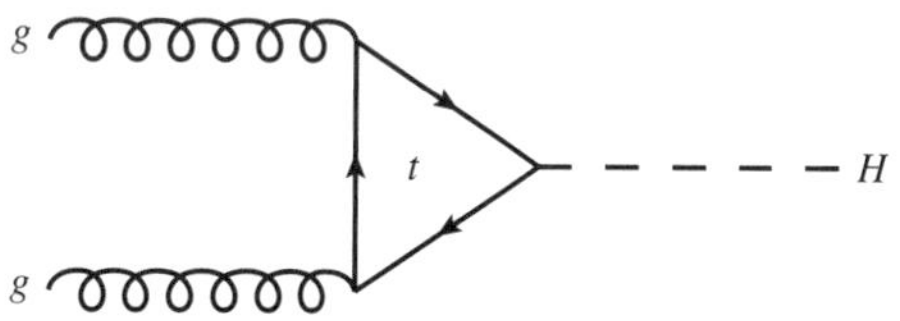


**FIGURE 11.5** The gluon fusion process responsible for Higgs boson production at the LHC. All quarks appear in the loop but the $H$ coupling to $t\bar{t}$ is much larger than to other $q\bar{q}$ pairs.

The leading order cross section for Higgs boson production through gluon fusion can be found from the matrix element $M$ for $H \to gg$ and it is related to the decay width as follows

$$\sigma_{g_1 g_2 \to H} = \frac{(2\pi)^4}{2m_H^2} \int \frac{d\mathbf{p}}{(2\pi)^3 2E} \delta(p_1 + p_2 - p)|M|^2 = \frac{8\pi^2 \Gamma_{H\to gg}}{n_g^2 m_H} \delta\left(p^2 - m_H^2\right) \tag{11.21}$$

where $n_g = 8$ is the number of gluons. The $H \to gg$ decay width follows from the $H \to \gamma\gamma$ decay width given in Equation 8.268 by ignoring the contribution from the $W^{\pm}$ loop and replacing the factor $\alpha n_c e_i^2$ by the color coupling $\alpha_s$ with a factor two included as the gluons are distinguishable giving

$$\Gamma_{H\to gg} = \frac{\alpha_s^2 g^2 m_H^3}{512\pi^3 m_w^2} |\Sigma_i F_i|^2. \tag{11.22}$$

Convolution with the gluon parton distribution functions permits computation of the cross section

$$\sigma_{pp\to H} = \int dx_1 dx_2\, x_1 g(x_1) x_2 g(x_2) \sigma_{g_1 g_2 \to H}. \tag{11.23}$$

Next-to-leading order corrections increase this expectation by about a factor 1.5.

The Higgs boson mass is not predicted by the standard model. The mass is an arbitrary parameter of the scalar potential and reflects the unknown physics of the Higgs sector. A naive expectation is that $m_H$ is comparable to $v$, $m_W$ and $m_Z$. As described in Chapter 8, the search for the $H$ produced in association with a $Z$ boson in $e^- e^+$ collisions at LEP excluded masses below about 114 GeV. Values in excess of 1 TeV are disfavored on theoretical grounds. The mass of the Higgs boson may be inferred from its participation as a virtual particle in standard processes analyzed in next to leading order, especially those involving heavy particles strongly coupled to the Higgs boson. As described in Chapter 9, the $W$ boson, $Z$ boson, and $t$-quark at next-to-leading order are effectively surrounded by a cloud of Higgs particles the density of which depends on $m_H$. The masses $m_W$ and $m_Z$ to leading order are fixed by symmetry breaking in the standard model but at next-to-leading order depend logarithmically on $m_H$ and $m_t$. Precise values of $m_W$ and $m_t$ with coupling constants determined from other standard model processes favor a Higgs mass near the lower bound from LEP.

The $p\bar{p}$ collider experiments at Fermilab searched for the Higgs boson produced by gluon fusion and in association with heavy weak bosons and top quarks and have excluded a Higgs mass in the interesting mass range near $2m_W$. In this range, the Higgs boson decays to pairs of weak bosons, a signature clouded little by other weak boson pair production processes. Luminosity, not energy, limited the ability of the Tevatron to explore the Higgs sector.

The cross section for Higgs boson production at the LHC running at design energy is shown in Figure 11.7 as a function of Higgs boson mass for $m_H$ as high as 1 TeV. The gluon fusion process is the dominant production mechanism although

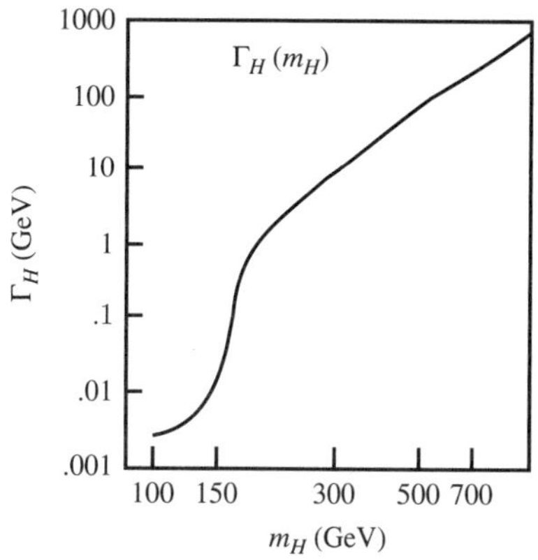


**FIGURE 11.6** Total decay width of the standard model Higgs boson as a function of Higgs boson mass. [A. Djouadi, J. Kalinowski and M. Spira, Comput. Phys. Commun. 108, 56 (1998)]

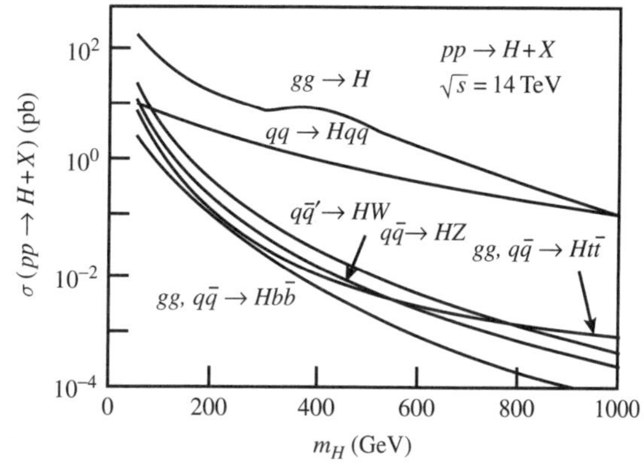


**FIGURE 11.7** Cross section for production of the Higgs boson at the LHC. [M. Spira, Fortschr. Phys. 46, 203 (1998)]

$H$ production in association with a $W$ boson, $Z$ boson, or $t$-quark contribute for low $m_H$. At the LHC, by virtue of the higher energy and luminosity, a Higgs boson of mass sufficient to decay to weak boson pairs is readily produced and detected. The cleanest signature is the case of

$$H \to ZZ \to (l^+l^-)(l^+l^-) \tag{11.24}$$

where a $l^\pm$ is one of the stable charged leptons $e^\pm$ or $\mu^\pm$ and one of the $Z$ bosons may be virtual. The CMS and ATLAS detectors offer seamless detection of electrons and muons over close to $4\pi$ of solid angle surrounding the interaction region. Reconstruction of events of the above form is tightly constrained. The invariant mass of at least one of the lepton pairs will be consistent with the $Z$ boson mass and the distribution of the invariant mass of the four leptons together will peak a unique value, $m_H$. This peak distinguishes Higgs boson production from background processes which source $Z$ boson pairs with a continuous distribution of invariant mass $m_{ZZ}$ at a comparable rate.

Identification of $H$ production is more difficult for $m_H$ values below threshold for decay to real or virtual weak bosons. A light Higgs boson decays predominantly to the heaviest quark permitted by energy and momentum conservations and hence

$$H \to b\bar{b} \tag{11.25}$$

is the dominant decay channel. The $b$-quark mass is $3 \times 10^{-4}$ of the LHC center of mass energy and $b\bar{b}$ quark pairs are produced in profusion. Both CMS and ATLAS are designed with elaborate silicon tracking systems capable of identifying the small displacements of tracks associated with the $b$-quark decays. However the resolution in measurement of the hadronic energy of $b$-quark initiated jets is much worse than the resolution in the measurement of the energy of a single lepton. The invariant mass of $m_{b\bar{b}}$ is consequently blurred, and the invariant mass peak associated with $H \to b\bar{b}$ is not distinguishable from the continuum production.

Consequently, both ATLAS and CMS will resort to a search based on the rare decay mode

$$H \to \gamma\gamma. \tag{11.26}$$

This decay takes place through a heavy particle loop in a fashion similar to the gluon fusion production process. Relative to the decay $H \to b\bar{b}$, the decay rate for $H \to \gamma\gamma$ is suppressed by the heavy fermion propagators, these factors obviated by the Higgs boson coupling being proportional to fermion mass, and by two factors of $\alpha$, and the branching fraction is of order one in ten thousand.

Despite the low branching fraction, the LHC luminosity permits production of sufficient numbers of Higgs bosons to enable a search for this rare decay mode. Both CMS and ATLAS are designed with finely segmented crystal electromagnetic calorimeters to detect isolated photons and to measure their energies with relative precisions approaching $10^{-3}$. Given high accuracy in locating of the collision point through tracking, the invariant mass of photon pairs is measured well enough that the peak in the distribution resulting from $H \to \gamma\gamma$ is distinguishable from the continuum background. The expected statistical significance of various Higgs boson decay signatures as a function of $m_H$ at the ATLAS experiment for an integral luminosity of 100 fb$^{-1}$ is shown in Figure 11.8. Even for $m_H$ in the low mass range near 120 GeV, a signal significance in excess of five standard deviations is accessible.

The LHC provides access to high statistics studies of a wide range of phenomena. Standard model studies will include multiple weak boson production, weak boson scattering, top quark production and decay, multiple jet production and fragmentation, and mixing and rare decays of heavy flavor mesons and baryons. At the kinematic limit, the LHC is capable of producing an object of invariant mass 14 TeV, some 80 times the mass of heaviest particle discovered to date (the top quark)

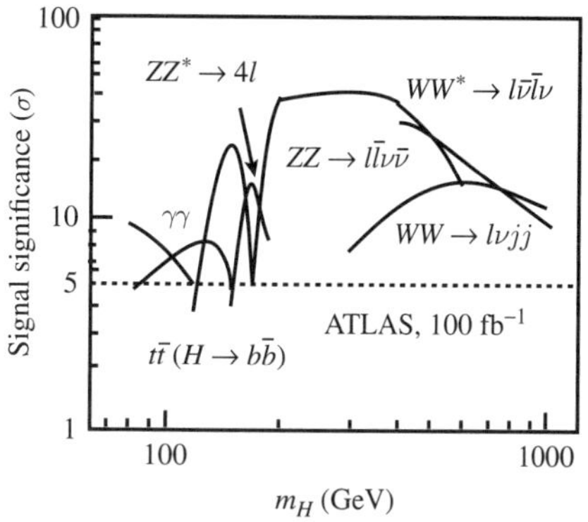


**FIGURE 11.8** LHC Higgs boson discovery potential. The expected statistical significance of various Higgs boson production and decay signatures is shown as a function of Higgs boson mass for an integral luminosity of 100 fb$^{-1}$ at the ATLAS experiment at the LHC. The dashed line indicates a nominal discovery level. [ATLAS design report]

and much larger than the scale of the spontaneous breaking of the electroweak symmetry. Analysis of the collisions will enable extension of present searches for new gauge bosons, super-particles, and other exotica. Of course, in the collision of the constituents of hadrons, the full energy is generally not available. As described by the parton distribution functions, the energy of a parton is on average but a fraction of the hadron energy. Yet the measured distributions extend to the full beam energy so the kinematic limit may be approached. The extreme intensity of the LHC and of an intensity upgrade already planned will for the first time allow exploration of physics in multiTeV range.

A possible follow-on to the LHC is a proposed International Linear Collider (ILC) - an $e^+e^-$ clashing beam machine operating at a nominal center of mass energy of 500 GeV with possible upgrade to 1 TeV (Figure 11.9). Other machines in the research and development stage include a high intensity $\mu^+\mu^-$ collider which would serve as a Higgs boson factory and, necessarily, as a high intensity neutrino source. Such lepton colliders have limited reach in energy but might be

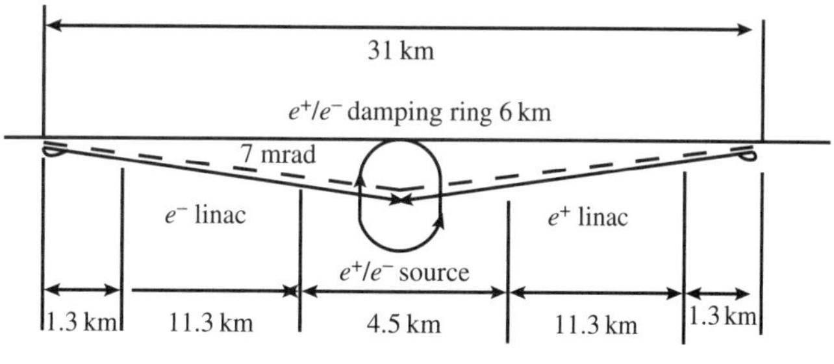


**FIGURE 11.9** Schematic layout of the International Linear Collider for 500 GeV center of mass energy. (ILC Reference Design Report Vol. 1, August 2007, linearcollider.org.)

dedicated to production of the standard model Higgs boson or of another neutral resonance that might be seen at the LHC. Precision measurements of the properties of such a particle would probe still higher energies through the effects of virtual particles. Another proposal is a 100 TeV center of mass energy $pp$ collider using advanced superconducting magnets in a synchrotron of order 60 miles in circumference. Such a machine is required to study physics beyond the reach of LHC experiments.

## 11.3 ■ MASSIVE NEUTRINOS

In the standard model, neutrinos are presumed to be massless. Only the left-handed fields appear while right-handed counterparts have vanishing coupling constants and are invisible in standard model interactions. Various evidence indicates that neutrinos have tiny masses, raising many questions about how the standard model must be extended to accommodate neutrino mass. The evidence all concerns oscillation of neutrino flavor as neutrinos propagate through matter and vacuum over macroscopic distances. In what follows, the essential data and the phenomenological description are presented. As we will see, there appears to be mixing of all three neutrino flavors described by a complex matrix the elements of which are being determined.

The existing data may be summarized as follows. Electron neutrinos sourced in solar fusion reactions in the core of the Sun disappear in route to Earth and it can be inferred they become both muon and tau neutrinos. The oscillation is substantially affected by interaction of the neutrinos with the material of the Sun. These low energy $\nu_e$ have insufficient energy to manifest muon or tau flavor so what they oscillate into is indirectly measured by weak neutral current interactions. High energy muon neutrinos from cosmic and artificial sources also oscillate over long baselines with a different frequency and apparently not often to electron neutrinos so by inference principally to tau neutrinos. Tau neutrino beams are not available for study. Anti-electron neutrinos from artificially induced fission reactions oscillate like electron neutrinos but again have insufficient energy to manifest muon or tau flavor. Long baseline accelerator experiments are planned to study in detail the appearance of $\nu_\tau$ and $\nu_e$ in $\nu_\mu$ beams and of $\nu_\mu$ and $\nu_\tau$ in $\nu_e$ beams and to study the charge conjugate oscillations and improved experiments will study matter effects in the Earth.

It is useful to have a two flavor oscillation model in mind as it characterizes well both the $\nu_e \leftrightarrow \nu_\mu, \nu_\tau$ oscillation and the $\nu_\mu \leftrightarrow \nu_\tau$ oscillation. We show later how it emerges from a three flavor oscillation model. Consider two neutrino energy eigenstates $\nu_1$ and $\nu_2$ with masses $m_1$ and $m_2$ related to the electron neutrino and another neutrino $\nu$ by a mixing angle $\theta$ with

$$\begin{aligned} \nu_e &= \nu_1 \sin\theta + \nu_2 \cos\theta \\ \nu &= \nu_1 \cos\theta - \nu_2 \sin\theta. \end{aligned} \tag{11.27}$$

The wave function of the energy eigenstate $\nu_i$ with momentum $p$ produced at $t = 0$ evolves as

$$\nu_i(t) = \nu_j^0 e^{-i(E_j t - px)} \tag{11.28}$$

where, for $p >> m_i$, we may make the approximation

$$p_i = \sqrt{E^2 + m_i^2} \simeq E - \frac{m_i^2}{2E}. \tag{11.29}$$

The probability that a $\nu_e$ produced at position $x = 0$ remains a $\nu_e$ at $x = L$ is given by

$$\begin{aligned} P(\nu_e \to \nu_e) &= | < \nu_e^0 \mid \nu_e(x = L) > |^2 \\ &= 1 - \sin^2(2\theta) \sin^2\left(\frac{\left(m_1^2 - m_2^2\right) L}{4E}\right) \\ &= 1 - \sin^2(2\theta) \sin^2\left(1.27 \frac{\Delta m^2}{1\ \mathrm{eV}^2} \frac{L}{1\ \mathrm{km}} \frac{1\ \mathrm{GeV}}{E_\nu}\right). \end{aligned} \tag{11.30}$$

The last expression shows in engineering units how, for fixed energy, the $\nu_e$ disappears and reappears as a function of propagation distance.

The earliest evidence for neutrino oscillation was a deficit of electron neutrinos produced in the solar core and bulk and observed on Earth in a pioneering experiment in 1968. To understand this and subsequent experiments observing solar neutrinos, we need to understand some of the details of the production processes. The standard model of solar structure and evolution includes a variety of nuclear reactions to intricate to review. The model is tested by observations of surface signatures of seismic oscillations which give confidence in the results of calculations of the spectrum of emitted flux of electron neutrinos. Electron neutrinos are by-products of the proton-proton fusion reaction chain and the so-called carbon-nitrogen-oxygen (CNO) cycle. The principal reactions which produce neutrinos are listed in Table 11.1 and the predicted energy spectrum is shown in Figure 11.10. Most of these neutrinos have energies below 0.5 MeV and result from the $pp$ reaction which is not confined to the solar core. The estimated error on their flux is a few percent. The core reactions that produce higher energy neutrinos are more sensitive to the core temperature. The uncertainties of the fluxes of the higher energy neutrinos are 15-50%. Of these, the high energy neutrinos from the B reaction have an uncertainty of about 15% and are important in what follows.

Cross sections for interactions of neutrinos with matter at the MeV energies characteristic of solar processes are well established (e.g. Equations 8.205) and may be used to precisely predict inverse beta decay nuclear interaction rates on Earth. The first observation of solar neutrinos was at the Homestake mine in Montana. It used 615 tons of tetrachloroethylene in which electron neutrinos of energy in excess of 0.81 MeV produce $^{37}$Ar nuclei through the reaction

$$\nu_e + {}^{37}\mathrm{Cl} \to {}^{37}\mathrm{Ar} + e^-. \tag{11.31}$$

| Reaction | $E_\nu$ |
|---|---|
| $p + p \to {}^2\text{H} + e^+ + \nu_e$ | $\leq 0.42$ MeV |
| $n + e^- + p \to {}^2\text{H} + \nu_e$ | $= 1.44$ MeV |
| ${}^3\text{He} + p \to {}^4\text{He} + \nu_e$ | $\leq 18.8$ MeV |
| ${}^7\text{Be} + e^- \to {}^7\text{Li} + \nu_e$ | $= 8.6$ MeV(90%), 0.38 MeV |
| ${}^8\text{B} \to {}^7\text{Be}^* + e^+ + \nu_e$ | $\leq 15$ MeV |
| ${}^{13}\text{N} \to {}^{13}\text{C} + e^+ + \nu_e$ | $\leq 1.2$ MeV |
| ${}^{15}\text{O} \to {}^{15}\text{N} + e^+ + \nu_e$ | $\leq 1.7$ MeV |
| ${}^{17}\text{F} \to {}^{17}\text{O} + e^+ + \nu_e$ | $\leq 1.7$ MeV |

**TABLE 11.1** Solar fusion reactions producing neutrinos.

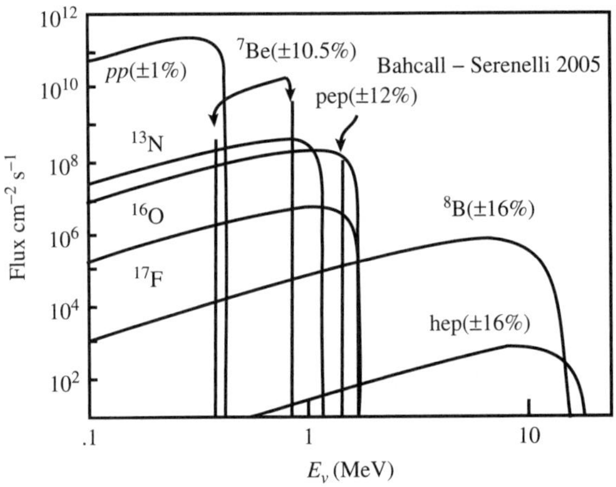


**FIGURE 11.10** Predicted flux of solar neutrinos from various reactions with estimates of their uncertainties. [John N. Bahcall, Aldo M. Serenelli, and Sarbani Basu, Astrophys. J. **621**, L85-L88 (2005)]

The signal is the 2.82 keV Auger electrons emitted by ${}^{37}$Ar with half-life 35 days. The signal results from the integral over all energies above threshold including intermediate energies for which the flux is relatively poorly predicted. This ground breaking experiment found a statistically significant signal deficit and spawned a number of oscillation searches at accelerators which, it turned out, were looking for flux changes over distances that were too short.

The Homestake experiment was followed by the 60 ton SAGE liquid gallium experiment and the GALLEX 100 ton gallium chloride experiment both of which made use of the reaction

$$\nu_e + {}^{71}\text{Ga} \to {}^{71}\text{Ge} + e^-. \tag{11.32}$$

The threshold energy for this reaction is only 0.23 MeV and therefore SAGE and GALLEX were sensitive to the majority of the solar neutrinos which originate in the *pp* reaction and for which the flux is best predicted. The SAGE experiment detected $71 \pm 6$ SNU in agreement with GALLEX which found $78 \pm 6$ or about 0.6 of the predicted value of $128 \pm 8$ SNU. The historical unit 1 SNU $= 10^{-36}$ captures per atom per second is used here as it is common in the field.

The 50 kton Super-Kamiokande imaging water cerenkov detector detects electron neutrinos through the charged and neutral current elastic scattering process

$$\nu + e^- \to \nu + e^- \tag{11.33}$$

with a threshold of about 5 MeV. It is therefore sensitive to the high energy part of the solar neutrino spectrum. In Figure 11.11, the correlation of the electron direction with the direction of the Sun is shown for candidate events. The solar component is clearly identified. The solar flux varies as expected with Earth-Sun distance over the course of the year and the recoil energy spectrum is close to model predictions. However the flux is about 0.4 of that predicted by the standard solar model without oscillation, a value different than that found by GALLEX and SAGE. It appears electron neutrinos have disappeared on their way from the Sun to the Earth at a rate that is energy dependent.

Other experiments dedicated to portions of the spectrum include the SNO 1000 ton heavy water imaging Cerenkov detector sensitive to neutrinos from the $^8$B reaction, and Borexino which uses a scintillating target to detect the 0.862 MeV line from the $^7$Be reaction [arXiv:0708.2251v2 [astro-ph]]. The SNO experiment

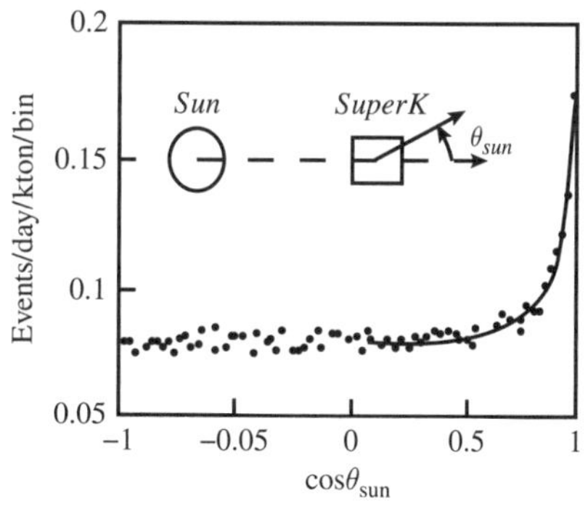


**FIGURE 11.11** The angular distribution of solar neutrino candidate events showing the correlation with the instantaneous direction of the Sun. The flat line seen under the peak in the solar direction represents background contributions. [J. P. Cravens *et al.*, Super-Kamiokande Collaboration, Phys. Rev. **D78**, 032002 (2008), arXiv:0803.4312v1 [hep-ex]]

detects neutrinos through the elastic scattering, charged current, and neutral current processes:

$$\begin{aligned} \nu_x + e^- &\rightarrow \nu_x + e^- \text{ (ES)} \\ \nu_e + d &\rightarrow p + p + e^- \text{ (CC)} \\ \nu_x + d &\rightarrow p + n + \nu_x' \text{ (NC)}. \end{aligned} \tag{11.34}$$

In elastic scattering and charged current events, the recoil electron is observed, while for neutral current events, the neutrons are detected by de-excitation gamma rays following neutron capture on another nucleus. The elastic scattering events result from charged and neutral currents for $\nu_e$ and from the neutral current for $\nu_\mu$ and $\nu_\tau$.

The SNO experiment measures the total flux of $^8B$ solar neutrinos with the NC reaction with a 4% uncertainty. The result agrees with the standard solar model with its 16% uncertainty. The flux inferred from the charged current reaction is 0.30 of the predicted $\nu_e$ flux over the range 5-11 MeV of recoil energies observed. The flux inferred from the elastic scattering rate is about 0.4 of that expected for $\nu_e$ but with some oscillatory structure. It therefore appears that the high energy $^8B$ electron neutrinos are oscillating into the other neutrino flavors such as $\nu_\mu$ or $\mu_\tau$ for which the neutral current reaction cross section is the same as for $\nu_e$ but for which the elastic scattering cross section is reduced - $\nu_e e$ elastic scattering can occur through both charged current and neutral current while $\nu_\mu e$ or $\nu_\tau e$ elastic scattering occurs only through the neutral current. That the electron fraction for the low energy neutrino differs from that for the high energy neutrinos suggests an energy dependent analysis is required to understand these results.

Terrestrial neutrino experiments have now clearly observed neutrino oscillations and have identified two $\Delta m^2$ values corresponding to oscillation lengths of $L/E \sim 10^6$ km at 1 MeV likely linking principally $\nu_\mu$ and $\nu_\tau$ and $L/E \simeq 32 \times 10^6$ km at 1 MeV that mixes $\nu_e$ into a mixture of $\nu_\mu$ and $\nu_\tau$. The oscillation lengths may be compared to 1 AU $= 1.5 \times 10^8$ km, to the difference between the minimum and maximum distance between the Earth and the Sun of $5 \times 10^6$ km, to the solar radius of $0.7 \times 10^6$ km, to the solar core radius which is roughly one quarter of the solar radius, and to the Earth radius 6371 km. As discussed below, we can consider $\nu_\mu$ and $\nu_\tau$ as tightly coupled. In a two flavor model in which $\nu_e$ mixes with one effective other flavor, the oscillation length for low energy solar neutrinos is long compared to the source dimensions or distance variations. For 10 MeV solar neutrinos, the oscillation length is still large compared to the (core) source size. Since the distance from the source to the detector on Earth is many oscillation lengths, the $\nu_e$ detection probability averaged over energies should be given by the energy independent quantity

$$< P(\nu_e \rightarrow \nu_e) > = 1 - \frac{1}{2} \sin^2(2\theta) \tag{11.35}$$

which is a measure of the mixing angle.

The explanation of the energy dependence of the $\nu_e$ disappearance probability for solar neutrinos appears to be the energy dependent coherent interaction of electron neutrinos with solar matter, called the MSW effect. In passing elastically through matter, electron neutrinos are distinguished from muon and tau neutrinos by their charge current exchange amplitude with target electrons. Consequently, the effective Hamiltonian for the mass eigenstates in the $\nu_e, \nu_\mu, \nu_\tau$ basis is

$$H = U \begin{pmatrix} 0 & 0 & 0 \\ 0 & \frac{\Delta m_{21}^2}{2E} & 0 \\ 0 & 0 & \frac{\Delta m_{31}^2}{2E} \end{pmatrix} U^\dagger + \begin{pmatrix} V & 0 & 0 \\ 0 & 0 & 0 \\ 0 & 0 & 0 \end{pmatrix}$$

where $U$ is the $3 \times 3$ mixing matrix and the effective potential $V$ derives from the amplitude for elastic scattering in the forward direction for $\nu_e$ by charge exchange and is given by

$$V = \sqrt{2} G_F n_e \tag{11.36}$$

with $n_e$ the electron number density. For solar core electron density $n_e = 6 \times 10^{25}$ cm$^{-3}$, the potential is $V = 0.75 \times 10^{-5}$ eV$^2$ MeV$^{-1}$. Since $\Delta m^2_{solar} = 7.6 \times 10^{-5}$ eV$^2$, the matter term dominates the vacuum evolution term for energies above about 4 MeV.

The effect of the matter term may be understood qualitatively as follows. For high energy neutrinos, we may neglect the vacuum oscillation portion of the preceding Hamiltonian and the electroweak neutrinos are the eigenstates. As the neutrino propagates to the surface of the Sun with adiabatic decrease in $n_e$ to zero, its state remains an energy eigenstate so it smoothly transitions to the higher mass vacuum eigenstate $\nu_2$ and then propagates without oscillation in this state from the solar surface to the Earth. The projection onto $\nu_e$ implies the high energy solar electron neutrino flux seen on Earth is the fraction $\sin^2 \theta \simeq 0.3$ of what would be expected without vacuum oscillation. In contrast, the low energy solar neutrinos are largely unaffected by the matter term, oscillate on their way out of the Sun and to Earth and the average low energy $\nu_e$ fraction seen on Earth is $< P >= 1 - \sin^2(2\theta)/2 \simeq 0.6$.

In the so-called MSW-LMA model which merges the calculation of matter mixing with large mixing angle vacuum mixing parameters, mixing effects in solar neutrino data are dominated by solar matter for energies above 3 MeV and by vacuum oscillations below 0.5 MeV. Regeneration in the Earth before detection is expected to add an additional 3-5% difference between upward going and downward going solar neutrinos. Borexino has identified the signal from the $^7$Be line in the intermediate energy region consistent with this model. Incidentally, Borexino has also observed a significant signal for $\bar{\nu}_e$ from geo-radioactivity.

We now turn to terrestrial studies of neutrinos. The seminal terrestrial observation of neutrino oscillation was of distance and energy dependent $\nu_\mu$ disappearance on a length scale of the Earth radius with $\nu_\mu$ of GeV energies. The discovery of the effect was made by the Super-Kamiokande experiment. The total mass of this experiment suffices to detect neutrinos produced in hadronic cosmic ray showers in the spherical shell of the Earth's atmosphere. These so-called atmospheric neutrinos originate predominantly in the decay chains

$$\begin{aligned}\pi^+ &\to \mu^+\nu_\mu \to \bar{\nu}_\mu e^+ \nu_e \nu_\mu \\ \pi^- &\to \mu^-\bar{\nu}_\mu \to \nu_\mu e^- \bar{\nu}_e \bar{\nu}_\mu\end{aligned} \tag{11.37}$$

and, since $\pi^+$ and $\pi^-$ are produced in roughly equal numbers, the number of $\nu_\mu$ is roughly twice the number of $\nu_e$. These neutrinos produce muons and electrons and hadronic showers in the Super-Kamiokande detector and, given the high energy, their direction is well correlated with the direction of the incident neutrino.

To first approximation, the flux of high energy cosmic rays incident upon the atmosphere is uniform. (The shadow of the Moon is observed.) Neutrinos are observed upward going with source distance simply related to polar angle as illustrated in Figure 11.12. The Super-Kamiokande collaboration observes a 50% deficit in muon neutrino flux as the travel distance increases to the maximum (one Earth diameter). [See Phys. Rev. **D81**, 092004 (2010).] The precise distance dependence of the $\nu_\mu$ disappearance is described by a two flavor mixing model as shown in Figure 11.13. The electron neutrino component appears constant on this length scale. The experiment has also observed evidence for the appearance of $\nu_\tau$ at a rate consistent with dominant oscillations between these two flavors but there with insufficient statistics to rule out a contribution from $\nu_e$ appearance.

Muon neutrino oscillation has been confirmed in studies of neutrinos produced at accelerators. The MINOS experiment uses a muon neutrino beam produced at Fermilab and observed both near the source and at a distance of 735 km in Minnesota. In Figure 11.14, the observed energy spectrum of charged current events producing muons in the far detector is compared to expectations based on the measurements with the near detector. (See http://www-numi.fnal.gov/pr_

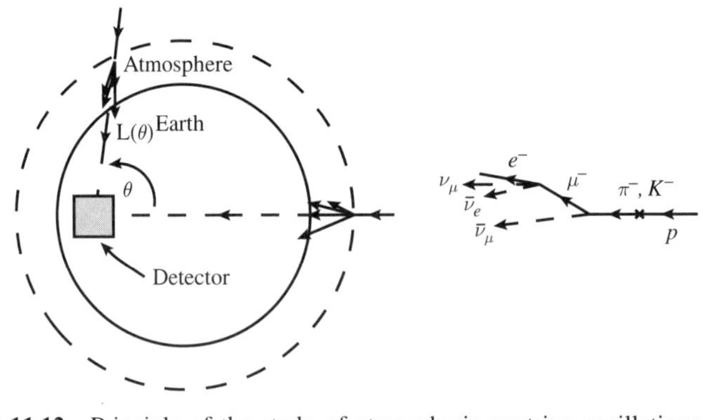


**FIGURE 11.12** Principle of the study of atmospheric neutrino oscillations. Neutrinos originate as the decay products of mesons in cosmic ray showers in the atmosphere and are detected with travel distance $L$ correlated with zenith angle $\theta$.

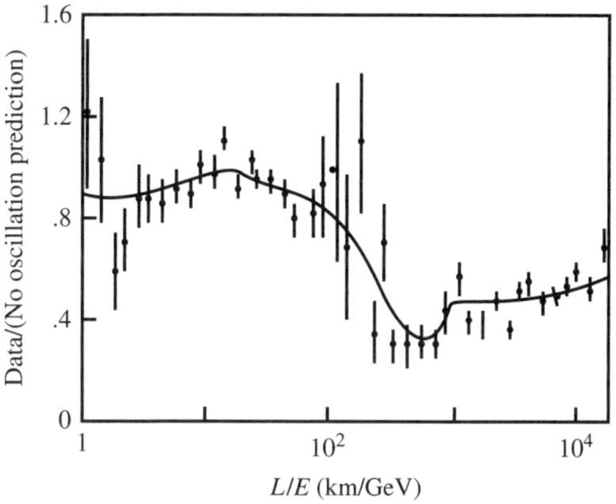


**FIGURE 11.13** Fit to oscillation model for neutrino flux as a function of distance divided by neutrino energy at the Super-Kamiokande experiment. [Super-Kamiokande Collaboration: Y. Ashie et al, Phys. Rev. Lett. 93, 101801(2004)]

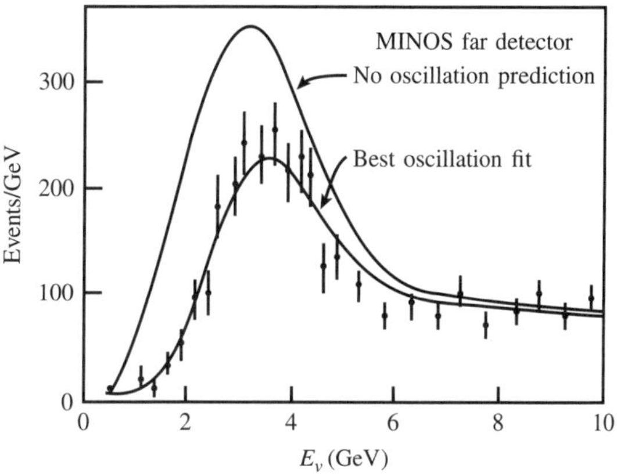


**FIGURE 11.14** Long baseline neutrino oscillations. Spectrum of charged current events in the MINOs far detector compared to expectation assuming the number of $\nu_\mu$ is conserved and compared to the oscillation hypothesis.

`plots/index.html`) A clear energy dependent $\nu_\mu$ deficit is found and is well described by a two flavor oscillation model. The neutral current energy spectrum (Figure 11.15) appears consistent with conservation of the total neutrino flux. A similar result was reported by the K2K long baseline experiment in Japan and additonal results will come from the T2K experiment. The $\nu_\mu$ disappearance experiments do not identify the flavor content ($\nu_e$ or $\nu_\tau$) of the neutrinos into which the $\nu_\mu$ is oscillating. Future experiments ICARUS, OPERA, and NOVA will accomplish this goal. ICARUS is a liquid argon time projection chamber capable of detailed reconstruction of tracks in $\tau^\pm$ production events from neutrino interactions. It will study atmospheric neutrinos and muon neutrinos directed from the

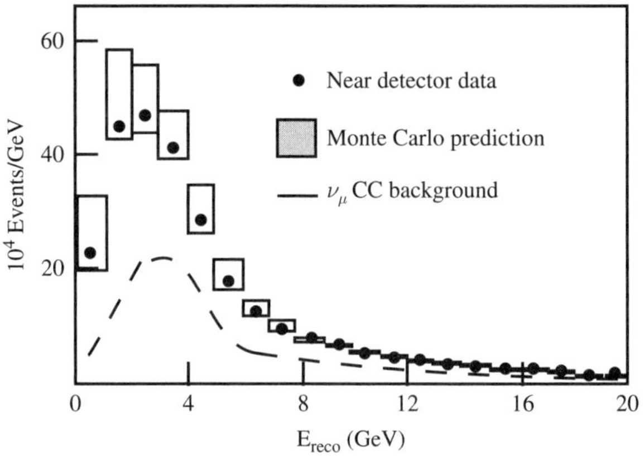


**FIGURE 11.15** Conservation of neutrinos. Spectrum of neutral current events in the MINOS far detector compared to expectation assuming the number of interacting neutrinos summed over flavor is conserved. The estimated background due to charged current events is also shown. [P. Adamson *et al.* (MINOS Collaboration), Phys. Rev. Lett.**107**,011802 (2011), Fermilab-Pub-11-183-E, 1104.3922v2 [hep-ex]]

CERN laboratory to the ICARUS location in the Gran Sasso, Italy. The OPERA experiment will use emulsion technology to search specifically for $\nu_\tau$ interactions and a $\nu_\tau$ component in neutrino beams from CERN. The NOVA experiment is a long baseline follow-on to the MINOS experiment focused on $\nu_e$ electron appearance and on comparing $\nu_\mu$ and $\bar{\nu}_\mu$ disappearance. These experiments also plan to study the oscillations of antineutrinos and search for $CP$-violation in the neutrino sector.

Additional information can be obtained by studies of antineutrinos ($\bar{\nu}_e$) produced in radioactive decay at nuclear power reactors. The first such experiments were CHOOZ and Palo Verde which operated with a 1 km baseline and placed limits on $\bar{\nu}_e$ disappearance. KamLAND is a 1000 ton liquid scintillator detector used to detect antineutrinos from distant reactors using the inverse beta decay reaction

$$\bar{\nu}_e + p \rightarrow e^+ + n \tag{11.38}$$

over an energy range 1-8 MeV. The prompt positron total energy essentially measures the neutrino energy (the neutron carries off only 10 keV) and neutron capture by free protons leading to delayed gamma ray production confirms the identification of the positron production. KamLAND is surrounded by 55 Japanese nuclear power reactor units, each an isotropic neutrino source. The reactor operation records are used to calculate the instantaneous fission rates using a reactor model, and the flux changes are used to identify the source distances. KamLand observes a clear signal for $\bar{\nu}_e$ disappearance (see The KamLand Collaboration, Phys. Rev. Lett. **100**, 221803 (2008)) well described by a two flavor oscillation model. The Daya Bay experiment (see arXiv:hep-ex/0701029v1) is a second generation experiment using eight detectors at three locations optimized to detect $\bar{\nu}_e$ from a cluster of six reactors in China and should significantly improve upon the pioneering experiments. This and other experiments planned or under construction hope to clarify the nature of the neutrino.

The results of the experiments described are consistent with the three flavor mixing model. One short baseline accelerator experiment LSND experiment found an excess of $\bar{\nu}_e$ in a $\bar{\nu}_\mu$ beam corresponding to a $\bar{\nu}_\mu \to \nu_e$ oscillation corresponding to a $\Delta m^2 \sim 1\ \text{eV}^2$ inconsistent with this model. The MiniBooNE experiment has searched for $\nu_\mu$ and $\bar{\nu}_\mu$ disappearance with a similar oscillation length and with inconclusive but tantalizing results suggesting that the oscillations of antineutrinos might differ from those of neutrinos.

The general phenomenological description of neutrino oscillations assumes that the three neutrino states $\nu_\alpha(\alpha = e, \mu, \tau)$ identified by standard model electroweak interactions are related to three mass eigenstates $\nu_i(i = 1, 2, 3)$ of mass $m_i$ by the unitary Maki-Nakagaw-Sakata-Pontecorvo (MNSC) matrix

$$V_{\alpha i} = < \nu_\alpha | \nu_i >, \;\; \nu_i = \Sigma_\alpha \nu_\alpha V_{\alpha i}, \;\; \nu_\alpha = \Sigma_i V^*_{\alpha i} \nu_i. \tag{11.39}$$

It should be born in mind that exactly three light neutrinos $\nu_\alpha$ are inferred from the width $\Gamma_Z$. If there are more than three light neutrinos, the additional ones must be of a sterile sort not coupling to the weak interaction and their inclusion requires an extended matrix.

Like the CKM matrix, the MNSC matrix may be parametrized as $V = UP$ with

$$\begin{aligned} U &= \begin{pmatrix} 1 & 0 & 0 \\ 0 & c_{23} & s_{23} \\ 0 & -s_{23} & c_{23} \end{pmatrix} \begin{pmatrix} c_{13} & 0 & s_{13}e^{-i\delta} \\ 0 & 1 & 0 \\ -s_{13}e^{i\delta} & 0 & c_{13} \end{pmatrix} \begin{pmatrix} c_{12} & s_{12} & 0 \\ -s_{12} & c_{12} & 0 \\ 0 & 0 & 1 \end{pmatrix} \\ &= \begin{pmatrix} c_{12}c_{13} & s_{12}c_{13} & s_{13}e^{-i\delta} \\ -s_{12}c_{23} - c_{12}s_{13}s_{23}e^{i\delta} & c_{12}c_{23} - s_{12}s_{13}s_{23}e^{i\delta} & c_{13}s_{23} \\ s_{12}s_{23} - c_{12}s_{13}c_{23}e^{i\delta} & -c_{12}s_{23} - s_{12}s_{13}c_{23}e^{i\delta} & c_{13}c_{23} \end{pmatrix} \end{aligned} \tag{11.40}$$

where $s_{ij} = \sin\theta_{ij}$ and $c_{ij} = \cos\theta_{ij}$ and the angles may be chosen to satisfy $0 \le \theta_{ij} \le \pi/2$. The matrix $U$ is unitary. The factor $P = \text{diag}(e^{i\alpha_1/2}, e^{i\alpha_2/2}, 1)$ applies if there are Majorana components.

A neutrino produced through the electroweak interaction after traveling a distance $x$ evolves into a quantum mechanical mixture

$$\nu_\alpha(x) = e^{iEx} \Sigma_j V^*_{\alpha j} e^{-im_j^2 x/(2E)} \nu_j^0. \tag{11.41}$$

The amplitude for observing flavor $\beta$ at distance $x = L$ is

$$< \nu_\beta | \nu_\alpha(x = L >= \Sigma_j V_{\beta j} \exp\left(-im_j^2 \frac{L}{2E}\right) V^*_{\alpha j}. \tag{11.42}$$

The probability that flavor $\nu_\beta$ appears through the electroweak interaction at distance $x$ is proportional to the square of this amplitude and has the form

$$P_{\alpha\to\beta} = \delta_{\alpha\beta} - 4\Sigma_{i>j}\text{Re}(U^*_{\alpha i}U_{\beta i}U_{\alpha j}U^*_{\beta j})\sin^2\frac{\Delta_{ij}}{2}$$
$$+2\Sigma_{i>j}\text{Im}\left(U^*_{\alpha i}U_{\beta i}U_{\alpha j}U^*_{\beta j}\right)\sin\Delta_{ij} \tag{11.43}$$

where we define $\Delta m^2_{ik} = m^2_k - m^2_i$ and

$$\Delta_{ij} \equiv \frac{\Delta m^2_{ij}L}{2E} = 2.534\frac{\Delta m^2_{ij}(\text{eV}^2)L(\text{km})}{E(\text{GeV})}. \tag{11.44}$$

Since $\Delta_{32} = \Delta_{31} - \Delta_{21}$, only two of these mass squared differences are independent and the transition probabilities for fixed energy are characterized by two spatial frequencies. Experiments identify two frequencies or mass differences: a high frequency associated with terrestrial length scale oscillations of GeV energy muon neutrinos made in the atmosphere, and a much smaller frequency associated with MeV energy solar electron neutrinos and reactor anti-electron neutrinos.

When two of the eigenstates have a mass difference small compared to their separation from the third, as appears to be the case, one may simplify the analysis. Assuming

$$|m^2_2 - m^2_1| << |m^2_3 - m^2_{1,2}|, \tag{11.45}$$

and $|\Delta_{21}| << 1$, examples of the disappearance probabilities expanded to first order in $\Delta_{21}$ (see Minako Honda, Yee Kao, Naotoshi Okamura, and Tatsu Takeuchi, arXiv:hep-ph/0602115v1.) are

$$P_{\nu_e\nu_e} = 1 - \sin^2(2\theta_{rct})\sin^2\left(\frac{\Delta_{31} - s^2_{12}\Delta_{21}}{2}\right)$$
$$P_{\nu_\mu\nu_\mu} = 1 - \sin^2(2\theta_{atm})\sin^2\left(\frac{\Delta_{31} - \kappa_{\mu\mu}\Delta_{21}}{2}\right)$$
$$P_{\nu_\mu\nu_e} = 4\left(\sin^2\theta_{rct}\sin^2\theta_{atm} - A\sin\delta\Delta_{21}\right)\sin^2\left(\frac{\Delta_{31} - \kappa_{\mu e}\Delta_{21}}{2}\right) \tag{11.46}$$

where atmospheric and reactor effective mixing angles are identified as

$$\sin\theta_{atm} = s_{23}c_{13} = \sin\theta_{23}\cos\theta_{13}$$
$$\sin\theta_{rct} = s_{13} = \sin\theta_{13} \tag{11.47}$$

and factors relevant to the next-to-leading terms are

$$A = \frac{1}{8}\sin(2\theta_{12})\sin(2\theta_{rct})\sin(2\theta_{atm})\sqrt{1 - \tan^2\theta_{rct}\tan^2\theta_{atm}}$$
$$\kappa_{\mu\mu} = c^2_{12} - (c^2_{12} - s^2_{12})\tan^2\theta_{rct}\tan^2_{atm} - \left(\frac{2A}{\cos^2\theta_{rct}\cos^2\theta_{atm}}\right)\cos\delta$$
$$\kappa_{\mu e} = s^2_{12} - \left(\frac{A}{\sin^2\theta_{rct}\sin^2\theta_{atm}}\right)\cos\delta. \tag{11.48}$$

To leading order ($\Delta_{21} = 0$), it is seen that disappearance experiments determine the effective mixing angles $\theta_{atm}$ and $\theta_{rct}$ and therefore $\theta_{23}$ and $\theta_{13}$. The preceding expressions apply for small $|\Delta_{21}|$ to the atmospheric neutrino disappearance $P_{\nu_\mu \nu_\mu}$ and to the short baseline CHOOZ experiment which measured $P_{\bar{\nu}_e \bar{\nu}_e} = P_{\nu_e \nu_e}$ and, in such cases, the appearance probabilities in this three flavor mixing model are:

$$\begin{aligned} P_{\nu_\mu \to \nu_\tau} &= \sin^2(2\theta_{23}) \cos^4(\theta_{13}) \sin^2 \Delta_{31} \\ P_{\nu_\mu \to \nu_e} &= \sin^2(2\theta_{13}) \sin^2(\theta_{23}) \sin^2 \Delta_{31} \\ P_{\nu_e \to \nu_\mu} &= \sin^2(2\theta_{13}) \sin^2(\theta_{23}) \sin^2 \Delta_{31} \\ P_{\nu_e \to \nu_\tau} &= \sin^2(2\theta_{13}) \cos^2(\theta_{23}) \sin^2 \Delta_{31}. \end{aligned} \tag{11.49}$$

The high frequency oscillation in atmospheric neutrinos yields parameters estimated by the Particle Data Group to be (PDG, 2005, 90% CL)

$$|\Delta m_{31}^2| = (2.43 \pm 0.13) \times 10^{-3} \text{ eV}^2 \text{ and } \sin^2(2\theta_{atm}) > 0.9 \tag{11.50}$$

from which it may be inferred that $|\cos\theta_{13}| \simeq 1$ and $|\sin\theta_{12}| \simeq 1$. From the CHOOZ experiment, the limit $\sin^2(\theta_{13}) < 0.19$ is derived while a global fit limit of about 0.05 has been quoted. It appears $\theta_{13} << 1$ so $\nu_\mu$ oscillates mainly to and from $\nu_\tau$ and $\nu_e$ is stable at this length scale.

When $|\Delta_{21}|$ is not small and $|\Delta_{31}| >> 1$, it is appropriate to average the general expression over the higher frequency. In this case, one finds

$$\begin{aligned} P_{\nu_e \nu_e} &= 1 - 2s_{13}^2(1 - s_{13}^2) - 4c_{12}^2 s_{12}^2 c_{13}^4 \sin^2 \frac{\Delta_{21}}{2} \\ &= 1 - \sin^2(2\theta_{12}) \sin^2 \frac{\Delta_{21}}{2} + O\left(|s_{13}|^2\right) \end{aligned} \tag{11.51}$$

where, in the last expression, $|s_{13}| << 1$ is assumed. This disappearance function applies to solar neutrinos (when corrected for the MSW effect) for which the effective mixing angle is $\theta_{sol} = \theta_{12}$.

Results from a global analysis are shown in Figure 11.16 and give the parameter values (with $1\sigma$ errors):

$$\begin{aligned} |\Delta m_{21}^2| &= (7.6 \pm 0.2) \times 10^{-5} \text{ eV}^2, \\ \sin^2(2\theta_{12}) &= 0.30 \pm 0.02, \\ |\Delta m_{31}^2| &= (2.4 \pm 0.1 \times 10^{-3} \text{ eV}^2, \\ \sin^2(2\theta_{23}) &= 0.50 \pm 0.07. \end{aligned} \tag{11.52}$$

This analysis also determines an upper bound ($3\,\sigma$) on the magnitude of the angle $\theta_{13}$ given by

$$\sin^2\theta_{13} < 0.056. \tag{11.53}$$

In sum, if we define $\nu_\pm = (\nu_\mu \pm \nu_\tau)/\sqrt{2}$, we have

$$\begin{aligned} \nu_3 &\simeq \nu_+ \\ \nu_2 &\simeq \cos\theta_{12}\nu_- - \sin\theta_{12}\nu_+ \\ \nu_1 &\simeq \sin\theta_{12}\nu_- + \cos\theta_{12}\nu_+ \end{aligned} \tag{11.54}$$

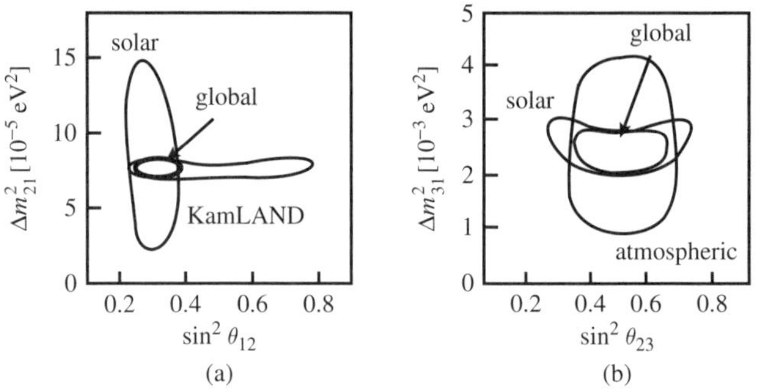


**FIGURE 11.16** Contours (99.73% confidence level) on neutrino mixing parameters from a global analysis of solar, atmospheric, accelerator, and reactor neutrino experiments. a) Leading "solar" oscillation parameters from solar neutrino oscillation measurements. Results from solar flux measurements and from KamLAND are compared to the global fit. b) Leading "atmospheric" oscillation parameters from data with natural and artificial neutrino sources. Results from MINOS and atmospheric neutrino experiments are compared to the global fit. [Thomas Schwetz, Mariam Tortola and Jose W. F. Valle, New J. Phys. **10**, 113011, 2008, arXiv:0808.2016v3 [hep-ph]]

and, if we assume the lowest mass vanishes, then $m_1 = 0$ eV, $m_2 \simeq 0.01$ eV, and $m_3 \simeq 0.05$ eV in the normal hierarchy or $m_3 \simeq 0$ eV, $m_1 \simeq 0.04$ eV, and $m_2 \simeq 0.05$ in the inverted hierarchy.

Notice that while $|\theta_{12}|$ appears to be small, the mixing angles $|\theta_{12}|$ and $|\theta_{23}|$ are large, unlike the quark flavor mixing described by the CKM matrix. Also, the factor $e^{i\delta}$ associated with $CP$-violating effects is multiplied by a factor $s_{13}$ which is small so $CP$-violation may be difficult to observe. It may be seen that the oscillation experiments determine the magnitudes of the differences between the squared masses, not the order. The case in which the nearly degenerate pair have masses below the third is called normal and the case in which the nearly degenerate pair have masses higher than the third is called inverted. Also, the scale of the masses themselves is still unknown. Mass limits of 3 eV, 190 keV, and 18 MeV for $\nu_e$, $\nu_\mu$, and $\nu_\tau$ are set by examining the endpoint energies of the weak decays. The masses must be as large as the mass differences so in the emerging picture at least one neutrino mass is larger than about $m > \sqrt{2\Delta m^2_{31}} = 0.07$ eV.

Neutrino mixing can be described in a simple extension of the standard model called $\nu$MSM introducing gauge invariant Higgs interactions to produce Dirac masses for the neutrinos $\nu_\alpha$ respecting the extremely tight constraints on conservation of lepton number. Because the neutrinos are neutral, the model entertains additional so-called Majorana mass terms linking neutrino to antineutrino in proportion to the Dirac masses. Such terms would not be consistent with gauge symmetry for charged leptons or quarks and imply lepton number violation proportional to the masses and consistent with present limits. Such a model introduces no interactions other than those implied by the Majorana mass terms. More general models attempt to incorporate neutrino mass and mixing in super-symmetric and grand unified theories and suggest generic reasons called the see-saw mechanism for a separation of the neutrino mass scale from the masses of other neutrinos. For details, the reader should consult the literature.

## 11.4 ■ PARTICLE PHYSICS AND COSMOLOGY

We now take a look at the role of particle physics in cosmology. To set the stage, we note that in cosmology the length scale of 1 Megaparsec (Mpc) is often used. One parsec is the distance at which one astronomical unit (the mean distance between the Earth and the Sun) subtends an angle of one second of arc. One second of arc is 1/3600 of one degree or $4.84 \times 10^{-6}$ radian or about the angular resolution of a diffraction limited lens of diameter 10 cm for visible light. One a.u. is $1.50 \times 10^{11}$ m and

$$1 \text{ pc} = 3.086 \times 10^{16} \text{ m} = 3.262 \text{ ly} \tag{11.55}$$

which is close to the distance between the Sun and its nearest neighbor star. The Milky Way galaxy is about 1000 ly thick and 50,000 ly or 15 kpc in radius. The Sun orbits the center at about one half the radius once every 250 million years or so and was last in it present position at the beginning of the Triassic period. The nearest comparable galaxy is the Andromeda galaxy at a distance of 2.2 million ly or 0.6 Mpc. The Earth is about 4.5 billion Earth years and 18 galactic years old. Its age is about one third of that of the Universe.

In the standard Big Bang model, the universe we see today originated in an explosion of energy. The expansion of the universe was first inferred from the recessional velocity of galaxies. The spectrum of visible light from a distant galaxy is measurably redshifted by its recessional velocity and Hubble discovered that the recessional velocity is proportional to distance, the proportionality expressed as

$$v = Hd \tag{11.56}$$

where $H$ is called the Hubble constant. By extrapolation backwards in time, it is inferred that the time of the explosion was about 14 billion years ago (roughly the value $H^{-1}$ which would apply if $H$ were constant over time) at which time the energy density and temperature were extremely high. The universe has since progressed through many stages which may be characterized by epochs of time or temperature as shown in Figure 11.17.

Einstein's general theory of relativity provides the framework for discussion of the cosmic expansion. In general relativity, the geometry of space and time are distorted by and evolves hand in hand with the energy content. In the generic homogeneous and isotropic model, the geometry is described by a time dependent curvature or scale factor. At very early times, cosmology theory is somewhat speculative. A landmark is the Planck time, around $10^{-43}$ s when it is supposed gravity and other forces were unified. Another landmark at $T \simeq 10^{17}$ GeV is suggested by extrapolation of coupling constants to what could be a grand unification scale. In the very early universe, it is postulated

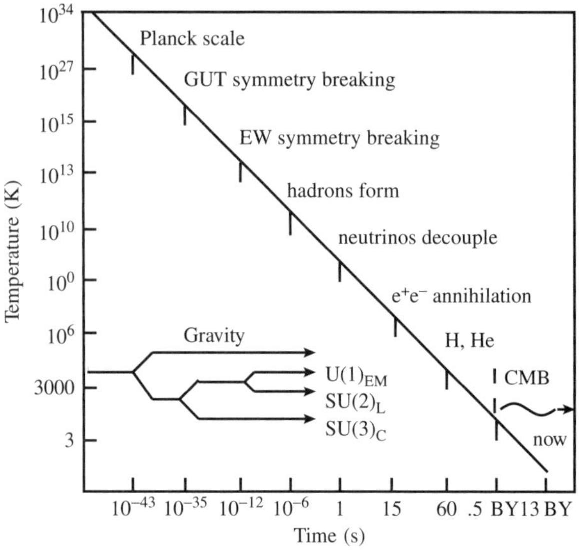


**FIGURE 11.17** Schematic timeline of events in the early universe.

that a cosmic inflation occurred - an exponential expansion due to a phase transition that erased local curvature fluctuations and produced locally the homogenous and isotropic region we find ourselves in today. A tiny imprint of gravitational and energy density fluctuations which seeded the formation of structure is observed in the cosmic microwave background and in galaxy distributions.

The universe emerged from the inflationary epoch as a hot plasma of relativistic particles. Some of these are the known quarks and leptons. Others such as SUSY particles may have been present. It is likely that most of such exotic particles are unstable and ultimately decayed leaving the photons and baryonic matter known today but some may be stable and constitute relic dark matter. This plasma was in thermal equilibrium and expanding. The expansion of the matter is linked dynamically to the geometry of the universe according to Einstein's general theory of relativity. The Einstein equations predict that under such initial conditions, spacetime expands, wavelengths expand, and particle energies drop. Consequently the temperature drops with time. In Newtonian terms, matter behaved as if thrown outwards in all directions with gravity slowing the expansion. At a temperature perhaps as low at 1 TeV, supersymmetry may have undergone a spontaneous phase transition with super particles emerging with masses of order 1 TeV. At a temperature corresponding to the weak scale, the Higgs field is presumed to have acquired a vacuum expectation value giving mass to quarks, leptons, and the $W$ and $Z$ boson.

After around $10^{-6}$ s, the temperature dropped to the scale of standard model physics and the evolution of the universe is better understood. The energy and mass scales we have encountered are the landmarks. The universe cooled and quarks and gluons combined to form hadrons when the temperature reached the hadronic scale. At about 1 s, the plasma was sufficiently dilute that neutrinos decoupled and thereafter have traveled freely. Relic neutrinos with small masses may be a small component of dark matter. As the temperature dropped below the mass of each known particle, particle-antiparticle pairs freely generated at higher temperatures could no longer be regenerated. Therefore the energy in that component of the particle spectrum was converted to lower mass particles and ultimately to photons through annihilation processes. A small excess of matter over antimatter was unable to annihilate and constitutes the stuff of the universe today.

At around three minutes, the temperature dropped to the nuclear scale and primordial nucleosynthesis began. Protons and neutrons combined forming deuterium and helium nuclei. It was a race against time for free neutrons decay to protons on this time scale. At around 20 minutes, the nucleon density had dropped sufficiently that nucleosynthesis ended, leaving about three times more deuterium nuclei than $^3$He nuclei by mass and essentially no higher mass nuclei. The abundances of light elements are sensitive to baryon number density $n_b$ usually expressed relative to the number density of photons $n_\gamma$ as $\eta = n_b/n_\gamma$ as illustrated in Figure 11.18.

A seminal point was the formation of atoms when the temperature dropped below the ionization potential of hydrogen. The universe changed from a florescent light bulb to a transparent gas of electrically neutral atoms. The scattering length for photons grew by an immense factor and, with expansion continuing, the photons essentially decoupled from matter. This decoupling occurred 380,000 years after

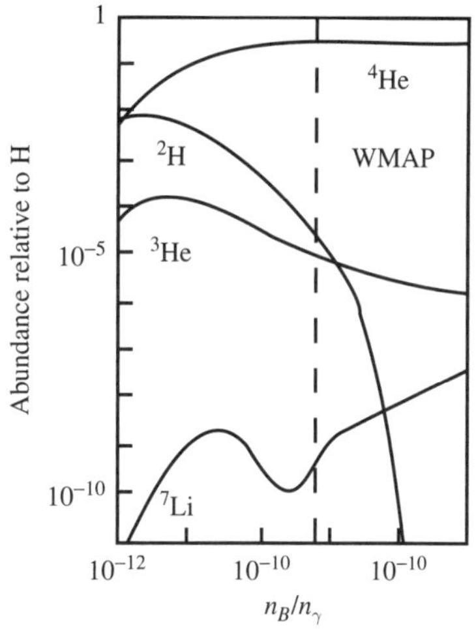


**FIGURE 11.18** Primordial nucleosynthesis. Relative abundances of light isotopes of H, He, and Li produced in the early universe are shown as functions of the baryon number density divided by the photon number density $\eta = n_b/n_\gamma$. The photon number density can be derived from the CMB intensity and the baryon density is related by the $\Lambda CDM$ model to the fluctuations in the CMB temperature. The value of $\eta$ from WMAP is shown.)

the explosion and the photons, cooled to a temperature of $2.725 \pm .002$ K and density $n_\gamma = 410.4$ cm$^{-3}$ today, are observable in the sky coming from in all directions exhibiting a thermal spectrum. This light is called the cosmic microwave background (CMB) radiation. An example CMB spectrum measured by the COBE satellite appears in Figure 11.19. The subsequent evolution of the universe entailed the gravitational collapse of dark matter and baryonic matter to form the vibrant collection of galaxies of compact objects and gas swarms in the structures seen today. The compact objects include main stream stars like the Sun and more exotic white dwarf stars, neutron stars, and black holes.

The life of a star has many stages and its fate depends on its initial mass. The star is supported against gravitational collapse by the release of energy in nuclear fusion. Nuclear fusion proceeds hesitantly through a sequence of steps in which light nuclei fuse, the fusion terminating in the formation of stable nuclei in the neighborhood of iron. At the end of the fusion process, a violent gravitational contraction can occur. An outer shell of matter is ejected and a compact object remains. Such events are observed as supernovae and in such stellar explosions, nuclei heavier than iron are formed. The aggregation of the star dust from stellar explosions along with primordial hydrogen and helium gas produced our solar system about five billion years ago. Supernova explosions release visible energy comparable to that of an entire galaxy for a short period of time. The intrinsic luminosity may be inferred from the shape of the emission spectrum. Supernovae therefore serve as standard candles in determining the distance to the host galaxies.

A remarkable feature of the CMB is the uniformity of the observed temperature. The largest anisotropic feature has a dipole form with a temperature difference between the extremes of 3.37 mK, interpreted as resulting from a peculiar solar velocity of $371 \pm 5$ km/s relative to the CMB, with 10% annual variation due to the Earth's orbital motion. Aside from this anisotropy, the temperature is found to be independent of direction to within of order one part in $10^5$. The temperature

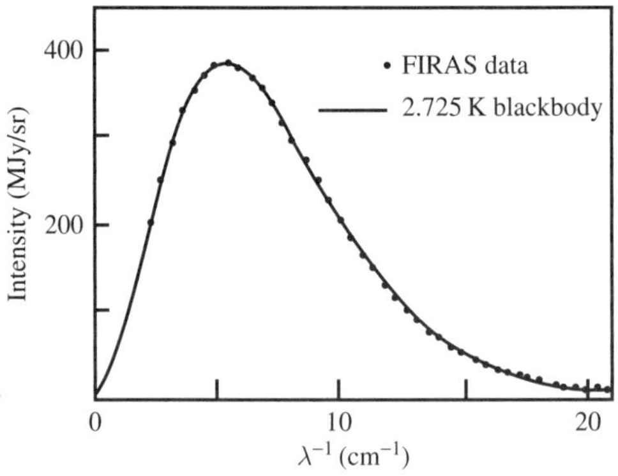


**FIGURE 11.19** Spectrum of cosmic microwave background radiation. The observed spectrum is that of a blackbody with a fitted temperature of 2.725 K.[ D.J. Fixen *el al*, Ap. J. **473**, 576 (1996)]

in a given direction reflects the temperature of the material in that direction at the time of photon decoupling. In reasonable models, temperature fluctuations are expected due to density fluctuations at the decoupling time, fluctuations seeded from still earlier invisible epochs. It is this uniformity that motivates the idea of an inflationary epoch. In inflationary models, a so-called inflaton field of unknown origin collapsed out of a metastable state injecting a vast energy into the dynamics of universe. The collapse is postulated to happen in such a way that a period of exponential expansion took place, erasing initial fluctuations in gravitational field energy and explaining the uniformity and isotropy of space. Examined more closely, the CMB temperature is found to fluctuate over angular scales of order one degree and smaller. Such fluctuations are shown in Figure 11.20 and can be connected to fluctuations in primordial energy components providing a window into opaque epochs that preceded the photon decoupling time.

Two mysterious components to the energy in the universe have emerged from cosmological and astrophysical studies. The first is dark matter associated with galaxies. The earliest evidence for dark matter was the rotational velocity distribution of stars in galaxies. A typical rotation velocity distribution is shown in Figure 11.21. Outside the core of a galaxy of visible mass $M$, the velocity of a star in a circular orbit at radius $r$ governed by Newtonian mechanics would be expected to be given by:

$$F = m\frac{v^2}{r} = \frac{mMG}{r^2} \rightarrow v(r) = \sqrt{\frac{MG}{r}}. \tag{11.57}$$

Observations show that $v(r)$ outside the visible core of a galaxy becomes independent of $r$ as would follow for orbits in a spherically symmetric distribution of dark matter with a density function $\rho_d \sim r^{-1}$. Observed rotation curves

**FIGURE 11.20** Temperature fluctuations in the cosmic microwave background radiation spectrum. Tiny deviations from the mean temperature, indicated by different colors, appear at different locations in the sky. [WMAP website gallery]

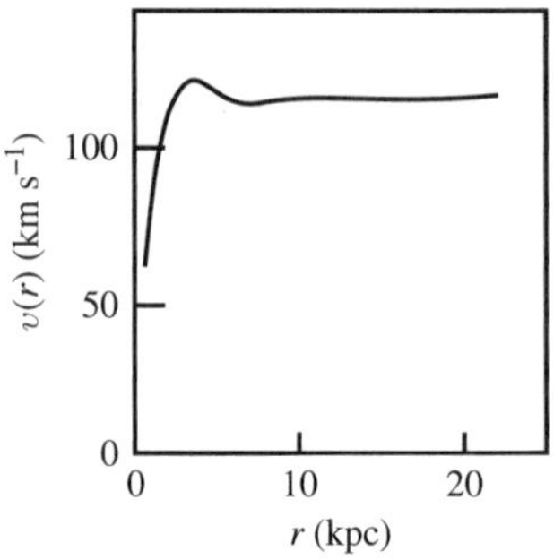


**FIGURE 11.21** Typical rotation curve for a galaxy. That the orbital speed $v$ of stars at radius $r$ outside the core is constant is explained by a spherical dark matter halo with density $\rho_d \sim r^{-1}$. [after van Albada *et al.*, Astrophysics Journal **295**, 305 (1985)]

can be fit by adding to the visible matter density a dark matter halo with mass distribution

$$\rho_d(r) = \frac{\rho_0}{\frac{r}{a}\left(1+\frac{r}{a}\right)^2} \tag{11.58}$$

where $\rho_0$ and $a$ are parameters. That an invisible source of gravity within galaxies exists was confirmed by the gravitational deflection of light passing near to a galaxy, the so-called gravitational lens effect predicted by Einstein. Observed deflections of light from far galaxies are stronger than predicted based on the mass inferred from the star light of deflecting nearer galaxies. Both kinds of observations are consistent with a dark matter component to galaxies with mass larger than that of the visible component.

Dark matter plays an important role in the standard cosmological model. In the standard model, gravitational aggregation of dark matter forms basins of attraction for the less consequential baryonic matter. Relativistic dark matter particles would free stream away from aggregation, smoothing out the gravitational potential wells into which matter falls. The standard cosmological model therefore favors cold nonrelativistic dark matter (CDM) particles. Cold dark matter particles are nonrelativistic at the time of their decoupling and the dividing line between hot and cold corresponds to a particle mass $m \simeq 1$ KeV. Observations of the tiny fluctuations in the CMB induced by early matter aggregation analyzed in the context of the $\Lambda CDM$ model described below indicate that relic neutrinos cannot account for more than a few percent of the dark matter content of the universe. Candidate dark matter particles include WIMPs (weakly interacting massive particles) and the lightest supersymmetric particle (LSP), perhaps the neutralino.

The second mysterious component is called dark energy. The first evidence for dark energy was the observation of an increase in the recessional velocity of supernovae in excess of that given by the Hubble relation. This acceleration was quite unexpected since gravity is expected to slow the expansion and, perhaps, even lead to a re-collapse. The magnitude versus redshift relationship of standard

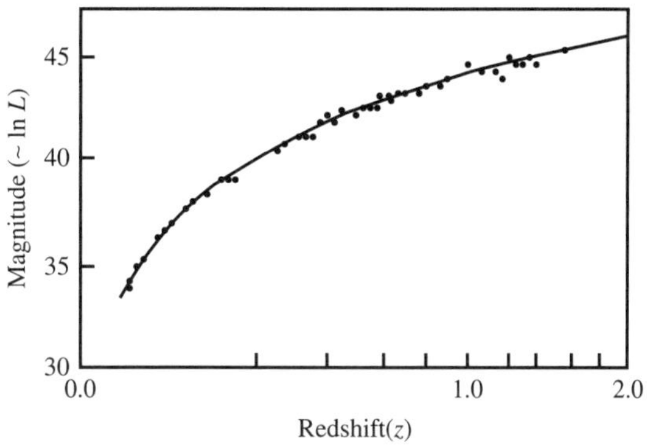


**FIGURE 11.22** Magnitudes of supernovae events versus redshift. Redshift is a measure of recessional velocity. Magnitude is a measure of distance. The relationship between the two is governed by the expansion history. A fit to the data in the $\Lambda CDM$ model is shown. [after M. Kowalski *et al.*, Astrophys. J. **686**, 749 (2008), arXiv:0804.4142 [astro-ph]]

candle supernovae out to redshifts beyond $z = 1$ is shown in Figure 11.22. The magnitude is the negative of the log of the observed luminosity and is a measure of apparent dimness. The redshift is defined by

$$z = \frac{\lambda_o - \lambda_e}{\lambda_e} \tag{11.59}$$

where $\lambda_e$ is the wavelength at emission and $\lambda_o$ is the wavelength observed. The largest galactic redshift value at the time of this writing is $z = 8.6$ ascribed to UDFy-38135539. The relationship between the magnitude and the redshift is model dependent. In a homogeneous and isotropic universe evolving with a time dependent scale factor $R(t)$ discussed in more detail below, the redshift is given by

$$z = \frac{R(t_o))}{R(t_e)} - 1 \tag{11.60}$$

where $t_o$ and $t_e$ and the proper times of observation and emission. There are several independent measures of distance. The luminosity distance $D_L$ of an object of absolute luminosity $L$ and apparent luminosity $l$ is defined by the inverse square relation

$$l = \frac{L}{4\pi D_L^2} \tag{11.61}$$

The apparent luminosity is reduced not just by luminosity distance but by a factor $(1+z)^{-1}$ due to photon energy shift, by another factor $(1+z)^{-1}$ due to time dilation of the emission rate, and by an flux factor $\lambda^{-2}$ which depends on the evolution of the geometry of the universe between emission and detection. In an homogeneous and isotropic universe,

$$\lambda = \frac{1}{\sqrt{k}} \sin\left(c\sqrt{k}\int_{t_e}^{t_o} dt\ R(t)^{-1}\right) \tag{11.62}$$

is the ratio of proper distances $ds(t_o)/ds(t_e)$ with $k = -1, 0$, or $+1$ the sign of the curvature. The relationship between absolute and apparent luminosity in such a model is then

$$l = \frac{LkR^2(t_e)}{4\pi\, R^4(t_o) \sin^2\left(c\sqrt{k}\int_{t_e}^{t_o} dt\; R(t)^{-1}\right)}. \tag{11.63}$$

Given $R(t)$, Equation (11.60) and Equation (11.63) provide a model for the observations of apparent magnitude verse redshift.

Dark energy is motivated by fits to the data shown in Figure 11.22. Dark energy is postulated as an effective pressure resulting in an excess positive acceleration of distant galaxies. It manifests itself in the luminosity versus redshift curve as a reduction in the dimming (larger negative magnitude) of distant (larger z) supernovae. It appears the dark energy effect can be modeled by a simple constant denoted by $\Lambda$ and called the cosmological constant. The cosmological constant was first introduced by Einstein to create a model of a static universe supported against gravitational collapse, and then discarded after the discovery of the present general expansion.

The standard cosmological so-called $\Lambda CDM$ model attempts to explain the entire history of the expanding universe leading from the inflationary epoch to present galaxies. Prominent parameters of the model include the present density of baryonic matter, the density of dark matter, and the value for the dark energy parameter $\Lambda$. The observables include fluctuations in the CMB spectrum, the distance versus redshift function for galaxies, the baryonic and dark matter densities in galaxies and gas estimated from a variety of observations, the patterns of aggregation of galaxies, and the relative abundances of the products of primordial nucleosynthesis. The stage for the model is Einstein's theory of the evolution of space, time, and energy.

## 11.5 ■ THE $\Lambda CDM$ MODEL

The evolution of the early universe in the $\Lambda CDM$ model assumes an expanding homogeneous and isotropic universe in thermal equilibrium governed by Einstein's theory of general relativity. Here we can provided but an overview of the governing equations.

Einstein's theory derives from the principle of equivalence according to which physics in a local free falling frame of reference is identical to that in a gravity-free inertial frame as described by the special theory of relativity. The theory of general relativity assumes space-time is non-Euclidean. The local invariant interval $ds^2$ can be written as

$$ds^2 = g_{\mu\nu}(x)dx^\mu dx^\nu = g_{\mu\nu}(y)\frac{\partial x^\mu}{\partial y^\alpha}\frac{\partial x^\nu}{\partial y^\beta}dy^\alpha dy^\beta \tag{11.64}$$

using a local metric tensor field $g_{\mu\nu}$ and general coordinates $x$ or $y$. A derivative covariant under a general coordinate transformation, reminiscent of the gauge covariant derivative, is defined by

$$A^{\nu}_{;\mu} \equiv \partial_{\mu} A^{\nu} + \Gamma^{\nu}{}_{\sigma\mu} A^{\sigma}. \tag{11.65}$$

where $A_{,\mu} \equiv \partial_{\mu} A^{\nu}$. The fundamental Christoffel symbols which appear here are

$$\Gamma^{\rho}{}_{\mu\nu} = \frac{1}{2} g^{\rho\sigma} (\partial_{\nu} g_{\mu\sigma} + \partial_{\mu} g_{\nu\sigma} - \partial_{\sigma} g_{muv}). \tag{11.66}$$

The curvature of space-time is represented by the covariant Ricci tensor field which can be expressed using the Christoffel symbols as

$$R_{\mu\nu} = \partial_{\nu} \Gamma^{\sigma}{}_{\mu\sigma} - \partial_{\sigma} \Gamma^{\sigma}{}_{\mu\nu} + \Gamma^{\sigma}{}_{\mu\rho} \Gamma^{\rho}{}_{\sigma\nu} - \Gamma^{\rho}{}_{\mu\nu} \Gamma^{\sigma}{}_{\rho\sigma}. \tag{11.67}$$

Einstein's equations relate the local curvature to the local energy-momentum tensor $T_{\mu\nu}$:

$$R_{\mu\nu} - \frac{1}{2} g_{\mu\nu} R + g_{\mu\nu} \Lambda = \frac{8\pi G_N}{c^4} T_{\mu\nu} \tag{11.68}$$

where $G_N$ is Newton's constant, $R = R^{\mu}_{\mu}$ is called the scalar curvature, and $\Lambda$ is the cosmological constant now associated with dark energy. The energy momentum tensor of a fluid collection of particles is

$$T^{\mu\nu} = -p g^{\mu\nu} + (p + \rho) u^{\mu} u^{\nu} \tag{11.69}$$

with $p$ the pressure, $\rho$ the mass density, and $u$ the flow 4-velocity field. It is implicit that the energy and momentum move locally as if free of gravity. Pressureless noninteracting dust-like particles which follow the so-called geodesic paths in curved space-time represent the simplest form of matter. Comparison of the term $g_{\mu\nu}\Lambda$ with $T_{\mu\nu}$ shows that dark energy behaves as a gas with pressure $p_{\Lambda} = \Lambda$. The equation of state parameter is defined as

$$w \equiv \frac{\rho}{p} \tag{11.70}$$

and, in this model of dark energy, $w_{\Lambda} \equiv \frac{\rho_{\Lambda}}{p_{\lambda}} = -1$.

The Ricci tensor is a nonlinear combination of $g_{\mu\nu}$ and its derivatives and the energy momentum tensor depends on $g_{\mu\nu}$. The nonlinearities present even in the absence of matter correspond to gravitational energy itself distorting space-time. The covariant energy and momentum tensor for a matter or radiation field is generally tied up with the metric tensor. For example, the energy momentum tensor for a free classical electromagnetic field is

$$T^{\mu\nu} = -\mu_0^{-1} \left( F^{\mu\rho} F_{\rho}{}^{\nu} + \frac{1}{4} g^{\mu\nu} F^2 \right) \tag{11.71}$$

where the field strength tensor is $F_{\mu\nu} = A_{\mu;\nu} - A_{\nu;\mu} = A_{\mu,\nu} - A_{\nu,\mu}$ with $A^\mu$ the vector potential. A locally linearized version of general relativity theory writes the metric tensor in the form

$$g_{\mu\nu} = \eta^0_{\mu\nu} + h_{\mu\nu} \tag{11.72}$$

where $h_{\mu\nu}(x)$ represents a small deviation from a flat metric represented by $\eta^0_{\mu\nu}$. One can construct the symmetric tensor field

$$\tilde{h}_{\mu\nu} = h_{\mu\nu} - \frac{1}{2}\eta_{\mu\nu}h \tag{11.73}$$

using the flat space-time metric tensor $\eta_{\mu\nu}$ with $h = \eta_{\mu\nu}h^{\mu\nu}$ and, in the linear approximation, find $\tilde{h}_{\mu\nu}$ satisfies the wave equation

$$\partial_\alpha \partial^\alpha \tilde{h}_{\mu\nu} = -16\pi G_N T_{\mu\nu} \tag{11.74}$$

where $T^{\mu\nu}$ is the energy-momentum tensor associated with matter. This equation has the structure of a wave equation for a massless spin-2 field with energy-momentum as a source and sink of gravitational waves.

Einstein's equations have solutions describing homogeneous and isotropic universes in which space coordinate intervals expand or contract as a function of co-moving proper time $t$ with a scale factor $R(t)$. The space-time interval can be written as

$$ds^2 = dt^2 - R^2(t)\left[\frac{dr^2}{1-kr^2} + r^2\left(d\theta^2 + \sin^2\theta\, d\phi^2\right)\right] \tag{11.75}$$

where $k = -1, 0, +1$ is a topological signature distinguishing open, flat, and closed universes respectively. The Einstein equations reduce to equations for the scale factor. For a fluid filled universe, the scale factor satisfies the Friedmann-Lemaitre equations that follow from the time and space components of the Einstein equations, namely

$$\begin{aligned} \left(\frac{\dot{R}}{R}\right)^2 &= \frac{8\pi G_N}{3}\rho - \frac{k}{R^2} + \frac{\Lambda}{3} \\ \frac{\ddot{R}}{R} &= -\frac{4\pi G_N}{3}(\rho + 3p) + \frac{\Lambda}{3} \end{aligned} \tag{11.76}$$

where the energy density $\rho$ and pressure $p$ are functions only of proper time. These equations imply

$$\dot{\rho} = -3H(\rho + p) \tag{11.77}$$

where the time dependent relative expansion rate $H \equiv \dot{R}/R$ is called the Hubble parameter. Its value today is usually written as

$$H \equiv \frac{\dot{R}}{R}|_{\text{today}} = 100\, h \text{ km s}^{-1} \text{ Mpc}^{-1} \tag{11.78}$$

where $h$ is close to 0.72. In solving the Friedmann-Lemaitre equations, the pressure may be eliminated in favor of temperature $T$ using the equation of state. Particle physics enters in the evolution of $\rho(T)$. For example, for a gas of photons, the number density is given by the Planck distribution.

A pedestrian derivation of the Friedmann-Lemaitre equations for $\Lambda = 0$ and nonrelativistic pressureless matter proceeds as follows. Consider a sphere of radius $R(t) = a(t)R_0$ containing particles of mass $m$ at zero pressure and presume the particles may be described by comoving coordinate $\mathbf{x} = a\mathbf{r}$ with $R_0$ and $\mathbf{r}$ dimensionless and $a(t)$ an arbitrary function. The relative velocity of two particles is

$$\mathbf{v}_{ij} = \dot{a}(\mathbf{r}_i - \mathbf{r}_j) \tag{11.79}$$

so each particle sees other particles receding with velocity proportional to distance. Mass conservation implies the mass density is proportional to $a(t)^{-3}$ which implies

$$\dot{\rho} + 3\frac{\dot{a}}{a}\rho = 0. \tag{11.80}$$

The gravitational potential energy of a particle due to particles at smaller radius is

$$U = -\frac{G_N m M(|\mathbf{x}|)}{|\mathbf{x}|}, \quad M(|\mathbf{x}|) = \frac{4}{3}\pi\rho|\mathbf{x}|^3. \tag{11.81}$$

The total energy is

$$E = \frac{1}{2}m\dot{\mathbf{x}}^2 + U = \frac{1}{2}\dot{a}^2|\mathbf{r}|^2 - \frac{4}{3}G_N\pi m\rho a^2|\mathbf{r}|^2 \tag{11.82}$$

and energy conservation implies

$$\left(\frac{\dot{a}}{a}\right)^2 = \frac{8\pi G_N}{3}\rho - \frac{kc^2}{a^2} \tag{11.83}$$

with $k = -2E/(mc^2r^2)$. The sign of $E$ determines the sign of $k$.

If $\Lambda = 0$, the solution to the Friedman-Lemaitre equations with $k = 0$ applies for the critical density

$$\rho_c = \frac{2H^2}{8\pi G_N} = 1.05 \times 10^{-5}\, h^2 \text{ GeV cm}^{-3}. \tag{11.84}$$

If $\Lambda = 0$ and $\rho > \rho_c$, then $k = +1$ and the universe is closed - the expansion is halted by gravity and the universe will ultimately collapse. If $\Lambda = 0$ and $\rho < \rho_c$ then $k = -1$ and the universe is open and expands forever. If $\Lambda \neq 0$, there can be a balance between the effective pressure $p_\Lambda$ and gravity.

The Friedmann-Lemaitre equations show that, in the absence of matter, a positive cosmological constant leads to exponential growth of the universe. Inflation models make use of this result and postulate a positive vacuum energy density associated with a generic scalar field and spontaneous symmetry breaking perhaps at the GUT scale. The exponential growth erases primordial inhomogeneities and dilutes relic objects like magnetic monopoles yielding a flat ($k = 0$) homogeneous and isotropic region within the visible horizon of a typical observer. The epoch of inflation is presumed to have ended by a phase transition to a lower vacuum energy density with vacuum energy materializing in the form of standard model particles and radiation. In the next stage of the expanding universe, the quantity $\Lambda$ which may be unrelated to inflation is small relative to the energy density associated with radiation and relativistic particles. Nonrelativistic matter appears at larger scale factors after expansion and cooling. Ultimately the matter and radiation energy density become negligible and the post phase transition value $\Lambda$ of the vacuum energy determines the evolution. In particular, were $\Lambda < 0$, the universe would ultimately collapse.

In many stages of the evolution of the universe, the interaction rate is large compared to the expansion rate and particles are in thermal equilibrium. In this case, their temperature and pressure distributions are described by thermodynamics and characterized by an evolving temperature. For a single component nonrelativistic baryonic gas with $p = \rho k_B T/m$, for example, we have $w = k_B T/m \simeq 0$ if $k_B T$ is much less than $m$ ~1 GeV. The mass density then satisfies

$$\dot{\rho} = -3(1+w)H\rho \rightarrow \rho(t) = \rho_0 \left(\frac{R(t)}{R_0}\right)^{-3(1+w)} \sim R^{-3} \tag{11.85}$$

which simply reflects the expansion of space. Given this expression, the Friedmann-Lemaitre equations for $k = \Lambda = 0$ imply $R(t) \propto t^{2/3}$ and $H = (2/3)t^{-1}$.

In the ideal gas approximation, the number density of particles with momentum between $p$ and $p + dp$ is

$$dn = \frac{g}{2\pi^3} \frac{p^2 dp}{e^{E/k_B T} \pm 1} \tag{11.86}$$

where $E = \sqrt{p^2 + m^2}$ is the energy and $g$ is the number of helicity states. The plus sign applies to fermions and the minus sign to bosons. For relativistic particles, $E \simeq p$ and the density and pressure that follow imply $w = -1/3$ and $\rho \sim R^{-4}$, an extra factor of $R^{-1}$ being associated with a redshift of wavelengths, the particle energy being proportional to $R^{-1}$. In this case, $R(t) \propto t^{1/2}$ and $H = (2t)^{-1}$. In particular, for photons, the energy density $\rho_\gamma$, pressure $p_\gamma$, number density $n_\gamma$, and entropy density $s_\gamma$ are given by

$$\rho_\gamma = g_\gamma \frac{\pi^2}{30} T^4 \; ; \; p_\gamma = \frac{1}{3}\rho_\gamma \; ; \; n_\gamma = \frac{2\chi(3)}{\pi^2} T^4 \; ; \; s_\gamma = \frac{1}{T}(\rho_\gamma + p_\gamma) \tag{11.87}$$

where $g_\gamma = 2$ in the number of photon spin states. Similar expressions hold for fermions with $g_\gamma \Rightarrow = (7/8)g_f$ where $g_f$ is the number of fermion degrees of freedom ($g_f = 2$ for the electron and $g_f = 1$ for the $\nu_e$ in the standard model). The temperature sets the scale of energy density and pressure and the Friedmann-Lemaitre equations then relate temperature to expansion rate and one finds $T \propto R^{-1}$ in an adiabatically expanding universe filled with relativistic particles.

In the early universe, when the temperature exceeded $m_e$, electrons and positrons were abundant. More generally, the number of relativistic degrees of freedom evolved from an effective $N = 427$ in the standard model when the temperature exceeded the weak scale to $N = 29$ for photons and $\nu$ and $\bar{\nu}$ when the temperature dropped below $m_e$. It can be estimated from neutrino collision cross sections and the evolving number densities that neutrinos went out of equilibrium at a temperature of about 3 MeV and thereafter have been free streaming. The density of photons took a slight jump due to $e^+e^-$ annihilation shortly afterwards and entropy conservation may be used to relate the relic neutrino temperature to the photon temperature:

$$T_\nu = (4/11)^{1/3} T_\gamma. \tag{11.88}$$

The entropy in photons and in the six decoupled neutrinos was then nearly the same and these have continued to fall together as the universe has expanded.

The post-decoupling expansion increases the wavelength of a particle by a factor $\alpha = R(T)/R(T_D)$ and a volume by a factor $\alpha^3$. When the temperature is $T$, the phase space distribution for the $i$-th neutrino species is

$$dn = \frac{g}{2\pi^3} \frac{p^2 dp}{e^{E/k_B T} \pm 1} \tag{11.89}$$

where $p = p_D/\alpha$, $T = T_D/\alpha$, and $E = \sqrt{p^2 + \tilde{m}_i^2}$. Here $\tilde{m}_i = m_i/\alpha << m_i$ has been neglected. Hence, for massive neutrinos, the momentum distribution still has the relativistic form even if the neutrinos become nonrelativistic. From the observations of the CMB photon temperature and density, one can estimate the present number density of neutrinos to be about 50 cm$^3$ per flavor with an equal number of antineutrinos. The mean relic neutrino momentum is

$$< p_\nu > \simeq 3.2 T_\nu = 5.2 \times 10^{-4} \text{ eV} \tag{11.90}$$

corresponding to a wavelength $\lambda_\nu = 2\pi / < p_\nu > = 2.3$ mm. Notice that if $m_\nu \simeq 5 \times 10^{-2}$ eV, the velocity is $v_\nu \simeq 0.01$ while if $m_\nu = 0.25$ eV the velocity is $v \simeq 2 \times 10^{-3}$. The velocity required to escape the Sun's gravity is $v_{\text{esc,Sun}} \sim 10^{-4}$ and the escape velocity for the Milky Way galaxy is $v_{\text{esc,MW}} \sim 10^{-3}$ so clustering of some neutrino flavors on a galactic scale is possible.

The present ratio of nucleon number density $n_b$ to photon number density $n_\gamma$ is estimated to be

$$\eta = \frac{n_b}{n_\gamma} \simeq 7 \times 10^{-9}. \tag{11.91}$$

Assuming $n_\gamma \propto T^3$ and $n_b \propto n_\gamma$, the temperature $T_m$ at which the radiation energy density became dominated by matter is given by $\rho_\gamma = m_N \eta n_\gamma$ implying $T_m \simeq 0.13$ eV. Residual electrons and protons plus hydrogen atoms and photons were in equilibrium and the temperature was above $T = 1.6 \times 10^5$ K corresponding to the ionization potential 13.6 eV. Hydrogen atoms were unable to absorb or scatter photons effectively and the universe became transparent at a decoupling temperature $T_d \simeq 3000$ K. The $\Lambda CDM$ model includes the evolution of the baryon number density, photon number density, and neutrino densities, models of reaction equilibrium governing what species of particles are present in thermal equilibrium, along with the evolution of noninteracting dark matter and dark energy.

The present day density of matter scaled to the critical density is called $\Omega_m = \rho/\rho_c$. The present energy density of relativisitic particles scaled to the critical density is called $\Omega_r$. The scaled vacuum energy density parameter is called $\Omega_\Lambda = \Lambda/(3H^2)$. The sum of these parameters determines the sign of the curvature and the Friedman-Lemaitre equation can be written as

$$\frac{k}{R^2} = H^2(\Omega_m + \Omega_r + \Omega_\Lambda - 1). \tag{11.92}$$

The $\Lambda CDM$ model has been used to predict the evolution of density and temperature fluctuations responsible for seeding the coalescence of matter into galactic structures. Consider an early spherical over-density fluctuation in a closely coupled baryon plus photon plasma together with weakly interacting dark matter. Pressure drives a radial acoustic pulse moving outwards at sound speed

$$c_s(\rho_b, \rho_\gamma) = c\frac{1}{\sqrt{3\left(1 + 3\frac{\rho_b}{\rho_\gamma}\right)}}. \tag{11.93}$$

At decoupling time $t_* \sim 4 \times 10^5$ yr, in the expanding universe, the pulse has reached a size

$$s = \int_0^{t_*} dt\, c_s(1+z) = \int_{z_*}^{\infty} dz\, \frac{c_s}{H(z)} \tag{11.94}$$

or, in the $\Lambda CDM$ model, $s \simeq 150$ Mpc. Density fluctuations translate to thermal fluctuations through the equation of state. After $t_*$, the photons stream away bearing the imprint and a temperature excursion appears at an angular scale of order one degree in the CMB. Not subject to pressure, the weakly interacting dark matter lags behind the baryonic matter. After condensation into atoms with the release of photons, the neutral gas falls back towards the dark matter in a so-called baryon acoustic oscillation (BAO). This implies a patch size for the growth of galactic structure correlated with the patch size observed in the CMB. The patch size is observable in the two point correlation function of galaxies.

Temperature fluctuations in the cosmic microwave background observed by the WMAP satellite are shown in Figure 11.20. The power spectrum of the angular distribution is shown in Figure 11.23 and is sensitive to the baryonic matter density, dark matter density, and cosmological constant which govern the rate of expansion, the sound speed, and streaming rates which smooth out fluctuations. Remarkably, the $\Lambda CDM$ model can be fit to the observed power spectrum and cosmological parameters can be extracted with high precision. Some of the important parameter

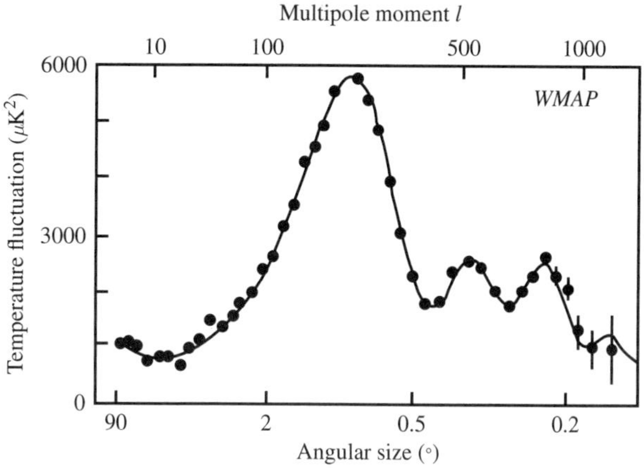


**FIGURE 11.23** Size spectrum of temperature fluctuations in the cosmic microwave background measured by the WMAP observatory [credit NASA / WMAP Science Team].

values are given in Table 11.2 and are consistent with the results obtained from other observations.

The convergence of observations of these signals of the early universe in the $\Lambda CDM$ is shown in Figure 11.24. The matter density and dark energy density values extracted from supernovae data, from the CMB, and from BAO data are all consistent. The values for $\Omega_m$ and $\Omega_\Lambda$ are consistent too with $\Omega_{tot} = 1$ as required by inflation. One of the interesting parameters is the number of neutrinos. Global fits to neutrino oscillation parameters found in terrestrial experiments indicate the sum of neutrino masses is $m_\nu^{tot} > 55.3$ meV while the present limit on the mass of the electron neutrino is 2.3 eV. The Katrin tritium beta decay experiment expects to reduce this limit to 200 meV. The influence of neutrinos on the angular distribution of the cosmic microwave background has been calculated (Barger *et al.*, hep-ph/0312065). Precise measurements with the PLANCK observatory are expected to be sensitive to $m_\nu^{tot}$ >260 meV.

The presence of relic neutrinos appears to be an inevitable prediction of the standard cosmological model but their detection is extremely difficult due to their low energy, low density, and weak interactions. A number of proposed methods have been considered. One method is zero energy absorption induced nuclear reactions such as

$$\nu_e + {}^3H \rightarrow e^{-3}He \tag{11.95}$$

which requires monitoring of order 1 kg of tritium. Incoherent scattering is governed by the cross section

$$\sigma \sim G_F^2 E_\nu^2 \sim 10^{-62}\ \text{cm}^2 \tag{11.96}$$

and a flux $j_\nu \sim 10^{12}$ cm$^{-2}$ s$^{-1}$ impinging on $10^{33}$ targets observed for $10^8$ s yields no events. However, since the wavelength $\lambda_\nu$ of relic neutrinos is of order

| Cosmological Parameter | Symbol | Value |
|---|---|---|
| Age of universe | $t_0$ | $13.72 \pm 0.12$ Gyr |
| Hubble constant | $H_0$ | $70.5 \pm 1.3$ km s$^{-1}$ Mpc$^{-1}$ |
| Baryon density | $\Omega_b$ | $0.0456 \pm 0.0015$ |
| Dark matter density | $\Omega_c$ | $0.228 \pm 0.013$ |
| Neutrino density | $\Omega_\nu h^2$ | $< 0.0071$ (95% CL) |
| Neutrino mass | $\Sigma m_\nu$ | $< 0.67$ eV (95% CL) |
| Number of light neutrino families | $N_{eff}$ | $4.4 \pm 1.5$ |
| Curvature fluctuation amplitude | $\Delta_R^2$ | $(2.445 \pm 0.096) \times 10^{-9}$ |
| Redshift of matter-radiation equality | $z_{eq}$ | $3253 \pm 89$ |
| Age at decoupling | $t_*$ | $377 \pm 3$ kyr |
| Redshift at decoupling | $z_*$ | $1090.88 \pm 0.72$ |
| Age at reionization | $t_{reion}$ | $432^{+90}_{-67}$ Myr |
| Total density | $\Omega_{tot}$ | $1.005 \pm .006$ |
| Equation of state index | $w$ | $-0.99 \pm 0.06$ |

**TABLE 11.2** Cosmological parameters derived from fits to cosmic microwave data in the $\Lambda$CDM model. [G. Hinshaw *et al.*, Astrophys. J. Suppl. **180**, 225 (2009), arXiv:0803.0732 [astro-ph]]

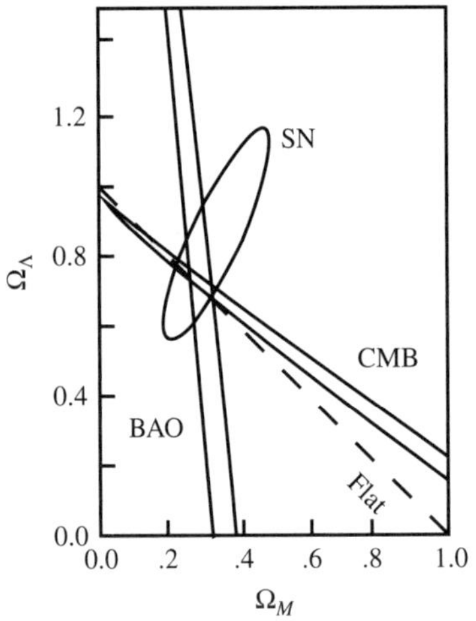


**FIGURE 11.24** Energy content of the universe. Correlated values for the dark energy density and matter density extracted from cosmic microwave background temperature fluctuations (CMB), supernovae luminosity and redshift data (SN), and baryon acoustic oscillation data (BAO) all point to values consistent with the critical density value $\Omega_\Lambda + \Omega_m = 1$ required by inflation. The observational contours at 68 % CL. [after R. Amanullah *et al.*, Astrophys. J. **716**, 712 (2010), arXiv:1004.1711 [astro-ph.CO]]

millimeters, coherent scattering (dominated by neutral current scattering from neutrons) is possible from a volume of order $\lambda^3$ leading to an enhancement in the cross section by factor of $N^2$ where $N \sim 10^{21}$ in the number of neutrons in the coherence volume. This has led to suggestions for detecting the net force due to the neutrino wind resulting from the Sun's proper motion ($v \sim 220$ km s$^{-1}$) modulated by the Earth's orbital motion through the neutrino sea using, for example, a torsion balance with balanced masses of unequal neutrino scattering rate. The long wavelengths have also suggested consideration of neutrino optics based on the refractive indices

$$n - 1 = \frac{2\pi}{p^2} \Sigma_i n_i f_{\nu i \to \nu i}(0) \tag{11.97}$$

computed from the weak interaction forward scattering amplitudes

$$f_{\nu i \to \nu i}(0) \simeq \frac{G_F E}{\sqrt{2}\pi} \left( g_V^i + g_A^i \boldsymbol{\sigma} \cdot \hat{\mathbf{p}} \right). \tag{11.98}$$

Polarized targets might be used, exploiting the sign difference between the amplitudes for neutrino and antineutrino scattering. Unfortunately, practical experiments based on these effects to not appear to have enough sensitivity to detect relic neutrinos.

## 11.6 ■ STRINGS AND OPEN QUESTIONS

Particle physics addresses basic questions about the universe. The standard model provides an accurate representation of the behavior of matter under even extreme conditions but it is not the end of the story. The list of open questions today includes:

- Why are there so many fundamental particles?
- Why are quarks and leptons governed by the peculiar mix of electroweak and color forces?
- Are there more fundamental particles?
- What is neutrino mixing telling us?
- What is dark matter?
- Why is there more matter than anti-matter?
- What is dark energy?

More fundamental perhaps are questions about the basic description of the universe in particle terms and the relationship between particles and space-time in the context of quantum mechanics. Kaluza-Klein models and string theory are two illustrative approaches to reformulations of physical theory that unify gravity with other fundamental interactions.

Einstein's theory of general relativity makes the geometry of space-time itself a full participant in physics. It is natural to attempt to treat the metric tensor of general relativity as a quantum field but the resulting theory is not renormalizable. This fact is related to the presence of a scale set by $G_N$. Extended matter models such as supergravity attempt to account for gravity through the introduction of a spin-2 particle in a way that respects renormalizability.

In a sense, the opposite approach to unification dates back to early in the 20th century and is to regard gauge interactions as a peculiar form of gravity in extra dimensions. In the simplest model, one new space-like dimension is presumed compact and forms a circle of microscopic radius providing a new degree of freedom at each point in space-time. The metric has the form

$$g_{AB} = \begin{pmatrix} g_{\mu\nu} + \kappa^2 A_\mu A_\nu & \kappa A_\nu \\ \kappa A_\mu & \phi \end{pmatrix} \tag{11.99}$$

where $\kappa A_\mu \equiv g_{5\nu}$. Application of general relativity to such a space-time with $\phi = g_{55}$ constant leads to a factorization into four dimensional general relativity plus an effective 4-vector field with all the properties of a $U(1)$ gauge field. This simple Kaluza-Klein model begs generalization to multiple compact dimensions which can set the stage for non-Abelian symmetry. By attaching 1+2+4 extra dimensional space-like manifolds of appropriate symmetry to ordinary space-time, one can accommodate the standard model gauge group $SU(3)_C \times SU(2)_L \times U(1)_Y$. Of course geometry is still only a theater for fermions and, even if gauge fields and the gravitation field can be unified by this artifice, nothing further can be said of the spectrum of matter.

String theory leaps over possible physics between accessible scales and the Planck scale and postulates a swarm of fundamental one dimensional things (strings) at the Planck scale endowed with fermions that interact in many flat dimensions. Particles are identified with string excitations. Particle interactions are extended and the infinities encountered in the renormalization of field theories of point particles are correspondingly ameliorated. The complexity of string interactions permits many strange dynamical universes in which compact and sheet structures emerge and different interactions may be confined to or tunnel between spaces with a reduced number of dimensions. Such a mechanism could explain the disparity the strength of gravity and the strengths of known gauge interactions. Research in this area proceeds at a great pace but no inescapable predictions for physics at the TeV scale and below have been discovered.

## 11.7 ■ FURTHER READING

The 2002 Nobel Prize in Physics 2002 was divided, one half jointly to Raymond Davis Jr. and Masatoshi Koshiba "for pioneering contributions to astrophysics, in particular for the detection of cosmic neutrinos."

Surveys of theories beyond the standard model include P. D. B. Collins, A. D. Martin, E. J. Squires, *Particle Physics and Cosmology*, Wiley (1989) and Graham G. Ross, *Grand Unified Theories*, Oxford (1985).

Cosmology and the particle physics in the early universe are the subject of many texts. See for example B. Ryden B, *Introduction to Cosmology*, Addison-Wesley (2003) and M. V. Berry, "Principles of Cosmology and Gravitation," IOP (1989), M. Rowan-Robinson, "Cosmology," Clarendon Press (1996).

Calculations of a wealth of particle physics processes in the standard model and beyond at colliding beam accelerators may be found in Vernon Barger and Roger Phillips, *Collider Physics*, Addison-Wesley (1997). For additional reading, consult "Resource Letter: The Standard Model and Beyond" Jonathan L. Rosner, Am. J. Phys. **71**, 302 (2003), arXiv:hep-ph/0206176v5.

An isospin triplet scalar field extension of the isospin doublet of the minimal model changes the ratio $\rho = m_W^2/(m_Z^2 \cos^2\theta_W)$ from the minimal model value of 1. See D. A. Rossand M. Veltman, Nucl. Phys. **B95**, 125-147 (1975), T. Blank, W. Hollik, Nucl. Phys. **B514** 113-134(1998), and J. R. Forshaw, B. E. White, D. A. Ross, J. High Energy Phys. **10**, 7 (2001).

A text devoted entirely to the phenomenology of Higgs bosons is John F. Gunion, Howard E. Haber, Gordon Kane, Sally Dawson, *The Higgs Hunter's Guide*, Addison-Wesley (1990). Updated perspectives appear in Gordon L. Kane (editor), *Perspectives on Higgs Physics II*, World Scientific (1997).

Physics at the LHC is described in the physics design reports of the CMS and ATLAS experiments, available at their websites. See also G. Kane and A. Pierce (eds.), *Perspectives on LHC Physics*, World Scientific (2010). The details of calculations of production processes at the LHC are reviewed in J. M. Campbell, J. W. Huston and W. J. Stirling,"Hard interactions of quarks and gluons: a primer for LHC physics," Rep. Prog. Phys. **70** 89 (2007), doi:10.1088/0034-4885/70/1/R02.

An introduction to supersymmetry is Stephen P. Martin, "A Supersymmetry Primer," arXiv:hep-ph/9709356v5, available at `http://zippy.physics.niu.edu/primer.html`. The MSSM is reviewed in H. P. Nilles, Phys. Rep. **110**, 1 (1984). A more current review is A. Djouadi *et. al.*,"The Minimal Supersymmetric Standard Model: Group Summary Report," arXiv:hep-ph/9901246v1. A program describing MSSM running masses and couplings is B.C. Allanach,"SOFTSUSY: a program for calculating supersymmetric spectra," Comput. Phys. Commun.**143**, 305 (2002).

The Feynman rules for the MSSM may be found in H.E. Haber and G.L. Kane, Phys. Rep. **117**, 75 (1985). An early discussion of SUSY Higgs boson signatures at *pp* colliders is J. F. Gunion and H. E. Haber, Nucl. Phys. **B307**, 445 (1988). One more current survey of non-standard Higgs physics is S. Kraml *et al.*, "Report of the CPNSH workshop, May 2004 - Dec 2005," CERN-2006-009, arXiv:hep-ph/0608079v1, available at `http://kraml.home.cern.ch/kraml/cpnsh/report.html`.

The International Linear Collider is described in design reports available at the project website `http://www.linearcollider.org`. The 2010

International Workshop on Linear Colliders reviews of physics at linear colliders and accelerator design progress are available at. `https://espace.cern.ch/LC2010/default.aspx`. Progress on the design of a $\mu^+\mu^-$ collider is described in Fermilab technical note TM2399 available at the project website `https://mctf.fnal.gov/`. Physics at a future muon collider is described in V. Barger, M.S. Berger, J.F. Gunion, and T. Han, Phys. Rep. **286**, 1(1997), Phys. Rev. **D56**, 1714 (1997), Phys. Rev. Lett. **78**, 3991 (1997), and Phys. Rev. **D55**, 142 (1997).

The Maki-Nakagaw-Sakata-Pontecorvo (MNSC) matrix was introduced in Z. Maki, M. Nakagawa and S. Sakata, Prog. Theor. Phys. **28**, 870 (1962).

The MSW effect is described in L. Wolfenstein, Phys. Rev. **D17**, 2369 (1978) and S. P. Mikheev and A. Yu. Smirnov, Sov. J. Nucl. Phys. **42**, 913 (1985), A. Smirnov, arXiv:hep-ph/0305106v1.

The standard solar model is described in Bahcall, Serenelli, and Basu, Astrophys. J. **L85**, 621 (2005) and **626**, 539 (2005), and J. N. Bahcall, S. Basu, M. H. Pinsonneault, Phys. Lett. **B433**, 1 (1998).

A website devoted to developments in neutrino physics with extensive references is "Neutrino Unbound," `http://www.nu.to.infn.it/`. Prominent recent experiments include MARE, Majorana, Cuore, ICARUS, ICECUBE, T2K, Daya Bay, BooNE, NOMAD, GALLEX, and SNO.

A extensive review of neutrino oscillations and their theoretical interpretation with 491 references is "Theory of neutrinos: a white paper," R. N. Mohapatra, *et. al* Rep. Prog. Phys. **70**, 1757 (2007). For a review of theoretical models of neutrino mass and mixing, see the Theory Working Group report of the APS sponsored "Joint Study on the Future of Neutrino Physics: The Neutrino Matrix," `http://www.aps.org/policy/reports/multidivisional/neutrino/index.cfm`. For SAGE results, see `http://ewiserver.npl.washington.edu/sage/SAGEResults.html`, and Phys. Rev. **C56**, 3391 (1997), hep-ph/9710491. For SNO results, see B. Aharmim *et al*, Phys. Rev. **C81**, 055504 (2010) .

The $\nu$MSM model is described in Takehiko Asaka, Steve Blanchet, Mikhail Shaposhnikov, Phys. Lett. **B631**,151-156 (2005), arXiv:hep-ph/0503065v1.

Two classic texts on particle physics and cosmology are S. Weinberg, "Gravitation and Cosmology", J. Wiley and Sons (1972), E. W. Kolb and M. S. Turner, The Early Universe, Addison-Wesley (1990). One more recent text is S. Weinberg "Cosmology," Oxford (2008).

For an introduction to the general theory of relativity, see L. D. Landau and E. M. Lifshitz, *Classical Theory of Fields*, Pergamon Press (1980) and S. Caroll, "Lecture notes on General Relativity," gr-qc/9712019.

Cosmological implications of fits to data from the Wikinson Microwave Asnistropy Probe (WMAP) are described in D. N. Spergel *et al*, Astrophys. J. Suppl.**170**, 377 (2007), arXiv:astro-ph/0603449v2. Further information can be found at the WMAP website http://map.gsfc.nasa.gov/.

Particle physics and inflationary cosmology are reviewed in Andrei Linde's text available online at arXiv:hep-th/0503203v1. A resource list on cosmology and astrophysics is Bharat Ratra and Michael S Vogeley, "Resource Letter BE-1:The Beginning and Evolution of the Universe," http://arxiv.org/abs/0706.1565v1.

A review of gravitational and cosmological evidence for dark matter and of experimental searches is "Deep Underground Science and Engineering Lab: S1 Dark Matter Working Group," D. S. Akerib *et. al.*, arXiv:astro-ph/0605719v3 (2007). See also the conference proceedings "Topics in Astroparticle and Underground Physics (TAUP 2009)," JOP COnference Series, `http://www.iop.org/EJ/toc/1742-6596/203/1`.

Baryon acoustic oscillations are described in Bruce A. Bassett, Renee Hlozek, arXiv:0910.5224 [astro-ph.CO].

For a list of US experiments and projects in high energy physics, see `http://www.fnal.gov/faw/experimentsprojects`. A list of CERN experiments is `http://greybook.cern.ch.html`. For US nonaccelerator particle physics related projects such as dark matter searches, see `http://www.er.doe.gov/hep/research/non\_accelerator.shtml`.

## 11.8 ■ PROBLEMS

### Problem 11.1. Proton lifetime estimate

Justify the factors in the crude estimate for the nucleon lifetime in a grand unified theory

$$\tau_p = \tau_\mu \left(\frac{m_\mu}{m_p}\right)^5 \left(\frac{m_X}{m_W}\right)^4$$

and calculate $\tau_p$ in years for $m_X = 10^{16}$ GeV. How does this estimate compare to the present experimental limit? [Ref: Pran Nath, Pavel Fileviez Perez, "Proton Stability in Grand Unified Theories, in Strings and in Branes," Phys. Rept. **441**, 191 (2007), arXiv:hep-ph/0601023v3]

### Problem 11.2. The decay $t \to H^+ b$

Charged Higgs particles $H^\pm$ are predicted in extensions of the minimal standard model and could provide a rapid decay mode for a heavy quark $Q$. The matrix element

$$M_{Q\to q+H} = \left(\frac{G_F}{\sqrt{2}}\right)^{1/2} m_Q \bar{q}[(1-\gamma_5)v_L + (1+\gamma_5)v_R]Q$$

implies the rest frame decay rate

$$\begin{aligned}\Gamma_{Q\to q+H} = &\frac{G_F m_Q^3}{8\pi\sqrt{2}}\frac{2k}{m_Q}\\ &\cdot \left\{\left(|v_L|^2 + |v_R|^2\right)\left[1 + (m_q/m_Q)^2 - (m_H/m_Q)^2\right]\right.\\ &\left. + \ 4Re(v_L v_R{}^*)m_q/m_Q\right\}\end{aligned}$$

where $k$ is the momentum of the Higgs boson. In a two Higgs doublet model, $g_L = \tan\beta$ and $g_R = (m_q/m_Q)\cot\beta$. The weak decay rate is

$$\Gamma_{Q\to q+W} = \frac{G_F m_Q^3}{8\pi\sqrt{2}}|V_{Qq}|^2 \frac{2k}{m_Q}\Big\{\left[1-(m_q/m_Q)^2\right]^2 + \left[1+(m_q/m_Q)^2\right](m_W/m_Q)^2 - 2(m_W/m_Q)^4\Big\}$$

where $k = [m_Q^2 - (m_W + m_q)^2]^{1/2}[m_Q^2 - (m_W - m_q)^2]^{1/2}/(2m_Q)$ is the $W$ boson momentum and $V_{Qq}$ is the CKM matrix element. Compute the weak decay width of the top quark for $m_q = 5$ GeV and show $\Gamma_{t\to Wb} \simeq 1.5$ GeV. Show that for $m_H = m_W$ and $v_L = 1$ and $v_R = 0$ the width for $\Gamma_{t\to H^+b}$ is comparable. [See I. Bigi et al, Phys. Lett. **B181**, 157 (1986).]

---

**Problem 11.3. Stoponium**

---

A variety of bound states can in principle form from supersymmetric and normal colored particles. The partial decay widths of the $\tilde{t}\bar{\tilde{t}}$ 1S $J^{PC} = 0^{++}$ ground state into gluon and photon final states are

$$\Gamma_{\eta_{\tilde{t}}\to gg} = \frac{4}{3}\alpha_s^2\frac{|R(0)|^2}{m_{\eta_{\tilde{t}}}^2}, \quad \Gamma_{\eta_{\tilde{t}}\to gg} = \frac{32}{27}\alpha^2\frac{|R(0)|^2}{m_{\eta_{\tilde{t}}}^2}$$

where $R(0) = \sqrt{4\pi}\psi(0)$ is the radial wave function at the origin. Potential models give

$$\frac{|R(0)|^2}{m_{\eta_{\tilde{t}}}^2} = \left(0.1290 + 0.0754L + 0.0199L^2 + 0.0010L^3\right)\ \text{GeV}$$

,

$$2m_{\tilde{t}} - m_{\eta_{\tilde{t}}} = (3.274 + 1.777L + 0.560L^2 + 0.081L^3)\ \text{GeV},$$

where $L = \ln(m_{\tilde{t}}/(250\,\text{GeV})$. For $m_{\eta_{\tilde{t}}} = 500$ GeV, show that the binding energy is about 3 GeV, the partial width into $gg$ is of order 2 MeV, and the di-photon fraction is about 0.005. [Ref: D. V Naanopoulos, Senji Ono, T. Yanagida, Phys. Letts **B137**, 363 (1984), S. P. Martin, Phys. Rev. **D77**, 075002 (2008), arXiv:0801.0237v2 [hep-ph]]

---

**Problem 11.4. Neutrino oscillations**

---

Consider two neutrinos with a Lagrangian mass term of the form

$$L = m_{aa}\bar{\nu}_a\nu_a + m_{bb}\bar{\nu}_b\nu_b + m_{ab}(\bar{\nu}_a\nu_b + \bar{\nu}_b\nu_a).$$

Show this may be diagonalized with the substitution

$$\begin{pmatrix} \nu_a \\ \nu_b \end{pmatrix} = \begin{pmatrix} \cos\theta & \sin\theta \\ -\sin\theta & \cos\theta \end{pmatrix} \begin{pmatrix} \nu_1 \\ \nu_2 \end{pmatrix}$$

with eigenvalues

$$m_{1,2} = \frac{m_{aa} + m_{bb}}{2} \pm \frac{1}{2}\sqrt{(m_{aa} - m_{bb})^2 + 4m_{ab}^2}$$

and mixing angle

$$\theta = \frac{1}{2}\tan^{-1}[2m_{ab}/(m_{bb} - m_{aa})].$$

Show that the probability $P(a \to a)$ that a type $a$ neutrino is observed to be type $a$ after traveling a distance $x$ can be cast in the engineering form

$$P(a \to a) = 1 - \sin^2 2\theta \sin^2(1.27\Delta m^2 x/E)$$

with $\Delta m^2 = m_1^2 - m_2^2$.

---

**Problem 11.5. Three flavor mixing**

---

Starting with the general expression for the mixing probability

$$\begin{aligned} P_{\alpha\to\beta} = {} & \delta_{\alpha\beta} - 4\Sigma_{i>j}\mathrm{Re}(U_{\alpha i}^* U_{\beta i} U_{\alpha j} U_{\beta j}^*) \sin^2\frac{\Delta_{ij}}{2} \\ & + 2\Sigma_{i>j}\mathrm{Im}(U_{\alpha i}^* U_{\beta i} U_{\alpha j} U_{\beta j}^*) \sin\Delta_{ij}, \end{aligned}$$

use $U^\dagger U = 1$ to eliminate $U_{\alpha 1}$ and $\Delta_{32} = \Delta_{31} - \Delta_{21}$ to eliminate $\Delta_{32}$ and show that

$$\begin{aligned} P_{\alpha\to\alpha} = {} & 1 - 4|U_{\alpha 2}|^2(1 - |U_{\alpha 2}|^2)\sin^2\frac{\Delta_{21}}{2} - 4|U_{\alpha 3}|^2(1 - |U_{\alpha 3}|^2)\sin^2\frac{\Delta_{31}}{2} \\ & + 2|U_{\alpha 2}|^2|U_{\alpha 3}|^2\left(4\sin^2\frac{\Delta_{21}}{2}\sin^2\frac{\Delta_{31}}{2} + \sin\Delta_{21}\sin\Delta_{31}\right). \end{aligned}$$

Expand to first order in $\Delta_{21}$ and show that

$$\begin{aligned} P_{\alpha\to\alpha} &= 1 - 4|U_{\alpha 3}|^2(1 - |U_{\alpha 3}|^2)\sin^2\frac{\Delta_{31}}{2} \\ &\quad + \left(2|U_{\alpha 2}|^2|U_{\alpha 3}|^2\sin\delta_{31}\right)\Delta_{21} + \ldots \\ &= 1 - 4|U_{\alpha 3}|^2(1 - |U_{\alpha 3}|^2)\sin^2\frac{\Delta_{31} - \kappa_{\alpha\alpha}\Delta_{21}}{2} + \ldots \end{aligned}$$

with $\kappa_{\alpha\alpha} = |U_{\alpha 2}|^2/(1 - |U_{\alpha 3}|^2)$, and, in terms of the standard mixing angles, the angle $\theta_{rct} = \theta_{13}$ is identified as the mixing angle used in two flavor analysis of reactor experiments. [G. Fogli, E. Lisi, and G. Scioscia, Phys. Rev. **D52**, 5334 (1995) and for higher order terms Minako Honda, Yee Kao, Naotoshi Okamura, and Tatsu Takeuchi, arXiv:hep-ph/0602115v1]

---

**Problem 11.6. Muon neutrino mass**

---

The momentum $p_\mu$ of the muon in the decay $\pi^+ \to \mu^+ \nu_\mu$ is sensitive to the neutrino mass. Express $p_\mu$ in terms of the pion, muon and neutrino masses. Given the Particle Data Group values and errors on the pion and muon masses, calculate the momentum for $m_\nu = 0$ and $m_\nu = 200$ keV and compare the difference to the error in the momentum for $m_\nu = 0$ resulting from the errors in the pion and muon masses. The muon mass is determined from the Zeeman splitting in muonium. The pion mass is determined from the spectrum of x-rays released when pions stop in matter forming pionic atoms. The error on the pion mass limits the neutrino mass limit from pion decay. Estimate the X-ray energy for a 4-3 transition in pionic magnesium ($Z = 12$) and the X-ray energy resolution in percent required to determine from this transition the mass of the $\pi^-$ to 400 eV. [Ref: K. Assamagan *et al*, "Upper limit of the muon-neutrino mass and charged-pion mass from momentum analysis of a surface muon beam," Phys. Rev. **D53**, 6065 (1996), S. Lenz *et al*, "A new determination of the mass of the charged pion," Phys. Letts. **B416**,50 (1998), Peter J. Mohr, Barry N. Taylor, and David B. Newell, "CODATA recommended values of the fundamental physical constants," Rev. Mod. Phys. **80**, 633 (2008)]

---

**Problem 11.7. Neutrino beam design**

---

MiniBooNE searched for $\nu_\mu \to \nu_e$ and $\bar{\nu}_\mu \to \bar{\nu}_e$ with neutrinos from $\pi^\pm$ and $K^\pm$ decay. Produced by 8 GeV kinetic energy protons hitting a 70 cm long Be target, the mesons were focused by a pulsed toroidal electromagnet called a horn - an inner cylindrical conductor of radius 2.2 cm and length 185 cm with a peak current of 174,000 A returned along a coaxial cylinder of radius 30 cm. Following the horn was decay region roughly 50 m in length terminated by an absorber. [See the MiniBooNE website and The MiniBooNE Collaboration, Phys. Rev. **D79**, 072002 (2009)]

a) Estimate the probability of interaction of an incident proton in the target. For an interaction in the middle of the target, what is the probability a produced meson escapes without interaction? b) Show that the (peak) field strength near the inner conductor of the horn is about 1.5 T. c) Show the radial momentum impulse to a relativistic $\pi^+$ meson traversing the horn axially near the inner conductor is about 1 GeV/c (comparable to the average meson transverse momentum). c) For a pion of 1 GeV total energy, what is the time dilation factor and what is the probability of a decay producing a $\nu_\mu$ while traversing the decay region? What is the probability for a $\pi^+ \to e^+\nu_e$ decay that produces a $\nu_e$? For a decay $\pi^+ \to \mu^+\nu_\mu$ halfway down the decay region, estimate the probability$\mu^+ \to \bar{\nu}_\mu e^+ \nu_e$ in the remainder of the decay region produces a $\nu_e$? Assume the muon energy is 0.5 GeV.

## Problem 11.8. High intensity neutrino beams

High intensity neutrino beams would be a byproduct of a high energy muon collider. The cross sections for inelastic charged current muon neutrino and antineutrino scattering from nucleons (average of $p$ and $n$) divided by the incident neutrino energy are

$$\sigma_{\nu_\mu N \to \mu^- X} \simeq 0.68 \times 10^{-38}\ \mathrm{cm}^2\ \frac{E_\nu}{1\ \mathrm{GeV}}$$
$$\sigma_{\bar{\nu}_\mu N \to \mu^+ X} \simeq 0.33 \times 10^{-38}\ \mathrm{cm}^2\ \frac{E_\nu}{1\ \mathrm{GeV}}.$$

The average mass density of the Earth is $\rho_{\mathrm{E}} = 5.52\ \mathrm{g\ cm}^{-3}$. a) For what muon neutrino energy is the mean free path for charged current interaction equal to the Earth's radius $R_E = 6.38 \times 10^6$ m? b) Show that for a neutrino beam energy of 70 TeV produced from muon decay, the beam radius at distance $R_E$ from the neutrino source is about

$$r \simeq \frac{m_\mu c^2}{E_\nu} R_E \simeq 10\ \mathrm{m}.$$

c) A rad is 0.01 joules/kg. Consider electron neutrinos of energy 70 TeV interacting and depositing energy in a concrete target of density $\rho_c = 2.5\ \mathrm{g\ cm}^{-3}$ over a depth of four hadronic interaction lengths or $L = 1.6$ m at distance $R_E$. Estimate the beam intensity $I$ in neutrinos per second required to produce one rad per second in the target. Neutral current interaction cross sections are about 0.30 and 0.36 times charged current cross sections and have been neglected. In addition, neutrino cross sections start to exhibit a logarithmic not linear dependence on beam energy for energies above $E_\nu \sim 10$ TeV. ["Destruction of Nuclear Bomb using Ultra-High Energy Neutrino Beam," Hirotake Sugawara, Hiroyuki Hagura, Toshiya Sanami, 29 June 2003, arXiv:hep-ph/0305062; "Ultrahigh energy neutrino-nucleon cross section and radiative correction," Gunter Sigl, Phys. Rev. **D57**, 3786 (1998)]

## Problem 11.9. Neutrinos in matter

Consider a two neutrino vacuum oscillation with

$$|\,\nu_1 > \; = |\,\nu_e > \cos\theta_v - |\,\nu_\mu > \sin\theta_v$$
$$|\,\nu_2 > \; = |\,\nu_e > \sin\theta_v + |\,\nu_\mu > \cos\theta_v$$

with masses $m_1 > m_2$. In matter with electron density $n_e$, the charged current weak interaction distinguishes the $\nu_e$ and the effective equation of motion in the $\nu_1 - \nu_2$ basis is

$$i\frac{d}{dt}\begin{pmatrix}\nu_1\\ \nu_2\end{pmatrix} = \begin{pmatrix}\frac{m_1^2}{2E} - V\cos^2\theta_v & -V\sin\theta_v\cos\theta_v \\ -V\sin\theta_v\cos\theta_v & \frac{m_2^2}{2E} - V\sin^2\theta_v\end{pmatrix}\begin{pmatrix}\nu_1\\ \nu_2\end{pmatrix}$$

with $V = G_F n_e$. Show that the eigenfrequencies are

$$\frac{1}{2}\left(\frac{m_1^2+m_2^2}{2E}-V\right) \pm \frac{1}{2}\sqrt{\left(\frac{m_1^2-m_2^2}{2E}-V\cos 2\theta_v\right)^2+V^2\sin^2 2\theta_v}$$

and show that the eigenstates for propagation in matter are

$$\begin{aligned}
\mid \nu_{1m}>&=\mid \nu_1>\cos\theta_1-\mid \nu_2>\sin\theta_1=\mid \nu_e>\cos\theta_m-\mid \nu_\mu>\sin\theta_m\\
\mid \nu_{2m}>&=\mid \nu_1>\sin\theta_1+\mid \nu_2>\cos\theta_1=\mid \nu_e>\sin\theta_m+\mid \nu_\mu>\cos\theta_m
\end{aligned}$$

where the mixing angles $\theta_1$ and $\theta_m$ are related to the vacuum mixing angle by $\theta_m = \theta_1 + \theta_v$ and given by

$$\begin{aligned}
\tan(2\theta_1) &= \frac{\tan(2\theta_v)}{1-(l_0/l_v)\sec(2\theta_v)}\\
\tan(2\theta_m) &= \frac{\tan(2\theta_v)}{1-(l_v/l_0)\sec(2\theta_v)}
\end{aligned}$$

with $l_0 = 2\pi/Gn_e$ and $l_v = (4\pi E)/(m_1^2 - m_2^2)$. Show the oscillation length in matter is

$$l_m = \frac{l_v}{\sqrt{1+(l_v/l_0)^2-2(l_v/l_0)\cos(2\theta_v)}}$$

and the transition probability for muon neutrino oscillations is

$$\mid<\nu_e\mid \nu_\mu(x)>\mid^2=\frac{1}{2}\sin^2(2\theta_v)(l_m/l_v)^2[1-\cos(2\pi x/l_m)].$$

For $l_0 >> l_v$, $l_m \simeq l_v$ and $\theta_m \simeq \theta_v$. For $l_v >> l_0$, $l_m \simeq l_0$. In between these extremes, there can be a resonant enhancement if $l_v/l_0 = 2\cos(2\theta_v)$. [L. Wolfenstein, Phys. Rev. **D17**, 2369 (1977)]

---

**Problem 11.10. Friedmann-Lemaitre equations**

---

Add a repulsive spring force counteracting the attractive spring force due to gravity to the pedestrian derivation of the Friedmann equation to find

$$\left(\frac{\dot a}{a}\right)^2 = \frac{8\pi G_N}{3}\rho - k\frac{c^2}{a^2} + \frac{\Lambda c^2}{3}.$$

Show that the Friedmann-Lemaitre equations

$$\begin{aligned}
\left(\frac{\dot R}{R}\right)^2 &= \frac{8\pi G_N}{3}\rho - \frac{k}{R^2} + \frac{\Lambda}{3}\\
\frac{\ddot R}{R} &= -\frac{4\pi G_N}{3}(\rho+3p) + \frac{\Lambda}{3}
\end{aligned}$$

imply $\dot{\rho} = -3H(\rho + p)$. See Jean-Philippe Uzan and Roland Lehoucq, Eur. J. Phys **22**, 371 (2001) for a discussion.

---

**Problem 11.11. Stau Catalyzed Fusion**

---

In one supersymmetric model, the lightest supersymmetric particle is a gravitino of mass $m_{3/2}$, a stable particle which could be a component of the dark matter in the universe. The next lightest supersymmetric particle could be the scalar partner of the tau lepton called the stau, $\tilde{\tau}$. The lifetime of the stau is

$$\tau \simeq 0.2 \text{ years} \left(\frac{m_{3/2}}{10 \text{ GeV}}\right)^2 \left(\frac{100 \text{ GeV}}{m_{\tilde{\tau}}}\right)^5 \left(1 - \frac{m_{3/2}^2}{m_{\tilde{\tau}}^2}\right)^{-4} .$$

a) What is the lifetime of a stau of mass 100 GeV if the gravitino mass is 20 GeV?
b) A heavy negatively charged stau brought to rest in a liquid deuterium target could form a sort of molecule. Estimate the size of $(\tilde{\tau}dd)_{\text{mol}}$ and compare to that of a muonic molecule $(\mu^- dd)_{\text{mol}}$.
c) The reactions

$$d + d \to {}^3\text{He} + \text{n} + 3.3 \text{ MeV}$$

$$d + d \to {}^3\text{H} + \text{p} + 4 \text{ MeV}$$

can occur with roughly equal probability by quantum tunneling through the repulsive Coulomb barrier. The nuclear fusion generally releases the stau which can catalyze another fusion until it becomes attached to one of the final state charged particles forming a stable atom. If the attachment probability is $2 \times 10^{-4}$, what total energy is released per stau? ["Stau-catalyzed Nuclear Fusion", K. Hamaguchi, T. Hatsuda, T. T. Yanagida, arXiv:hep-ph/0607256v3]

---

**Problem 11.12. WIMP elastic scattering**

---

If dark matter in the Milky Way galaxy is comprised of weakly interacting massive particles (WIMPS) in a non-rotating spherical distribution, a terrestrial detector orbiting the Milky Way galactic center at a radius of eight kpc would be expected to encounter a flux of dark matter particles with a density of 0.3 GeV cm$^{-3}$ and typical velocity $v \sim$300 km s$^{-1}$ with an annual modulation of 30 km s$^{-1}$ due to the Earth's velocity around the Sun. Show that a nucleus of mass $m_A$ recoiling from a head-on collision with a dark matter particle of mass $mc^2 = 10 - 1000$ GeV will have a recoil energy

$$K_A = \frac{4m_A/m}{(1 + m_A/m)^2} \frac{1}{2} mv^2$$

and that for $m_A c^2 \sim$50 GeV that the recoil energy is of order tens of keV. Such recoils may be detected through telltale ionization trails.

## Problem 11.13. Neutrinos from supernovae

In 1987, nineteen antineutrinos were detected via the reaction $\bar{\nu}_e p \to n e^+$ in the IMB and Kamiokande detectors and five neutrinos in the Baskan detector were detected over an interval of 13 s. The burst was associated with Supernova 1987a in the Large Magellanic Cloud at a distance $d_{1987} = 168,000$ LY $\sim 51$ kpc. The energies of such neutrinos are expected to be near 10 MeV typical of nuclear reactions and the detector thresholds were also a few MeV. Show that the arrival time difference of two relativistic neutrinos with masses $m_1$ and $m_2$ produced simultaneously at a distance $d$ is

$$\Delta t = t_1 - t_2 = \frac{d}{2}\left[\left(\frac{m_1}{E_1}\right)^2 - \left(\frac{m_2}{E_2}\right)^2\right].$$

To use the arrival times to determine the mass differences, the arrival times must be large compared the neutrino production time scale $\tau$. For $\tau = 1$ s, $E_1 = E_2 \simeq 10$ MeV, $\Delta m^2 = m_1^2 - m_2^2 = 8 \times 10^{-5}$ eV$^2$, show that the minimum distance is much larger than $d_{1987}$ and, since the number $N$ of neutrinos that would be detected scales as $N \propto 1/d^2$, that the neutrino spectrum cannot be deduced by this method without a lower threshold. What detector energy threshold would be required for a time delay sensitive to this value of $\Delta m^2$ at 51 kpc? [Ref: B.Jegerlehner, F.Neubig, G.Raffelt, Phys. Rev. **D54**, 1194(1996), arXiv:astro-ph/9601111]

## Problem 11.14. Relic density and average energy

The relic particle phase space distribution at decoupling temperature $T$ is

$$dn = \frac{1}{\exp\left(\frac{E}{k_B T} - \mu\right) \pm 1} \frac{dV\,d\mathbf{p}}{(2\pi)^3}$$

where $E = \sqrt{\mathbf{p}^2 + m^2}$ and the plus sign applies to a fermion helicity state and the minus sign to a boson helicity state. For photons, $m_\gamma = 0$ and $\mu_\gamma = 0$ and $E_\gamma = |\mathbf{p}_\gamma|$. Since it is believed that $m_\nu << k_B T \sim 1$ MeV, we can assume $E_\nu = |\mathbf{p}_\nu|$. In thermodynamics, one encounters integrals of the form

$$\int_0^\infty dz\, z^n (e^{z-\mu} \pm 1)^{-1} = \mp\Gamma(n+1)\mathrm{Li}_{n+1}(\mp e^\mu)$$

where $\Gamma(n+1) = n!$ and the polylogarithm is $\mathrm{Li}_s(z) = \Sigma_{k=1}^\infty k^{-s} z^k$. The chemical potential $\mu_\nu$ for neutrinos vanishes if there is lepton number symmetry and the density of neutrinos and antineutrinos are equal. In this case, we can use the simpler results

$$\int_0^\infty dz\, \frac{z^n}{e^z - 1} = n!\zeta(n+1); \int_0^\infty dz\, \frac{z^n}{(e^z + 1} = (1 - 2^{-n})n!\zeta(n+1)$$

where the Riemann zeta function values we require are $\zeta(2) = \pi^2/6 = 1.645$, $\zeta(3) = 1.202$ and $\zeta(4) = \pi^4/90 = 1.082$. Show that the number density $n_b$ for a relativistic boson helicity state and the number density $n_f$ for a fermion helicity state are

$$n_b = \frac{4}{3} n_f = \frac{2.404}{2\pi^3} \left( \frac{k_B T}{\hbar c} \right)^3$$

and the energy densities are

$$\rho_b c^2 = \frac{1}{2} \sigma T^4, \ \rho_f c^2 = \frac{7}{16} \sigma T^4$$

where $\sigma = \pi^2 k_B^2 / (15 c^3 \hbar^3) = 7.56 \times 10^{-16}$ J m$^{-3}$ K$^{-4}$ is the Stefan-Boltzmann constant. Use the CMB temperature and $T_\nu = (4/11)^{1/3} T_\gamma$ to compute the number of photons and electron neutrinos (equal to the number of electron antineutrinos) per cubic centimeter today.

## Problem 11.15. Oopsie bursts

Cosmic neutrino absorption spectroscopy has been proposed as a method of detecting relic neutrinos. The spectrum of ultra high energy cosmic protons is cut off at high energy by collisions with cosmic microwave background photons potentially exposing a cosmic neutrino spectrum from primordial sources. Consider a collision of a neutrino $\nu_i$ of flavor $i$ energy $E_\nu$ with a nearly stationary relic antineutrino $\bar{\nu}_i$ of mass $m_i \simeq 1$ eV. a) Compute the resonant energy in GeV at which $\nu_i \bar{\nu}_i \Rightarrow Z \to X$ will create an absorption feature ($Z$ burst) in the cosmic neutrino spectrum. (Off resonance, $\nu\bar{\nu} \Rightarrow f\bar{f}$ and $\nu\bar{\nu} \to W^+W^-$ provide a featureless attenuation. For lighter neutrino masses, the absorption feature may be blurred by target neutrino motion.) What are the corresponding energies for $\nu_i \bar{\nu}_i \Rightarrow \Upsilon(1S) \to X$ and $\nu_i \bar{\nu}_i \Rightarrow J/\psi \to X$? b) Compute the threshold energy $E_p$ at which inelastic collisions beginning with $p + \gamma \to \Delta(1232) \to X$ will attenuate the cosmic proton spectrum assuming $E_\gamma \simeq k_B T_{CMB}$. Above this "GZK" cutoff, it is expected that the primordial cosmic neutrino spectrum might be observable without background from the decay products of cosmic proton interactions. c) Use the formula

$$\sigma_{ab \to cd} = \frac{4\pi(2s_X + 1)}{(2s_a + 1)(2s_b + 1) p_{ab}^2} \frac{m_X \Gamma_{X \to ab} m_X \Gamma_{X \to cd}}{(s - m_X^2)^2 + m_X^2 \Gamma_X^2}$$

with $\Gamma_{\Delta \to p\gamma} / \Gamma_\Delta \simeq 0.5\%$ and the present number density of CMB photons to estimate the attenuation length in Mpc for protons at the "GZK" cutoff energy. d) Use the same formula and $\Sigma_i \Gamma_{Z \to \nu_i \bar{\nu}_i} / \Gamma_Z = 20\%$ and the relic neutrino density today to estimate the attenuation length for cosmic neutrinos and show that it is comparable to the age of the universe (the light speed travel time for a distance of

$10^4$ Mpc is 30 billion years.) so cosmic expansion must be considered. e) Use the estimates

$$\frac{\Gamma_{J/\psi \to \Sigma_i \nu_i \bar{\nu}_i}}{\Gamma_{J/\psi \to e^+e^-}} = \frac{27 G_F^2 m_\psi^4}{256\pi^2\alpha^2}\left(1 - \frac{8}{3}\sin^2\theta_W\right) = 4.5 \times 10^{-7}$$

with $\Gamma_{J/\psi \to e^+e^-}/\Gamma_{J/\psi} = 5.94\%$ and

$$\frac{\Gamma_{\Upsilon rightarrow \Sigma_i \nu_i \bar{\nu}_i}}{\Gamma_{\Upsilon \to e^+e^-}} = \frac{27 G_F^2 m_\psi^4}{64\pi^2\alpha^2}\left(-1 + \frac{4}{3}\sin^2\theta_W\right) = 4.1 \times 10^{-4}$$

with $\Gamma_{\Upsilon \to e^+e^-}/\Gamma_{\Upsilon} = 2.38\%$ to estimate the attenuation length for these resonances. The expansion implies increased neutrino density and more attenuation at early times but a cosmic smearing of the absorption feature by an amount equal to the redshift to reach significant absorption. This smearing is much larger for the lighter resonances compared to that for the $Z$. See G. Barenboim, O. M. Requeljo, C. Quigg, Phys. Rev. **D71**, 083002 (2005), arXiv:hep-ph/0412122v1.

---

**Problem 11.16. Relic neutrino detection**

---

Suppose a natural radioactive decay $N \to N' e_1^- \bar{\nu}_e$ converts nucleus $N$ into $N'$ and that the electron energy spectrum for $m_\nu = 0$ extends to $E_0$. Then $N'$ can capture a relic neutrino of essentially zero energy resulting in

$$\nu_e N \to N' + e_2^-.$$

If $m_\nu > 0$, the energy spectrum of $e_1^-$ vanishes at $E_1 = E_0 - m_\nu$ while the energy of $e_2^-$ has a fixed value $E_2 = E_0 + m_\nu$. Given an energy resolution of order $m_\nu$, these cases can be distinguished and the relic neutrino density in principle observed. The capture cross section is proportional to the transition rate for the decay in the region of phase space for which the neutrino is at rest. That transition rate is roughly the total decay transition rate times a suppression factor due to the Coulomb interaction of the electron and the nucleus N' which is exponentially sensitive to the nuclear charge. Hence short half-life, high Q, low Z radioactive nuclei such as $^3H$ are preferred. The tritium capture cross section is

$$\sigma_{^3\mathrm{H}\ \nu_e \to {}^3\mathrm{He}\ e^-} = \frac{7.8e \times 10^{-49}\ \mathrm{m}^2}{v_\nu / c}$$

for neutrino velocity $v_\nu$. The mean relic neutrino number density is $n_\nu \simeq 50\ \mathrm{cm}^{-3}$ and the temperature $T_\nu = (4/11)T_\gamma = 1.7 \times 10^{-4}$ eV from the Big Bang model. The mean kinetic energy is $\bar{E}_\nu = 6.5 T_\nu^2/m_\nu$ for nonrelativistic neutrinos and $\bar{E}_\nu = 3.15 T_\nu$ for relativistic neutrinos. For $m_\nu = 0.6$ eV, estimate the number of captures per year for 0.1 kg of $^3H$. [Alfredo G. Cocco, Gianpiero Mangano, Marcello Messina, J. Phys. Conf. Ser. **110**, 082014 (2008), arXiv:hep-ph/0703075v2]

## Problem 11.17. Black hole evaporation

A black hole appears when the escape velocity defined by $mv^2/2 = GMm/r$ equals light speed. The mass $M$ must be contained within the Schwarzchild radius $r_s = 2GM/c^2$. A black hole plausibly radiates as a blackbody with temperature $T = \hbar c^3/(8\pi GMk_B)$. a) Using the Stefan-Boltzmann constant $\sigma = \pi^2 k_B^4/(60\hbar^2 c^2)$ and the area $A = 4\pi r_s^2$, show that the energy loss rate is

$$P = -\frac{dE}{dt} = -c^2\dot{M} = \frac{K}{M^2}$$

where $K = \hbar c^6/(15360\pi G^2) = 3.56 \times 10^{32}$ W kg$^2$. Neglecting any radiated particles other than photons, integrate to show that the evaporation time is

$$\tau = \frac{c^2 M^3}{3K} = \frac{5120\pi G^2}{\hbar c^4} M^3 \sim 1.3 \times 10^{-17} \left(\frac{M}{1\ \text{kg}}\right)^3 \text{ s}.$$

b) Evaporation today requires a temperature in excess of the temperature 2.7 K of the cosmic microwave background. What maximum mass, in units of the lunar mass $7.36 \times 10^{22}$ kg, evaporates and what is the evaporation time for such a mass neglecting absorption of radiation? c) What is the maximum mass with an evaporation time shorter than of the age $13.7 \times 10^9$ yr of the universe?

# Index

NOTE: An *f* following a page number indicates a figure; a *t* following a page number indicates a table.


Acceleration, relativistic, 73, 98
Accelerator-based experiments, present and future, 517–522
Accelerators. *See also* Fermi National Accelerator Laboratory Tevatron
  linear, 74, 102
  radio frequency quadrupole (RFQ), 92, 103
  relativistic mechanics and, 72–76
  ß-function, 81
  Van der Graaf, 74, 99
ALEPH, 437, 446
ALICE, 89
Angular momentum
  combining two spin 1/2 wave functions, 23
  orbital angular momentum operator, 14
  vector addition of, 504–505
Anomalous electromagnetic couplings, 243
Anomalous magnetic moment
  of electron, 428
  of muon, 429
Anti-hypertritium in heavy ion collisions, 315f
Anti-matter and charge conjugation (*C*), 268–270
Anti-screening in gauge theory, 443–444, 443f
Antiparticle, 4
Antiproton source, 103–104
Askaryan effect, 140
ATHENA, 51
ATLAS, 89, 127, 129f, 130–131, 520–521
Atmospheric neutrinos, 528, 529f
Atoms, exotic, 55. *See also* Muonic atom; Muonium; Positronium
  $(\mu^+e^-)$ atom, 33
  $(\pi^{\pm}\mu^{\pm})$ atom, 204
Axions, 246–247

B-meson decay, 494, 508
Barn, 84
Baryons, 4, 463
  acoustic oscillation, 548
  $\Delta^+$, 463–464
  decuplet, 463
  $\Lambda$, 42, 463
  $\Lambda_b$, 46
  $\Lambda_c$, 46
  magnetic moment values, 467, 467t
  mass, 369
  number, 52
    quarks and, 52
    violation, 512
  *p*, 55, 70–71, 76, 99, 103, 133, 506
  $\Sigma^-$, 463
  $\Sigma_b^-$, 46
  $\Omega$, 39, 463
  $\Omega_b$, 46
  $\Omega_{sss}$ (1672), 46
  $\Xi^0$, 46, 463
  $\Xi^-$, 463
  $\Xi_b$, 46
$B_c$, 43
$B_d$, 43, 491, 493, 495
Beam gas lifetime, 105–106
Beam stability and emittance, 77–83
BELLE, 493
Berkeley Bevatron, 71
Berm design, 138
Betatron, 100
  oscillations, 80–81, 81f, 100
Black hole evaporation, 565
Bohr radius, 14
BOREXINO, 525
Boson decays, *W* and *Z*, 373–377, 375f
Boson interactions. *See also* Electroweak boson interactions, form factor in chromodynamics, 327, 327f
Boson masses, prediction of weak, 410–411
Boson production
  at collider, 390–397
  multiple weak, 397–399
Boson(s), 1–2, 3f, 23, 117f, 320, 352f, 393, 393f, 394, 394f, 397, 397f. *See also* Higgs boson; *W* boson
  decays to a virtual *W*, 412–413
  and standard model interactions, 47–48
Bound state resonances, 182f
BRAHMNS, 344
Branching fraction, 31